INDIVIDUAL RIGHTS AND
THE LAW IN BRITAIN

Individual Rights and the Law in Britain

Edited by

CHRISTOPHER McCRUDDEN
Fellow of Lincoln College
Oxford

and

GERALD CHAMBERS
Research and Policy Planning Unit
the Law Society

THE LAW SOCIETY

CLARENDON PRESS · OXFORD

1994

Oxford University Press, Walton Street, Oxford OX2 6DP
Oxford New York Toronto
Delhi Bombay Calcutta Madras Karachi
Kuala Lumpur Singapore Hong Kong Tokyo
Nairobi Dar es Salaam Cape Town
Melbourne Auckland Madrid
and associated companies in
Berlin Ibadan

Oxford is a trade mark of Oxford University Press

Published in the United States
by Oxford University Press Inc., New York

British Library Cataloguing in Publication Data
Data available

Library of Congress Cataloging in Publication Data
Individual rights and the law in Britain /
[edited by] Christopher McCrudden and
Gerald Chambers.
p.   cm.
Includes bibliographical references.
1. Civil rights—Great Britain—History. I. McCrudden,
Christopher. II. Chambers, Gerald.
KD4080.R54 1993
342.73'085—dc20
[347.30285] 93-4899
ISBN 0-19-825741-4

1 3 5 7 9 10 8 6 4 2

Typeset by Best-set Typesetter Ltd., Hong Kong
Printed in Great Britain
on acid-free paper by
Biddles Ltd., Guildford and King's Lynn

Foreword

Rights are matters about which most people feel very strongly. The upholding or enforcement of a right is sometimes concerned with the recovery of money. The rights which give rise to the strongest feelings, however, usually concern a principle or a cause, or involve a real or perceived oppression or abuse of power, either by the state or by a person or corporation which is more powerful and influential than the injured citizen. It is in these circumstances that the public needs independent lawyers to ensure that justice is achieved. As lawyers, therefore, solicitors have a particular duty to uphold the rule of law and the rights of the citizen.

The Law Society's long-term strategic plans (*Succeeding in the 90s*, Law Society, 1991) stress the commitment of the Society to 'serve law and justice and to help people gain access to justice'. Thus the Society sees itself as having a particular role to play in supporting solicitors who deliver legal advice and assistance related to clients' rights and in campaigning to preserve those rights. It is therefore particularly appropriate that the Law Society should demonstrate this concern for human rights and the rule of law by developing within its research programme this current study. The idea for the study arose from Council members (notably Sir Richard Gaskell, the President of the Law Society in 1988–9), following the recognition that a great deal of legislation affecting rights had been enacted over the preceding ten years. It was further recognized that the Society could make an important contribution to the corpus of knowledge on domestic human rights issues by documenting these changes clearly. It was therefore decided that the subject should be researched in a rigorous manner by the Society's Research and Policy Planning Unit.

The study was developed in consultation with the Society's International Human Rights Working Party and the International Human Rights Consultative Group for lawyers. Both these groups helped to define the remit of the project and to set the study in an international context. In order to keep the project within manageable limits, it was decided to concentrate on changes in civil and political rights since 1950. In addition, it was felt to be important to deepen the content by bringing together experts in various areas of law rather than by commissioning one individual authority to present an analysis. It is indeed an achievement that Christopher McCrudden and Gerry Chambers have been able to gather contributions for this book from such an eminent group of experts on human rights law.

The topics covered in this volume are diverse as are the opinions and views of the contributing authors. The editors have tried, however, to ensure that there was a consistency of treatment in each contribution. Each author was therefore asked to address a set of core questions which were laid out in a commissioning Schema. In addition, authors attended several discussion meetings, sponsored by the Law Society, during the life of the project. At these meetings agreement was reached on the terms of the Schema and draft chapters were discussed around the table in a collegiale atmosphere. As well as making an individual contribution, each author has therefore worked with a set of common aims, objectives, and guidelines which ensured that broadly the same key issues were addressed within each chapter.

Whilst this book is an important achievement, it must be recognized that it portrays only one part of a much larger picture: an equivalent examination of the wider social, economic and environmental rights of individuals and groups has not been tackled in this book. Furthermore, the picture is constantly shifting: the book covers developments at least up to January 1993, although some contributors have been able to take some developments after that date into account. What is clear however is that human rights law is likely to be an area of increasing importance in the future both in Britain and in Europe. We hope that this book, by documenting clearly what has happened in Britain to date, will make a useful contribution to these developments.

CAROLE F. WILLIS

Research and Policy Planning Unit
The Law Society

Editors' Preface

We wish first to express our gratitude to the Law Society for making the production of this book possible and for supporting the project from its inception. However, any views and opinions expressed or policies proposed do not necessarily accord with those of the Law Society. Our co-authors have taken part in the project with good grace, intellectual commitment, and in a co-operative spirit for which we are most grateful. As editors we have not attempted to impose any ideological or theoretical direction on contributing authors. Any opinion expressed in the book belongs to its author and is not necessarily in accordance with the views other authors might have on the topic concerned. Nor are our introductory and concluding chapters attempts to arrive at a consensus view. Evelyn Collins is writing in a personal capacity.

We were most fortunate to be able to draw on the research skills of Julia Black who assisted in the preparation of the tables of cases and statutes and the bibliography. Emma Powell of the Law Society contributed to the preparation of the Appendices. Brendan O'Leary read and commented extensively and helpfully on drafts of Chapters 1 and 16, as did Peter Ashman and Laurence Helfer for Chapter 15. Lastly our grateful thanks to Richard Hart and Jane Williams at Oxford University Press for their encouragement, patience, and wisdom.

We are grateful to the following for permission to draw on previously published material which we have updated and adapted: the Council of Europe for use of the Figure in Chapter 1; R. R. Churchill for permission to draw on material from an article forthcoming in J. P. Gardner (ed.), *The European Convention on Human Rights: Aspects of Incorporation* which appears as part of Appendix D; and Liberty for permission to use and amend table 1 from *A People's Charter* which also appears as part of Appendix D.

CHRISTOPHER MCCRUDDEN
GERALD CHAMBERS

Contents

List of Contributors

GERALD CHAMBERS: Senior Research Officer in the Research and Policy Planning Unit of the Law Society.

EVELYN COLLINS: Chief Equality Officer at the Equal Opportunities Commission for Northern Ireland.

ANN DUMMETT: Adviser to the Commission for Racial Equality.

JOHN EEKELAAR: Reader in Law at Pembroke College, Oxford.

KEITH EWING: Professor of Law at King's College, London.

JOHN GARDNER: Fellow and Tutor in Law at Brasenose College, Oxford.

CONOR GEARTY: Reader in the Faculty of Laws at King's College, London.

JOHN JACKSON: Reader in the Faculty of Law at the University of Sheffield.

LEONARD LEIGH: Professor of Law at the London School of Economics and Political Science.

CHRISTOPHER MCCRUDDEN: Fellow and Tutor in Law at Lincoln College, Oxford.

ELIZABETH MEEHAN: Professor of Politics at Queen's University, Belfast.

JAMES MICHAEL: Lecturer in the Faculty of Law at University College, London.

SEBASTIAN POULTER: Senior Lecturer in the Faculty of Law at the University of Southampton.

ROBERT REINER: Professor of Law at the London School of Economics and Political Science.

GENEVRA RICHARDSON: Reader in the Faculty of Law at Queen Mary and Westfield College, London.

ROBERT WINTEMUTE: Lecturer in the Faculty of Law at King's College, London.

Table of Cases

Table of United Kingdom Statutes

Table of International Instruments, European Community Legislation, and Legislation from Other Jurisdictions

1

Introduction: Human Rights in British Law

CHRISTOPHER McCRUDDEN AND
GERALD CHAMBERS

INTRODUCTION

The aims of this book are twofold: first, and most importantly, to describe
and discuss those changes which have been introduced into English law
since 1950 which have had an impact on the civil and political rights of the
individual, and, secondly, to set those changes in the context of European
and international human rights law. The book concentrates on civil and
political rights, broadly speaking those rights defined in the Universal
Declaration of Human Rights, the International Covenant on Civil and
Political Rights (ICCPR), and the European Convention on Human Rights
and Fundamental Freedoms (ECHR). The substantive areas discussed
include: police powers, due process, privacy, prisoners' rights, freedom of
association, freedom of assembly, freedom of speech, minority rights,
immigration and nationality, racial discrimination, women's rights, the
rights of children and rights in the family, sexual orientation, and civil
liberties in the context of political violence. The period since 1950 provides a convenient starting-point for assessing the effect of the ECHR,
which the United Kingdom ratified in 1951, coming into force on 3
September 1953. The ICCPR came into effect in 1976, the year the
United Kingdom ratified the Covenant.

The book provides an overview which addresses the questions: where
are we now? what changes have there been since 1950? why have these
changes come about? what are the ideals which the United Kingdom has
committed itself to in international and regional human rights agreements?
how does the United Kingdom give effect to the ideals propounded in
these agreements? The study draws on a number of different approaches
to the issues considered in order to set the domestic human rights debate
in an international context and to address the issue in its legal, political,
and theoretical context; and to provide a scholarly basis for further work
in the area. The choice of topics is grounded in the perspective of the
international and European conventions and not determined solely by
current political debate or topicality, though it clearly reflects current
domestic concerns.

The geographical focus of the book is primarily on developments in

England and Wales, and the law as stated in the following chapters relates to that jurisdiction as at 1 January 1993 unless otherwise stated. Several chapters do, however, draw on developments in Scotland (and, to a lesser extent, Northern Ireland), to add to a better understanding of the position in England and Wales. We have frequently used the terms 'Britain' and 'British' to describe this area of coverage.

This introductory chapter begins with a sketch of some of the major features of the political and social context which have shaped the law relating to individual rights in Britain since 1950. We argue that this law can be better appreciated when seen as arising from, and intimately related to important political, social and intellectual trends, both national and international. The chapter then turns to a consideration of one of the most important of these developments: the growth since the Second World War of an international law of human rights. We describe briefly those features of this law which are of most importance for an under-standing of the specific rights discussed in the subsequent chapters. The introduction ends with an explanation of the structure of the book and a brief account of the chapters which follow.

POLITICAL AND SOCIAL CONTEXT

The protection of civil liberties and the advancement of human rights do not take place in a political or legal vacuum. An important theme of this book is that political, economic, and social contexts significantly affect the extent and type of protection promised and delivered by a particular legal system. Indeed, it can be argued that human rights law is required to be more responsive and sensitive to context than other areas of the law. In seeking to understand human rights law in Britain since 1950 it is therefore necessary to have an historical awareness of the political frame-work and of the immense changes that have been taking place both on the national and international scenes. In order to set legal develop-ments in context we have therefore provided in Appendix A an outline of the main events in British and international politics since 1950 alongside key statutory developments in the domestic law of individual rights during the same period.

European and international developments

The reconstruction of Western Europe after the Second World War led France and Germany to leave behind their role of old adversaries and take on their current role as joint leaders of the new Europe. Britain lost its empire, declining from world power to its current role as regional

player in the European Community which it joined in 1973. These decades saw the rise of the United States as the dominant world power, the
ideological and military confrontation between the United States and the
former Soviet Union in the 'cold war', the rise and ultimate collapse of
the Soviet Union, and with it its empire in Eastern Europe. New independent states in Asia and Africa have emerged from the former colonies of
the Western powers, often coupled with bloody, long-running wars of
national liberation against the former colonial power (or its replacement)
as in Vietnam and Algeria. There have been numerous other localized
wars such as in Korea, Nigeria, the Falklands, and in the Gulf. Britain
has witnessed substantial international migration movements, between
poor Third World countries in Africa, the Americas, and Asia, and
between them and the countries of Western Europe and North America.
Political violence became a threat on an international scale, particularly
from the Middle East, arising from the long-standing conflict involving
Israel, the Palestinians, and the Arab world. In addition, a handful of
groups dedicated to the violent overthrow of Western European liberal
democratic structures appeared in the 1970s.

These developments were brought to public knowledge in Britain with
speed and immediacy, first with the expansion of cinema newsreel, supplementing radio, and then through the massive extension of television
broadcasting as the main medium of mass communication. With the
advent of television, the post-war world began to seem a much smaller
place, and one in which ideas and events from abroad were increasingly
available to a wide audience. However, television is but one of many
major technological and scientific innovations to which we are now ac-
customed, which previously would have seemed like ideas from science
fiction. The development of nuclear weapons, space exploration, the
exponential expansion in data collection methods, the exploitation of the
microchip, the discovery of DNA, the development of the contraceptive
pill, and so on, make the world not only a smaller but also a more
complex and technical place, and one where choices seem to be required
of us more and more. Indeed, the development of technocratic societies
where access to power is increasingly determined by access to expert
knowledge systems seems to exclude many ordinary citizens from effective participation in decision-making.

The period between the 1950s and the 1990s has seen important shifts
in the dominant ideologies. The collapse of communism in Eastern Europe
has already been mentioned, but prior to this, there has been since the
1960s decreasing acceptance of traditional authority, as exemplified for
instance by the growth of important movements opposing entrenched
patterns of racial discrimination in the United States and in South Africa,
and the flourishing of the women's movement. More recently, the re-

surgence of Islam has had important implications for world political developments, and in the Western hemisphere a move from socialist or social democratic towards market-oriented economic principles has significantly altered the political landscapes of some of these countries. Recent years have also seen a world-wide upsurge in the identification of many peoples with their particular ethnic, religious, linguistic, or national communities, shattering the illusion that these issues had long since been resolved.

British politics was affected by these international movements and events in ways addressed in subsequent chapters. Most importantly for our purposes, a strong current has been carrying Britain in the direction of an increasing acceptance of supra-national, and for the most part European, laws and institutions which may in practice limit domestic freedom of action and circumscribe national sovereignty. The international treaties, covenants, agreements, and conventions to which the United Kingdom has subscribed since the Second World War will be discussed subsequently, together with their specific relevance to domestic legal developments. They demonstrate the profound impact international agreements can potentially have on governmental autonomy in both the domestic and international spheres. European institutions and laws will continue to impinge on and shape United Kingdom domestic policy over the next decade.

Domestic political developments

However, domestic issues also played an important role in affecting developments in human rights protection. The Labour Government of Clement Atlee was elected in 1945 and continued in office until 1951, when it was replaced by a Conservative Government, first under the premiership of Winston Churchill, then under Anthony Eden, Harold Macmillan, and Alec Douglas-Home. This in turn gave way in 1964 to the first Labour Government of Harold Wilson, re-elected in 1966. This was replaced by a Conservative Government under Edward Heath in 1970 which lasted until 1974. The two elections in 1974 allowed back a Labour Government, first under Harold Wilson and subsequently under James Callaghan. The Conservative Government under Margaret Thatcher replaced the Callaghan administration in the 1979 election, being returned again with substantial majorities in 1983 and 1987. John Major replaced Margaret Thatcher as leader of the Conservative Party and as Prime Minister in 1990, leading the Conservatives to their fourth successive general election victory in April 1992.

Despite changes in government, from a Labour to a Conservative administration in 1951 and from Conservative to Labour in 1964, there was a remarkable political consensus for at least the first twenty-five years after the Second World War. This consensus comprised significant agreement between the two major parties on the appropriate mix between government and private enterprise in the economy, the appropriate provision of social welfare, the role of trade unions in political and economic life, and the place of local government. The effect of this consensus was a significant expansion of social provision during the 1950s and 1960s, in such areas as health care, welfare provision, education, and housing.

Rising expectations of personal autonomy and freedom during the 1950s and 1960s challenged and transformed power relations between citizen and state, child and parent, employee and employer, and men and women. The 1950s was a period of rising affluence and in Britain was the era characterized by Harold Macmillan's quip that 'you never had it so good'. Increased affluence brought the possibility of independence for young people from their parents at a much earlier stage in their life. The decline of the extended family as a means of social support was to some extent ameliorated by the development of the welfare state of which the national health service was the most potent post-war example. The period since 1950 has seen a massive entry of women (and especially married women) into the work-force.

The 1960s saw this increasing economic well-being and security being translated into demands for greater freedom on the part of young people in Britain and in the other affluent countries of Western Europe and North America. In Britain the Labour Government of Harold Wilson was committed to economic modernization and the promotion of greater equality. The first legislation promising equality to ethnic minorities across a range of issues, and to women in pay, was enacted in 1965 and 1970 respectively. The growth of demands for increased sexual autonomy contributed to the partial decriminalization of male homosexuality and a liberalization of the abortion laws in 1967. Britain, which once prided itself on its cultural homogeneity, saw the growth of alternative and underground movements, a flourishing of nationalism on its Celtic fringe, the development of a civil rights movement among the minority Catholic population in Northern Ireland, and immigration from the Caribbean, the Asian sub-continent, and Africa. As youth culture became increasingly internationalized (or perhaps Americanized) it was possible for widespread international dissent and protest significantly to affect domestic politics, as happened with protests over the involvement of the United States in Vietnam and with the student protests in 1968.

A key feature of the British political scene since the Second World War has been popular extra-parliamentary political protest. In the 1950s, 1960s, and 1980s there were demonstrations protesting against nuclear weapons, in the 1960s against American policy in Vietnam, in the 1970s and 1980s against industrial relations conditions (most notably the Grunwick dispute in 1977, and the miners' strike during 1984 and 1985), and at the end of the 1980s against the community charge (or 'poll tax'), a form of personal taxation which replaced property tax as a method of funding local government. During the 1980s there was an increase in rioting. Disturbances on a scale unseen in Britain for generations occurred during 1981 in Brixton and Southall in London, in Toxteth in Liverpool, in the Moss Side area of Manchester, and in parts of the West Midlands. Similar outbreaks of disorder occurred later in the decade in Handsworth and Tottenham in 1985 and in Bristol in 1987.

These outbreaks of disorder in Britain were much less serious than the civil disturbances and conflict in Northern Ireland which re-emerged after 1968. The subsequent descent into political violence became a daily feature of life, occasionally spilling over into Great Britain with horrific results, such as the IRA bombings in Birmingham and Guildford during 1974, and the attack on Margaret Thatcher and senior Conservative Party politicians in Brighton in October 1984. The political and social context of policing has thus become increasingly fraught and controversy-ridden from the late 1950s, but especially so after the early 1970s. Not surprisingly, the role of the security services of the state has, at least in public perception, grown in importance.

The years since 1970 have seen also the collapse of the post-war political consensus. The collapse is difficult to pin-point precisely but the election of Margaret Thatcher to the leadership of the Conservative Party in 1975 is an important marker. All of these issues, and the pervasive question of Britain's role in Europe, became areas of substantial and increasingly bitter disagreement between the political parties, and indeed within the political parties during the 1970s and 1980s. In particular, the governments of Margaret Thatcher had a number of primary political and economic goals which distinguished them from previous governments of either political party allegiance: 'rolling back the frontiers of the state', deunionization, the reduction of inflation and public expenditures, the curbing of powers of local government, tax reform (including decreases in direct taxes on income), and an increase in the extent to which the market played a distributional role. A growing disillusionment with state intervention was particularly apparent in several of these policies.

Constitutional reform

Successive British governments were criticized as elective dictatorships and there arose a growing movement in favour of administrative and constitutional reforms in the United Kingdom. Many features of the British Constitution have been subject to the most acute criticism, and proposals were made for significant reforms. Several reforms were en-acted, including the introduction of the Parliamentary Commissioner for Administration (the Ombudsman) in 1967, followed later by ombudsmen for local government and the health service. The committee system of the House of Commons was modified to permit greater scrutiny by MPs of the administration. Judicial review by way of administrative law increased considerably in its scope from the mid-1960s. Several more far-reaching reforms were, however, advocated unsuccessfully, including the introduc-tion of proportional representation instead of the current 'first-past-the-post' electoral system; the reform of the House of Lords; parliamentary assemblies (or complete independence) for Scotland and Wales; devolu-tion of power to the English regions.[1] Many of these proposals were supported, at least in part, because civil liberties and human rights would be better protected as a result of these reforms.

The reform seen by some as most likely to improve the provision and protection of human rights in Britain, was the adoption of a domestic and entrenched Bill of Rights. Though Bills of Rights had earlier been proposed in the House of Lords as early as 1947, the 1968 Fabian pamphlet by Anthony Lester,[2] followed by the 1974 Hamlyn Lectures by Sir Leslie Scarman (as he then was),[3] mark the beginning of a campaign which continued throughout the period under study.[4] The debate con-centrated for much of the period on the value of giving effect to the European Convention on Human Rights through its incorporation into domestic law by legislation. Support for incorporation has been particu-larly apparent from groups and individuals, including several prominent judges,[5] who are outside the traditional conflict between the two major political parties. Several bills incorporating the Convention were success-ful in the House of Lords, but failed in the House of Commons indicating the scepticism which important sections of political and legal opinion have of proposals for a domestic Bill of Rights.[6]

More recently, however, two major sets of proposals for a domestic Bill of Rights advocated a mixture of ECHR and ICCPR provisions,[7]

---

[1] See, in general, R. Brazier, *Constitutional Reform: Reshaping the British Political System* (1991).

[2] A. Lester, *Democracy and Individual Rights* (1968). See further, A. Lester, 'Funda-mental Rights in the United Kingdom: The Law and the British Constitution', 125 *U. Penn. LR* 337 (1976).

[3] Sir L. Scarman, *English Law: The New Dimension* (1974).

[4] For a useful history of the debate, see M. Zander, *A Bill of Rights?* 3rd edn. (1985).

[5] In 1992, for example, the Master of the Rolls and the Lord Chief Justice both called for a domestic Bill of Rights, see *The Independent*, 1 December 1992.

[6] See e.g. K. D. Ewing and C. A. Gearty, *Freedom under Thatcher: Civil Liberties in Modern Britain* (1990).

[7] Institute for Public Policy Research, *A British Bill of Rights* (1990). For a more jurisprudential discussion, see R. Dworkin, *A Bill of Rights for Britain* (1990).

or a mixture of these together with other relevant international human rights provisions,[8] because of a belief in the inadequacy of the ECHR provisions standing alone. Advocates of a Bill of Rights also diverge significantly over which of several different approaches to judicial review is preferable, and in particular what effect the Bill should have on parliamentary sovereignty. Some emphasize a judicial role which is decisive, requiring that judges be responsible for enforcing rights against all legislative and government action.[9] Others would give the courts the power only to send back legislation for reconsideration, allowing Parliament to violate human rights if they chose to do so explicitly by ordinary legislation.[10] Yet others would effectively limit judicial review under the Bill of Rights only to striking down administrative action on the grounds of violation of human rights.[11] Perhaps the most complex set of proposals is that advocated by Liberty (formerly the National Council for Civil Liberties, NCCL)[12] which has proposed that while some rights such as freedom from torture would never be capable of being overridden by Parliament, other rights could be abrogated by legislation for a limited period of time only, and such an abrogation could be re-enacted only where a general election had intervened. A judicial decision that an Act of Parliament had infringed the Bill of Rights could be reversed by a two-thirds majority of a special cross-party parliamentary committee.

The emergence of rights' consciousness

A common feature of many of the international and domestic developments discussed above is the increased role which rights' consciousness played, particularly in providing the intellectual framework in which the developments and proposals were presented. Thus we see increasing use being made of the right to self-determination, the right not to be discriminated against, the right to leave or enter one's country, the right to freedom of expression, and so on. In tandem with the expanding use of rights-talk came the expansion of the nature of rights at issue. In Western Europe and America, traditional rights such as freedom of expression have often been interpreted as requiring government to desist from acting. New rights emerged, however, which required government to act positively, by providing the means by which these traditional rights could be

---

[8] Liberty, *A People's Charter: Liberty's Bill of Rights: A Consultation Document* (1991).

[9] IPPR, *British Bill of Rights*, though with the caveat that a special parliamentary majority might amend the Bill of Rights.

[10] Zander, *Bill of Rights?* 68–74. S. 33 of the Canadian Charter of Rights and Freedoms, instituting such a legislative override power, has been particularly influential in this regard.

[11] G. Robertson, *Freedom, the Individual and the Law*, 6th edn. (1989), 397 (discussing a Bill drafted by Lord Scarman).

[12] Liberty, *A People's Charter*.

made a reality, such as the right to work, the right to an adequate standard of living, the right to an adequate standard of nutrition, the right to be adequately housed, the right to a basic level of income, and so on. Government was expected at least to take a role in, and in some cases to become the primary mechanism for, the positive promotion of rights. In subsequent chapters, therefore, we do not restrict discussion to rights against the state and other public bodies. Where relevant, contributors discuss when and how the state may intervene to provide protection of the rights of individuals against other individuals. We consider, in other words, not just negative liberty, liberty from state interference, but also positive liberty, the positive role of state intervention in helping to further human rights.

The increase in rights consciousness not only contributed to the development of international and European human rights conventions, it also led to the enhancement of human rights provisions within the constitutions of several countries to which Britain felt particularly close. The most far-reaching development was the new interpretation by the United States Supreme Court of the Bill of Rights in the United States Constitution after the Second World War. Partly because of the perceived impotence of the federal and state legislatures and state courts, partly because of the odium which certain practices were bringing to the United States in international affairs, the United States Supreme Court exercised considerable judicial activism in the area of civil liberties and human rights. Particularly during the period when the Supreme Court was presided over by Chief Justice Earl Warren, case after case under the Bill of Rights invalidated laws and practices in the areas of racial discrimination, police powers, due process, freedom of speech, freedom of assembly, voting rights, freedom of religion, privacy, and women's rights. The Warren Court became identified by some as the role model for appropriate judicial activism in the cause of human rights, although more recently disillusionment with the Court's approach to human rights has set in among its erstwhile admirers. This judicial activism, coupled with the public and academic controversy which it engendered, meant that developments in the United States were followed closely in many countries of the world.

The perceived success of the Warren Court and the developments in international law discussed above, prompted several countries, which had previously followed the Westminster model of parliamentary democracy, to consider whether they should develop a Bill of Rights enforced by judicial review. Canada enacted a Bill of Rights in 1960 which proved something of a non-event. However, in 1982, as part of a larger constitutional settlement, the Canadian *Charter* of Rights and Freedoms came into force. New Zealand enacted a Bill of Rights in 1990 and

Australia almost did,[13] and may yet.[14] From 1945 onwards all of the newly emerging states which had formerly been under British control (except Israel) adopted a constitution with a Bill of Rights attached.

Each subsequent chapter considers how particular human rights have developed domestically. The major changes in legislation, the major case-law developments, and the principal methods adopted to protect, advance, or retard the particular liberty or right at issue are all considered. To the extent that there have been changes, the contributors consider what the primary methods have been by which these changes have occurred and the extent to which the major economic and social developments discussed above account for the changes which have taken place. We shall consider also how central to mainstream party political debate the issue has been, the role of pressure groups, the extent to which developments in other countries have been seen as particularly relevant, and with what effect. Many of the chapters also consider what major reform proposals have been put forward, by whom and with what results. Later, we consider to what extent comparisons and developments in these and other countries are seen in Britain as particularly relevant, and with what effect.

DEVELOPING AN INTERNATIONAL LAW OF HUMAN RIGHTS

The post-war growth in rights consciousness has contributed to the development of a veritable explosion of international agreements relating to the protection of human rights.[15] The Second World War marked a significant point of departure for a very clear reason. The sense of outrage, disgust, and guilt at the systematic practice of genocide committed during the war caused the victorious allies to prosecute Nazi leaders for war crimes before the International Military Tribunal at Nuremberg. But it also led to a desire that the international community should ensure that such gross violations of human rights would not occur again. These human rights have been elaborated under the auspices of several international organizations: the United Nations, the International Labour Organization, the Council of Europe, the European Community, and the Conference on Security and Co-operation in Europe.

---

[13] New Zealand Bill of Rights, 1990. See New Zealand Dept. of Justice, *A Bill of Rights for New Zealand: A White Paper* (1985).
[14] Australian Senate, Standing Committee on Constitutional and Legal Affairs, *A Bill of Rights for Australia?* (1985).
[15] P. Sieghart, *The International Law of Human Rights* (1983).

Developments under United Nations auspices

In 1945, the newly established United Nations was designed to be an important actor in protecting human rights.[16] The UN Charter provided that its purposes included 'promoting and encouraging respect for human rights and . . . fundamental freedoms for all'.[17] All UN members pledged themselves 'to take joint and separate action in co-operation with the Organization for the achievement of . . . universal respect for, and observance of, human rights'.[18] In 1946, after establishing a Commission on Human Rights, the UN's Economic and Social Council (ECOSOC) announced that the drafting of an international bill of rights would be the Commission's first priority. On 10 December 1948, the General Assembly adopted the non-binding, but morally and legally significant, Universal Declaration of Human Rights by a vote of forty-eight to none with eight abstentions. The day before, the Convention on the Prevention and Punishment of the Crime of Genocide had been approved by the General Assembly.

The Universal Declaration on Human Rights has little binding force on states, except to the extent that provisions in it may be considered part of customary international law. Attempts to go further, however, and provide a Convention under which those states which ratified it would be legally bound, together with effective enforcement machinery, soon became embroiled in the cold war confrontation which was to engulf much of international relations until 1989. The cold war also facilitated states' claims to sovereignty and non-interference in their internal affairs. Therefore work towards an international Bill of Rights proceeded slowly. No international Bill of Rights was agreed until 1966. All that could be achieved until then was agreement on specific, limited human rights conventions in particular areas where consensus could be achieved. For example, in 1951, the important Convention Relating to the Status of Refugees was concluded, later amended by a Protocol in 1966. The protection accorded by the Convention was supplemented by the establishment of the office of the United Nations High Commissioner for Refugees (UNHCR) by the General Assembly in 1951. The UNHCR's role is to provide protection to refugees falling within the scope of the 1951 Convention and the 1966 Protocol, as well as refugees in countries not party to the Convention or Protocol, and those not yet recognized as refugees by a state.

The institutional vacuum created by the absence of an international

---

[16] For an assessment, see P. Alston (ed.), *The United Nations and Human Rights: A Critical Approach* (1992).

[17] Art. 1(3).

[18] Art. 56 with Art. 55.

Bill of Rights covering the broad range of human rights issues prompted the institutions set up by the Charter to develop a human rights role. The ECOSOC and the Commission on Human Rights established by ECOSOC predated the Universal Declaration on Human Rights. The Human Rights Commission set up working groups and commissioned special rapporteurs on many issues, but the principal subsidiary of the Commission is the Sub-Commission on the Prevention of Discrimination and the Protection of Minorities, which in its turn has established numerous working groups and special rapporteurs. From the late 1960s, a mechanism was established to receive communications or petitions from individuals alleging gross violations of human rights, under the provisions of ECOSOC Resolution 1235 of 1967 and ECOSOC Resolution 1503 of 1970.[19]

These bodies, however, have not lived up to the hopes which some held for them. ECOSOC has been described as 'a quintessential political body' with its members, elected by the General Assembly, being formal representatives of UN member states. So too are the members of the Commission on Human Rights. Writing in 1988, Rosalyn Higgins declared that the Human Rights Commission 'is a body in which supporting one's friends' record often assumes priority along with attacking the human rights of those who are not one's friends'.[20] '[The Sub-Commission] too', according to Higgins, 'is a fairly political body in a rather robust sense, while still putting out some very serious and worthwhile studies'.[21] Farer, among others, has criticized the petition procedure in biting terms.[22] These institutions have, however, been highly marginal in their impact on domestic human rights developments in the United Kingdom and little more will be said of them.

Instead of one convention being agreed in 1966 covering the broad range of human rights, two conventions were drafted, one (the ICCPR) dealing with civil and political rights (to which the Western bloc was attracted),[23] and another (the International Covenant on Economic, Social, and Cultural Rights, ICESCR) covering economic, social, and cultural rights (to which the Eastern bloc was attracted). States were able to ratify either convention or both. The United Kingdom has ratified both Covenants. In this book, however, we concentrate only on rights derived

---

[19] See further Ton J. M. Zuijdwijk, *Petitioning the United Nations: A Study in Human Rights* (1982).
[20] R. Higgins, 'United Nations Human Rights Committee', in R. Blackburn and J. Taylor, *Human Rights for the 1990s: Legal, Political and Ethical Issues* (1991), 67.
[21] Ibid.
[22] T. J. Farer, 'The UN and Human Rights: More than a Whimper, Less than a Roar', in A. Roberts and B. Kingsbury, *United Nations, Divided World: The UN's Roles in International Relations* (1988), 95, at 119–33.
[23] See L. Henkin (ed.), *The International Bill of Rights: The Covenant on Civil and Political Rights* (1981).

from the ICCPR, and not those derived from the ICESCR, except where they overlap.[24]

The circumstances in which the Covenants were agreed means that the enforcement procedures for both Covenants are weak. Although the ICCPR and the ICESCR are binding on ratifying states in international law, these states have pledged themselves to nothing more onerous than providing a periodic report on national compliance. In addition to these reporting procedures, states ratifying the ICCPR have two further options regarding enforcement procedures which they may exercise. First, there is the option of recognizing the jurisdiction of the Covenant's Human Rights Committee (HRC) to hear complaints from other states that have also accepted this procedure. The Committee may hold hearings and promote a friendly settlement. Second, the state may recognize the Committee's authority to hear petitions from individuals alleging violations of the Covenant by that state and adjudicating upon that petition. This second option is available by way of an optional protocol. The United Kingdom has accepted the first (inter-state) procedure, though in practice it has yet to be put into operation as regards any state. The United Kingdom, however, has neither signed nor ratified the optional protocol permitting the HRC to deal with individual communications regarding the United Kingdom, on the grounds that 'in some respects it compares unfavourably from the individual's standpoint' with the equivalent procedure under the European Convention on Human Rights, to which the United Kingdom has acceded.[25] As regards the United Kingdom, therefore, the implementation of the ICCPR is in practice confined to periodic reports which the United Kingdom must make to, and be questioned about, by the Committee.

Reports under the ICCPR are transmitted through the Secretary-General to the Human Rights Committee, which is authorized to study and thereafter transmit reports to the states parties and to ECOSOC, along with 'such general comments as it may consider appropriate'.[26] The Human Rights Committee, composed of independent experts, has been in operation since 1977.[27] In contrast, the practice for much of the period since the ICESCR came into force was for reports under the Covenant to be submitted to ECOSOC for consideration in accordance with the provisions of the Covenant. ECOSOC in turn transmitted them to the Human Rights Commission (set up under the Charter) 'for study and

---

[24] For a perspective on economic, social, and cultural rights, see R. Beddard and D. M. Hill (eds.), *Economic, Social and Cultural Rights: Progress and Achievement* (1992).

[25] Hansard HC Debs. 962, written answers, col. 262, 8 Feb. 1979. This view was subsequently affirmed, see HL Debs. 495, col. 591, 28 Mar. 1989.

[26] Art. 40.

[27] For a discussion of its operation, see D. McGoldrick, *The Human Rights Committee* (1991).

general recommendation'.[28] In May 1986, however, a new committee of independent experts was established to assist the ECOSOC in monitoring states parties' compliance with their obligations under the ICESCR.[29]

The United Kingdom has submitted three reports to the Human Rights Committee under the ICCPR, in 1977, 1984, and 1989. The 1977 report,[30] the first to be submitted by the UK government, is notable for the explanation given of the legal framework within which rights are protected and remedies sought in the UK. The introduction to the report by the UK representative to the Committee relies heavily on the traditional exposition of the methods by which civil liberties are said to be protected in Britain.

1. Sir James Bottomley (United Kingdom) . . . Since his country's law had developed so long ago, the means by which it ensured the protection of fundamental rights differed from those adopted in many other States parties to the Covenant. The United Kingdom had no written constitution. It had an omnicompetent Parliament with absolute power to enact any law and to change any previous law. The courts have not, at least in recent times, recognized any higher legal order by reference to which acts of Parliament could be held to be void.

2. The constitutions of many other countries contained a bill of rights, which could be altered only by some special constitutional procedure. In his country's law, however, there was no similar code of rights, but there were specific sets of reciprocal rights and duties and civil remedies or criminal prohibitions. Some of the rights guaranteed by the Covenant, such as the right to life recognized in article 6 of the Covenant, were implied in United Kingdom law, which made any act that would interfere with those rights unlawful. Other rights, such as the right to freedom of expression recognized in article 19 of the Covenant, were secured by the absence of any legal inhibition on freedom of action or by the limitation of such action to specific and defined situations.

3. Another difference between his country and many other countries was that it recognized no distinction between public law governing the actions of the State and private law governing relations between citizens. In addition, the United Kingdom had no separate and systematized code of administrative law, although there were arrangements for dealing with individual grievances against the administration through the office of the Parliamentary Commissioner for Administration and through commissioners who exercised similar functions in relation to the acts of local authorities and the National Health Service.

4. In the United Kingdom, there was no principle by which international treaties and conventions automatically became part of domestic law. His country's practice was to consider, before ratification of an instrument, whether its domestic law adequately fulfilled the obligations it was about to assume and, if it did not, to alter the law so that it conformed to those obligations. Consequently . . . the

---

[28] Art. 19.
[29] See 'UN Committee on Economic, Social and Cultural Rights', (1989) 42 *ICJ Review* 33.
[30] CCPR/C/1/Add. 17, Add. 35, Add. 37, and Add. 39.

Covenant did not of itself have the force of law in the United Kingdom, whose ability to ratify the Covenant had rested upon the fact that the rights recognized in the Covenant were already guaranteed by law, subject to the reservations which had been made upon signature or ratification.[31]

An important element of the reporting procedure is the scrutiny of the report by the Committee and the questions put to the Government's representative. Many Committee members took the view that the 1977 UK report espoused general principles and statements of intent but fell short on detail, on elucidating the legal status of the Covenant and on the mechanisms for redress.[32] Interestingly, in 1977 the Government was able to take some pride in announcing to the Committee its record under the European Convention, thus indicating that this was a method of validating its record which was in favour at that time.

The report provoked close questioning on various issues which have been of concern to civil liberties groups in Britain.[33] Questioning on discrimination and on equality of opportunity figured prominently: one member, deprecating the lack of detail in the report, requested more information on the participation of women in public life and in the judiciary, and queried whether higher unemployment rates among Catholics in Northern Ireland were evidence of discrimination against them. There was questioning on whether the UK government's immigration policy was racially discriminatory, on the conditions of proof in race discrimination cases, on prison conditions in the wake of protests by prisoners, and on the status and conditions of prisoners held on remand.

By the time of the 1984 Report[34] the Government was able to report on several important pieces of legislation which had a bearing on civil rights and liberties, including the British Nationality Act (discussed in detail by Dummett in Chapter 11), the Police and Criminal Evidence Act (see Reiner and Leigh, Chapter 3), the Data Protection Act (see Michael, Chapter 9), and the various acts affecting employment rights and trade unions affairs (see Ewing, Chapter 8) which had become a hallmark of the early years of the Thatcher administrations. In addition, the record of the UK government under the European Convention was by this time less favourable. By 1985 the European Court had issued judgments in several cases. Four issues dominated the discussion in the 1984 report: due process concerns and police powers; prisoners' correspondence; freedom

---

[31] CCPR/C/SR 67, p. 2.

[32] *Yearbook of the Human Rights Committee*, 1977–8, vol. i. *United Nations* (1986), 237–51.

[33] For consideration by the Committee, see CCPR/C/SR 67, 69, and 70; SR 147–9; SR 161–2; SR 164.

[34] See CCPR/C/32/Add. 5, Add. 14, and Add. 15. For its consideration by the Committee, see CCPR/C/SR 593–8; and SR 855–7.

of association concerns with regard to trade unions, including issues surrounding closed shop agreements; and citizenship rights and entitlements to enter the UK.

The 1989 report[35] is a document of a different order from the previous two. Leaving aside its greater length, it differed in several important ways. It acknowledged the development of several important additional methods of protection which supplemented the (impliedly) inadequate traditional approaches, in particular the increasing influence of the European Commission and Court of Human Rights, and of judicial review by the domestic courts of administrative action. In addition, perhaps as a result of its experience before the European Commission and Court, the Government appeared to be less complacent in its approach and the report acknowledged four areas in which 'human rights and personal freedom are at present a matter of particular concern'.

1. On the prevention of terrorism the report states that 'the rights of the individual have to be balanced by the protection and safety of the public'.

2. Regarding immigration and asylum claims it is stated that it is 'necessary to operate tight controls and apply rigorous investigative procedures which may involve inconvenience or distress for the individuals concerned'.

3. Direct and indirect discrimination on grounds of race and sex 'is still practised . . . in ways which are difficult to identify and even more difficult to correct'.

4. There are problems about conducting fair and impartial investigations into alleged miscarriages of justice in the above three areas and into 'complaints of misconduct by public officials'.

The acknowledgement that there were dilemmas and shortcomings in these areas is a departure from the more self-satisfied approach of earlier reports.

The 1989 report also varies from the two previous reports in its much greater reliance on and use of statistical and research material, particularly material produced by government departments. This may have been a response to criticisms of the lack of detail in the 1977 report. Thus, comparative statistical data on the following issues is presented for the years intervening since the 1985 report: use of firearms by the police; cases referred to the Police Complaints Authority; detentions under deportation orders and of illegal entrants; removals under deportation orders and under illegal entry powers. In addition, there is statistical data on the size of the ethnic minority population, on ethnic minorities in the criminal justice system, and in employment; on the prison population; on

---

[35] CCPR/C/58/Add. 6.

cases of discrimination handled by the Commission for Racial Equality; and on women in public life. These two new features in the 1989 report are of course linked: the publication of such statistical material was designed to back up the recognition that problems such as race and gender discrimination were more entrenched than had previously been acknowledged and to show that the government was initiating research as part of a serious effort to address the issues.

When considering a government's periodic report, the members of the Human Rights Committee may receive information from non-governmental organizations and pressure groups who take an interest in human rights issues. This information may be taken into consideration by members of the Committee when it questions the government representatives, although it is rare for the Committee to refer to the organization by which it was submitted during the deliberations.[36] Information was supplied to the Committee concerning the 1989 UK government report by several non-governmental organizations concerned with human rights, among others Liberty, Amnesty International, Article 19, Charter 88, the Howard League for Penal Reform, JUSTICE, the Joint Council for the Welfare of Immigrants, and the Campaign for the Administration of Justice.

In addition to the general international covenants just discussed, numerous other agreements have been drawn up under the auspices of the United Nations. The most important, listed chronologically, are included in Appendix B, including several which will be mentioned in subsequent chapters, together with the date signed, the date the treaty came into force, and whether the United Kingdom has ratified the Treaty. It should be noted that similar reporting procedures to those of the ICCPR exist for parties to six other major international human rights agreements, concluded under the auspices of the United Nations.[37] Of these, we can note in particular the reporting procedures of the International Convention on the Elimination of All Forms of Racial Discrimination,[38] the

---

[36] McGoldrick states: '[t]he general, though not invariable, practice of members . . . has been not to refer directly to the source of their information . . . This general practice makes it difficult to determine the sources and the extent of the use made of outside information.' However, he goes on to say, '[t]he situation above has changed significantly since 1986. The effects of glasnost have reached the HRC and many members specifically cite sources, have good relations with non-governmental organisations, and receive a wide range of material . . . The question of the use of sources by HRC members is no longer an issue in the HRC.' McGoldrick, *Human Rights Committee*, 78–9.

[37] These are the International Covenant on Economic, Social, and Cultural Rights, the International Convention on the Elimination of all Forms of Racial Discrimination, the International Convention on the Supression and Punishment of the Crime of Apartheid, the Convention on the Elimination of All Forms of Discrimination Against Women, and the Convention Against Torture and Cruel, Inhuman or Degrading Treatment or Punishment. See further, J. L. Gomez del Prado, 'United Nations Conventions on Human Rights' (1991) 7 *HRQ* 492–513; T. Meron, *Human Rights Law-Making in the United Nations* (1986).

[38] See N. Lerner, *The UN Convention on the Elimination of all Forms of Racial Discrimination*, 2nd edn. (1980).

Convention on the Elimination of All Forms of Discrimination Against Women, and the Convention on the Rights of the Child.[39]

Notwithstanding limitations in its methods of operation and bearing in mind that the first periodic report by the UK under the ICCPR was made as recently as 1977, the work of the Human Rights Committee (and of other equivalent supervisory bodies) is clearly of relevance in furnishing information on the basis of which it might be possible to assess progress and advancement in the protection of civil rights and liberties in the UK. Such information is essential if we are to be able to assess the extent to which human rights and civil liberties are protected. The periodic government reports and the accompanying documentation from non-governmental organizations provide an accumulating body of evidence which will assist scholarship in this area in the future. Consideration is given in the final chapter of this volume to further sources of data that might be available for carrying out an assessment of the UK's human rights record and to the pitfalls and problems that exist in measuring human rights infringements and in detecting and classifying progress in this area.

Under the auspices of the International Labour Organization[40]

Several subsequent chapters consider human rights standards set by the International Labour Organization, as well as by the United Nations. The ILO was originally established in 1919, as an international organization with a wide mandate in the area of employment welfare and workers' rights.[41] Since the Second World War it has become an authoritative international human rights body, though in a limited area.[42] Over 150 states are currently members of the ILO, including the United Kingdom.[43] One of the ILO's main organs is the General Conference, which attempts to influence employment standards by preparing and adopting conventions and recommendations as well as providing a more general forum for international discussions of such issues. The ILO conventions most relevant for subsequent chapters, listed chronologically, are included in Appendix B, together with the date signed, the date the treaty came into force, and whether the United Kingdom has ratified. The Conference in turn elects members of the Governing Body which has itself two standing committees which consider human rights matters: the ILO Freedom of Association Committee (on issues relating to infringement of trade union

[39] See M. S. Pais, 'The Committee on the Rights of the Child', 47 *ICJ Review* (1991), 36.
[40] This discussion draws on E. Lawson, *Encyclopedia of Human Rights* (1991), 964–5.
[41] A. Alcock, *History of the International Labour Organisation* (1971).
[42] See C. W. Jenks, *Human Rights and International Labour* (1960).
[43] K. D. Ewing, *Britain and the ILO* (1989).

rights) and the ILO Committee of Experts on the Application of Conventions and Recommendations (which supervises adherence to several ILO instruments which concern human rights issues). In addition, the Governing Body has power to establish a committee to examine complaints relating to discrimination in employment.

Under the auspices of the Council of Europe

The Council of Europe was founded in May 1949, reflecting the desire for closer co-operation among European states. By June 1992 it comprised twenty-seven member states, including the twelve member states of the European Communities and four of the former communist Eastern and Central European states (Hungary, Poland, Bulgaria, and Czechoslovakia).[44] One of its most important accomplishments has been its role in formulating a series of treaties open for ratification by members of the Council of Europe (and sometimes beyond). Of these treaties by far the most significant is the European Convention for the Protection of Human Rights and Fundamental Freedoms, usually abbreviated to the European Convention on Human Rights (ECHR), which, as we have seen, was agreed in 1950, ratified by the United Kingdom in 1951,[45] and came into force on 3 September 1953.[46] By March 1992 twenty-four of the twenty-seven member states of the Council of Europe had ratified the ECHR (Hungary, Poland, and Bulgaria had not ratified).

The Convention drew on the Universal Declaration of Human Rights, but the rights protected were less extensive. The Convention includes within its protection the classic civil and political rights, including the right to life, freedom from torture, freedom from slavery, due process rights, freedom of association, freedom of assembly, freedom of speech, protection of privacy and the family, and freedom from the discriminatory application of these rights and freedoms. In contrast with the ICCPR, however, it does not specifically include protection of minority group rights, and the protection from discrimination provided under the European Convention is considerably weaker in its scope, issues to which we return in subsequent chapters.

---

[44] For discussion of further geographical enlargement of the Council of Europe in the future, see Parliamentary Assembly of the Council of Europe, *The Geographical Enlargement of the Council of Europe: Policy Options and Consequences*, repr. in (1992) 13/5–6 *HRLJ* 230.

[45] For a discussion of the Cabinet debate on whether to sign, see A. Lester, 'Fundamental Rights: The United Kingdom Isolated?' (1984) *PL* 46.

[46] For general discussions of the substance and procedure, see J. E. S. Fawcett, *The Application of the European Convention on Human Rights*, 2nd edn. (1987); P. van Dijk and G. J. H. van Hoof, *Theory and Practice of the European Convention on Human Rights*, 2nd edn. (1990).

The typical method by which the European Convention proceeds is to state the substantive right in apparently absolute terms in the first paragraph of the relevant article, and then in the second paragraph to include several important limitations on that right. The right to freedom of expression is a fairly typical example of this approach.

ARTICLE 10

1. Everyone has the right to freedom of expression. This right shall include freedom to hold opinions and to receive and impart information and ideas without interference by public authority and regardless of frontiers. This article shall not prevent States from requiring the licensing of broadcasting, television or cinema enterprises.

2. The exercise of these freedoms, since it carries with it duties and responsibilities, may be subject to such formalities, conditions, restrictions or penalties as are prescribed by law and are necessary in a democratic society, in the interests of national security, territorial integrity or public safety, for the prevention of disorder or crime, for the protection of health or of morals, for the protection of the reputation or the rights of others, for preventing the disclosure of information received in confidence, or for maintaining the authority and impartiality of the judiciary.

The framing of the basic right and the paragraph 2 exceptions thus gives considerable latitude to both the European Commission on Human Rights and the European Court of Human Rights in their interpretation of the Convention. Not surprisingly, perhaps, these institutions were at first noticeably cautious in upholding complaints by applicants, usually allowing a considerable 'margin of appreciation' to the state in deciding the appropriate extent of the right and of the limitations. The institutions tended to accept states' justifications, partly in order to increase the confidence of those states which were still highly sensitive to considerations of national sovereignty in the reasonableness of the Convention's institutions. This 'confidence building' period has since given way to a somewhat more robust approach, as can be seen from a reading of Chapter 6 on prisoners rights and Chapter 15 on sexual orientation discrimination.

In addition to the so-called 'paragraph 2 exceptions', there are two general exceptions which may be drawn on by a member state. The first general exception is set out in Article 15, which provides that in certain emergency situations, a state may derogate from its obligations. Such a derogation is only permitted, however, with regard to certain articles. No such derogation is permitted, for example, with regard to the right to be free from torture. The second general exception is provided by Article 17 which provides that the protections guaranteed in the Convention may not be used by groups or individuals in such a way as to undermine the liberal democracy which the Convention seeks to safeguard.

The Convention further provides that the list of rights and freedoms may be added to by means of additional protocols, which in turn may be adhered to by a member state. There is no requirement that a state which adheres to the main Convention should adhere to these additional rights and freedoms. Ten additional protocols have been agreed which supplement the original text. All but Protocols No. 9 and No. 10 were in force by the beginning of 1993. These protocols are of two types. The first type adds to the list of rights protected by including, for example, a right to peaceful enjoyment of one's possessions (Protocol No. 1), a right to education (Protocol No. 1), a right to enter and leave one's country (Protocol No. 4), and the abolition of the death penalty (Protocol No. 6). The second type of protocol modified the procedural and institutional arrangements in the original text in particular respects; the most important will be mentioned below.

As regards enforcement mechanisms, the Convention provides for the establishment of a European Commission of Human Rights and a European Court of Human Rights which are located in Strasbourg. A simplified diagram of the procedure is shown in the figure below. The Convention provides that complaints of breaches of the Convention by a state party may be made to the European Commission of Human Rights either by another state party (as in the case brought by Ireland against the United Kingdom concerning Northern Ireland, see Chapter 5), or by an individual victim, if the state complained against has expressly recognized the right of individual petition. The United Kingdom originally recognized the competence of the Commission to receive individual petitions in 1966. This acceptance was renewed for a further period of five years commencing on 14 January 1991.

In 1991, a total of 1,648 applications were registered by the Commission. After registering a complaint, the Commission must first decide whether an application is admissible. One important consideration at this stage is whether the applicant, who must not be anonymous, has exhausted the domestic remedies available under the state's domestic law, a point to which we return below. In addition, the application must have been made within six months of a final decision by the courts or authorities of that state. Currently, of all the applications investigated by the Commission, about 10 per cent are declared admissible. In 1991, a total of 217 applications were declared admissible by the Commission. If the Commission (or, in certain circumstances since the coming into force in January 1990 of Protocol No. 8, a *Chamber* of the Commission) finds the application admissible, it must then attempt to establish the facts and attempt to secure a friendly settlement on the basis of respect for the human rights guaranteed in the Convention between the applicant and the state. If no friendly settlement is reached, then a report is prepared, setting out the facts as found by the Commission and containing the Commission's

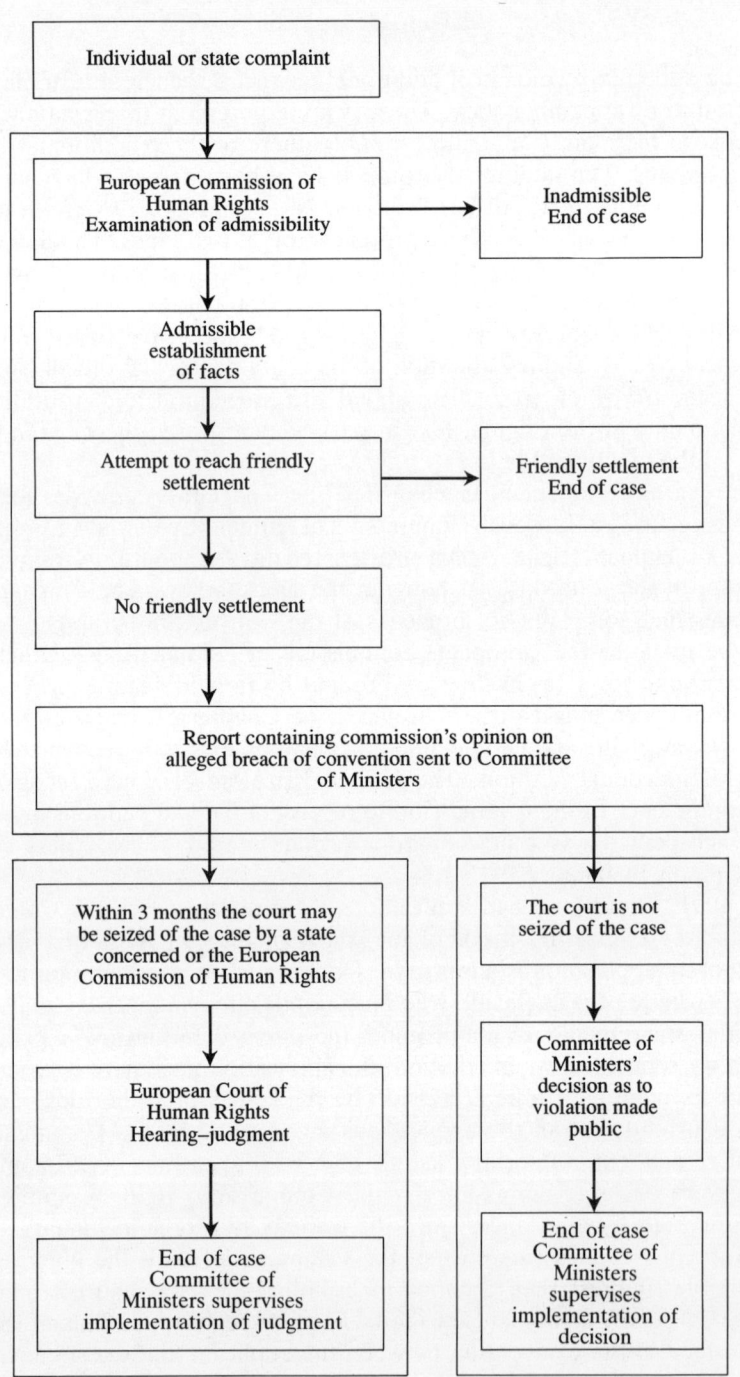

How the European Convention human rights institutions work
Source: *The Council of Europe and the Protection of Human Rights* (Strasbourg, 1990), 15.

opinion on whether the Convention has been violated. The report is initially sent only to the state or states involved and to the Committee of Ministers of the Council of Europe, a political body.

The case may be referred, within three months, to the European Court of Human Rights either by the Commission, or by a state party, if the defendant state has accepted the compulsory jurisdiction of the Court. (The United Kingdom originally recognized the competence of the Court in 1966. The jurisdiction of the Court was recognized for a further period of five years commencing on 14 January 1991.) If the case has not been referred to the Court within three months of the Committee of Ministers first receiving the report of the Commission, the Committee decides whether there has been a violation. Article 32 provides that the Committee decides by a majority of two-thirds of the members entitled to sit on the Committee. (Protocol No. 10 would amend Article 32 to enable the Committee to take such decisions by means of a simple majority.)

Between 1959 (when the Court was established) and 1990, the Court has had 252 cases referred to it. In the proceedings before the Court, the Commission does not act as a party to the case but presents its opinion. The Convention does not allow individual applicants to refer a case to the Court. The Rules of Court do, however, permit an applicant to indicate that he wishes to take part in proceedings once a case has been referred by a state or the Commission. Protocol No. 9, which as we have seen was not in force by the beginning of 1993, will enable individuals to refer their case to the Court after it has been examined by the Commission. The Rules of the Court permit another person or body to submit arguments as *amicus curiae*. The Court, normally after a public hearing, gives a reasoned judgment in open court which is transmitted to the Committee of Ministers which supervises its execution.

When the Convention institutions may legitimately take up a case is set out in Article 26, which provides that the Commission may only deal with individual complaints 'after all domestic remedies have been exhausted, according to the generally recognised rules of international law'. Article 13 provides that everyone whose rights and freedoms set out in the Convention are violated 'shall have an effective remedy before a national authority notwithstanding that the violation has been committed by persons acting in an official capacity'. Subsequent chapters consider the effects of the United Kingdom's membership of the Council of Europe in general, and its ratification of the European Convention on Human Rights in particular. We shall examine the use made in the domestic courts of the Convention. Particular attention is paid to recourse to the European Convention institutions, and the effects of such recourse on British law.

In addition to the European Convention on Human Rights, several other important agreements have been concluded under the auspices of the Council of Europe. Some of the more important of these (as well as the ECHR and its various protocols) are listed in Appendix B, together with details of ratification by the United Kingdom, and the date when the Conventions and their protocols came into force. Two are considered in depth in subsequent chapters.[47] First, in 1961, the European Social Charter was signed, coming into force in 1965. The Charter and an additional protocol set out several social and economic principles covering such areas as the right to work, the right to bargain collectively, and to strike, for example.[48] States adhering to the Charter (twenty had done so by 1992) commit themselves to comply with the Charter. Their progress in doing so is monitored through a rather inadequate supervisory machinery including, *inter alia*, a Committee of Independent Experts.[49] A new protocol amending the Charter's provisions for supervision was agreed in October 1991 but had not come into force by 1992.[50] Second, in 1989, the European Convention for the Prevention of Torture and Inhuman or Degrading Treatment or Punishment came into force. This Convention provides additional safeguards for those detained in prison or elsewhere. A Committee has been established which is empowered to visit places of detention and to make recommendations for improving the conditions of those held therein.[51]

Under the auspices of the European Communities

The Treaty of Rome established the European Economic Community in 1957. Originally the members of the Community were France, West Germany, Italy, Belgium, the Netherlands, and Luxembourg. The United Kingdom, Ireland, and Denmark acceded to the Community in 1972. Greece became a member in 1981, and Portugal and Spain in 1986. The Treaty of Rome had both a political and an economic rationale. It was aimed both at increasing closer political co-operation in Europe in the longer term ('to lay the foundation of an ever closer union among the peoples of Europe'), in part, by means of achieving a common market for goods and services, and by co-operation in economic policy. As part of

[47] The European Commission for Democracy through Law has proposed an additional European Convention for the Protection of Minorities, repr. in (1991) 12 *HRLJ* 269.

[48] D. J. Harris, *The European Social Charter* (1984).

[49] D. J. Harris, 'The System of Supervision of the European Social Charter: Problems and Options for the Future', in L. Betten, D. Harris, and T. Jaspers (eds.), *The Future of European Social Policy* (1989). See further, P. O'Higgins, 'The European Social Charter', in R. Blackburn and J. Taylor (eds.), *Human Rights for the 1990s* (1991), 121.

[50] For text see (1992) 13 *HRLJ* 179.

[51] See *First General Report of the European Committee for the Prevention of Torture and Inhuman and Degrading Treatment or Punishment*, Jan. 1991, repr. in (1991) 12 *HRLJ* 206.

the attempt to construct a common market, the Treaty provided for the free movement of workers of the member states within the common market.[52] This provision has subsequently been supplemented with several regulations and directives.[53]

There was an economic rationale for another Treaty provision, which provided for equal pay for men and women. This provision is one of several in the Treaty in the area of social policy, and these other articles set a context in which Article 119 operates. Article 117, for example, specifies that member states are agreed on the need 'to promote improved working conditions and an improved standard of living for workers'. Article 118 gives the Commission the task of promoting close co-operation between member states in the social field, particularly in matters relating to employment, labour law and working conditions, vocational training, social security, prevention of occupational accidents and diseases, occupational hygiene, and the right of association. The equal pay provision was inserted because without it some countries with national equal pay laws feared that they would be at a competitive disadvantage to those member states without such laws.[54] Several significant directives have also expanded the right to non-discrimination on grounds of sex beyond pay, into other areas connected with employment such as access to employment, working conditions, dismissal, and pensions.[55] Despite the primarily economic rationale for the equal pay provision, the non-discrimination requirement has become part of what might be described as the 'fundamental rights' aspect of the Community, an aspect which has become more prominent because of the activities of Community institutions, institutions in the member states, and concerned individuals.[56]

---

[52] Article 48 EEC.

[53] See especially Regulation 1612/68 on freedom of movement for workers within the Community, OJ L257, 19 Oct. 1968, 2; and Council Directive 68/360/EEC on the abolition of restrictions on movement and residence within the Community for Workers of member states and their families, OJ L257, 19 Oct. 1968, 13.

[54] C. Docksey, 'The European Community and the Promotion of Equality', in C. McCrudden (ed.), *Women, Employment and European Equality Law* (1987), 3.

[55] Council Directive 75/117/EEC of 10 Feb. 1975 on the approximation of the laws of the member states relating to the application of the principle of equal pay for men and women; Council Directive 76/207/EEC of 9 Feb. 1976 on the implementation of the principle of equal treatment for men and women as regards access to employment, vocational training and promotion and working conditions; Council Directive 79/7/EEC of 19 Dec. 1978 on the progressive implementation of the principle of equal treatment for men and women in matters of social security; Council Directive 86/378/EEC of 24 July 1986 on the implementation of the principle of equal treatment for men and women in occupational social-security schemes; Council Directive 86/613/EEC of 11 Dec. 1986 on the application of the principle of equal treatment between men and women engaged in an activity, including agriculture, in a self-employed capacity, and on the protection of self-employed women during pregnancy and motherhood.

[56] An ambitious study of the position of human rights in the Community has been carried out at the European University Institute, see A. Cassese, A. Clapham, and J. Weiler (eds.),

The Single European Act, a treaty concluded between the member states in 1986 and in force from 1 July 1987, introduced further provisions to achieve a more effective internal market 'comprising an area without internal frontiers in which the free movement of goods, persons, services and capital is ensured'. In conjunction with these efforts to achieve a more effective internal market, all but the United Kingdom agreed to a Community Charter of the fundamental social rights of workers on 9 December 1989 to accompany the economic measures. We shall consider these developments in greater detail in several of the following chapters, particularly those concerning women's rights, and immigration and nationality.

The institutions of the Community comprise the Commission, the Council, the European Parliament, the European Court of Justice (ECJ), and the Court of First Instance. The ECJ has consistently held that valid Community law has supremacy over any national law which conflicts with it. Community law comprises the law embodied in the Treaty (and amending Treaties), together with the secondary legislation promulgated by the Commission and Council. Generally speaking, the Council legislates on a proposal from the Commission. Secondary legislation of this type may take the form of regulations or directives. Regulations are directly applicable in all member states and have effect automatically in the member states irrespective of whether the member state has taken any action to implement it. Nor may member states purport to implement them in different terms. Directives are binding 'as to the result to be achieved upon each Member State to which it is addressed, but shall leave to the national authorities the choice of form and method', using appropriate methods of national implementation.

Under Article 169 of the EEC Treaty, the European Commission has the right to initiate infringement proceedings in its own name against a member state before the ECJ, alleging that the member state is not complying with European legal requirements. In practice, however, the main mechanism by which issues of European law raised in the domestic context get to European institutions is by the reference procedure under Article 177 of the Treaty which enables any court or tribunal in a member state to refer to the ECJ questions as to the interpretation of Community law. If the domestic court is the final court of appeal on an issue, the court must refer the European law question to the ECJ, unless the European law is clear.

The question of when Community law gives rise to rights and obligations that can be enforced by individuals in national courts is an issue of

Human Rights and the European Community: Methods of Protection (1991); A. Cassese, A. Clapham, and J. Weiler (eds.), *Human Rights and the European Community: The Substantive Law* (1991); A. Clapham, *Human Rights and the European Community: A Critical Overview* (1991).

Community law. The Court has held that Treaty Articles and Regulations may be 'directly effective' in certain circumstances. Where Community law is directly effective it can be used in the domestic courts by an individual, to establish a self-standing right, enforceable by a domestic court, against other individuals as well as against member states. Article 119 of the Treaty, providing for equal pay for men and women, has such direct effect. To the extent that the ECJ gives Article 119 a wide interpretation, the prospects of using EC law as directly effective against other individuals is thereby increased. However, much of the EC law to be found in subsequent chapters is to be found in the form of directives. Such directives have been held by the Court only to have 'vertical direct effect'; that is, they may only be relied upon against an organ of the state.[57] The potential use of such directives has been broadened by the expansive approach taken by the ECJ as to what constitutes a state body. The ECJ has held that the test is whether the body was 'made responsible, pursuant to a measure adopted by the State, for providing a public service under the control of the State and had for that purpose special powers beyond those which result from the normal rules applicable in relations between individuals'.[58]

These limitations do not mean, however, that directives can be ignored in domestic litigation in circumstances where a public body in this sense is not a party. Directives have, even in these circumstances, what might be called 'indirect effect'. The ECJ has held that national legislation which is implementing an EC directive has to be interpreted by domestic courts in light of that directive. A national court is obliged to interpret national law as far as possible in accordance with rules laid down by Community law, whether prior or subsequent.[59] Any ambiguities or uncertainties (terms which have a very broad meaning for the ECJ in this context) in domestic legislation, irrespective of when it was enacted, have to be resolved in favour of European law, if the directive indicates that European law would come to a different answer.

The Court has proven itself innovative not only in the context of the enforcement of Community law, but also in holding in several cases that the general principles of Community law include protection for the human rights which are included in the European Convention on Human Rights and are part of the common constitutional traditions of the states.[60] Community law, then, has two different concepts relevant for a

---

[57] *Marshall* v. *Southampton and SW Hampshire Area Health Authority*, Case 152/84, [1986] QB 401 (ECJ); [1986] 2 ECR 723.

[58] *Foster* v. *British Gas, plc.*, Case C188/89, [1991] 1 QB 405 (ECJ).

[59] *Marleasing* v. *La Comercial Internacional de Amnetacion*, Case C106/89 [1990] ECRI 4135.

[60] See M. A. Dauses, 'The Protection of Fundamental Rights in the Community Legal Order' (1985) 10 *European LR* 398.

discussion of the place of human rights in the Community. We have seen earlier that certain rights (such as equal pay) have been considered 'fundamental rights' and these have supremacy over national law. In addition, we have now introduced a second concept of human rights as part of the general principles of Community law. This second principle does not mean that any human rights violation by a member state is in breach of Community law. Rather, the human rights content of these general principles is drawn on by the ECJ for the specific purpose of limiting the activities of Community institutions, and as principles of interpretation for Community laws.

For the past two decades there has also been intermittent consideration given to whether the Community as a whole should accede to the European Convention on Human Rights,[61] but so far without this occurring. The European Parliament adopted a Declaration of Fundamental Rights and Freedoms in April 1989 'with the goal of securing its eventual adoption by other Community institutions and inclusion in the Treaties and in due course in a Treaty of European Union'.[62]

In addition to the formal decision-making institutions and structures of the original Treaty, the member states of the Community meet regularly outside the context of the formal structures. One mechanism is the European Council (not the Council of Ministers). Member state co-operation and joint decision-making outside the context of the Treaty have also led to several initiatives. For our purposes, two of the most important are the Schengen Accords and the activities of the *TREVI* group, considered further in Chapters 5 and 11. The Schengen agreement, signed in June 1985, deals with such matters as extradition, political asylum, hot pursuit, and a common police information system. Although it is restricted to the Benelux countries, Germany, and France, it has formed the model on which similar European wide co-operation might develop further. TREVI was established in 1976 as an intergovernmental forum, with meetings every six months at ministerial and official level, at which member states of the EC discuss practical, operational co-operation against terrorism, drugs trafficking, and other serious crime and public-order problems.

In 1990, initiatives were taken by the European Council to increase political and economic integration. They resulted in a new treaty, concluded on 13 December 1991 at Maastricht in the Netherlands, which provides for the establishment of a new supra-national body, the European Union.[63] The Treaty had not been ratified by all member states by 1

---

[61] See Commission Communication on Community accession to the European Convention for the Protection of Human Rights and Fundamental Freedoms and some of its Protocols, SEC (90) 2087 final, 19 Nov. 1990.

[62] C. Turpin, *British Government and the Constitution*, 2nd edn. (1990), 326.

[63] Treaty on European Union, 7 Feb. 1992.

January 1993. When this happens the European Economic Community becomes the European Community. The Community in turn becomes part of the European Union.[64] One of the objectives of the European Union is 'to strengthen the protection of the rights and interests of the nationals of its Member States through the introduction of a citizenship of the Union'.[65] The Treaty further provides that the Union 'shall respect fundamental rights as guaranteed by the European Convention for the Protection of Human Rights and Fundamental Freedoms and as they result from the constitutional traditions common to the Member States as general principles of Community law' (Article F(2)). Article E provides that 'the European Parliament, the Council, the Commission and the Court of Justice shall exercise their powers under the conditions and for the purposes provided for, on the one hand, by the provisions of the Treaties establishing the European Communities', and on the other hand, 'by the other provisions of this Treaty', including (presumably) Article F(2). The Court of Justice does not, however, have jurisdiction over any provision relating to the European Union, unless the Treaty specifically states that it does.

Citizenship of the Union guarantees the right to move and reside freely within the territory of the member states, the right of non-nationals to vote and to stand as a candidate in municipal elections in the member state in which he or she resides, the right to petition the European Parliament, and the right to apply to the (newly established) European Ombudsman. The Council of Ministers, acting unanimously, may adopt provisions to strengthen or to add to the rights laid down in the Treaty.

There are important provisions on co-operation in the sphere of justice and home affairs which echo those included in the Schengen and TREVI arrangements. Article A of these provisions provides that 'for the purpose of achieving the objectives of the Union, in particular the free movement of persons, and without prejudice to the powers of the European Community', the following areas are now regarded by the member states 'as matters of common interest': asylum policy, entry into member states, immigration policy and policy regarding nationals of third countries (such as conditions of residence by nationals of third countries in the territory of member states, including family reunion and access to employment), judicial co-operation in civil and criminal matters, customs co-operation, police co-operation for the purposes of preventing fraud, unlawful drug trafficking, and other serious forms of international crime. These matters are to be dealt with 'in compliance with' the European Convention on Human Rights and the Convention Relating to the Status of Refugees

---

[64] Art. A.
[65] Art. B.

1951 and its 1967 Protocol, 'and having regard to the protection afforded by Member States to persons persecuted on political grounds'.

These provisions received the signatures of all the member states of the European Community. In addition, a further agreement was concluded between the member states of the European Community, with the exception of the United Kingdom, on social policy. This amends the Treaty of Rome in several respects, notably Article 119 (equal pay). New provisions are included giving greater competence to the Community in social affairs.

Subsequent chapters consider the effects of the United Kingdom's membership of the Community on human rights and civil liberties. We examine the use made in the domestic courts of the Community law, action taken by the European Commission, recourse to the ECJ, and the effects of such action on domestic law.

Under the auspices of the Conference on Security and Co-operation in Europe

The Conference on Security and Co-operation in Europe (CSCE) was a product of the cold war. It was established as a mechanism for discussion between the then communist countries of Eastern and Central Europe, and the democracies of Western Europe and North America. Preliminary consultations among these states resulted in the first meeting of foreign ministers in Helsinki in 1973. In July 1975 a summit meeting, again in Helsinki, agreed the declaration known as the Helsinki Final Act which set out the basic principles guiding relations between the participating states. Between then and the end of the cold war in the late 1980s, the CSCE provided a process by which the Soviet and Western blocs could discuss measures which would lessen tension and build a degree of confidence between them. Among these issues was the question of human rights.

An understanding of certain features of the CSCE process is necessary for subsequent discussion. Its primary rationale is the building of international security through resolving conflict between states which may endanger that security. It is concerned with human rights issues where those endanger peace and security in Europe. The CSCE is not a legal mechanism for conflict prevention or resolution and it has no legal status as an international institution, nor are any agreements concluded under its auspices legally binding. It is a political mechanism which relies on political pressure and reciprocation for its ability to achieve change. It has expanded considerably since its establishment in 1973 with thirty-five participating states, and now covers much of the Northern hemisphere. One-third of these states joined during 1992.

From the beginning of the process, decisions have been taken on the basis of consensus, meaning that all participating states must at least acquiesce in a proposal in order for it to be adopted. This consensus method of decision-making has only been altered in a limited number of circumstances. First, the Council of the CSCE decided in Berlin in June 1991 to adopt an emergency mechanism for consultation and co-operation in conflict situations. This emergency mechanism can be put into operation as soon as one state requests it, where it is supported by twelve or more participating states. In that case, a meeting of the Committee of Senior Officials (CSO) will be held within 24 hours, and it will act as a mediator in the dispute. Any other decisions, however, must be taken on the basis of consensus. Second, the Concluding Document of the Moscow Meeting[66] agreed in October 1991 that each participating state is allowed, when supported by nine other states, to send observers to another state where it considers that gross violations of human rights are taking place. Each participating state can name three experts who can be sent as CSCE observers into another state at short notice. Third, where clear, gross, and uncorrected violations of human rights occur, the Council of Ministers meeting in Prague in January 1992 agreed that a consensus-minus-one method may be used to adopt political measures against the state in violation.[67]

Until recently, the CSCE had little institutional manifestation. It existed only in periodic summits, preparatory meetings, follow-up meetings, and meetings of experts. These were government-to-government negotiations. Since 1990, however, the CSCE structures have been strengthened in several respects. Ministers for Foreign Affairs of the participating states now meet regularly as a Council, and at least once a year. The CSO is in charge of the preparation of the Council meetings, and for the overview, management, and co-ordination of the process, as well as acting as the Council's agent in taking appropriate decisions. There is a chairman-in-office who is responsible on behalf of the Council for the co-ordination of consultation on, and communication of, CSCE business and decisions. In addition, three small offices have been established: a secretariat in Prague (to provide administrative support for the negotiations), a Conflict Prevention Centre in Vienna (to assist in reducing the risks of conflict), and an Office for Democratic Institutions and Human Rights in Warsaw (to serve as the institutional framework for CSCE activities in the area of human rights). The Helsinki Summit in July 1992 approved a further institutional innovation: the creation of a CSCE High Commissioner for National Minorities. Lest too much be read into these institu-

---

[66] (Nov. 1991) 30 *International Legal Materials* 1670.
[67] 'Prague Document on Further Development of CSCE Institutions and Structures', repr. in (1992) 13/4 *HRLJ* 174–6.

tional developments, however, it is important to realize that the budget and staffing for these offices have been kept very low indeed. Only nine diplomats altogether staff the three offices. As important, these diplomats do not constitute a permanent CSCE bureaucracy, being merely detached from the participating states' foreign offices.

One of the important concerns of the CSCE process is with what it terms 'the human dimension'.[68] The 'human dimension' of the CSCE is the term used in CSCE documents to describe matters relating to 'all human rights and fundamental freedoms, human contacts and other issues of a related humanitarian character'.[69] The concept of the human dimension has been expanded since it was originally coined, to include also the issues of democratization, the rule of law, and the protection of minorities. Prior to the Vienna Follow-up Meeting in 1989, advancing the human dimension was largely restricted to the setting of standards. At the Vienna meeting a new mechanism was established and this was further elaborated at the meetings in Copenhagen and Moscow during 1990 and 1991 on the human dimension. This Vienna mechanism consists of four stages: first, an exchange of information in which each state is entitled to require another state to provide information on alleged violations of human rights; second, bilateral meetings may be convened by one state with another to consider the question with a view to resolving it; third, the initiating state may inform all other CSCE states about its concern; and fourth, the question may be raised at CSCE meetings. These are in addition to the Moscow mechanism, and the High Commissioner for National Minorities, discussed above.

From the perspective of human rights, one of the most noteworthy effects of the CSCE process was the stimulus it gave to the creation of non-governmental human rights organizations in several of the then communist states of Eastern and Central Europe. These Helsinki monitoring groups stimulated in turn the foundation of several such organizations in several Western countries. Their common *raison d'être* was that the commitments which their governments entered into in the field of human rights should not be forgotten.

THE STRUCTURE OF THE BOOK

The sequence of chapters which follows this introduction follows roughly the sequence of issues in the European Convention on Human Rights.

---

[68] A. Bloed and P. van Dijk, *Essays on Human Rights in the Helsinki Process* (1985); A. Bloed and P. van Dijk, *The Human Dimension of the Helsinki Process: The Vienna Follow-up Meeting and its Aftermath* (1991); R. Brett, *The Development of the Human Dimension Mechanism of the Conference on Security and Co-operation in Europe* (1991).
[69] Vienna Concluding Document, 28 *ILM* 527, Mar. 1989.

As we shall see, however, there is by no means an exact fit. Several topics will require consideration of more than one Article. In arranging the chapters as we have, we have merely sought to identify an area with the Article with which it might reasonably be thought to be most closely associated.

We concentrate on many of the applicable civil and political rights, but do not attempt to include economic and social rights, as laid down, for example, in the ICESCR.[70] We have not, however, interpreted this separation as watertight, nor indeed as particularly convincing. Thus, we have not regarded ourselves as restricted by this distinction from discussing what might be thought the economic and social aspects of particular rights; for example, the right to freedom of association, or freedom from race and sex discrimination.

While attempting to be wide-ranging in its coverage and approach, this book does not claim to be comprehensive. Several substantive issues which are not included may briefly be mentioned. We do not discuss the right to vote, or the right to property. These rights have been discussed extensively elsewhere and might make an already large book unmanageable.

More debatably, we have not included discussion of discrimination on grounds of mental and physical disability. This has become an important area for discussion in recent years. In 1983 the International Labour Organization drafted a recommendation (No. 168) and a Convention (No. 159) concerning the vocational rehabilitation and employment of disabled people. The United Nations has issued a report on national legislation for the equalization of opportunities for people with disabilities.[71] The EC has shown some interest in becoming involved in the area,[72] and there is now extensive legislation against discrimination on these grounds in the United States.[73] In the United Kingdom, however, the relevant legal provisions are few and far between, and those that do exist are largely unenforced. Reformers have advocated new legislation but so far to no avail.[74]

In Chapter 2, the first of two contributions to this volume, Gearty argues that in Britain the right to freedom of assembly is best seen as a negative right, since the right is nowhere enshrined in law. Assembly is permissible to the extent that, in assembling, existing laws are not thereby

---

[70] For an overview, see B. Bercusson, 'Fundamental Social Rights and Economic Rights in the European Community', in A. Cassese. A. Clapham, and J. Weiler (eds.), *Human Rights and the European Community: Methods of Protection* (1991), 195.

[71] United Nations, *Report on National Legislation for the Equalization of Opportunities for People with Disabilities: Examples from 22 Countries and Areas* (1989).

[72] W. Albeda, *Disabled People and their Employment* (1984).

[73] Americans with Disabilities Act 1990.

[74] See, generally, I. Bynoe, M. Oliver, and C. Barnes, *Equal Rights for Disabled People: The Case for a New Law* (1991).

infringed. In practice, two distinct legal attitudes to assembly can be identified; on the one hand, freedom to assemble is an aspect of the common-law presumption in favour of freedom of speech which is to be cherished and preserved as a basic freedom; on the other hand, assembly is regarded as necessarily a potential threat to public order and tranquillity and for this reason requires control and regulation. Gearty argues that a presumption of freedom of assembly has in practice been eroded in favour of the preservation of public order. He considers in particular whether those concerned about the preservation of the right of freedom of assembly should question the degree of police discretion in this area and the extent to which greater police accountability in public-order operations is necessary.

Reiner and Leigh examine, in Chapter 3, the rights of persons under suspicion of criminal offences and consider the question of police powers, locating the problem of the liberties of suspects in the wider context of the controversy about police discretion. They discuss the origin of police discretion, which has implications for how civil liberties in this area may be protected: is discretion inevitable because most policing is low-level and invisible or is the existence of discretion a result of a failure on the part of the legislature to formulate laws which control it? The chapter considers in detail the Police and Criminal Evidence Act 1984, which they see as an attempt to regulate low-level policing and they review evidence which indicates whether or not the new principles introduced by the Act were successful.

While Reiner and Leigh focus on pre-trial issues, in Chapter 4 Jackson considers due-process issues connected with prosecution and trial. He argues that a model of criminal justice which sees the ultimate goal of the criminal process as the conviction of criminals does not sit easily with an individual rights perspective. Rather, it is argued that the criminal process requires that those accused of crimes be treated as individual moral agents. This view lends itself well to the appropriate consideration of the place of individual liberty in the criminal process. However, although English law may give recognition to due-process values, this does not mean that these values are necessarily reflected in the workings of the system.

In Chapter 5, his second contribution, Gearty considers the developments affecting civil liberties which have arisen from the increase in political violence, including that arising in the context of Northern Ireland. He argues that in attempting to respond to political violence the state has made serious incursions into civil liberties. He considers whether such legislation has been successful in controlling political violence, and, indeed, the extent to which the continued existence of 'temporary' provisions is justified by empirical evidence as to its effect in ameliorating terrorist threats. He argues that the reasons for the introduction and continued existence of legislation such as the Prevention of Terrorism Act lies both in its symbolic importance, that is, as a sign of public revulsion and fear, and in the absence of serious political discussion about its necessity.

Richardson considers, in Chapter 6, the rights of prisoners who have been deprived of their liberty by the state through incarceration. She argues that the difficulty of defining prisoners' rights stems from a confusion over what the function of imprisonment is—punishment, deterrence, or treatment. Prisoners can, however, be thought to have two different sorts of rights: first the residue of the rights that accrue to every citizen (that is, those that have not been taken away by the basic deprivation of liberty inherent in imprisonment), but in addition special rights which accrue from the fact that, by virtue of the deprivation of freedom inherent in imprisonment, prisoners enter into a special relationship with the state and thereby become dependent on the state for the provision of basic needs. Private law has not been as effective a vehicle for change as public law and the intervention of the European Court of Human Rights has been crucial on certain issues.

Gardner argues, in Chapter 7, that there exists a moral right to freedom of expression. This freedom seeks to protect the ability to express one's lifestyle and preferences. In this context freedom of expression is more than just free speech. It includes a whole variety of activities which give expression to a lifestyle. Censorship can be seen from this perspective as any activity which seeks to invalidate a lifestyle or cultural activity. Within this perspective Gardner considers developments in domestic law in the following areas: obscenity, racial hatred, blasphemy, defamation, confidentiality, contempt of court, and official secrets; consideration is also given to the way in which the law regulates the media (press, cinema, theatre).

In Chapter 8, Ewing characterizes freedom of association as a two-dimensional freedom: it promotes the individual's freedom to combine with others in protecting his or her own interests and it consists of what might be seen as a 'collective right' to act together to promote a common interest. Although there is no general protection in British law for the right to freedom of association as such, there has in practice been state support in various ways for the right to associate, as well as and alongside various restrictions affecting for example civil servants and members of outlawed organizations. Ewing considers whether there is also a right not to associate and discusses how this has been considered in the courts and legislated for by Parliament.

Michael's discussion, in Chapter 9, focuses on privacy and in particular on issues connected with the protection of private information or data.

The chapter reviews the most significant developments in the UK with respect to informational privacy. The Council of Europe is seen as having had considerable importance, both in terms of Britain's obligations under the ECHR and in the requirements imposed on the UK by way of adherence to the Council of Europe Convention on Data Protection. However, a wider definition of privacy is possible which would include the physical aspects of privacy and would encompass such issues as the legal regulation of sexual conduct, abortion, and the privacy of family or marital life, issues which are dealt with by Eekelaar.

In Chapter 10, on families, children, and the law, Eekelaar first considers rights embodied in the legal relationships between adults, including the legal rights of both married men and married women, cohabitation, divorce, and the right to marry. A second theme of Eekelaar's chapter concerns the various rights embodied in the relationships between parents and children, including both the rights of parents regarding their children and the rights of children themselves. Eekelaar argues that developments in family law in the recent past can be characterized as embodying a shift from a welfarist to a rights-based ideology.

Dummett shows, in Chapter 11, that British domestic law on nationality since 1945 has been about the regulation, control, and restriction of, or entitlement to, citizenship and residence. She argues that the right to residence in Britain and to British nationality is complicated by the absence of any clear concept of citizenship in the United Kingdom. To understand developments regarding the law on citizenship in the UK it is necessary to understand the evolution of policies on immigration control both with respect to foreigners who were seen as potentially politically subversive, and to non-white entrants who were seen as raising particular problems of cultural assimilation. Dummett argues that the debate about citizenship has in the United Kingdom become inextricably linked to the debate about immigration of non-whites from the New Commonwealth.

Collins and Meehan outline, in Chapter 12, how recent legislation in the UK has sought to combat and outlaw gender discrimination at work, and related areas. They argue that confusion over what is meant by equality, and controversy in the women's movement about strategies for achieving equality, have hampered progress in this area. European Community directives and decisions in the European Court of Justice have been crucial in advancing women's rights at work.

In Chapter 13, McCrudden notes that English common law had very little to say about discrimination or about a right to equal treatment irrespective of racial origin, and that this could be explained in part by the dominance in common law of the concept of freedom of contract. Despite the absence of common-law protection in this area, Britain has since the middle of the 1960s developed detailed legislation and enforcement mechanisms in the area of racial discrimination. McCrudden documents how a growing awareness of the reality of racial discrimination led to the development of legislation in the 1960s and 1970s. He considers the ambiguity of the broader aims and objectives of anti-discrimination legislation and points to the 1976 Act as reflective of both an individual justice model of rights and of a group justice model.

Poulter examines the status of the rights and interests recognized as applicable to ethnic, religious, or cultural minorities in Chapter 14. He contrasts assimilationist and pluralist approaches to minority rights. Assimilationists argue that minorities should seek to adapt themselves to the predominant cultural values of English society while pluralists advocate respect for the distinctive cultural traditions and identities of minorities. Poulter shows that recent legislation has endorsed both the pluralist and assimilationist approaches, often within the same Act.

In Chapter 15, Wintemute deals with discrimination against gay, lesbian, and bisexual men and women, and same-sex emotional-sexual conduct. He examines the concepts of 'sexual orientation' and 'sexual orientation discrimination', and several ways in which such discrimination can be viewed as presenting a human rights issue. He goes on to describe how English law discriminates against persons who engage in same-sex emotional-sexual conduct in the criminal law, in employment, housing, and service provision, in the regulation of partnerships, and by imposing legal or other barriers to parenthood. He considers the actual and potential impact of the ECHR and of European Community law in surmounting discrimination and reviews various proposals for law reform.

The final chapter, Chapter 16, attempts to extract common themes from the prevous chapters, and to bring together some of the conclusions which can be drawn from them.

2

Freedom of Assembly and Public Order

CONOR GEARTY

INTRODUCTION: THEORETICAL CONSIDERATIONS

The title of this chapter is aptly Janus-headed. It reflects the schizophrenia that afflicts the treatment of the subject by both politicians and members of the public. The law manifests the same confusion, being rooted simultaneously in two opposites. On the one hand, there is freedom of assembly; on the other there is a concern for public order. As regards the first of these, a presumption in favour of freedom of assembly is firmly based in traditional liberalism's concern for individual liberty. In this context, the assembling that is protected by such a preference is not the sort that is a familiar part of most of our daily routines: meeting the family at home, getting the bus to work, queuing for the cinema, partying with friends. These gatherings normally have no political dimension, and the restrictions imposed on them (barring orders, bus lanes, health by-laws, noise-abatement zones) are compelled by public-interest considerations unrelated to the content of what is being uttered. The freedom of assembly involved here, in contrast, is the liberty to join with others in order to communicate a political message. Such a gathering exists not as an accidental consequence of birth (the family), nor as a necessary condition of getting somewhere else (the bus or the cinema queue), but solely so as to be able more forcibly to relay a point of view about some aspect of the way that society organizes its affairs.

If it is correct to describe such assemblies as being distinguished by their having political purposes, then 'political' must be understood in a broad sense. It might be a march calling for the resignation of the government or for the withdrawal of troops from Northern Ireland. It might be a meeting demanding an end to the rates or the poll tax. But it could just as easily be a vigil outside a private laboratory involved in animal experiments, a protest in front of a building society engaged in repossessions, or a trade union picket at the entrance to a factory or an office-block. Viewed in this way, freedom of assembly can be seen as a particular application of a broader constitutional and legal presumption in favour of freedom of speech; it is the simultaneous exercise of the latter freedom by more than one person acting in concert. Its importance lies in the belief that there is truth as well as safety in numbers. As the revolutions in Eastern and Central Europe in 1989 demonstrated, thousands of people gathered together in a public place will more powerfully advocate a shared point of view than the same number each chanting privately at home. In a society where political expression is generally mediated through controllable conduits (in particular, television, radio, and the national press), such freedom of public assembly represents the ultimate safeguard for democracy.

The case which most powerfully illustrates the operation of this common-law presumption is the famous nineteenth-century decision of *Beatty* v. *Gillbanks*.[1] The appellant and others were members of the Salvation Army who were bound over by local magistrates after a peaceful procession which they had organized encountered violent opposition from a rival group who were antagonistic to their aims. The key question for the divisional court, hearing the case on appeal, was whether the violence of their opponents (the 'Skeleton Army') could effectively veto the Salvationists' freedom of assembly. The answer was unanimously in the negative:

What has happened here is that an unlawful organization has assumed to itself the right to prevent the appellants and others from lawfully assembling together, and the finding of the justices amounts to this, that a man may be convicted for doing a lawful act if he knows that his doing it may cause another to do an unlawful act. There is no authority for such a proposition.[2]

This is as close to an unequivocal statement of principle that English judges have achieved in this field.

More usual are balancing remarks such as those of Otton J. in a recent case that the 'courts have long recognized the right to free speech to protest on matters of public concern and to demonstrate on the one hand and the need for peace and good order on the other'.[3] In his influential report into the Red Lion Square disorders of 1974, Lord Scarman went so far as to erect the language of rights around both sides of this judicial compromise, when he suggested that 'amongst our fundamental human rights there are, without doubt, the rights of peaceful assembly and public protest and the right to public order and tranquillity'.[4] The tone here is rather less certain than was the court in *Beatty*. In fact that case has not escaped criticism. It has been judicially described as 'somewhat unsatisfactory' by a later Chief Justice,[5] and an Irish Lord Chancellor has 'frankly' admitted that he could 'not understand' it.[6] Dicta such as these

---

[1] (1882) 9 QBD 308.

[2] Ibid., at p. 314 per Field J.

[3] *Hirst and Agu* v. *Chief Constable of West Yorks.* (1987) 85 Cr. App. R. 143, at p. 151.

[4] The Rt. Hon. Lord Justice Scarman, *The Red Lion Square Disorders of 15 June 1974. Report of an Inquiry by the Rt. Hon. Lord Justice Scarman, OBE*, Cmnd. 5919 (1975), para. 5.

[5] *Duncan* v. *Jones* [1936] 1 KB 218, at p. 222 per Hewart LCJ.

[6] *O'Kelly* v. *Harvey* (1883) 15 Cox CC 435, at p. 446 per Law LC.

reflect a position at the opposite end of the legal spectrum to that occupied by *Beatty*. Here, we find a public-order perspective on public assembly. This places the emphasis not on individual liberty but rather on tranquillity and security. The maintenance of the Queen's peace is seen as the primary function of the law, and this must necessarily entail keeping a watchful eye on all large gatherings of people.

The law on public order has to be wide enough to catch the rioters, the hooligans, the disorderly football fans, and the aggressive drunkards. But it necessarily also extends to people who have gathered together in pursuit of their freedom to assemble as understood in the civil liberties sense, since there is no guarantee that such groups are inherently non-violent merely because their initial purpose in gathering together is political; large protests have often degenerated into looting and criminal damage, and even well-organized pickets have on occasion become pitched battles between strikers and workers. Indeed, on this point of view, there are compelling public-order justifications for giving the authorities the power to restrict or ban assemblies in advance where this would prevent violence or disorder from breaking out in the first place. Thus, in a case decided within a year of *Beatty* v. *Gillbanks*, the Irish appellate courts upheld a decision by a magistrate to disperse a peaceful meeting in order to prevent an attack on it by its opponents. The officer's 'paramount duty was to preserve the peace unbroken, and that by whatever means were available for the purpose':

Accordingly, in the present case, even assuming that the danger to the public peace arose altogether from the threatened attack of another body on the plaintiff and his friends, still, if the defendant believed and had just grounds for believing that the peace could only be preserved by withdrawing the plaintiff and his friends from the attack with which they were threatened, it was I think the duty of the defendant to take that course.[7]

From this perspective, freedom of assembly can more often resemble a gamble with the Queen's peace than the exercise of a fundamental liberty.

This clash of judicial attitudes should not surprise us. The principles at stake involve problematic issues for every liberal democracy. An absolute guarantee of assembly would be chaotic and anarchic whilst a total pre-occupation with peace would be mind-numbing and authoritarian. In most of Western Europe, the legal solution is found by guaranteeing a right of assembly to which exceptions are then made. As we shall see, this is the method adopted by the European Convention on Human Rights and Fundamental Freedoms. In Britain, in contrast, the law does not favour assembly as a positive right. The emphasis is instead on the freedom to assemble as a negative liberty, as something that we are free

---

[7] Ibid., at pp. 445–6.

to do to the extent that we are not restricted by any law, whether statute or judge-made. In the words of the Victorian jurist, A. V. Dicey, the 'right of assembling is nothing more than a result of the view taken by the courts as to individual liberty of person and individual liberty of speech'.[8] Dicey was confident that such an approach produced a much more concrete protection for freedom than the broad commitments often to be found in continental constitutions. *Beatty* v. *Gillbanks* was for him the key case which 'established, or rather illustrated' the validity of his point of view.[9]

Is there any difference in substance between the European and the English approaches? This depends on the one hand on the number of laws that restrict freedom of assembly in the 'Diceyite' United Kingdom, and on the other on the breadth and range of the exceptions to the right that are to be found in the European system. In this chapter, we will look first at British law up to 1950, and secondly at the various developments there have been in statute and common law over the past forty years. This will lead us to an examination of how the law is applied in practice in the United Kingdom. In this regard, we will consider recent moves towards a more militarized police establishment, a development which has already had very significant implications for freedom of assembly. We will then analyse why our domestic law has developed in the way that it has, paying particular attention to the role of political parties, pressure groups, and law reformers. Finally, we will compare the present-day situation in this country with the position in the United States and with the text and case-law of the European Convention on Human Rights and determine whether an approach rooted in rights would have made or could still make a substantial difference.

THE LAW IN 1950

Even Dicey admitted that some restriction on freedom of assembly was required, 'grounded on the absolute necessity for preserving the Queen's peace'.[10] *Beatty* v. *Gillbanks* accepted that a person was liable where a 'disturbance of the peace was the natural consequence'[11] of his or her acts. In fact, the law has always gone much further than either of these statements would suggest. Property rights in particular have traditionally been protected in such a way that they have invariably prevailed over

[8] A. V. Dicey, *An Introduction to the Study of the Law of the Constitution*, 10th edn. (1959), 271.

[9] Ibid., 274.

[10] Ibid., 277.

[11] (1882) 9 QBD 308, at p. 314 per Field J.

conflicting civil liberties. This has been more to do with the structure of the common law than the bias or predilections of particular judges, and the rules that applied in this early period, on trespass[12] and nuisance[13] for example, remain with us today. The civil law apart, by 1950 Dicey's vision of the common law, and *Beatty* v. *Gillbanks* itself, had taken further batterings from both Parliament and the judges.

Even before the emergence of Dicey's leading text in 1885, section 7 of the Conspiracy and Protection of Property Act 1875 had criminalized a wide variety of behaviour in a way that could, in some contexts at least, be characterized as restrictions on the liberty to assemble. The section was drafted in very broad terms and included the intimidation and persistent following of any person where such behaviour was undertaken with a 'view to compel [him or her] to abstain from doing or to do any act which such other person has a legal right to do or abstain from doing'. The Public Order Act 1936, passed partly at least as a response to the threat to the peace posed by the followers of Sir Oswald Mosley, carried the legislative attack into the heartland of freedom of assembly, namely the liberty to participate in a protest march on the highway. Section 3(1) permitted the chief officer of police in a local government area a wide discretion to impose a variety of conditions on public processions where these appeared 'necessary for the preservation of public order'. Section 3(2) went further, permitting the imposition of bans on 'all public processions or of any class of public procession' where it was thought that the power to impose conditions would not be sufficient to prevent 'serious public disorder'. Clearly these powers could and were exercised in a way which stifled legitimate as well as potentially violent demonstrations.

As regards the judges, two cases were of particular importance in defining the law by 1950. In the first, *Thomas* v. *Sawkins*,[14] a public meeting was held in a library to protest against an incitement to disaffection bill then before Parliament and to demand the dismissal of the local chief constable. Between 500 and 700 people were present, including three police officers sitting in the front row. The organizers did not want any members of the local constabulary present, and these three were asked to leave. When they refused to do so, an attempt was made forcibly to eject them but this was thwarted by the sudden arrival of large numbers of police reinforcements. The question for the Divisional Court was whether the police were entitled to 'enter on private premises and to remain there against the will of those who, as hirers or otherwise, are for the time being in possession of the premises'.[15] The Lord Chief Justice, Lord

---

[12] *Harrison* v. *Duke of Rutland* [1893] 1 QB 142.
[13] *Lyons and Sons* v. *Wilkins* [1899] 1 Ch. 255. Cf. *Tynan* v. *Balmer* [1967] 1 QB 91.
[14] [1935] 2 KB 249.
[15] Ibid., at p. 254.

Hewart, thought such a right existed not only where a crime has been, or is being, committed, but also where an officer has 'reasonable ground for believing that an offence is imminent or is likely to be committed'.[16] Avory J. and Lawrence J. emphasized the duty of the police to maintain order and preserve the peace. The case was important because of the wide discretionary power it reposed in police officers, as long as they could say that they reasonably anticipated a breach of the peace—and it is this part of the court's ruling which was preserved when the Police and Criminal Evidence Act was enacted in 1984.[17]

The second case, *Duncan* v. *Jones*,[18] illustrated a different aspect of the same police power. A group of about thirty people had gathered on the roadway opposite the entrance to a training centre for the unemployed. The idea was to hold a protest meeting but when the appellant gave the appearance of being about to address the crowd, she was told by the chief constable of the district who happened to be present that she could only continue by moving to another street some 175 yards away. This would have defused the impact of her protest and the appellant refused, where-upon she was arrested by an Inspector Jones and charged with obstruction of a police officer in the execution of his duty. Her subsequent conviction in the magistrates' court was upheld by the King's Bench division. As in *Thomas* v. *Sawkins*, there was no suggestion that the appellant or any other person present had either committed, incited, or provoked any breach of the peace. Fourteen months previously, however, a disturbance had followed a similiar meeting also addressed by the appellant, and on this basis the police claimed that their action now was designed to prevent a reasonably apprehended breach of the peace. Once again, this reasoning met with the approval of the high court judges. It was a 'plain case' having 'nothing to do with the law of unlawful assembly'.[19] Lord Hewart began his judgment with the declaration that there had 'been moments during the argument in this case when it appeared to be suggested that the Court had to do with a grave case involving what is called the right of public meeting',[20] but, quoting Dicey, his Lordship was emphatic that 'English law does not recognize any special right of public meeting for political or other purposes'.[21]

DEVELOPMENTS IN THE LAW SINCE 1950

It is clear that by 1950 the law on freedom of assembly had already drifted firmly in the direction of public order. This early period remains

---

[16] [1935] 2 KB 249, at p. 255.
[17] Police and Criminal Evidence Act 1984, s. 17(6).
[18] [1936] 1 KB 218.
[19] Ibid., at p. 223 per Humphreys J.
[20] Ibid., at pp. 221–2.
[21] Ibid., at p. 222.

important because it set the tone for the development of the law in recent years. There have been no dramatic changes or breaches with this restrictive tradition. Rather, what we have seen has been a deepening in a legal edifice already in place. There have been new laws, and new applications of old laws, but there has been little that has been innovatory in a civil libertarian sense. The Dicey model remains in place and *Beatty* continues to have pride of place in the textbooks, but the theoretical framework underpinning both the distinguished jurist and the divisional court decision has been sharply undermined. The legislative and judicial activism of the post-war years has been almost wholly restrictive in impact. We will look first at developments in three areas in particular: the Public Order Act 1936, the breach of the peace/obstruction power, and the offence of obstruction of the highway. Then we shall consider the impact of the only major legislative initiative in this area since the Second World War, the Public Order Act 1986.

The application of the Public Order Act 1936

As we have seen, this Act provided a framework for the banning of public processions where it was considered that there was a risk of serious public disorder. Soon after its enactment, the power was invoked for a six-week period in a part of the East End of London. In 1960, it was applied to St Pancras in the same city in the context of rent disturbances there. Nuclear disarmament protests in 1961 led to the imposition of a ban covering Trafalgar Square and its immediate vicinity. Planned fascist marches occasionally led to bans both in London and elsewhere during the early part of the 1960s.[22] But until 1981, the overall impact of this potentially restrictive provision was slight. During the period 1936–80, banning orders were employed on no more than eleven distinct occasions.[23] The atmosphere changed after the summer riots of 1981. In that year alone, forty-two banning orders were issued. The figures for the following years were thirteen in 1982, nine in 1983, and eleven in 1984.[24] In 1981, the Campaign for Nuclear Disarmament mounted a legal challenge to a twenty-eight-day ban which had been imposed on almost all processions in London.

Neither the High Court nor the Court of Appeal were willing to overturn the Metropolitan Police Commissioner's opinion about the risk of an outbreak of serious public disorder.[25] This was despite the fact that

---

[22] The details are in D. G. T. Williams, *Keeping the Peace: The Police and Public Order* (1967), 60.

[23] *Review of the Public Order Act 1936 and Related Legislation*, Cmnd. 7891 (1980), para. 31.

[24] *Review of Public Order Law*, Cmnd. 9510 (1985), para. 4.7.

[25] *Kent* v. *Metropolitan Police Commissioner*, *The Times*, 15 May 1981.

one member of the Court of Appeal, Sir Denys Buckley, thought the reasons for the order to be 'meagre'. All the judges emphasized that the danger of disorder came not from the marchers but rather from the fact that '[h]ooligans and others might attack the police who were doing their duty in escorting a peaceful procession'.[26] Ackner LJ suggested that the demonstrators might become a 'target for their viciousness'.[27] This was significant because it took the case fully into *Beatty* v. *Gillbanks* territory. Perhaps the older decision is distinguishable. Such violence as was anticipated here would have been directed largely against the police rather than the procession; there was no 'battle of ideas', merely a fear that an excuse for thuggery would not be turned down. Nevertheless, this rather tenuous point apart, it would now appear that the banning procedure in the Public Order Act, as interpreted by the courts, permits (without, of course, compelling) senior police officers to set aside the principle in *Beatty* v. *Gillbanks*.

Section 5 of the 1936 Act provided a further platform for police discretion, and the application of this law had an occasional impact on freedom of assembly through the years. As amended in 1965, the section covered any person who in any public place or at any public meeting used threatening, abusive, or insulting words or behaviour, or distributed or displayed any writing, sign, or other visible representation which was threatening, abusive, or insulting. None of these actions was criminal in itself. The conduct had also to be intended to provoke a breach of the peace or to have been of such a nature as to have been likely to have occasioned such a breach. Amongst those convicted under the section have been protestors who shouted 'Remember Biafra' during the two minutes' silence at the 1969 Remembrance Day ceremony in Whitehall,[28] and a man who in 1968 handed to US servicemen leaflets which opposed the war in Vietnam.[29] Pat Arrowsmith, however, was acquitted when she gave similar anti-war material to British troops, this time relating to Northern Ireland. The prosecution of an anti-apartheid demonstrator for 'insulting' behaviour in disrupting a Wimbledon tennis match involving a South African player was also unsuccessful.[30]

From the point of view of principle, the most important case decided under the section was *Jordan* v. *Burgoyne*.[31] The defendant, a person of extreme right-wing views, addressed a crowd of some 5,000 people at a public meeting in Trafalgar Square. A group of 200–300 young people, mainly communists, members of the Jewish faith, and advocates of nuclear

---

[26] *Kent* v. *Metropolitan Police Commissioner*, per Lord Denning MR.
[27] Ibid.
[28] S. H. Bailey, D. J. Harris, and B. L. Jones, *Civil Liberties, Cases and Materials*, 2nd edn. (1985), 154.
[29] *Williams* v. *DPP* [1968] Crim. LR 563.
[30] *Brutus* v. *Cozens* [1973] AC 854 (HL).
[31] [1963] 2 QB 744.

disarmament, were gathered immediately in front of the speakers' platform. They were aggressively hostile to all those who were participating in the meeting. When the defendant used words in his speech which were particularly offensive to this group, there was complete disorder, an outcry, and a general surge forward by the crowd. The defendant was afterwards convicted under section 5, and this was upheld in the divisional court. He had engaged in no acts of violence himself; his offence lay in the fact that what he had chosen to say caused others to react criminally. Speaking for a unanimous court, Lord Parker CJ held that every such speaker was required to take his or her audience as he or she found it:

[I]f in fact it is apparent that a body of persons are present—and let me assume in the defendant's favour that they are a body of hooligans—yet if words are used which threaten, abuse or insult—all very strong words—then that person must take his audience as he finds them, and if those words to that audience or that part of the audience are likely to provoke a breach of the peace, then the speaker is guilty of an offence.[32]

It may well be that the result here was justified on the facts of the case. Racist and anti-Semitic speech may be regarded as on a uniquely offensive plane, justifying strong restrictive measures, even at some cost to our traditional liberties. Moreover, the defendant here was engaged in provocation rather than communication.[33] However, the judgment of the Lord Chief Justice appears to go rather wider than this. In particular, it seems to extend to all political speech which has this violent impact, regardless of the content of what is being said. A meeting to call for the withdrawal of troops from Northern Ireland would be vulnerable if members of the Territorial Army decided deliberately to disrupt it. So would a demonstration in favour of republicanism if it found that part of the crowd was composed of devoted and affronted royalists. The difficulty is that it is often precisely where speech is most inflammatory that it is most required. A framework of laws that protects only preaching to the converted is hardly likely often to expose itself to the shock of the new—and possibly true. *Jordan* v. *Burgoyne* takes us perilously close to the 'heckler's veto' which Field J. in *Beatty* v. *Gillbanks* was determined to avoid. As with the earlier case on banning marches, however, the effect of the decision is to give the police the discretion to decide whether to protect or to arrest a controversial speaker.

The breach of the peace/obstruction power

As we have seen, *Duncan* v. *Jones* and *Thomas* v. *Sawkins* confirmed the existence of a broadly based police power to act so as to prevent reason-

---

[32] Ibid., at p. 749.
[33] Cf. *Wise* v. *Dunning* [1902] 1 KB 167.

ably apprehended breaches of the peace. The statutory offence of ob-struction of a constable in the execution of his duty[34] is available for those who flout police instruction in this area, but the crucial phrase, 'breach of the peace' has been left undefined by Parliament and its meaning is hard to pin down even after close scrutiny of the case-law. One recent Court of Appeal decision has stressed that there must be 'an act done or threatened to be done which either actually harms a person, or in his presence his property, or is likely to cause such harm, or which puts someone in fear of such harm being done'.[35] But Lord Denning MR in a different case in the same year suggested (though neither judge sitting with him agreed) that a breach of the peace occurred 'whenever a person who is lawfully carrying out his work is unlawfully and physically prevented by another from doing it'.[36] This would seem to include many forms of peaceful protest, including sit-ins, certain public meetings, and even some pickets.

Of more importance than strict legal definitions has been the way in which the police have utilized this vague power. The courts have main-tained their reluctance retrospectively to overrule a police officer's judg-ment, not only in relation to whether a breach of the peace is reasonably to be anticipated but also on the secondary question of what is required to prevent it. Thus, the case-law reveals a wide variety of legitimate police action, ranging from the simple removal of a provocative emblem from a person's clothing[37] to the banning of entire meetings.[38] The decis-ions that have arisen in the context of industrial disputes are particularly instructive. The leading authority is *Piddington* v. *Bates*.[39] During an industrial dispute at a printers' works in London, the police decided to allow only two pickets at each of the two entrances to the works. The defendant disagreed with this decision. He felt that this number was not enough to communicate with workers entering and leaving the premises (usually at the same time) or with transport workers as and when they delivered goods to the place. When he attempted to join one of the pairs of pickets, however, he was arrested and, although there had been no blockage of the highway, disorder, or violence, he was later charged with obstruction of a police officer in the execution of his duty. The police argued that they had reasonably apprehended that a breach of the peace might occur if there had been more than two pickets at each entrance.

The Divisional Court, presided over by the Lord Chief Justice, Lord

---

[34] Now to be found in the Police Act 1964, s. 51(3).

[35] *R.* v. *Howell* [1982] QB 416 (CA), at p. 426 per Watkins LJ.

[36] *R.* v. *Chief Constable of the Devon and Cornwall Constabulary, ex parte Central Electricity Generating Board* [1982] QB 458, at p. 471.

[37] *Humphries* v. *Connor* (1864) 17 ICLR 1.

[38] *O'Kelly* v. *Harvey* (1883) 15 Cox CC 435. The action in this case was taken by a magistrate rather than the local constable, but the ruling does not hinge on this fact.

[39] [1961] 1 WLR 162.

Parker, emphasized that there had to be a 'real possibility' of a breach of the peace, but went on to find that just such a situation of menace existed here: eighteen people 'milling about' when there were only eight people in the works created a 'real danger of something more than mere picketing'.[40] The authorities therefore were entitled to act as they did. The exact number to be allowed was entirely a matter for the constable who 'must be left to take such steps as on the evidence before him he thinks are proper'.[41] This case provided the legal basis for the police power to limit numbers and to form cordons to allow lorries and employees through to work. The requirement that there be a real possibility of a breach of the peace is only very rarely subject to any form of judicial scrutiny. A stark indication of the breadth of this power was given by May J. in a later divisional court decision:

Where . . . a police officer reasonably anticipates that in the circumstances obtaining in the particular case the consequence of any peaceful picketing may well be a breach of the peace, either by the pickets or by spectators, whether supporters of the pickets or not, then it is the duty of that police officer to take such steps as are reasonably necessary to prevent that anticipated breach of the peace. Those steps may include requiring would-be pickets to desist, for so long as the police officer may reasonably deem necessary to prevent the breach of that peace, from any attempt to picket at that place.[42]

This power, which began life as the exercise of discretion by an individual constable, has in practice hardened into a set of rules for the management of industrial disputes. The employment legislation of the 1980s imposed new restrictions on picketing by increasingly exposing it to civil liability, but it never made the activity of picketing in itself criminal. The breach of the peace power as interpreted in *Piddington* v. *Bates* has often had this effect, however, particularly when allied to the statutory offence of obstruction. During the miners' strike in 1984–5, an important new dimension became apparent. In *Moss* v. *McLachlan*,[43] a convoy of motor vehicles containing striking miners was prevented from leaving a motorway at a junction which was within several miles of four Nottinghamshire collieries. The police correctly assumed that the miners were on their way to one or other of these pits in order to demonstrate and picket, and deduced from this a reasonable apprehension of a breach of the peace which, they argued, justified their action in blocking the motorway. The divisional court agreed and upheld the conviction on obstruction charges of four of the striking miners. The police had to 'honestly and reasonably

---

[40] Ibid., at p. 170 per Parker LCJ.
[41] Ibid.
[42] *Kavanagh* v. *Hiscock* [1974] QB 600, at p. 612.
[43] [1985] IRLR 76.

form the opinion' that there was 'a real risk of a breach of the peace in the sense that it [was] in close proximity both in place and time',[44] but in coming to this decision they were entitled to take into account not only the facts before them in the case but also 'knowledge gleaned from . . . colleagues and from the widespread public dissemination of the news that there had been severe disruptions of the peace, including many incidents of violence, at collieries within the . . . area'.[45] Furthermore, 'anyone with knowledge of the . . . strike would realise that there was a substantial risk of an outbreak of violence'.[46]

This road-blocking power has since been extended beyond situations of industrial conflict. It was used in 1985 to prevent anti-nuclear protestors from reaching the vicinity of a Royal Air Force base in Cambridgeshire. There have been press reports of it being employed as a matter of course in order to regulate the movement of fans on their way to and from football matches. On one occasion in Northern Ireland, the Royal Ulster Constabulary stopped a bus containing members of a Protestant apprentice boys band, 30 miles from the largely Catholic town where it was their intention to march. After confiscating their musical instruments, the police allowed them to proceed, and they went on to parade. The lawfulness of this action was afterwards upheld in court. Even though the fear of violence may have been justified here, particularly in the context of sectarian strife in Northern Ireland, the 30-mile control zone must surely be at the outer reaches of police discretion in this area.

Obstruction of the highway

Whilst it is true that this breach of the peace power is vague and indefinite at its edges, it is fairly clear that, dicta from Lord Denning notwithstanding, some connection with violence, whether actual or apprehended, is required. Where no such situation exists, however, the police are not powerless to act. The statutory offence of obstruction of the highway has emerged as a powerful supplementary tool in the control of popular protest. The offence, which has a long history,[47] has been codified on two occasions since 1950, most recently in section 137 of the Highways Act 1980: '(1) If a person, without lawful authority or excuse, in any way wilfully obstructs the free passage along a highway, he shall be guilty of an offence.'[48] In an interpretation of an earlier identical provision which has since been accepted as authoritative, Lord Parker CJ laid down three elements to this offence. First, 'any occupation of part of a road . . . interfering with

---

[44] Ibid., at p. 78 per Skinner J.
[45] Ibid.
[46] Ibid.
[47] See *Lowdens* v. *Keaveney* [1903] 2 IR 82.
[48] See also Highways Act 1959, s. 121.

people having the use of the whole of the road is an obstruction'.[49] Secondly, wilful obstruction means no more than that 'the obstruction is caused purposely or deliberately'.[50] Thirdly, there must be proof that the use in question is 'an unreasonable use'. This depends 'upon all the circumstances, including the length of time the obstruction continues, the place where it occurs, the purpose for which it is done, and ... whether it does in fact cause an actual obstruction as opposed to a potential obstruction'.[51]

This offence is broad enough to catch many forms of otherwise legitimate public protest, and the fact that there is a power of arrest without warrant in relation to it[52] only serves to emphasize the section's potentially mischievous impact on liberty. The cases that have reached the higher courts seem to bear out the suspicion that the police have occasionally succumbed to the temptation to utilize it in a way which has undermined peaceful protest. In *Arrowsmith* v. *Jenkins*,[53] the defendant was a well-known campaigner for a variety of radical causes. Her public meeting on a street in Bootle in April 1962 caused a total obstruction of the highway for five minutes and a partial block for one quarter of an hour. Her conviction for obstruction was unanimously upheld in the Divisional Court in a short extempore judgment delivered by Lord Parker CJ. In *Broome* v. *Director of Public Prosecutions*,[54] the House of Lords unanimously held that a strike picket who stood in front of a lorry urging its driver not to proceed in a certain direction could validly be convicted of obstruction of the highway, notwithstanding that the then industrial relations law gave such demonstrators certain rights in relation to the peaceful persuasion of non-striking workers. In a similar case in 1983,[55] the divisional court allowed an appeal against the acquittal of hospital employees who had blocked an entrance to a hospital in pursuit of their claim for higher wages. It is hard to see how, in any of these cases, the courts could have done other than confirm the breadth of the law.[56]

The Public Order Act 1986

In view of this plethora of controls on political protest, it is hardly surprising that a government review in 1985 should have 'revealed no

---

[49] *Nagy* v. *Weston* [1965] 1 WLR 280, at p. 284.
[50] Ibid.
[51] Ibid.
[52] See PACE 1984, s. 25(3)(*d*)(v).
[53] [1963] 2 QB 561 (DC).
[54] [1974] AC 587 (HL).
[55] *Jones* v. *Bescoby*, unreported 8 July 1983 (Divisional Court) quoted in *Hirst and Agu* v. *Chief Constable of West Yorks.* (see n. 3), at p. 148 by Glidewell LJ.
[56] For a case where convictions were successfully challenged, see *Hirst and Agu* v. *Chief Constable of West Yorks.*

yawning gaps' in the law.[57] The same White Paper accepted that no 'amount of tightening of the law, short of draconian measures which would be quite unacceptable, can guarantee the prevention of all disorder'[58] and drew attention to the fact that, when disorder had broken out, the problem was 'essentially one of enforcement'.[59] Nevertheless, and somewhat paradoxically, reform was required to provide 'the police with adequate powers to deal with disorder, or where possible to prevent it before it occurs, in order to protect the rights and freedoms of the wider community'.[60] The law was 'complex and fragmented'[61] and it was necessary to 'bring up to date the age-old balance between fundamental but sometimes competing rights in our society'.[62] This was the intellectual case for the enactment of a new Public Order Act in 1986. In fact the legislation did not codify the law on public order or even repeal all of the 1936 Act which it was presented as superseding. The old section 5 was re-enacted in a tighter form and a new section was added which criminalized 'threatening, abusive or insulting words or behaviour or disorderly behaviour' which was 'within the hearing or sight of a person likely to be caused harassment, alarm or distress thereby'.[63] This has proved a controversial section, with prosecutions being brought under it against types of political speech.[64] Part three of the Act extended the law on the control of racial hatred, but it has not been enough to prevent the increasing number of racial attacks that have been occurring in recent years, particularly in London. These have led to calls for more and better policing of inner city areas, demands that the police have attempted to respond to positively, albeit without making the commitment that many community leaders feel to be essential.

The main importance of the Act for our present purposes, however, lies in the way that it extended the law on meetings and processions. The power to control and ban processions where there is a reasonable apprehension of serious public disorder is preserved and the new Act continues to require such a fear as a precondition for a total ban. But it expands the rest of this part of the law in three important respects. First, subject to a very few exceptions, the police must now be given written notice of all public processions at least six days before they are due to take place.[65]

---

[57] *Review of Public Order Law*, para. 1.9.
[58] Ibid.
[59] Ibid., para. 1.10.
[60] Ibid., para. 1.8.
[61] *Review of the Public Order Act 1936*, para. 21.
[62] 79 HC Debs. 508 (16 May 1985) (Mr Leon Brittan).
[63] Public Order Act 1986, s. 5.
[64] K. D. Ewing and C. A. Gearty, *Freedom under Thatcher: Civil Liberties in Modern Britain* (1990), 121–5.
[65] s. 11.

Secondly, the Act permits the imposition of conditions on stationary meetings as well as on marches. Such restrictions may be imposed on 'an assembly of 20 or more persons in a public place which is wholly or partly open to the air'.[66] *Duncan v. Jones* clearly demonstrated that the police already had this power, regardless of how many people were at a meeting, as long as they reasonably anticipated a breach of the peace. This takes us to the third change. The criteria on the basis of which conditions may be imposed on either meetings or processions include, but extend beyond, the apprehension of serious public disorder. They now embrace a judgement by the police that the purpose of a protest is the intimidation of others 'with a view to compelling them not to do an act they have a right to do, or to do an act they have a right not to do'. They also extend to a reasonable belief by the police that a march or meeting may cause 'serious damage to property' or 'serious disruption to the life of the community'.[67]

If one of these criteria is fulfilled to the police officer's satisfaction, then such conditions on a procession may be imposed 'as appear to him necessary to prevent such disorder, damage, disruption or intimidation, including conditions as to the route of the procession'. The officer may prohibit it 'from entering any public place specified in the directions'.[68] If it is a meeting, then the conditions may relate 'to the place at which the assembly may be (or continue to be) held, its maximum duration, or the maximum number of persons who may constitute it'.[69] These powers may be exercised in advance or on the spot, by a police constable who happens upon the preparations for a protest. There is a power of arrest without warrant and refusal to obey police instructions may result in a gaol sentence. It is quite clear (and has already been demonstrated in *Duncan v. Jones*) that the imposition of conditions may be as effective as an outright ban. An animal rights march around an empty common is less effective, if causing less inconvenience, than one down a high street. A meeting about South Africa is less meaningful at Waterloo Bridge than in front of the country's embassy in Trafalgar Square, especially if the number is cut to 20 after 2,000 turn up.[70]

The breadth of these powers serves to focus attention on the controlling phrases, particularly 'intimidation' and 'serious disruption to the life of the community'. Neither is defined in the Act. The Government has suggested that, apart from the 'obvious example' of picketing,[71] intimi-

---

[66] s. 16.
[67] See ss. 12 and 14.
[68] s. 12.
[69] s. 14.
[70] See *Police* v. *Reid (Lorna)* [1987] Crim. LR 702.
[71] *Review of Public Order Law*, para. 5.10.

dation would include a 'National Front march through Asian districts' and also 'animal rights protestors, who on occasion have marched on furriers' shops or food factories with the intention of preventing the employees from working'.[72] It has also argued that 'a march may be coercive simply by reason of the number of marchers compared with its objective (for example, 1,000 people marching on the home of a local councillor, or an inquiry inspector)'.[73] As far as 'serious disruption to the life of the community' is concerned, the Government has referred to 'the disruption caused by demonstrations outside neighbouring embassies'[74] and to 'marches being held through shopping centres on Saturdays or through city centres in the rush hour'.[75] All these examples make quite clear that the Act has dispensed with the overarching principle which used to govern the law in this area, namely that freedom of assembly may only be restricted when there is some fear (however attenuated in practice) of an outbreak of violence. In its stead are vague formulae, the effect of which is to place the police more firmly than ever in their now increasingly prevalent role as the arbiters of acceptable protest.

THE LAW IN PRACTICE

Our recounting of the law has gone well beyond *Beatty* v. *Gillbanks*. The breadth and range of the restrictions we have identified leaves little of Dicey's residual liberty intact. The goals behind these various developments have been many, including the preservation of the public peace; the protection of property rights; the securing of free passage along the highway; the reduction of inconvenience; and the protection of the individual's contractual right to work. Freedom of assembly may occasionally have been a factor in the judicial or legislative mind when these matters were in issue but it has never in itself been the basis for any of the laws we have discussed.[76] The freedom is trapped by its complacent classification as a residual liberty, and without legislation or positive common-law rules to act as its defender, it is being slowly squeezed into extinction.

But this is not to say that the freedom to assemble has disappeared off the British legal scene. Such an assertion would be to fly in the face of the facts. All of us have seen animal rights' activists distributing literature in our local shopping precincts and watched (at least on television) as large-scale protests have brought London to a halt. If we are not white, some

[72] *Review of Public Order Law*, para. 4.24.
[73] Ibid.
[74] Ibid., para. 5.9.
[75] Ibid., para. 4.22.
[76] For the Education (No. 2) Act 1986, s. 43, see below.

of us may even have bolted our doors as a heavily protected fascist group has processed through our area. Whether directly or indirectly, we are witnesses to popular protest all the time. The law may not compel adherence to the spirit of *Beatty* v. *Gillbanks*, but neither does it force a departure from it. The matter is left to the police. The courts will back their judgement, regardless of the direction in which it leads them. Thus, in one case,[77] a potential site for a nuclear power station was occupied by protestors. The electricity board wanted the police to remove them, but the local constabulary refused. The Court of Appeal was asked to compel the police to act but refused to do so. Lord Denning MR thought that they had the power to intervene if they chose to, but that the matter was 'a policy decision with which . . . the courts should not interfere'.[78] Lawton LJ emphasized the discretion of the officers on the ground: '[p]olice constables are no one's lackeys'.[79] Templeman LJ was emphatic that the 'police on the spot must decide when to intervene'.[80] Earlier cases showed us the courts backing police action; here we see support for inaction.

It is not unreasonable to conclude that in this branch of civil liberties we have freedom under the police rather than freedom under the law. Generally speaking, the authorities have used their discretion in a way which has been reasonably true to the legacy of *Beatty* v. *Gillbanks*. Of course we have seen that there was large-scale banning in the early 1980s, but this may be viewed as somewhat exceptional, a reaction to a period of particular disorder rather than a conscious change of policy. Of course there are exceptions. A march planned by the Manchester branch of the Apprentice Boys of Derry was banned in 1987, as was a procession organized by the Manchester Martyrs' Memorial Committee planned for the following day.[81] The controversy generated by Salman Rushdie's novel, *The Satanic Verses*, led to numerous public order measures, including a nine-day ban on marches in Dewsbury, West Yorkshire.[82] More recently, the Metropolitan Police Commissioner has called for a re-examination of the right of provocative groups to march through multi-racial areas.[83] But the very fact that the matter was put in those terms indicates a presumption, at the highest level, against the use of the breach of the peace law or the statutory power to impose conditions as ways of debilitating unpopular protest.

---

[77] R. v. *Chief Constable of the Devon and Cornwall Constabulary, ex parte Central Electricity Generating Board* [1982] QB 458.
[78] Ibid., at p. 472.
[79] Ibid., at p. 476.
[80] Ibid., at p. 480.
[81] 176 HC Debs. (WA) 204 (11 July 1990) (Mr Peter Lloyd).
[82] *Independent*, 24 June 1989.
[83] *Guardian*, 13 June 1991.

If the freedom of assembly perspective has survived in decisions about whether to permit popular demonstrations, then it is public order concerns that have come recently to dominate the way in which such protests have been handled. Indeed, the transformation of the way in which the police deal with public disorder in general, and marches and demonstrations in particular, has been one of the most marked and disquieting developments of recent years. Consideration of popular protest in the 1960s evokes nostalgic memories of an era in which unprotected, unarmed, and good-humoured constables managed crowd control through the peaceful tactic of strength in numbers. To some extent, this rosy picture is not altogether accurate. At a nuclear disarmament rally in Trafalgar Square in 1961, 1,314 people were arrested. In March 1968, mounted police were used to end a protest in front of the United States embassy in London, and there were 280 arrests. However, a much larger rally later the same year went off peacefully, and its management by the police was more typical of that generation of popular protest.[84]

The 1970s were rather more violent. There was an increase in industrial tension, such as at Grunwick where, during a long dispute, more than 300 police and civilians were injured and over 500 arrests were made. At times, the police deployment outside the works exceeded 4,500 officers.[85] Another source of conflict were the extreme right-wing groups which mushroomed during the decade. They were often opposed by anti-fascist fronts and the result was violence in which demonstrators, counter-demonstrators, and police became embroiled. At one such occasion in Lewisham in 1977, 270 police officers and 57 members of the public were injured.[86] Occasionally there were fatalities, as in Red Lion Square in 1974[87] and Southall in 1979,[88] and this added to the pressure on the police. Despite such disorder, however, it was still possible for a Home Office Green Paper to be written in the following terms as late as 1980:

The police are the servants of the community. This applies to the policing of demonstrations as to other aspects of policing. The British police do not have sophisticated riot equipment—such as tear gas or water cannon—to handle demonstrations. Their traditional approach is to deploy large numbers of officers in ordinary uniform in the passive containment of a crowd. Neither the Government nor the police wish to see this approach abandoned in favour of more aggressive methods.[89]

---

[84] See D. Waddington, K. Jones, and C. Critcher, *Flashpoints: Studies in Public Disorder* (1989), 56–8; A. Sherr, *Freedom of Protest, Public Order and the Law* (1989), 41–9.

[85] *Review of the Public Order Act 1936*, para. 7.

[86] Ibid.

[87] See Scarman, *Red Lion Square Disorders*.

[88] See National Council for Civil Liberties, *Southall, 23 April 1979: The Report of the Unofficial Committee of Enquiry* (1980).

[89] *Review of the Public Order Act 1936*, para. 15.

Great changes began early in the 1980s. The catalyst was the rioting that took place in the spring and summer of 1981. Disturbances on a scale unseen in Britain for generations occurred in Brixton and Southall in London, in Toxteth in Liverpool, in the Moss Side area of Manchester, and in parts of the West Midlands. Similar outbreaks of disorder occurred later in the decade in Handsworth and Tottenham in 1985 and in Bristol in 1987, but it was the unexpectedness and ferocity of the disorders of 1981, and the fact that much of the anger was directed squarely at individual constables, which so shocked and frightened senior police officers. Shortly afterwards came the bitterest and most divisive industrial action since the general strike in 1926. The miners' dispute of 1984–5 involved injuries to 1,486 police officers in the United Kingdom, with casualties amongst members of the public being at least at this level. The police worked over 14 million hours' overtime, at an estimated cost of £140 million.[90] Accompanying these events, and partly in reaction to them, three significant changes in the way the police dealt with public disorder became apparent.

The first change related to the national co-ordination of policing strategy. Police forces in this country are local forces accountable to local police authorities, except in London where the police authority is the Home Secretary. Despite the absence of a UK-wide police force, a National Reporting Centre was set up in Scotland Yard in 1972. This body, which is only activated in times of crisis, played a key role during the miners' strike in co-ordinating police strategy and in the deployment of officers on a national basis in consultation with the Home Office. It is also believed that the Centre gave policing orders to local forces, thereby overruling the operational discretion of chief constables in the regions, and setting a precedent for a covertly centralized police force. The second change to emerge was the revival of élite groups of police officers within each police force. These groups are trained for an aggressive role in crowd control, and represent a significant move away from traditional ideas of minimum force and containment. The Special Patrol Group had been particularly controversial in the 1970s, with Lord Scarman drawing attention in his Brixton report to the dangers of 'too inward-looking and self-conscious an *ésprit de corps* developing in the Group'.[91] Nevertheless such bodies, whether the Territorial Support Group (TSG), the Tactical Aid Group (TAG), or the Police Support Units (PSU), enjoyed a renaissance during the miners' strike and have remained a key part of public order policing ever since.

Thirdly, and most dramatically of all, the rioting in 1981 led to a radical

---

[90] See P. Wallington, 'Policing the Miners' Strike' (1985) 14 *Industrial Law Journal* 145.
[91] Lord Scarman, *The Scarman Report: The Brixton Disorders, 10–12 April 1981* (1986), 144–5.

reorganization in the way in which the police deal with disorder. A *Public Order Manual of Tactical Options and Related Matters* was prepared by the Association of Chief Police Officers and its contents were put into effect during the miners' dispute. The document was drafted in secret, allegedly after consultation with the Hong Kong police, with part of its contents only becoming known when the manual was referred to in the course of a criminal trial arising out of the strike. It contains references to long-shield formations, short-shield units, and horses. The first are designed to present a formidable appearance in front of a crowd; the second are sent into a crowd to disperse and/or incapacitate people with truncheons; and the third are used to create a fear and scatter effect.[92] The manual is also alleged to deal with the situations in which CS gas and baton rounds may be used on British streets.[93] To date, twenty-six police forces have been issued with CS gas, a practice which has been upheld in the Court of Appeal.[94] There have been recent allegations that the manual also says that armed police using conventional tactics of isolation or containment would have problems dealing with a gunman hiding in a crowd or in a building protected by a mob. In this situation, the manual is reported as providing that armed officers would be protected by an inner cordon of police with plastic bullets and CS gas and an outer ring of riot squad officers. These would provide a 'sterile' area enabling police to 'neutralise' a gunman firing at them.[95]

These dramatic alterations in the approach to public order policing are all the more important when the breadth and range of the restrictive law available to the police is remembered. Thus, during the miners' strike, over 11,000 arrests were made, on suspicion of offences ranging right across the available spectrum, from obstruction of the highway to riot. Procedural laws were also employed such as binding over orders and bail conditions.[96] After the strike, the police faced a variety of civil actions in relation to alleged false imprisonment and assaults against pickets. One Yorkshire miner received £60,000 damages after being beaten unconscious,[97] and in a highly controversial out of court settlement South Yorkshire police agreed to pay £500,000 in damages and costs to 39 miners injured during violent clashes outside the Orgreave coking plant near Sheffield. Charges against 95 miners in relation to events on the

---

[92] See generally G. Northam, *Shooting in the Dark* (1988).

[93] The relevant guidelines are now set out at 174 HC Debs. (WA) 429–30 (18 June 1990) (Mr Peter Lloyd).

[94] *R. v. Secretary of State for the Home Dept., ex parte Northumbria Police Authority* [1989] QB 26 (CA).

[95] *Independent*, 9 Apr. 1990.

[96] *R. v. Mansfield Justices, ex parte Sharkey* [1985] QB 613. On the miners' dispute in general, see Wallington, 'Policing'.

[97] *Guardian*, 19 Dec. 1990.

same day had earlier collapsed and the plaintiffs in the civil actions had a variety of complaints, ranging from broken legs to multiple injuries.[98]

It would be astonishing if these dramatic changes in the approach to public order policing had not spilt over into the non-industrial arena and this is in fact what has happened. Since 1985, popular protest, public processions, and assemblies have generally been allowed to proceed, but they have involved much more aggressive levels of confrontation with the police than in earlier eras, with far more allegations and counter-allegations about disproportionate police action and official brutality. The annual movement of large numbers of travellers to celebrate the summer solstice in the vicinity of Stonehenge has led to repeated violent clashes with the police, with the most dramatic of these (in 1985) having since come to be immortalized as the 'Battle of the Beanfield'. In 1989, 800 officers, two helicopters with searchlights, snatch squads, mounted police, and a private security firm were deployed to deal with the 500 travellers who had made the trip that year.[99] Also in 1989, a protest march about *The Satanic Verses* attracted 20,000 demonstrators to Hyde Park, but the occasion degenerated into one of serious disorder, with running battles between the police and protestors continuing for over three hours.[100] Severe disturbances also broke out in central London during and after a protest march and meeting about the community charge or 'poll tax' on 31 March 1990. The police decision to disperse the demonstrators using special equipment and forty-nine mounted officers was particularly controversial.

Student groups have been the focus of similar police action on two occasions. In 1985, considerable disquiet was caused by the methods used by Greater Manchester Police to deal with protestors during the visit of the then Home Secretary, Leon Brittan, to Manchester University. The way in which the police had cleared a way through to the front door of the building at which Mr Brittan was to speak was the focus of allegation and counter-allegation, and a postgraduate student was afterwards awarded £50,000 damages for assault, false imprisonment, and malicious prosecution.[101] In 1988, part of a large student protest against education loans was broken up by the police after it had gathered on the south side of Westminster Bridge, away from the agreed route of the procession. To police this occasion, the authorities used 36 vans, 1,500 officers, 2 boats, 1 helicopter, and 47 mounted officers in what the Home

---

[98] *Independent*, 20 June 1991.

[99] See Ewing and Gearty, *Freedom under Thatcher*, 125–8.

[100] Metropolitan Police Commissioner, *Annual Report 1989*, Cm. 1070 (1990), 20.

[101] *Independent*, 31 July 1990. See generally, Manchester City Council, *Leon Brittan's Visit to Manchester University Students' Union: Report of the Independent Inquiry Panel* (1985).

Secretary afterwards described as 'a controlled action . . . not a charge'.[102] The National Union of Students estimated the number of injured at over 100 and made a series of allegations about police brutality. The Police Complaints Authority appointed the Cambridgeshire Chief Constable to investigate the day's events, and a year later his report unequivocally exonerated the police from any wrongdoing. His conclusions included 'praise for all officers involved in policing the event'.[103]

There have been great difficulties in using the police complaints procedure to resolve questions of alleged excessive force by the police. After serious disturbances outside News International's Wapping plant in 1987, complaints involving more than 100 officers were referred to the Authority, but a large number of prosecutions have since failed due to delay in bringing charges. Many police officers were angry that the cases had been brought at all, since the methods employed by the defendant constables were said to have involved no more than what was set out in the Home-Office-approved *Tactical Options Manual*. This is the nub of the difficulty. The police have over the years developed a perception of how to deal with popular protest which is at variance with how the general public and the demonstrators expect to be treated. It is understandable that, after the riots of 1981, the police should have sought to learn more efficient tactics of crowd control and self-protection. It is clear from the miners' strike and subsequent events, however, that what was learnt then has become a standard rather than an exceptional response to disorder. Of particular concern is the fact that the whole process has taken place in secret, without any opportunity for democratic input or control. The controversy that surrounds public order policing today is largely a consequence of this lack of accountability.

POLITICAL PARTIES, PRESSURE GROUPS, AND LAW REFORM

An important point to be made in defence of the police is to recognize that they can hardly be held responsible for the increasingly violent society which they are being asked to protect. The rioting and industrial disorder of the 1980s was more serious than at any time since the war. There has, however, been no economic or social rights perspective in this area. The extent to which major economic and social developments account for the changes that we have seen in the frequency and intensity of public disorder and popular protest is hotly contested. There are many

---

[102] 142 HC Debs. (WA) 174–6 (29 Nov. 1988) (Mr Douglas Hurd).
[103] Metropolitan Police Commissioner, *Annual Report 1989*, 20.

within the Conservative Party who view such disturbances as manifestations of straightforward criminality, to be dealt with by the police without regard to context or motive. In this way the Tories have tried—often successfully—to turn the issue of freedom of assembly into a subset of their general commitment to law and order, thus embarrassing the Labour Party which they can then seek to present as 'soft on crime'. In this regard, the following exchange during the Public Order Bill's second reading between the then Home Secretary, Mr Hurd, and his Labour shadow is instructive:

MR HURD. Some will speak, quite naturally, as they have begun to do from the Opposition Benches, this afternoon, for those who define liberty in terms of the right to march, picket and demonstrate, to go unhindered to a football match, to do what they like on housing estates and in shopping precincts without police interference.
MR KAUFMAN. Nonsense.
MR HURD. Well we shall judge what happens in debate.[104]

Labour are less comfortable than the Conservatives in their handling of issues relating to freedom of assembly and public order. On the one hand, they reject what they regard as an overly simplistic approach which equates all disorder (other than police disorder) with criminality. Labour's inclination is to look for root causes. On the other hand, however, they are very wary of being laid open to the accusation from their political opponents that they support lawlessness. The Labour position, therefore, is often a rather complex one which does not lend itself to easy media sound bites, and this in turn compounds its problems of presentation. This was seen in its most acute form during the miners' strike when the Labour leader had to engage in a series of complex verbal manœuvres so as to avoid appearing to back violence without at the same time giving the impression that he was supporting the Government.

The politics of such situations apart, however, it is difficult to avoid the conclusion that the disorder that has surrounded industrial disputes in the 1980s was closely bound up with political and economic factors such as a rising unemployment rate and a perception that management, government, and police have been engaged in co-ordinated action against workers. As regards disorder unrelated to industrial action, the simple law and order equation becomes even more difficult to maintain. There are some who share the view of Labour MP Mr Clive Soley, that until 'there is some understanding of the Socialist principle of more equity between the distribution of wealth and power, the riots that have broken out on our streets will continue to distort the fabric of our society'.[105] As far as the inner

---

[104] 89 HC Debs. 800 (13 Jan. 1986).
[105] 89 HC Debs. 857 (13 Jan. 1986).

city riots are concerned, it is difficult not to agree with Lord Scarman that the 'core of the problem' is:

[a] decaying urban structure, with its attendant evils of bad quality and inadequate housing, and a lack of job opportunities, with its inevitable evil of high unemployment. These depressing conditions coexist with the crucial social fact that these areas have a high proportion of ethnic minority groups—blacks and Asians. And these groups believe and feel, with considerable justification, that it is the colour of their skins, and their first or second generation immigrant origins which count against them in their bid for a fair share in our society. We cannot avoid facing this important racial dimension.[106]

As the inner-city riots, the miners' dispute, and the interchange between Mr Hurd and Mr Kaufman strikingly demonstrate, the concerns of freedom of assembly and public order have regularly both reflected and generated national political controversies. This has always been the case. The Public Order Act 1936 was a response to the fascist threat then facing not only Britain but also the rest of Europe. In the 1960s, opposition to nuclear arms and the Vietnam war, two of the key political issues of the day, led to many protests. In the 1970s, disputes such as those at Saltley Gate and at Grunwick became part of political folklore. None of this is at all surprising. The freedom to assemble is most likely to be powerfully utilized when some matter of deep political concern is at issue. A correlation between political passion and popular protest is therefore to some extent inevitable. We should expect to see the most controversial topics of the day translated on to the streets. That has been the case in the 1980s, when large public protests have reflected the electorate's concern over issues such as education, the National Health Service, the Gulf war, and, as we have already seen, the community charge.

Reform in the law in this area has generally been reactive to events and incidents. The Public Order Act 1986, for example, owes its existence first to disturbances at Southall in 1979[107] which led to a Green Paper on possible changes in 1980,[108] and secondly to the miners' strike in 1984–5 which resuscitated a drive for reform which appeared in the intervening period to have lost momentum. Some laws reach the statute-book as responses to isolated mischiefs which for various reasons manage to become central items on the political agenda for short periods of time. Into this category falls section 39 of the Public Order Act, introduced late in the bill's passage through Parliament to deal with the 'sense of outrage at the degradations' caused to a farmer in the south of England by 'the mass invasion' of his land by the 'hippy convoy'.[109] Section 43 of the

---

[106] _The Scarman Report_, p. xiv.
[107] 969 HC Debs. 441 (27 June 1979) (Mr Whitelaw).
[108] _Review of the Public Order Act 1936._
[109] 103 HC Debs. 830 (4 Nov. 1986) (Mr Hogg).

Education (No. 2) Act 1986, concerning freedom of speech in universities, polytechnics, and colleges, was a response to the difficulties right-wing speakers like John Carlisle MP and Ray Honeyford were experiencing in getting a hearing before student gatherings. In the words of the then Prime Minister, Mrs Thatcher, '[h]igher education requires freedom of thought and freedom of speech. It is not for people who expect to have that for themselves to deny it to others'.[110] The Entertainment (Increased Penalties) Act and Part IV of the Public Order Act 1986 were introduced to deal with acid house parties[111] and football hooliganism[112] respectively. The danger in *ad hoc* legislation of this type is that the drafting may not be tight enough to achieve the desired ends or, alternatively, that the law that is produced is so broad that it comes to be applied to situations which were not anticipated by Parliament.

The centrality of public-order/freedom-of-assembly issues to mainstream political debate is a fact with which pressure groups and law reformers have each had to grapple. Apart from the police, there are few broadly based pressure groups in this area and there is no evidence of the higher courts being used to further the aims of such organizations. Proposals for reform in the law have come from three sources: the Law Commission, the House of Commons Select Committee on Home Affairs, and Lord Scarman. Each has enjoyed some degree of influence. The Law Commission's proposals on the rationalization of a large part of the law on public order were broadly accepted by the government.[113] The Select Committee's 1980 report on public order recommended an advance notice requirement for public processions, the introduction of a power to impose conditions on assemblies, and the extension of the criteria for imposing conditions on processions.[114] Variations on all these proposals were later to emerge as part of the Public Order Act 1986. Lord Scarman's criticisms of the race relations legislation in his Red Lion Square report led to a change in the law[115] and his recommendation in his Brixton report that a specific offence of racially prejudiced conduct be included in the police's discipline code has been adopted.[116]

The risk with all law reform proposals, of course, is that the authorities will discount certain recommendations whilst enthusiastically adopting

---

[110] 94 HC Debs. 159 (18 Mar. 1986). A list of the meetings disrupted is at 100 HC Debs. (WA) 189–92 (25 June 1986) (Mr Walden).

[111] See the short debate at 177 HC Debs. 67–9 (23 July 1990).

[112] Home Affairs Committee of the House of Commons, *Policing Football Hooliganism* (HC 1 of 1990–1).

[113] Law Commission, *Criminal Law: Offences Relating to Public Order* (Report no. 123, 1983) (HC 85); Public Order Act 1986, pt. 1.

[114] Home Affairs Committee of the House of Commons, *The Law Relating to Public Order* (HC 756 of 1979–80).

[115] Scarman, *Red Lion Square Disorders*, para. 125; Race Relations Act 1976, s. 70.

[116] *The Scarman Report*, p. xviii.

those that conform to their policy preferences. Thus, the police have sought to heed Lord Scarman's call for better training in the handling of disorder, but, writing five years after his Report, his Lordship felt compelled to deplore the fact that the 'government . . . ha[d] not tackled the core of the inner city problem—the factor of racial disadvantage'.[117] Just as serious in his view was the disregard of another key proposal relating to the establishment of 'a system of independent investigation' of allegations of police wrongdoing; despite Lord Scarman's view that 'the police must not investigate themselves', no change has occurred in this critical area, though the level of independent supervision of police investigations of themselves has been improved.[118] Nor have other suggestions by Lord Scarman, such as a new power to ban individual marches, found their way into law. Thus we may conclude that, whilst independent bodies and individuals have had an important influence on how the law has developed in this field, the initiative has remained firmly with the government, something which is hardly surprising given the political sensitivity of the issues involved.

AMERICAN AND EUROPEAN COMPARISONS

The United States Constitution declares in the First Amendment that Congress 'shall make no law . . . abridging the freedom of speech . . . or the right of the people peaceably to assemble'. After many decades of neglect, this guarantee was successfully invoked in a number of important decisions by the Supreme Court around the turn of the 1960s, the most famous of which was the *Pentagon Papers* case decided in 1971.[119] This new judicial emphasis on freedom meant that no state could 'forbid or proscribe advocacy of the use of force or of law violation except where such advocacy [was] directed to inciting or producing imminent lawless action and [was] likely to incite or produce such action'.[120] The implications of this for freedom of assembly were brought home in the Supreme Court of Illinois decision in *Skokie* v. *National Socialist Party of America*.[121] A neo-Nazi group proposed to march through a predominantly Jewish village, a substantial minority of whose inhabitants were survivors of Hitler's concentration camps. Against the objection of these locals, the Illinois court not only upheld the group's right to march but

---

[117] *The Scarman Report*, p. xvii.
[118] PACE 1984, part ix.
[119] *New York Times* v. *US* 403 US 713 (1971).
[120] *Brandenburg* v. *Ohio* 395 US 444, at p. 447 (1969); see also *Cohen* v. *California* 403 US 15 (1971) and *Coates* v. *City of Cinncinnati* 402 US 611 (1971).
[121] 373 NE 2d 21 (1978).

also permitted them to wear their National Socialist uniform and display their party emblem, the swastika. It was 'firmly settled' that 'the public expression of ideas may not be prohibited merely because the ideas are themselves offensive to some of their hearers'.[122] The state had 'no right to cleanse public debate to the point where it [was] grammatically palatable to the most squeamish amongst us'.[123] The court, moreover, could 'not indulge the facile assumption that one can forbid particular words without also running a substantial risk of suppressing ideas in the process'.[124]

The question whether a similar line of reasoning would be followed in Britain so as to emasculate our present controls on racist speech is a very controversial one, particularly in the context of the present Bill of Rights debate. What is clear is that the European Convention on Human Rights is nothing like so unequivocal as its American counterpart. Article 11 provides that '[e]veryone has the right to freedom of peaceful assembly' but then goes on to say:

No restrictions shall be placed on the exercise of these rights other than such as are necessary prescribed by law and are necessary in a democratic society in the interests of national security or public safety, for the prevention of disorder or crime, for the protection of health or morals or for the protection of the rights and freedoms of others.

The Convention has not often been argued before British courts in this context. It has occasionally been referred to in the political debate, sometimes in the context of justifying a proposed restriction.[125] In 1981, the European Commission of Human Rights held that a two-month ban on marches in London did not infringe Article 11.[126] Neither a neo-Nazi meeting in Austria[127] nor a demonstration by Laplanders in Oslo[128] found any protection in its provisions when they were banned or broken up by the authorities. In only one case does the European Court of Human Rights suggest that 'participants must . . . be able to hold [a] demonstration without having to fear that they will be subjected to physical violence by their opponents'.[129] Given the breadth of Article 11 and the deferential record shown by British judges in this area, it is unlikely that incorporation of the Convention into domestic law would presage a wholesale revision of our public order law, though the banning of racist marches

---

[122] Ibid., at p. 23, quoting *Bachellar* v. *Maryland*, 397 US 564 (1970), at p. 567 (1970).
[123] Ibid., at p. 24.
[124] Ibid.
[125] *Review of Public Order Law*, para. 1.8.
[126] *Christians against Racism and Fascism* v. *UK* (1981) 24 YBECHR 178.
[127] *App. 9905/82* v. *Austria* (1985) 7 EHRR 137.
[128] *Apps. 9278/81 and 9415/81* v. *Norway* (1983) 6 EHRR 357.
[129] *Plattform 'Ärzte für das Leben'* v. *Austria* (1988) 13 EHRR 204.

might be vulnerable to the classic liberalism which protected the neo-Nazis in *Skokie*.

CONCLUSION: REFORM

In this chapter we started with the proposition that the law in this area of civil liberties was driven by two different objects, one that sought to promote freedom of assembly, the other that strove to preserve public order. From our examination of how the law has developed, we may conclude that neither has achieved pre-eminence, but that the question of which to prefer has become a matter for the police rather than for Parliament or the courts. The law is designed in such a way as to enable the police to choose either alternative without fear of acting beyond their legal powers. Judicial deference is the order of the day. This is the context in which the drift towards a more public-order-oriented police force should be viewed. The trend is manifested most clearly of all in the quasi-military way in which protest is now managed. It is submitted that the most pressing reform in this area may well be the restructuring of police discretion so as to draw the authorities firmly back from this quickly formed habit of aggressive policing. Bound up with such a change, and helping to achieve it, would be a renewed emphasis on accountability.

The situation at the moment is most unsatisfactory. Police forces are local and therefore not accountable to Parliament, yet many decisions on public order are taken in a way that appears so co-ordinated that it suggests centralized control. This was most evident during the miners' strike. At the same time chief constables habitually resist local accountability on 'operational' matters. The net effect of this system is that large areas of extremely controversial policing are beyond the reach of the democratic process. The revelations about the previously unheard of *Tactical Options Manual* make this point clearly. It is suggested that the most pressing area for empirical research in this field lies in the examination of discretion in the area of public order policing. Valuable work has already been on the miners' dispute[130] and on the Stonehenge travellers,[131] but much more needs to be accomplished at this practical rather than theoretical or legal level.

Apart from tackling these questions of police actions, secrecy, and lack of accountability, is there any legal change which should be made? The European Community, the Social Charter, and most international instru-

---

[130] See Wallington, 'Policing'.
[131] NCCL, *Stonehenge: A Report into the Civil Liberties Implications of the Events Relating to the Convoys of Summer 1985 and 1986* (1986).

ments have no relevance in this part of the law. As we have seen, the European Convention on Human Rights and Fundamental Freedoms presents a superficially attractive alternative, but its effect might well be very slight in practice. Perhaps a more promising route would be to enact a statutory right to demonstrate, setting out in detail the circumstances in which, and the extent to which, this liberty could be exercised. Such a law, which would have the advantage of being drafted by elected representatives rather than manufactured out of vague language by unaccountable judges, would begin to reclaim freedom of assembly for a statute and common law that has almost forgotten its importance.

3

Police Power

ROBERT REINER and LEONARD LEIGH

> Much power is vested in a police constable, and many opportunities are given him to be hard and oppressive, especially to those in his custody. Pray avoid harshness and oppression; be firm but not brutal, make only discreet use of your powers . . . You are not absolutely *bound* to arrest. You ought to exercise your discretion, having regard to the nature of the crime, the surrounding circumstances, and the condition and character of the accuser and the accused.[1]

INTRODUCTION: THE PROBLEM OF POLICE POWER

The public standing of the police is now at its lowest ebb since they came to be established in their modern form in the first half of the nineteenth century. This is attested to by many opinion polls, which since 1989 have regularly recorded much lower levels of general public confidence in the police than at any time since such polls began. In addition, there is the more solid evidence of a number of systematic research surveys which have charted this decline, notably the Home Office British Crime Surveys.[2]

The critical turning point is 1989, when a seeming Pandora's Box of scandals was opened up by the Court of Appeal's October decision to release the 'Guildford Four' (the three men and a woman sentenced to life imprisonment in 1974 for the Guildford and Woolwich pub bombings). Much of the damage to the criminal justice system's reputation focused on the police, as the appeal decision was based on evidence (gathered by the Avon and Somerset Constabulary) showing that some of the Surrey police officers in the 1974 case 'must have lied' (in the words of Lord Lane LCJ). Since 1989 similar scandals have burst on to the public scene with numbing regularity: the associated case of the Maguire family, and the 'Birmingham Six', whose convictions were also overturned in 1990 and 1991 respectively, were two other long-running *causes célèbres* whose reverberations will continue for many years. So great was the damage

---

[1] Preface by Sir Henry Hawkins, 'one of Her Majesty's Judges', to C. E. Howard Vincent, *The Police Code and General Manual of the Criminal Law for the British Empire* (Francis Edwards & Simkin, London, 1893).

[2] See in particular W. G. Skogan, *The Police and the Public in England and Wales: A British Crime Survey Report* (1990).

done to confidence in the system of criminal justice, that the Home Secretary had to establish a Royal Commission on Criminal Justice under the chairmanship of Lord Runciman, the first to be appointed since the Conservatives took office in 1979, and the first since the Royal Commission on Criminal Procedure reported in 1981.

The signs of a coming crisis in public confidence in the police long predate this latest wave of scandals, of course. For the last two decades there have been clear indications of an erosion of the massive public esteem, indeed affection, which the British police enjoyed, especially in the middle years of this century. However, until relatively recently, this withdrawal of support and greater questioning of the police was largely confined to particular groups: notably the young, economically and socially dispossessed, and especially black population of the inner cities, and the liberal intelligentsia, the '*Guardian*-reading classes'. This decline in police legitimacy has many roots and symptoms.[3] However, a central concern has been the question of how the police use (or perhaps abuse) their legal powers.

Anxiety about this was the main ideological basis of the opposition to the initial establishment of modern professional police forces in the early nineteenth century. Many—at both ends of the spectrum of social status—felt that a professional police force was alien to British traditions of civil liberty, and necessarily inimical to individual rights and freedom. The slow, unsteady, uneven, but ultimately successful achievement of Sir Robert Peel and those who were appointed to lead the nascent police force has often been told.[4] Clearly central to acceptance of the force by the majority of the public was the belief that the police were subordinate to the rule of law, and that they lacked either legal powers or the coercive capacity to police other than by the consent of the populace. This is the partly mythical concept of the police as 'citizens in uniform', which played a focal part in the reports of both the 1929 and 1962 Royal Commissions on the Police.[5] This concept was vitiated already in 1929, let alone 1962, by the accretion of power based on bureaucratic organization, technology, and training, and it has been further undermined since by a steady accretion of formal legal powers. The myth of the police as 'citizens in uniform', whatever its attractions as an ideal, conceals the true nature of the growth of police power and powers, and the consequential problem of regulating their exercise. When translated into the further myth that the

---

[3] These are analysed more fully in R. Reiner, *The Politics of the Police* (1992), ch. 2; and R. Reiner, 'Policing A Postmodern Society' 55 *Modern Law Review* 6 (1992).

[4] T. A. Critchley, *A History of Police in England and Wales* (1978); Reiner, *Politics of Police*, chs. 1 and 2; C. Emsley, *The English Police: A Political and Social History* (1991).

[5] See the *Report of the Royal Commission on the Police*, Cmnd. 1728 (1962), para. 30, pp. 10–11.

chief constable is in constitutional essence just a constable, it has become the basis of the doctrine of constabulary independence, and acts as a bulwark against effective legal or democratic accountability.[6]

The Royal Commission on Criminal Procedure Report of 1981, and the Police and Criminal Evidence Act 1984 which it generated, constitute a significant watershed, consolidating and clarifying the changes in formal powers and concrete practice which had built up since the Second World War, separating the police officer from the ordinary citizen. The new settlement is based on a new ideological rationale. It is recognized that the police do have powers significantly different from the citizen. However, these are formulated so that they fundamentally balance the tension between police effectiveness and individual liberty.[7] Furthermore, an adequate machinery of safeguards is constructed to regulate the exercise of police powers.

This ideological rationale is compatible with the provisions both of the United Nations Covenant on Civil and Political Rights 1966 and the European Convention on Human Rights. Although the terms of the latter document in particular were strongly influenced by English law, they are drafted in terms sufficiently wide to accommodate continental systems of procedure as well.[8] They are thus compatible with modern English procedures which, during the police phase at least, increasingly resemble procedures used elsewhere in Europe.[9] Certain well-publicized exceptions such as arrest for terrorism and orders for surveillance apart, English criminal procedure appears to be in conformity with the European Convention as it has been interpreted by the European Commission and the Court.[10] There are at least two variables at work. The first is that English law changes, and any such change may produce a collision with the Convention and indeed with other international instruments. The second is that, in the long term, the European Court in particular may so interpret the Convention as to reduce the flexibility of disposition available to states and may, in consequence, produce incompatibilities between the Convention and English law and practice. There is, however, a further

---

[6] Cf. T. Jefferson and R. Grimshaw, *Controlling the Constable* (1984); L. Lustgarten, *The Governance of the Police* (1986); R. Reiner, *Chief Constables: Bobbies, Bosses or Bureaucrats* (1991), chs. 1, 2, 11, 13.

[7] This is the concept of 'the fundamental balance', which underpinned the deliberations of the Royal Commission on Criminal Procedure, cf. their 1981 *Report*, Cmnd. 8092 (1981), paras. 1.11 and 1.12, pp. 4–5.

[8] See further D. Poncet, *La Protection de l'accusé par la Convention européenne des droits de l'homme* (1977).

[9] Indeed, English practices in detention and interrogation are superior in respect of the provision of legal advice and the tape-recording of interviews to procedures in e.g. France and Germany.

[10] See further L. H. Leigh, 'La Procédure pénale anglaise à la lumière de la Convention européenne des droits de l'homme' (1988) 3 *Rev. Sci. Crim.* 453–68.

point worth noting. A comprehensive analysis of English law and practice may disclose that cherished practices are by no means invulnerable to challenge. We deal with some of these matters below.

The influence of the European Convention on police practices and the law which regulates them is actually and potentially significant. It is not pervasive simply because certain cherished rights, while prominent in common-law systems, either do not fall within the Convention or do so only to a limited extent. This is true, for example, of the right to silence and the privilege against self-incrimination which are affected only to the extent that Article 3 of the European Convention forbids the use of torture, inhuman or degrading treatment or punishment, and therefore, as is well known, forbids the use of interrogation techniques involving sensory deprivation.[11] Other techniques such as the obtaining of intimate and non-intimate samples are only marginally affected by provisions against torture, and the integrity of other procedures such as identification parades may be affected by fair trial guarantees such as Article 6 of the European Convention, provided that these are largely and liberally construed in favour of the defence.

The influence of the European Convention is indirect. As an international instrument it does not have direct effect in English domestic law, but because the United Kingdom permits individuals to petition Strasbourg, it provides a forum for the redress of individual grievances which, in turn, provokes changes in domestic law to bring English law and practices into conformity with the Convention. It thus provides an impetus to reform of an immediate character. The European Convention, like other international instruments, will in any event be taken account of in the formulation of new legislation.[12] In litigation courts and litigants increasingly refer to the Convention, but this seldom influences the case in hand. While the Convention can be invoked in the interpretation of legislation where ambiguity appears, and as an aid in the interpretation of Codes of Practice, and therefore in the regulation of discretion, it only seldom has more than rhetorical significance, outside the particular context of terrorism.[13] Care has been taken to ensure that English law and practice will not be subject to challenge in Strasbourg, at any rate where the police procedures in question are relatively visible and so likely to be the subject of effective challenge. The European Convention has, however, a potential impact in the regulation of discretion which should not be

[11] *Ireland* v. *UK* (1978) 2 EHRR 25. Art. 7 of the UN International Covenant on Civil and Political Rights is somewhat wider.

[12] For a striking example, see *Report of the Commission to Consider Legal Procedures to Deal with Terrorist Activities in Northern Ireland*, Cmnd. 5185 (1972).

[13] As e.g. in *Re Chinoy* [1991] COD 207. By contrast, Lord Diplock in *Gleaves* v. *Deakin* [1980] AC 477 proposed that a prosecutor might take account of Art. 10 in deciding whether to permit a prosecution for criminal defamation to proceed.

overlooked, for the terms of the Convention could well influence advice to police officers concerning how to operate procedures.

These notions will be elaborated and explored in the rest of this chapter. The next section will develop a theoretical analysis of the concept of police discretion, which lies at the root of the problems of understanding and regulating police power. Following this, we chart and attempt to explain the development of police powers in Britain since 1950, taking due account of the actual and potential influence of international and European law. Finally, we will review the implications of empirical research on police practice for an appreciation of the impact of legal change on policing.

'BLUE-LETTER LAW':
PRINCIPLES, RULES, AND ACTION

Roscoe Pound's celebrated distinction between 'the law in the books' and 'the law in action' has been the theoretical underpinning for much of the subsequent development of legal realism, sociological jurisprudence, and socio-legal studies. Research and debate on the police, which has flourished in the last thirty years in Britain, North America, and increasingly in other parts of the world, has been a paradigmatic case of this realist approach. Its primary thread has been the revelation and analysis of a gap between legality and police practice, and the development of policies to close that gap.

Much of the early American work on the police focused on the 'gap' issue, stemming from a recognition of the inevitable discretion enjoyed in practice by the police, and especially by the lowest but most significant patrol ranks. A clutch of influential articles and books in the early 1960s (the era of *Mapp* and *Miranda*, the seminal Supreme Court decisions affirming a strong exclusionary rule for evidence obtained in violation of suspects' rights) homed in on the implications for principles of due process of law and civil rights.[14] By the mid-1960s the fact of police discretion, and its practical departure from legality, were taken for granted by empirical researchers. Thus the first major observational study of police work in the USA, Jerome Skolnick's *Justice Without Trial*, begins by observing that his purpose was 'not to reveal that police violate rules

---

[14] The key examples are J. Goldstein, 'Police Discretion not to Invoke the Criminal Process: Low Visibility Decisions in the Administration of Justice' (1960) 69 *Yale Law Journal* 543; W. LaFave, 'The Police and Non-Enforcement of the Law' (1962) *Wisconsin Law Review*; H. Goldstein, 'Police Discretion: The Ideal vs. the Real' (1963) 23 *Public Admin. Rev.*; W. LaFave, *Arrest* (1965); K. C. Davis, *Discretionary Justice* (1969); and *Police Discretion* (1977).

and regulations. That much is assumed.'[15] Instead, he was concerned to advance the understanding of the determinants of police deviation from the rule of law, so as to provide an informed basis for reform. This has been the flavour of most subsequent American research, which has been concerned to analyse and to structure police discretion, taking for granted its existence and indeed inevitability.[16]

In Britain the wisdom (never mind the inescapable fact) of police discretion has always been officially accepted, as the quotation from Sir Henry Hawkins which heads this chapter illustrates. Nevertheless, when empirical research on the police began in the 1960s, the leading early examples all emphasized the extent to which the law in action was shaped by the exercise of police discretion.[17] This is summed up in the title of one piece from the early period of British police research: 'The Police Can Choose.'[18]

There have always been two broadly distinct interpretations of discretion: *de facto* discretion, and *de iure* discretion. The former concept assimilates the ideas of discretion and choice. It is illustrated by Kenneth Culp Davis's frequently cited definition: 'A public officer has discretion whenever the effective limits on his power leave him free to make a choice among possible courses of action.'[19] It is this usage which tends to inform empirical research literature, in which the idea of discretion sometimes becomes virtually coterminous with all decision-making.[20]

The problem with the *de facto* interpretation of police discretion is that in legal discourse (and indeed in ordinary language) the term 'discretion' has an inescapably normative dimension, which is eclipsed by an essentially positivist, behaviourial account. We would not speak of the police officer's discretion to take a bribe for not enforcing the law, even if this was one of his routine decisions and within his effective power. Implicit in the concept of discretion is the idea of background principles against which the propriety of the exercise of discretion can be reviewed. As Fletcher put it: 'There is a contingent quality to discretionary decisions.

---

[15] J. Skolnick, *Justice without Trial* (1966), 22.

[16] In the USA there *has* been recent debate about the legal and constitutional propriety of police discretion. Davis, *Discretionary Justice* and *Police Discretion*; R. Allen, 'The Police and Substantive Rule-Making: Reconciling Principle and Expediency' (1976) *Pennsylvania Law Review* 62; R. Allen (ed.), 'Discretion in Law Enforcement' (1984) 47 *Law and Contemp. Problems* 4; G. H. Williams, *The Law and Politics of Police Discretion* (1984).

[17] M. Banton, *The Policeman in the Community* (1964), ch. 5; J. Lambert, *Crime, Police and Race Relations* (1970), ch. 5; J. Young, 'The Role of the Police as Amplifiers of Deviancy', and M. Cain, 'On the Beat', both in S. Cohen, (ed.), *Images of Deviance* (1971).

[18] J. Lambert, 'The Police Can Choose' (1969) 14 *New Soc.* 352.

[19] Davis, *Discretionary Justice*.

[20] e.g. M. Brown, *Working the Street: Police Discretion and the Dilemmas of Reform* (1981), 3–4.

There is in the background a sense of a higher authority's approving and tolerating the discretion.'[21] Or in Dworkin's pithier formulation: 'Discretion, like the hole in a doughnut, does not exist except as an area left open by a surrounding belt of restriction.'[22]

The tightness of the belt of restriction can vary. Discretion may be more or less restrictively bounded and structured by the higher authority which approves it. In Dworkin's analysis, it may be more or less 'weak' or 'strong', depending on whether judgement must be exercised according to definite standards or objectives with reference to which discretion is structured.

However, in addition to variation according to the 'weakness' or 'strength' of the discretion given, there is a further dimension of the possibilities for reviewing actions. 'Weak' discretion decisions may in effect be as difficult to review as 'strong' discretion. This sort of situation often occurs in policing. The exercise of legal power by constables requires their prior observation of specific factual conditions which give the constable reasonable grounds to suspect an offence.[23] These facts can usually only be ascertained by the constable observing specific actions, of which his superior officers cannot have direct knowledge. For this reason they cannot give unqualified orders for the constable to use his powers. In the words of Lawton LJ in *R.* v. *Chief Constable of Devon and Cornwall, ex parte CEGB*:[24] 'Chief constables . . . cannot give an officer under command an order to do acts which can only lawfully be done if the officer himself with reasonable cause suspects that a breach of the peace has occurred or is imminently likely to occur or an arrestable offence has been committed.' Thus even 'weak' discretion, i.e. discretion which must be exercised according to specific standards, may be effectively unreviewable, either within the police service or before the courts.

It is this problem which has particularly vexed analysts of police discretion. The issue has not been seen generally as the strength of the discretion enjoyed by the police. Rather attention has focused on a significant peculiarity of police work, its 'low visibility' as Goldstein expressed it more than three decades ago.[25] The bulk of police work—uniform patrol, and also most detective work—consists of dispersed individual or small-group operations, out of sight of organizational supervisors. The

---

[21] G. P. Fletcher, 'Some Unwise Reflections About Discretion', in R. J. Allen (ed.), *Discretion in Law Enforcement* (1984) 47 *Law and Contemp. Problems* 282. Discretion is here distinguished from 'prerogatives', which 'are intrinsically free of regulation'.

[22] R. Dworkin, 'The Model of Rules', in *Taking Rights Seriously* (1977), 31.

[23] Lustgarten, *Governance*, 13–14.

[24] [1980] 1 All ER 797 at p. 826.

[25] Goldstein, 'Police Discretion'.

consequence is that: 'The police department has the special property that within it discretion increases as one moves *down* the hierarchy.'[26]

It is important to emphasize that this invisibility is primarily social rather than physical. Almost all exercises of police discretion will be visible to someone other than the police officer concerned. Even if never reported to police supervisors, it will be known to citizens in their capacity as suspects, victims, or witnesses. If any of these citizens are disgruntled by the exercise of police discretion and fail to report it, this is largely because of the anticipated difficulties of establishing a case. The root of the 'invisibility' of the operation of police discretion lies in what has been called 'the politics of discreditability'.[27] The police are accepted as more creditable witnesses than most of the predominantly low-status people they come into adversarial contact with, and their testimony is likely to prevail in any review forum, apart from the relatively rare case where there is independent evidence.

The 'invisibility' of street-level police discretion derives from this social superiority of the police in situations where their decisions are reviewed. The anticipation of this in turn keeps many complaints from ever being raised. This socially constructed 'invisibility' of rank-and-file police decisions gives them a considerable measure of *de facto* discretion. The focus of most empirical research on the police has been to chart and explain the operation of low-level discretion. In effect researchers have claimed that the real policies of the police are determined at ground level, because of the practical autonomy of rank-and-file behaviour from review.

Police discretion has remained an important issue because of the pattern of its exercise which research has revealed. Discretion is officially justified as inevitable because of scarce police resources, and beneficial in permitting non-enforcement of the law when enforcement would violate common-sense notions of justice in particular cases ('obsolete' laws, 'stale' cases, and so on). Problems arise because the criteria invoked for non-enforcement may be less consensual than the police pretend (e.g. the view of domestic disputes as essentially 'private' matters), and because of overwhelming evidence that the outcome of the exercise of police discretion is unjustly discriminatory. This has been especially well established in the area of racial discrimination in the exercise of police powers.[28] This is an inevitable concomitant of police organizations built around a core

---

[26] J. Q. Wilson, *Varieties of Police Behaviour* (1968), 7.

[27] S. Box and K. Russell, 'The Politics of Discreditability: Disarming Complaints against the Police' (1975) 23 *Sociol. Rev.* 2.

[28] The evidence is summarized in T. Jefferson, 'Race, Crime and Policing: Empirical, Theoretical and Methodological Issues', (1988) 16 *Int. J. Sociol. of Law* 521–39; and R. Reiner, 'Race and Criminal Justice' (1989) 16 *New Community: Special Issue on Race, Criminal Justice and the Legal System* 5–22.

activity of uniformed patrol of public space. The police culture which many studies have described, and which has been characterized as conservative, sexist, and prejudiced against racial and other minorities, mediates police discrimination. But 'cop culture' is the consequence of rather than the cause of the police mission.[29] It is because policing in a hierarchical society mainly involves processing the least powerful social groups that police culture develops and encourages negative stereotypes of these groups (despite official policies to counteract this). These groups have been evocatively referred to as 'police property'.[30]

In the last decade there has developed a powerful structuralist critique of the earlier tradition in police research, which emphasized police culture as the effective determinant of police practice, given the difficulties of reviewing the low visibility of discretion. The inspiration for this critique was largely the work of Doreen McBarnet, though the structural analysis has been advanced subsequently by several others. McBarnet argued that police discretion arises out of the permissive formulation of rules governing police practice by legislatures and courts. Researchers have made the low-level police 'the "fall-guys" of the legal system taking the blame for any injustices'.[31] Responsibility ought to be placed on 'the judicial and political elites' who make rules of such elasticity as to accommodate unjust police practices. So what most researchers have characterized as *de facto* police discretion violating the rule of law is seen by McBarnet as *de iure* discretion arising from inadequately and loosely formulated law. Police behaviour departs from the rhetoric but not the

---

[29] For analyses of the characteristics and determinants of 'police culture', see S. Holdaway, *Inside the British Police* (1983); and Reiner, *Politics of Police*, ch. 3.

[30] The term comes from Ed Cray, *The Enemy in the Streets* (1972), 11. See also J. A. Lee, 'Some Structural Aspects of Police Deviance in Relations with Minority Groups', in C. D. Shearing (ed.), *Organisational Police Deviance* (1981); and M. Brogden, T. Jefferson, and S. Walklate, *Introducing Policework* (1988), ch. 6. Striking empirical confirmation comes from a recent study of custody in police stations. It was found that the overwhelming majority of people detained were drawn from the 'police property' group of the economically and socially marginal. Over half (55%) had no paid employment, almost all being unemployed young men. Most of the rest (a third overall) were in manual unskilled working-class jobs. Only 6% of the sample had non-manual occupations, and of these only one-third (i.e. 2% overall) were in professional or managerial occupations. Most were young, 87% were men, and 12% were black. In short, the weight of adversarial policing falls disproportionally on young men in lower socio-economic groups; cf. R. Morgan, R. Reiner, and I. McKenzie, *Police Powers and Policy: A Study of Custody Officers* (1990).

[31] D. McBarnet, 'False Dichotomies in Criminal Justice Research', in J. Baldwin and A. K. Bottomley (eds.) *Criminal Justice* (1978); 'Arrest: The Legal Context of Policing', in S. Holdaway (ed.), *The British Police* (1979), 39. See also D. McBarnet, *Conviction* (1981). Subsequent work developing this structuralist perspective include M. Brogden, *The Police: Autonomy and Consent* (1982), ch. 5; C. D. Shearing (ed.), *Organisational Police Deviance*; R. Ericson, *Making Crime: A Study of Detective Work* (1981) and *Reproducing Order: A Study of Police Patrol Work* (1981); R. Grimshaw and T. Jefferson, *Interpreting Policework* (1987).

substance of legality. This analysis accepts the extent of discretion the police enjoy, but shifts the account of its origins. While in principle discretion could be controlled by non-permissive rule formulation, on this structuralist analysis it is not easy to achieve because of the mystificatory function vague rules perform for state elites, in deflecting attention away from their responsibility for police practice. Presumably if the elite was concerned about the outcome of police discretion it could act to restrict it.

The structuralist critique is undoubtedly a powerful and influential corrective to early socio-legal research on the police, which assumed too readily that the 'law in action' could float entirely free of the 'law in the books'. However, in so far as it implies that changing the 'law in the books' is the key to reforming police practice, it is throwing out the baby with the bathwater. The 'low visibility' of routine police work remains a problem for effective regulation of policing. Nor does the structuralist analysis displace the concept of police culture as one factor in police practice, even if it is not the prime mover. To say that the laws governing police behaviour are 'permissive' leaves considerable leeway for police culture to shape police practice (although the culture is itself patterned by the role of the police in the social structure, as argued above).

Analysing the relationship between law and police practice requires a synthesis of the 'culturalist' and 'structuralist' positions. It is far from certain that changes in the black-letter 'law in the books' will be translated into corresponding changes in the 'law in action', above all because of the low visibility of routine police work. On the other hand, the autonomy of police culture is itself a relative one, and can be influenced by changes in organization and training for instance. Furthermore, culture is not automatically translated into action. Police officers may be inhibited by effectively policed rules from acting out cultural values of a prejudicial kind, for example.

Legal rules, then, are neither the complete determinants of police practice (as naïve black-letter lawyers might believe), but nor are they are irrelevant to action, as rule-sceptical realists might claim. What effective policing of the police requires is an analysis of 'blue-letter law': the implicit rules of working practice which are a product of the black-letter rules of formal law, mediated by the police organization and culture. This is well brought out in the distinction between three types of rules made in the celebrated Policy Studies Institute (PSI) research on policing London, one of the key sources for an understanding of the pre-PACE law in action.[32] The PSI study distinguishes between: (*a*) 'working rules', the ones which police officers actually follow in practice, not necessarily

---

[32] D. Smith *et al.*, *Police and People in London*, 4 vols. (1983).

consciously; (*b*) 'inhibiting rules', which are formal legal or disciplinary rules which have a deterrent effect on practice, because they are perceived as likely to be sanctioned; (*c*) 'presentation rules', which are used to impart an acceptable gloss to accounts of actions which are actually informed by different 'working rules'. The relationship between any of these sets of rules and the law is problematic. Legal rules may well be used presentationally, rather than being effective inhibitors let alone working rules. They then act as an ideological facade constructing a cosy canopy over the messy and profane realities of policing.

The art of successfully regulating policing practice is dependent on understanding the complex relationship between formal rules and procedures, the sub-cultural rules of the police themselves, the structure of the police organization, and the practical exigencies of the tasks of policing. In short, from a policy-maker's perspective, the issue is to ensure that formal rules do not become merely presentational, but become working rules, or at least inhibiting ones. Unfortunately, although much research has demonstrated the 'gap' between these types of rules, we understand very little about the determinants of the size of the gap. What makes rules more or less presentational, inhibiting, or actually incorporated into the working rules of police culture? Research has not advanced very far towards an understanding of this vital issue.

In the next two sections we will turn to an account of the key developments in black-letter law domestically and from Europe concerning police powers. The final section will review what empirical research has shown about the gap between legal rules and policing practice, and will be concluded by a consideration of the implications for future policy.

THE DEVELOPMENT OF POLICE POWERS IN ENGLAND AND WALES SINCE 1950

There can be no doubt that the powers of the police have expanded enormously since the post-war period. This is not just a question of legal powers. The manpower and material resources available to the police have increased considerably[33] (although in the last couple of years public expenditure restraint and the search for efficiency in the use of public

---

[33] In 1959 the strength of the police service was 70,000 (this was 6,000 under authorized establishment). By 1969 this had gone up to nearly 98,000, and by 1979 just over 113,000. In 1989–90 police numbers were 126,204, a 15% increase since 1979. During the 1980s civilian staff also increased by 17%, and in the 1970s they had gone up by 46%. (*Statistics of the Criminal Justice System 1969–79* (1980), table 8.1; *A Digest of Information on the Criminal Justice System* (1991), 66.) Expenditure on the police rose from £1,036m. in 1973 to £1,222m. in 1979, and had reached £3,731m. by 1989. (Ibid. 74.)

resources have begun to take their toll on the police, in common with all government services).[34] Their formal legal powers have also been enhanced, both by case-law and statute. At the same time, the mechanisms for attempting to regulate the exercise of police powers have been bolstered in a variety of ways, although how adequately this has been accomplished is very much open to debate. In all these respects the first few years of the 1980s were a watershed of unparalleled significance. The increase in police manpower and resources accelerated dramatically, while the Police and Criminal Evidence Act 1984 (PACE), the Prosecution of Offences Act 1985, and the Public Order Act 1986 constitute decisive landmarks in the development of police powers. (The latter two statutes are mainly be dealt with in chapters 4 and 2 respectively.)

This part of the chapter will concentrate primarily on the regime instituted by PACE. It will begin, however, with a brief consideration of the trends in police powers before PACE, and the background to the Act.

Post-war developments in police powers up to PACE

PACE is recognized by both its friends and foes as signifying a formal extension of police powers. What is debated is the extent to which this is adequately balanced by an effective system of safeguards over the use of police powers. The extension of formal powers marked by PACE must also be qualified by recognition of the degree to which both case-law and police practice had anticipated powers which had not been clearly stated by statute until PACE.[35] Although the direction and trend of legal development only become really unambiguous in the decade before PACE, the general thrust of change since 1950 is clearly to enhance the powers of the police.

Key statutory developments 1950–1984

There are few major statutory developments in police powers in the postwar period prior to PACE. However, these do point in the direction which was consolidated by PACE.

(i) *The Police Act 1964.* This Act grew out of the Royal Commission on the Police Report of 1962, and is the most important post-war statute bearing

---

[34] *Report of HM Chief Inspector of Constabulary 1989* (1990), 12; the Audit Commission, *Police Powers*, 6 and 8 (1990).

[35] The police evidence to the Royal Commission on Criminal Procedure emphasized that the powers they were seeking were already part of police practice to a considerable extent, so that PACE partly constitutes a ratification of existing *sub rosa* practices (cf. *Written Evidence of the Commissioner of Police of the Metropolis to the Royal Commission on Criminal Procedure* (1979), pt. 1, p. 2).

on the police, prior to PACE. The Royal Commission had been announced in 1959 in the wake of a series of controversies concerning alleged abuses of police powers, and the accountability of chief constables to local police authorities.[36]

The main theme of the Act is police accountability to government, central and local. The 1964 Police Act consolidates the tripartite system for the governance of the police, which has remained the nominal framework ever since. The tripartite structure divides responsibility for policing between the chief constable, the Home Secretary, and the local police authority (comprised of two-thirds councillors and one-third JPs) for all forces outside London. The precise division of roles is unclear and disputed but all commentators agree that the first two parties increased their power relative to the latter due to the Act. Subsequent case-law,[37] statute,[38] and organizational change (notably the amalgamation of the 121 forces of the mid-1960s into today's 43 large forces) have rendered the tripartite structure unbalanced in the extreme, with the power of the local police authorities atrophying to virtual insignificance. It is clear that chief constables are now independent of any form of effective accountability of their decisions about how to enforce the law (though they are subject to numerous influences of course), while at the same time their policies are increasingly based on circulars and guidance from the Home Office and their own Association of Chief Police Officers (ACPO, which is an informal body without any basis in statute or case-law).[39]

[36] The scandals included corruption allegations against three chief constables, accusations that a boy in Thurso had been beaten by a constable, conflict between the Watch Committee and chief constable in Nottingham, and an altercation over a traffic incident in Hyde Park which involved Brian Rix, the famous farceur. See S. Stevenson and A. Bottoms, 'The Politics of the Police: A Royal Commission in a Decade of Transition', in R. Morgan (ed.), *Policing, Organised Crime and Crime Prevention* (1990). The most important contemporary exposition and critique of the Royal Commission and the Act is G. Marshall's seminal *Police and Government* (1965).

[37] Most significantly, *R. v. Metropolitan Police Commissioner, ex parte Blackburn* [1968] 2 QB 118; *R. v. Metropolitan Police Commissioner, ex parte Blackburn (no. 3)* [1973] QB 241; *R. v. Chief Constable of Devon and Cornwall, ex parte CEGB* [1981] 3 WLR 867; *R. v. Oxford, ex parte Levey*, The Times, 1 Nov. 1986; and *R. v. Secretary of State for the Home Dept., ex parte Northumbria Police Authority* [1988] 2 WLR 590. The *Blackburn*, *CEGB*, and *Levey* cases confirm a strong doctrine of constabulary independence from direction by police authorities or the courts over decisions about how to enforce the law. The *Northumbria* case establishes that both as a result of the Police Act and the Royal Prerogative the views of the Home Secretary, on the basis of professional advice from HM Inspectors of Constabulary, override those of local police authorities on the means necessary for maintaining order.

[38] The Local Government Act 1985 abolished the Metropolitan Authorities, replaced their police authorities by joint boards, increased the direct share of central government in local police expenditure to 51%, and strengthened its power to influence the rest by its support grant towards local costs. B. Loveday, 'The New Police Authorities in the Metropolitan Counties' (1991) 1/3 *Policing and Soc.* 193–212.

[39] Lustgarten, *Governance*; Reiner, *Chief Constables*. The proposals announced to Parliament on 23 March 1993 by the then Home Secretary Kenneth Clarke have the net effect

Apart from consolidating the constitutional independence of the police in law-enforcement matters, the Police Act 1964 affects police powers in a number of other ways. It required chief officers for the first time to record and investigate all complaints against the police, to appoint investigators from outside forces in serious cases, and to refer all cases where a criminal offence may have been committed to the Director of Public Prosecutions (section 49). While these procedures were already supposed to be in place in most forces, they were frequently neglected until the Act placed them on a statutory footing. The result was a sharp rise in the number of recorded complaints following the Act.[40]

The Act also made chief police officers liable for torts committed by constables in their forces, reversing the common-law position that there was not a master–servant relationship for tort purposes between chief constables and their subordinates.[41] This is very significant in terms of enhancing the practical possibilities for suing police officers for wrongful exercise of their powers. Civil actions against the police have begun to increase considerably in recent years, making this at least as important an avenue for redress of grievances as the formal complaints system.[42] It may also have an unavowed effect on regulating practice. Many more cases are brought against the police than a perusal of reported decisions might suggest. A substantial number of these are settled and, where particular patterns of police misconduct emerge, one would expect forces to take remedial action. The fact that such civil complaints are handled in many forces by force solicitors such as the Metropolitan Police Solicitors ought to conduce to this result. Unfortunately in practice it seems not to!

The Act gave statutory form to the common-law offences of assaulting or obstructing a constable in the execution of his duty (section 51). The celebrated case of *Rice* v. *Connolly*[43] indicated an initial judicial tendency to interpret obstruction strictly, as it was held that not answering a constable's reasonable questions, while being an obstruction in fact, was not a 'wilful' one in that a citizen had no legal duty to answer police questions. However, some later cases suggest a greater readiness to find uncooperative behaviour an obstruction.[44]

---

of further diminishing the accountability of the police to elected local authorities, and enhancing the power of central government. For a critique see R. Reiner and S. Spencer (eds.) *Accountable Policing: Effectiveness, Empowerment and Equity* (1993).

[40] M. Maguire, 'Complaints against the Police: The British Experience; in A. Goldsmith (ed.), *Complaints against the Police* (1991), 180–1.

[41] The *Fisher* v. *Oldham* [1930] 2 KB 364 judgment, the locus classicus for the modern constabulary independence doctrine is thus reversed on this specific point. However, the doctrine in general as it concerns the governance of the police is reinforced in the rest of the Act.

[42] R. Clayton and H. Tomlinson, *Civil Actions against the Police* (1987).

[43] [1966] 3 WLR 17.

[44] e.g. *Ricketts* v. *Cox* (1981) 74 *Cr. App. R.* 298, where silence coupled with a hostile manner was held to be obstruction by the Divisional Court.

(ii) *The Criminal Law Act 1967 and arrest powers.* This placed on a statutory basis the general power to arrest without a warrant,[45] which had been developed previously in the common law. Until the 1967 Act, the common-law powers of arrest without warrant had been restricted to felonies. The Act abolished the felony/misdemeanour distinction, but instead intro-duced the concept of the 'arrestable offence': one for which the sentence was either fixed by law, or for which a hitherto unconvicted person might be sentenced by statute to a prison term of five years or more (section 2). In their details these provisions so closely resemble those now in force under section 24 of the Police and Criminal Evidence Act 1984 that it is unnecessary to rehearse them in detail.[46]

(iii) *Statutory powers to search with warrant.* At common law there tradition-ally was no general power to search a person before arrest without consent or a warrant. However, a plethora of local statutes did confer such powers. The prototype was the Metropolitan Police Act 1839, section 66, which gave a constable power to stop, search, or detain any person who was reasonably suspected of carrying anything stolen or unlawfully obtained. In addition, a variety of statutes dating back to the nineteeth century gave search powers for specific offences or circumstances, ranging from the Badgers Act 1973, section 10, to the Firearms Act 1968, section 47(3) (both are re-enactments of earlier statutory powers).[47]

The old local statutes did have a wide impact in the places they covered, but the other statutes were of limited scope. The first nationally applicable statute which conferred a stop and search power which had wide national applicability was the Misuse of Drugs Act 1971. This conferred a power for any constable to stop and search any person reasonably suspected of being in unlawful possession of a controlled drug (section 23(2)). A constable is also empowered to search any vehicle or vessel where he has reasonable grounds to suspect drugs may be found, and to seize anything found for the purpose of proceedings under the Act. This was a significant extension of police powers to stop and search, and particularly controversial because of evidence of its extensive use, particularly against young males from ethnic minorities.[48] It was

---

[45] In addition, numerous Acts of Parliament confer specific powers to arrest without warrant. The bewildering variety of these is conveyed by app. 9 of the Royal Commission on Criminal Procedure, *The Investigation and Prosecution of Criminal Offences in England and Wales: The Law and Procedure* Cmnd. 8091–2.

[46] This consolidated in statutory form the growing gulf between the powers of a constable and a citizen, which had already appeared at common law, in such cases as *Walters* v. *W. H. Smith* [1914] 1 KB 595.

[47] A full list of such special search powers is given in the Royal Comm. on Criminal Procedure, *Investigation*, app. 1.

[48] The Royal Commission showed in most years of the 1970s that more than 10,000 searches under the Act were made in London alone (Royal Commission on Criminal Procedure: *The Investigation and Prosecution of Criminal Offences in England and Wales: The Law and Procedure* (1981), app. 2). Evidence of its unequal impact is given in C. Willis,

widely argued and indeed ultimately conceded that, despite Home Office guidelines to the contrary, the power was triggered by police stereotyping, sometimes unconsciously, of young blacks and those of unconventional appearance as likely suspects.

Developments in case-law 1950–1984

In this section we will indicate how the trend in case-law in the decades immediately before PACE was to advance police powers in the key areas involved in the investigation of crime, in ways which anticipate the PACE package.

(i) *Stop and search*. The patchwork quilt of statutory stop and search powers prior to PACE has already been noted. In a number of cases in the years before PACE the courts held that some of the statutes also extended a power to question briefly those who were stopped.[49]

(ii) *Arrest*. Until fairly recently it appeared that the common law would not countenance arrest for the purpose of questioning a suspect. The wartime case of *Dumbell* v. *Roberts*,[50] for example, upheld the view that although the reasonable suspicion which justified an arrest could fall short of 'a *prima facie* case for conviction', nevertheless the police were duty-bound to make all practicable and reasonable inquiries into a case *before* arrest. As recently as 1978, the Court of Appeal ruled that 'police officers can only arrest for offences', and had no power to arrest a person 'so that they can make inquiries about him'.[51] In the same case, however, Lawton L. J. took the view that, having arrested someone for a specific offence, the police could hold the arrested person in custody while they made inquiries, although as soon as they had enough evidence to charge they should do so without delay.[52] In an even earlier case, Lord Devlin had also indicated that questioning after arrest was permitted at common law.[53]

By the time PACE took effect, the common law had come to accept fully the power of the police to arrest in order to obtain evidence by questioning the suspect. This is shown by the House of Lord's ruling in

The Use, Effectiveness and Impact of Police Stop and Search Powers (1983); and Smith *et al.*, *Police and People*.

[49] e.g. this was held in *Daniel* v. *Morrison* [1980] Crim. LR 181 with regard to the 1839 Metropolitan Police Act power under s. 66, and in *Green* [1982] *Crim LR* 604, in relation to the 1972 Misuse of Drugs Act.

[50] [1944] 1 All ER 327.

[51] *R.* v. *Houghton and Franciosy* [1978] 68 Cr. App. R. 197.

[52] At p. 205.

[53] *Shaaban Bin Hussein* v. *Chong Fook Kam* [1969] All ER 1626 at 1630.

Mohammed-Holgate v. *Duke*[54] that, although the motive for arrest was to put the defendant under greater pressure in custody than if she had been questioned without arrest, it was legitimate for the police to do so in order to dispel or confirm reasonable suspicion. Even so, no such arrest is valid unless the constable acts on reasonable suspicion, the existence of which was, in the above case, perhaps unwisely conceded by counsel.

Other extensions of arrest powers at common law before PACE include the relaxation of the principle that an arrested person had to be brought before a court as soon as reasonable, and could not be detained while further evidence was sought or by being taken to a place where such evidence might be found.[55] For example, in *Dallison* v. *Caffery*,[56] Lord Denning declared it permissible for a constable to make reasonable investigations after arrest, including taking the suspect to his house to look for stolen goods. (This anticipates section 30(10) of PACE.)

(iii) *Detention and suspects' rights.* Until PACE the only statutory time limit on police detention was that a person in custody had to 'be brought before a magistrates' court as soon as practicable'.[57] In *Hudson* this was interpreted by the Court of Appeal as meaning generally within 24 hours, and certainly within 48.[58] This 48-hour limit was reiterated in other cases in the late 1970s and early 1980s.[59]

Although there are exceptions to this, the decade before PACE saw numerous cases in which evidence was held admissible despite violations of the suspect's rights under the Judges' Rules, for example, access to legal advice. In such cases as *Lemsatef*[60] the Court of Appeal held admissible statements made by the suspect, despite clear breaches of the Judges' Rules, and stated that it was a matter of discretion for the trial judge.[61] Finally in *R.* v. *Sang*,[62] the House of Lords stated that 'it is not part of the judge's function to exercise disciplinary powers over the police or prosecution as respects the way in which evidence to be used at the trial is obtained by them'.[63] The Lords accepted that trial judges had discretion to exclude confession evidence if it had been obtained in violation

[54] [1984] AC 437.

[55] e.g. *R.* v. *Lemsatef* [1977] 2 All ER 835.

[56] [1965] 1 QB 348.

[57] Magistrates' Court Act 1980, s. 43(4).

[58] *R.* v. *Hudson* [1980] 72 Cr. App. R. 163. Hudson's conviction was overturned after he had confessed at the end of five days in custody.

[59] Notably *R.* v. *Houghton and Franciosy*, and *Re Sherman and Apps* [1981] 2 All ER 612.

[60] See n. 55.

[61] Judges generally did allow such evidence, as in *Elliott* [1977] *Crim. LR* 551, though there were exceptions, e.g. *Allen* [1977] *Crim. LR* 431.

[62] [1980] AC 402.

[63] Per Lord Diplock.

of the rules, but contrary to previous case-law declared that all other sorts of evidence (such as that obtained as a result of *agent provocateur* tactics as in *Sang*'s case itself) were admissible, even if gathered by unfair or illegal means.

(iv) Search and seizure. Another area in which there were indications of judicial readiness prior to PACE to extend police powers permissively was search and seizure. Dicta in *Jeffrey* v. *Black*,[64] *Chic Fashions*,[65] and *Ghani* v. *Jones*,[66] indicated that a search of an arrested person's home for evidence relating to that offence, and the taking of goods found there which were reasonably believed to be evidence in relation to that crime, would both be acceptable. However, these statements were all *obiter*, and in *McLorie* v. *Oxford*[67] the Divisional Court rejected the view that an arrested person's premises could be searched without warrant.

Police complaints before PACE

This selective review of key areas of police powers suggests a tendency before PACE for statute and case-law to advance powers in an increasingly permissive direction, although the trend was neither unilinear nor uncontradictory. One significant counter-trend was the further development after the 1964 Police Act of the procedures for dealing with complaints against the police.

The Police Act 1976 established for the first time an independent element in the process, in the shape of the Police Complaints Board (PCB), which had the power to review the papers produced by the internal police investigation and recommend, or if necessary direct, that disciplinary charges be brought. Although resented by the police (it prompted Sir Robert Mark's resignation in 1976) the PCB was universally seen as a toothless body. Indeed its 1980 Triennial Report acknowledged its own virtual impotence, and recommended a number of changes. In 1981 the Board's new chairman, Sir Cyril Philips (hot from chairing the Royal Commission on Criminal Procedure) declared that 'the existing Board had kept so low a profile that it has climbed into a ditch'.

During the early 1980s a chorus of official reports (notably the Scarman Report into the Brixton disorders, and the Select Committee on Home Affairs) argued strongly for an independent element in the investigation as well as adjudication of complaints. The Police Federation, in a remarkable volte-face, came out in support of a fully independent system, as did some chief officers. In an important decision, *R.* v. *Police Complaints*

---

[64] [1978] 1 All ER 555 (per Lord Widgery).
[65] [1968] 2 QB 299.
[66] [1969] 3 All ER 1700 (per Lord Denning).
[67] [1982] 3 WLR 423.

Board, ex parte Madden,[68] it was held wrong for the PCB to follow a double-jeopardy rule whereby if the DPP decided not to launch criminal proceedings as a result of a complaint, the officer accused would also be shielded from disciplinary action. This enhancement of the complaints systems by statute and case-law before PACE prefigures the idea of balancing increasing police powers and safeguards over their use (albeit not very satisfactorily).

In conclusion, developments in statute and case-law in several respects already pointed in the direction of PACE. None the less, PACE is a significant landmark, the importance of which as a statutory rationalization of police powers is impossible to overstate. We turn to this in the next section.

The Police and Criminal Evidence Act 1984: Origins, content, and interpretation

The origins of PACE

The broad background for understanding the emergence of PACE is the increasingly fraught and controversy-ridden political and social context of policing, since the late 1950s, but especially since the early 1970s.[69] There are two contradictory facets of this which are specially relevant to PACE. On the one hand, the problems facing the police multiplied. Crime rates in particular began to climb seemingly inexorably from the mid-1950s, and after the mid-1970s the rate of increase accelerated, in what some criminologists have described as a 'hyper-crisis'.[70] At the same time, the clear-up rate, the most commonly used indicator of police effectiveness in dealing with crime, steadily fell.[71]

The police were not only faced with more crime work, but were apparently doing it less well. At the same time, the social climate became increasingly less deferential and recalcitrant to police authority. Police

---

[68] [1983] 1 WLR 447.

[69] The wider problems of policing are discussed in Reiner, *Politics of Police*.

[70] R. Kinsey, J. Lea, and J. Young, *Losing the Fight against Crime* (1986), 12. The pitfalls of interpreting official crime rates are well known: cf. K. Bottomley and K. Pease, *Crime and Punishment: Interpreting the Data* (1985). However, surveys indicate that victimization is rising, albeit not as fast as recorded crime, cf. P. Mayhew, D. Elliott, and L. Dowds, *The 1988 British Crime Survey* (1989); P. Mayhew and N. A. Maung: *Surveying Crime: Findings From the 1992 British Crime Survey* (Home Office Research and Statistics Department, 1992). In any event, a higher reported crime rate does mean more crimes which the police must deal with.

[71] Indeed, the best explanation of falling clear-up rates is the increasing amount of crime police officers had to handle. Clear-ups per officer—a better productivity measure—actually rose. cf. Audit Commission, Police Paper 8, *Effective Policing: Performance Review in Police Forces* (1990).

have complained for the last three decades that people were becoming generally more aware of their rights, and consequently harder to police. However welcome this may be in itself, what it means is that in many respects the police now seek to accomplish through the exercise of legal powers what they used to be able to achieve through 'voluntary' compliance with police requests. In addition, the profane reality which lay behind the 'Dixon of Dock Green' myth of policing in the 1930s, 1940s, and 1950s, was that regular abuses of powers took place against the socially powerless 'police property' groups, but was less often questioned by the recipients, and less often believed by the respectable classes when it was.[72] The decline in social deference, which constitutes social progress for liberals, poses severe adjustment problems for the police.

These contradictory background pressures induced changes in policing which made it far more controversial. The coercive tactics and powers of the police were used more frequently, and they were reorganized in ways which relied increasingly on technology and specialized serious crime units of a variety of kinds, from Special Patrol Groups to Regional Crime Squads. These measures, which were responses to a more questioning social context, in turn exacerbated the problem of declining public confidence and support.

The concrete result of these pressures which ultimately produced PACE (as well as the developments in the law before PACE which have already been described) was the growth of two conflicting pressure groups, campaigning for contradictory changes in police powers. On the one hand, there emerged a 'law and order' lobby, in which the police themselves were prominent.[73] This was manifest in some victories in the late 1960s and early 1970s, such as the introduction of majority verdicts and the pretrial disclosure of alibi defences.[74] It was also reflected in the fierce debate about the abortive Eleventh Report of the Criminal Law Review Committee (CLRC) on evidence in criminal cases.[75] The 'law and order' lobby was prominent in support of the CLRC proposal to end the right of silence, which has remained one of its key demands ever since. At the time, it also opposed the proposal to tape-record police interviews, although the police have more recently come to welcome tape-recording. The police and 'law and order' lobby were also very active in the 1979

---

[72] See e.g. the anecdotes in R. Mark, *In the Office of Constable* (1978), ch. 2; and M. Brogden, *On the Mersey Beat: Policing Liverpool between the Wars* (1991), chs. 4, 5. For earlier evidence see: C. Emsley, '"The Thump of Wood on a Swede Turnip": Police Violence in 19th Century England', in L. A. Knafla (ed.), *Crime, Police and the Courts in British History* (1990).

[73] The emergence of the 'law and order' lobby is charted in R. Reiner, 'Fuzzy Thoughts: The Police and "Law and Order" Politics' (1980) 28 *Sociol. Rev.* 2.

[74] Robert Mark's autobiography, *In the Office*, explains his role in this; cf. pp. 68–70.

[75] Cmnd. 4991, June 1972.

general election campaign, in which the issue was a very significant factor in the Conservative victory.[76]

Opposing the 'law and order' lobby was an increasingly active array of civil liberties and penal reform groups. Some were long-existing bodies concerned with civil liberties in a broad sense, such as the National Council for Civil Liberties, the Howard League, Justice, the Haldane Society, and so on. However, there has also proliferated since the 1960s a variety of single-issue groups mobilized around specific alleged miscarriages of justice.[77]

The *cause célèbre* which had the greatest impact as a precursor of PACE was the *Confait* case in 1972, in which three teenage boys were wrongly convicted of murder. An official inquiry chaired by Sir Henry Fisher reported in 1977, and revealed routine avoidance of the Judges' Rules.[78] Fisher's suggestion that reform required a broader inquiry, 'something like a Royal Commission', is usually regarded as the immediate trigger for the announcement of the Royal Commission on Criminal Procedure (RCCP).

The Royal Commission was subject to intensive pressure from both the 'law and order' and civil liberties lobbies, as well as the legal profession's representative bodies. The RCCP also commissioned an extensive body of research into how the system actually functioned. When it reported in January 1981[79] it recommended what it argued was a balanced package, giving due weight to the need for the police to have adequate powers to investigate crime, but with workable safeguards for suspects to prevent abuses. This reflected also the contradictory pressures from the diverse lobbies. The report was broadly welcomed by the Conservative government and the police, but widely (though not universally) condemned by the civil liberties groups and the left. Harriet Harman of the NCCL described it as a 'triumph of the "law and order" lobby', presaging a chorus of critical condemnation.[80] In one respect the Royal Commission and its critics and, indeed ultimately the Government, were strangely

[76] I. Taylor, 'The Law and Order Issue in the British General Election and Canadian Federal Election of 1979' (1980) 5 *Canadian J. of Sociol.* 3; and R. Reiner and M. Cross (ed.), *Beyond Law and Order* (1991), 2–3.

[77] Examples can be found in P. Scraton and P. Gordon (eds.), *Causes for Concern* (1984); and B. Woffinden, *Miscarriages of Justice* (1989).

[78] A fuller account of the *Confait* case and the Fisher report can be found in J. Baxter and L. Koffman, 'The Confait Case: Forgotten Lessons?' (1983) 14 *Cambrian LR* 11.

[79] Cmnd. 8092 (1981).

[80] Some examples of the debate provoked by the RCCP are the *Crim. LR*, special issue July 1981; *Politics and Power*, 4. *Law, Politics and Justice* (1981), articles by Jones, Hillyard, and Reiner; D. McBarnet, 'The Royal Commission and the Judges' Rules' (1981) 8 *Brit. J. Law and Soc.*; (1982) 10 *Int. J. Sociol. of Law*, articles by Kinsey and Baldwin, and McConville and Baldwin; L. H. Leigh, 'The Royal Commission on Criminal Procedure' (1981) *MLR*; M. Zander: 'Police Powers' (1982) 53 *Political Quarterly*.

silent. That was whether the proposed reforms were in conformity in all respects with the European Convention on Human Rights. The Royal Commission noted the point in respect of electronic surveillance, and proposed a proper warrant procedure to bring English law into conformity with the Convention.[81] Electronic surveillance was, not, however, reformed until the Interception of Communications Act 1985 was passed, and then, as James Michael notes in chapter 9, imperfectly. Other discrepancies between police powers and the European Convention were not noted. This is unfortunate.

The Government launched Mark I of the Police and Criminal Evidence Bill in October 1982. This was widely perceived even by erstwhile liberal supporters as having tipped the balance heavily towards bolstering police powers rather than safeguards.[82] After an especially rough parliamentary ride, the bill fell with the announcement of the June 1983 general election.[83] A revised Mark II of the bill was introduced in October 1983, and seemed to redress the balance in the more even-handed style of the RCCP. On the whole it regained the qualified support of the mainstream professional legal bodies, though it continued to attract the ire of civil liberties groups. The police representative bodies began to express some misgivings, seeing Mark II as tilting too far to appease the legal and civil liberties lobbies.[84] The Act finally passed through Parliament on 31 October 1984, and came fully into force in January 1986. The next section will give a summary account of its contents.[85]

The content of PACE

As noted above, the RCCP Report was based on the principle of striking a 'fundamental balance' between giving the police the powers they required to investigate crime, while also providing adequate safeguards over how they are exercised. This principle continued to inform the

[81] Cmnd. 8094, paras 3.53–3.60.

[82] e.g. W. Merricks (a member of the RCCP) in *The Times* 19 Nov. 1982, and M. Zander, 'Police and Criminal Evidence Bill' (1983) 133 *New LJ*.

[83] The parliamentary history of PACE and its forerunner is told in more detail in R. Reiner, 'The Politics of the Act' (1985) *Public Law*, and L. Leigh, 'Some Observations on the Parliamentary History of the Police and Criminal Evidence Act 1984', in C. Harlow (ed.), *Public Law and Politics* (1986).

[84] Cf. *Police* (Jan. 1984), 4.

[85] Fuller accounts may be found in a number of books including: Leigh, *Police Powers in England and Wales* (1985, 2nd ed.); M. Zander, *The Police and Criminal Evidence Act 1984* (1990); M. Freeman, *The Police and Criminal Evidence Act 1984* (1985); V. Bevan and K. Lidstone, *A Guide to the Police and Criminal Evidence Act 1984* (1985, 2nd ed. 1992). Useful collections presenting diverse viewpoints on the Act are: J. Baxter and L. Koffman (eds.), *Police: The Constitution and the Community* (1985); J. Benyon and C. Bourne (eds.), *The Police: Powers, Procedures and Proprieties* (1986); *Public Law* Symposium on PACE, Autumn 1985; (Sept. 1985) *Crim. LR*, Special Issue. See also M. Freeman, 'Law and Order in 1984' (1984) 37 *Current Legal Problems* 175, for a stimulating critique.

Government's presentation of PACE (although the consensus view was that Mark I had tilted too much to the police side, and the civil liberties groups felt this about the RCCP Report itself as well as PACE in its final form).

The idea of balance is inscribed into the structure of the Act, although how adequately it is realized is debatable. On the one hand, the Act gives the police a variety of powers which they did not enjoy on a statutory basis before (though to an extent these merely rationalize the hitherto existing statute and common-law position, and/or legitimate what was already police practice). On the other hand, the exercise of each of these is governed by requirements which are set out partly in the Act itself, and partly in the accompanying Codes of Practice.[86] These Codes are buttressed by section 67 of PACE, which makes failure to comply with them a disciplinary offence, and makes a breach admissible as evidence in any criminal or civil proceedings if it is thought relevant by the judge(s). To aid review of police actions, the Act also implements through the Codes of Practice the RCCP's solution to the problem of the 'low visibility' of routine police work. The adoption of this device has been criticized in that practices and entitlements laid down in the Codes, albeit buttressed by exclusionary rules of evidence and a police discipline code which may bear harshly upon a constable who transgresses the rules, do not confer rights upon suspects. It may conversely be argued that the Government was prepared to give a more liberal content to the Codes than it would have done had they been put into law.

A variety of recording requirements are laid down for all exercises of powers, giving reasons for what is done, and there is to be contemporaneous recording of interviews. In addition to these particular safeguards for specific powers, the package as a whole includes sections purporting to enhance police accountability more generally. Section 106 obliges police authorities to make arrangements for obtaining the views of the community about policing in their area. Part IX of the Act institutes the Police Complaints Authority to enhance the independent element in the complaints system. We will briefly review some key aspects of PACE to amplify these points.

(i) *Stop and search.* Police powers to stop and search are clearly boosted by section 1 of the Act. The power to search for stolen goods, which was

---

[86] There are five of these: (A) on stop and search; (B) search or premises and seizure of property; (C) on detention, treatment and questioning of suspects; (D) on identification procedures; (E) on tape-recording of interviews. These each contain detailed guidance about procedures in each of these areas. They were recently revised on the basis of the experience of PACE so far. The revised codes were approved by Parliament in Dec. 1990, and came into effect on 1st April 1991. They can be found in Zander, *PACE 1984*, 2nd edn.

previously given by local legislation in many places (e.g. the Metropolitan Police Act 1839), is extended nationally. New powers are given to stop and search for articles made or adapted for use or intended to be used for burglary, theft, obtaining property by deception, or taking a motor vehicle without authority. Finally, a power to stop and search for offensive weapons is given, defined as either weapons made or adapted for use to cause injury, or intended by the person carrying it for such use.

There are two main safeguards over the new stop and search powers. First, there is a record-keeping requirement. The constable must make a detailed record of the search and the reasons for it, and inform the suspect that he has a right to request a copy within 12 months. Second, Code of Practice A on stop and search warns against irresponsible or overfrequent use of the power. Specifically, it lays down that there must be an objective basis to the idea of reasonable suspicion, connected to the individual searched. It cannot arise simply from the person's membership in a category stereotyped as more likely to offend, e.g. black or young people, or those with long hair.

(ii) *Arrest.* Arrest powers are also clearly extended, not only compared to the existing law, but by contrast with the RCCP recommendations. Section 24, which gives the power to arrest for arrestable offences, is essentially modelled on section 2(1) of the Criminal Law Act 1967. However, it goes beyond it in some respects. First, while the Criminal Law Act's definition of arrestable offence included only *statutory* offences carrying a possible penalty of five or more years imprisonment, PACE extends this also to common-law offences. Second, PACE classes as arrestable offences a number of statutory offences which were previously non-arrestable (though some did carry a power of arrest from their specific statutes). They are listed in section 24(2) of PACE, and include some Official Secrets Act offences, indecent assault on a woman, taking a motor vehicle without authority, and going equipped for stealing.

However, the major extension of arrest powers under PACE comes in section 25. This gives a power for a constable to arrest for a non-arrestable offence where service of a summons would be impracticable because certain 'general arrest conditions' apply. These are that the officer has reasonable grounds to suspect that arrest is necessary to prevent (i) the suspect causing physical harm to the suspect or any other person; (ii) the suspect suffering physical injury; (iii) loss or damage to property; (iv) an offence against public decency which ordinary members of the public could not reasonably avoid; (v) to protect a child or other vulnerable person.

The RCCP had seen these sorts of conditions as part of its 'necessity principle': as additional considerations for deciding to use arrest rather than summons for *arrestable* offences. There was indeed common-law authority for the proposition that arrest ought not to be used where a summons would suffice.[87] The Act erodes the necessity principle by providing that a constable may exercise his discretion to arrest a suspect when any one of the arrest conditions appears. In effect, the presence of any one condition presumptively brings the case into the category where arrest is a proper means of proceeding. Not only does the Act extend arrest powers beyond the existing law and the RCCP's proposed increases. It does so without establishing any specific safeguards for the use of arrest powers.

The very structure of powers, which was intended to simplify the law, has become ever more complicated. Powers under local and private Acts have disappeared but fresh legislation such as the Public Order Act 1986 have added new powers of arrest as offences are created, the maximum penalty for which would not bring them within the category of arrestable offences.

There are at least three cases in which the law of arrest may conflict with the European Convention. The first concerns terrorism and we allude to it only briefly. Article 5(1)(c) specifies that the purpose of arrest is to bring a suspect before a competent legal authority on reasonable suspicion of having committed an offence. Arrest for terrorism is not, however, arrest for a specific offence since terrorism is a generic term which describes activities which fall under several different offences. In *Brogan*, the European Court held that an arrest for terrorism was, however, an arrest for an offence since terrorism as a term contained specific offences.[88] This reasoning has been stringently, and we believe rightly, criticized.[89] The case also raised an issue of general importance: whether an arrest for the purpose of discovering evidence of criminal involvement is proper.[90] This was an issue which could not be proved against the authorities in *Brogan* but which could potentially arise generally in police arrest practices.

Most encounters intended to elicit evidence fall under stop and search powers and while the definition of arrest taken at its widest could certainly comprehend stops made for the purpose of searching they are unlikely to be considered arrest for the purposes of Article 5 of the European Convention.[91] For this there are two reasons. First, English domestic law and indeed the laws of other European systems distinguish between stops

---

[87] Per Goddard LJ (as he then was) in *Dumbell* v. *Roberts* (1944) 133 JKB 185.

[88] *Brogan and other* v. *UK* (1988) Series A, No. 145B.

[89] In addition to articles already cited see P. M. Roche, 'The United Kingdom's Obligation to Balance Human Rights and its Anti-Terrorism Legislation: The Case of *Brogan and Others*' (1989–90) 13 *Fordham Int. LJ* 328.

[90] See further W. Finnie, 52 *MLR* 703; Livingstone, 40 *NILQ* 288.

[91] On the definition of arrest generally, see *Lewis and Another* v. *Chief Constable of the South Wales Constabulary* [1991] 1 All ER 206; *Murray* v. *Ministry of Defence* [1988] 1 WLR 692.

and arrest. The European Convention must have been intended to reflect this distinction. Secondly, Article 5(4) of the Convention which provides for proceedings to test the legality of detention presumes a more formal and longer detention than stop and search typically involves.

Secondly, a problem of want of conformity could arise under section 24, subsections (4), (5), and (7) of PACE. These provisions permit arrest where a person is committing an offence, or is guilty of an offence, or is about to commit an offence, but where the arrester has no reasonable cause for suspecting him of doing so. This raises the problem of the arresting officer's hunch or, as Professor Glanville Williams once put it, second sight.[92] Before the Criminal Law Act 1967, there was doubt whether an officer could arrest where he lacked reasonable cause to suspect the offender of, for example, committing an offence. The case of *Dadson* suggested that he could not, at any rate in the absence of specific statutory powers.[93] The 1967 Act, the provisions of which have been carried forward into PACE with amendments, deliberately jettisoned the rule in *Dadson*.[94]

Article 5.1(c) of the European Convention, however, states the principle of reasonable cause in an unqualified form and thus hearkens back to the former rule of English law. This provides an opportunity for conflict between English law and the Convention, albeit one which will arise but rarely. The issue of which rule is to prevail may well have to be faced. In favour of the drafting of the European Convention it may be urged that the Convention, in specifying reasonable cause, requires that circumstances of excuse be known to whoever relies upon them, and that, in the interests of the liberty of the subject, an executive officer should not be entitled to exercise powers to which a condition attaches without first being aware that that condition exists.[95]

(iii) *Entry, search, and seizure.* PACE places on a statutory basis powers to enter premises for purposes of search in connection with suspected offences, and to seize property as evidence, which had hitherto existed only in an uncertain way in the common law, and in a variety of specific statutes.[96] Whilst thus extending and rationalizing the powers, PACE also constructs an elaborate scheme of safeguards. There can be no doubt that

[92] G. L. Williams, 'Statutory Powers of Arrest without Warrant', [1958] *Crim. LR* 72.

[93] (1850) 3 Car. & Kir. 148, 2 Den. 35.

[94] See Criminal Law Revision Committee 7th Report, *Felonies and Misdemeanours*, Cmnd. 2659 (1965), para. 13.

[95] Professor Williams saw the issue differently, as one on which if e.g. a crime were taking place, there would be no unlawful act by a constable sufficient to found criminal or tortious liability on his part. But if a necessary condition precedent is not made out the act must surely be tortious, however socially desirable the arrest of a criminal may be.

[96] These are listed in the RCCP *Law and Procedure* vol., apps. 5 and 6.

this scheme conforms fully to Article 8 of the European Convention on Human Rights.

Warrants may be issued under section 8 by magistrates to search premises for material which may be of evidentiary or other substantial value to the investigation of a serious arrestable offence (SAO).[97] However, certain types of material are given extra protection. 'Special procedure material' is material created or acquired for journalism, other than records or documents, or material acquired in any occupation or office which is held under an obligation of confidence, or subject to a statutory restriction on disclosure. If such material is likely to be relevant evidence or of substantial value to investigating an SAO, an order for production or access may be applied for from a circuit judge in an *inter partes* hearing (section 9). Excluded material is an even more tightly protected category. It includes personal records, human tissue taken for medical purposes, or journalistic material consisting of documents or other records (section 11). Here an order for production will only be granted under legislation which existed before PACE. Finally, legally privileged material cannot be the subject of a search warrant or an order for production or access at all.

Apart from consolidating powers to search with warrant, PACE clarifies the common-law situation with regard to powers to search without warrant, and to seize items found, invariably in ways which enhance police powers. In sections 17–19 the powers of search and seizure which had been indicated by such cases as *Jeffrey* v. *Black*, *Chic Fashions* v. *Jones*, and *Ghani* v. *Jones*, are placed on a clear statutory footing. The contrary 1982 case of *McLorie* v. *Oxford*,[98] in which it was held that the police had no right to enter the premises of a person arrested for attempted murder to search for the motor car which was the alleged murder weapon, is now overruled by statute. Thus the police now have a clear power to enter and search premises occupied or controlled by an arrested person, if they have reasonable grounds for suspecting they will find evidence relating to that or a similar (but not an unconnected) offence. This is subject to an independent written authorization by an officer of the rank of inspector or above. They may seize anything which they believe reasonably to be evidence relating to that offence or any other offence, and which they

---

[97] The concept of the 'serious arrestable offence' is the trigger throughout PACE for the operation of more intrusive police powers. It is defined in s. 116 of the Act. It encompasses a number of what the RCCP listed as 'grave' offences. In addition, any arrestable offence may be deemed serious if it has led or is likely to lead to any of certain consequences listed in s. 6: (*a*) serious harm to state security or public order; (*b*) serious interference with the administration of justice or the investigation of offences; (*c*) anyone's death; (*d*) serious injury to any person; (*e*) substantial financial gains; (*f*) serious financial loss, judged by the circumstances of the victim. The last element in particular attracted controversy because of the definitional scope it affords to the investigating officer.

[98] *McLorie* v. *Oxford* [1982] QB 1290.

believe it is necessary to seize in order to prevent it being concealed, lost, altered, or destroyed, or which they believe has been obtained through commission of an offence.

(iv) *Detention and questioning.* PACE consolidates the trend in common law which had moved towards acceptance of the power to detain for questioning, but structures this with an elaborate network of safeguards. Section 37(2) permits a suspect's detention without charge if 'the custody officer has reasonable grounds for believing that his detention without being charged is necessary to secure or preserve evidence relating to an offence for which he is under arrest or to obtain such evidence by questioning him'. The custody officer is the officer appointed (under section 36) whose function it is to supervise and deal with the process and conditions of detention in designated police stations. He is supposed to be a sergeant (unless one is not readily available), and must not be involved in the investigation of the offences of which the detainees he is responsible for are suspected.

The custody officer is thus the linchpin of the Act's supposed safeguards for detained prisoners. He is responsible for informing the prisoner of his rights (to see a solicitor (section 58), to have someone informed of his arrest (section 56), and to consult the Codes of Practice), regulating access to the prisoner of investigation officers (section 39), ensuring the conditions of detention are met (set out in Code C), and deciding on charge, bail, and release (sections 46 and 47). The custody officer is also obliged to open and keep a custody record, containing details of the detention process. For the first time PACE sets statutory limits to the time which may be spent in detention before charge. The necessity for continued detention must be reviewed by an officer of the rank of inspector or above after six hours, and again not more than nine hours later (but the reviews may be postponed if 'impracticable'). For detention beyond 24 hours, the offence must be an SAO, and the authorization must come from a superintendent or above who declares it necessary for effective investigation (section 42). After 36 hours further detention for an SAO is possible, but only on the warrant of a magistrates' court after an *inter partes* hearing (section 43). This warrant can be for a maximum of 36 hours. After 72 hours one further extension up to an absolute maximum of 96 hours is possible, again on a magistrates' warrant after an inter partes hearing. Even if he was held incommunicado before,[99] the suspect

[99] These rights may be withheld, if the offence is an SAO, on the authority of a superintendent or above, if he has reasonable grounds for believing that the exercise of the right will lead to interference with evidence connected with an SAO or interference with or physical injury to other people, or to the alerting of other persons suspected of having committed such an offence, but not yet arrested, or will hinder the recovery of any property obtained through that offence: see PACE 1984, ss. 56 and 58 as amended.

must have access to a solicitor at the time of the first magistrates' court sitting (i.e. after 36 hours).

It is clear that PACE extends police powers in all the key areas of the investigative process. In addition to the specific safeguards laid down for some of these, however, some general balancing factors are introduced.

(v) *Exclusion of evidence.* One of the main criticisms made by civil libertarians of the RCCP Report, as well as the initial versions of the Police and Criminal Evidence Bill, was their failure to include a strong exclusionary rule, rendering inadmissible evidence gathered in violation of suspects' rights.[100] Indeed, in some respects the intention was to dilute further the common-law position. As seen above, this rejected a disciplinary role for the courts in relation to police practice. However, it was long established that confessions were inadmissible if they were not voluntary because they had been induced 'by fear of prejudice or hope of advantage held out by a person in authority',[101] or obtained through oppression.[102] The RCCP rejected the voluntariness principle as unrealistic, on the basis of research demonstrating the inherently coercive nature of being in custody in a police station.[103] It proposed its replacement by the test of unreliability, coupled with a continued rejection of confessions based on violence or inhuman and degrading treatment. The value invoked is that of Article 3 of the European Convention and the wording reflects its terms.[104] It follows that those techniques of sensory deprivation which the European Court condemned in *Ireland* v. *United Kingdom* are unlawful in themselves and must give rise to the exclusion of evidence as having been obtained by torture, or inhuman and degrading treatment.[105] Admissions so obtained would, of course, also be excluded under the veracity principle. The reliability test leaves judges with more discretion than the old voluntariness one. Inducements and threats do not automatically mean inadmissibility.[106]

PACE incorporates these changes in section 76. A confession is to be admitted only if the prosecution can prove beyond reasonable doubt that

---

[100] Cf. e.g. P. Sieghart, 'Sanctions against Abuse of Police Powers' (1985) *PL* 440.

[101] *R.* v. *Ibrahim* [1914] AC 599.

[102] *R.* v. *Prager* [1972] 1 All ER 1114 and earlier cases.

[103] B. Irving, *Police Interrogation: A Case Study of Current Practice* (1980).

[104] See the Royal Commission on Criminal Procedure, *Report*, Cmnd. 8092 (1981), para. 4.132.

[105] This is a distinct advance upon the guarded terms of the Parker Report which concluded that such techniques might be illegal: see *Report of the Committee of Privy Councillors Appointed to Consider Procedures for the Interrogation of Persons Suspected of Terrorism* (Chairman, Lord Parker of Waddingham), Cmnd. 4901 (1972).

[106] See further, L. Leigh, 'Le Royaume Uni', in M. Delmas-Marty (ed.), *Raisonner la raison d'état* (1989), 329–55.

it was not obtained by oppression or by any means likely to make it unreliable, once such a claim has been raised by the defence. Section 77 requires a judge to warn the jury of the need for special caution before convicting a mentally handicapped accused on the basis of a confession.

An important departure, moving further towards a general exclusionary rule, was incorporated into the Act as a result of an amendment introduced by Lord Scarman at the report stage of the bill.[107] Scarman's amendment was not accepted by the Government, but instead a somewhat modified version was incorporated into the Act as section 78.[108] This provides that:

In any proceedings the court may refuse to allow the evidence on which the prosecution proposes to rely to be given if it appears to the court that, having regard to all the circumstances in which the evidence was obtained the admission of the evidence would have such an adverse effect on the fairness of the proceedings that the court ought not to admit it.

This seems on the face of it to advance the power to exclude evidence beyond the common law discretion found in *Sang* and other cases. (The latter is preserved in section 82(3).) It gives an ampler measure of protection to suspects than that found either in the European Convention on Human Rights or the United Nations International Covenant on Civil and Political Rights, 1966.

(vi) *Complaints against the police.* Part IX of PACE introduces a number of reforms in the system for dealing with complaints against the police, in recognition of a growing critical consensus. The main one is the replacement of the Police Complaints Board by the Police Complaints Authority (PCA). The PCA has several powers and duties which the PCB lacked. The key one is the supervision of the police investigations of certain serious complaints, the first time there has been any independent element in the investigation process. Under section 87 some complaints (those alleging conduct resulting in death or serious injury) *must* be referred to the PCA, while others may be, by the Home Secretary, a chief constable, or a local police authority, and the PCA may itself select certain cases for supervision.

Another important innovation is the establishment of procedures for informal resolution of minor complaints (section 85). This is a response to the criticism that the system was overloaded with minor complaints, for which the full apparatus of investigation was inappropriate.

---

[107] This was a somewhat weaker version of an earlier defeated amendment by Scarman which would have introduced a 'reverse onus' exclusionary rule for all evidence whose lawfulness was challenged.

[108] The full legislative history is discussed in Sieghart, 'Sanctions'.

(vii) Consultation with the community. Section 106 requires arrangements to be made in each police area for 'obtaining the views of people in that area about matters concerning the policing of the area, and for obtaining their co-operation with the police'. In provincial forces the responsibility lies on the police authority after consultation with the chief constable. In the Metropolitan area it lies on the Secretary of State to issue guidance to the Commissioner. Nothing is laid down in the Act itself about the form of such consultation. However, Home Office guidelines in Circular 2/1985 have now been followed throughout the country, and the uniform means of implementing section 106 is through the establishment of police consultative committees on lines suggested by the Home Office.

(viii) The effect of PACE on the common law. One of the key fears of the critics of PACE was that too much scope was left for judicial discretion in policing the Codes of Practice through not having a strong exclusionary rule, despite the recommendation of the Royal Commission on Criminal Procedure. It was felt that the pre-PACE attitude of extending police powers by permissive responses to breaches would continue. Although clearly the judicial reaction has not been uniform, most commentators concur in the view that such fears have not been confirmed.[109] With exceptions, the typical judicial response seems to be that, as PACE reflects a balanced package in which adequate powers are given to the police, they must stick to them and the corresponding safeguards must be followed. As David Feldman concludes, after a comprehensive and incisive review of post-PACE case-law on detention, it 'is surprising in the light of . . . the pre-PACE record of English judges in exercising their common law exclusionary discretion. Yet it shows that the judges now see themselves as having a disciplinary and regulatory role in maintaining the balance between the powers of the police and the protection of suspects.'[110] How far this might go is shown in *R. v. Mason*.[111] In this case a reliable confession to fire-bombing a car was held inadmissible by the Court of Appeal because it had been obtained by a trick on Mason and his solicitor, who had falsely been told that Mason's fingerprints were found on fragments of glass from the bomb. This shows the lengths to

---

[109] Detailed reviews include: A. A. S. Zuckerman, 'Illegally Obtained Evidence: Discretion as a Guardian of Legitimacy', (1987) *Current Legal Problems* 55–70; R. May, 'Fair Play at Trial: An Interim Assessment of s. 78 of the Police and Criminal Evidence Act 1984' (1988) *Crim. LR* 722–30; D. Birch, 'The PACE Hots up: Confessions and Confusions under the 1984 Act' (1989) *Crim. LR* 95–116; A. Choo, 'Improperly Obtained Evidence: A Reconsideration' (1989) *Legal Studies* 261–83; D. Feldman, 'Regulating Treatment of Suspects in Police Stations: Judicial Interpretations of Detention Provisions in the Police and Criminal Evidence Act 1984' (1990) *Crim. LR* 452–71.

[110] Feldman, 'Regulating Treatment', 469.

[111] [1987] 3 ALL ER 481.

which some judges are prepared to go in allowing suspects to be acquitted even where strong evidence of guilt exists, in order to safeguard the 'fairness of the proceedings' of which section 78 speaks.

The seminal case has, however, been *R. v. Samuel*.[112] Here a Court of Appeal ruling held inadmissible statements made after access to a solicitor had been denied on a superintendent's authority because there was said to be a 'likelihood of other suspects to be arrested being inadvertently warned'. The Court held that access to legal advice should only be denied in rare circumstances. Mistrust of a solicitor had to be based on very specific tangible facts. *Samuel* sent shock waves through the police as it came to be realized painfully that PACE would be interpreted much more stringently than the Judges' Rules. A number of cases soon after *Samuel* reinforced the message.[113] However, comfort for the police arrived in later cases in which the Court of Appeal led by Lord Lane CJ held that confessions obtained after wrongful denial of access to legal advice was not necessarily inadmissible. In these cases it was held that statements thus obtained may be admitted if this was still fair in all the circumstances, for example, because the defendant knew his rights anyway.[114]

Even more significantly, Lord Lane made clear his view that PACE had increased the defendant's rights to a point where the balance needed some tilting in the opposite direction, notably by allowing adverse comment on the suspect's exercise of the right of silence. In raising this issue once again, after the RCCP had expressly decided to preserve the right of silence as a part of its package, Lord Lane gave important support for a growing movement in the criminal justice establishment. In July 1987 the then Home Secretary, Douglas Hurd, had floated the issue at a Police Foundation lecture; in November 1988 the Criminal Evidence (NI) Order 1988 abolished the right for Northern Ireland; and in July 1989 a Home Office Working Group was set up to look into it. The matter is thus firmly back on the political agenda.[115]

In a later case, Lord Lane made even more clear his continued support for a less strict response to breaches of the Act.[116] He states that a 'breach of the Act or the Code did not mean that any statement would necessarily be ruled out. Every case had to be determined on its own particular facts.' The judicial attitude to PACE is thus variable. However, a string of cases since *Alladice* in which police infractions have rendered

---

[112] [1988] 2 WLR 920.

[113] e.g. *Dawson* [1988] *Crim. LR* 442, *Cochrane* [1988] *Crim. LR* 449, *Chung* [1991] 92 Cr. App. R. 314. The effect continues, cf. L. Wallace, 'A Change in PACE' (1991) 155 *Justice of the Peace* 325.

[114] *R. v. Alladice*, (1988) 87 Cr. App. R. 380; *R. v. Dunford* (1990) 91 Cr. App. R. 150.

[115] Cf. S. Greer and R. Morgan (eds.), *The Right to Silence Debate* (1990).

[116] *R. v. Parris* [1989] Crim. LR 214.

evidence inadmissible[117] suggests that the courts are taking a much less permissive line on police powers than before PACE, or than anyone anticipated during the pre-PACE debate.[118] Both under the Act and Code C, as under the old Judges' Rules, the courts conclude that a breach of the rules does not as such render confessions and admissions inadmissible, but the capital difference lies in the emphasis which courts give to such breaches. Bad faith is relevant to the issue of exclusion. So, too, is overbearing conduct.[119] The effects of this on police practice are less clear, and will be considered in the following section which looks at empirical evaluations of the impact of PACE.

PACE IN PRACTICE: EMPIRICAL RESEARCH ON POLICE POWERS

Pre-PACE empirical research

It was noted at the beginning of this chapter that much of the empirical research on policing before PACE demonstrated that a gulf existed between the 'law in the books' and the 'law in action'. More precisely, as Doreen McBarnet emphasized, the permissive and elastic formulation of the legal powers of the police meant that they enjoyed considerable discretion, so that decisions and actions which were not in reality informed by principles of legality were nevertheless not sanctionable violations of specific legal rules. The working rules of the police differed from formal rules of law, but the latter were used presentationally to give acceptable accounts of incidents and practices when necessary.

Empirical research based on observations of police work found repeatedly that decisions to stop, arrest, charge, or use other powers were informed by criteria derived from police culture rather than the Judges'

---

[117] e.g. *Fenneley* [1989] Crim. LR 142, *Absolam* [1988] Crim. LR 748, *Waters* [1989] Crim. LR 62.

[118] The continuation of this trend is indicated by a recent case, *R.* v. *Beckford, The Guardian*, 25 June 1991, p. 39. The Court of Appeal quashed the conviction of Beckford who was arrested for possession of heroin after police entered and searched his house having arrested a man who had just left it. The entry was under s. 32(2)(*b*) of PACE, giving a constable power to enter and search premises where a person was when or immediately before he was arrested, to search for evidence connected to that offence. The trial judge ruled the evidence admissible, directing the jury that the police were 'perfectly entitled' to enter. The Court of Appeal held that he should have left to the jury the question of fact whether the police intended to enter to search for evidence of the other man's offence, or to find evidence against Beckford. The judge effectively pre-empted this issue by his direction, so the conviction was quashed.

[119] See e.g. *R. Dunford* (1990) 91 Cr. App. R. 150; *R.* v. *Maguire (Jason)* (1990) 90 Cr. App. R. 115 (questioning juvenile without an adult present).

Rules. They were governed by the 'Ways and Means Act', not Act of Parliament or the common law. The view that police work could not be done effectively if legal procedures were properly adhered to had often been expressed by chief officers,[120] and was a part of the police evidence to the RCCP. Similar attitudes were common amongst the rank and file.[121] It is still often argued now by the less progressive chief officers, who claim that PACE has hampered their work further.[122]

It had often been shown by research before PACE that these views were routinely translated into practice. An observational study by a former police sergeant described a variety of tactics for controlling suspects which 'distance . . . officers from the constraints of legal rules and force directives', such as 'verballing' or 'working the oracle' (i.e. fabricating statements), or using physical force.[123] The PSI study also found that while 'outright fabrication of evidence is probably rare . . . departure from rules and procedure affecting evidence are far more common . . . There will be no fundamental change as long as many police officers believe that the job cannot be done effectively within the rules.'[124] Police deviation was shielded by the low visibility of most police work, buttressed by a variety of devices for constructing acceptable presentational accounts to supervisors or courts. How to 'cover your ass' was a central aspect of the craft skills of police culture.[125] 'Reasonable suspicion' and other legal criteria only enter police thinking in after-the-event accounting, not when making decisions.

A particular concern, which was highlighted especially by more quantitative research on the outcomes of police decision-making, was the apparently socially discriminatory character of policing. It was noted above that the brunt of coercive policing is borne by the 'police property' groups of the most socially and economically marginal. As the PSI concluded after a wide-ranging, large-scale study of Londoners and the Metropolitan Police, including extensive observation: 'The weight of police activity bears much more heavily on sections of the lower working class and others whom the police tend to lump in with them.'[126]

This is confirmed by numerous studies of each stage of the investigative process. Being young, male, black, unemployed, and economically dis-

---

[120] e.g. Mark, *In the Office*, 58; D. McNee, *McNee's Law* (1983), 180–3.
[121] Cf. R. Reiner, *The Blue-Coated Worker* (1978), 77–81, 221–3.
[122] Reiner, *Chief Constables*, 144–60.
[123] Holdaway, *Inside the British Police*, 101.
[124] Smith *et al.*, *Police and People*, iv. 228–30.
[125] See e.g. M. Chatterton, 'The Supervision of Patrol Work under the Fixed Points System', and P. Manning, 'The Social Control of Police Work', both in S. Holdaway (ed.), *The British Police* (1979).
[126] Smith *et al.*, *Police and People*, iv. 166.

advantaged, are all associated with a higher probability of being stopped, searched, arrested, charged, making a complaint against the police (especially one of assault), and failing to have these complaints substantiated.[127] There was some debate about the extent to which this reflected 'pure' discrimination, as distinct from differential offending, or at any rate detectability, in different social groups.[128] However, the consensus was that at least a good deal was due to discrimination.

It is unsurprising, therefore, that civil liberties groups responded to PACE (and the RCCP Report before it) with anxiety. As Ole Hansen of the Legal Action Group put it: 'If they exceed their present powers why should they not exceed wider powers?'[128a] The Act's vaunted safeguards lay not in precise, specific, and mandatory legal remedies,[129] but in a bureaucratic regime of form-filling, backed up by the police disciplinary code more than any external sanctions. Given the often jaundiced view of senior officers towards the legislation, and the evident disregard of the rank and file for suspects' rights and for cumbersome paperwork,[130] what chance was there that the Act would bite as well as bark?

---

[127] On stop and search see A. Brogden, 'Sus is Dead, but what about Sas?' (1981) 9 *New Community*; M. Tuck and P. Southgate, *Ethnic Minorities, Crime and Policing* (1981); S. Field and P. Southgate, *Public Disorder* (1982); C. Willis, *The Use, Effectiveness and Impact of Police Stop and Search Powers* (1982); M. McConville, 'Search of Persons and Premises: New Data From London' (1983) *Crim. LR* 605–14; M. Brogden and A. Brogden, 'From Henry VIII to Liverpool 8: The Complex Unity of Police Street Powers' (1984) *International Journal of Sociology of Law* 37; Smith *et al.*, *Police and People*, i. 95–102, iii. 96–7; P. Southgate and P. Ekblom, *Police: Public Encounters* (1984). On search of premises: McConville, 'Search'; K. Lidstone, 'Magistrates, the Police and Search Warrants' (1984) *Crim. LR* 454. On arrest: P. Stevens and C. Willis, *Race, Crime and Arrests* (1979); Field and Southgate, *Public Disorder*; Smith *et al.*, *Police and People*; M. Cain and S. Sadigh, 'Racism, the Police and Community Policing' (1982) *Journal of Law and Society* 87. On charging, cautioning, and prosecution: T. Bennett, 'The Social Distribution of Criminal Labels' (1979) 19 *British Journal of Criminology* 134; C. Fisher and R. Mawby, 'Juvenile Delinquency and Police Discretion in an Inner-City Area' (1982) 22 *Brit. J. Criminology*; S. Landau and G. Nathan: 'Selecting Delinquents for Cautioning in the London Metropolitan Area' (1983) 23 *Brit. J. Criminology* 128; A. Sanders: 'Class Bias in Prosecutions' (1985) 24 *The Howard J.* 176. On complaints, see P. Stevens and C. Willis, *Ethnic Minorities and Complaints against the Police* (1981); M. Tuck and P. Southgate, *Ethnic Minorities, Crime and Policing*; S. Box and K. Russell, 'The Politics of Discreditability: Disarming Complaints against the Police' (1975) 23 *Sociological Review* 2.

[128] P. Stevens and C. Willis, *Race, Crime and Arrests*; J. Lea and J. Young, *What is to be Done about Law and Order?* (1984), ch. 4; Jefferson 1988, op. cit; R. Reiner, 'Race and Criminal Justice' (1989) 16 *New Community* 5.

[128a] *New Society*, 22 January 1981, 161.

[129] A. Sanders, 'Rights, Remedies and the Police and Criminal Evidence Act' (1988) *Crim. LR* 802–12.

[130] It had been shown e.g. that about half of all stops were not recorded, and those that were tended to be justified in a Delphic or formulaic way. Cf. C. Willis, *The Use, Effectiveness and Impact of Police Stop and Search Powers* (1983); D. J. Smith, *Police and People in London* (1983), i. 114. Rights such as access to legal advice were often side-stepped: cf. M. Zander, 'Access to a Solicitor in the Police Station' (1979) *Crim. LR* 342.

Post-PACE empirical research

Although an extensive amount of evaluative empirical research has been commissioned by a variety of bodies[131] on the effects of PACE, the jury is still out. It is somewhat premature to come to final conclusions, especially as the Codes of Practice have been revised recently (taking effect on 1 April 1991), in the light of early research and practitioner reaction. Research is under way to assess the new Codes.

What is known so far suggests a rather complex picture, which throws up more questions for further research and analysis rather than definite conclusions. PACE has certainly had a profound effect on the nature and outcomes of the police handling of suspects. Routine practice has incorporated the rituals and procedures of the Codes of Practice, and many indices of suspects' access to rights indicate improvement. On the other hand, there has been an uneven and incomplete incorporation of the PACE rules and Codes into police working practices. Much of the change is presentational and affects little of substance in the experience of suspects. Furthermore, there is some indication that early changes were an impact effect of new procedures, and that old working practices are reviving.

On the positive side, the research indicates that:

1. Suspects are almost invariably informed of their rights on reception at the police station.[132]

2. As a result, the proportion receiving legal advice has increased

---

[131] These are mainly: (*a*) the Economic and Social Research Council, which launched an initiative to monitor PACE in 1986, supporting a number of studies including: D. Dixon, A. K. Bottomley, C. A. Coleman, M. Gill, and D. Wall, 'Reality and Rules in the Construction and Regulation of Police Suspicion' (1989) 17 *Int. J. Sociol. of Law* 185–206, D. Dixon *et al.*, 'Safeguarding the Rights of Suspects in Police Custody' (1990) 1/2 *Policing and Society* 115–40, and D. Dixon *et al.*, 'Consent and the Legal Regulation of Policing' *J. of Law and Soc*; I. McKenzie, R. Morgan, and R. Reiner, 'Helping the Police with their Inquiries: The Necessity Principle and Voluntary Attendance at the Police Station' (1990) *Crim. LR* 22–33; and R. Morgan, R. Reiner, and I. McKenzie, *Police Powers and Policy: A Study of the Work of Custody Officers* (1990); K. Lidstone, 'Powers of Entry, Search and Seizure' (1989) 40/4 *Northern Ireland Legal Quarterly* 333–62; B. Irving and I. McKenzie, *Police Interrogation* (1989) and 'Interrogating in a Legal Framework, in R. Morgan and D. Smith (eds.) *Coming to Terms with Policing* (1989). (*b*) The Home Office: C. Willis, J. Macleod, and P. Naish, *The Tape Recording of Police Interviews with Suspects: A Second Interim Report* (1988); M. Maguire, 'Effects of the PACE Provisions on Detention and Questioning: Some Preliminary Findings' (1988) 28/1 *Brit. J. Criminology* 19–43; D. Brown, *Detention at the Police Station under the Police and Criminal Evidence Act* (1989); M. Maguire, 'Complaints against the Police: The British Experience', in A. Goldsmith (ed.), *Complaints against the Police* (1991), and M. Maguire and C. Corbett, *Complaints against the Police* (1991); (*c*) The Lord Chancellor's Dept.: A. Sanders, L. Bridges, A. Mulvaney, and G. Crozier, *Advice and Assistance at Police Stations and the 24 Hour Duty Solicitor Scheme* (1989), and A. Sanders and L. Bridges, 'Access to Legal Advice and Police Malpractice' (1990) *Crim. LR* 494–509.

[132] Sanders and Bridges, 'Access'; Morgan, Reiner, and McKenzie *Police Powers*.

somewhere between two to four times, and is now about a quarter of all suspects.[133]

3. The extended powers available for SAOs on a superintendent's authority are obtained relatively infrequently, in about 2 per cent of cases.[134]

4. The extent of use of 'tactics' to extract confessions or incriminating statements in interrogation has declined.[135]

5. Tape-recording of interviews seems to have reduced arguments about what occurred, and to be welcomed now by the police who were previously suspicious.[136]

6. The average period of detention in police stations has declined (from about 6 hours and 20 minutes to about 5 hours and 20 minutes).[137]

7. The PCA supervision of complaints is often vigorous and active, especially in serious cases.[138]

There is, however, also much evidence on the negative side:

1. Detention is authorized almost automatically and invariably. The idea of the custody officer as an independent filter on this has proved unrealistic.[139]

2. The information to suspects about their rights is often given in a ritualistic or meaningless way. This may account for the overwhelming majority who do not take them up. In addition, it appears that 'ploys' may be used to dissuade suspects from taking up rights, though the extent to which they are used is debatable. Conversely, it appears from recent research that the right to silence may benefit some serious offenders disproportionately.[140]

---

[133] Estimates vary between studies both pre- and post-PACE, and each study shows a considerable unexplained variance between police stations: cf. D. Brown, *Detention at the Police Station under the Police and Criminal Evidence Act* (1989), Sanders and Bridges, 'Access', Morgan, Reiner, and McKenzie, *Police Powers*. There does not seem too much decline in the proportion of suspects seeking legal advice who fail to receive it, however (around one-fifth). The revised codes require custody offices to tell suspects legal advice is available free of charge. This has further boosted the take-up rate. D. Brown, T. Ellis and K. Larcombe: *Changing the Code: Police Detention under the Revised PACE Codes of Practice* (HMSO, London, 1992).

[134] Brown op. cit., pp. 48–50. Also Home Office Statistical Bulletin (15 July 1991) 14/91 *Statistics on the Operation of Certain Police Powers under the Police and Criminal Evidence Act.*

[135] Irving and McKenzie, *Police Interrogation*, contrasted with B. Irving, *Police Interrogation* (1980) for pre-PACE evidence.

[136] Willis, Macleod and Naish, *Tape Recording.*

[137] Morgan, Reiner and McKenzie, *Police Powers.*

[138] Maguire, 'Complaints'; Maguire and Corbett, *Complaints.*

[139] McKenzie, Morgan, and Reiner, 'Helping the Police'.

[140] On access to lawyers cf. Sanders *et al.*, *Advice and Assistance of Police Stations* (1989); Morgan Reiner, and McKenzie, *Police Powers*. For the exercise of the right to silence, see T. Williamson and S. Moston: 'The Extent of Silence in Police Interviews', in Greer and Morgan (eds.), *The Right to Silence Debate.*

3. Later stages in the detention process (such as reviews, or regulating access to suspects by investigating officers) are less punctiliously followed than the reception rituals. Custody officers are also less scrupulous about recording pre-detention events (such as delay between arrest and arrival at police stations).[141]

4. PACE procedures can frequently be side-stepped by securing 'voluntary' compliance by suspects with police requests. Such 'consent' is especially important for the stop and search power, which is often circumvented.[142]

5. The use of 'tactics' in interrogation is increasing again, though not to pre-PACE levels.[143]

6. The average length of detention for more minor cases has increased since PACE.[144]

7. However adequate PCA supervision of complaints investigations may be, public confidence and complainant satisfaction remain disastrously low, and the police have also become alienated.[145]

8. Consultative committees work more to legitimate police authority than impose public priorities on the police.[146]

9. The social-discriminatory pattern of use of police powers seems as marked as before.[147] This is because the social role of the police remains essentially unchanged, and PACE has done little to alter the culture and organization of policing.[148]

CONCLUSION

There can be no doubt that the powers of the police have expanded considerably in the last two decades. PACE is obviously a watershed, placing new powers on a more systematic basis than ever before. At the same time it attempts to erect a network of safeguards to prevent the abuses which have been revealed both by a series of *causes célèbres* and by empirical research. Contrary to most expectations, the judiciary have

[141] Morgan, McKenzie and Reiner op. cit.

[142] Dixon *et al.*, 'Consent and Legal Regulation'. For 'voluntary' attendance at police stations as an alternative to detention, see McKenzie, Morgan and Reiner, 'Helping the Police'. Lidstone op. cit. notes a decline in searches with relative to without warrants, another means of avoiding external scrutiny procedures.

[143] Irving and McKenzie, *Police Interrogation*.

[144] Morgan, Reiner, and McKenzie, *Police Powers*.

[145] Maguire and Corbett, *Complaints*.

[146] R. Morgan, 'Policing by Consent: Legitimating the Doctrine', in R. Morgan and D. Smith (eds.), *Coming to Terms with Policing* (1989).

[147] Morgan, Reiner and McKenzie, *Police Powers*.

[148] Dixon *et al.*, 'Consent and Legal Regulation'. McConville, A. Sanders and R. Leng: *The Case for the Prosecution* (Routledge, London, 1991).

shown greater willingness on the whole to police the new settlement than they did the earlier Judges' Rules.

The combination of changes have changed police practices profoundly. However, they have done so unevenly and patchily. There remains much evidence of deviation, and of a constant substance of socially discriminatory policing under new formalities. The analytic and research agenda for the future ought to try and penetrate the factors which account for when legal change does alter practice, when this change is merely presentational, and when it is ignored.

A working hypothesis, synthesized from the research cited above, would suggest the following. Deterrence, symbolism, police culture, organization, and training are all important. However, of fundamental importance is the outside social and political context.

Deterrence is important. If powers are precisely rather than permissively formulated, procedures to render visible occasions of usage are constructed,[149] and supervisors and the courts determined to police them, change can occur. Thus, the booking-in rituals, which are precise, relatively visible, clearly enjoined in training, and policed by supervisors, are followed religiously. However, the danger of precisely formulated rules is also evident here. They can be satisfied by ritualistic observance with little meaning, defeating their intended objectives.

If discretion is then to be allowed (and of course some is inevitable in any case) the key changes must be in police working rules, in the informal culture of the police. These can be penetrated by official policy, through symbolism, training, and organization. Symbolism is very important in the reception of PACE. The fact that the Codes were issued with a legitimacy derived from an Act of Parliament undoubtedly gave them greater weight in police culture than the Judges' Rules had. Organization and training are clearly important too. Thus the procedures are more faithfully followed (in substance as well as form) when custody officers are encouraged to identify with the role by serving in it for extended periods of time, performance in the job is relevant to career prospects, and special training is devoted to specialist custody officers, than if they are ordinary operational sergeants serving a turn because they drew the short straw. They become dedicated custody officers in every sense.

However, even if organizational changes do succeed in achieving a

---

[149] Some of the key aspects of PACE's safeguards, such as record-keeping and tape-recording, are examples of what has been called the 'transparent stewardship' model of accountability. R. Morgan, 'Policing by Consent: Legitimating the Doctrine', in R. Morgan and D. Smith (eds.), *Coming to Terms with Policing* (1989). More could be done along these lines, and experimentation is occurring with e.g. the video-taping of interviews (successfully tried in some parts of North America, cf. A. Grant, 'Video-Taping Police Interviews' (1987) *Crim. LR* 375, and computerization of custody records.

greater congruence between black-letter and blue-letter law, the substance of policing will remain essentially unchanged. This means that the sources of controversy and dissatisfaction, such as the socially discriminatory use of police powers, will continue. Policing in a hierarchical and divided society will never be even in its impact, and thus legal regulation will always be inadequate by itself to secure legitimacy and consent.

4

Due Process

JOHN JACKSON

INTRODUCTION: DUE PROCESS AND CRIME CONTROL

This chapter is concerned with due process in the context of criminal procedure but due process is not only of significance in criminal procedure. There is now a general acceptance of the need for due process or fair procedures where decisions are taken affecting the rights of individuals, although there remains the difficult question of determining the precise application of this principle.[1] It might be thought that due process should extend to decisions affecting as vital a matter as the liberty of the subject but this is not always recognized. Chapter 11 makes it clear that little recognition has been given to the procedural rights of persons detained under the immigration laws or to those subject to deportation orders made by the Secretary of State. Although the courts have recognized that prisoners facing disciplinary charges have a right to a fair hearing, Chapter 6 discusses the continuing resistance to the idea that prisoners have a right to be heard before decisions are made on segregation and release for parole. The European Court of Human Rights has ruled that Article 5(4) of the European Convention on Human Rights requires periodic review of the detention of psychiatric patients,[2] but there have been doubts about whether mental health review tribunals are an appropriate 'court' of review.[3]

Discussions of due process in the context of criminal procedure often begin by contrasting Packer's two models of criminal justice, the crime-control model and the due-process model.[4] The crime-control model places a premium on the effective processing of cases through the system, an assembly line producing conveyor-belt justice. The due-process model in contrast operates as a constraint on the way in which the crime-control model works, an obstacle course designed to present formidable obstacles to carrying accused persons along the assembly line, placing value on the primacy of the individual and the limitation of state power. Packer

---

[1] D. J. Galligan, *Discretionary Powers: A Legal Study of Official Discretion* (1986), ch. 7.

[2] *X. v. UK* (1982) 4 EHRR 188.

[3] J. Peay, *Tribunals on Trial: A Study of Decision-Making under the Mental Health Act* (1989).

[4] H. Packer, *The Limits of the Criminal Sanction* (1968), 150–72.

seemed to envisage two value systems, the crime-control model holding that the repression of crime is the most important function of the criminal process and the due-process model holding that the individual accused must not be aggrieved in the interests of repressing crime.

This does not mean that a crime-control model dedicated to the repression of crime may not allow for some due process in its pursuit of the guilty. As no system can guarantee a correct outcome, procedures have to be designed which allocate the risk of errors according to the magnitude of harm that will be caused if a particular error is made.[5] A cost-benefit analysis of a system designed to convict criminals may not necessarily put equal value on the conviction of the guilty and the acquittal of the innocent.[6] Such a system requires that the public has confidence in the accuracy of the process. It may be that a greater loss of confidence arises when it transpires that innocent persons are convicted, because this may be perceived as a greater injustice. Even if the two risks of acquittal of the guilty and conviction of the innocent are regarded as equally harmful, a system may wish to place a greater burden of proof on the prosecution on the grounds that there is such a disparity of resources between the prosecution and defence that the defence should be favoured in the allocation of proof.[7]

The problem with such a cost-benefit analysis is that no independent weighting is attached to due process. The case for due process is that it gives expression to certain values that should be recognized within the criminal process even if this means reducing efficiency in terms of repressing crime. Above all, due process gives expression to the Kantian injunction that persons should be treated as ends in themselves and not merely as means to an end. But what exactly does due process entail? A distinction is often made between procedures which safeguard values that are related to outcomes and procedures which safeguard values that are independent of outcomes.[8] Dealing first with outcomes, it can be argued that individuals are entitled to procedures which help to prevent mistaken decisions concerning matters that have grave consequences for them. Dworkin has argued that an individual who is falsely convicted does not just suffer the bare harm of the pain and penalty of punishment but suffers also a 'moral' harm in the sense of a loss of dignity and respect.[9] He goes on to argue in favour of two procedural rights: a right to have the risk of this moral harm reflected in criminal procedures and the right

---

[5] See A. A. S. Zuckerman, *The Principles of Criminal Evidence* (1989), 129.

[6] For a discussion of the costs and benefits of procedures, see Galligan, *Discretionary Powers*, 326–9.

[7] A. Ashworth, 'Concepts of Criminal Justice' (1979) *Crim. LR* 412, 414–15.

[8] See e.g. D. Resnick, 'Due Process and Procedural Justice', in J. R. Pennock, and J. W. Chapman (eds.), *Due Process* (1977), 206–28.

[9] R. Dworkin, 'Policy, Principle and Procedure', in C. Tapper (ed.), *Crime, Proof and Punishment: Essays in Memory of Sir Rupert Cross* (1981), 193–225.

to a consistent weighting of the importance of this risk in all cases. This does not mean that defendants have a right to the most accurate procedures whatever the cost in providing them, but it does entitle them to require a balance to be carried out between the gravity of the harm of false conviction and the costs involved in avoiding them, which is a very different balancing exercise from one which starts from the premiss that the aim of the criminal process is simply to convict criminals. Special weighting is then given to the right not to be falsely convicted, although a sense of proportion has to be maintained between the importance of the right and the costs incurred in safeguarding it, both in terms of resources and in terms of accuracy, because protecting the innocent may involve acquitting the guilty.

This right entails that individuals are given a fair trial but it would also seem to entail that accurate procedures are observed throughout the criminal process and not merely at trial. There is an intimate connection between what happens in the investigatory process and what happens at trial, in the sense that a failure to observe accurate standards of investigation before trial runs the risk of evidence being skewed at trial and the accused being falsely convicted.

Apart from the need to sift out the innocent at an early stage of the process, there are a variety of other decisions affecting both defendants and victims which police, prosecutors, and magistrates make before trial, such as whether to investigate, arrest, caution, prosecute, or negotiate, what the mode of trial should be, and whether bail should be granted. Defendants and victims can claim a right to be treated in the same way as others in the same position. But one problem with the pre-trial stage, and the sentencing stage as well, is that there is a lack of clarity about which outcomes are desirable and this can be explained by the fact that there are differing views about what the aims of the criminal process are.[10] For example, there has been increasing concern in recent years about the plight of victims in the criminal justice system,[11] but until we have clarified what role victims should have in the criminal process, we cannot decide what interest, if any, they should have in decisions to prosecute and in sentencing decisions and what rights, if any, they should be accorded in the decision-making process. Again there is a lack of agreement about whether sentencing be geared towards the prevention of criminal acts or whether it should merely reflect the seriousness of the offence. The aims of the process and the interests at stake have to be clarified before procedures can be devised to facilitate them.[12]

[10] See D. J. Galligan, 'Regulating Pre-Trial Decisions' and N. Lacey, 'Discretion and Due Process at Post-Conviction Stage', in I. H. Dennis (ed.), *Criminal Law and Justice* (1987).

[11] See e.g. M. Maguire and J. Pointing, *Victims of Crime: A New Deal?* (1988).

[12] Galligan, 'Regulating', 191.

It can also be argued that individuals are entitled to have intrinsic process rights which are unrelated to instrumental goals such as avoiding false convictions. An accused who is denied the right to a fair trial, for example, is arguably subjected to moral harm in the sense in which Dworkin defines it, simply because he has not been treated as a rational, autonomous person but rather as an object to be dealt with.[13] A public process of proof at which the defendant is present, is entitled to legal representation, and knows the case against him no doubt aids the process of truth finding. It has been argued that the interests of truth finding require a participatory model of adjudication.[14] But a fair trial can be viewed not only as an inquiry into the defendant's guilt but also as a demonstration of the defendant's guilt which calls a defendant to answer for his actions.[15]

If individuals are to be treated with respect before trial as well as at trial, then this would seem to require that coercive action be taken against them only where there is reasonable cause for it and that they be brought to trial within a reasonable time. It would also seem to require the pre-trial process to be viewed, like the trial, as a communicative process in which allegations are put to suspects and suspects are able to respond. As at trial, however, the right to participate ought to mean the right to participate in an informed manner and this may entail the need for legal advice so that informed choices can be made.

Once due-process requirements have been given concrete expression, there is the problem of ensuring that they are realized. First it is necessary to ensure that they are recognized in rules of procedure. Over the years the American Supreme Court has interpreted the due-process clause in the US Constitution as requiring that an accused person be put on fair notice of the nature of the acts he is charged with, that he be given adequate opportunity to present his case through counsel before a fair and impartial tribunal free from prejudicial influences, that he be entitled to be continuously present at the trial and to confront his accusers, and that he be free from the damaging effect of coerced confessions.[16] Although there has been a tendency in Anglo-American literature to believe that only adversarial systems are capable of upholding the presumption of innocence,[17] many European countries which operate so-called inquisitorial systems of criminal justice have ratified the European

---

[13] See L. H. Tribe, *American Constitutional Law* (1978), 503.

[14] J. Jackson, 'Two Methods of Proof in Criminal Procedure' (1988) *MLR* 549–68.

[15] See R. A. Duff, *Trials and Punishments* (1986), ch. 4.

[16] See S. H. Kadish, 'Methodology and Criteria in Due Process Adjudication: A Survey and Criticism' (1956) *YLJ* 319–63.

[17] For criticism of this 'Manichaean dichotomy', see M. Damaska, 'Evidentiary Barriers to Conviction and Two Models of Criminal Procedure: A Comparative Study' (1973) *Univ. Penn. LR* 506, 569–70.

Convention on Human Rights, Article 6 of which specifically guarantees the presumption of innocence. Article 6 also requires accused persons to be brought before a public hearing presided over by an independent and impartial tribunal and guaranteed the right to be informed promptly of the case against them, time and facilities to prepare for their defence, free legal assistance, and the right to examine and have examined witnesses against them and to obtain the attendance and examination of witnesses on their behalf under the same conditions as witnesses against them.

One of the problems with adversarial procedures, with their emphasis on the trial as the focal point of the system, is that the pre-trial stage of the process is left largely unregulated until the point at which a defendant is charged. By contrast, in inquisitorial systems, the police and investigatory agencies are regulated more by the courts and by supervising judges.[18] Rather than the process of proof highlighting one particular event—the trial—inquisitorial systems represent a continuous process of proof with perhaps a number of phases of investigation in serious cases, the trial representing merely the final phase. The lack of control over the police in adversarial systems, coupled with the willingness to permit parties to negotiate over the plea, means that it is especially important to provide safeguards, such as the availability of independent legal advice and the right of silence, to prevent suspects confessing or pleading unnecessarily.

Once due-process requirements are recognized, it is important to ensure that remedies are available when due-process requirements have not been met. Since there is such an important link between fairness before trial and fairness at trial it can be argued that the courts need to be prepared to exercise a supervisory role at trial over pre-trial procedure, particularly in adversarial systems where they have little direct control over the pre-trial process. One way of doing this is for the courts to exclude, at the trial, evidence obtained unfairly before trial. Opinions differ as to how extensive this exclusionary jurisdiction should be. Three different principles have commonly been advocated for guiding the courts on this question.[19] The reliability principle holds that the courts are only justified in excluding evidence which is unreliable. The protective principle holds that evidence should be excluded where the unfairness has caused the defendant to be put at some disadvantage. Finally, the disciplinary principle holds that evidence may be excluded to punish those who have caused unfairness and to deter breaches in the future. Since so many defendants plead guilty in the adversarial system it can be argued

[18] See discussion in M. Damaska, *The Faces of Justice and State Authority: A Comparative Approach to the Legal Process* (1986), 47–56.

[19] See A. Ashworth, 'Excluding Evidence as Protecting Rights', (1977) *Crim. LR* 723–35.

that the courts' supervisory role ought to extend beyond ensuring fairness in the individual contested case and include a disciplinary function, at least to the extent of excluding evidence obtained in breach of rules that are designed to prevent false convictions.[20]

Finally, it is necessary to appreciate that procedural rules and enforcement mechanisms are not enough in themselves to ensure that due process is realized. Empirical assessments are necessary to see how the rules are operating and to gauge the extent to which they are being observed by officials in their everyday practices.

It is now time to consider how these issues have been dealt with in Britain since 1950. The following sections review the structure of the British criminal process; the steps that have been taken both to restrict and advance the due-process interests of defendants; what particular contribution the courts have made to safeguard and enforce these interests; and the impact of international standards. The final section considers empirical assessments.

THE STRUCTURE OF BRITISH CRIMINAL JUSTICE

The English, Scottish, and Northern Irish systems of criminal justice are basically adversarial in structure, with the result that the prosecution has to make a case against the defendant and if necessary prove guilt beyond reasonable doubt in a contested trial. The structure of the three systems has remained essentially unchanged in the post-war period, although this may change in the wake of the report of the Royal Commission on Criminal Justice which at the beginning of 1993 was conducting a radical review of the workings of the entire English system in the light of the miscarriages of justice in the cases of the Guildford Four, the Birmingham Six, and the Maguire Seven.

The most important structural change in England and Wales has been the institution of a Crown Prosecution Service, headed by the Director of Public Prosecutions, under the Prosecution of Offences Act 1985. The service which is charged with the prosecution of most criminal offences is divided into thirty-one areas throughout England and Wales, each of which is headed by a Chief Crown Prosecutor accountable to the Director.[21] In Northern Ireland an independent Director of Public Prosecutions was instituted in 1972 with powers to prosecute serious cases.[22] In Scotland the decision to prosecute has long been

---

[20] For further argument on this point see J. Jackson, 'In Defence of a Voluntariness Doctrine for Confessions: *The Queen* v. *Johnston* Revisited' (1986) *Irish Jurist* 208–46.

[21] F. Bennion, 'The Crown Prosecution Service' (1986) *Crim. LR* 3–15.

[22] Prosecution of Offences (NI) Order 1972 (SI No. 538). See *Report of the Working Party on Public Prosecutions*, Cmnd. 552 (1971).

in the hands of legally qualified procurators fiscal accountable to the Lord Advocate.[23]

Once prosecuted there are two different modes of trial. The vast majority of defendants are tried summarily, which means trial in a magistrates' court in England and Northern Ireland or in a district court or sheriff court in Scotland. Magistrates' courts account for 97 per cent of all criminal business in England and Wales. The alternative is trial on indictment in a Crown Court in England and Northern Ireland (Crown Courts replaced the old quarter sessions and assize courts in 1971 in England and in 1978 in Northern Ireland) or trial by solemn procedure in the sheriff court or High Court of Justiciary in Scotland. The main difference between the two modes of trial is that in trials on indictment or in solemn procedure defendants are entitled to trial by jury if they contest their case. (In Northern Ireland an exception is made when persons are tried for offences linked to terrorism.)[24]

There is a striking contrast in all three jurisdictions between the quality of justice that is dispensed in the lower courts and the higher courts, what one writer has described as 'two tiers of justice'.[25] One noticeable difference is that many defendants are unrepresented in summary trials. Things have improved in the post-war period with the advent of legal aid, but legal aid is subject to a means test and will not be granted unless it is considered desirable to do so in the interests of justice. The Widgery Committee on legal aid in England and Wales did not believe that legal aid was necessary in many summary cases and it laid down non-statutory criteria (known as the Widgery criteria) to guide magistrates on when it is in the interests of justice to grant legal aid.[26] The result has been that, while legal aid is almost always granted in the Crown Court, it is restricted much more in the magistrates' court and, despite the guidance, the chances of obtaining legal aid still seem to vary considerably depending on which particular magistrates' court the defendants appear in.[27] The Royal Commission on Legal Services argued for an extension of the legal aid scheme but its proposals were not implemented.[28] The Commission's recommendation that duty solicitor schemes be extended to provide legal advice in all busy magistrates' courts was, however, taken up, and duty solicitor schemes were placed on a statutory basis in 1982 to help ensure that legal advice was more readily available than before. But there has been concern in recent years about the low levels of

---

[23] See S. Moody and J. Tombs, *Prosecution in the Public Interest* (1982).

[24] See Ch. 5.

[25] D. McBarnet, *Conviction. Law, the State and the Construction of Justice* (1981), ch. 7.

[26] *Report of the Departmental Committee on Legal Aid in Criminal Court Proceedings*, Cmnd. 2934 (1966). See now s. 22 of the Legal Aid Act 1988.

[27] See R. C. White, *The Administration of Justice*, 2nd edn. (1991), 79.

[28] *Final Report of the Royal Commission on Legal Services*, Cmnd. 7648 (1979).

remuneration available for duty solicitors and under the legal aid scheme; there is evidence that solicitors are giving up legal aid work in criminal cases.[29]

When defendants are represented at court, the vast majority, particularly in the magistrates' court, plead guilty. Fewer than two out of ten defendants plead not guilty in the English magistrates' courts compared to one in three in the Crown Court. Defendants plead guilty for a variety of reasons.[30] Apart from the fact that they may perceive themselves to be guilty, they may believe that a guilty plea will result in a discount in a sentence and that their case will be dealt with more quickly, a considerable inducement to a defendant in custody. Under the Bail Act 1976 there is a presumption that a defendant should be granted bail unless there are compelling reasons for believing that if released he or she will abscond, commit an offence, or obstruct justice, but in practice it appears to be difficult to be granted bail when prosecutors object to it.[31] Although the proportion of defendants remanded on bail has increased, large numbers of persons are still remanded in custody. A cautious attempt has recently been made in England to follow the Scottish example of time limits in criminal cases. Under section 22 of the Prosecution of Offences Act 1985, custody limits of 70 days between the defendant's first court appearance and committal or trial for indictable or either-way offences and a maximum period of 112 days between committal and trial have been introduced in most areas, but the courts may extend the period if satisfied that there is a good and sufficient cause and the prosecution has acted with all due expedition.[32] The most probable reason why defendants plead guilty is because very few defendants have exercised their rights of legal advice and their right of silence in the police station, with the result that they have incriminated themselves before they arrive in court. By the time their cases reach court the police and prosecutors have a monopoly of information about them and defence lawyers are under pressure to reach settlements and advise a guilty plea.[33]

The greater proportion of contests in the Crown Court than in the magistrates' court is explained by the fact that defendants who want to plead not guilty see the Crown Court as offering a better chance of

---

[29] See the editorials in the *New LJ* 15 Sept. 1989, 11 Oct. 1991. Recent government proposals for a system of fixed fees for legal aid work in magistrates' courts have attracted particular criticism. See F. Gibb, 'Mass Rally for Justice', *The Times*, 11 February 1992.

[30] See the discussion in M. Zander, *A Matter of Justice*, rev. edn. (1988), 188–90.

[31] White, *Administration*, 81.

[32] Prosecution of Offences (Custody Time Limits) Regulations (SI 1987 No. 299).

[33] See J. R. Baldwin, *Pre-Trial Justice: A Study of Case Settlement in Magistrates Courts* (1985); M. McConville, A. Sanders, and R. Leng, *The Case for the Prosecution* (1991), ch. 8.

acquittal.[34] The appointment of lay magistrates by local committees which advise the Lord Chancellor is shrouded in secrecy. There has been a clear over-representation of middle classes on the magistrates' bench and an under-representation of women and members of ethnic minorities. One study has concluded that advisory committees operate racist criteria, such as 'being assimilated into the English way of life' as a criterion of suitability for the selection of black magistrates.[35] Another barrier to representativeness, apart from the selection procedure, is the demands of the job which require persons to 'have time to carry out the full range of magisterial duties'.[36] Apart from concerns about the representativeness of magistrates, there have also been concerns about the close ties between the magistracy and the police and about magistrates becoming easily case-hardened.[37] Research has shown that magistrates have a tendency to accept police evidence.[38] Case-hardening is perhaps a particular danger in the case of stipendiary (or professional) magistrates who have been given an increasingly important role in a number of magistrates' courts in England and Wales.[39]

Doubts about the representativeness, independence, and impartiality of the tribunal of fact do not arise to the same extent in the case of jury trial. Although less than 2 per cent of cases are tried by jury, jury trial has long played a vital symbolic role in characterizing the fairness of the English criminal process. The institution of the jury has been variously heralded as trial by one's peers, as an example of participatory democracy, and as a safeguard against the abuse of state power and oppressive laws and prosecutions.[40] The historical justification for these claims is dubious but there are clear advantages for defendants in trial by jury as compared with trial by magistrates. In 1972 the property qualification for jurors was abolished and this marked a major step in the direction of the principle of trial by random selection. Although this principle does not guarantee trial by a representative cross-section of the community and can cause difficulties in individual cases where black defendants, for example, can

[34] D. Riley and J. Vennard, *Triable Either-Way Cases: Crown Court or Magistrates' Court* (1988).

[35] M. King and C. May, *Black Magistrates: A Study of Selection and Appointment* (1985).

[36] Lord Chancellor's Dept., 'The Qualities Looked for in a Justice of the Peace' (1988) *The Magistrate* 78. See J. W. Raine, *Local Justice: Ideals and Realities* (1989).

[37] S. Enright and J. Morton, *Taking Liberties: The Criminal Jury in the 1990s* (1990), 96–9.

[38] J. Vennard, *Contested Trials in Magistrates' Courts* (1982).

[39] For discussion of their history see J. R. Spencer (ed.), *Jackson's Machinery of Justice in England* (1989), 412–15. For figures see H. C. Debs, vol. 200, cols. 419–20, 11 December 1991. Resident magistrates sit exclusively in the Northern Ireland Magistrates' courts.

[40] For a recent critique of these claims see, P. Darbishire, 'The Lamp that Shows that Freedom Lives: Is it Worth the Candle?' (1991) *Crim. LR* 740–52.

end up with an all-white jury, the principle is arguably a much better foundation for impartiality and independence than the principle of selection by unaccountable advisory committees appointed by the Lord Chancellor. Independence and impartiality are also encouraged by the fact that jurors come to each case fresh, unsullied by the case-hardening syndrome and without predisposition to one side or the other.

Another advantage in jury trial is that jurors are not privy to inadmissible evidence which may prejudice any tribunal of fact. Many of the rules of procedure and evidence in the English contested trial can be explained against the background of a right not to be falsely convicted— for example the presumption of innocence, the burden of proof on the prosecution, the standard of proof beyond reasonable doubt, the rules relating to similar facts, character, corroboration, restrictions of cross-examination of the accused, and the rules concerning the admissibility of confessions.[41] The difficulty with tribunals like the magistracy which act as the tribunal of fact as well as law is that they are privy to the evidence which they hold to be inadmissible and therefore have to go through the mental gymnastics of disregarding what they have already been told.

PRESSURES FOR CHANGE

The pressure groups described in Chapter 3 which have campaigned in contradictory ways for changes in police powers have also been active in calling for changes throughout the criminal process as a whole. The crime-control or law-and-order lobby, which has claimed that the process leans too heavily in favour of the defendant, particularly the sophisticated, professional criminal, has focused on issues such as the right of silence and the jury system, as well as on the punishment of offenders. The due-process lobby, which has argued that the process does not go far enough to protect inarticulate, young, and inexperienced defendants, has tended to focus on miscarriages of justice that have caused innocent persons to be wrongly convicted.

The political complexion of respective governments has played some part in helping or hindering the ability of crime-control and due-process adherents to effect change. The Conservative Party has traditionally tended to present itself as the party of 'law and order' and a strong crime-control lobby within the party has been particularly effective in calling for tough measures against offenders. In 1983, for example, the Conservative government radically altered the parole system by announcing that serious offenders would not be granted parole until the final months of

[41] D. J. Galligan, 'More Scepticism about Scepticism' (1988) *Oxf. J. Legal Studies* 249–65.

their sentence, with the result that the paroling rate in longer sentence cases was sharply reduced.[42] But the importance of party politics in effecting change can be exaggerated. The Police and Criminal Evidence Act 1984 (PACE), which introduced important safeguards for suspects in custody (discussed in Chapter 3) and the Prosecution of Offences Act 1985 were both the products of Conservative, not Labour, administrations. Many due-process adherents have lamented the poor record of Labour governments in the area of due process.[43]

It can be argued that the key to many of the legislative changes in the post-war period has been not so much the particular political complexion of the party in power as the extent to which proponents of reform have managed to create a consensus of opinion across the crime-control–due-process spectrum that something ought to be done to safeguard crime-control or due-process values. This can be illustrated by examining how successful each lobby has been in achieving reform on the issues highlighted.

The right of silence

The right of silence has assumed great symbolic importance for both due-process and crime-control adherents. To the due-process lobby it is the cornerstone of British justice, underpinning the principle that it is for the prosecution to prove the accused's guilt.[44] In recent years crime-control adherents have seen the right as a clear demonstration of their claim that criminals can use the rules to escape justice. As discussed in Chapter 3 the issue has constantly surfaced in the context of police powers. One of the problems for the crime-control lobby, however, has been to justify the claim that the exercise of the right before and at trial has actually impeded the conviction of criminals. Recent empirical evidence does not show that exercise of the right results in a failure to charge defendants or in their acquittal.[45]

In Northern Ireland the claim that large numbers of terrorist suspects were exercising the right provided the justification for radically curtailing it under the Criminal Evidence (NI) Order 1988.[46] In Scotland concern

---

[42] For discussion of the impact of this change, see *Report of the Review Committee: The Parole System in England and Wales*, Cm. 532 (1988).

[43] See e.g. T. Sargant and P. Hill, *Criminal Trials: The Search for Truth* (1986).

[44] See e.g. J. Wood and A. Crawford, *The Right of Silence: The Case for Retention* (1989).

[45] See S. Greer, 'The Right to Silence: A Review of the Current Debate' (1990) *MLR* 709, 711. See also now R. Leng, *The Right to Silence in Police Investigation: A Study of Some of the Issues Underlying the Debate* (1993), Research Study No. 10 for the Royal Commission on Criminal Justice.

[46] See Ch. 5 and J. Jackson, 'Recent Developments in Criminal Evidence' (1988) *NILQ* 105–18.

about the number of accused persons who went back on allegedly voluntary confessions prompted the Thomson Committee on Criminal Procedure to recommend the revival of the judicial examination procedure.[47] From 1980 accused persons may be brought before the sheriff and asked about any relevant matter by the prosecutor. A failure to mention matters later relied on may be commented on at the trial. In England and Wales the only major incursion into the right of silence to date has been in fraud cases. Concern over the ability of perpetrators of serious fraud to escape justice led to the establishment of the Roskill Committee on Fraud Trials which made a number of proposals for the policing and prosecution of fraud.[48] As a result the Criminal Justice Act 1987 established a Serious Fraud Office to investigate serious fraud and to take over prosecutions of such cases and it gave the Office power to require anyone under investigation or believed to have relevant information to answer questions and provide information or documents.[49] In addition, the Act provided for the abolition of committal proceedings in serious fraud cases and their replacement by a preparatory hearing conducted by a Crown Court judge who may require defendants to disclose their defence under penalty of adverse comment at the trial.

Although little opposition was mounted against the Roskill Committee's proposals for the policing and prosecution of fraud, other proposals affecting criminal procedure more generally were not so favourably received. The Committee believed that the hearsay rule was preventing the admission of important documentary evidence against fraudsters and it recommended the virtual abolition of the hearsay rule for documents. Sweeping changes to the hearsay rule in all criminal cases were proposed in the Bill preceding the 1987 Act but these met with serious opposition in the House of Lords and a much watered down version of the reforms was enacted the next year in the Criminal Justice Act 1988. The Act only provided for the admission of first-hand documentary hearsay in particular situations where it was impossible or unreasonable to call the maker.[50]

Serious consideration has recently been given to the idea that the disclosure provisions now imposed on the defence in fraud cases should be extended to all cases tried in the crown court. Although there continues to be much opposition to any encoachment on the right of silence

[47] *Criminal Procedure in Scotland*, Cmnd. 6218 (1975).

[48] *Fraud Trials Committee Report* (1986). For discussion see M. Levi, 'Reforming the Criminal Fraud Trial' (1986) *J. Law and Soc.* 117–30.

[49] A recent House of Lords decision has held that this statutory power does not come to an end even when the person under investigation is charged with an offence (*Smith* v. *Director of Serious Fraud Office* [1992] 3 All ER 456).

[50] For discussion of these reforms see D. Birch, 'The Criminal Justice Act 1988: Documentary Evidence' (1989) *Crim. LR* 15–31.

at the police investigation stage, it is considered more justified to require the defence to make some disclosure of its case after the prosecution has disclosed its case to the defence.[51] To date the defence is only required to disclose details of alibi defences and expert evidence it intends to call in the crown court.[52] The Royal Commission on Criminal Justice, at the time of writing, was considering whether greater defence disclosure is desirable. The most appropriate stage at which disclosure should take place would seem to be at preliminary hearings or at pretrial reviews which have proved popular both in the magistrates' court and in the Crown Court as a means of clarifying the contested issues and obtaining estimates of how long any trial will last.[53] Despite approval given to these hearings, fears have been expressed that the occasions may be used by prosecutors to force deals and put pressure on the defence to plead guilty.[54]

Trial by jury

Another issue that has sparked major controversy has concerned the place and role of the jury in the criminal process. We have seen that the institution of the jury is still of great significance as a symbol of due process and few have advocated outright abolition of the jury. The debate has rather centred around whether the jury needs to be reformed and in what direction reform should go.

Crime-control adherents have highlighted a number of concerns about the operation of the jury system. The competence of jurors has been a persistent theme ever since the property qualification was abolished in 1972. In 1986 the Roskill Committee on Fraud Trials recommended that jury trial should be replaced in cases of serious fraud by a Fraud Trials Tribunal, on the ground that jurors were not capable of understanding the evidence in fraud trials. In 1988 a highly publicized survey conducted by *The Times* claimed that professional people were able to evade jury duty easily with the result that there was a heavy bias in favour of unemployed and manual members on juries.[55] Another concern has been that jurors are sometimes intimidated or bribed into acquitting

[51] In its evidence to the Royal Commission on Criminal Justice, the Law Society has recommended that after committal it should be open to the prosecution to apply to the defence for disclosure of a list of witness whom they intend to call at trial and in serious cases to apply to the judge to direct the defence to answer a list of questions to determine the contested issues.

[52] S. 11 of the Criminal Justice Act 1967; s. 81 of the Police and Criminal Evidence Act 1984.

[53] Pre-trial reviews are not regulated by statute except under s. 11 of the Criminal Justice Act 1989. See Baldwin, *Pre-Trial Justice*.

[54] McConville, Sanders, and Leng, *Prosecution*, 168–70.

[55] 'The Jury on Trial', *The Times*, 24–7 Oct. 1988.

professional criminals. This was the theme of Sir Robert Mark's famous Dimbleby Memorial lecture on BBC Television in 1973 and it has been a constant theme of jury critics ever since.[56] Critics are also outraged when juries appear to acquit clearly guilty people, for example when Clive Ponting was acquitted in 1985 of charges under the Official Secrets Act and when Messrs. Randle and Pottle were acquitted in 1991 of assisting the escape of a convicted spy in the 1960s.

Jury supporters have responded to these claims in a number of ways. In Lord Devlin's view criticism of the 'perversity' of juries misses the point as it is the very ability of juries to be perverse which characterizes their strength.[57] On this view the function of the jury is not just to apply the law to a set of facts but to mediate between the law and community values and sentiment. It has also been easy to point to the lack of empirical evidence to substantiate the claims made. One piece of jury research directly contradicted Mark's claims about the acquittal of professional criminals.[58] The Roskill Committee's argument for the abolition of the jury in complex fraud trials was founded on little actual evidence of jury incomprehension.[59] Criticisms were also made about the lack of evidence for *The Times*' claims.[60]

Despite this, jury critics have managed to achieve a number of reforms aimed at checking the jury, such as the abolition of the need for unanimous verdicts, disqualification for jurors who have committed offences themselves, jury vetting in particular kinds of cases, restrictions on the defence right to challenge without cause, and restrictions on the defence right to elect for jury trial by making certain offences purely summary. Findlay and Duff have argued that the key to the enactment of these reforms was that they were proposed not purely on crime-control grounds but in terms of values which due-process adherents could not quarrel with, such as the need to preserve the impartiality and independence of juries.[61] So, for example, the argument for amending the unanimity rule enacted in the Criminal Justice Act 1968 was based on fears of jury 'nobbling' in cases involving professional criminals. The need to preserve impartiality was also the basis for the enactment of the Juries (Disqualification) Act 1984 which prevented persons serving as jurors if

---

[56] Sir Robert Mark, 'Minority Verdict' (1973), Dimbleby Lecture, BBC Publications, London. See also S. Davis and T. Rayment, 'Rough Justice? Jury System Goes on Trial', *Sunday Times*, 5 June 1988.

[57] P. Devlin, *The Judge* (1979).

[58] M. Zander, 'Are Too Many Criminals Avoiding Conviction? A Study in Britain's Two Busiest Courts' (1974) *MLR* 28–61.

[59] See the dissenting opinion of W. Merricks, 190–9 and R. Harding, 'Jury Performance in Complex Cases', in M. Findlay and P. Duff, *The Jury under Attack* (1988), 74–94.

[60] See B. Raymond, 'Not Guilty Verdict on the Jury', *Guardian*, 18 Nov. 1988.

[61] Findlay and Duff, *Jury under Attack*.

they had been sentenced within the previous ten years to a term of custody, a suspended sentence, or a community service order, or within the previous five years to a probation order. More controversially, the case for jury vetting and the abolition of the peremptory challenge was also argued for on grounds of impartiality. In 1978 the Attorney-General issued guidelines permitting jurors to be stood by in serious cases involving 'strong political overtones', on the ground that jurors with strong political beliefs might put improper pressure on other jurors.[62] Revised guidelines in 1980 narrowed the class of cases for which full vetting would be available to cases involving national security and terrorist cases, but the danger of extreme jurors continued to provide a justification for vetting in these cases.[63] Then in 1986 the Roskill Committee recommended that the right of peremptory challenge be abolished on the ground that it enabled defendants to rig juries in their favour. Despite evidence to show that challenging had little effect on defendants' chances of acquittal,[64] this recommendation was enacted in the Criminal Justice Act 1988. The result is that the defence's powers of challenge have been significantly diminished.[65] Although these reforms may be seen as victories for the crime-control lobby, increasing as they do the prosecution's power to influence the composition of the jury at the expense of the defence, they appear to have been achieved by couching the arguments in terms of values that can claim widespread support.

This gradual shift in the balance of power, from the defence towards the prosecution, is also evident in the restrictions that have been placed on the defence right to elect for jury trial. The pressures here have come not so much from the crime-control lobby as from what may be called an administrative or bureaucratic lobby, consisting of the Home Office, successive Lord Chancellors, members of the judiciary, the magistracy, prosecutors, and justices' clerks, all of whom have been concerned about the increase in the number of cases committed to the Crown Court with the consequent rise in cost and delay.[66] In 1976 the James Committee was set up to 'consider within the existing legal framework what should be the distribution of criminal business between the Crown Court and the magistrates' court'.[67] The Committee recommended that a number of

[62] See *Note* [1980] 2 All ER 457.

[63] See *Note* [1980] 3 All ER 785. For discussion see White, *Administration*, 117–21; Enright and Morton, *Taking Liberties*, ch. 3.

[64] J. Vennard and D. Riley, 'The Use of Peremptory Challenge and Standby of Jurors and their Relationship to Trial Outcome' (1988) *Crim. LR* 731–8.

[65] The Government has, however, also subsequently restricted the prosecution's power to stand by jurors, see (1989) 88 Cr. App. R. 123.

[66] For discussion see Enright and Morton, *Taking Liberties*, ch. 6.

[67] *Report of the Interdepartmental Committee on the Distribution of Criminal Business between the Crown Court and Magistrates' Court*, Cm. 6323 (1975).

offences be transferred to the sole jurisdiction of magistrates and the Criminal Law Act 1977 enacted many of these recommendations by shifting offences from the category of indictment only to the category of either-way (for example, most forms of burglary) and from either-way to purely summary (for example, various public-order offences and drink-driving offences). Only a spirited opposition in the House of Lords prevented the classification of thefts under £20 becoming purely summary. Despite these changes, pressure for further restrictions on jury trial on grounds of cost and delay continued into the 1980s and in 1988 the Criminal Justice Act made the offences of taking a motor vehicle without consent, driving whilst disqualified, and common assault purely summary.

Miscarriages of justice

One of the most effective ways of illustrating the need to introduce safeguards for accused persons has been to highlight particular mis-carriages of justice. Indeed it can be argued that two of the most important pieces of legislation affecting the rights of suspects and defend-ants in the post-war period, PACE which is discussed in Chapter 3 and the Prosecution of Offences Act 1985, both owe their existence to the exposure of particular miscarriages. Miscarriages of justice are able to create a consensus across the crime-control–due-process spectrum that something has to be done. To adherents of crime control they highlight inefficiencies in the system in terms of convicting the guilty and to due-process adherents they highlight what has been argued to be one of the fundamental due-process values, the need to safeguard against false convictions.

Appeals

Often it has been left to individual campaigners to highlight miscarriages of justice in the first place, but the organization Justice founded in 1959 has played a leading role in exposing miscarriages of justice and therefore in creating a momentum for change. So, for example, in 1961 Justice began a full-scale investigation into the operation of the Court of Appeal after the Court had proved unable to halt miscarriages of justice in the Rowland and Timothy Evans cases. This led the Government to set up the Donovan Committee on the Court of Appeal and as a result of its recommendations a number of changes were made and consolidated in the Criminal Appeal Act 1968. In contrast to appeals against summary convictions which involve a complete rehearing of the evidence, the sanctity of the jury verdict has traditionally been a considerable barrier to any effective review of findings of fact on which verdicts have been based.

Until 1966 appeals on grounds of miscarriage of justice could only be successful where the verdict was 'unreasonable or could not be supported having regard to the evidence'. The Criminal Appeal Act 1966 substituted a wider ground of appeal, namely where the conviction is, in all the circumstances of the case, unsafe or unsatisfactory. Two years earlier the Court of Criminal Appeal was granted the power to order a retrial instead of merely upholding or quashing the verdict where fresh evidence had become available since the trial on indictment.[68] Section 23 of the Criminal Appeal Act 1968 also widened the grounds on which fresh evidence could be admitted. The Court had previously refused to admit fresh evidence unless it was unavailable at the original trial.[69] Section 23 required the Court to receive admissible evidence when it was likely to be credible and when there was a reasonable explanation for the failure to adduce it at the time of the original trial. The Court was also given a discretion to admit the evidence of any competent witness when it was in the interests of justice to do so.

Prosecutions

The new appellate procedures under the 1968 Act did not alleviate concern about miscarriages of justice and in 1970 Justice published a report which argued that it was wrong in principle for the police both to investigate and prosecute criminal cases.[70] The report recommended that England and Wales should follow the Scottish system where the decision to prosecute is taken by independent Procurators-Fiscal appointed by the Lord Advocate. Concern about the prosecution system in England heightened as a result of the *Confait* case discussed in Chapter 3. *Confait* was one of the rare cases which was serious enough to merit prosecution by the Director of Public Prosecutions under the Prosecution of Offences Act 1879. Despite this, the Fisher Report concluded that the case showed that it was possible 'for a prosecution to proceed through all stages up to the start of the trial based on a time of death outside the brackets given by medical witnesses, without any attempt to clarify the medical evidence or to discover whether it was consistent with the case to be presented' and 'for a prosecution to be based almost wholly on uncorroborated confessions without proper steps having been taken to seek evidence to support or contradict the evidence of the confessions'.[71] The report made a number of recommendations aimed at preventing this occurrence in the

---

[68] *R. v. Parks* (1962) 46 Cr. App. R. 29. See M. Knight, *Criminal Appeals* (1975), ch. 4.

[69] S. 43 of the Criminal Justice Act 1988 has since given the Court of Appeal a discretion to order a retrial in any case where it is in the interests of justice to do so.

[70] Justice, *The Prosecution Process in England and Wales* (1970).

[71] *Report of an Inquiry by the Hon. Sir Henry Fisher into the Circumstances Leading to the Trial of Three Persons on Charges Arising out of the Death of Maxwell Confait and the Fire at 27 Doggett Road, London SE 5* (1977), para. 2.29.

future, including the exclusion of all uncorroborated confessions of children, the mentally retarded, and anyone whose confession was either not tape-recorded or obtained in breach of the Judges' Rules. It was also sympathetic to the idea of involving an independent, legally trained official, as in Scotland, at an early stage of criminal proceedings to evaluate the weaknesses of the prosecution case.

The Royal Commission on Criminal Procedure which was established after the report of the Fisher Inquiry was less enthusiastic than Fisher about the idea of involving legally trained prosecutors at an early stage of criminal proceedings. It doubted whether it was desirable to separate completely the functions of investigation and prosecution and recommended instead that the police retain complete responsibility for investigating offences and for making decisions to charge and to caution but that a system of locally based Crown prosecutors should be set up to decide whether to proceed on the charges brought or to withdraw them. In 1983 the government issued a White Paper in which the idea of Crown prosecutors who would be accountable to local supervisory bodies was rejected in favour of a national prosecution service headed by the Director of Public of Prosecutions and under the superintendence of the Attorney-General.[72] But the paper envisaged the same basic framework recommended by the Royal Commission with the police deciding first whether to bring a prosecution and the Crown prosecutor then deciding whether to continue it. The Prosecution of Offences Act 1985 requires the Director of Public Prosecutions who heads the new Crown Prosecution Service to take over the conduct of all but the most minor of criminal proceedings that have been instituted by the police. According to the Code for Crown Prosecutors that has been issued under the Act in order to promote consistent decision-making, prosecutors must first decide whether there is a realistic prospect of a conviction, and then whether the public interest requires a prosecution, before they can take over a prosecution.[73] Factors such as the likely penalty that will be imposed, the attitude of the complainant, the staleness of the offence, and the youth, age, or infirmity of the offender are relevant in determining the public interest. The Code also provides guidelines on other decisions that are taken by prosecutors, including the choice of charge, the mode of trial, and whether to accept a guilty plea to a lesser offence or to fewer offences than originally charged.

Disclosure to the defence

Another issue highlighted by the *Confait* case was the failure on the part of the prosecution to disclose alibi evidence helpful to the defence. At

---

[72] *An Independent Prosecution Service for England and Wales*, Cmnd. 9074 (1983).
[73] For discussion of the Code, see A. Ashworth, 'The "Public Interest" Element in Prosecutions' (1987) *Crim. LR* 595–607.

that time the defence was not entitled to any disclosure in summary proceedings. In indictable cases the defence was informed of the nature of the prosecution case at committal but it was not entitled to see statements which might be helpful to the defence.[74] The Royal Commission recommended that the prosecution should be required to disclose a summary of the case on request by the defence in summary proceedings. It ruled out any routine disclosure of witness statements in indictable cases but urged prosecutors to disclose to the defence on request any documents which had 'some bearing on the offences charged or the surrounding circumstances of the case', subject to an exception for sensitive material.[75] In 1982 the Attorney-General responded by issuing guidelines which upheld the principle of disclosing 'unused material' to the defence in indictable cases but recognized a discretion not to make disclosure where there were grounds for fearing intimidation or for believing a statement to be untrue or for fearing that a witness may give an untrue statement to the defence, as well as where the statement is 'sensitive'.[76] The government also subsequently laid down rules entitling the defence to disclosure of the prosecution case in summary cases, provided the offences charged are triable either way.[77]

Doubts about the post-war reforms

Despite the reforms enacted by PACE and the Prosecution of Offences Act a number of miscarriages of justice continued to be revealed throughout the 1980s, largely due to the efforts of Justice and a BBC Television series called *Rough Justice*.[78] The publicity given to the Irish cases mentioned earlier led to the establishment of the Royal Commission on Criminal Justice in 1991 and since its establishment several wrongful convictions have been quashed by the Court of Appeal.[79] These cases have raised serious questions about the effectiveness of the post-war reforms.

First of all, they raise questions about the PACE reforms which were discussed in chapter 3. Despite PACE a number of cases have come to light where individuals have confessed falsely in the police station. As

---

[74] See *R.* v. *Bryant and Dickson* (1946) 31 Cr. App. R. 146; *Dallison* v. *Caffrey* [1965] 1 QB 348.

[75] *Report of the Royal Commission on Criminal Procedure*, Cmnd. 8092 (1980), para. 8.19. The Commission adopted the recommendations of a Home Office working party set up in the wake of the Fisher Report. See *Report of the Working Party on Disclosure of Information on Trials on Indictment* (1979).

[76] *Practice Note* [1982] 1 All ER 734.

[77] Magistrates' Courts (Advance Information) Rules (SI 1985 No. 601). See C. J. Emmins, 'Why No Advance Disclosure for Summary Offences?' (1987) *Crim. LR* 608–14.

[78] P. Hill and M. Young, *Rough Justice* (1983), and P. Hill, M. Young, and T. Sargant, *More Rough Justice* (1985).

[79] See e.g. the cases of John McGranaghan (October 1991), Raghip, Silcott and Braithwaite (December 1991), Jacqueline Fletcher, Stefan Kiszko (February 1992), Judith Ward (June 1992), Wayne and Paul Darvell (August 1992).

already mentioned the Fisher Report recommended that confessions obtained in certain circumstances should be corroborated. This idea was rejected by the Royal Commission on Criminal Procedure but at the beginning of 1993 was being considered afresh by the Royal Commission on Criminal Justice.[80]

Secondly, the cases question the ability of the prosecution system to screen out weak cases. Although the Irish cases were tried before the institution of the Crown Prosecution Service, they were, like the *Confait* case, prosecuted by the DPP and the new service has no greater role in the prosecution of cases than the DPP had when he prosecuted in these cases.[81] Unlike the Scottish system it merely has the power to take over or discontinue prosecutions already begun by the police and unlike the Scottish and Northern Irish systems it has no power to direct the police to make investigations.

The cases have also raised questions about the adequacy of the procedures for disclosing evidence to the defence. The precise status of the Attorney-General's guidelines has been unclear, although failure to comply with them has formed the basis of a number of successful appeals.[82] The problem with the guidelines is that they leave vital decisions about what to disclose and what not to disclose to the police and the prosecuting authorities, and do not give the defence a right to view all the relevant evidence. A recent attempt by the Court of Appeal to clarify the position has not inspired confidence. The Court said that material witness statements should be provided to the defence 'unless there are good reasons for not doing so', but the Court gave no explanation of what were 'good reasons'.[83] The cases have also illustrated weaknesses in the disclosure mechanisms after conviction. During the Birmingham Six appeal the Court of Appeal expressed concern about the failure of the office of the DPP to disclose two vital pieces of scientific evidence to the defence which were in its possession at least a month before the 1987 appeal. One of these was a statement showing that two people travelling on the same ferry as the Birmingham Six on the night of the bombings had the same positive readings for explosives as the Six but that it was decided that the result was caused by an innocent substance.

Above all, the cases have illustrated defects in the appellate procedures provided for under the Criminal Appeal Act 1968. All the Irish cases

[80] See M. McConville, *Corroboration and Confessions: The Impact of A Rule Requiring that No Conviction Can be Sustained on the basis of Confession Evidence Alone* (1993). Research Study No. 13 for the Royal Commission on Criminal Justice.

[81] A. Sanders, 'An Independent Prosecution Service?' (1986) *Crim. LR* 16–27.

[82] See e.g. *R.* v. *Phillipson* (1989) 91 Cr. App. R. 226, *R.* v. *Sanson* (1991) 92 Cr. App. R. 115, *R.* v. *Maguire* [1992] 2 All ER 433, *R.* v. *Ward* [1993] 1 WLR 619.

[83] *R.* v. *Ward* [1993] 1 WLR 619, 680. See P. O'Connor, 'Prosecution Disclosure: Principle, Practice and Justice' (1992) *Crim. LR* 464–77.

were originally appealed unsuccessfully to the Court of Appeal after conviction. Once a defendant has appealed unsuccessfully to the Court of Appeal, it appears that he has no right to a further appeal and must instead petition the Home Secretary either to recommend the royal prerogative of mercy or to refer the case back to the Court of Appeal under section 17 of the 1968 Act.[84] The royal prerogative of mercy can be invoked on grounds of compassion as well as miscarriage and is therefore an unsatisfactory basis for clearing a defendant's name. The time it took for the Home Secretary to refer the Irish cases back to the Court of Appeal, together with criticism of the Court's refusal to quash the convictions of the Birmingham Six when the case was referred back to it in 1987, highlighted the claim that successive Home Secretaries have been reluctant to exercise their powers of referral and that the Court of Appeal is extremely reluctant to reverse its earlier findings.[85] In 1982 the House of Commons Select Committee concluded that the Court of Appeal was not an appropriate body to conduct an inquiry of fact and recommended that section 17 be repealed in favour of an extra-judicial review body which would investigate cases of alleged miscarriages of justice.[86] At the time the government argued against the idea on the ground that miscarriages of justice that occur within the judicial system should be corrected by that system but the recent revelations have increased pressure for more effective ways of remedying miscarriages of justice.[87]

THE ROLE OF THE COURTS

Two particular questions can be examined to determine what contribution the courts have made to safeguard the due-process interests of accused persons in the post-war period. First there is the question how far they have tried to avert the risk of miscarriages of justice in particular cases and then there is the question of the extent to which they have tried to enforce the various rules that have been designed to protect defendants throughout the criminal process.

The risk of false conviction

We have mentioned that the English courts recognize a number of rules and principles in order to safeguard against false convictions. The most

---

[84] This is the effect of *R. v. Pinfold* [1988] 2 All ER 217.

[85] M. Tregilgas-Marcus, 'Miscarriages of Justice within the English Legal System' (1991) *New LJ* 608, 715.

[86] House of Commons Select Committee on Home Affairs, *Sixth Report* (1982).

[87] 'Justice Must Come First', leader, *Independent*, 28 Mar. 1991.

important is the principle laid down in *Woolmington* v. *DPP*[88] that it is
for the prosecution to prove the accused's guilt beyond reasonable doubt.
This principle was stated by Viscount Sankey to be subject to the defence
of insanity and subject also to any statutory exceptions. In *R.* v. *Hunt*[89]
the House of Lords refused to limit the latter to express exceptions but it
indicated that the question whether Parliament has impliedly shifted the
burden of proof could not be settled, as had earlier been thought,[90] by
the form of the statutory provision, and it proceeded to lay down guidelines of statutory construction to determine whether the burden has
shifted. Apart from the wording of the provision, these require the courts
to take account of the 'substance and effect' of the provision.[91] The factors mentioned include taking account of the ease with which a defendant
could prove his innocence, the extent to which the prosecution of offences
would be particularly difficult or burdensome if the burden of proof
were on the prosecution, and the seriousness of the offence in question.
Opinions differ on the extent to which this decision has given proper
weight to the *Woolmington* principle.[92]

Another principle that is also important in safeguarding against a false
conviction is what has been called the principle of confining the verdict to
the charge.[93] This imposes a duty on the courts to ensure that juries in
particular do not behave prejudicially against accused persons. The courts
have not proved very receptive to the argument that the particular
composition of a jury is likely to prejudice a defendant. Challenges for
cause have been made particularly difficult because of the reluctance of
the courts to permit the questioning of potential jurors to establish
cause.[94] There was a period in the 1970s and 1980s when some judges
were prepared to stand by jurors to ensure a multi-racial jury in racially
sensitive cases,[95] but in *R.* v. *Royston Ford*[96] the Court of Appeal held
that the power of the judge to stand by jurors was limited to removing
jurors who were not competent to serve on a jury and that any attempt to
influence the composition of a particular jury detracted from the principle
of random selection.

The courts have proved more responsive to developing mechanisms for
dealing with particularly prejudicial evidence. The House of Lords has

---

[88] [1935] AC 462.
[89] [1987] AC 352.
[90] See *R.* v. *Edwards* [1975] QB 27.
[91] [1987] AC 352, 372, per Lord Griffiths.
[92] Compare P. Mirfield, 'The Legacy of *Hunt*' (1989) *Crim. LR* 19–30 and D. Birch,
'Hunting the Snark: The Elusive Exception' (1988) *Crim. LR* 221–32.
[93] Zuckerman, *Principles*, 222–3.
[94] See R. Buxton, 'Challenging and Discharging Jurors: 1', (1990) *Crim. LR* 225–35.
[95] S. Enright, 'Multi-racial Juries' (1991) *New LJ* 992–6.
[96] (1989) 89 Cr. App. R. 278.

continued to affirm the existence of a common law discretion to exclude evidence whose probative value is outweighed by its prejudicial effect,[97] and this discretion has been used, albeit erratically, to exclude character evidence which is otherwise admissible under the Criminal Evidence Act 1898.[98] The courts have generally restricted the need for corroboration warnings but they have endorsed the need for the judge to warn the jury to treat evidence which is challenged by the defence with caution.[99] The Court of Appeal has also laid down very specific guidelines governing the way a judge should direct a jury when the evidence against an accused rests substantially on identification evidence based on personal impression.[100] These guidelines known as the Turnbull guidelines were laid down after a committee was set up to examine two cases of mis-identification which led to miscarriages of justice.[101] Apart from warning juries of the special need for caution before convicting on identification evidence, the guidelines require judges to mention the reason for the warning and the circumstances of the identification and to withdraw cases altogether where the quality of the evidence is poor and there is an absence of supporting evidence. Although it can be argued that these guidelines do not go far enough to protect defendants against the risk of identification evidence, principally because they do not address the problem of how to detect mistaken identifications,[102] they do at least show that the courts are sensitive to the dangers of this kind of evidence.

The courts would appear to have been less successful in dealing with dubious forensic evidence. First of all there has been concern about the independence of the scientific experts who have given evidence in the courts and their failure to disclose evidence helpful to the defence in court. In *Preece* v. *HM Advocate* the Scottish High Court found that in a murder case a forensic scientist, Dr Alan Clift, had failed to disclose to the court the fact that the dead victim had the same group A secretor as the defendant, with the result that the seminal grouping result could have been due to contamination by vaginal secretion.[103] In his first interim report into the Maguire case Sir John May also raised serious questions about the independence of the Forensic Science Services by revealing that

---

[97] *Selvey* v. *DPP* [1970] AC 304, *R.* v. *Sang* [1980] AC 402.

[98] See e.g. *R.* v. *Watts* [1983] 3 All ER 101. The later cases of *R.* v. *Burke* (1986) 82 Cr. App. R. 152 and *R.* v. *Powell* [1985] 1 WLR 1364 make the discretion appear rather limp.

[99] *R.* v. *Beck* [1982] 1 WLR 461, *R.* v. *Spencer*; *R.* v. *Smails* [1986] 2 All ER 928.

[100] *R.* v. *Turnbull* [1977] QB 224.

[101] *Report to the Secretary of State for the Home Dept. of the Departmental Committee on Evidence of Identification in Criminal Cases* (1976).

[102] See e.g. I. H. Dennis, 'Corroboration Requirements Reconsidered' (1984) *Crim. LR* 316, 332–6; J. Jackson, 'The Insufficiency of Identification Evidence Based on Personal Impression' (1986) *Crim. LR* 203–14.

[103] [1981] *Crim. LR* 783.

the forensic scientist involved in the case had failed to disclose that the tests carried out did not differentiate between nitroglycerine and another explosive substance and had also failed to disclose other tests that were negative.[104] In the later Judith Ward appeal it transpired that notebooks in the possession of the government scientists contained results which showed that tests carried out for nitroglycerine were misleading as everyday household dyestuff could be enfused with nitroglycerine. The Court of Appeal chose to put the blame for this non-disclosure squarely on the scientists involved, but these cases also raise questions about the appropriateness of the adversarial system for dealing with scientific evidence.[105] Experts are called for a particular side and they may develop particular feelings of loyalty towards that side. Once they have made a statement for a particular side, moreover, they may feel a need to maintain that position. The Court of Appeal has made it clear that expert witnesses owe a duty to disclose material which has some bearing on the offence to the defence,[106] but as with more general disclosure provisions this depends on the good will of the scientists involved. It is too early to say whether the recent change in the status of the Home Office Forensic Science Service to that of an independent agency will command greater confidence in the forensic services.[107] A recent report has suggested that the new arrangements which require police forces to pay for forensic science services out of their own budget may discourage the police from submitting items which could exculpate the defendant.[108] Even if evidence relevant to the defence is discovered and disclosed, the use which the defence makes of it will depend on its ability to call upon forensic experts to assist it.

The other concern relates to the increasing use of forensic evidence in the courts and anxiety about the quality of the techniques that are adopted by the scientific community. In the Maguire case Sir John May found that tests which at the time of the trial were said to be almost foolproof methods of detecting nitroglycerine proved on later discovery to be defective. DNA profiling is the latest technique to be hailed as the solution for identifying individuals with crime and is being increasingly used in criminal investigations.[109] But experience in the United States suggests that although there is general acceptance of the molecular

---

[104] *Return to an Address of the Honourable House of Commons dated 12 July 1990 for the Inquiry into the Circumstances Surrounding the Convictions Arising out of the Bomb Attacks in Guildford and Woolwich in 1974* (1990).

[105] For discussion see D. J. Gee and J. K. Mason, *The Courts and the Doctor* (1990), ch. 10.

[106] *R. v. Maguire* [1992] 2 All ER 433.

[107] See House of Commons, Home Affairs Committee, *Report on the Forensic Science Service* (HC 26, 1987) and government reply (Cmnd. 699, 1989; Cmnd. 1549, 1991).

[108] P. Robert, C. Willmore and G. Davis, *The Role of Forensic Science Evidence in Criminal Proceedings* (1993). Research Study No. 11 for the Royal Commission on Criminal Justice.

and genetic principles involved in DNA analysis, the application of the technology to forensic samples can be problematic and there is a need for quality assurance to ensure that the samples match and to calculate the statistical probabilities of the match being random.[110] Unlike the US courts, however, which require scientific techniques to have general acceptance in the relevant scientific field,[111] the courts in the UK have no established procedure for ensuring the reliability of novel scientific evidence. The problem came to a head in December 1992 when DNA evidence submitted in an armed robbery case was not accepted by the trial judge following evidence from an American forensic scientist. This prompted a meeting of forensic scientists, police officers and prosecutors to try to reach agreement on the use of DNA evidence.[112] A recent report has called on the courts to establish their own criteria of admissibility.[113]

Turning to confessional evidence which has been said to constitute one of the main sources of miscarriages of justice,[114] the courts have considered that they have a common law discretion to exclude unreliable confessions, aside from any inadmissibility test.[115] But the problem here is that judges need to be able to recognize unreliable confessions for this discretion to work effectively. Experts now claim that there are two types of false confession, what have been called the 'coerced compliant' confession and the 'coerced-internalized' confession.[116] The former are usually the result of forceful or persistent questioning, whilst the latter are the result of a belief on the part of the suspect that he or she actually committed the crime. But the restrictive rules on expert evidence, which prevent experts testifying on the likelihood of a person making a false confession unless he or she is suffering from a recognized mental illness or is educationally subnormal, mean that the courts are less able to identify false confessions in particular cases.[117] So far as the new admissibility test under section 76 of PACE is concerned, the courts have made it clear that the unreliability head of the test does not require that the police have to be guilty of any impropriety before a confession can be

---

[109] It has been estimated that the test had been performed in around 2,000 criminal investigations in the UK by Feb. 1991, see J. Sufian, 'DNA in the Courtroom' (Feb. 1991) *Legal Action* 7–9.

[110] Ibid.

[111] *Frye* v. *US* 293 F.2d 1013 (DC Cir. 1923).

[112] D. Campbell, 'Police Seek Agreement on DNA Crisis', *The Guardian*, 15 Jan. 1993.

[113] B. Stevenson, *The Ability to Challenge DNA Evidence* (1993), Research Study No. 9 for the Royal Commission on Criminal Justice.

[114] Justice, *Miscarriages of Justice* (1989).

[115] See e.g. *R.* v. *Miller* [1986] 1 WLR 1191. For discussion of this discretion see P. Mirfield, *Confessions* (1985), 118 ff.

[116] For discussion see R. Pattenden, 'Should Confessions be Corroborated?' (1991) *LQR* 317, 319.

[117] Ibid. 326–7.

excluded.[118] But they have tended to interpret the provision as requiring that something is said or done in the course of questioning which is out of the ordinary, such as a failure to comply with the rule that an independent adult is present during interviews with juveniles, mentally disordered, and mentally handicapped persons.[119]

The other danger with confessional evidence is that confessions may have been concocted. The revelations of police misconduct in the Birmingham Six case and the disbanding of the serious crimes squad in the West Midlands has led to fears that the police routinely manufacture admissions as a short cut to convictions. One problem here is that defendants are discouraged from alleging concoction because the effect of such an allegation is to open them up to cross-examination on their criminal record.[120] The courts have excluded confessions under sections 76 and 78 of PACE where there have been breaches of the duty to record interviews contemporaneously.[121] Apart from enforcing these so-called 'verballing provisions', the Court of Appeal has also recognized that the presence of a solicitor can have the effect of helping to authenticate a police account,[122] but it has not always been prepared to exclude confessions obtained after wrongful denial of access to legal advice.[123]

As well as protecting defendants from prejudicial evidence in court, the courts have been concerned to protect them from prejudicial press comment about the case out of court. The discussion of the Contempt of Court Act 1981 in Chapter 7 mentions that it is a contempt to publish material which constitutes a substantial risk of prejudice. It is clear, for example, that publications referring to an accused's past are a contempt of court.[124] Although there is no contempt liability for accurate reports of legal proceedings or for public-interest discussions which contain a risk of prejudice, the courts have been quite active in shielding defendants from the prejudicial effects of court reporting and public discussion. The Court of Appeal recently held that public discussion may be so pointed and prejudicial that discharge of the jury may be necessary to avoid the risk of injustice to an accused.[125] It also seems that judges have quite commonly exercised their power under section 4(2) of the Contempt of Court Act 1981 to postpone the reporting of trial proceedings to avoid a substantial

---

[118] *R.* v. *Fulling* [1987] QB 426; [1987] 2 All ER 65, *R.* v. *Harvey* [1988] Crim. LR 241.

[119] See *R.* v. *Goldenberg* (1988) 88 Cr. App. R. 285, *R.* v. *Crampton*, *The Times*, 22 Nov. 1990.

[120] See *R.* v. *Britzman* [1983] 1 WLR 350, [1983] 1 All ER 369 (CA).

[121] See *R.* v. *Doolan* [1988] Crim. LR 747; *R.* v. *Delaney* (1989) 88 Cr. App. R. 338 (CA); *R.* v. *Keenan* [1990] 2 QB 54, [1989] 3 All ER 598; *R.* v. *Canale* [1990] 2 All ER 187.

[122] *R.* v. *Dunn* (1990) 91 Cr. App. R. 237 (CA).

[123] *R.* v. *Alladice* (1988) 138 *New LJ* 347 (CA), (1988) 87 Cr. App. R. 380, *R.* v. *Dunford* (1990) 91 Cr. App. R. 150 (CA).

[124] C. J. Miller, *Contempt of Court* (1990), 203–6.

[125] *R.* v. *McCann, Cullen and Shanahan* [1991] Crim. LR 136.

risk of prejudice.[126] The reporting of committal proceedings, which carries a particular danger of the publishing of evidence later held to be inadmissible, was severely curtailed by Parliament under section 8 of the Magistrates' Courts Act 1980, although there are no restrictions on publishing the names of accused persons.[127]

While the courts have made genuine efforts to avert the risk of miscarriages of justice, they have been less able to remedy miscarriages of justice once defendants have been convicted. Reference has already been made to the reluctance of the Court of Appeal to reverse earlier findings when cases are referred back to it by the Home Secretary under section 17 of the Criminal Appeal Act. A number of criticisms have also been made of the way the courts have interpreted this Act. The use of practice directions empowering the Court to order that time spent waiting for an appeal does not count towards the custodial sentence can have the effect of discouraging meritorious as well as unmeritorious appeals.[128] There has also been criticism of the way the courts have interpreted the power to admit fresh evidence under section 23 and to deal with fresh evidence once it has been admitted.[129] In *Stafford and Luvaglio* v. *DPP*[130] the House of Lords held that once fresh evidence is admitted the question the Court of Appeal should ask itself was not whether the new evidence might have led the original jury to have a reasonable doubt but whether the Court thought that in the light of the new evidence the conviction was unsafe or unsatisfactory. This has required the Court of Appeal to blend the fresh evidence with the rest of the evidence adduced in the original trial. The Law Lords did not suggest how the Court should do this but there has been a tendency for the Court's view of the fresh evidence to be coloured by accepting the truth of the rest of the evidence.[131] The judgment also had the effect of inhibiting the Court from making use of its power to order a retrial.[132]

Procedural irregularity

It has already been argued that it is particularly important in adversarial proceedings, where investigating authorities such as the police have not

---

[126] Miller, *Contempt*, 334.

[127] The Sexual Offences (Amendment) Act 1976, s. 6, which gave persons accused but not convicted of rape a right of anonymity, was repealed by s. 158(5) of the Criminal Justice Act 1988. For background see J. Temkin, *Rape and the Legal Process* (1987), 193–4. For discussion of the whole question of affording anonymity to defendants, see R. Munday, 'Name Suppression: An Adjunct to the Presumption of Innocence and to Mitigation of Sentence' (1991) *Crim. LR* 680–8, 753–62.

[128] *Practice Direction*, [1970] 1 All ER 1119, *Practice Direction*, [1981] 1 All ER 555.

[129] See P. O'Connor, 'The Court of Appeal: Trials and Tribulations' (1990) *Crim. LR* 615–28.

[130] [1974] AC 878 (HL).

[131] O'Connor, 'Court of Appeal', 618.

[132] Tregilgas-Marcus, 'Miscarriages of Justice'.

traditionally been accountable to judicial authorities, to ensure that individuals have some form of redress when due-process requirements have not been met. One way of providing redress is to allow individuals to sue the police for a breach of duty owed towards them. There are a number of torts available to remedy police misconduct, such as assault and battery, false imprisonment, and trespass to property or goods. Defendants are also able to sue the police for malicious prosecution when the police knew the defendants did not commit the offence. Defendants have no cause of action for wrongful conviction, however, although they do now have a statutory right to compensation which is determined by the Home Secretary when they have been pardoned as a result of a miscarriage of justice or when their conviction has been quashed out of time or on a reference to the Court of Appeal by the Home Secretary.[133]

Although civil actions against the police are becoming more common,[134] there are difficulties with civil claims. For one thing not every breach of the rules is subject to a civil claim. A breach of the police codes of conduct under PACE or of the Code for Crown Prosecutors under the Prosecution of Offences Act or of the Attorney-General's Guidelines on Disclosure of Evidence does not give rise to a civil claim. Another problem is that civil claims are generally made after criminal proceedings are over, with the result they are not able to provide instant relief. Habeas corpus is the most famous example of a remedy that does provide instant relief as it permits a means by which a prisoner can test the legality of his detention and secure his release. In the past this has proved a useful remedy against unlawful detention in the police station but the wide powers given to the police to detain arrestees under PACE have limited the use that is made of this remedy in criminal procedure. It has, however, been used as a remedy in Northern Ireland where unlawful force was used against a person who had been detained lawfully.[135] The remedy still has a useful function in other contexts, especially in view of a House of Lords ruling in an immigration case that where an executive officer's decision to restrict a person's liberty is dependent on the existence of certain facts, the court must be satisfied to the civil standard of proof that the facts did exist at the time the power was exercised.[136]

Apart from habeas corpus the courts have developed three particular remedies to deal with procedural irregularities whilst criminal proceedings are still pending. These are the use of judicial review to quash convictions,

---

[133] S. 133 of the Criminal Justice Act 1988.

[134] See R. Clayton and H. Tomlinson, *Civil Actions against the Police* (1987).

[135] *In re Gillen's Application* [1988] 1 NIJB 47.

[136] *R. v. Secretary of State for the Home Dept. ex parte Khawaja* [1984] AC 74 (HL). In practice, though, the courts have been very reluctant to grant habeas corpus in illegal entry cases, see ch. 11.

the use of the abuse of process doctrine to stay proceedings, and the exclusion of evidence unfairly obtained before trial.

In a number of cases the Divisional Court has granted applications for judicial review on grounds of breach of natural justice or on grounds of fraud, collusion, or perjury to remedy possible miscarriages of justice. In *R. v. Leyland Justices ex parte Hawthorne*,[137] for example, the Court held that the failure of the police to inform the defence about a potential witness who had made a statement to them which was helpful to the defence case amounted to a breach of the rules of natural justice. The Court has also granted applications in situations where, through no fault of the prosecution, the court has been misled by the state of the evidence. In a recent case it quashed a conviction for drunken-driving after a plea of guilty when it was revealed that the doctor who had taken blood samples had used contaminated cleaning swabs. The Court held that, although no dishonesty had been involved, the conduct of the prosecution was analogous to fraud, collusion, or perjury as it gave the appellant no opportunity to decide whether to plead guilty or not.[138]

Although individual prosecutorial decisions and the Code for Crown Prosecutors have been quite well insulated from review by the courts,[139] judicial review has also been used to remedy an abuse of process. The doctrine of abuse of process was virtually unknown in criminal procedure until 1964 when a majority in the House of Lords in *Connelly* v. *DPP*[140] took the view that, aside from the special pleas of *autrefois acquit* and *convict*, the courts could stay criminal proceedings to prevent multiple prosecutions of the accused for essentially the same conduct. Since then the doctrine of abuse of process has been recognized as a remedy to stay proceedings in a variety of other situations where it has been alleged that the prosecution has taken advantage of the defendant unfairly.[141] One of the commonest uses of the doctrine has been in situations where prosecutions have been delayed unjustifiably.[142] The courts have taken quite a strict view of their power to extend time limits between charge and committal or trial,[143] but these rules do not, of course, deal with delays up to charge. It is generally necessary to show that the prosecution

---

[137] [1979] 1 All ER 209.

[138] *R. v. Bolton Justices ex parte Scally* [1991] 2 All ER 619.

[139] See Ashworth, 'The "Public Interest" Element'. But see *R. v. Chief Constable of Kent ex parte C*; *R. v. Crown Prosecution Service ex parte B.*, *The Times*, 17 Apr. 1991.

[140] [1964] AC 1254, [1964] 2 All ER 401 (HL).

[141] See e.g. *R. v. Horsham Justices ex parte Reeves* (1982) 75 Cr. App. R. 236, *R. v. Derby Crown Court ex parte Brooks* (1985) 80 Cr. App. R. 164, *R. v. Liverpool City Justices ex parte Ellison* [1989] Crim. LR 369.

[142] See R. Pattenden, 'Abuse of Process in Criminal Litigation', (1990) *J. Crim. Law* 341–57.

[143] *R. v. Southampton Crown Court ex parte Roddie*, *The Times*, 11 Feb. 1991.

was responsible for the delay, that the delay was unjustifiable, and that the defendant would be prejudiced in the preparation or conduct of the defence.[144] Although the House of Lords has yet to rule on the matter, it now seems settled that magistrates' justices may stay proceedings on grounds of abuse of process as well as Crown Court judges.[145] But there is some doubt whether an application for judicial review founded on abuse of process may be made to the Divisional Court before the inferior court has been asked to stay the proceedings.[146]

So far the remedy of staying proceedings on the ground of abuse of process has been limited to situations governing the institution or continuation of prosecutions and it has not generally been used to deal with unfairness by the police before the prosecution.[147] Entrapment, for example, has been ruled out as a ground for staying proceedings,[148] as has the tampering or destruction of evidence.[149] The remedy for unfairly obtained evidence is to ask the court to exclude the evidence as a result of unfairness. Discussion of this issue in Chapter 3 revealed that while the courts have traditionally been reluctant to exclude unfairly obtained evidence, they have been prepared in a number of decisions to exclude evidence under section 78 of PACE on grounds of unfairness to the proceedings where there has been a breach of PACE or the Code on Police Detention and Questioning. Although Chapter 3 argues that the judges now see themselves as having a disciplinary and regulatory role in maintaining the balance between the powers of the police and the protection of suspects, certain decisions have continued to disclaim any disciplinary role.[150] Decisions which stress the bad faith of the police do suggest a disciplinary approach,[151] but decisions which stress the need for a causal link between the breach and the obtaining of evidence point more towards a protective approach and no clear principle seems to emerge from the cases.[152] The courts have arguably not paid enough

---

[144] It seems now that prejudice may be presumed from a substantial delay: *R.* v. *Bow Street Stipendiary Magistrate ex parte DPP and Cherry* [1990] Cr. App. R. 283, [1990] Crim. LR 318.

[145] *R.* v. *Brentford Justices ex parte Wong* [1981] QR 445 (summary proceedings); *R.* v. *Telford Justices ex parte Badhan* [1991] 2QB 78 (committal proceedings).

[146] Pattenden, 'Abuse of Process', 352–3.

[147] See *R.* v. *Bow Street Metropolitan Stipendiary Magistrate ex parte DPP* [1992] Crim. LR 790.

[148] *R.* v. *Sang* [1980] AC 402 (HL).

[149] *R.* v. *Sadridin* unreported judgment, 11 July 1985, discussed by Pattenden, 'Abuse of Process', 343–4.

[150] *R.* v. *Mason* [1988] 1 WLR 139, [1987] 3 All ER 481, 484, per Watkins, LJ.

[151] See e.g. *R.* v. *Fulling* [1987] QB 426, [1987] 2 All ER 65 (CA), *R.* v. *Mason* [1987] 3 All ER 481 (CA), *R.* v. *Alladice* (1988) 87 Cr. App. R. 380, (1988) 138 New LJ 347 (CA).

[152] *R.* v. *Samuel* [1988] QB 615, [1988] 2 All ER 135 (CA), *R.* v. *Alladice* (1988) 87 Cr. App. R. 380. For discussion see D. Birch, 'The PACE Hots Up: Confessions and Confusions under the 1984 Act' (1989) *Crim. LR* 95–116.

attention to what their role should be. If, as has been argued, the courts have a responsibility above all to ensure that persons are not innocently convicted, then in so far as rules such as the verballing provisions and rights such as the right of access to legal advice assist in preventing wrongful convictions, the exclusion of evidence obtained in breach of them can be justified, however reliable the evidence in an individual case, because it illustrates the courts' commitment to protecting the innocent.

INTERNATIONAL STANDARDS

Although the United Kingdom is a signatory to both the European Convention on Human Rights and the International Covenant on Civil and Political Rights, there have been few signs of these treaties influencing the direction of the English criminal process. Article 3 of the European Convention influenced the decision of the Royal Commission on Criminal Procedure to recommend an admissibility test for confessions that excluded confessions obtained by torture, inhuman, or degrading treatment. The government was clearly embarrassed when the European Court held that section 14 of the Prevention of Terrorism Act 1989 was in breach of Article 5 of the Convention because it permitted terrorist suspects to be detained for up to seven days before being brought to court but, instead of altering the law, it entered a derogation in respect of Article 5 of the Convention.[153]

Until recently, Article 6 and the corresponding article relating to a fair trial in the Covenant, Article 14, have had little impact on criminal justice debates. Obviously, one of the reasons for this is the fact that these articles have not been incorporated into English law. Although defendants have taken cases to the European Commission alleging a breach of Article 6, one difficulty apart from the need to exhaust local remedies is that Article 6 is designed to deal with procedural irregularities in the administration of justice and is not concerned with whether the domestic courts have correctly assessed the evidence.[154] The Article does not therefore provide a remedy for miscarriages of justice *per se*. Recent concern about miscarriages of justice and a feeling that these are attributable to defects in procedure has, however, stimulated increasing interest in whether English procedures are compatible with the provisions of Article 6.

The jurisprudence of the European Commission and Court in relation to Article 6 provides considerable scope for opening up aspects of the

---

[153] *Brogan and others* v. *UK* (1988), Series A, No. 145B. See discussion in Ch. 5.
[154] F. Jacobs, *The European Convention on Human Rights* (1980), 106–7.

English criminal process to scrutiny. First of all, the Commission and Court have given a wide meaning to the term 'criminal charge', with the result that Article 6 has been considered to apply to proceedings before appeal courts,[155] to proceedings on sentence,[156] and to proceedings where defendants plead guilty.[157] There is more doubt about whether the protections provided for in Article 6 apply to pre-trial proceedings. The European Court has concluded that compliance with the law when taking evidence does not affect the fairness of the criminal trial.[158]

Secondly, the Court has accepted that a criminal trial may not conform to the required standard of fairness even though the minimum requirements laid down in Article 6(2) and 6(3) have been complied with.[159] One of the principles of fairness that the Commission and Court have recognized as endemic in the notion of a fair hearing is the concept of 'equality of arms' which requires that the prosecution and defence are put on an equal procedural footing so far as influencing the tribunal in the case is concerned. This concept would seem to provide greater protection in adversarial systems where the parties themselves collect and present the evidence. It has been argued that the factual inequality of resources between prosecution and defence to vet jurors, and the legal inequality in their rights to challenge jurors, raises the question whether the prosecution has too much influence on the composition of the jury.[160] There is also scope for applying the concept of equality of arms to the present English rules on disclosure of evidence collected by the police to the defence and to rules governing access to forensic evidence.[161] If the parties should be on an equal footing so far as influencing the tribunal of fact is concerned, then arguably this requires that they have equal access to the information which influences the tribunal's decision.

The impact of the International Covenant on Civil and Political Rights has been even more remote but one question which has emerged recently is whether restrictions on the right of silence are compatible with Article 14(2)(g) of the Covenant which states that 'no one shall be compelled to testify against himself or to confess guilt'. No such guarantee is expressly provided for in Article 6 of the Convention but the Northern Ireland Order which curtailed the right of silence in Northern Ireland is, at the time of writing, being challenged before the Commission on the ground that the rights contained in Article 14(2) of the Covenant represent accepted international standards of fairness which must be applied in

---

[155] *Application 5006/71* v. *UK*, 39 Collection of Decisions 91.
[156] *Application 4623/70* v. *UK*, 39 Collection of Decisions 63.
[157] *Application 5076/71* v. *UK*, 40 Collection of Decisions 64.
[158] The *Schenk case*, ECHR 1988, Series A, No. 140.
[159] See the Pfunders case, *Austria* v. *Italy* (1963) 6 *YBEC* HR 740, 792.
[160] T. Gallivan and C. Warbrick, 'Jury Vetting and the European Convention on Human Rights' (1979) *Human Rights Rev.* 176–201.
[161] Amnesty International, *United Kingdom: Human Rights Concerns* (1991), 13–14.

interpreting Article 6. Another question that has been raised is whether the use of uncorroborated confessions at trial is in breach of Article 6. The European Commission was recently asked to rule on this issue in a case which resulted in the conviction of three defendants for the murder of a police officer in disturbances at Broadwater Farm in London, but the application was ruled inadmissible on the ground that the local remedy of referral by the Home Secretary to the Court of Appeal had not been exhausted.[162]

EMPIRICAL ASSESSMENTS

There has been an increasing flow of empirical evidence on the day-to-day workings of many aspects of the criminal process in this country. Apart from the numerous studies concerning the police that have been discussed in Chapter 3, other aspects of the criminal process that have been examined include prosecution decision-making,[163] decisions on mode of trial and decisions to plead guilty,[164] pre-trial reviews,[165] the trial process,[166] jury decision-making,[167] sentencing practices,[168] and the workings of the Court of Appeal.[169] These studies have helped assessments to be made about the extent to which there is compliance with the rules, the extent to which defendants avail themselves of the safeguards, and the effect of the rules on the attitudes of participants. Psychological research has also highlighted the dangers that there are in confessional evidence and identification evidence and the extent to which lay persons recognize such dangers.[170] This has helped to inform debate about what should be done to protect defendants from such evidence.

There is a problem, however, in assessing the extent to which present

---

[162] Ibid. 18. The Court of Appeal subsequently quashed the convictions on a referral by the Home Secretary (see *R. v. Raghip and others*, *The Times*, 9 Dec. 1991).

[163] A. Sanders, 'Constructing the Case for the Prosecution' (1987) *J. Law and Soc.* 229–53; McConville, Sanders, and Leng, *Prosecution*.

[164] J. Baldwin and M. McConville, *Negotiated Justice* (1977); J. Baldwin, *Pre-Trial Justice* (1985); D. Riley and J. Vennard, *Triable Either-Way Cases: Crown Court or Magistrates' Court* (1988).

[165] Baldwin, *Pre-Trial Justice*.

[166] P. Carlen, *Magistrates' Justice* (1976); McBarnet, *Conviction*.

[167] S. McCabe and P. Purves, *The Jury at Work* (1972); *The Shadow Jury at Work* (1974); W. R. Cornish and A. P. Sealy, 'Jurors and the Rules of Evidence' (1973) *Crim. LR* 208; J. Baldwin and M. McConville, *Jury Trials* (1979); J. Vennard, 'The Outcome of Contested Trials', in D. Moxon, *Managing Criminal Justice* (1985).

[168] D. Thomas, *Current Sentencing Practice* (1982).

[169] Knight, *Criminal Appeals*; K. Malleson, 'Miscarriages of Justice: The Accessibility of the Court of Appeal' (1991) *Crim. LR* 323–32.

[170] See S. Lloyd-Bostock, *Evaluating Witness Evidence* (1983), *Law in Practice* (1988); G. H. Gudjonsson, *The Psychology of Interrogations, Confessions and Testimony* (1992).

procedures contribute to miscarriages of justice. The difficulty is that there is no independent criterion which is capable of measuring the correctness of convictions.[171] This has posed particular difficulties for jury research. Researchers have realized that they cannot validate or invalidate verdicts but they have tried to assess the performance of the jury by observing shadow juries and by comparing verdicts with the views of professional participants. But the problem with these methods is that they assume there are agreed criteria of competency and rationality.[172] There is a similar problem in trying to assess the performance of the Crown Prosecution Service. In its assessment of the cost-effectiveness of the Service in 1989 the National Audit Office revealed significant regional variations in terms of discontinuance/prosecution rates, conviction, acquittal, and case dismissal rates, processing delays, and prosecution costs.[173] While these performance indicators may say much about cost-effectiveness, they are less conclusive in measuring 'correct' decision-making. Indicators such as conviction and acquittal rates in particular assume the validity of the way cases are disposed of in the courts. Similarly, research attempts to identify weak cases that are prosecuted assume a notion of what a 'weak case' is.[174]

In view of the impossibility of assessing performance against some independent correct outcome, it would seem necessary for there to be more research into measuring outcomes against the views of participants including jurors and those most directly involved, victims and defendants.[175] Studies of victim and defendant attitudes enable assessments to be made about satisfaction with the system generally and these can have policy implications in terms of what can be done to improve satisfaction. Research elsewhere has found that defendants express more satisfaction with results which they believe are procedurally fair, even if personally unfavourable, than with results that are procedurally unfair.[176]

Although there is now widespread acceptance of the need for empirical assessment of the way the criminal justice system operates, there has been much less acceptance of the need to look at the way other systems operate. Despite Britain's membership of the European Community,

---

[171] See Z. Bankowski, 'The Value of Truth' (1981) *Legal Studies* 257–66.

[172] M. D. A. Freeman, 'The Jury on Trial' (1981) *Current Legal Problems* 65–111.

[173] *Review of the Crown Prosecution Service* (1989).

[174] A. Sanders, 'Some Dangers of Policy Oriented Research: The Case of Prosecutions', in Dennis, *Criminal Law*.

[175] See A. Bottoms and J. D. McClean, *Defendants in the Criminal Process* (1976); J. Shapland *et al.*, *Victims in the Criminal Justice System* (1985); J. Morgan and L. Zedner, *Child Victims: Crime, Impact and Criminal Justice* (1992). There has been virtually no research in the United Kingdom on the attitudes of jurors. For an exploratory survey, see J. Jackson *et al.*, *Called to Court: A Public View of Criminal Justice in Northern Ireland* (1991).

[176] See E. A. Lind and T. R. Tyler, *The Social Psychology of Procedural Justice* (1988).

there has been little interest in the criminal justice systems of other community countries and often a great degree of scepticism about the fairness of such systems. This lack of interest has even extended to systems within the United Kingdom such as Scotland. The Royal Commission on Criminal Procedure, for example, believed that comparisons with the Scottish prosecution system were undesirable because of the differences between the two systems.

There are good reasons to be cautious about utilizing foreign experience but this is no reason for not examining the procedures of other countries to see what can be learnt.[177] Attitudes towards foreign systems are now changing in the light of recent concern about miscarriages of justice and increasing doubts about the ability of the adversarial system to avoid them.[178] There is now a growing interest in whether the non-adversarial procedures of European countries are better able to find the truth than adversarial procedures.[179] The Royal Commission on Criminal Justice has to consider within its terms of reference the possibility of the courts 'having an investigative role both before and during the trial'. The Home Secretary has said that the Royal Commission should consider whether England and Wales ought to move towards continental-style law and that it must judge between the present adversarial system and the inquisitorial system in France.[180] Two reforms have attracted particular interest. First it has been suggested that judicial examiners should supervise all aspects of pre-trial preparation including interrogation.[181] We have seen that judicial examination is provided for in Scotland and the idea is not as foreign to English procedure as is sometimes believed, but the example that is most frequently cited is the *juge d'instruction* in France. The other suggestion is the establishment of an independent forensic science service which would work for the courts and which would undertake research and laboratory work requested by the parties.[182] One danger with these ideas is that they seem to detract from the defence's right to prepare and present its own case and to be provided with sufficient resources to call on

---

[177] L. H. Leigh, 'Liberty and Efficiency in the Criminal Process: The Significance of Models' (1977) *Int. and Comparative LQ* 516–30.

[178] Some have been expressing doubts for some time, see D. McBarnet, 'Fisher Report on the Confait Case: Four Issues' (1978) *MLR* 455.

[179] For the difficulties in resolving this question, see J. Jackson, 'Theories of Truth Finding in Criminal Procedure: An Evolutionary Approach' (1988) *Cardozo LR* 475, 481–6.

[180] *Sunday Times*, 29 Sept. 1991, p. 2.

[181] See Devlin, *The Judge*, 74–8; Lord Scarman, 'Justice in the Balance', *The Times*, 5 Mar. 1991.

[182] For discussion of the relative advantages and disadvantages of court appointed experts, compare M. N. Howard, 'The Neutral Expert: A Plausible Threat to Justice' (1991) *Crim. LR* 98–105, and J. Spencer, 'The Neutral Expert: An Implausible Bogey' (1991) *Crim. LR* 106–10.

expertise of its choice.[183] It is ironic that at a time of increasing interest in European procedures, other European countries are moving towards more adversarial procedures.[184]

CONCLUSION

Although the increasing attention that is being paid to international standards, empirical assessments, and the procedures of other countries is a welcome development, there remains a lack of consensus amongst participants in and outside the criminal justice system about the importance to be attached to due-process values. There has been a general consensus on the need for accuracy and impartiality in decision-making but these values have often been exploited to justify changes which have advantaged the prosecution more than the defence. Only the consistent exposure of celebrated miscarriages of justice has produced changes which have increased safeguards for defendants in the criminal process. What has been lacking is a clear statement of rights that can command widespread support within and outside the criminal justice system but this is unlikely to materialize until due-process values are given greater recognition by both the police and the courts.

---

[183] R. Stockdale, 'Running with the Hounds' (1991) *New LJ* 772–5.

[184] See e.g. E. Amodio and E. Selvaggi, 'An Accusatorial System in a Civil Law Country: The Italian Code of Criminal Procedure' (1989) *Temple LR* 1211–24.

5

Political Violence and Civil Liberties

CONOR GEARTY

INTRODUCTION

This chapter is not exclusively about a particular freedom or right, such as expression, assembly, association, or due process. Nor is its prime focus a defined group within society, like prisoners, the family, or a specific minority. Rather it is concerned with a particular problem and the impact on all our freedoms of trying to counter it: politically motivated violence, or 'terrorism' to accord it its contemporary label. Though invariably condemned today, political violence has a long and often honourable history in Britain. The institutions of the state were built on the successful subversion of Oliver Cromwell and the unabashed violence of William III. In the nineteenth and early twentieth centuries, some chartists, suffragettes, trade unionists, and Irish nationalists were condemned for using violence in the pursuit of political ends that have since become the orthodoxies of today. The Irish question in particular presented the authorities with the issue of 'terrorism' for the first time, with occasional bombings in Britain and regular expressions of the nationalists' physical force tradition in their home country. By 1922 the threat posed by it had, however, largely disappeared from Britain when the partition of Ireland in that year ushered in a period of relative calm.[1] The decades from the 1920s through to the end of the 1960s were almost totally free of domestic subversive violence. There were problems of public disorder and acute confrontations between government and organized labour, but the violence that these episodes generated was not systematically political in its intent. The Irish Republican Army (IRA) mounted a bombing campaign in England at the start of the Second World War but its much more ambitious 'border campaign' of 1956–62 was barely noticed in Britain since its violence was restricted to Northern Ireland.[2] So relaxed were the times that a leading civil liberties lawyer was able to write a best-selling book in 1963 entitled *Freedom, the Individual and the Law* without

---

[1] Northern Ireland and the then Irish Free state remained in turmoil for a considerable period after the agreement between the Government and the revolutionary Irish administration in Dec. 1921.

[2] For the parliamentary debates on Northern Ireland during this period, see 562 HC Debs. 1265–7 (19 Dec. 1956); 620 HC Debs. 1402–67 (30 Mar. 1960); 656 HC Debs. 1711–1817 (30 Mar. 1962).

making any reference to terrorism, or even to Northern Ireland,[3] and terrorism as a subject did not appear in the index to the Commons' debates until 1971.

The resurgence of political violence at the end of the 1960s was all the more dramatic for this lack of precedents in the remembered past. The reasons for the change were partly international and partly local. The student revolts in France and the United States in 1968 created a new vocabulary of rebellion which was fuelled both by the campus popularity of Maoism and by the controversial involvement of the United States in Vietnam. Out of this radical atmosphere emerged a handful of Western European groups dedicated to the violent overthrow of liberal democratic structures of government, the most prominent being the Baader-Meinhof gang in West Germany[4] and the Red Brigades in Italy.[5] The killings and kidnappings in which these groups engaged were few but the publicity they secured was immense and the label 'terrorist' re-entered the popular parlance as a useful way of describing them. Britain had its own rather puny version in the form of the Angry Brigade but it soon died out whereas the 'euroterrorists' have persisted through several generations as minor irritants on the body politic. Middle-Eastern-inspired violence has been just as enduring and rather more serious since its upsurge after the humiliation of the Arab cause in the six-day war in 1967. The Palestinian movement's lurch towards extremism was reflected in a wave of aircraft hijacks, and attacks on innocent civilians and undefended targets in European cities. The high point of this form of 'international terrorism' was reached with the assault by Black September on the Munich Olympics in 1972, when nine Israeli athletes, five Palestinian gunmen, and a German police officer died in the shoot-out that eventually ended the crisis. After the mainstream Palestinian leadership drew back from such actions in late 1974, the pace of terrorism slackened but renegade extremists remained capable of mounting extremely bloody attacks, with the hijacks at Entebbe and Mogadishu being the best remembered. In the 1980s, the situation was further complicated by the involvement of countries like Syria, Libya, and Iran in political violence in Europe, in what came to be known as 'state-sponsored terrorism', the most serious examples being the gun attacks on Rome and Vienna airports which claimed twenty lives in December 1985 and the blowing-up of a packed Pan Am jumbo jet over Lockerbie in Scotland in December 1988.

This was the international context which from 1968 created and maintained a sense of insecurity in the United Kingdom and an intense

---

[3] H. Street, *Freedom, the Individual and the Law* (1963).
[4] See S. Aust, *The Baader-Meinhof Group: The Inside Story of a Phenomenon* (1987).
[5] See A. Jamieson, *The Heart Attacked: Terrorism and Conflict in the Italian State* (1989).

nervousness about terrorism and terrorists. British aircraft have been hijacked, Englishmen have been taken hostage in Beirut, and events like Lockerbie, the shooting of the Israeli ambassador in London in 1982, and the death outside the Libyan People's Bureau of WPC Fletcher in 1985 have suggested that not even the streets of Britain are entirely safe. But the greatest contributor to tension over terrorism has been the indigenous IRA. The group re-emerged in Northern Ireland in the two years after the then Prime Minister Harold Wilson ordered troops into the Province in August 1969 to preserve the peace and protect Catholic communities from sectarian attack. The army soon incurred the enmity of the IRA, rejuvenated by the prospect of fighting again for the united Ireland that they believed had been wrongfully denied the Irish majority fifty years before. The battle has been continuing with remarkable consistency ever since. Over 3,000 people have been killed in the Province; one-third of these have been members of the security forces, half have been civilians, and the rest have been from one or other of the paramilitary groups that have multiplied since the outbreak of the disturbances. The death rate was most severe in the years 1972–6, when an average of 285 people lost their lives annually. Since then the yearly toll has levelled out at rather less than one hundred, with the direct cost of the violence now being estimated as in the region of £410 million every year.[6] In the financial year 1991–2, the amounts paid for criminal damage and criminal injuries resulting from terrorist activity were £29,765,000 and £10,878,000 respectively.[7]

The IRA's campaign has not been restricted to Northern Ireland. From the start it has also had its sights on Britain. In February 1972 an explosion at the Aldershot army base killed five waitresses, a gardener, and an army chaplain. In 1973, there were 86 explosions in Britain, involving one fatality and 380 other casualties. In February 1974 11 died and 14 were injured when a huge bomb blew up a bus full of British service personnel and their families as they were travelling through Yorkshire. Later that year 21 people were killed and 162 injured in the infamous Birmingham pub bombings. As in Northern Ireland, the level of atrocities has been much less severe since the 1970s, and there have been successive years when no incidents have occurred. However, the IRA's capacity to strike has remained, as evidenced by the Knightsbridge car bomb in 1983 which killed five, and the assaults on army bandsmen in Regent's Park and the Royal Marine School of Music at Deal, both of which incidents cost the lives of service personnel. The Brighton bomb in

---

[6] D. McKittrick, ' "Troubles" Cost UK and Ireland £410 Million a Year, Report Says', *Independent*, 22 May 1990.
[7] 212 HC Debs. 602 (WA) (27 Oct. 1992).

1984 showed the IRA's ability to take the fight directly to their enemy, as did the murders of leading political figures like Airey Neave and Ian Gow, both of whom were close advisers to Mrs Margaret Thatcher. Any one of the IRA's attacks during summer 1990 on the London stock exchange, the Conservative Carlton Club, and a party at a territorial army barracks could have caused loss of life on a scale reminiscent of the early 1970s. In 1990, there were 3 fatalities and 56 injuries in Britain— more IRA-related incidents than at any time since 1975, and this does not count the IRA's campaign against the British forces based in Europe. In 1991 there were three further fatalities and forty persons injured, when the IRA strikes included those against Victoria Station and No. 10 Downing Street. In 1992, there were 48 separate such incidents in Britain causing injuries to 222 people and killing six.[8]

This is the domestic background against which to view the law that has grown up over twenty years to counter political violence both in Britain and in Northern Ireland. There can be no doubt that in terms of breadth, comprehensiveness, and impact, the anti-terrorism legislation enacted by successive UK governments, and the ministerial orders and police powers that have either accompanied or been authorized by these Acts, represent the most dramatic incursion into civil liberties that has taken place since the Second World War. The issues raised by these laws are not easy to resolve, for a dilemma confronts all democratic societies with terrorist problems. Unlike the totalitarian or authoritarian government, which need brook no dissent whatsoever, the liberal democratic ideal of scrupulous adherence to the rule of law and generous tolerance of opposition helps make a terrorist problem possible in the first place. There is freedom of movement, little police surveillance, and a news media driven by ratings waiting to have its attention grabbed by the right story. The violent subversive that emerges from this atmosphere of relative tolerance is more than willing to use these liberal sentiments in order better to destroy the very society that boasts of them. The temptation therefore is to clamp down, but by turning to repression democratic society hands the terrorists a victory of sorts: they have been noticed; they are on an equal footing with the enemy; above all, their assertion that society was not really free, previously scoffed at, appears to carry more weight. Before rushing to repression, furthermore, it is important correctly to identify why dissent has turned to violence and also to assess the level of danger posed by a terrorist group. For, despite its noisy impact, terror is the weapon of the weak[9] and no purely terrorist group (as opposed to guerrilla army) has ever succeeded in achieving its

---

[8] See the comprehensive list of terrorist incidents in Britain since 1974 at 220 HC Debs. 282–8 (WA) (4 Mar. 1993).

[9] W. Laqueur, *The Age of Terrorism* (1987).

main aims.[10] The violence may be an ideological experiment or it might be a desperate throw from a group which has been excluded from the political process by an unfair distortion of democracy. In the latter case particularly, the dangers of an attempt at a purely military or police solution may outweigh the threat posed by the state's opponents. A victory achievable only by descending to the level of the terrorist is a defeat disguised by violence.

In this chapter, we summarize the laws that have been enacted in Britain and examine their impact upon our traditional freedoms. Then we analyse the political context which has made these changes possible. This will take us to a discussion of the role of pressure groups and reformers, after which we will consider the extent to which the courts have become involved in this highly sensitive area. Finally, we will assess the impact on the law of international instruments, in particular the European Convention on Human Rights. That document neatly captures the challenge that politically motivated violence poses for liberal democracies when it balances its guarantee of various traditional civil liberties with the caveat set out in Article 17: 'Nothing in this Convention may be interpreted as implying for any State, group or person any right to engage in any activity or perform any act aimed at the destruction of any of the rights and freedoms set forth herein or at their limitation to a greater extent than is provided for in the Convention.'

DEVELOPMENTS IN THE LAW

Personal liberty

In both Northern Ireland[11] and the Republic of Ireland[12]—but not in Great Britain—a provision authorizing the introduction of internment remains on the statute book. After every atrocity, there are always calls from Unionists north of the Border for its immediate introduction, but the British government has yet to yield to this pressure, mindful perhaps of the consequences of its unsuccessful application by the then Unionist government in the early 1970s.[13] In the absence of internment, the most serious consequences for individual liberty have flown from the definition of 'terrorism' in the anti-terrorism legislation. The word covers 'the use of

---

[10] This is of course a complicated question which involves far closer attention to definitions than can possibly be accorded here. See generally C. Gearty, *Terror* (1991).

[11] Northern Ireland (Emergency Provisions) Act 1991, s. 34.

[12] Offences Against the State (Amendment) Act 1940, last used during the IRA's 'border campaign' 1956–62.

[13] Lord Gardiner, *Report of a Committee to Consider, in the Context of Civil Liberties and Human Rights, Measures to Deal with Terrorism in Northern Ireland*, Cmnd. 5847 (1975).

violence for political ends', including 'any use of violence for the purpose of putting the public or any section of the public in fear'.[14] The first part of this formulation makes clear that no element of terror is required; every violent act for a political purpose is terrorist under UK law. Surprisingly perhaps, there is no separate crime of 'terrorism'. Rather, the importance of this definition lies in the various administrative and police powers that it unlocks. Thus the authorities have wide powers of arrest and detention where a person is suspected of being 'concerned in the commission, preparation or instigation of acts of terrorism'.[15] These may be exercised in two distinct situations. First, any person arriving from Northern Ireland may be stopped and 'examined' at the point of entry, 'for the purpose of determining' whether they appear to be involved in such activities. There is no need for a reasonable suspicion unless the person is held for longer than twelve hours. Indeed, if the examination lasts less than one hour it does not appear in the government's statistics and it would seem that a large number of travellers are processed in this way every year. The maximum length of such 'examinations' is twenty-four hours, after which the individual must be freed or detained for a further twenty-four hours (making a total of forty-eight hours in all). The second situation in which these powers are exercisable is broader than the port power in that it allows arrest and detention on the same basis anywhere in Britain but it is narrower in that it requires the presence of a reasonable suspicion from the start. Like the port power, the maximum time a suspect can be held is forty-eight hours, but both types of detention may be extended for a further five days by order of the Home Secretary.[16] The criteria that guide the Secretary of State are not set out in the Act and there is no provision for any judicial or even quasi-judicial oversight.

In 1977–90, more than 4,300 people were detained in Britain under these powers. The busiest period was the late 1970s, and in 1980 alone there were 537 detentions. In later years, however, the pace slackened somewhat. Since 1983, the figure has hovered around the 170 mark, with highs of 193 in 1985 and 147 in 1986. In 1990, the number detained in Britain was 169, more than half of whom were arrested in just three police areas: London, Merseyside, and Dumfries and Galloway. In 1991 and 1992, the figures were respectively 121 and 140. Fewer people are now being detained at airports and ports than used to be the case—as many as 73 per cent of detentions were of this sort in 1987, but the corresponding statistic for 1990 was 31 per cent and for 1992, 25 per cent. It should not be forgotten, however, that a large number of stops at points of entry count as examinations rather than detentions—and there

---

[14] Prevention of Terrorism (Temporary Provisions) Act 1989, s. 20(1).
[15] Ibid., s. 14(1)(*b*); sch. 5, para. 2(1).
[16] s. 14(5); sch. 5, para. 6(3). See *Forbes* v. *HM Advocate* [1990] SCCR 69.

were 254 of these relating to Northern Ireland and lasting over one hour in 1992, a figure one-third higher than for 1989. As far as the length of detention proper is concerned, the majority of suspects are released within the forty-eight hour deadline. However, a handful of detainees suspected of terrorism connected with Northern Ireland are held without charge for up to seven days every year. In 1990, there were 13 such cases, rather more than in 1989 (2), 1988 (4), and 1987 (7), but rather less than in both 1986 (15) and 1985 (29). In 1991 and 1992 the figures were down again, to zero and 5 respectively.

The main purpose behind these arrests and detentions is said to be questioning with a view to the institution of criminal proceedings. But the vast majority of those held are neither charged with criminal offences nor removed from Great Britain. In 1990, for example, of the 169 persons detained only 25 were charged with offences, and of these no more than 6 faced prosecution under the terrorism legislation. Lord Shackleton, who reviewed the operation of the law in 1978, felt that it was not 'simply a question of arresting people who can promptly be charged with offences',[17] and as one leading authority has put it, the 'rate of charging betokens a dominant police interest in gathering background information through questioning a detainee about his political views, friends and colleagues'.[18] The same author refers to the occurrence of police sweeps during which 'everyone connected with a person against whom there is firm evidence' is arrested, and this has happened on at least three occasions before Christmas in 1979, 1981, and 1983.[19] Despite Home Office advice to the contrary, the police have sometimes allowed the impression to develop that those arrested have been chosen more for the nature of their politics than for their connections with political violence.[20] Indeed, the suspicion is held by some that the arrest, examination, and detention powers in the prevention of terrorism legislation have become the basis of a vast but legally uncontrolled surveillance exercise on the Irish in Britain.

Freedom of movement

In 1990, fifteen of those detained under the terrorism laws were excluded from Great Britain. The total number of 'exclusion orders' in force at the end of the year was ninety-seven. By the end of 1992, this figure had

[17] Lord Shackleton, *Review of the Operation of the Prevention of Terrorism (Temporary Provisions) Acts 1974 and 1976*, Cmnd. 7324 (1978), para. 135.
[18] C. Walker, *The Prevention of Terrorism in British Law* (1992), 181.
[19] Ibid. 184.
[20] C. Scorer and P. Hewitt, *The Prevention of Terrorism Act: The Case for Repeal* (1981), 39–45.

been reduced to eighty-one. This power is one of the most controversial and hotly debated in the whole of the counter-terrorism arsenal. Under what is now section 5 of the Prevention of Terrorism Act 1989, if the Secretary of State is satisfied that any person (*a*) is or has been concerned in the commission, preparation, or instigation of acts of terrorism connected with Northern Ireland; or (*b*) is attempting or may attempt to enter Great Britain with a view to being concerned in the commission, preparation, or instigation of such acts of terrorism, then the Secretary of State may make an exclusion order against him or her. 'Terrorism' here has the same meaning assigned to it as we saw in the context of arrest and detention. An exclusion order, which prohibits a person from 'being in, or entering, Great Britain', lasts for a maximum of three years, though the authorities are free to make fresh orders. A British citizen who 'is at the time ordinarily resident in Great Britain and has then been ordinarily resident in Great Britain throughout the last three years' is exempt. Breaching an exclusion order, or helping to circumvent one, is an offence carrying a maximum penalty, on indictment, of five years in jail. As with the detention power, exclusion orders were widely employed when the terrorism legislation was first put in place in 1974. The three-year ordinary residency exemption was not then part of the law and many Irish families who had lived in Britain for years found that they were being forcibly uprooted. The time limit, first introduced in 1984, precipitated a sharp decline in the number of orders in force, from 235–50 in each of the years 1980–4 to an average of just over 100 since 1986.

Over the years there have been allegations that threats of exclusion orders have been used to harass the Irish community in Britain, particularly those of its members who hold pronounced republican views. In 1982, the then Home Secretary used the power to prevent a visit to Britain by three leading members of Provisional Sinn Féin, a lawful political party closely associated with the IRA. (One of the excluded triumvirate, Mr Gerry Adams, subsequently won a seat in the House of Commons, at which point the exclusion order against him was lifted.) A person who objects to his or her proposed banishment may exercise a power under the Act to write to the Home Secretary setting out the grounds of the complaint and requesting a personal interview in order to discuss them. The Secretary of State then refers the matter 'for the advice of one or more persons nominated by' him. The Minister has an absolute discretion in choosing these people. There is no requirement that they be lawyers or that they have any familiarity with the Northern Irish situation. They could be police officers or civil servants. What usually happens is that one of the advisers interviews the aggrieved suspect. Legal representation may be allowed but no witnesses are called. The supplicant is not helped by the fact that he or she is not informed of the reasons why the exclusion order has been made: suspicions that cannot be seen are not ones that can be easily allayed. The Home Secretary's final decision is based on the view of the advisers, together with an objective report on the interview and the suspect's original written representations. A substantial proportion of exclusion orders are now being made subject to this appeal procedure,[21] though as is to be expected, given the limited terms of reference, only a very few of these are successful. Since 1982, 144 exclusion orders have been applied for and all but 18 of these have been granted by the Secretary of State.

Due process

As is to be expected, the most dramatic changes in the requirements of due process have been in Northern Ireland. The wholesale suspension of the judicial system which was tried in the form of internment in 1971 turned out to be a disaster, with its inept introduction alone being responsible for greatly increasing IRA support.[22] Since its phasing out in 1975, the emphasis has been on dealing with suspected terrorists in the criminal courts, but various changes have been made to meet what is perceived to be the vulnerability of the ordinary process to paramilitary manipulation. First, the emergency law passed in a series of enactments since 1973 has fundamentally transformed the relationship between the individual and the state, with the security forces now having wide powers of stop and search, entry, arrest, and detention which would be unthinkable in the rest of the United Kingdom and which extend well beyond the anti-terrorism legislation. The Northern Ireland (Emergency Provisions) Act 1991, for example, added new offences such as that of directing, 'at any level, the activities of an organization which is concerned in the commission of acts of terrorism' (maximum sentence of life imprisonment)[23] and possessing something 'in circumstances giving rise to a reasonable suspicion that the item is . . . for a purpose connected with the commission, preparation or instigation of acts of terrorism'.[24] The latter offence comes with a power of entry and arrest for the police, based on reasonable suspicion,[25] and the 'item' caught by the section could in itself be quite harmless. It is the officer's reasonable suspicion as to the reason for having the item that is the offence—a type of double thought crime.

Secondly, apart from changes in the substantive law, there have been

---

[21] Lord Colville, *Review of the Operation of the Prevention of Terrorism (Temporary Provisions) Act 1984*, Cm. 264 (1987), para. 11.3.

[22] Gardiner, *Report*, Cmnd. 5847.

[23] s. 27.

[24] s. 30.

[25] s. 17(2).

important modifications made to the criminal process. Trial by jury has been sharply reduced in the Province with the vast majority of serious crimes now being heard before a single judge, with a right of appeal to the three-judge court of appeal.[26] The theory initially was that only terrorist-related crimes should be dealt with in this way as a matter of course, but whether by conscious design, excessive caution, or the operation of financial constraints, such courts have quickly become the norm. During the passage through Parliament of the Emergency Provisions Act 1991, the Government resisted strong pressure from the Opposition to shift the presumption back to jury trials. The right to silence has been similarly emasculated.[27] Suspects who fail to mention something when they are questioned, or who omit to explain 'any object, substance or mark' on their person, or who refuse to tell the police why they are in a certain place at a certain time, may find the judge at their later trial prepared to infer guilt from their reluctance to co-operate. The judge may also use this silence as corroboration of other evidence for the purpose of securing a conviction. Thus, in place of the traditional police caution about the right to silence, there are now a variety of complex messages, one of which may be given by way of illustration:

You do not have to say anything unless you wish to do so but I must warn you that if you fail to mention any fact which you rely on in your defence in court, your failure to take this opportunity to mention it may be treated in court as supporting any relevant evidence against you. If you wish to say anything, what you say may be given in evidence.

The court may also now draw adverse inferences from an accused's failure to give evidence in court on his or her own behalf. The law requires that a silent accused be publicly called to give evidence, notwithstanding his or her declared intention not to do so. It is hard to see what the purpose behind this piece of contrived theatricality could be; its effect is to prejudice further the standing of the defendant in the eyes of the court.

Finally, there have also been a number of modifications to the rules governing inquests in Northern Ireland. This is a matter of importance because of the number of people killed in incidents involving the security forces. Between 1969 and 1 March 1984, 233 people were killed by the police or army on duty in the Province.[28] An Amnesty International Report published in 1988 drew attention to the fact that, between

---

[26] For the background to the change, see Lord Diplock, *Report of the Commission to Consider Legal Procedures to Deal with Terrorist Activities in Northern Ireland*, Cmnd. 5185 (1972). The current legal basis for non-jury courts is to be found in Part 1 of the Northern Ireland (Emergency Provisions) Act 1991.

[27] The Criminal Evidence (Northern Ireland) Order 1988 (1988 No. 1987 (NI 20)).

[28] 56 HC Debs. 494 (WA) (21 Mar. 1984).

November 1982 and April 1988, 49 people were killed by the security forces and in 19 of the 32 incidents identified the victims were unarmed. The vast majority of those killed were Catholic.[29] In view of the difficulty of obtaining criminal convictions against members of the police and army, the inquest has over the years assumed importance as an additional (and sometimes the only) forum for sorting out what are often remarkably different versions of the same fatal incident. Traditionally, with a few specific and limited exceptions, every person is a competent and compellable witness at an inquest, though, to quote from the English rule, '[n]o witness . . . shall be obliged to answer any question tending to incriminate himself' and if 'it appears to the coroner that a witness has been asked such a question, the coroner shall inform the witness that he may refuse to answer'.[30] In Northern Ireland, however, paragraph 9 of the relevant rules states that:

9(2) Where a person is suspected of causing the death, or has been charged or is likely to be charged with an offence relating to the death, he shall not be compelled to give evidence at the inquest.

9(3) Where a person mentioned in paragraph (2) offers to give evidence the coroner shall inform him that he is not obliged to do so, and that such evidence may be subject to cross-examination.[31]

Thus, those responsible for a killing in the Province, being persons 'suspected of causing the death', are not required to explain themselves even in a partial way to this limited body. This leaves the inquest in Northern Ireland with what is often the perfunctory task of stating the obvious—the identity of the victim and the manner of his or her death. Relatives and friends of deceased persons have on occasion refused to accept what they have seen as a manipulation of the process in the authorities' favour. This has resulted in sporadic outbursts of disorder. More often, it has led to elaborate verbal jousts as lawyers for the families of those deceased have tried to transform the inquest into the inquiry that the rules have contrived to forbid.

None of these dramatic modifications of the rule of law has occurred in Britain, though there have been calls for an end to the right to silence and a Home Office working party recommended changes in 1989.[32] Access to

---

[29] Amnesty International, *Northern Ireland: Killings by the Security Forces and 'Supergrass' Trials* (1988). For recent allegations of security forces' collusion in paramilitary action see the Channel 4 'Dispatches' programme transmitted in October 1991, on which see the debate in the House of Commons at 216 HC Debs. 569–79 (17 Dec. 1992).

[30] Coroners' Rules 1984, rule 22.

[31] Rule 9 of the Coroners (Practice and Procedure) Rules (Northern Ireland) 1963, as amended by the Coroners (Practice and Procedure) (Amendment) Rules (Northern Ireland) 1980.

[32] See the summary of the Working Party's recommendations at [1989] *Crim. LR* 677. For a recent indication of the continuing importance of the right to silence in the context of 'terrorist' trials, see *R. v. Cullen, Shanahan and McCann* [1991] *Crim LR* 136 (CA).

a lawyer when in detention may however be delayed in terrorist cases across the United Kingdom for up to forty-eight hours, rather than for the thirty-six hours which is the outer limit for ordinary crime,[33] and there have been occasional allegations that the threat of exclusion orders has been used to coerce recalcitrant detainees into conversation. While jury trial is still required, it is now standard practice in Britain for vetting to take place in 'terrorist' cases. This is to protect against the danger that

a juror's political beliefs are so biased as to go beyond normally reflecting the broad spectrum of views and interests in the community to reflect the extreme views of sectarian interest or pressure group to a degree which might interfere with his fair assessment of the facts of the case or lead him to exert improper pressure on his fellow jurors.[34]

The scrutiny is normally conducted by the police special branch but the Attorney-General may authorize an inquiry beyond perusal of criminal records and police files. No information on these vetting exercises needs to be revealed to defence counsel, the court, or the wider public, so it is not possible to know the extent to which the practice has become routine. The jury, however, has not been the most culpable element in the succession of wrongful convictions in terrorist cases in Britain which Lord Devlin has described as 'the greatest disasters that have shaken British justice in [his] time'.[35] The cases of the Birmingham Six, the Guildford Four, Judith Ward, and the Maguire Seven raise questions concerning the role of the police, the prosecuting authorities, the senior judiciary, and the forensic science service which strike right to the core of the integrity of the whole criminal process in this country.[36]

Freedom of association

Other mainstream civil liberties have been affected by the legislative response to Northern Irish terrorism. One of the least contentious is the prohibition on membership of the IRA. The ban was introduced in 1974 because it was felt that 'the public should no longer have to endure the affront of public demonstrations in support of' the IRA.[37] Another

---

[33] Police and Criminal Evidence Act 1984, s. 58(13). For Northern Ireland, where the law is similar, see Northern Ireland (Emergency Provisions) Act 1991, s. 45(1).

[34] See Practice Note, Attorney-General's Guidelines on Jury Checks, reproduced at [1988] 3 All ER 1086.

[35] Lord Devlin, 'The Conscience of the Jury' (1991) 107 *LQR* 398.

[36] See Return to an Address of the Honourable House of Commons dated 12 July 1990 for an inquiry into the Circumstances Surrounding the Convictions Arising out of the Bomb Attacks in Guildford and Woolwich in 1974 (HC Paper 556, 1990).

[37] 882 HC Debs. 636 (28 Nov. 1974) (Mr Roy Jenkins). In Northern Ireland, both the IRA and the INLA are banned as are the following groups: Cumann na mBan; Fianna na hEireann, the Red Hand Commando; Saor Eire; the Ulster Freedom Fighters; the Ulster Volunteer Force; the Ulster Defence Association; and the Irish People's Liberation Organization.

group, the Irish National Liberation Army (INLA) was later also banned. Membership of either group is an offence punishable by up to ten years in gaol. It is also a crime to arrange, manage, or address any meeting of three or more persons (whether private or public) if it is known that the meeting is to support or to further the activities of the IRA or INLA or is to be addressed by someone belonging to one or other of them. The 1989 Act has added to this by introducing new laws aimed at the financing of proscribed organizations.[38] An offence will also be committed where any person in a public place either (*a*) wears any item of dress or (*b*) wears, carries, or displays any article, 'in such a way or in such circumstances as to arouse reasonable apprehension that he is a member or supporter of a proscribed organization'.[39] It is claimed that these provisions achieve 'the avoidance of public outrage, but also the averting of any danger of this outrage being expressed in disorder'.[40] There have, however, been very few prosecutions in Britain for membership or organizing meetings and only a handful under the unlawful dress provision. On the other hand, there have been suggestions that legitimate public debate has been hampered, with groups like the Troops Out Movement having difficulty getting permission to hold rallies and book halls for their public meetings.[41] In his official review of the Act, Earl Jellicoe admitted that it was 'asking a lot of the police service to apply these provisions fully in relation to proscribed organizations themselves, while not affecting the free expression of views about Northern Ireland'.[42]

Freedom of expression

One of the liberties that has been most severely affected by the disorder in Northern Ireland has been freedom of expression. To a large extent, this has been the result of self-censorship on the part of the broadcasting and print media. In particular, coverage of Northern Ireland on television has become the subject of elaborate systems of regulation by senior management, with programmes being withdrawn or modified if they are judged by executives to be too sensitive. The most famous incident was the *Real Lives* affair in 1985, when a BBC programme was first delayed and then altered after pressure from the Home Office. The film involved a portrait of one of Sinn Féin's leaders, Martin McGuinness. In September 1988, Channel 4 dropped a programme in the *After Dark* series when it became apparent that Gerry Adams was to be one of the guests. In early October the same year, a Panorama documentary on the SAS was trans-

[38] See generally the 1989 Act, pts. 1 and 3.
[39] s. 3.
[40] Earl Jellicoe, *Review of the Operation of the Prevention of Terrorism (Temporary Provisions) Act 1976*, Cmnd. 8803 (1983), para. 207.
[41] Scorer and Hewitt, *Prevention of Terrorism*, 18–20.
[42] Jellicoe, *Review*, para. 212.

mitted in an edited form after the BBC, having received the advice of the
D-notice Committee, had refused to show the original programme.[43] The
law has played something of a background role in all of this. Since 1976,
it has been an offence where a person 'fails without reasonable excuse' to
disclose to the police or army information 'which he knows or believes
might be of material assistance—(a) in preventing the commission by any
other person of an act of terrorism connected with the affairs of Northern
Ireland; or (b) in securing the apprehension, prosecution or conviction of
any other person for an offence involving the commission, preparation or
instigation of such an act'.[44] There have been few arrests under this
section and only very rarely any convictions, but its main utility has been
as a background threat against journalists whose contacts with subversives
in the Province have earned them the enmity of the security forces.[45]
There have been public warnings to distinguished journalists from both
Britain and the United States, but as yet the media has not had to cover a
dramatic trial of one of its own.[46] Following the transmission in October
1991 of a Channel 4 'Dispatches' programme alleging collusion between
the security forces and paramilitary organizations in Northern Ireland, an
order was obtained under the Prevention of Terrorism Act requiring
both Channel 4 and the independent production company involved to
produce certain documents. Both refused to do so and were fined £75,000
by the Divisional Court for contempt of court. The researcher involved in
making the programme was later charged with perjury but these proceedings were discontinued in November 1992.

The most important official constraint on freedom of expression is the
media ban introduced by the then Home Secretary Mr Hurd in October
1988. The exact words of the prohibition require the BBC and the IBA to:

refrain from broadcasting any matter which consists of or includes—any words
spoken, whether in the course of an interview or discussion or otherwise, by a
person who appears or is heard on the programme in which the matter is broadcast
where—(a) the person speaking the words represents or purports to represent an
organization specified in paragraph 2 below, or (b) the words support or solicit
or invite support for such an organization, other than any matter specified in
paragraph 3 below.

Paragraph 2 makes clear that all proscribed organizations under either
the Prevention of Terrorism or Emergency Provisions Acts are caught by

---

[43] *Guardian*, 13 Oct. 1988.

[44] The words quoted here are taken from s. 18 of the 1989 Act.

[45] Between Mar. 1984 and Dec. 1990, eight people were detained in Britain on suspicion
of committing an offence under this section. Four were subsequently charged but only a
single conviction was obtained.

[46] For further information, see K. D. Ewing and C. A. Gearty, *Freedom under Thatcher*
(1990), 241–3.

the ban. But a legal organization, Sinn Féin, is also included. The matter specified in paragraph 3 as being exempt from the ban is: 'any words spoken (a) in the course of proceedings in Parliament, or (b) by or in support of a candidate at a parliamentary, European Parliamentary or local election pending that election'.

The ban has remained controversial since its introduction, for three reasons in particular. First, it is extremely wide, catching not only the interview with a Sinn Féin or IRA representative, but also the reporting by a journalist of news information about the IRA or Sinn Féin. The transmitting of archive material on former Irish leaders who were nationalist fighters in their youth is also caught, as are comments by Sinn Féin councillors on aspects of local government services in their area. Shortly after its introduction, the Government indicated that it did not intend the ban to reach as far as its words seemed to suggest, and it is now clear that Sinn Féin representatives are allowed to speak in their capacity as local councillors and that even political matters can be addressed as long as subtitles or an actor's voice are employed on screen when the 'banned' person is speaking. The second objection to the ban is that this sharp infringement of civil liberties has been permitted for no very clearly articulated reason. For its part, the Government has pointed to the 'widespread offence' caused by the appearance of terrorist sympathizers on radio and television, and has also alleged that both media have been used by the paramilitaries 'to deliver indirect threats'.[47]

The main point in its favour from the Government's point of view, however, is that the 'terrorists themselves draw support and sustenance from access to radio and television' and that the ban has severely inhibited their ability to secure the 'oxygen of publicity' from this source. It is undoubtedly the case that the whole republican movement in Northern Ireland has suffered from a marked decline in media interest since the ban came into force. This has been true of foreign as well as local journalists and the result has been that major publicity campaigns, such as that to commemorate the twentieth anniversary of the deployment of troops on the streets of Northern Ireland, have largely failed to get off the ground. In these terms, therefore, the ban has been a success—not so much because of what it says as because of its impact as a further stimulus to self-censorship. But this leads to the third objection to the ban which is that it is an illegitimate intrusion into the political process because it inhibits the expression of political views not only by 'terrorists' from banned organizations but also by members of Sinn Féin, a lawful political

---

[47] The Home Secretary's original statement about the ban is at 138 HC Debs. 893–903 (19 Oct. 1988). The Commons debate on the banning Order is at 139 HC Debs. 1073–1155 (2 Nov. 1988).

party which has had a sitting Member of Parliament and has many locally elected councillors. Sinn Féin may have declined somewhat from its electoral peak in its post-hunger-strike heyday, but it remains a considerable political force, capable of securing 11 per cent of the votes cast at recent Province-wide elections. Opponents of the ban say that it is misguided to make it possible to forget that Sinn Féin are popular in some parts of Northern Ireland and that it is a dangerous precedent to allow such direct censorship of a lawful group which, though it may have a shade of criminality about it, nevertheless has a political message to communicate.

Other freedoms

The impact of political violence on freedom is not restricted to subversion of the IRA variety. We mentioned earlier that international and state-sponsored terrorism have also been a cause of some deaths and much anxiety over the years. In 1984, Parliament widened the detention power under the terrorism legislation so as to embrace 'international terrorism', by which was meant 'the use of violence for political ends' of any description 'except acts connected solely with the affairs of the United Kingdom or any part of the United Kingdom other than Northern Ireland'.[48] This clause has given rise to the embarrassing oddity that PLO and ANC representatives, including such senior figures as Mr Yasser Arafat and Mr Nelson Mandela, have been liable to arrest as suspected 'terrorists' during their stays in Britain, whereas active members of groups like the Animal Liberation Front and other indigenous subversive groups have not been vulnerable in the same way. Perhaps because of such considerations, the power has not been utilized as often as might have been expected. In 1990, a total of twenty-four suspects were held, an increase on the two previous years when eighteen (in 1989) and sixteen (in 1988) were picked-up. In 1991 and 1992, the figures were thirty-two and twenty respectively. The highest number of suspected 'international terrorists' detained in any given year was seventy-three in 1985, a twelve-month period during which there was a large number of well-publicised terrorist incidents around the world. Whilst it is disturbing that nearly two-thirds of those held under this provision have been neither charged with any offence nor deported, removed, or excluded, the real risk from the point of view of civil liberties is that a future government will be misled by a false analogy with it into extending the power to cover all forms of domestic subversion involving some dimension of violence. If the Irish

[48] See now ss. 14(2)(*b*) and 20(1) of the 1989 Act. The definition of 'terrorism' is the same for both types of political violence; the text gives the first and inclusive part of the definition in s. 20(1).

example were to be repeated, such a development would have the effect of seriously inhibiting legitimate dissent.

In contrast to detention, the exclusion power is not generally relevant to international terrorism since other mechanisms are available for securing the removal of unwanted dissidents. Extradition law allows the return of persons convicted or suspected of serious offences in foreign countries as long as the necessary international arrangements are in place. The law has traditionally allowed fugitives to remain in this country where the offence of which they are accused is 'one of a political character',[49] though over the years the courts have responded to the increase in terrorism by narrowing the meaning of the phrase so as to exclude many politically motivated crimes involving violence.[50] This judicial move has been matched by an international drive towards greater political co-operation, and Britain now has agreements with many European nations which specifically exclude the political defence.[51] In its place is to be found a new emphasis on the motives of the requesting state. A fugitive will not be extradited if it can be established that

(i) the requisition for his surrender has in fact been made with a view to try or punish him on account of his race, religion, nationality or political opinions; or (ii) that he might, if surrendered, be prejudiced at his trial or punished, detained or restricted in his personal liberty by reason of his race, religion, nationality or political opinions.[52]

It is unlikely that these provisions will come into play where a Western European state is the requesting authority.[53]

Apart from extradition, the ordinary law on immigration is also relevant in so far as it governs the situation where an asylum application is made by a foreign national. Such a person need not necessarily have been engaged in political violence at home, but an assertion of an involvement of this sort may be a part of their claim. Once again, there has been a hardening of the position in recent years. The situation is governed by the United Nations Convention relating to the Status of

[49] Extradition Act 1870, s. 3(1); see now Extradition Act 1989, s. 6.

[50] See *Cheng* v. *Governor of Pentonville Prison* [1973] AC 931 (HL).

[51] See the Suppression of Terrorism Act 1978, giving effect to the European Convention on the Suppression of Terrorism, Cmnd. 7031 (1977); see further an *Agreement Concerning the Application of the European Convention on the Suppression of Terrorism among Member States of the European Communities*, Cmnd. 7823 (1980). For the complicated position with regard to extradition from the Republic of Ireland, see the two statutes enacted by that country in 1987: the Extradition (European Convention on the Suppression of Terrorism) Act 1987 and the Extradition (Amendment) Act 1987.

[52] Extradition Act 1989, s. 24.

[53] See also ss. 11(3), 12(2) which protect a fugitive from a return which would be 'unjust or oppressive', a provision that is less unlikely to be employed against a liberal democratic state: see *Sinclair* v. *DPP* [1991] 2 All ER 366 (HL).

Refugees[54] which has been incorporated into British law so as to exempt from ordinary immigration rules a person who has only one country to go to but it is 'one to which he is unwilling to go owing to well-founded fear of being persecuted for reasons of race, religion, nationality, membership of a particular social group or political opinion'.[55] In 1987, the House of Lords interpreted the requirement that the applicant's fear of persecution should be well-founded as meaning 'that there has to be demonstrated a reasonable degree of likelihood that he will be persecuted for a Convention reason if returned to his own country'[56] and that it was for the authorities to decide this on an objective rather than subjective basis. In reaching their decision, their Lordships rejected an interpretation adopted by the Court of Appeal and an alternative approach argued for by counsel briefed to appear on behalf of the United Nations High Commissioner for Refugees, who had presented 'a careful and thoughtful argument largely based on the travaux préparatoires of the Convention and Protocol'.[57] Apart from this contribution from the judges, the Government has progressively made it more difficult for asylum seekers to substantiate their claims, and in summer 1991 the Home Secretary announced an end to the availability of independent legal aid for applicants for asylum, a decision deplored by the then chairman of the Bar Council as 'an action of which the Government should be ashamed'.[58]

EXPLAINING THE CHANGES

As we have seen, an array of new laws has emanated from Parliament in the past twenty years, empowering a wide range of officials to take action to guard society from 'terrorists' and from others seeking to escape conflict in foreign countries. The authorities responsible for the enforcement of these provisions include Home Office officials and the police. Particularly important is the specialist counter-terrorism squad based in London under the auspices of the Metropolitan Police. Customs and immigration officers also exercise responsibility in this area, reflecting the importance to counter-terrorism of border and immigration controls. Bubbling just beneath the surface is the secret world of counter-terror. The terms of reference of the security service (better known as MI5)

---

[54] The Convention, dated 28 July 1951, is published as Cmd. 9171.

[55] Statement of Changes in Immigration Rules (HC 169 of 1983), rule 73, made under the Immigration Act 1971, s. 3(2).

[56] *R. v. Secretary of State for the Home Dept., ex parte Sivakumaran and others* [1988] 2 WLR 92 (HL), at p. 98 per Lord Keith.

[57] Ibid., at p. 99.

[58] A. Scrivener, 'Barrier to Law in the Land of the Free', *Observer*, 18 Aug. 1991.

include 'the protection of national security and, in particular, its protection against threats from . . . terrorism'.[59] This is the legal basis on which the executive may authorize MI5 to engage in activities which would otherwise be unlawful, such as house-breaking or the illicit removal of documents.[60] Telephone tapping and the interception of communications generally is covered by different legislation which permits such conduct where it is 'in the interests of national security'.[61] The safeguards against the abuse of any of these powers are vague and inadequate, and the Official Secrets Act 1989 criminalizes all unauthorized disclosures,[62] so there is no way of knowing exactly how much covert action is engaged in under the protective umbrella of counter-terrorism.[63] MI5 were recently given the lead role in the official fight against terrorism.[64]

Describing the edifice that has grown up around the battle against subversion does not explain how it came to be built. Economic and social developments do not play a big part here, except to the extent that the most important terrorist group, the IRA, has grown and survived in an atmosphere of local economic depression, spiced with nationalist idealism. The comparative poverty of their hinterland is therefore relevant but not decisive: terrorism is a Belfast phenomenon, not a Liverpudlian or Glaswegian one and this is more for reasons of history, culture, and national identity than because of especially grotesque economic degradation. As regards the response of the state, the explanation for the breadth of counter-terrorism laws does not lie in the level of violence, particularly where Britain is concerned. Even during its worst periods, terrorism has never killed many people in England and there have been sequences of years during which very few incidents have been reported. The likelihood of death is infinitely greater from a whole range of activities like smoking, matrimonial quarrelling, and car accidents, none of which has provoked the same legislative response. The defeat of terrorism is also not a rationale that can be pointed to with any confidence; if this were the prime goal, then it has manifestly been a failure. While it is impossible to tell how much the existence of these laws has prevented even greater levels of violence, it is clear that the law and order responses of successive governments have on occasion, such as with the introduction of internment in 1971 and the 'shoot to kill' controversy in the early 1980s, increased alienation rather than reduced violence. It is

[59] Security Service Act 1989, s. 1(2).
[60] s. 3.
[61] Interception of Communication Act 1985, s. 2(2)(*a*).
[62] See in particular s. 1.
[63] See Ewing and Gearty, *Freedom under Thatcher*, chs. 3 and 6.
[64] See the statement by the Home Secretary, Mr Kenneth Clarke, at 207 HC Debs. 297–306 (8 May 1992).

now widely accepted that a lasting peace can only be achieved by political rather than by purely military means.[65]

An explanation for the legislative reaction to terrorism lies in the nature of its violence. It is considered to be uniquely threatening on account of its perceived indiscriminateness, its tendency to strike anywhere, and its destructive political motive.[66] Thus in the first months of 1991, the IRA's mortar attack on Downing Street and its bombing of Victoria Station in London caused few casualties relative to the reaction of horror and shock that they drew from the general public, the police, and the politicians. These emotions were evident again in subsequent attacks on Bishopsgate and Warrington. It is in such understandable though emotional responses that we see why so many counter-terrorism laws have come to be passed. In no other field are our laws so much the consequence of a desire to be seen to be doing something. They owe their existence not to any strategic policy goal or covert masterplan, but rather to the 'something must be done' syndrome that afflicts all politicians trying to assuage public indignation. To borrow a famous phrase, 'the politics of the last atrocity' lie at the root of many of these laws. The continued violence and the bureaucratic determination of the lead counter-terrorist agencies then makes unravelling the legislation seem to many to be well-nigh impossible. A good example of this process is the prevention-of-terrorism legislation, first passed in 1974. On the day after the Birmingham pub bombings in November of that year, the then Home Secretary Mr Jenkins told Parliament that there were 'certain security measures which would . . . justify emergency legislation in the present circumstances'.[67] The Act, described by Mr Jenkins as 'draconian' and 'unprecedented in peacetime'[68] passed through all stages in both Houses in forty-two hours. There was no division on second reading in the Commons because there was no member willing to act as teller for the noes.[69] The comments of one participant capture the atmosphere of the debate:

MR LITTERICK: Here we are, a panic striken mob—[HON. MEMBERS: 'Rubbish']—
   because we have had avalanches of letters from panic-striken, shocked people . . .
   That is the kind of pressure we are supposed to resist. We are not supposed to
   hurry into these judgments like frightened people.[70]

Nine years later, Clare Short, now a Labour MP but then a junior Home Office official, recalled an encounter she had in Parliament on the night the Bill was being debated:

[65] The clearest recognition of this fact is the Anglo-Irish Agreement concluded between the UK and the Republic of Ireland in 1985.
[66] See Gearty, *Terror*, ch. 1.
[67] 881 HC Debs. 1671 (22 Nov. 1974).
[68] 882 HC Debs. 35–6 (25 Nov. 1974).
[69] 882 HC Debs. 752 (28 Nov. 1974).
[70] Ibid. 673.

Late that night I was sitting in the Ministerial box listening to the debate . . . Next to me was the man who drafted the Bill. Repeatedly, I turned to him to ask in what way the different provisions would have prevented the Birmingham bombings. Eventually he replied, 'You know very well that is not what it is about. We have to appease them. They are after capital punishment. They have to be given something'.[71]

The subsequent history of the legislation gives an insight into how it has survived. When he introduced the first bill, Mr Jenkins said that he did 'not think that anyone would wish these exceptional powers to remain in force a moment longer than [was] necessary'.[72] Accordingly, the Act provided for its own expiry after six months unless renewed by affirmative parliamentary approval of an order of the Secretary of State. This duly occurred in May 1975 by a vote of 161 to 10, with the Home Secretary declaring that the 'security situation continue[d] to give cause for concern', though he admitted that there had been 'no terrorist incidents in Britain since the end of January'.[73] Perhaps a little embarrassed by this, Mr Jenkins promised that 'unless in November I feel able to recommend the dropping of substantial parts of the Act, I shall not ask the House to proceed by order'.[74] However, this undertaking led to a new Act rather than merely the end of the old one, with the death of eight people and the injury of over 170 at the hands of terrorists during the intervening period making it appear to the Home Secretary that it would be 'inconceivable within the House and incomprehensible outside' to drop the legislation.[75]

This set the pattern for all subsequent debates, with each renewal of the legislation (done on an annual basis after 1976) being accompanied by statistics of violence presented as arguments for retention. The Government (regardless of which Party happens to be in power) has usually been able to point to various episodes of terrorist activity, but occasionally the figures have not been so supportive. In 1977 the new Home Secretary Mr Merlyn Rees secured the Act's retention despite the fact that there had been no terrorist attacks in Britain at all in the last seven months of 1976 and only two incidents (neither involving casualties) in the first two months of 1977.[76] The following year, Mr Rees recognized that the 'past year has been free of terrorist attacks' but argued to retain the legislation because this was 'only part of the picture'.[77] The Act was renewed

[71] 47 HC Debs. 90 (24 Oct. 1983).
[72] 882 HC Debs. 642 (28 Nov. 1974).
[73] 892 HC Debs. 1083 (19 May 1975).
[74] Ibid. 1091.
[75] 901 HC Debs. 877 (26 Nov. 1975). The details of the IRA actions in the preceding three months are at ibid. 879. The vote on second reading for the new measure was 183 in favour to 14 against.
[76] 927 HC Debs. 1473 (9 Mar. 1977).
[77] 946 HC Debs. 547 (15 Mar. 1978).

annually throughout the mid-1980s despite the fact that there were no IRA actions causing casualties in Britain in 1985, 1986, or 1987. The executive's argument has it both ways: if there is terrorism we need the Act to counter it; if there is no terrorism, we need to keep the Act because it is doing such a fine preventive job. The empirical evidence that underpins these assertions is shrouded in secrecy and it is therefore impossible to confirm their accuracy.

The issue of terrorism was not a controversial party-political matter for the first seven years of the operation of the terrorism legislation. It was a Labour government that introduced the 1974 Act and guided it through its 1976 consolidation and annual renewal until 1979, supported throughout, as would be expected, by the Conservative Party. Indeed, the cross-party consensus was such that the annual renewal debates were poorly attended and rarely contentious, with Mr Rees beginning his 1977 speech for retention with the following revealing remarks: 'I must confess that I am tempted to move the motion formally so that we can all go home, because there does not appear to be anyone present who is interested in the business. I see that one hon. Member has now entered the Chamber, so I shall have to stay.'[78] The number of Labour MPs opposing the legislation grew very slowly and by 1980 there were no more than twenty-six votes against renewal.[79] By this time, however, Labour were out of power and were adopting a variety of new policies in a number of areas. This led to a change of approach in 1981, when the new Shadow Home Secretary Mr Hattersley called for a review of the legislation and then, after this proposal was rejected by the Government, Labour abstained on the vote for renewal.[80] The following year, this policy of abstention was repeated but Mr Hattersley voiced his 'deepest distaste for the Prevention of Terrorism Act and the thought of its being extended for another year'.[81] In 1983, the politicization of the issue was confirmed when Labour finally decided to vote against renewal.[82] Since then, the Party has consistently opposed the legislation, voting against every renewal of its provisions.[83] Its stance on the two full re-enactments of the legislation in 1984 and 1989 has been more complicated. On each occasion

[78] 927 HC Debs. 1472 (9 Mar. 1977).

[79] See the debate at 980 HC Debs. 405–38 (4 Mar. 1980).

[80] The debate is at 1 HC Debs. 336–74 (18 Mar. 1981). The debate on the renewal order is at ibid. 375–95.

[81] 20 HC Debs. 154 (15 Mar. 1982). The vote for renewal was 138 in favour, 53 against; see ibid. 169.

[82] 38 HC Debs. 639 (7 Mar. 1983).

[83] The debates are recorded at: 73 HC Debs. 1299–1320 (21 Feb. 1985); 92 HC Debs. 415–37 (19 Feb. 1986); 110 HC Debs. 263–86 (10 Feb. 1987); 127 HC Debs. 925–49 (16 Feb. 1988); 168 HC Debs. 819–44 (6 Mar. 1990); 187 HC Debs. 21–71 (4 Mar. 1991); 204 HC Debs. 689–736 (24 Feb. 1992).

Labour abstained on the second reading but in 1989 it voted against the bill at third reading in the Commons.[84]

It may be that Labour's position is a laudable one from the point of view of civil liberties, but it has not been without its political costs. There have been allegations of inconsistency and of lack of principle, with critics pointing out that the Party was happy to maintain the Act when it was in power and that it only began actively to oppose it after having been in opposition for four years. It is not illogical for Labour to answer this by pointing to the level of violence and arguing that whereas the legislation was required in the past the point has now come when it is no longer essential, though this still leaves the problem of the Party's support in the late 1970s, when violence was at a lower level in Britain than it has been during a number of years in the 1980s when they have opposed renewal. Of equal seriousness has been Labour's difficulty in persuading the general public that opposition to the prevention of terrorism legislation, however principled, is not the same as support for terrorism in general and the IRA in particular. They have not been helped in this by the support openly shown to Provisional Sinn Féin by some of their MPs, and since 1983 the Conservatives have sought to make political capital out of Labour's position. In that year, the then Home Secretary expressed his regret that the Opposition were 'departing from the position that they followed so wisely for nine years in the national interest'.[85] Two years later, the Minister of State at the Home Office Mr Waddington described Labour as having 'taken leave of its senses' and being 'demonstrably composed of people . . . who are simply not fitted to exercise the power necessary for the protection of the people of this country'.[86] As late as 1991, the Home Secretary Mr Baker felt able to equate Labour's position with support for terrorism when he called upon the Opposition not to 'give the IRA the chance to claim a victory in the House'.[87] When asked in Parliament 'to condemn the official opposition' for voting against the renewal order in 1987, the then Prime Minister Mrs Thatcher declared it to have been 'a bad day for law and order'.[88] Throughout her period in office, terrorism was at or near the top of Mrs Thatcher's personal agenda, and the Labour Party was under sustained pressure, just as were other bodies, such as the media, the Irish Government, and the European

---

[84] For the second reading in the Commons on the 1984 Act, see 47 HC Debs. 52–116 (24 Oct. 1983). For the second reading in the Commons on the 1989 Act, see 143 HC Debs. 207–73 (6 Dec. 1988). The debate on the third reading is at 146 HC Debs. 29–108 (30 Jan. 1989).

[85] 47 HC Debs. 60 (24 Oct. 1983) (Mr Brittan).

[86] 73 HC Debs. 1318 (21 Feb. 1985).

[87] 187 HC Debs. 29 (4 Mar. 1991).

[88] 110 HC Debs. 765 (17 Feb. 1987).

Court of Human Rights, whom she perceived from time to time as being 'soft on terrorism'.[89] It is difficult to gauge precisely how damaging Labour's position has been in electoral terms, or even if it has been damaging at all, but the Conservative Party has always judged it to be an issue worth concentrating on.

PRESSURE GROUPS

The emotions generated by terrorism, and the ways in which changes in the law can be introduced very speedily, make the position of civil liberties pressure groups extremely difficult. The most powerful influence of all is the general public, particularly in the aftermath of a serious atrocity when calls for action are hardest to resist. The security forces can always be expected to argue consistently both for the maintenance of the law and order status quo and for the addition of new powers. On the other side of the political spectrum, Amnesty International and Liberty (formerly the National Council for Civil Liberties) have conducted various inquiries, but without achieving much impact, one Amnesty investigation being condemned by Mrs Thatcher as a 'stunt without status'.[90] In Northern Ireland, the Committee for the Administration of Justice (CAJ) briefs Members of Parliament and others on its preferred changes to the law. During the passage of the Northern Ireland (Emergency Provisions) Act 1991, for example, it provided background briefing papers for opposition legislators and it has produced its own draft Bill of Rights for the Province. Its position, which is critical of much of the Province's emergency provisions, has not endeared it to the authorities. In contrast to the CAJ, the Standing Advisory Commission for Human Rights (SACHR) is a statutory body of long standing in Northern Ireland, having been set up under section 20 of the Northern Ireland Constitution Act 1973, with terms of reference that include a watching brief over civil liberties developments.[91] However, its authority has been diluted in recent years by the apparent failure of Government to consult it over many recent changes such as the modification of the right to silence and the media ban, and by the refusal of the authorities to countenance a statutory role for the Commission in the monitoring of emergency provisions law.

[89] For two of the most famous episodes, the 'Death on the Rock' and Fr. Ryan affairs, see Ewing and Gearty, *Freedom under Thatcher*, ch. 7.

[90] The phrase is that of the late Mr Ian Gow MP, but Mrs Thatcher agreed with his use of it: see 130 HC Debs. 1274 (31 Mar. 1988). The report was into the shooting by the security forces of three members of the IRA in Gibraltar.

[91] SACHR's first annual report is at HC 632 of 1974–5.

Inquiries into counter-terrorist law and policies have enjoyed most influence when they have been appointed by the government, and have been targeted to specific issues. There have been allegations that such inquiries are inevitably compromised and partisan and this has been particularly the case where the reports have been into specific incidents involving alleged abuses of civil liberties by the authorities, such as Lord Widgery's investigation into the 'Bloody Sunday' killings by the British army in 1972[92] and Sir Edmund Compton's consideration of the allegations of physical brutality that accompanied the introduction of internment in 1971.[93] It is true and unsurprising that reports which have been more involved in making recommendations than allocating blame have been most successful when their conclusions have fitted in with the official policy of the day. Into this category fall the reports by Lord Diplock into 'legal procedures to deal with terrorist activity in Northern Ireland'[94] and by Lord Gardiner[95] into 'measures to deal with terrorism', with the first of these providing the case for the abolition of the jury in terrorist-related trials, and the second taking a civil liberties line on internment for which the government of the day was not ungrateful. However, it is fair to say that other inquiries with unattractive findings have also had an impact, most notably the 1979 report by H. F. Bennett into police interrogation practices which led to a number of changes in procedure.[96]

Apart from these *ad hoc* inquiries, there have been three general reviews of the terrorism legislation, all of them conducted by members of the upper House who were not part of the Government when they conducted their reviews. Lord Shackleton reported in 1978,[97] Earl Jellicoe in 1983,[98] and Lord Colville in 1987.[99] In addition, there have been annual reviews of the workings of the Act since 1984, with Sir Cyril Philips reporting on the years 1984 and 1985 and Lord Colville conducting the same exercise every year since then. All these inquiries have been profoundly affected by terms of reference which have involved an assumption about the need for the legislation under review. And whilst certain concessions have been wrested from the Government on points of

[92] Lord Widgery, *Report of the Tribunal Appointed to Inquire into the Events on Sunday 30 January 1972 which Led to Loss of Life in Connection with the Procession in Londonderry on that day* (HL 101; HC 220 of 1972).

[93] Sir Edmund Compton, *Report of the Enquiry into Allegations against the Security Forces of Physical Brutality in Northern Ireland Arising out of the Events on 9 August 1971*, Cmnd. 4823 (1971).

[94] Diplock, *Report*, Cmnd. 5185.

[95] Gardiner, *Report*, Cmnd. 5847.

[96] His Honour Judge H. G. Bennett, *Report of the Committee of Inquiry into Police Interrogation Procedures in Northern Ireland*, Cmnd. 7497 (1979).

[97] Shackleton, *Review*, Cmnd. 7324.

[98] Jellicoe, *Review*, Cmnd. 8803.

[99] Colville, *Review*, Cm. 264.

detail, there has been a consistent refusal to countenance any changes of substance. Four examples may be given by way of illustration. As early as 1978, Lord Shackleton reported that the 'longer the exclusion order power continues, the stronger will be the case . . . that the Government should reconsider this issue'.[100] In his annual report for 1985, Sir Cyril Philips condemned such orders as 'objectionable in principle',[101] and the following year Lord Colville recommended an end to them altogether, an approach which he thought 'would be the correct one both in terms of civil rights in the United Kingdom and this country's reputation in that respect among the international community'.[102] Despite all this pressure, the Government has refused to budge, although it has recognized that exclusion orders 'very considerably infringe ordinary civil liberties'.[103] Secondly, and in a similar vein, the suggestion from both Lord Colville and Lord Shackleton that the offence of withholding information should 'be allowed to lapse'[104] has been rejected.[105] Thirdly, a similar fate has been meted out to Lord Colville's recommendation, contained in his separate 1990 report into emergency powers in the Province, that the internment power should be removed.[106] Finally, when Lord Colville suggested in his annual review for 1990 that a tribunal should be set up to oversee the seven-day detention power, the Home Secretary thanked him for the review but went on to tell the Commons that the 'Government's answer [was] no'.[107]

THE COURTS

There are few ways in which the exercise of counter-terrorist powers can be reviewed. The ordinary police complaints procedure is available and in Northern Ireland a system for monitoring dissatisfaction with the army has recently been set up with an independent assessor having been appointed by the Government to keep under review the operation of the Army Complaints procedure. Sir Louis Blom-Cooper has been appointed as independent commissioner to monitor conditions and procedures in those centres where suspects are held for questioning.[108] Codes of practice

---

[100] Shackleton, *Review*, Cmnd. 7324, para. 128.
[101] Sir Cyril Philips, *The Prevention of Terrorism (Temporary Provisions) Act 1984: Review of the Year 1985* (1986), para. 20.
[102] Colville, *Review*, Cm. 264, para. 11.6.1.
[103] Comment made by the Home Secretary in his letter of reply to Philips, op. cit. For further discussion on the Philips report see the debate at 92 HC Debs. 415–37 (19 Feb. 1986).
[104] Shackleton, *Review*, Cmnd. 7324, para. 133.
[105] See 964 HC Debs. 1505–1624 (21 Mar. 1979).
[106] Lord Colville, *Review of the Northern Ireland (Emergency Provisions) Acts 1978 and 1987*, Cm. 1115 (1990), para. 11.9.
[107] 187 HC Debs. 24 (4 Mar. 1991) (Mr Baker).
[108] Northern Ireland (Emergency Provisions) Act 1991, s. 60.

are also expected soon to be introduced to regulate the activity of the security forces in the Province.[109] As we have seen, an individual subject to a proposed exclusion order may make representations to advisers appointed by the Home Secretary. None of these mechanisms is entirely satisfactory and the overall picture is of a system in which self-regulation plays an important role. In practice this will often mean deference to the security services and specialist anti-terrorist police officers. In view of the potential for abuse in such an arrangement, we might have expected a prominent role for the courts as the guardians of liberty, and the Northern Irish judges have indeed tried to improve procedural safeguards for suspects and defendants. But as far as the British courts are concerned, the record has been remarkable for its total and consistent deference to executive authority. The English judges' only impact in this field has been as legitimators of governmental activity.

Since the start of the Northern Irish disorders in 1969, six major challenges to executive action have reached the House of Lords and all have been unsuccessful. In *McEldowney* v. *Forde*,[110] regulations under Northern Ireland's Special Powers Act were in issue. Made in 1967, they provided that all 'republican clubs' or 'any like organization howsoever described' were to be unlawful associations.[111] The legal basis for this was claimed to be a power vested in Northern Ireland's Minister for Home Affairs 'to take all such steps and issue all such orders as may be necessary for preserving the peace and maintaining order' and furthermore 'to make regulations . . . for making further provision for the preservation of the peace and maintenance of order'.[112] The appellant was prosecuted for membership of a republican club which was accepted to be at no time 'a threat to peace, law and order. There was nothing seditious in its pursuits or those of its members'.[113] Nevertheless a divided House of Lords upheld the validity of the regulation. The court 'should not substitute its judgment for that of the Minister'.[114] Lord Hodson thought that 'the use of the words "any like organization howsoever described" [lent] some support to the contention that the regulation [was] vague and for that reason invalid, but on consideration [he did] not accept the argument based on vagueness'.[115] For his part Lord Guest—writing in 1971—did not 'know what significance the word "republican" [had] in Northern Ireland. It may well be that it will bear a different construction in Northern Ireland from what it might bear in another context. These,

---

[109] Ibid., ss. 61–2.
[110] [1971] AC 632 (HL).
[111] Civil Authorities (Special Powers) Acts (Amending) (No. 1) Regulations (Northern Ireland) 1967.
[112] Civil Authorities (Special Powers) Act (Northern Ireland) 1922, s. 1.
[113] *McEldowney* v. *Forde*, op. cit., at p. 647 per Lord Guest.
[114] Ibid., at p. 645 per Lord Hodson.
[115] Ibid..

however, [were] matters for the Minister'.[116] The Minister concerned was Mr William Craig, who was a somewhat controversial figure in the Province, and especially noted for his sectarian views, having on one occasion justified discrimination against Catholic lawyers on the ground that 'they were educationally and socially inferior'.[117]

In the later case of *Murray* v. *Ministry of Defence*,[118] the army arrived at the plaintiff's home at 7 a.m., and it was obvious from this moment that her movements were restricted and that she was under arrest. However, she was not informed of this until thirty minutes later, apparently in breach of the statutory requirement that to be lawful the fact of the arrest had to be communicated to the arrestee.[119] In the plaintiff's action for false imprisonment, the House of Lords unanimously found for the army. The delay in speaking the words of arrest was justified by the threatening nature of the situation, and the army's practice was 'sensible, reasonable and designed to bring about the arrest with the minimum of danger and distress to all concerned'.[120] As for pedantically applying the words of the section, 'it would be to close one's eyes to the obvious not to appreciate the risk that the arrest may be forcibly resisted'.[121] In *R.* v. *Secretary of State for the Home Department, ex parte Brind*,[122] a likewise united House of Lords upheld the Sinn Féin media ban.[123] Lord Bridge thought it 'perhaps surprising . . . that the restriction imposed [was] of such limited scope'.[124] Lord Templeman also considered that the 'interference with freedom of expression [was] minimal and the reasons given by the Home Secretary [were] compelling'.[125] In an odd remark aimed presumably at the appellants and other civil libertarians who had opposed the ban, Lord Ackner said that 'the vehemence of the criticism of the Secretary of State's decision [was] perhaps a clear indication of the strength of the impact of the terrorist message when he is seen or heard expressing his views'.[126]

Their Lordships have also had three opportunities to involve themselves in the law surrounding deaths caused by the security forces, some of which fatalities, as we have seen, have given rise to allegations of a 'shoot to kill' policy in the Province.[127] In *Attorney-General for Northern*

---

[116] *McEldowney* v. *Forde*, at p. 649.
[117] J. J. Lee, *Ireland 1912–85: Politics and Society* (1989), 418.
[118] [1988] 1 WLR 692.
[119] See now the Northern Ireland (Emergency Provisions) Act 1991, s. 18(2).
[120] *Murray* v. *Ministry of Defence*, op. cit., at pp. 700–1 per Lord Griffiths.
[121] Ibid., at p. 700.
[122] [1991] 1 AC 696.
[123] On which, see above text to n. 43 ff.
[124] [1991] AC 696, at p. 749.
[125] Ibid., at p. 751.
[126] Ibid., at p. 759.
[127] See above text to n. 28 ff.

Ireland's Reference (No. 1 of 1975),[128] a soldier shot and killed an unarmed man who had run away when challenged. It was accepted in court that he was an 'entirely innocent person who was in no way involved in terrorist activity'. Nevertheless, the House accepted that at the time the soldier could reasonably have apprehended an 'imminent danger to himself and other members of the patrol if the deceased were allowed to get away and join armed fellow members of the Provisional IRA who might be lurking in the neighbourhood', since this could lead to 'the killing or wounding of members of the patrol in ambush', thereby 'encouraging the continuance of the armed insurrection'.[129] In *Farrell* v. *Secretary of State for Defence*,[130] the killing of three men by the army in an incident at a bank in Newry was held to be reasonable force in the prevention of crime, notwithstanding evidence that the army planning had been such that the only possible way of stopping the men was by shooting them. In *McKerr* v. *Armagh Coroner*,[131] the House was again unanimous, this time in upholding Northern Irish rules that allow members of the security forces to refuse to attend or give evidence at inquests into the deaths of civilians in which they are known to have been involved. On this occasion, the House overturned a unanimous decision of the Province's Court of Appeal which had held the rules to be *ultra vires*.

Apart from these six decisions, the GCHQ case is also important in this area. It will be seen in chapter 8 that, in the context of the banning of trade unions without consultation at a state communications centre, the House held that it was for the government of the day to determine what was required in the interests of national security and that this could not be questioned in any court, even where the conduct of the Government was otherwise unlawful.[132] Citing this decision, the Divisional Court has upheld the refusal by the Home Secretary to give reasons to an applicant for an exclusion order that was made against him.[133] The courts have also refused to question the factual basis for deportation orders said to be 'conducive to the public good for reasons of national security'.[134] Thus during the crisis provoked by the Iraqi invasion of Kuwait and by the subsequent dispatch of a large American-led multinational force to the Gulf, a large number of British residents of Palestinian and Iraqi origin were detained pending deportation on the ground of 'national security'.

---

[128] [1976] 3 WLR 235 (HL).

[129] Ibid., at p. 247 per Lord Diplock.

[130] [1980] 1 WLR 172, [1980] 1 All ER 166 (HL).

[131] [1990] 1 WLR 649 (HL). For a very good analysis of the decision, see the note by A. Jennings at (1990) 140 *NLJ* 633.

[132] *Council of Civil Service Unions* v. *Minister for the Civil Service* [1985] AC 374 (HL).

[133] *R.* v. *Secretary of State for the Home Dept., ex parte Stitt*, The Times, 3 Feb. 1987.

[134] *R.* v. *Secretary of State for the Home Dept., ex parte Hosenball* [1977] 1 WLR 766; *R.* v. *Secretary of State for the Home Dept., ex parte B*, Independent, 29 Jan. 1991.

They were not told the reasons for their intended removal but the Home Office asserted that their involvement with terrorist groups made their presence in the country a security risk. No further details were proferred. The courts refused to intervene on their behalf, with the Master of the Rolls repeating that national security was 'the exclusive responsibility of the executive' and denying that any reasons needed to be given to those affected.[135] A number of the detainees were released after successful political campaigns on their behalf and there were subsequent allegations that many mistakes had been made by the security services when the lists of detainees had been drawn up.

INTERNATIONAL COMPARISONS

Political violence and civil liberties is an area that has attracted its share of international attention. We have already seen that the European Convention on the Suppression of Terrorism has remodelled the political defence to extradition where certain European countries are concerned. The United Nations Universal Declaration of Human Rights adopted in 1948 has had a more general political influence and has been important, notwithstanding that its provisions are not part of British domestic law and cannot be relied upon in court. The UK government has found itself on more than one occasion forced publicly to defend its civil liberties record in Northern Ireland before UN committees in New York, and the organization's representatives visiting the Province have filed adverse reports on the state of the law.[136] The greatest international pressure, however, is from the need to comply with the European Convention on Human Rights and Fundamental Freedoms, not least because its enforcement mechanism is far more sophisticated than that to be found at the UN. The sixth protocol, concerning the abolition of the death penalty has yet to be ratified by the United Kingdom, a fact that remains irrelevant as long as Parliament continues to vote to keep the punishment off the statute book; in the appropriate case, however, the Convention's prohibition on 'inhuman or degrading treatment or punishment' may be used to prevent the return of subversives to face the death penalty in third countries.[137] It is in other areas of the United Kingdom's counter-terrorism laws that the Convention has had its most dramatic effect. In what is perhaps its most famous case, the European Court ruled in 1978 that the techniques of sensory deprivation employed by the authorities

---

[135] *R.* v. *Secretary of State for the Home Dept., ex parte Cheblak* [1991] 2 All ER 319.
[136] For a recent example, see *Independent*, 15 Feb. 1989.
[137] *Soering* v. *UK* (1989) 11 EHRR 439.

against detainees who had been interned in 1971 amounted to 'a practice of inhuman and degrading treatment' in breach of article 3.[138] The case had been brought by the Republic of Ireland rather than by any of the individuals concerned and was a major embarrassment to successive UK governments throughout the 1970s.

The practices impugned by the Court were no longer being engaged in when the decision was handed down, so the Government was faced with no dilemma over whether or not to comply. Such a situation did arise in 1989 when in *Brogan* v. *United Kingdom*,[139] the Court found by a vote of twelve to seven that the seven-day detention period permitted under the terrorism legislation was in breach of the requirement in Article 5(3) that all detainees should 'be brought promptly before a judge or other officer authorized by law to exercise the judicial power'. Of the four detention periods declared too long by the Court, the shortest was for four days and six hours. The Government responded by derogating from the Convention to the extent required to avoid this decision. To achieve this, reliance was placed on the formula in Article 15 which requires that there be a 'public emergency threatening the life of the nation', and the assertion to this effect was under challenge before the Strasbourg authorities at the time of writing. A succession of notices of derogation covering security measures in Northern Ireland had been filed since 1968,[140] but the Government had been confident that these were no longer necessary after certain amendments were made to the law in 1987.[141] The *Brogan* decision was therefore, to put it mildly, something of a shock. In a later case also involving the UK's terrorism laws, but not covered by the derogation issued after *Brogan*, the European Court refused to accept that a power of arrest was 'reasonable' as required by Article 5(1)(*c*) where the authorities declined to reveal the evidence on which they based the exercise of their power and where the only evidence to which they were prepared to point was the criminal record of the detainees.[142] The Convention's capacity to intrude into UK law and practice on behalf of the individual is not exhausted by these decisions, and the erosion of the right to silence is thought by some to be vulnerable. The challenge to the

---

[138] *Ireland* v. *UK* (1978) 2 EHRR 25. The Court disagreed with the Commission that the practices amounted to torture.

[139] (1989) 11 EHRR 117.

[140] See (1971) 14 YBEC 32; (1973) 16 YBEC 24; (1975) 18 YBEC 18; (1978) 21 YBEC 22.

[141] Northern Ireland (Emergency Provisions) (Amendment) Act 1987. See further on derogations from international instruments to which the UK is a party the various reports of Northern Ireland's Standing Advisory Commission on Human Rights. For the Republic of Ireland's successful reliance on Art. 15(3) in respect of a challenge to its use of the power of internment during the IRA's 'border campaign' 1956–62, see *Lawless* v. *Ireland* (1961) 1 EHRR 15.

[142] *Fox, Campbell and Hartley* v. *UK* (1990) 13 EHRR 157. See W. Finnie at (1991) 54 *MLR* 288.

media ban, however, fell at the first hurdle, being declared to be manifestly ill-founded by the European Commission. The time taken by each application severely diminishes its impact and this has led to calls for the Convention to be incorporated into UK law. It is not at all certain, however, that the British courts and in particular the House of Lords would interpret the Convention in the same spirit as the Strasbourg court in this most sensitive of areas.

The jurisdiction of the European Court embraces countries other than Britain with 'terrorist' problems. Italy has had the Red Brigades; Germany has had the Red Army Faction; and the Basque nationalist group ETA has fought for independence from Spain for over thirty years. It is tempting but unfair to draw unflattering comparisons between Britain's handling of the IRA in Northern Ireland and how these countries have dealt with their domestic subversives. Germany and Italy confront groups whose dedication to violence is for ideological reasons which are lost on their own communities and—with the collapse of communism—probably also now lost on the perpetrators as well. Members of such groups live isolated lives, constantly on the run from the authorities. They are able only occasionally to mount desultory attacks and do not even have the comfort of being able to point to anywhere in the world where their approach has reaped any rewards. In such a context, the main responsibility of the governments that oppose them has been to avoid overreacting (as the German authorities did in the early 1970s) and to develop policies which allow for the reintegration into society of convicted terrorists. Both Italy and Germany now have such schemes in place.[143] In Spain, the subversive threat has been more serious because of the avowedly nationalist aims of ETA, though the Madrid authorities have been greatly assisted by the fact that the Basque region is not divided along sectarian or cultural lines like Northern Ireland. This has enabled the authorities to grant autonomy to the region without provoking an anti-ETA terrorist backlash and this in turn has greatly diminished ETA's influence as local Basque moderates have been seen to take power.[144] The group remains an irritant but has lost the capacity fatally to wound.

Neither the European Social Charter nor the European Community are directly relevant to the law on political violence. The Schengen agreement signed in June 1985, and further developed through a fresh treaty in 1990, deals with such matters as extradition, political asylum, hot pursuit, and a common police information system. Restricted to the Benelux countries, Germany, and France, it is regarded by some as a blueprint for the ending of border controls. The TREVI group is also, strictly speaking, outside the terms of reference of the European Community

---

[143] Gearty, *Terror*, 132–9.
[144] Ibid. 106–11.

since it is concerned with policing matters in which the Community has no competence. Established in 1976, it is an intergovernmental forum with meetings every six months at ministerial and official level at which the member states of the EC can discuss 'co-operation at a practical, operational level against terrorism, drugs trafficking and other serious crime and public order problems'.[145] It is 'essentially an informal body whose objective is to advance co-operation on the ground',[146] and to this end it has four working groups, one of which is devoted to terrorism. A permanent secretariat for TREVI was agreed at its December 1990 meeting and there are further plans to develop its remit and powers. The growth of TREVI in this informal and *ad hoc* way has given rise to concern in the European and UK parliaments about the extent to which it is operating in secret and outside the normal requirements of democratic accountability. For its part, the British government has emphasized the answerabilty in Westminster of all those of its ministers who participate in TREVI.[147]

CONCLUSION

It is clear that the political violence arising out of the conflict in Northern Ireland has posed a challenge to the British authorities which greatly exceeds that which has confronted other European governments (with the exception of the Republic of Ireland which has been dealing with the same problem). Without in any way minimizing the gravity of the threat posed by the IRA and other subversive groups, it is necessary to draw attention to three strands in the development of our law which are a source of particular concern to civil libertarians. First, over the past twenty-five years, we have witnessed a marked normalization of emergency powers. What was thought of as uniquely draconian in 1974 is now rarely commented upon or controversial. The agony that used to accompany the deliberate erosion of freedom in the face of the terrorist threat is now largely gone. This is most evident in the way in which our terrorism legislation has become effectively permanent whilst remaining technically 'temporary' in its title if in nothing else. But the same process is also illustrated by the way in which recent changes like the media ban and the limitations on the right to silence in Northern Ireland have been introduced without any time limit on their operation. Even if the IRA

---

[145] Home Affairs Committee of the House of Commons, Practical Police Co-operation in the European Community (7th Report, HC 363, Session 1989–90), para. 48 quoting from a Home Office memorandum submitted to the Committee.
[146] Ibid.
[147] Ibid. The Government's reply is at Cm. 1367 (1991).

were to suspend their campaign of violence tomorrow, it is by no means obvious that the Government would—or would be expected to—bring the media ban to an end.

The second concern is that as the anti-terrorism laws have remained on the statute books, so also have they grown in size as more and more powers have been added to meet what appears at times to be a bewildering variety of terrorist threats. The Prevention of Terrorism Act now on the books is a far cry from its much slimmer 1974 predecessor, and as we have seen, the 1991 version of Northern Ireland's Emergency Provisions Act contains enormous extra powers for the police and army. Whilst each new provision may be difficult for opponents of terrorism to gainsay, the overall effect is of a corpus of security laws which has the potential dramatically to affect the relationship between the individual and the state. Accompanying this formal growth has been a similar expansion in the ranks of the counter-terrorists, and though the details are shrouded in secrecy, we may be quite certain that these factions are unlikely to be persuaded easily to give up the power that they have accumulated and exercised for two decades with the minimum of democratic control. A related anxiety concerns the way in which the threat of terrorism has infected other areas of law, justifying changes which are inimical to freedom and which could normally only have been made at a much higher political cost. Thus, jury vetting, the refusal to prosecute police officers for murder, the wide powers given by statute to MI5, and various other changes have all been justified on national-security/terrorism grounds. Blank cheques are always dangerous, especially when made out to a state's secret service.

Finally, in relation to Northern Ireland, the great number of changes in the rule of law have occurred because of the anxiety of government ministers both to limit the activities of the IRA and to secure as many convictions as possible through the ordinary courts. Both are laudable objectives, but if the tactics employed to achieve them undermine the perceived impartiality of the legal process, then the whole approach quickly becomes self-defeating. The rule of law has a vital role in fostering a culture of reconciliation in the Province by being demonstrably fair, even-handed, and responsive to the community in which it is applied. On the other hand, a policy of criminalization which separates 'law and order' and security issues from politics, and changes the law to suit its ends, hinders the ending of sectarian divisions by appearing to confirm the partisanship of British law. In such circumstances, measures aimed at tackling subversion in fact compound it.

6

Prisoners and the Law: Beyond Rights

GENEVRA RICHARDSON

INTRODUCTION: PUNISHMENT AND RIGHTS

This chapter is primarily concerned with the interests and legal status of the imprisoned. In the debate surrounding the present state of British prisons much has been made of the notion of prisoners' rights: the claim to a right can be a powerful political device. However, if the debate is to penetrate beyond rhetoric it is essential to examine more closely the nature of and justification for any claim to the existence of prisoners' rights.

A claim to the possession of a right can be made at a variety of levels. It can, as suggested, be merely rhetorical, or it can refer to specific moral or legal entitlements. An individual asserts a right when she makes a claim to performance, either action or forbearance, on the part of another.[1] The right is legal when the correlative duty is owed at law and moral when the duty is morally enforceable. This chapter will be based on the assumption that a core of basic or moral rights exists, that it is possessed by all human actors, that it should inform the relationship between the individual and the state, and that it should be reflected in law.[2] Commentators who accept the existence of such rights, however, differ significantly as to their contents.[3] This uncertainty is only exacerbated when the relationship between punishment and rights is considered. While most who accept the existence of basic rights are prepared also to accept their partial loss through punishment, the extent of that loss is rarely specified. Thus the problems associated with the identification and specification of basic rights are greatly compounded when the individual possessor of those rights is the subject of punishment by the state.

In England and Wales imprisonment is the most severe form of state punishment. Individuals who are convicted of serious breaches of the

---

[1] A. L. Corbin, 'Legal Analysis and Terminology' (1919) 29 *YLJ* 163, 167. The definition expresses the Hohfeldian notion of a claim-right: W. N. Hohfeld, *Fundamental Legal Conceptions* (1991).

[2] See A. Gerwith, 'The Epistemology of Human Rights' (1984) 1/2, *Social Philosophy and Policy* 1–24, and J. R. Pennock and J. W. Chapman, *Human Rights: Nomos 23* (1981).

[3] See e.g. R. Nozick, *Anarchy, State and Utopia* (1974); R, Dworkin, *Taking Rights Seriously* (1977); and T. D. Campbell, *The Left and Rights: A Conceptual Analysis of the Idea of Socialist Rights* (1983).

criminal law are removed from the rest of society, deprived of their liberty, and subjected to numerous other restraints frequently amounting to an alteration in their legal, and possibly basic, rights and obligations. If it is accepted that punishment may properly entail a loss of basic rights, then the precise extent of that loss must be dependent on the justification underlying the imposition of the punishment. Thus, in principle, when the punishment is imprisonment, any alteration in a prisoner's rights, whether legal, basic, or both, can only be justified by reference to the recognized aims of imprisonment. Unfortunately the force of this statement is significantly reduced by the absence of any clear agreement as to the proper purposes of imprisonment. Indeed it seems likely that no single purpose will ever be agreed. In the late nineteenth century retribution and deterrence gave way to deterrence, rehabilitation, and social protection as the main aims of imprisonment.[4] By the 1970s sufficient scepticism had emerged concerning the feasibility and acceptability of both rehabilitation and social protection that they were rejected by many in favour of a new retributivism or just deserts model.[5] Now in the 1990s the early rigour of the new retributivism has been somewhat softened by a more pragmatic approach.[6]

This confusion surrounding the proper purpose of imprisonment is perhaps unremarkable. Just as punishment cannot be regarded as singular,[7] so the purposes of imprisonment are likely to vary according to the character of the offender, the nature of the offence, and the attitudes of society. Nevertheless, despite this inevitable plurality, it is possible to argue that no 'acceptable' prison regime can justify the denial of all basic rights in the imprisoned: the prisoner must always retain some residue, whether the purpose of imprisonment is expressed in terms of retribution, rehabilitation, social protection, or deterrence.[8]

Rather than favour a particular justification for imprisonment, this chapter will adopt a minimalist approach. It will argue that, whatever the underlying purpose of imprisonment, that purpose can be met by the mere segregation of the prisoner from society. In relation to social protection such a position is easy to justify. With regard to retribution, although it is clearly possible to be more punitive, there is now widespread acceptance of the claim that offenders are sent to prison as punishment,

---

[4] For an account of the evolution of modern penal policy see L. Radzinowicz and R. Hood, *The Emergence of Penal Policy in Victorian and Edwardian England* (1990).

[5] A. Von Hirsch, *Doing Justice: The Choice of Punishments* (1976).

[6] See e.g. the discussion of the role of the prison service in the Woolf Inquiry Report, *Prison Disturbances April 1990*, Cmnd. 1456 (1991), paras. 10.04–10.44.

[7] D. Garland and P. Young (eds.), *The Power to Punish* (1983), and N. Lacey, *State Punishment, Political Principles and Community Values* (1988).

[8] This question is pursued further in G. Richardson, 'The Case for Prisoners' Rights', in M. Maguire, J. Vagg, and R. Morgan, (eds.), *Accountability and Prisons* (1985).

not for punishment.[9] The loss of liberty is therefore enough. This capping of the claims of retribution carries inevitable consequences for deterrence, since deterrence is now regarded as properly limited by the demands of retribution. No more can therefore be demanded in the name of deterrence than can be justified by the claims of retribution.[10] Potentially, treatment could prove more difficult to accommodate within this minimalist stance but, owing to a now general recognition of the dangers inherent in the idea of coerced treatment, treatment has ceased to be regarded as a primary justification for the use of imprisonment.

Imprisonment, therefore, will be regarded here as justifying only the removal of those rights necessarily affected by the need to segregate. All other rights, whether legal or basic, should be retained. This approach accords well with the views of those who advocate adherence to the principles of minimal interference and normalization within prison regimes.[11] Arguably, however, the claim to rights in prisoners should not end with the recognition of a significant residue. Society, for whatever reason, has deprived prisoners of their liberty and has thus increased their dependence. Society thus has a duty to protect prisoners and to provide them with certain facilities be they recreational, educational, medical, or merely the 'physical necessities of life'. Whatever the merits of the debate concerning society's duty to provide such facilities to non-prisoners, it can be argued that prisoners, by virtue of their exceptional and involuntary dependence, are in a special position and are therefore entitled to special rights against the authorities. To some degree this extension of the notion of prisoners' rights beyond the residue into the realm of special rights blurs the distinction between 'civil and political', on the one hand, and 'social and economic' rights on the other. However, there is a strong argument that any concept of prisoners' rights which is restricted to a residue of the rights generally possessed by non-prisoners against the state, most of which are best classified as 'civil and political', and ignores the prisoners' claims to additional and often 'social and economic' rights, would be seriously deficient.

The discussion so far has centred on the notion of 'basic' rather than legal rights. In relation to prisoners there is a particularly strong case to argue that such basic rights should be reflected in the law. Legal recognition is essential to enable prisoners realistically to assert their rights. There is also an argument that the retention by prisoners of the legal rights ordinarily possessed by non-prisoners is a basic or moral right, encapsulated in notions of equal treatment before the law, and should be

---

[9] See, most recently, the Woolf Inquiry Report.

[10] H. L. A. Hart, *Punishment and Responsibility: Essays in the Philosophy of Law* (1968).

[11] R. D. King and R. Morgan, *The Furture of the Prison Service* (1980).

recognized as such by the law. Recognition by the law of the existence of a right does not, of course, imply that the formal legal structure should be the primary mechanism for the enforcement of that right, and in the discussion that follows thought will be given to alternative enforcement mechanisms and their proper relationship with the courts.

THE DEVELOPMENT OF PRISONERS' RIGHTS IN ENGLAND AND WALES

The statutory framework

Imprisonment in England and Wales is primarily governed by the Prison Act 1952. The Act, however, provides only the barest framework and certainly contains no comprehensive statement of the rights and duties of either the prison authorities or the inmates. It vests the Home Secretary with constitutional responsibility for the administration of the prison system in England and Wales and authorizes him to issue rules for 'the regulation and management of prisons . . . and for the classification, treatment, employment, discipline, and control of persons required to be detained therein'.[12] This he has done in the form of the Prison Rules 1964 and subsequent amendments.[13]

The Prison Rules are considerably more detailed than the Act, specifying, for example, the duties of boards of visitors and prison chaplains, listing the various offences against discipline, and laying down basic rules concerning letters and visits. They do not, however, as will be explained below, create rights in prisoners directly enforceable at private law. Indeed, although some are relatively specific, others vest considerable discretion in the governor or the Secretary of State. The Rules have therefore been supplemented by Standing Orders and Circular Instructions. The former are designed as formal statements of a prisoner's privileges and obligations, while Circular Instructions provide internal administrative guidance on specific issues and procedures. The Prison Rules in the form of statutory instruments are public documents and have obtained parliamentary approval. Both Standing Orders and Circular Instructions are issued internally by the prison service and, until recently, have been regarded as classified. Following comments by the European Commission of Human Rights in *Silver* v. *The United Kingdom*,[14] however, the practice with regard to Standing Orders has been modified. Many are now published and should, in theory, be readily available to prisoners. Circular

[12] Prison Act 1952, ss. 1 and 47.
[13] SI 1964 No. 388.
[14] (1980) 3 EHRR 475.

Instructions, by contrast, remain in general inaccessible, but they are not classified and copies are sent on request to interested parties.

The formal rules governing the management of prisons, whatever their constitutional status however, do not, either singly or in combination, purport to provide a code of directly legally enforceable rights in prisoners. Indeed it is not profitable at present to search for such a code. In conformity with its traditional residual approach to such matters, the law in England and Wales chooses merely to regard the prisoner as retaining the rights possessed by non-prisoners, provided those rights are not removed by imprisonment. For example, while a prisoner is now allowed to marry in prison, there is no right to conjugal visits.[15]

In *Raymond* v. *Honey*, in a speech now widely quoted to illustrate the legal status of prisoners, Lord Wilberforce stated that a prisoner 'retains all civil rights which are not taken away expressly or by necessary implication'.[16] At first sight this statement seems encouraging to those seeking to enhance the rights of prisoners, particularly since many of the express legal disabilities which used to attach to the prisoner have been removed. The Forfeiture Act 1870 imposed disabilities on convicted felons in both private and public law. In 1948 section 70 of the Criminal Justice Act removed the private-law disabilities and, with the abolition of the distinction between felonies and misdemeanours by the Criminal Law Act in 1967, most public-law disabilities were removed. Some remain. Perhaps most notably, section 4 of the Representation of the People Act effectively disenfranchises convicted prisoners.

Nevertheless, even on its face, Lord Wilberforce's statement reveals some fundamental difficulties. In the first place, there is the incorrigible problem of identifying exactly what 'civil rights' are possessed by non-prisoners, which must be faced before it is possible even to start speculating on the impact of imprisonment on those rights. Secondly, the notion of 'necessary implication' allows extensive scope for judicial discretion in a sensitive area of public policy. Further, when set against the model for prisoners' rights discussed above, it is clear that Lord Wilberforce's statement is concerned exclusively with the retention of existing rights: no reference is made to the possibility of additional rights attaching to prisoners by virtue of their exceptional dependence.

With Lord Wilberforce's statement and the statutory background in mind, it is necessary to consider briefly the decided cases in order both to obtain some idea of how the assertion of rights by prisoners has been

---

[15] It is significant to note that it was only after the intervention of the European Commission of Human Rights that this bare right to marry while in prison was properly recognized, *Hamer* v. *UK* (1982) 4 EHRR 139. Originally marriages could not take place in prison and prisoners could be denied leave to enable them to marry elsewhere.

[16] [1983] 1 AC 1 (HL), at p. 10.

dealt with in practice by the courts, and to assess the impact of court judgments on the development of prisoners' rights. It is convenient to start with the position in private law.

Private law

Duty of care

It is now well established that the prison authorities owe a duty of care to the prisoners in their custody and that prisoners are therefore entitled to compensation for injury incurred through any breach of that duty.[17] A prisoner like any other citizen has a right not to be injured through the negligence of another. In some respects the common law, in recognition of the exceptional vulnerability and dependence of prisoners, has gone further and has refined this general right in the prisoner's favour. It is now established that the authority's duty of care extends to providing one prisoner with reasonable protection against the assaults of other prisoners. Thus prisoners have a right to reasonable protection from the authorities against the intentional violence of other prisoners. Such rights of protection against intentional violence on the part of third parties are rare in the common law. Reported cases suggest, however, that the enforcement of this right by prisoners is not easy: courts appear reluctant to find a breach of duty on the part of the authorities.[18]

By contrast to the case of inter-prisoner violence, the fact of imprisonment has been used, presumably 'by necessary implication', to reduce the duty imposed on the prison authorities with regard to the provision of medical care. The prisoner's right to medical care, it seems, is inferior to that of non-prisoners. In *Knight*, a case involving the suicide of a prisoner awaiting transfer to a psychiatric unit, the High Court was unable to accept that a prison hospital was to be judged according to the standards appropriate to an ordinary psychiatric unit, despite the fact that the prison regularly contained a number of potential suicides.[19]

Outside prison, those injured at work receive special protection under the law. The law recognizes that the relationship between employer and employee gives rise to special rights and duties. Within prison convicted prisoners can be required to work in prison workshops. A prisoner who is injured in the course of such work is not, however, offered the same protection as is available to other victims of 'industrial injury'. The courts

---

[17] *Ellis* v. *Home Office* [1953] 2 QB 135 (CA) and *Christofi* v. *Home Office*, *The Times*, 31 July 1975.

[18] *Ellis* (see n. 17) and *Egerton* v. *Home Office* (1978) *Crim. LR* 494. But see *H* v. *Home Office*, *The Independent*, 6 May 1992

[19] *Knight* v. *Home Office* [1990] 3 All ER 237.

have denied that the Factories Act applies to prison workshops[20] and have further denied the existence of the common law relationship of master and servant.[21] The prisoner injured in the prison workshops has, therefore, to rely on the general principles of negligence. On the face of it, the law deprives the prisoner of rights which apply to non-prisoners and does so with little justification, since there is no apparent conflict between the purposes of imprisonment and the objectives of either the Factories Act or the common law rules. The willingness of the Home Office to provide *ex gratia* payments fails to alter that fact.

Prison conditions

The appalling physical conditions which prevail in a large number of British prisons are now well documented. Thousands of prisoners, both convicted and unconvicted, are locked in overcrowded cells with no integral sanitation for up to twenty-three hours a day. According to the notion of rights suggested above, a prisoner, who by definition is entirely dependent on the prison authorities for the conditions in which he or she lives, should be entitled to special rights guaranteeing an agreed minimum. Prisoners who have attempted to assert such rights in private law have, however, met with little success. The Prison Act does not define imprisonment nor does it specify the conditions which should prevail within a prison. Section 14 merely specifies that no cell may be used as sleeping accommodation unless it has been certified as adequate, and requires the certificate to specify the maximum number of prisoners who should be housed in each cell. Rule 23 of the Prison Rules, however, allows the maximum number to be exceeded with leave of the Secretary of State. The legal force of section 14 is thus effectively neutralized. In October 1991 the total prison population in England and Wales exceeded the certified number by 2,050 and by March 1992 9,160 prisoners were sharing two to a cell designed for one.[22] Although the prison population has since fallen it is significant to note that overcrowding still tends to be at its worst in the local prisons where large numbers of remand prisoners are held, a population which according to the International Covenant should be accorded special status.[23] Remand prisoners by definition have been convicted of no offence and yet they often suffer the worst conditions: 'we found consistently that the physical conditions for remand

---

[20] *Pullen* v. *Prison Commissioners* [1957] 1 WLR 1186.

[21] *Davis* v. *Prison Commissioners*, *The Times*, 21 Nov. 1963. The Home Office does have a duty to train prisoners adequately in the use of machinery, a duty that derives from the 'neighbour principle', *Ferguson* v. *Home Office*, *The Times*, 8 Oct. 1977.

[22] *Report on the Work of the Prison Service April 1991–March 1992*, Cm. 2087 (1992).

[23] *Ibid.*, p. 6. And see The International Covenant on Civil and Political Rights, Art. 10(2)(*a*).

prisoners was particularly poor'.[24] In 1991 the Woolf Report stressed the need to improve conditions for the remand population, and in its subsequent White Paper the government has undertaken to reflect the special status of remand prisoners.[25]

The failure of the primary legislation to provide any enforceable rights to minimum physical conditions is compounded by the courts' attitude to the Prison Rules, which provide the most comprehensive legislative statement of the obligations of the prison authorities. While many of the Rules, such as rule 23(2) itself, provide the authorities with considerable discretion, some are phrased in mandatory and relatively unambiguous terms: rule 24 states 'each prisoner shall be provided with a separate bed and with separate bedding adequate for warmth and health'. None the less the courts have denied that the Prison Rules vest prisoners with any special rights. An action for breach of statutory duty will not lie with respect to injury flowing from non-compliance with the Rules.[26] In any context the court's decision whether or not to allow an action for breach of statutory duty will be a matter of statutory interpretation and is likely to involve an assessment of the relevant policy implications. In relation to the Prison Rules those policy implications have been assessed overwhelmingly in favour of the authorities: 'it would be fatal to all discipline in prisons if governors and wardens had to perform their duties always with the fear of action before their eyes if they in any way deviated from the rules'.[27] Thus even if the conditions of a prisoner's confinement are in clear breach of the Prison Rules no action for breach of statutory duty can arise.

Thwarted by the courts' attitude to the Prison Rules, prisoners have sought to use false imprisonment to challenge the lawfulness of the conditions within which they are being held. However, in *Williams* v. *The Home Office (No. 2)* Tudor Evans J. held that imprisonment is justified in law by the order of the court and by section 12 of the Prison Act, and that it cannot become unlawful merely by reason of the conditions in which it is served.[28] Until recently it might have been thought that subsequent case-law concerning both prisoners and detained mental patients had slightly softened the impact of this ruling. It seemed that conditions could deteriorate so far as to render the detention unlawful either because the conditions constituted 'cruele and unusuale' punishment under the terms

---

[24] Report of HM Chief Inspector of Prisons 1988 (HC Paper 491, Session 1988/89), para. 2.06.

[25] *Custody, Care and Justice*, Cm. 1647 (1991) p. 70.

[26] *Arbon* v. *Anderson* [1943] KB 252. This position has been confirmed by the House of Lords in *R.* v. *Deputy Governor of Parkhurst Prison, ex parte Hague* [1992] 1 AC 58.

[27] Lord Goddard, *Arbon* v. *Anderson*, supra, p. 255. For further discussion see G. Richardson, *Law, Process and Custody* (1993) chs 5 and 6.

[28] [1981] 1 All ER 1211.

of the Bill of Rights 1688, or because they had become 'intolerable'.[29] Conditions which sank too low, it seemed, could be said to negate the statutory defence supplied by section 12 and thus to constitute false imprisonment.[30] However, the threshold of lawfulness was never particularly demanding. Ackner LJ suggested that a prisoner would cease to be lawfully detained if conditions were such as to be seriously prejudicial to health 'e.g. because it [the cell] became and remained seriously flooded, or contained a fractured gas pipe allowing gas to escape into the cell'.[31] However, the House of Lords in *Hague* has now put a stop to all such speculation. An action in false imprisonment is not available to a prisoner to challenge the conditions of his or her imprisonment.[32]

It must be concluded from the above that the private law has generally been reluctant to uphold rights in prisoners. Residual rights have typically been restricted by a generous interpretation of the 'necessary implication[s]' of imprisonment, and additional special rights have been denied. Thus the assertion of rights in private law has provided little direct pressure for improvement in either the treatment of prisoners or the conditions of imprisonment. The only 'pressure' may come from the shock engendered by the law's formal acceptance, in cases such as *Knight*, of lower standards where prisoners are involved. Indeed there is now a growing body of opinion in favour of radical reform of the provision of medical care in prison. Dating from the early anxieties expressed concerning the 'liquid cosh', the overuse of drugs to tranquillize troublesome prisoners, and more recently the increase in suicide and self-harm in prisoners, the pressure for the reform of the prison medical service has been growing. The prison medical service, it is argued, should no longer be part of prison management but should become part of the National Health Service; prisoners are entitled to the same level of care as non-prisoners.[33] In response a series of official bodies, including the House of Commons Select Committee on the Social Services, Her Majesty's Chief Inspector of Prisons and a departmental efficiency scrutiny, have looked at the issue.[34] In the event the integration of the prison medical service

---

[29] *R. v. Deputy Governor of Parkhurst Prison, ex parte Hague* [1991] 3 WLR 1210, [1990] 3 All ER 687 (CA).

[30] Ibid. 707.

[31] *Middleweek* v. *Chief Constable of Merseyside Police* [1990] 3 All ER 662, 668.

[32] See n. 26. In appropriate cases an action in negligence or assault would still lie, and an action in false imprisonment could lie against an officer acting in bad faith.

[33] See e.g. the section on 'Medicine and Control' in M. R. Fitzgerald and J. Sim, *British Prisons* (1979). For a history of medicine in prison see J. Sim, *Medical Power in Prisons* (1990). In Apr. 1991 the Health Care of Prisoners Bill was introduced as a private member's bill, sponsored by Inquest and the National Association of Probation Officers.

[34] See e.g. the Third Report from the Social Services Committee, Session 1985–6, *Prison Medical Service* (1986), HM Chief Inspector of Prisons' Report, *Prison Suicides*, Cm. 1383 (1990), and the *Report of an Efficiency Scrutiny of the Prison Medical Service* (1990).

within the NHS has not been approved, but extensive contracting out of prison medical care is now being encouraged and health care standards are being devised. Whatever solution is found the present unjustifiable discrimination against prisoners cannot be allowed to continue.[35]

Public law

The dearth of support offered to prisoners by the private law has led to an increased reliance on public law. In *O'Reilly v. Mackman* Lord Diplock held that, since a prisoner's private rights were unaffected by the imposition of a disciplinary award in breach of natural justice, his only remedy lay in public law.[36] In certain areas of prison life the courts have been willing to recognize and uphold public law rights in prisoners and on occasions such intervention, perhaps assisted by the European Court and/or Commission of Human Rights, has led to changes in government policy. In this respect penal and mental-health pressure groups, notably Liberty (the National Council for Civil Liberties as was), PROP, Justice, and MIND, have played an important role in identifying suitable cases and supporting them through the courts.

Prison discipline

The modern era of domestic judicial intervention began in the context of disciplinary hearings. The Prison Act empowers the Secretary of State to make rules for discipline within prisons.[37] The Prison Rules in their original 1964 form provided a code of disciplinary offences and corresponding penalties. Charges were heard by either the governor or the board of visitors, a lay body appointed for each prison by the Secretary of State and, despite the fact that the penalties involved could include loss of remission amounting to the equivalent of several years additional detention, there were few procedural safeguards. In 1979 the Court of Appeal held that boards of visitors acting in their disciplinary capacity were susceptible to certiorari, even though loss of remission did not technically involve interference with a prisoner's rights, and were bound by the rules of natural justice.[38] A series of applications for judicial review then followed in which prisoners successfully alleged breaches of the board of visitors' duty to act fairly.[39] In 1984 the Divisional Court,

---

[35] The attitude of the European Torture Committee to the psychiatric wing in Brixton is described below.

[36] [1983] 2 AC 237 (HL).

[37] s. 47.

[38] *R. v. Board of Visitors of Hull Prison, ex parte St. Germain* [1979] QB 425 (CA).

[39] *R. v. Board of Visitors of Blundeston Prison, ex parte Fox-Taylor* [1982] 1 All ER 646; *R. v. Board of Visitors of Gartree Prison, ex parte Mealy, The Times*, 14 Nov. 1981; and *R. v. Board of Visitors of Highpoint Prison, ex parte McConkey, The Times*, 23 Sept. 1982.

distinguishing an earlier Court of Appeal decision, held that boards of visitors had a discretion to allow a prisoner to be legally represented and laid down the factors which might influence the exercise of that discretion in favour of representation.[40] Finally in 1988 the House of Lords, in *Leech*, reversing an earlier Court of Appeal decision, held that the governor was susceptible to judicial review when hearing disciplinary charges and was subject to the duty to act fairly.[41] Interestingly, the House of Lords' decision in *Leech* was clearly influenced by the decision of the Northern Irish Court of Appeal in *R. v. Governor of the Maze Prison, ex parte McKiernan*, where the Irish court had refused to follow the earlier restrictive ruling of the English Court of Appeal.[42]

Thus through the assertion of the public law right to a fair hearing prisoners managed to open up a previously hidden system of justice to court scrutiny and encouraged adherence to the basic requirements of natural justice. It was no mean achievement but, arguably, the largely procedural improvements that followed did not go far enough. Prisoners were still subject to an internal code of discipline which rendered them vulnerable to additional loss of liberty, but which was administered by a lay body, appointed by the Secretary of State and bound by procedural requirements which were inferior to those imposed on ordinary criminal courts. In addition to the activities of the domestic courts there were other pressures, both European and domestic, for change in relation to prison discipline. At one stage it seemed likely that the European Convention would play a significant role when the Commission suggested that the board of visitors was insufficiently independent to constitute a 'court' under Article 6.[43] In the event the Court failed to endorse the Commission's view and pressure from Strasbourg accordingly diminished.[44]

In 1985 non-judicial domestic pressure for change emerged with the Report of the Prior Committee into Prison Discipline.[45] Prior urged the removal of the board of visitors' disciplinary role and their replacement by a new Prison Disciplinary Tribunal. After an initially positive response, however, the Government rejected that proposal. Apparently the resource implications were too great and the anticipated pressure from Europe had failed to materialize. In the event the reforms which finally emerged in 1989 amounted to little more than a revision and reclassification of the disciplinary charges and a reduction in the penalties available to the

---

[40] *R. v. Secretary of State for the Home Dept., ex parte Tarrant*, [1985] QB 251, distinguishing *Fraser* v. *Mudge* [1975] 1 WLR 1132.
[41] *Leech* v. *Deputy Governor, Parkhurst Prison* [1988] AC 533, overruling *R. v. Deputy Governor, Camphill Prison, ex parte King* [1985] QB 735 (CA).
[42] (1985) 6 *NIJB* 6. See Lord Bridge in *Leech* (see n. 41) at p. 561.
[43] *Campbell and Fell* v. *UK* (1982) 5 EHRR 207.
[44] (1984) 7 EHRR 165.
[45] *Report of the Committee on the Prison Disciplinary System*, Cmnd. 9641 (1985).

board of visitors.[46] Also, as an experiment, magistrates' clerks were introduced into certain establishments to clerk board adjudications.

Despite all the consultations, discussions and pressures for change, therefore, the government decided to retain the basic structure of the prison disciplinary system. No independent, legally qualified panel was established, and the system remained internal, with a jurisdiction overlapping that of the criminal courts. Without doubt the procedures before both boards of visitors and governors did improve enormously as a result of the courts' willingness to ensure adherence to the duty to act fairly, but it is perhaps instructive to speculate as to what might have occurred had the courts failed to intervene. If faced with the shockingly oppressive system that prevailed before *St Germain* would the European Court of Human Rights have reacted more strongly? Would the Government have been forced to accept Prior despite the cost? Did the recognition of the prisoner's right to a fair hearing serve merely to prop up an essentially unacceptable system? The answer to all three questions is probably 'no', but they need to be asked.

In 1992 structural reform was finally achieved when the Woolf Report succeeded in inspiring real change. The Report recommended the abolition of the two-tier disciplinary system. It proposed that all disciplinary charges be heard by the governor, with a maximum of 28 days loss of remission, and that all other offences be dealt with by the ordinary criminal courts. The Report's proposals mirrored the situation in Scotland and had much to recommend them in terms of both principle and practicality.[47] They were simple and involved the creation of no extra adjudicatory mechanism. They also respected the properly limited objectives of an internal disciplinary system. While a disciplinary system is necessary in order to provide the Prison Service with the means to effect the safe segregation of prisoners, its scope should not extend beyond the demands of that limited objective. The Woolf proposals might have overstepped those limits to some degree by reluctantly accepting the continued use of loss of remission as a disciplinary penalty, but their whole thrust was towards a limited system and, being carefully designed to minimise the financial and administrative implications of change, they finally succeeded in effecting reform where many others had failed. In April 1992 the boards of visitors lost their disciplinary functions. All disciplinary charges are now heard by the governors, who may award a maximum of 28 additional days, and all criminal offences deemed worthy

---

[46] These reforms were achieved by amendment of the Prison Rules and relevant Standing Orders and by the publication of a new Manual on the Conduct of Adjudications (1989).

[47] See Woolf Inquiry Report (n. 6) section 14 and for further discussion see Richardson (n. 27) ch. 7.

of higher penalties are referred to the police for prosecution through the normal channels.[48]

Prisoners' correspondence and access to lawyers

For a person confined to prison the ability to maintain contact with the outside world is of considerable importance. Further, the ability to obtain independent legal advice is essential to the full protection of the prisoner's rights and interests. Perhaps in reflection of the importance of the issues involved, the Prison Rules are relatively detailed on the subject of correspondence, but nevertheless require supplementing by Standing Order 5, running to 29 pages.

The evolution of this present policy on correspondence has been significantly influenced by court action, both domestic and European. Arising out of an incident in 1969, Sydney Golder, a sentenced prisoner, wished to consult a solicitor about possible proceedings in libel against a prison officer. Under the regulations then in force a prisoner wishing to seek legal advice had first, unless he or she was already a party to legal proceedings, to petition the Secretary of State. Further, if the proposed action concerned treatment by the prison authorities, the prisoner was required first to make a formal complaint through the internal channels: the prior ventilation rule. When Mr Golder's request to contact his solicitor was refused he petitioned the European Commission of Human Rights and his complaint was ultimately upheld by the Court.[49] The Court held that the Home Secretary's action was in breach of both Article 6, guaranteeing a right of access to court, and Article 8, guaranteeing a right of respect for correspondence. In grudging recognition of the ruling the Government amended the relevant Prison Rules to enable prisoners to contact lawyers concerning actions to which they might become a party, but such access was made 'subject to any direction of the Secretary of State'. Significantly, the relevant directions, in the form of a Circular Instruction, maintained a prior ventilation rule.

In 1980 the whole question of prisoners' correspondence came before Europe again in *Silver* v. *United Kingdom*.[50] In addition to access to legal advice this case involved the whole question of the censorship and regulation of prisoners' correspondence. In *Silver* the European Commission doubted the propriety of the prior ventilation rule and, following their report, the British government amended the relevant Standing Order, replacing prior with simultaneous ventilation. At this point the

---

[48] Prison (Amendment) Rule 1992, SI 1992/514. Under Criminal Justice Act 1991 s. 42, additional days have replaced loss of remission.

[49] *Golder* v. *UK* (1975) 1 EHRR 524.

[50] (1980) 3 EHRR 475 (European Commission), (1983) 5 EHRR 347 (European Court).

domestic courts finally intervened. Emboldened by the case of *Raymond* v. *Honey*,[51] in which the House of Lords had upheld a prisoner's right of access to court, the Divisional Court held that the simultaneous ventilation rule constituted an impediment to the prisoner's right of access to the courts, and was thus *ultra vires* the Prison Rules.[52] The Standing Orders have been amended once more and the simultaneous ventilation rule has finally disappeared. Rule 37A and Standing Order 5 combine now to provide that letters to and from lawyers concerning current legal proceedings will not be read or stopped, unless the governor has reason to doubt that the letter is what it purports to be.[53]

In relation to censorship in general *Silver* has had a significant impact. Not only did it lead to the publication of Standing Orders, as described above, but it forced a revision of Order 5, leading to a significant reduction in the restrictions on all prisoners' correspondence.

The recent liberalization of the practice regarding a prisoner's access to the outside world has not, however, been a response to judicial activity alone. In 1984 the Home Secretary set up the Control Review Committee (CRC) 'to review the maintenance of control in the prison system'. In its report the CRC expressed concern that the system sent the wrong signals to long-term prisoners.[54] Most of the identifiable privileges were concentrated in the high-security training prisons where overcrowding was less intense and regimes more liberal, thus individual prisoners had no incentive to seek transfer to a less secure prison. In the years following the CRC report the prison service has responded to this concern by withdrawing routine reading of correspondence first in open prisons and then, in March 1988, in category C prisons. A similar reduction in censorship was initiated in a sample of more secure establishments in February 1990. While encouraged by these moves, the Woolf Inquiry advocated the end of routine censorship and the government has responded positively.[55] Also following the CRC report, cardphones were installed in all open prisons and low security establishments, and are now being extended to all category B prisons.

Thus over the last 20 years the prisoners' rights of correspondence with the outside world have been greatly extended. Certainly the initial pressure for reform came in the form of a clear recognition of 'rights' by the ECHR, which was eventually echoed by the domestic courts. More

---

[51] See n. 16.

[52] *R. v. Secretary of State for the Home Dep., ex parte Anderson* [1984] QB 778.

[53] The Court of Appeal has now decided that Prison Rule 33 is *ultra vires* the Prison Act in so far as it purports to allow a prisoner's letters to his or her solicitor to be stopped or read to ensure that they comply with the general restrictions on correspondence: *R v. Secretary of State, Ex p. Leech (No. 2)*, *The Times*, 20 May 1993.

[54] *Managing the Long-Term Prison System* (1984).

[55] *Custody, Care and Justice* (n. 25) p. 83.

recently, however, the liberalizing steps taken by the prison service seem to have been inspired by the desire to rationalize the system and to facilitate the smooth management of inmates. The Woolf Report, while adopting this management focus by avoiding the rhetoric of rights and concentrating on the value to management of the maintenance of justice within prisons, none the less recognized the fundamental significance of the issues involved.[56]

Segregation, transfer, and release

As was described above, the prisoner's right to a fair hearing was first taken seriously by the courts in the context of prison discipline. In relation to segregation, transfer, and release decisions the judicial approach had, until recently, been more restrained. Rule 43 enables a governor to remove a prisoner from association in the interests of 'good order or discipline'. This initial removal was originally limited to 24 hours and was renewable by either a member of the board of visitors or the Secretary of State, for a period not exceeding 28 days and for further 28-day periods. By Circular Instruction 10/74 arrangements were made to enable governors of dispersal prisons to transfer a particularly disruptive prisoner under rule 43 to segregation in a local prison, a practice known as 'ghosting'. In theory, segregation and transfer under these provisions are managerial and not punitive decisions, none the less their impact on individual prisoners can be severe.[57] Attempts have hence been made to assert some of the procedural rights which have been recognized in the formal disciplinary context. In *Ex parte Hague*, however, the Court of Appeal refused to recognize a right to be heard either before the initial segregation or before the renewal after 24 hours.[58] The context was distinguished from that of a formal disciplinary charge where the Prison Rules specifically require a hearing, and it is evident that the Court of Appeal felt there were sound policy reasons against extending such procedural rigour to the decision to segregate. The Court further refused to recognize a legal right in the prisoner to be given the reasons for his or her segregation, although Taylor LJ states that '[no] doubt in many cases the governor will be able, as here, to give reasons at the time of the decision or shortly after'.[59]

In 1990 the Circular Instructions regulating the implementation of both rule 43 itself and ghosting were revised. Interestingly CI 26/1990 concerning rule 43 is most grudging in its recognition of procedural rights in

[56] Woolf Inquiry Report (n. 6) para 14.274.
[57] HM Chief Inspector of Prisons, *A Review of the Segregation of Prisoners under Rule 43* (1986).
[58] See n. 29.
[59] Ibid. 698.

prisoners. While it encourages the giving of reasons to prisoners, it remarks that 'inmates have no entitlement to be given reasons for their segregation under rule 43'. The new Circular concerned with ghosting, CI 37/1990, on the other hand, although it still denies any absolute right to reasons, declares that the prisoner should be told the reasons for his transfer, before it is effected if possible, and, if segregated at the receiving prison, should again be given reasons. Although some reservations remain, the procedural safeguards demanded by CI 37/1990 represent a significant improvement and it must be hoped that the government's commitment to the giving of reasons for simple rule 43 segregation, expressed in the 1991 White Paper, is also fully reflected in practice.[60] Quite how far, if at all, the courts can be credited with influencing these procedural reforms is hard to assess. The timing alone is inconclusive.[61]

The selective early release of fixed term prisoners at the discretion of the executive was introduced in England and Wales by the Criminal Justice Act 1967. The Act created a dual key system whereby the Secretary of State was empowered to release if recommended to do so by the parole board. The decision-making structure was seen as administrative rather than judicial; the primary recommendation emanated from a panel of experts in the form of the parole board and the final decision rested with the executive.[62] The Act and accompanying regulations imposed few procedural requirements: there was no obligation to provide a hearing, the prisoner had no access to the dossier on which decisions were reached, and there was no obligation to give reasons, save after recall. The procedure for the release on licence of life-sentence prisoners was identical in many relevant respects, but life-sentence prisoners had no right of reference to the parole board. Their first review date would be set by the executive in consultation with the judiciary and was related to the period deemed necessary to reflect the gravity of the offence, the tariff period.[63]

As in the context of transfer and segregation, prisoners have sought to persuade the courts to recognize their procedural rights with regard to release. Again until recently they have met with little success. In 1981 the Court of Appeal in *Payne* took an unsympathetic line which dominated the case-law for a decade.[64] Payne was a mandatory life-sentence prisoner whose application for release on licence was refused. He sought a declaration that, among other things, he was entitled to reasons for that

[60] *Custody, Care and Justice* (n. 25) para. 5.32.

[61] The evolution of the new Circulars is discussed further in Richardson (n. 27) chs 5 and 6.

[62] For an account of the history see *Report of the Parole System in England and Wales*, Cmnd. 532 (1988).

[63] A clear account of the relevant procedures can be found in *Report of the Select Committee on Murder and Life Imprisonment* (1989).

[64] *Payne* v. *Lord Harris* [1981] 1 WLR 754.

refusal. Much of the argument in the Court of Appeal centred on the precise nature of the statutory requirements and much significance was attached to the fact that, although reasons were required on recall, no such requirement was imposed in the case of an initial refusal to re-commend release. While accepting that the common law requirements of natural justice could be used to supplement the statutory structure, the Court of Appeal unanimously declined to put it to such a use in the context of parole. It was, as Lord Denning freely admitted, a question of policy: 'what does public policy demand as best to be done? To give reasons or to withhold them?'[65] Their Lordships clearly believed that the weight of argument was against the giving of reasons. Parole was a privilege; the parole board was an expert and advisory body which should not be required to provide reasons. Indeed any obligation to do so would apparently hamper its work, provide ammunition for litigious prisoners, and inhibit the production of candid reports.

A similar judicial attitude was evident in *Ex parte Gunnell* when a recalled life-sentence prisoner sought to challenge the process of his recall. He argued that he was entitled to a written statement of reasons and a hearing before the parole board, involving at the very least the disclosure of police and medical reports. He was unsuccessful before both the Divisional Court and the Court of Appeal.[66] Neither court was pre-pared to read into the statutory requirements more than was absolutely necessary. The circumstances of the case were such, the court felt, that the prisoner must have understood the reasons and the statute did not specifically require a written statement to be given. Nor did the statute demand an oral hearing or full disclosure of adverse reports. The disclo-sure point was taken up again in *Ex parte Benson* where the Divisional Court, in ordering the disclosure of medical reports prior to an application for judicial review, reasserted that there was no general right to disclo-sure where parole was refused.[67] Disclosure would only be ordered where the reports were shown to be central to an application for judicial review.

The disinclination of the courts to recognize rights in prisoners in the context of release decisions was not limited to strictly procedural matters. At the Conservative party conference in 1983 Leon Brittan, the then Home Secretary, announced changes to parole policy which effec-tively withdrew parole from certain categories of prisoner serving over five years and ensured that certain categories of murderer would serve at least twenty years. As a result a number of long-term prisoners found their release dates recede literally overnight.

---

[65] Ibid. 759.

[66] *R. v. Secretary of State for the Home Dept., ex parte Gunnell* (1984) *Crim. LR* 170 and (1985) *Crim. LR* 105 (CA).

[67] *R. v. Secretary of State for the Home Dept., ex parte Benson, The Times,* 8 Nov. 1988.

In *Re Findlay* two of the appellants were lifers who had been transferred to open prisons in the expectation of fairly imminent release.[68] On the day of the party conference they were transferred back to closed conditions and, in light of the new policy, could not then anticipate release before 1989 and 1993 respectively. The two men argued that, whatever the legality of the overall policy, their cases were exceptional and should be recognized as such. The Home Office never alleged that the return to closed conditions was related to the appellants' behaviour and, indeed, the court seemed to accept that it resulted exclusively from the change in policy. However, neither the Court of Appeal nor the House of Lords would accept that the transfer to open conditions, with all its implications, constituted exceptional circumstances. Further, Lord Scarman, in the only speech provided by the House of Lords, dismissed any suggestion that the appellants had a legitimate expectation of early release. According to his lordship the only thing a convicted prisoner can legitimately expect 'is that his case will be examined in the light of whatever policy the Secretary of State sees fit to adopt, provided always that the adopted policy is a lawful exercise of the discretion conferred on him by the statute'.[69]

In the context of release, therefore, the domestic courts were reluctant to recognize any 'rights' in prisoners beyond those specifically included in the relevant legislation and accordingly provided little direct pressure for change.[70] Change has, however, occurred. Since the mid-1970s penal reform groups have expressed disquiet concerning the discretionary nature of parole and the procedural inadequacies of the decision-making structure, but official policy, being influenced primarily by a desire to control the size of the prison population, appeared unreceptive to radical reform. However, in the mid-1980s small adjustments to the parole structure, introduced to relieve prison overcrowding, drew fierce opposition from the judiciary, who feared an erosion of differentials where short sentences were concerned. This judicial opposition, combined with the disquiet caused by the overtly political manner in which Leon Brittan's new policy had been introduced, finally obliged the Government to initiate a major review. In 1987 the Carlisle committee was set up to examine the whole parole system. Its 1988 report was followed in 1990 with a government White Paper accepting the majority of its recommendations, including fairly substantial procedural reforms.[71] This in turn has been followed by

---

[68] *Re Findlay* [1985] AC 318.

[69] Ibid. 338.

[70] An exception to this general trend may be found in *R. v. Secretary of State for the Home Dept., ex parte Handscomb* (1987) 86 Cr. App. R. 59, but arguably the courts were there motivated primarily by the desire to protect their own standing rather than the interests of the prisoners.

[71] For the Carlisle Report see *Report of the Parole System* (n. 62) and for the White Paper see *Crime Justice and Protecting the Public*, Cmnd. 965 (1990).

the Criminal Justice Act 1991. Parole has been abolished for all sentences of under four years and for those serving four years or more the procedures surrounding discretionary early release have been significantly improved. Although the decision-making process will remain an essentially paper exercise, the prisoner will have access to his or her parole dossier, will be interviewed by a parole board member and will be given reasons for the decision, whether positive or negative.[72]

In the case of life-sentence release procedures, domestic pressure for change has been strengthened by the judgments of the European Court of Human Rights in relation to both prisoners and psychiatric patients. In 1981 the Court held that Articles 5(1) and 5(4) of the Convention guaranteed a mentally disordered offender, compulsorily detained under a restricted hospital order, regular access to a 'court' to determine the lawfulness of his detention.[73] Under the legislation then in force in England and Wales the release of such a patient was in the discretion of the Secretary of State. Under the Mental Health Act 1959 mental health review tribunals were advisory only in the case of restricted patients. In the opinion of the ECHR there was, therefore, no 'court' with relevant powers to which the patient had access. The tribunal, which might have been sufficiently independent, had no power to order release; the review jurisdiction of the High Court was not capable of testing the substantive lawfulness of the detention; and the Secretary of State, who had the necessary powers, lacked the necessary independence. The United Kingdom was, therefore, in breach of its obligations under the Convention.

As a result of this ruling the new mental health legislation, which was then before Parliament, had to be revised to provide the required 'court' with the relevant powers. The solution chosen was to empower the mental health review tribunal to order the discharge of restricted patients in certain circumstances.[74] It was, arguably, the least that was required by the Convention, but it is inconceivable that even this much would have been achieved without the impetus of *X* v. *United Kingdom*. Significantly, perhaps, the mental health pressure group MIND, which had promoted the case to Strasbourg, had shied away from advocating the removal of executive discretion in its official recommendations prior to the 1982 reforms.[75] An unambiguous ruling from the ECHR was clearly thought likely to achieve more than any amount of reasoned argument within the domestic political process.

The principles enunciated in *X* have more recently been applied in relation to discretionary life sentences. In 1987 the ECHR found the

[72] The early release structure is contained in Criminal Justice Act 1991 Part II. The procedural requirements are non-statutory and are found in CI 26/1992.

[73] *X.* v. *UK* (1982) 4 EHRR 188.

[74] Mental Health Act 1983, ss. 72 and 73.

[75] L. O. Gostin, *A Human Condition*, ii (1977), ch. 11.

United Kingdom government in breach of Article 5(4) in respect of its treatment of a discretionary life-sentence prisoner.[76] In 1966, when 17 years old, Weeks had been sentenced to life imprisonment for armed robbery. He had held up a pet shop with a starting pistol loaded with blanks, and had stolen 7s. When the sentence was affirmed on appeal the Court of Appeal remarked that it had been imposed 'in mercy to the boy'. Between 1976 and 1987 Weeks was released on licence three times and on each occasion his licence was revoked following some relatively minor incident.

The ECHR felt that the indeterminate sentence had been imposed for the purposes of social protection because of Weeks's mental instability and potential dangerousness. Both of these conditions were susceptible to change over time and, as in the case of *X*, Article 5(4) guaranteed Weeks regular access to a 'court' in order to test the continued lawfulness of his detention. As neither the parole board nor the High Court nor the Secretary of State could constitute such a 'court', the United Kingdom was in breach of the Convention. Given a generous interpretation, this judgment carried broad implications for the release structure provided for all discretionary life sentences. The government, however, chose to take no action, presumably hoping that *Weeks* v. *United Kingdom* was special to its facts.

In 1988, following disquiet expressed in the House of Lords concerning the structure of release from life sentences, a House of Lords Select Committee was set up to consider the crime of murder, the penalty it should attract and the arrangements for determining release. Having taken evidence as to the requirements of the Convention with regard to release procedures, the Committee finally recommended the removal of executive discretion and the creation of an independent tribunal to make all decisions concerning the release and recall of life-sentence prisoners.[77] Still the government failed to act.

In 1990 the ECHR published its decision in *Thynne, Wilson, and Gunnell* v. *United Kingdom*.[78] This judgment makes it clear that the principles expressed in *Weeks* cannot be restricted to the facts of that case alone. According to *Thynne, Wilson, and Gunnell* the requirements of Article 5(4) apply to all discretionary life-sentence prisoners who are detained, on the basis of their mental instability and dangerousness, after the expiry of the punitive, or tariff, period of their sentence. The government sought to distinguish the cases of Thynne, Wilson, and Gunnell on the grounds that, since they, unlike Weeks, had all committed serious offences, no clear dividing line could be drawn between the punitive and the social protective elements in their life sentences. At no clearly ident-

---

[76] *Weeks* v. *UK* (1988) 10 EHRR 293.
[77] *Report on Murder* (n. 62) (1989).
[78] Vol. 190 of Series A, European Court of Human Rights.

ifiable point did social protection become the single justification for continued detention. The ECHR was unconvinced. While the Court agreed that it might be hard to distinguish the punitive from the protective elements in a discretionary life sentence, it felt that at some point, however heinous the offence, the punitive element must expire, leaving the continued detention to be justified solely on the grounds of the prisoner's potential dangerousness. A person's capacity to be dangerous may change over time and thus, on the principles of *Weeks*, anyone detained solely on the grounds of their dangerousness must be entitled regularly to challenge the lawfulness of that detention before a 'court'. When the Criminal Justice Bill was first published in November 1990, no concessions were made to the European ruling. Following wide-spread revolt in the House of Lords over the retention of the mandatory life sentence for murder, however, the government finally introduced provisions, contained within the Criminal Justice Act 1991, to enable the parole board to order the release of discretionary life-sentence prisoners. Section 34 empowers the board to direct the release of a discretionary life-sentence prisoner if it is satisfied 'that it is no longer necessary for the protection of the public that the prisoner should be confined'. The new scheme provides for the specification of the relevant (or tariff) period by the trial judge in open court, and empowers the board to release the prisoner after the expiry of that period. It came into operation in October 1992, and in August rules were made specifying the procedures to be followed by the board when hearing individual cases.[79] Under the transitional arrangements provided by the Act the scheme has also been extended to apply to discretionary life-sentence prisoners sentenced before October 1992 who have reached their tariff date. Since October 1992, therefore, the release procedures for discretionary lifers have been radically changed. The tariff date will be declared in open court, the release decision itself is taken by an independent tribunal following an oral hearing and the executive veto has been removed. These are major improvements which could scarcely have been achieved without the intervention of the ECHR. Inevitably, however, some problems remain. It is particularly unfortunate that the government has followed the pattern adopted by the Mental Health Act 1983, and has required the board to be 'satisfied that it is no longer necessary for the protection of the public that the prisoner should be confined'. Such statutory wording effectively creates a presumption against release, despite the expiry of the tariff period, which must be rebutted if the prisoner is to obtain release.[80]

---

[79] The Parole Board Rules 1992. These rules are not statutory instruments.

[80] For an account of the interpretation of the release provisions of the Mental Health Act 1983 see J. Peay, *Tribunals on Trial: A Study of Decision Making under the Mental Health Act 1983* (1989), and for further discussion in the context of life sentences and mentally disordered offenders see Richardson (n. 27).

It is also important to realise that the new procedures apply only to discretionary life sentences certified under section 34. Thus it is possible that some offenders sentenced to life for offences for which that sentence is not mandatory will not receive the benefits of section 34. More significantly the new procedures do not apply to murderers, all of whom receive a mandatory life sentence. The decision whether or not to release a mandatory lifer still rests with the Secretary of State.[81]

Finally since the intervention of the ECHR in *Thynne, Wilson and Gunnell* the domestic courts have at last begun to shake off the deadening influence of *Payne*. In 1992, emboldened by the judgment in *Thynne, Wilson and Gunnell* and the impending reforms contained in the Criminal Justice Act, the Court of Appeal agreed to depart from *Payne* and to order the disclosure of the parole dossier in the case of a discretionary life-sentence prisoner whose case was considered prior to the implementation of section 34.[82] While it is still too early to predict how far this new enthusiasm for the procedural rights of prisoners will extend, it is evident that the judiciary were prepared directly to depart from *Payne* in the case of mandatory life sentences as well. In the event such intervention was rendered unnecessary by the Secretary of State's decision to allow mandatory lifers access to their dossiers and reasons for their release decisions.

The impact of international legal obligations

In the context of prison discipline, the ability of the domestic courts to achieve improvements in formal procedures was recognized, but reservations were expressed as to the possible impact of judicial intervention on the achievement of more radical change. In relation to the European Commission and Court no such reservations are appropriate. A decision of the European Commission persuaded the United Kingdom to recognize a prisoner's right to marry.[83] With regard to prisoners' correspondence, the decisions of the Court, enforcing rights to privacy and access to the courts, eventually forced a reluctant government to act. It was only after the basic reforms had been achieved that a change in management philosophy encouraged greater liberalization. In the context of release procedures, again, the ECHR has played a central role. The enforcement of Article 5(4) in *X* v. *United Kingdom*, led directly to the introduction of a new structure for the release of mentally disordered offenders. Most recently, with *Weeks* and *Thynne, Wilson, and Gunnell*, the ECHR has

---

[81] The position of mandatory lifers was, at the time of writing, still under challenge in Europe, but meanwhile the Secretary of State had promised to review the structure for the release of mandatory lifers.

[82] *R* v. *Parole Board, Ex p. Wilson* [1992] 2 All ER 576.

[83] *Hamer* v. *UK* (n. 15).

finally achieved reform of the system of release from discretionary life sentences.

By comparison with the direct impact of the European Convention, the effect of other international agreements has been remote. The International Covenant on Civil and Political Rights, although covering many similar rights to those guaranteed under the European Convention, does not as far as the United Kingdom is concerned carry the right of individual petition. Other relevant international provisions include the European Standard Minimum Rules for the Treatment of Prisoners and the Body of Principles for the Protection of All Persons under any form of Detention or Imprisonment, emanating from the United Nations, but neither is of binding force, although the European Rules can be influential. Only the European Convention for the Prevention of Torture and Inhuman or Degrading Treatment appears likely to have any real impact.[84] It establishes a committee of individuals who can visit any place within the jurisdiction of a party to the Convention where people are deprived of their liberty by a public authority. Following a visit the committee reports to the state authorities, and may publish a statement if the state refuses to introduce the necessary improvements. In 1990 the committee visited five prisons and five police stations in the UK, including the psychiatric wing at Brixton. While it found no incidents of ill-treatment, it considered the overcrowding and sanitation in three establishments so outrageous as to amount to 'inhuman and degrading treatment'. In 1991 the government allowed publication of the committee's report and published its own response.[85] While the government denied that the conditions found amounted to 'inhuman and degrading treatment', it has promised improvements and it is significant to note that the notorious F Wing at Brixton has subsequently been closed.

Conclusions on the role of the domestic courts

It should be evident from the above that the record of the domestic courts as a protector of the 'rights' of prisoners has been variable. The ability of the private law to provide any real protection is greatly inhibited by the court's attitude to the Prison Rules. There are no effective private law rights to sensible minimum standards. In relation to negligence the court's assessment of the implications of imprisonment has varied. With

---

[84] (1987) 9 EHRR 161. The UK ratified the Convention in June 1988 and it came into force on 1 Feb. 1989.

[85] *Report to the United Kingdom Government on the Visit to the United Kingdom by the European Committee for the Prevention Torture and Inhuman or Degrading Treatment or Punishment* (1991) and the *Response of the United Kingdom Government* (1991). For a useful account of the record of the Convention see M. Evans and R. Morgan, 'The European Convention for the Prevention of Torture: Operational Practice' (1992) 41 *ICLQ* 590.

regard to inter-prisoner violence courts have been prepared to recognize special duties of protection on the part of the authorities, although they appear reluctant to find those duties breached. In the context of medical care, by contrast, the fact of imprisonment has led the courts to accept a lower standard of care from the authorities.

The inadequacies of the private law have led to greater reliance on public law, and in certain circumstances prisoners have met with considerable success. Most notably as far as domestic courts are concerned, the judiciary has intervened in the disciplinary system and has imposed fairly stringent procedural requirements on adjudicators. Other areas of judicial activity have, as emphasized above, owed much to the involvement of the European Court and Commission of Human Rights. In relation to prisoners' correspondence the domestic courts did eventually add support and recognized a right of access to the court, but in the context of early release and transfer the domestic judiciary until recently has been reluctant to recognize rights beyond those few expressly stated in the legislation. Without doubt, however, what success there has been before the courts, both domestic and European, owes much to the persistence of penal and mental health pressure groups and advocates and their astute case selection.

It has been suggested elsewhere that the approach of the domestic courts to prisoners' claims in both public and private law displays considerable sympathy for the arguments of administrative convenience, typically offered by the authorities in response to any attempts to enforce rights in prisoners.[86] The judicial interpretation of the 'necessary implication' referred to by Lord Wilberforce tends in practice, therefore, to favour the authorities. The areas where the courts have been prepared to intervene in order to resolve uncertainties in favour of prisoners have typically been either those, such as prison discipline, where the context is 'quasi-judicial' and the courts feel confident and appropriately equipped, or those where rights which they might hold particularly dear, such as access to justice, are concerned. In the more 'administrative' areas of transfer, and the conditions of imprisonment, the courts of England and Wales have shown little enthusiasm for acting as a mechanism for the enforcement of prisoners' interests. With reference to Lord Wilberforce's statement, therefore, the case-law suggests that the interpretation of 'necessary implication' typically adopted by the judiciary will tend to lead to the recognition of only a restricted residue of rights in prisoners. Further, it is clear that the courts are reluctant to step beyond Wilberforce and to recognize additional special rights in prisoners.

---

[86] This argument is addressed in G. Richardson, 'Judicial Intervention in Prison Life', ch. 3 in Maguire, Vagg, and Morgan (eds.), *Accountability* (n. 8).

These conclusions, if accepted, would suggest that if the courts are to be used more widely as direct enforcers of prisoners' rights, the relevant legislation must be more specific, and in particular it must expressly create special rights in prisoners enforceable at law. A relatively simple move in this direction could be achieved by the express application to prisons of specific regulatory schemes. The remaining areas of Crown immunity could be lifted and schemes such as that contained in the Factory Act applied to prisons. It is encouraging in this context to note that the Race Relations Act has been held to apply to prisons.[87] The extension of otherwise applicable regulations in this way would be entirely in line with the principles of normalization and with the model of prisoners' rights advocated earlier. It would enable prisoners formally to retain the same rights as non-prisoners, their residue of rights, and would bring to an end their reliance on *ex gratia* payments from the Home Office. The creation of special rights in prisoners, additional to those possessed by non-prisoners, is likely to prove much more difficult.

EXTRA-JUDICIAL ENFORCEMENT

Any account of prisoners' rights that was limited to a discussion of the activities of the courts would, however, be seriously deficient both as a representation of present practice and as a basis for considering avenues for reform. Even if suitably empowered, the ordinary courts are unlikely to provide an effective mechanism for the routine oversight of prisoners' interests.[88] It is clear that at present the courts play only a marginal role in most areas of daily prison life. This lack of routine involvement is not, however, necessarily to be decried. There is little reason to suppose that the courts are the most appropriate body to provide regular oversight. A specialized, but truly independent, lay body could be more accessible, could possess more relevant expertise and could be in a better position to achieve improved standards through conciliation. The court's role could then be limited to the provision of an ultimate means of enforcement. At present, however, the extra-judicial safeguards available to prisoners are sadly inadequate.

Many aspects of a prisoner's existence are governed by administrative rules and guidelines. The types of letters that may be written and received, for example, or the 'privileges', now 'facilities', that may be expected, are the subject of detailed Standing Orders. A prisoner who feels that he or

---

[87] *Alexander* v. *Home Office* [1988] 1 WLR 968.

[88] R. Morgan, and A. J. Bronstein, 'Prisoners and the Courts: The U.S. Experience', ch. 18 in Maguire, Vagg, and Morgan (eds.), *Accountability* (n. 8).

she has suffered as a result of non-compliance with these Orders, or who wishes to complain for any other reason has two main mechanisms available. In the first place she may raise the matter internally within her own establishment and then, via the area manager, with the Secretary of State. Evidence gathered by Austin and Ditchfield and Her Majesty's Chief Inspector suggests that inmates are broadly satisfied with the management of 'applications' at wing and governor level.[89] With regard to petitioning the Secretary of State, however, dissatisfaction was widespread among prisoners and frequently echoed by staff.[90] In September 1990 the procedures were amended to encourage greater confidentiality, speedier resolution, and the provision of reasoned decisions but, while it is to be hoped that these new procedures will meet most of the criticisms, the system of applications and petitions must inevitably remain internal.

In addition to the internal system of applications, a prisoner may raise a grievance with the board of visitors which currently provides the most accessible mechanism for external review. Much has been written concerning the procedural vagaries of some boards, their lack of conspicuous independence, their strong tendency to identify with management, and, until April 1992 at least, the difficulties inherent in their dual role as complaints mechanism and disciplinary tribunal.[91] In the light of these criticisms, Woolf recommended a strengthening of the boards' complaints role and an end to their involvement in the disciplinary system. While the latter recommendation has now been acted upon, it remains to be seen whether local boards of visitors can ever possess sufficient status and independence to provide an effective external safeguard of the rights and interests of prisoners.

If it be accepted, as suggested above, that the court's involvement in the oversight of prisoners' interests should be limited to the provision of last resort enforcement, then there is a strong case for the introduction of a truly independent and appropriately empowered national body to whom prisoners would have access for the resolution of grievances. Following the recommendations of the Woolf Report that an independent complaints adjudicator for prisons be created, the government is in the process of appointing a prison ombudsman to provide both a central independent element within the grievance procedure and an independent 'appellate' tier for the disciplinary system. Whether this new figure, who

---

[89] C. Austin and J. Ditchfield, 'Internal Ventilation of Grievances: Applications and Petitions', ch. 10 in Maguire, Vagg, and Morgan (eds.), *Accountability* (n. 8), and HM Chief Inspector of Prisons, *A Review of Prisoners' Complaints* (1987).

[90] See also the reservations expressed by Lord Bridge in *Leech* v. *Deputy Governor, Parkhurst Prison* [1988] AC 533, 567, where he suggested that any investigating civil servant, if faced with a dispute between a prisoner and a governor, 'is likely simply to accept the governor's account'.

[91] M. Maguire, 'Prisoner's Grievances: The Role of the Board of Visitors', ch. 9 in Maguire, Vagg, and Morgan (eds.), *Accountability* (n. 8).

for the present will enjoy no statutory status, will possess sufficient standing, resources or powers to be truly effective in either capacity remains to be seen.

Whatever the nature of the extra-judicial enforcement mechanisms, however, their efficacy will be strictly limited until a set of minimum standards can be agreed. In the absence of such standards no bench mark exists against which the regime provided in any prison can be assessed and no objective statement of a prisoner's entitlements can be found. It is encouraging, therefore, that in accordance with the recommendations of the Woolf Report, the prison service is now devising a code of standards for application throughout the service.[92] After many years of debate the authorities have finally accepted that the creation of standards covering all aspects of the prison regime is both feasible and desirable. While it is clear that these standards will not be designed to create legally enforceable substantive rights in individual prisoners, establishments will be expected to comply and extra-judicial mechanisms to encourage compliance will be introduced. In addition, these standards might be held to constitute a statement of a prisoner's legitimate expectations, particularly if they are to be reflected in the 'compacts' between prisoners and their individual establishments. The precise implications flowing from the court's recognition of a legitimate expectation are, of course, hard to predict with any certainty, but judicial recognition of the expectation created by the standards could be used to spur on the efforts of any extra-judicial enforcement. Given the nature of the existing statutory framework, the development of such an indirect mechanism for the protection of the interests of prisoners might prove both more attractive to the judiciary and more effective in improving prison conditions than any direct evocation of substantive rights. It is also possible that the notion of a legitimate expectation could be used by the inmate of a contracted out prison as a means of attracting judicial attention to non-compliance by the contractor with the regime specifications set out in the contract with the Home Office.

CONCLUSION

This chapter has argued that a claim to prisoners' rights can be justified and has suggested a model for the identification of those rights. It has also suggested that such rights are inadequately recognized by the present legal structure, both statutory and common law. Whatever the formal legal position, there is also ample evidence that prisoners' rights, in a broader sense, are widely neglected in practice.

---

[92] Woolf Inquiry Report, para. 12.129. See A Code of Standards for the Prison Service: A Discussion Document (1992).

There is currently almost universal agreement that the state of British prisons is entirely unacceptable.[93] In such prisons it is unrealistic to suppose that any but the most essential of a prisoner's 'rights' can be respected. Over the last few years extensive evidence of poor physical conditions has emerged[94] and this has been supplemented by equally strong evidence of less tangible inadequacies within the system. The Chief Inspector has provided reports on grievance procedures, removal from association under rule 43, and prison suicides;[95] independent research has been undertaken on boards of visitors and race relations;[96] and inquiries have been conducted into parole and life-licence procedures.[97] There is no absence of empirical evidence. Data abound confirming an overall failure in practice to respect the rights of prisoners, both civil and political and economic and social. The Woolf Report, published in 1991, voiced grave concern about the conditions in prison and warned of the serious consequences that could flow from a failure to act. In some important respects, prison discipline and minimum standards for example, the government has heeded that warning and has acted upon the recommendations contained within the Report. However, the government has yet to display the strong commitment necessary to achieve real and lasting change.[98] Prisons may be news when prisoners sit on their roofs, burn them down, or dramatically escape from them. For the majority of the voting public, however, they are marginal and, despite the prison governors' warning that 'dreadful conditions, broken promises and dashed expectations provide fertile conditions for those intent on trouble',[99] the government has priorities more pressing than prisoners' rights. Real change it seems will only occur when it is no longer politically acceptable to delay reform.

While the concept of rights might possess some political potential to encourage pressure for reform, experience suggests that the courts are unlikely to exploit that potential under the present legislative structure.

---

[93] R. D. King and K. McDermott, 'British Prisons 1970–87: The Ever Deepening Crisis' (1989) 29 *Brit. J. Criminology* 107–28.

[94] The Report of HM Chief Inspector of Prisons 1988 concludes 'that in prison men were still having to exist in conditions which offend against any reasonable standard of decency' (HC paper, no. 491, Session 1988/89, HMSO, London), para. 205.

[95] See nn. 89, 57, and 34 respectively.

[96] M. Maguire and J. Vagg, 'Who are the Prison Watchdogs? The Membership and Appointment of Boards of Visitors' (1983) *Crim. LR* 238, and E. Genders and E. Player, *Race Relations in Prisons* (1989).

[97] See *Report on Parole System* (1988) (n. 62) and *Report on Murder* (1989) (n. 63) respectively.

[98] For an assessment see Prison Reform Trust, 'Implementing Woolf: The Prison System One Year On' (1992).

[99] Evidence of the Prison Governors' Association to the Woolf Inquiry, 1990.

Further, although the energetic development of the concept of legitimate expectations in relation to prisoners might provide some interim protection and oversight, it is no substitute for radical reform. Real change must come and must come via Parliament. It is now over 40 years since our major democratic institution expressed its views and a new legislative framework is urgently needed. The nature of imprisonment and the treatment of prisoners are matters of real public interest and cannot be left to administrative discretion. Parliament must be required to debate the issues, and must be invited to create a framework which recognises the legitimate entitlements of prisoners and provides for the articulation of those entitlements with sufficient clarity and openness to render them effective and enforceable.

7

Freedom of Expression

JOHN GARDNER

INTRODUCTION

This discussion starts from the assumption that we have a moral right to freedom of expression, whatever the law may say about it. The first section records some distinguishing features of this right, and outlines the main arguments which support and shape it. The next section surveys some aspects of English law which bear on freedom of expression, and asks whether the moral right already sketched has been respected by our legal system in recent years. Finally, the focus shifts to some international legal orders in which the United Kingdom has become involved. Apart from the obvious question of whether recent English law has been meeting its international law obligations concerning the protection of free expression, there is also the widely neglected question of whether meeting those obligations would bring respect for the moral right to freedom of expression in its wake.

THE MORAL RIGHT

The moral right to freedom of expression confers its powerful protection on all activities which are conventionally understood to have a primarily expressive or communicative function. Some activities—talking, writing, publishing, broadcasting—are always protected. Others are protected only in certain contexts. Lighting a candle for a departed soul is a protected activity, for example, but lighting a candle to see during a power cut is not. The fact that a particular activity is protected by the moral right to freedom of expression means that one has a special moral weapon with which to repel arguments for restricting its pursuit. Indeed, like all other rights, the moral right to freedom of expression fires two different calibres of normative ammunition at hostile arguments. Some arguments for restricting the pursuit of expressive or communicative activities simply fall to be outweighed by the right, whereas others are excluded from consideration altogether. But it does not follow that one can force every argument for restricting expression or communication into retreat. Both outweighing and excluding have their limitations, and some hostile arguments survive them both. After all, a right is only as powerful as its justification can make it.

The right is justified in part by the contribution it makes, when it is respected, to maintaining a free flow of information. Writers who notice that the right makes this contribution sometimes assume that the contribution is valuable simply because we live in a democracy, under a system of government which cannot function without wide public access to information about politicians and their policies.[1] This assumption tends to encourage the view that the right to freedom of expression is primarily a right to engage in political analysis and debate. No doubt the importance of democracy does lend extra weight to the right to freedom of expression in some contexts. But the value of its contribution to maintaining a free flow of information is much more pervasive.[2] There are plenty of other public goods which, like democracy, can function adequately only if a good deal of information is freely available. Think about labour markets, for example, or public transport systems. And there is a more general point as well. People cannot thrive in a highly complex and mobile pluralist society unless they know what their options are for work, recreation, and education; how their lives are likely to be affected by present and future laws, government policies, and business practices, and so on. One needs to be able to make plans and commitments with some idea of the viable alternatives and likely repercussions. Now the right to freedom of expression is not, of course, a right to *receive* all the information required for these various purposes.[3] It is merely a right to supply the information. But granting suppliers a right to supply is a good way to make sure that, under normal conditions of social life, receivers often receive. This is why the value of wide public access to information helps to justify our right to freedom of expression.

The right's protection extends, however, well beyond informative communications. It covers avant-garde films and prayer books as well as job advertisements and bus timetables. Those who think otherwise are overlooking or underrating what Joseph Raz calls the 'validating' role of expressive activities and the related 'invalidating' role of certain types of state censorship.[4] Expressive activities function not only as sources of

---

[1] See e.g. A. Meiklejohn, 'Free Speech and its Relation to Self-Government', in his collection of essays entitled *Political Freedom: The Constitutional Powers of the People* (1965).

[2] T. M. Scanlon, 'A Theory of Freedom of Expression' (1972) 1 *Philosophy and Public Affairs* 204.

[3] Receipt of information is protected by the right to freedom of information, a right which is not treated separately in this volume. See P. Birkinshaw, *Freedom of Information: The Law, the Practice and the Ideal* (1989).

[4] J. Raz, 'Free Expression and Personal Identification' (1991) 11 *Oxf. J. Legal Studies* 303 at 311.

information, but also as reflections and portrayals of people's experiences and ways of life. There are magazines about bodybuilding and television plays dealing with disability, newspapers for political activists and commercials featuring harassed mothers. However questionable in other respects, these share the valuable feature that they give the experiences and ways of life with which they are concerned a place in public culture, and thus some kind of public recognition. This public recognition, which can only be secured through expression, plays a special role in developing people's pride in their ways of life and identification with their own experiences, and hence their well-being. Associated with this is the unambiguous social meaning of those prohibitions on expression which are imposed by the state authorities because the expression portrays or reflects a certain way of life or its characteristic experiences. Such 'life-style censorship' is always understood as an authoritative condemnation of the whole way of life in question. And while it is one thing not to have a voice in public culture, it is quite another to have one's life written off by one's society. If the former detracts from the possibilities for pride and personal identification, the latter strikes at the heart of one's membership of society, and deprives one of the sense of ease with one's environment which is essential to a fulfilling life. Both points can be illustrated from the recent history of Britain's gay community. The regular Gay Pride festivals during the 1980s have powerfully illustrated the validating impact of expressive activities, the positive value of having a distinctive public voice. Meanwhile the controversy surrounding the notorious 'Section 28'—a 1988 statutory provision banning local authorities from promoting the teaching of 'the acceptability of homosexuality as a pretended family arrangement'—perfectly illustrates the unambiguous social meaning of life-style censorship, and its profound negative impact upon those who are singled out for condemnation by it.[5]

So these—the pervasive need for information, and the positive and negative impact of 'validation' and 'invalidation' respectively—are the principal considerations which conspire to justify the moral right to freedom of expression, endowing it with its often formidable power to outweigh and exclude arguments in favour of restricting expression. As far as outweighing is concerned, the right obviously weighs more heavily against arguments for restricting expressive or communicative activities the more the restriction would tend to inhibit the flow of information or limit the scope for validation of people's lives. The right is strong in defence of accurate news reporting, for example, but weak in defence of sensationalist fabrications. It puts its weight behind those who would publish leaked government documents, but not behind those who deal in sinister disin-

---

[5] Local Government Act 1988, s. 28. See chapter 15.

formation. It stands strong against attempts to restrict broadcasters to mainstream programming (say by requiring that every programme pay for itself in advertising revenues), but puts up much less resistance against regulations which push in the opposite direction, requiring some minority programming, public access, community broadcasting, or the like. All these variations in weight follow from the fact that the right exists to assist the flow of information and the availability of validation. The right's power to exclude hostile arguments altogether also follows from its justi- fication. What the right to freedom of expression excludes, above all else, are arguments in favour of life-style censorship. The exclusion covers arguments based not only on the intrinsic merits of the concerns and attitudes conveyed in the expression, but also on the fact that they may influence others to cultivate similar concerns and attitudes, with or without harmful side-effects.[6] It may be asked why such arguments are excluded and not merely subject to powerful outweighing. The answer is that the government of a pluralistic society, while not bound to see that every way of life or experience is validated, and entitled to condemn particular actions such as rape and theft, is generally bound to avoid condemning anybody's *whole* way of life outright. The burden of estab- lishing that a way of life is so devoid of redeeming features that nothing in it should stand uncondemned is an extremely difficult one to discharge, and it seems unlikely that it can be discharged in relation to any of the general groups (racists, IRA supporters, travellers, sado-masochists, etc.) who are the predictable targets of life-style censorship in our society. And yet life-style censorship cannot, because of its social meaning, be anything other than total condemnation of the whole way of life associated with the expression. That is why the right to freedom of expression has ex- clusionary force against arguments for life-style censorship, whereas it wields a variable outweighing power against other arguments for restrict- ing expressive activities.

This, in sum, is how the moral right to freedom of expression defends expressive and communicative activities. It is frequently underestimated by government committees and judges, and even by some civil liberties campaigners. Among other things, it is apparently quite often confused with the famous 'harm principle', which bans governments from restricting any activity except in order to prevent harm.[7] The harm principle is justified by the need to limit the damage inflicted on people's lives by coercion and the use of invasive sanctions. As such, its protection extends well beyond expressive and communicative activities. Every use of the law, from planning enforcement notices to murder convictions, has to be

[6] Compare what T. M. Scanlon calls the 'Millian Principle' in 'Theory of Freedom'.
[7] The best-known modern account is H. L. A. Hart, *Law, Liberty, and Morality* (1963).

justified on harm-prevention grounds. But only expressive and communicative activities benefit from the extra protection of the right to freedom of expression. Some restrictions on expression are indefensible simply because the expression is harmless—in which case there is no need to plead the right to freedom of expression, because the harm principle will do the trick by itself—but many other restrictions on expression are indefensible because the harms which the expression causes are outweighed or excluded from consideration by the moral right to freedom of expression. These harms, if they cannot be averted by other means, are simply part of the price that must be paid for living in a free society.

FREE EXPRESSION UNDER ENGLISH LAW

One way to ensure widespread respect for the moral right to freedom of expression, now favoured in many countries, is to create a general *legal* right to freedom of expression. In domestic English law, however, there is no such thing, except in a few peculiar contexts. The first peculiar context is that of parliamentary debate. Parliament's Committee on Privileges imposes some limited restraints on MP's speeches, but under our 1689 Bill of Rights other bodies (including the courts) are precluded from restraining or attaching sanctions to what MPs say in the course of proceedings in either House, on pain of being held in contempt of Parliament.[8] There is also a legal right to freedom of expression, of sorts, for participants in court proceedings. The remarks of judges, witnesses, and jurors cannot be the subject of civil actions, although criminal liability is apparently not ruled out.[9] The third context in which there is something like a legal right to freedom of expression, a much more recent addition, is the university or college campus. Under 1986 legislation, higher and further education institutions have been required to issue their own 'codes of practice' on free speech, providing for disciplinary measures against students and staff members in case of infringement.[10] Speech which is prohibited or actionable in law is excluded from protection under these 'codes'. The arrangements were introduced, not to override legal restrictions on expression, but to eradicate extra-legal restrictions on expression, such as student union 'no platform' policies.

The fact that it grants no legal right to freedom of expression outside these contexts does not mean, however, that English law is generally blind to the existence of the moral right. On the contrary, modern judges

[8] See P. Leopold, 'Freedom of Speech in Parliament: Its Misuse and Proposals for Reform' (1981) *PL* 30.
[9] See e.g. *Marrinan* v. *Vibart* [1962] 3 All ER 380.
[10] Education (No. 2) Act 1986, s. 43.

have tried to give the moral right some legal recognition by establishing a legal *principle* of free speech. This serves to justify narrow readings of statutes and common-law rules which threaten to inhibit expression. It also serves to outweigh (although not to exclude) various countervailing considerations which would otherwise justify censorship. In theory such a principle could yield good legal protection for freedom of expression, every bit as good as a fully-fledged legal right. But in fact the common-law principle has major limitations. In the first place, it has never been used by the English courts except to defend the expressive activities of individuals. Local authorities and charities, for example, are denied freedom of expression on important questions of public policy.[11] There is no principled support for this distinction; it rests on the mistaken and over-individualistic assumption that the right to freedom of expression is justified by the interests of the right-holder alone, rather than by public interests in free-flowing information and wide-ranging validation. In the second place, the principle of free speech is not used by the courts except to protect verbal expression and communication: hence 'free *speech*'. This is an entirely arbitrary stopping-point. A picture or gesture can often convey an attitude or piece of information more effectively than a whole book. Words have no special magic. And finally, the principle has been relied upon less in relatively black-and-white areas of law, like criminal law, than in areas of law which involve explicit 'balancing' tests, like family law.[12] One might imagine, on this basis, that the astonishing tidal wave of judicial review proceedings against public bodies during the 1970s and 1980s, bringing many opportunities for creative 'balancing' operations, would have enhanced the usefulness of the principle of free speech. But unfortunately that is not the case. In 1991, the House of Lords confirmed, without having to overrule any previous decisions in the process, that the rules of judicial review which apply to exercises of official discretion, unlike the rules of statutory and common-law construction, do not include any principle of free speech.[13]

Obscenity

Even more alarming, perhaps, is the widespread disregard for the principle of free speech in some areas of English law which are all about the control of expression. Obscene publications law is a good example.

---

[11] *R.* v. *Inner London Education Authority ex parte London Borough of Westminster* [1986] 1 WLR 28, [1986] 1 All ER 19; *McGovern* v. *Attorney-General* [1982] Ch 321, [1981] 3 All ER 493 (on Amnesty International).
[12] Compare *Attorney-General* v. *Able* [1984] QB 795, [1984] 1 All ER 277 with *Re X (a minor)* [1975] 1 All ER 697 (CA).
[13] *Brind* v. *Secretary of State for the Home Dept.* [1991] AC 696 (HL).

The common-law offence of 'obscene libel' was sustained by generations of judges as a general response to the publication of literary and pictorial representations of human sexual activity or anything remotely connected with it. For a publication to count as obscenely libellous, it was enough that some part of it would have a 'tendency to deprave and corrupt' the most sexually naïve person who might chance upon it.[14] The bare mention of a frowned-upon sexual practice was frequently enough to secure a conviction. A spate of prosecutions for obscene libel in the 1950s—to which *The Kinsey Report* and *The Decameron* both fell prey—led to a vigorous reform campaign orchestrated by the Society of Authors. The result was the Obscene Publications Act 1959. This preserved the 'tendency to deprave and corrupt' test for obscenity, which is still with us to this day. What the Act added was that the effect of the publication *taken as a whole* on its *likely* readership was now to be assessed, and subject to a brand new 'public good' defence for material with literary, scientific, or other merits.

As expected, it was the public-good defence that made the biggest splash. Expert witnesses could now be called to identify the merits of the publication, but not to comment on its alleged tendency to deprave and corrupt. For a new, more tolerant, generation of jurors in the late 1960s and early 1970s, the expert evidence of literary figures and sociologists commending a work's positive contributions was apparently more compelling than a prosecuting counsel's inexpert observations about its role in the decline of civilization. Moreover, it was accepted that almost any kind of redeeming social value in a publication, including its value as an *exposé* of depraved lifestyles, could be relied upon to support a public-good defence.[15] So from the mid-1970s, the Director of Public Prosecutions (to whom all prosecution decisions in this area are referred for the sake of consistency) largely gave up trying to prosecute the publishers of works with any pretensions to be more than crudely pornographic. Along with the resulting slump in prosecutions, it has to be said, there came a corresponding increase in the use of the 1959 Act's distinctive 'forfeiture' procedure—also supervised by the DPP—which allowed magistrates to grant an order for confiscation and destruction of obscene publications. The public-good defence, everyone knew, cut less ice with magistrates than with jurors. The procedure is still widely used. All the same, there has been a genuine liberalization since the mid-1970s. That prosecutions and forfeitures have not been reduced to a trickle reflects the fact that

---

[14] *R. v. Hicklin* (1868) LR3 QB 360 at 371 (HL).

[15] Indeed, only arguments based on the supposed therapeutic or educational value of pornography as such have ever been explicitly ruled out, and those much later on. See *DPP v. Jordan* [1976] 3 WLR 887, [1976] 3 All ER 725 (HL) and *Attorney-General's Reference (No. 3 of 1977)*, [1978] 1 WLR 112, [1978] 3 All ER 1166 (CA).

pornographic production lines are ever more productive and their products ever more odious, not that the authorities have maintained their earlier repressive zeal. Reputable publishers can now get away with just about anything. Even a novel like *American Psycho*, a catalogue of revolting sexual violence disguised as social comment, can now be published without running a serious risk of official action. Compare this with 1967, when the far less gratuitous book *Last Exit to Brooklyn*—recently turned into a mainstream film—sparked an enthusiastic, if unsuccessful, prosecution.

But perhaps *American Psycho* would have escaped an obscenity prosecution even without the liberalizing influence of the public-good defence. One thing that the *Last Exit* case established firmly was that material does not tend to 'deprave and corrupt' when it is more likely to put its readership off the life-style portrayed than to attract them to it.[16] This shows that obscenity has nothing to do with offensiveness. Extremely offensive matter will be all the more likely to benefit from this 'aversion' argument. It is a virtue of English obscene publications law that it has avoided the 'offensiveness' approach favoured in many other countries (including the usually free-expression-conscious United States), and has tried to work with a 'harm' model instead. The problem is that it is far from clear what the harm is supposed to be. The words 'deprave' and 'corrupt' are both ambiguous. They may simply refer to an undesirable alteration of outlook or attitude, or they may refer to an undesirable alteration of outlook or attitude brought about by insidious means, such as manipulation or addiction, or they may refer to an undesirable alteration of outlook or attitude leading to undesirable behaviour. The courts should surely have applied their famous principle of free speech to resolve the ambiguity here in favour of a narrower interpretation. But the principle of free speech is never mentioned in this context. Lord Reid did insist, back in 1972, that 'deprave and corrupt' means more than 'lead morally astray', but unfortunately he did not say how much more.[17] So it was not clear, even after the 1970s liberalization, whether English obscene publications law met the requirements of the harm principle, let alone the more demanding requirements of the moral right to freedom of expression.

It was against this background that the Committee on Obscenity and Film Censorship, chaired by Bernard Williams, was appointed by the Home Office in 1977. The Committee's 1979 report proposed the harm principle as the correct principle by which to fix the limits of obscene publications law, and neglected the additional power of the right to freedom of expression.[18] But at the same time it invoked an extremely narrow concept of 'harm'. The result was an argument which took account

---

[16] *R.* v. *Calder and Boyars Ltd.* [1969] 1 QB 151, [1968] 3 All ER 644 at 648 (CA).
[17] In *Knuller* v. *DPP* [1973] AC 435, [1972] 2 All ER 898 at 904.
[18] *Report of the Committee on Obscenity and Film Censorship*, Cmnd. 7772 (1979).

of too much and too little. It took account of too much by focusing on harms which the rule against life-style censorship eliminates from the argument altogether (namely the violence which may be inflicted upon women because pornography influences its readers to have barbaric attitudes to women). It took account of too little because it ignored the wide range of unexcluded harms (like manipulation of readers, harassment of women unwillingly exposed to the material, exploitation of models) which can often be quite plausibly laid at the door of pornography. But the result of all this confusion was strikingly simple. Finding the link between pornography and violence unproven, as most studies do, the Committee recommended the virtual abolition of obscene publications offences. Following Denmark's lead of a decade earlier, outright bans were to be replaced with display and distribution restrictions. Needless to say, the recommendation has not been implemented. The 1980s saw the introduction of the new display and distribution restrictions, but *in addition* to the outright bans of the 1959 Act, not replacing them.[19] There were also various unsuccessful attempts to reform or displace the 1959 Act by way of private members' bills. But in a period during which the right to freedom of expression and the harm principle both seem to have become unfashionable across the whole party-political spectrum, they were all a far cry from the Williams Committee recommendations. The sponsors and promoters of the various unsuccessful bills all shared an enthusiasm for further state censorship of sexual images on 'offensiveness' grounds.

A shift to 'offensiveness' would be a radical departure for obscene publications law, but not a radical departure for English law as a whole. There are wide common law offences of 'outraging public decency' and 'conspiracy to corrupt public morals' which have persisted alongside obscene publications law. The House of Lords ruled in 1961 that these were not obscenity-based offences, and so had survived the 1959 reforms intact.[20] Not only are they not obscenity-based; they are not harm-based. Their use typically infringes the harm principle, to say nothing of the right to freedom of expression. Fortunately, they are rarely used. But when they are used there are none of the various protections which the 1959 Act affords to artists and publishers, such as the 'public good' defence and the 'aversion' argument. Soon after the common-law offences were 'rediscovered' in 1961, the Solicitor-General gave an undertaking in the House of Commons that they would not be used to circumvent the 1959 Act's defences. Perhaps that undertaking has not strictly been violated

---

[19] Local Government (Miscellaneous Provisions Act) 1982; Indecent Displays (Control) Act 1981.

[20] *Shaw* v. *DPP* [1962] AC 220, [1961] 2 All ER 446 (CA).

in the intervening years; nor, however, has its spirit been honourably upheld.[21]

Racial hatred

Unlike expression dealing with sex or violence, racist expression has never been treated as a threat to public morals or public decency at common law. The traditional weapon against political extremists, rather, was the crime of 'seditious libel'. This once extended to any expression intended to promote discontent or ill-feeling, but it was narrowed down, by 1947, to cases where violence or public disorder was intended as well.[22] Although the Board of Deputies of British Jews exerted pressure throughout the 1950s for a specific race-hate offence to supplement seditious libel, it was growing racial tension in the late 1950s and early 1960s that really put the issue high on the political agenda. Honouring a manifesto pledge, the new Labour government of 1964 included a provision in the Race Relations Act 1965 making it a crime to use threatening, abusive, or insulting words or behaviour with the intention of stirring up racial hatred and likely to stir up racial hatred. The requirement that both intention and likelihood be proved reflected the concern, voiced by the influential and enlightened Campaign Against Racial Discrimination among others, that a more inclusive offence would pose an excessive threat to freedom of expression.

Difficulties of prosecution combined with continuing pressure to prosecute have, however, led to successive relaxations of the offence's definition since 1965.[23] A new version in the Race Relations Act 1976 dropped the requirement of an intention to stir up racial hatred, because juries could easily be persuaded, apparently, to see the most insidious pseudo-educational racist material as intended to prompt discussion or encourage compassion rather than stir up hatred. But the new version cast in terms of likelihood alone also faced problems. Among other things, a kind of debased 'aversion' argument surfaced: the most hate-filled material was represented as unlikely to stir up racial hatred because it would put most people off rather than attracting them to racism. So with all-party agreement there was yet another relaxation of the wording—let us hope it will be the last—in the Public Order Act 1986. Now an intention to stir up racial hatred and a likelihood of its being stirred up have become *alternative* grounds for conviction. Throughout

---

[21] The undertaking is at (1964) HC Debs. 695, col. 1212. A recent departure from the spirit of the undertaking is *R.* v. *Gibson* [1991] 1 All ER 439.

[22] *R.* v. *Caunt, The Times*, 18 Nov. 1947.

[23] For part of the story see P. Leopold, 'Incitement to Hatred: The History of a Controversial Criminal Offence' (1976) *PL* 389.

these changes the requirement that the words or behaviour be 'threatening, abusive or insulting' has remained constant. Prosecutions, moreover, have all along required the consent of the Attorney-General. That consent has not been forthcoming very often. For comparison: between 1976 and 1981, a period during which over a thousand obscene publications prosecutions and twice as many forfeiture proceedings were launched, there were only around twenty prosecutions for incitement to racial hatred.

Embarrassed liberals sometimes make misguided calls for more zealous prosecution of racist speakers just because of their racist attitudes or racist influence. What they should be calling for are stronger laws to deal with racial harassment and intimidation. The incitement-to-racial-hatred laws, as drafted, carry an alarming element of life-style censorship, in spite of the nominal emphasis on the immediate impact of the expression ('threatening, abusive or insulting'). For all that there have been campaigns, particularly in the late 1980s, for there to be a analogous offence of incitement to *religious* hatred in England, as there already is, for obvious reasons, in Northern Ireland. Fortunately the campaigns have been unsuccessful. The common law offence of 'blasphemous libel' remains in place, however, giving anomalous and wide-ranging legal protection to the religious sensibilities of Christians. After more than fifty years of welcome desuetude, the offence was revived in a much-publicized private prosecution against Gay News in 1977.[24] In 1990, Muslims hoping to bring Salman Rushdie to trial in connection with *The Satanic Verses* made an unsuccessful attempt to persuade the Divisional Court that blasphemy extends to protect religions other than Christianity.[25] Even the government of the day, not otherwise known for its civil libertarian leanings, cited Rushdie's right to freedom of expression in refusing to introduce legislation extending blasphemy to protect other religions. It is a shame that this unexpected enthusiasm for free expression did not extend to implementing the Law Commission's sound 1985 advice that the crime of blasphemy should be abolished altogether.[26]

Defamation

Some writers liken racist or sectarian agitation to defamation of character. But the analogue is of limited value, because so little racist agitation belongs to the category, that of purely empirical assertion, with which defamation law is properly concerned. The moral right to freedom of

[24] *Whitehouse* v. *Lemon* [1979] AC 617, [1979] 1 All ER 898 (HL).
[25] *R.* v. *Chief Metropolitan Stipendiary Magistrate ex parte Choudhury* [1991] 1 All ER 306. On the *Satanic Verses* affair, see chapter 14.
[26] *Offences against Religion and Public Worship* (1985).

expression gives virtually no protection to communicators of false information, but because of the structure of its justification it swings back into
action to protect the expression of attitudes and concerns, however base
or misconceived. The history of English defamation law can be seen as an
attempt to capture that distinction. The basic rule is that any statement
which lowers somebody in the estimation of 'right-thinking people' is
actionable in the civil courts. This already puts vulgar abuse and crude
generalization out of the range of liability, since no 'right-thinking person'
would be swayed by such things. But there are also various distinct
defences. Most significantly, one can escape liability by proving the truth
of the facts stated, or by showing that the statement was fair comment.
The 'fair comment' defence has been interpreted broadly, to cover virtually every assertion which is not purely empirical, and in modern cases
the law's elusive principle of free speech has helpfully been invoked
where the limits of the defence have been in doubt.[27]

In spite of these well-conceived defences, however, the threat of defamation writs has seriously inhibited writers and editors in England during
the twentieth century. Not least among their worries has been the common
law's regime of strict or 'no-fault' liability for defamation. It is generally no defence, according to judicial decisions in the early part of the
century, that the publisher of the alleged defamation could not have
known that the plaintiff's reputation would be called in question by the
publication.[28] The 'terror to authorship' created by this strict liability was
one of the main concerns of the Committee on the Law of Defamation,
chaired by Lord Porter, which reported in 1948.[29] The Government took
no action on the Committee's report, but a private member succeeded in
having some of its recommendations enacted as the Defamation Act
1952. Among other things, the Act created a new immunity (or 'privilege')
for certain kinds of verbatim reports in newspapers and broadcasts, and
offered a new defence against strict liability. The new defences were
conditional, however, in the sense that they were available only to those
who had offered to publish corrections or retractions of the statements
said to be defamatory. The Porter Committee had rightly seen this 'conditional defence' model, which it analogized to the more extensive French
droit de réponse, as an ideal way to deal with two apparently competing
concerns at once. One could mitigate the harm to those defamed, while at
the same time reducing the 'chilling effect' of defamation law, its tendency

---

[27] *Kemsley* v. *Foot* [1952] AC 345, [1951] 1 All ER 331 at 342 (CA); *Slim* v. *Daily Telegraph Ltd.* [1968] 2 QB 157, [1968] 1 All ER 497 at 503 (CA).

[28] *Hulton & Co.* v. *Jones* [1908–10] All ER Rep 29; *Cassidy* v. *Daily Mirror Newspapers* [1929] All ER Rep 117.

[29] *Report of the Committee on the Law of Defamation*, Cmnd. 7536 (1948). The reference to the 'terror of authorship' is in *Knupffer* v. *London Express Newspaper Ltd.* [1942] 2 All ER 555 at 561 per Goddard LJ.

to deter the publication of true information as well as false information.

Unfortunately, the 1952 Act did not really make much practical difference to the 'terror' of publication. The protection for publishers not at fault, in particular, was quickly found to be too narrowly circumscribed to be of much use to anyone. As a result, another Committee on Defamation was appointed in 1971, this time with Faulks J. in the chair. This Committee was also enthusiastic about the 'conditional defence' model, and in 1975 it recommended, among other things, considerable strengthening of the 1952 Act's provisions on newspaper and broadcasting privilege and innocent publication.[30] But there has been no further legislation in the wake of these recommendations. The corrections and withdrawals that appear regularly in the pages of newspapers and magazines nowadays are of no special legal significance. They are there simply because some of those who are defamed will settle for a retraction instead of serving a writ, or as the price for withdrawing a writ. The alternative is long and costly litigation. The result can be a huge windfall for the plaintiff and a crippling outlay for the defendant, even if the offending story was well checked at the time. The problem is partly that a jury is used, not only to decide the case—itself an anomaly in the civil law—but also to fix the damages award. Some absurdly high awards in the 1980s generated renewed pressure for a major overhaul of the system. Recent legislation has provided for increased appellate controls on damages awards in defamation cases.[31] But an implementation of the Faulks Committee recommendations remains equally urgent.

The seriousness of the situation is made clear, if more clarity be needed, by a decision of the New York Supreme Court in 1992. Faced with a request for local enforcement of a British libel award, a New York judge held that he was entitled to decline the request—an exceptional step to take—on the ground that the cause of action on which the award was based was 'repugnant to the public policy of the state.' The repugnance in question, he said, stemmed from the fact that defamation liability in England is strict, and from the fact that it places the burden of proving the truth of a defamatory statement on the defendant. Because of their serious 'chilling effect' on the dissemination of true information, either of these facts would have sufficed by itself to put English libel law in violation of the free speech guarantee in the First Amendment of the United States Constitution. English standards in this area are, said the judge, 'antithetical' to the protections offered to the press in the United States.[32] Coincidentally, the English Court of Appeal had recog-

---

[30] *Report of the Committee on Defamation*, Cmnd. 5909 (1975).

[31] Courts and Legal Services Act 1990, s. 8(1); RSC Ord. 59 R. 11(4).

[32] *Bachchan* v. *India Abroad Publications Inc.* (S. Ct. NY, transcript 28692/91, 13 April 1992).

nised the same antithesis a month or two earlier, in a landmark decision denying the right to sue in defamation to public bodies, partly supported by reference to the free speech jurisprudence of the US Constitution.[33] The House of Lords recently upheld this decision, taking the more disingenuous line that the arguments used in the American constitutional cases are already a recognised part of English law, captured in its free speech principle.[34] Certainly the decision to deny the defamation action to public bodies is a step in the right direction, but in countless other respects, as the New York humiliation reminds us, the English law of defamation still has a very long way to go before it falls into line with the First Amendment, let alone honours the (in some respects still tougher) moral right to freedom of expression.

Confidentiality

The civil law of defamation exists to prevent and remedy certain kinds of harmful misinformation, and its adverse impact on the dissemination of accurate information is just a side-effect, albeit a devastating one. The civil law of breach of confidence, by contrast, exists precisely in order to inhibit the free flow of perfectly accurate information. In recent years there has been a huge expansion of this head of liability beyond its traditional role of protecting trade secrets from disclosure by disgruntled employees and unscrupulous commercial collaborators. Communications between husband and wife, between litigants, and between Cabinet Ministers, have all been held to carry obligations of confidence.[35] There is still no general test for deciding whether information has been imparted in confidence. And yet once a duty of confidentiality is in place it is enforceable not only against the original recipient of the information, but also against third parties to whom it has subsequently been passed, at least once they realize that it was passed to them in breach of confidence. The breach of confidence action virtually confers a proprietary right in information. Only the fact that information is already 'in the public domain'—as in the final stages of the infamous *Spycatcher* litigation in 1988—will put an end to the duty to keep it secret.

Where defamation law has its 'fair comment' defence, the law relating to breach of confidence provides a defence of 'disclosure in the public interest'. The view earlier in the century was that there was no public

---

[33] *Derbyshire County Council* v. *Times Newspapers Ltd.* [1992] 1 QB 770, [1992] 3 All ER 65.

[34] [1993] 1 All ER 1011.

[35] Respectively: *Duchess of Argyll* v. *Duke of Argyll* [1967] Ch 302, [1965] 1 All ER 611; *Distillers Co. (Biochemicals) Ltd.* v. *Times Newspapers Ltd.* [1975] QB 613, [1975] 1 All ER 41; *Attorney-General* v. *Jonathan Cape Ltd.* [1976] QB 752, [1975] 3 All ER 484.

interest in the disclosure of 'private' wrongs or immoralities, but only in the reporting of crimes.[36] In a line of Court of Appeal cases in the 1960s and 1970s, however, the public-interest defence was developed to permit public disclosures which would lend the lie to attractive public images or would draw attention to questionable business methods, with much emphasis placed on the right to freedom of expression.[37] The 1980s seem to have brought something of a retreat, or at any rate an equivocation, beginning with a disturbing remark, in a 1980 decision by the House of Lords, that the exposure of mismanagement in a nationalized industry would not be in the public interest in the relevant sense.[38] More recently, in the *Spycatcher* case, the House of Lords has accepted that the public interest might extend beyond revelations of wrongdoing in some cases, but also endorsed the view that the public interest would often warrant disclosure to the relevant authorities only, and not to the public at large.[39] The Law Lords apparently thought that the right to freedom of expression could be adequately respected here by shifting the burden of proof on the public-interest question from defendant to plaintiff in cases brought by the Government, although not in cases brought by 'private citizens'.[40] In a far from radical report on the subject in 1981, by contrast, the Law Commission had suggested that the shift to the plaintiff of the main burden of proof relating to public interest was required in all cases, public or private, to reflect the importance of the 'free circulation of information'.[41]

But divining the precise details of the public-interest element in liability for breach of confidence may be somewhat academic. Since topical *exposés* tend to go stale, the decision which matters for practical purposes in most confidentiality cases is the decision to grant or withhold an interlocutory injunction in the early stages of the litigation, not the ultimate ruling at the trial. The possibility of protecting freedom of expression in this area therefore depends, in practice, on the extent to which courts are prepared to depart from the standard 'balance of convenience' test for interlocutory injunctions, a test which would naturally militate against the publication of information, since information cannot be made secret again once disclosed. In defamation cases the courts depart from the balance of convenience test by refusing to issue interlocutory injunctions if a remotely

---

[36] *Weld-Blundell* v. *Stephens* [1919] 1 KB 520.
[37] *Initial Services* v. *Putterill* [1968] 1 QB 396, [1967] 3 All ER 145; *Hubbard* v. *Vosper* [1972] 2 QB 84, [1972] 1 All ER 1023; *Woodward* v. *Hutchins* [1977] 1 WLR 760, [1977] 2 All ER 751 (CA). A notable common factor in all of these cases was Lord Denning.
[38] *British Steel* v. *Granada Television* [1981] 1 All ER 417 (HL) at 455 per Lord Wilberforce.
[39] *Attorney-General* v. *Guardian Newspapers Ltd. (No. 2)* [1988] 3 All ER 545 at 657 (Lord Griffiths) and 659–60 (Lord Goff).
[40] Ibid. at 640 (Lord Keith) and 660 (Lord Goff).
[41] Law Commission, *Breach of Confidence*, Cmnd. 8388 (1981), 140.

arguable defence of justification or fair comment will be entered.[42] But in 1984 the Court of Appeal rejected the view that a similar rule might exist where a public-interest argument will be made in a breach of confidence action.[43] On the other hand, after a period of considerable doubt in the 1960s and 1970s, it now seems to be accepted that the ordinary balance of convenience test must be modified in breach of confidence cases to allow considerations of the public interest to be given some weight even at the interlocutory stage.[44] Interlocutory injunctions in breach of confidence cases nevertheless represent a serious risk to the free flow of information. Nor are interlocutory injunctions the only threat to freedom of expression encountered in the pre-trial process. In civil actions the 'discovery' procedure can be used to ascertain the sources of journalists' information, so long as this is in the interests of justice, national security, or crime prevention, and while newspapers and broadcast media are exempt from the duty to identify their informants in this way in defamation actions, they are not so exempt in breach of confidence actions.[45] The risks involved in leaking incriminating or embarrassing material to the press are therefore high, if one values one's career, and informants will often be deterred even where an overwhelming public-interest argument could ultimately be made. This intolerable situation prompted a notable show of journalistic courage in a much-publicized 1989 case, in which a young writer was prepared to go to prison rather than give in to a discovery order which would have revealed the source of his information.[46]

Official secrets

Both attacks on reputation and disclosures of confidential information can ground criminal as well as civil liability. The criminal law counterpart of defamation is the common-law offence of 'defamatory libel', occasionally invoked to deal with extreme cases of character assassination. The criminal law counterpart of breach of confidence is not quite so general, but it is more frequently used. It only affects those who disclose *official* secrets.

The Official Secrets Act 1911 made it an offence to disclose or receive any government information without authorization. There was no need for the information to be about national security or anything remotely

---

[42] *J. Trevor & Sons* v. *Solomon*, *The Times*, 16 Dec. 1977.

[43] *Lion Laboratories Ltd.* v. *Evans* [1985] QB 526, [1984] 2 All ER 417.

[44] *Attorney-General* v. *Guardian Newspapers Ltd.* [1987] 3 All ER 316 (*Spycatcher* at the interlocutory stage).

[45] Compare *Lyle-Samuel* v. *Odhams Ltd.* [1920] 1 KB 135 with *British Steel* v. *Granada* [1981] 1 All ER 417. The general rule limiting disclosure to matters of justice, national security, or crime prevention is in s. 10 of the Contempt of Court Act 1981.

[46] *X Ltd.* v. *Morgan-Grampian (Publishers) Ltd.* [1990] 2 All ER 1.

sensitive; indeed in this context it was no defence that the information was already in the public domain.[47] The section did allow disclosures to a person to whom the discloser owed a duty of disclosure 'in the interests of the State'. But a 1962 House of Lords decision suggested that 'the interests of the State' fell to be determined by the policies of the government of the day, at least as far as defence matters were concerned.[48] So this was clearly not a general public-interest defence. Nevertheless, in a highly controversial 1970 prosecution arising out of the disclosure of documents relating to the Biafran war, the jury appear to have insisted on treating it as a public-interest defence, spurred on by a judge who made no attempt to disguise his contempt for the 1911 Act.[49] The Government responded by appointing a committee to examine the operation of the section, chaired by Lord Franks, which agreed that section 2 was a 'mess' and recommended limiting the liability to disclosures of information about defence, the economy, international relations, and a few other matters.[50] There were various abortive moves towards statutory reform between 1978 and 1988. Over the same period there was a steady stream of prosecutions under section 2, including a pair of infamous cases in 1984 against civil servants Sarah Tisdall and Clive Ponting, both of whom had leaked documents not so much jeopardising the national interest as embarrassing the Government. The adverse publicity surrounding these cases, and the evident willingness of juries to go on treating the 'interests of the State' proviso as a public-interest defence, finally got too much for a government increasingly obsessed with secrecy. A new White Paper in 1988 set out proposals for what was advertised as a significantly liberalizing reform, but which was in reality nothing of the kind.[51]

The Official Secrets Act 1989 adopts the Franks Committee approach of limiting criminal liability to those who disclose information falling into certain specified categories, namely information about security and intelligence, about defence, about international relations, and about certain law-enforcement matters. There are also new defences for unwitting disclosers and those whose disclosures do no damage, but the main burden of establishing these lies on the defence, an extraordinary aberration in a criminal law context. In many cases involving security and intelligence matters, moreover, there is an irrebuttable presumption of damage. The old 'interests of the State' proviso has gone altogether, depriving juries of the chance to read it as a public-interest defence. With security and

---

[47] *R.* v. *Crisp* (1919) 83 JP 121.
[48] *Chandler* v. *DPP* [1964] AC 763, [1962] 3 All ER 142 (HL).
[49] *R.* v. *Aitken* (1971), documented in Jonathan Aitken, *Officially Secret* (1971).
[50] *Report of the Departmental Committee on Section 2 of the Official Secrets Act 1911*, Cmnd. 5104 (1972).
[51] *Reform of Section 2 of the Official Secrets Act 1911*, Cm. 408 (1988).

intelligence matters, in effect, the Government is always the arbiter of what is to be disclosed without criminal sanction. This position is backed up by a provision confirming the *Spycatcher* doctrine that members and former members of the security services owe a lifelong duty of confidentiality to the Crown, but shorn of the *Spycatcher* allowance for a public-interest defence: so, in the civil courts also, the Government will now have the final say in relation to this category of information. Identification of a journalist's source of information can still be ordered in the interests of justice, national security, or crime prevention, and not only as part of the civil procedure, but also to enable criminal prosecution.[52] And the draconian powers of search and seizure available under the 1911 Act—memorably used to impound the tapes of the *Secret Society* series at BBC Scotland in 1987—are left entirely intact under the new legislation.

Any relaxation of the law in this new Act is more or less cosmetic, since the old section 2 was never used in practice for disclosures falling outside the new protected categories. The main aim of the changes, rather, has been to reinforce the seal on the least appetizing areas of British governmental activity, eliminating all concessions to freedom of expression. It is easy to exaggerate the role which concessions to freedom of expression could play in making our government less secretive, however. Remember that the right to freedom of expression is a right to convey information, not to receive it. The right contributes to the free flow of information because people are often willing to pass information on. The Government's 'D-notice' system, under which the media co-operate in keeping certain official information under wraps, shows that such willingness cannot always be relied upon even among those whose job it is to pass on information. No doubt we can expect less willingness still from senior civil servants whose job is apparently precisely the opposite.

Contempt of court

It is perhaps rarer to find such secrecy in the judicial arm of government. Private hearings in chambers are used for some proceedings, and with these there will occasionally be a 'conspiracy of silence' among the parties and their lawyers. But normally the only way to stop information about judicial proceedings from getting out in our system of largely open-court trials is to prohibit people from passing it on, restricting their freedom of expression. This has always been one role of the law relating to contempt of court.

In prominent cases in the 1950s it was confirmed that contempt of court

---

[52] *Secretary of State for Defence* v. *Guardian Newspapers Ltd.* [1985] AC 339, [1984] 3 All ER 601 (HL).

was a crime of strict liability at common law, extending to any activity which created any risk of prejudice to pending or imminent criminal or civil proceedings, or which tended to interfere with the course of justice in such proceedings.[53] There was much criticism of these decisions in the press, and a committee of the influential organization Justice recommended that exceptions be carved out of the strict-liability rule.[54] The Government adopted this recommendation in the Administration of Justice Act 1960, providing a cluster of 'innocent publication' defences. Then it was brought home, in the 1960s, that even a remote risk of prejudice, arising from publications long before a trial, would be sufficient for contempt liability.[55] At least it seemed clear, at that point, what kind of reports would be covered and why. It was contempt to publish previous convictions, or alleged confession or identification evidence, or bare assertions of guilt, all of which might give jurors and witnesses preconceptions about the case before it started; and of course attempts to sway a judge, however futile, would be interferences with the course of justice. It was easy to see the point of these restrictions on freedom of expression: the established authoritative process for deciding whether certain information is accurate, so that its dissemination is strongly protected by the right to freedom of expression, should not have its conclusions corrupted by premature exercises of that right.

But the 1973 decision of the House of Lords in a widely publicized contempt case against the *Sunday Times* took the restrictions a good deal further.[56] It was a contempt, because an interference with the course of justice, to publish material encouraging a litigant to give up defending a civil action by denying that the defence had any merit, even if there was no intention or likelihood of swaying the court itself. What was crucial was not prejudice to the trial, but prejudgment of the issues, exposing the litigant to the pressure of public disapproval. This was another case, like many in the defamation and breach of confidence contexts, in which a company was protected from embarrassing public criticism of its immoral activities by a law which was conveniently diverted from its proper function. Decisions to settle, unlike the carefully circumscribed decisions of judges and juries, are ordinary decisions to be made on ordinary grounds; they are not established authoritative processes for establishing the truth of anything and there is no more reason to protect them from outside influences than there is to protect the making of government

---

[53] *R.* v. *Evening Standard Co. Ltd. ex parte Attorney-General* [1954] 1 QB 778, [1954] 1 All ER 1026, *R.* v. *Odhams Press Ltd. ex parte Attorney-General* [1957] 1 QB 73, [1956] 3 All ER 494, *R.* v. *Griffiths ex parte Attorney-General* [1957] 2 QB 192, [1957] 2 All ER 379.

[54] Justice, *Contempt of Court: A Report by Justice* (1959).

[55] *R.* v. *Thomson Newspapers Ltd. ex parte Attorney-General* [1968] 1 WLR 1, [1968] 1 All ER 268, *R.* v. *Savundranayagan* [1968] 1 WLR 1761, [1968] 3 All ER 439 (CA).

[56] *Attorney-General* v. *Times Newspapers Ltd.* [1974] AC 273 [1973] 3 All ER 54 (HL).

policy from outside influences. It is hardly surprising, then, that the European Court of Human Rights found in 1979 that there had been a violation of the right to freedom of expression here.[57] An attempt was made, in the Contempt of Court Act 1981, to accommodate the objections of the European Court, and also to implement recommendations made by a Committee on Contempt of Court which had been set up, coincidentally, just before the *Sunday Times* case.[58] The 1981 Act does not cover those who act with the intention of prejudicing or interfering in trials, who still fall under the common law rules. Liability for unintentional contempts, on the other hand, is subject to a new range of statutory controls. Most important, there must now be a substantial risk of serious prejudice and there is a defence for 'public interest' discussions with merely incidental prejudicial effect. For the most part, the courts seem to have taken the new restrictions seriously, explicitly recognizing their importance as contributions to freedom of expression.[59] But one ominous sign was the hint, in a 1983 decision, that cases exactly like *Sunday Times* might still be caught by the law on unintentional contempt.[60] And the subsidiary *Spycatcher* actions have given real cause for concern that, with their usual incapacity to understand the simple concept of intention, the courts will treat merely *knowing* prejudice to trials as intentional, thereby circumventing the 1981 Act's controls in a wide range of 'public interest' cases, and not merely in the 'trial by media' cases which the residual common law of intentional contempt was presumably designed to catch.[61]

There remain some aspects of judicial proceedings which it is automatically a contempt, or in some cases a statutory offence, to report. These include the details of committal proceedings, the deliberations of the jury, and matter validly placed under reporting restrictions by the judge.[62] The limits and force of such reporting restrictions are far from clear, but their imposition where they are not essential to the administration of justice has recently twice met with the disapproval of the Divisional Court.[63] Nor should we forget 'contempt in the face of the court', which is committed by those who refuse to be silent in court, insult the jury, or such like, and 'scandalizing the court', committed by those who go too far in their attacks on the judiciary. Fortunately, it now seems that it is virtually

---

[57] *Sunday Times* v. *UK* (1979) 2 EHRR 245.

[58] *Report of the Committee on Contempt of Court*, Cmnd. 5794 (1974).

[59] *Attorney-General* v. *Times Newspapers Ltd.*, *The Times*, 12 Feb. 1983, *Attorney-General* v. *News Group Newspapers Ltd.* [1987] QB 1, [1986] 2 All ER 833 (CA).

[60] *Attorney-General* v. *English* [1983] 1 AC 116, [1982] 2 All ER 903 (HL).

[61] See *Attorney-General* v. *Newspaper Publishing Plc* [1988] Ch 33, [1987] 3 All ER 276 (CA); *Attorney-General* v. *Observer Ltd.* [1988] 1 All ER 385.

[62] Magistrates Courts Act 1980, s. 8; Contempt of Court Act 1981, ss. 4, 8 and 11.

[63] *R.* v. *Evesham Justices ex parte McDonagh and Berrows Newspapers* [1988] QB 540, [1988] 1 All ER 371; *Attorney-General* v. *Guardian Newspapers Ltd* (No. 3) [1992] 1 WLR 874 [1992] 3 All ER 38.

impossible to go too far, with a 1968 Court of Appeal decision citing the right to freedom of expression as an insuperable obstacle to putting intemperate critics of the judiciary behind bars.[64]

Media controls

It would be easy to imagine, from what has been said so far, that judges and juries have had, between them, a virtual monopoly on the regulation of expression in our country since the second world war. The courtroom seems to have been the cutting-room for everything from pornographic magazines to leaked military secrets. But in fact a good deal of control over expression in Britain has been and continues to be carried out by officials with specific delegated powers, either to censor material before it goes out or to censure its producers when the complaints come in.

Of these delegated censorship powers the oldest to survive into the post-war period was that of the Lord Chamberlain, under the Theatres Act 1843, to ban the production of any play or part of a play which he considered hostile to 'the preservation of good manners, decorum or the public peace'. It was with the work of the new dramatists of the 1950s and 1960s that serious tensions started to emerge. A Joint Committee on Theatre Censorship was set up, and recommended complete abolition of the Lord Chamberlain's powers. In 1968 this recommendation was implemented by statute, and the Government took the excellent precaution of removing liability for common law public decency and public morals offences from theatre productions at the same time. In return, producers and directors became subject to prosecution (with the consent of the Attorney-General only) according to the test of obscenity in the Obscene Publications Act 1959, a limited form of control which has stood them in reasonably good stead ever since. A homosexual rape scene in a 1981 National Theatre production of Howard Brenton's morality play *The Romans in Britain* gave rise to pressure for prosecution from the usual quarters, but the Attorney-General sensibly declined.

A much more recent creation than the Lord Chamberlain's jurisdiction is that of the British Board of Film Classification, previously and less euphemistically known as the British Board of Film Censors. It was originally set up in 1912 by the cinema industry itself, hoping to present a responsible image to suspicious prophets of moral decline. Local authorities, who had a statutory power to control cinemas by licence, increasingly included requirements of BBFC certification in cinema licence conditions. The Cinematograph Act 1952 gave implicit statutory recognition to the

---

[64] *R. v. Metropolitan Police Commissioner ex parte Blackburn (No. 2)* [1968] 2 All ER 319 (CA).

system by placing local authorities under a duty to certify films as fit for children, adding that they could rely on the certification of 'such other body as may be specified in the [cinema] licence', an oblique reference to the BBFC. But certification for adults was still in the discretion of local authorities, and although many used BBFC certificates here as well, some devised their own conditions instead. The Greater London Council, for example, attempted to limit licensees only by the standards of the general law, requiring them not to show material that was obscene or likely to stir up racial hatred. In 1976 judicial review proceedings, it was held that the GLC had erred too far on the side of tolerance. The Obscene Publications Act 1959 did not at that stage extend to films, but the *recherché* common law of 'indecent exhibitions' did, so that the GLC, sanctioning the display of films that were not obscene but merely indecent, was sanctioning the commission of criminal offences and acting beyond its legal powers.[65]

The irony was that the GLC could in practice have achieved the desired result with impunity by relying on BBFC certificates. The BBFC was already prepared to grant 'X' (adults only) certificates to films with content that would doubtless have counted as indecent by ordinary legal standards, and it was clear both that criminal prosecutions would not in practice be mounted by the DPP against cinema films with BBFC certificates, and that a prima-facie reliance on BBFC certificates would not have been held to be beyond the council's powers. In any case, one upshot of the litigation was that the Government included a provision in the Criminal Law Act 1977 replacing the 'indecent exhibitions' liability of cinemas with liability under the Obscene Publications Act 1959. Liability for incitement to racial hatred was extended to cinema films under the Public Order Act 1986, but in practice BBFC certification continues to provide significant *de facto* protection against criminal prosecutions, the odd private prosecution apart. And the BBFC has on occasion taken successful steps to discourage private prosecutions by mediation and powerful advocacy, as with *The Last Temptation of Christ* in 1988. For a non-statutory body, the BBFC has remarkable control over the regulation of cinema. Set against the background of our inhospitable criminal law, it is hard to deny that the organization serves rather than retards the positive values underlying the right to freedom of expression, a fact which perhaps helps to explain the Williams Committee's enthusiasm for most of the film censorship system.[66] The only question can be whether the BBFC is sufficiently identified with the state for its influence to carry the social meaning of life-style censorship. Perhaps the Government's

[65] *R. v. Greater London Council ex parte Blackburn* [1976] 1 WLR 550, [1976] 3 All ER 184 (CA).
[66] *Report on Obscenity*, 158.

decision to adopt the BBFC as statutory censor of video films for home use under the Video Recordings Act 1984 has already changed it from being a bulwark against official interference to being an official interferer in its own right. Certainly, impending litigation in respect of the BBFC's refusal to certify the film *Visions of Ecstacy* for video distribution suggests that the organisation has crossed that fatal line. The film, rejected on grounds of religious insensitivity, represents if anything less of a challenge to Christian doctrine than *Last Temptation*, so adeptly steered through the film certification process only a couple of years earlier.

If the BBFC has an ambiguous status *vis-à-vis* the state, then the governing bodies of national broadcasting are in an even more complex position. The British Broadcasting Corporation has since its inception in 1926 consisted of a Board of Governors, which supervises broadcasting according to the terms of the Corporation's royal charter and its special 'Licence and Agreement'. Independent broadcasting, meanwhile, was inaugurated in 1954 with the creation of the Independent Television Authority (later the Independent Broadcasting Authority) as a licensing and supervisory body. The Television Act 1954 required the ITA to ensure that, among other things, programmes under its auspices were not tasteless, indecent, or offensive, did not encourage crime or disorder, preserved 'due impartiality' on controversial current affairs issues, and did not express the opinions of the Authority or its members on public policy questions. The BBC already had in its licence a requirement not to broadcast its own opinions on public policy, and in 1964 provided its own undertakings to the Government concerning taste, decency, the setting of bad examples, and due impartiality. Exempt from the law on obscene publications and incitement to racial hatred, both bodies emerged as anomalous self-regulators, in a special relationship of trust with the Government. Both the BBC board and the IBA have always been made up entirely of government appointees, with no requirement of 'due impartiality' in their selection. There have been many instances of the BBC board cravenly complying with government demands to withdraw or cut specific programmes, or simply taking the easy way out with potentially controversial items. The IBA, likewise, has regularly stepped in to prevent the broadcast of politically sensitive or otherwise injudicious material, especially in the wake of a Court of Appeal decision in 1973 taking it to task for delegating decisions too readily and failing to engage in enough pretransmission scrutiny.[67] On the other hand, there have been notable instances of fortitude in the face of government pressure, as with the IBA's admirable decision to stand by the documentary *Death on the Rock*

[67] *Attorney-General ex rel. McWhirter* v. *Independent Broadcasting Authority* [1973] QB 629, [1973] 1 All ER 689 (CA).

in 1988, or the BBC board's sturdy defences against facile allegations of left-wing bias. So it is just not clear whether these bodies should be seen as the government's agents in the control of broadcasting or as embattled broadcasters struggling against a hostile state apparatus.

In a sense this tension has almost reached breaking-point during the 1980s. On the one hand, the Government has made no bones about packing the BBC board and IBA with its loyal supporters, and expecting them to tow the official line. On the other hand, it has lashed out against both sides of the industry with growing hostility. It created governing bodies it could trust, only to inflict ever-increasing damage on the whole relationship of trust on which broadcasting had thrived since the early days. The re-regulation of broadcasting started innocently enough with the creation in 1980 of the Broadcasting Complaints Commission, set up on the recommendation of the Annan Committee on Broadcasting to adjudicate public complaints of unfair treatment and unwarranted invasions of privacy.[68] But then came the Broadcasting Standards Council, set up in 1988 amid gloomy predictions of its likely influence on creativity, and put on a statutory footing under the Broadcasting Act 1990. It exists to monitor the standards of taste and decency of television and radio, to provide its own codes, and to adjudicate related complaints. The BSC's bark has so far been worse than its bite—much of its energy has been devoted to harmless, although intellectually dubious, 'research'—but its creation was a sign of growing hostility towards the broadcasters themselves. And there was more to come. Late in 1988, as discussed in chapter 5, the Home Secretary announced that there would be a general ban on the broadcasting of direct statements by representatives of Sinn Fein and a range of other organizations. It was presented as being the broadcasters' fault that this blatant life-style censorship was called for: after all, they were the ones who had insisted on giving the 'oxygen of publicity' to supporters of violence. And as if broadcasters had not by now been sniped at enough, there came the final volley of shots. In 1990 the IBA was to be closed; there was to be a new Independent Television Commission and a new Independent Radio Authority; they were to issue brand new 'Impartiality Codes' and 'Programme Codes' with severe penalties for uncooperative broadcasters instead of routine pretransmission scrutiny; and, subject to an apparently undemanding 'quality threshold', the independent broadcasting franchises were to be sold to the highest bidder. At the same time both the BBC and the independent sector were to be brought under the law relating to obscene publications and incitement to racial hatred. The Government described this, in all seriousness, as 'broadcasting deregulation'.

---

[68] *Report of the Committee on the Future of Broadcasting*, Cmnd. 6753 (1977); Broadcasting Act 1980, s. 17.

I pointed out, at the outset, that the right to freedom of expression puts more weight into defeating some kinds of media regulation than others. It is weak against regulation that will require programming diversity, will grant public access, and so on. These enhance validation. The right is strong against regulation that will damage minority programming, such as regulation that forces broadcasters into purely mainstream programming by setting financial qualifications. The 1980s began with welcome regulation of the first kind—the creation of Channel Four—but they ended with regulation of the second kind, which is likely to pose serious threats to freedom of expression once the treadmill of regular franchise bidding is fully under way. The first auction, in 1991, did not go exactly as expected, since some bids were rejected, not for being too small, but for being so large as to threaten the commercial viability of the bidder. Nevertheless it would be hard to pretend that the overall results have defied expectations: independent television has certainly drifted closer and closer to a mass-market homogeneity in the wake of the franchise changes, and the BBC has been pulled in the same direction through its efforts to maintain impressive ratings in the run up to its Charter renewal date in 1996. And there has been regulation in other unwelcome categories. There are the bulging rules on 'impartiality', for example, an ideal which is quite inconsistent with any kind of complexity or creativity in programming. Of course, there is the familiar argument that television and radio are run by government bodies, the BBC board and the ITC, who therefore necessarily impose the stamp of official authority on whatever they transmit, and should not be partisan for that reason. But that claim cuts both ways. If the BBC Board and the ITC are indeed thought of as agents of the Government then any editorial control they exercise over their producers and editors itself carries the social meaning of official censorship. That would mean that a BBC board could not legitimately ban or curtail programmes because of the attitudes or concerns they convey. If they did it would be an act in the same class as the notorious 'section 28' ban on the promotion of homosexuality by local authorities—in the same class, indeed, as the very Sinn Fein ban which caused so much wringing of hands at the BBC.

INTERNATIONAL STANDARDS

The legal challenge to the Secretary of State's broadcasting ban, rejected in the House of Lords, was taken (unsuccessfully, as discussed in chapter 5) to the European Commission on Human Rights. So the regulation of broadcasting joins the long list of civil rights issues, many of them related to freedom of expression, on which Britain's performance has been called into question before international legal tribunals.

Britain's international commitments concerning freedom of expression go back as far as 1948, to the adoption of the Universal Declaration of Human Rights by the then fledgling United Nations General Assembly. Article 19 of the Declaration recorded, with brief elaboration, that 'everyone has the right to freedom of opinion and expression'. It took the United Nations another twenty-eight years to carry this moral assertion over into international law, by means of its International Covenant on Civil and Political Rights. Article 19 of the Covenant enacted that 'everyone shall have the right to freedom of expression', with the new word 'shall' emphasizing that this would now be a legal and not merely a moral right. A few sentences of elaboration, altered only slightly from the 1948 Declaration, made clear that any expressive or communicative activity would be covered by the right (indeed a right to receive information was thrown in for good measure), but added that any expressive or communicative activity would be open to domestic-law restrictions if such restrictions were necessary to protect the rights or reputations of others, or to protect national security, public order, public health, or public morals.

It is not hard to see how English law's main restrictions on expression might be squeezed into these categories. As Britain ratified the Covenant in 1976 without agreeing to let individuals petition the UN Human Rights Committee, however, its legal arrangements have never been authoritatively tested against the Covenant's requirements. Moreover, the arrangements of other nations that have been tested do not present very useful analogues. Nor have either of the two Article 19s been much cited in British courts. They have featured from time to time in political debate, and the British government did make a notable (and solitary) stand in defence of Article 19 of the Declaration when the UN General Assembly itself was considering draconian measures to stamp out racist ideas and ideologies world-wide.[69] But it is fair to say that Article 19 matters more for what it symbolizes than for the use to which it is put.

The same cannot be said of Article 10 of the European Convention on Human Rights, to which Britain has been bound since 1953, with a right of individual petition since 1966. Like Article 19 of the Covenant, Article 10 of the European Convention creates a right to freedom of expression in international law. As in the Covenant, a right to receive information is appended to the right to freedom of expression as if it were part of the same right. And as in the Covenant, there is a list of permissible restrictions on the right's exercise. This list is a bit longer: it adds confidentiality and the authority of the judiciary to the catalogue of interests which may be protected at the expense of free expression, so long as the protection is

---

[69] Cited in E. Schwelb, 'The International Convention on the Elimination of All Forms of Racial Discrimination' (1966) 15 *Int. and Comparative LQ* 996 at 1025.

by law and 'necessary in a democratic society'. Of the major cases which have reached Strasbourg from British shores, several have concerned Article 10 and this extended list of permissible restrictions. The *Sunday Times* case, in which the European Court of Human Rights disapproved aspects of the law relating to contempt of court as going beyond what was necessary for preserving the authority of the judiciary, has already been mentioned.[70] More recently, the Government's embarrassment over the *Spycatcher* affair was compounded by a defeat in the same forum.[71] Other cases were less successful. In a 1975 case about the use of the Obscene Publications Act the Court held that governments have a 'margin of appreciation' as to the level of regulation needed 'for the protection of morals', but rightly added that prohibition on 'offensiveness' grounds would not meet the requirements of the Convention.[72] And among the countless cases rejected by the European Commission on Human Rights before even reaching the Court, we find an attempt to defeat the *Gay News* blasphemy conviction and an attempt by a right-wing politician to establish his right to buy broadcasting time.[73] Although Britain has had more than its fair share of cases in this area, countless cases originating in other European countries offer useful guidance on whether English law conforms to Convention standards: for example, the Court's ruling that Austrian defamation law must be made to distinguish more sharply between false information and misguided opinion, and the Commission's ruling that Dutch racist agitators could not claim Article 10 protection in respect of their agitation.[74]

From this miscellany of cases it is hard to extract any general account of the development of Article 10 jurisprudence. There are remarkable contrasts in tolerance, even between cases decided very close together. Two trends can be identified, however. First, reliance on the 'margin of appreciation' has become an increasingly predictable feature of Article 10 cases.[75] Second, there has been a perceptible move towards the idea that the interests enumerated in the catalogue of permissible restrictions are to be balanced against the interests of freedom of expression, rather than treated as specific narrow exceptions to its supremacy.[76] While the first

[70] *Sunday Times* v. *UK* (1979) 2 EHRR 245, dealt with above at n. 57.

[71] *The Observer and The Guardian* v. *UK* (1992) 14 EHRR 153; *The Sunday Times* v. *UK* (No. 2) (1992) 14 EHRR 229.

[72] *Handyside* v. *UK* (1976) 1 EHRR 737. But c.f. *Müller* v. *Switzerland* (1991) 13 EHRR 212 at 227.

[73] *Gay News* v. *UK* (1983) 5 EHRR 123; *X* v. *UK* (1972) 38 Coll Dec. 86.

[74] *Lingens* v. *Austria* (1986) 8 EHRR 407; *Glimmerveen* v. *Netherlands* (1980) Dec. & Rep. 187.

[75] For an extreme use, see the Court's decision in *Markt Intern Verlag* v. *Federal Republic of Germany* (1990) 12 EHRR 161.

[76] See, e.g. *X* v. *Germany* (1989) 11 EHRR 101. With *Purcell* v. *Republic of Ireland* [1991] 2 H.R. Case Digest 72, the language of the Convention itself seems to have become irrelevant.

trend makes the Convention less useful in practical terms, the second actually threatens its claim to create a *right* to freedom of expression at all. If deprived of all exclusionary force, and simply turned into a consideration to weigh in the balance, freedom of expression is no longer a right; it is simply a value or principle. In any case, the increasingly deferential and compromise-based approaches of both Court and Commission have made the Convention an increasingly safe instrument for our domestic courts to invoke. It is now common to see Article 10 invoked in place of the common-law principle of free speech, precisely because it allows impressive references to grand ideals without actually importing much substance. In both the *Spycatcher* case and the case dealing with the Sinn Fein broadcasting ban, Law Lords were able to exploit the growing vacuity of the Convention to give an already foregone conclusion an aura of respectability.[77] In its recent foray into defamation law, too, the House of Lords felt able to boast that the common law of England, with its principle of free speech, already accepts the standards of Article 10.[78] Admittedly, this may not be far from the truth. But to present it as a virtue is to trade on the widespread but mistaken view that accommodating international law human rights jurisprudence is the same thing as respecting human rights. There could come a point, indeed, where it would be better to have no Convention on Human Rights at all than one that can be used to legitimate and congratulate any decision under the sun.

The European Convention is also invoked, from time to time, in the case-law of the European Community, but its appearances here tend to be no less cosmetic. The European Court of Justice's one achievement for freedom of expression in England was not an application of Article 10 or any other freedom of expression rule; it was an application of the Community's laws concerning trade barriers. Before 1986 English customs officers were empowered by law to seize 'indecent or obscene' matter at port of entry, whereas English police officers were only empowered to seize 'obscene' material under the Obscene Publications Act. The result was that material published abroad was liable to confiscation when it would not have been so liable had it been home-produced. The Luxembourg Court held that this put continental European pornographers at an unfair trading disadvantage, and required that English customs officers seize obscene articles only.[79] But the time is coming when the European

---

[77] *Attorney-General* v. *Guardian (No. 2)* [1988] 3 WLR 776, [1988] 3 All ER 545 (HL) at 660 (Lord Goff); *Brind* v. *Secretary of State for the Home Dept.* [1991] 1 All ER 720 (HL) at 725–6 (Lord Templeman).

[78] *Derbyshire County Council* v. *Times Newspapers* [1993] 1 All ER 1011.

[79] *Conegate Ltd.* v. *Customs and Excise Commissioners*, Case 121/85, [1987] QB 284, [1986] 2 All ER 688. Unfortunately, a rerun of the same argument did not pay off in a recent Irish application concerning the censorship of abortion information: see *Society for the Protection of Unborn Children Ireland Ltd.* v. *Grogan*, Case 159/90, [1991] 2 ECR 539.

Community will be much more involved in freedom of expression as such. We now have a Directive on Broadcasting and a Code of Practice on Sexual Harassment, which have major freedom of expression implications. There is also proposed legislation on official secrecy. It is to be hoped that the Community institutions will not insist on treating these as primarily economic or administrative issues. They are about maintaining the free flow of information and facilitating, so far as possible, the validation of people's diverse ways of life.

CONCLUSION

It is hard to evaluate the state of freedom of expression in Britain, now or at any other time. The extent of free expression is not crudely quantifiable; as our discussion has revealed, the right is respected and violated in many incomparable dimensions. Some general trends in recent English law may be identified, however. I mentioned above that the judiciary have been inclined to read their principle of free speech in an excessively individualistic way, focusing mainly on the interests of the person or body claiming the freedom and not the public interest in their enjoyment of that freedom. And yet the public interest in free expression—stemming primarily from its contribution to free-flowing information and wide--ranging validation—nevertheless saw increasing legal recognition after 1950. Sometimes the interest was recognized by statute, as with the 'public good' defence in the Obscene Publications Act 1959 or the new 'privilege' defences in the Defamation Act 1952. On rarer occasions, as with the development of a more generous 'public interest' defence to breach of confidence actions during the 1960s and 1970s, the recognition was judicial. In the 1980s, however, we can detect a retreat from this communitarian interpretation of the right. This is not to say that the Government ignored free expression issues altogether. The Government appreciated Salman Rushdie's right to freedom of expression, for example, because it was his right exercised, as the government saw it, largely in his own interests. But there was no such appreciation of the importance of free expression by the BBC, say, or by local authorities, or by those employees who set out to blow the whistle on their employers' sharp practices, or by campaigning charities. The Government was evidently baffled by the idea that these people hold their right to freedom of expression primarily because of the value their holding it and exercising it has for others, the value of the information and the validation their expression supplies. Moreover, the tremendous zeal for privatization in the 1980s led to increased gagging of employees (for example in the health service and in educational institutions), whose criticisms of essentially public services were no longer tolerated by commercially minded managements. Some imagine that the development of this more individualistic climate had some kind of resonance with the liberal values on which freedom of expression itself is based. One aim of this discussion has been to suggest that freedom of expression is rightly associated with liberal concerns, but that those concerns are more communitarian than they are individualistic. Respecting people and their diverse ways of life requires that freedom of expression, and many other things besides, be viewed more as public goods than as private assets.

8

Freedom of Association

K. D. EWING

INTRODUCTION

Freedom of association is rightly regarded as one of the cornerstones of liberal democracy. It was an essential feature of political life identified by Alexis de Tocqueville in his *Democracy in America* where he argued that the 'most natural privilege of man, next to the right of acting for himself, is that of combining his exertions with those of his fellow creatures and of acting in common with them'. For him the right of association was 'almost as inalienable in its nature as the right of personal liberty'.[1] Freedom of association is in fact indispensable to democratic and accountable government for it provides the constitutional basis of the right to form and join political parties, to take part in the activities of pressure groups, and to meet with others to discuss matters of common concern. In view of the importance of freedom of association, it is unsurprising that it should be protected by a variety of international legal documents, most notably the European Convention on Human Rights. By Article 11(1) this provides that 'Everyone has the right to freedom of peaceful assembly and to freedom of association with others, including the right to form and to join trade unions for the protection of his interests.' By the same token, it is unsurprising that protection for freedom of association should be found expressly in many constitutional documents, including the Canadian Charter of Rights and Freedoms of 1982 and the rather differently enforced New Zealand Bill of Rights Act 1990.[2]

But freedom of association is important not only to facilitate effective participation in civil and political society. It is equally important in the fields of social and economic activity and is particularly significant as a basis for securing trade union freedom from interference by the state on the one hand and from employers on the other. Consequently, freedom of association occupies a prominent role in international legal instruments promoting social and economic rights. A particularly vivid example of this is Convention 87 of the International Labour Organization which provides by Article 2 that 'Workers and employers, without distinction whatso-

---

[1] A. de Tocqueville, *Democracy in America* (1945 edn.), 196.
[2] It is also expressly guaranteed in a number of European constitutions, including Germany, Italy, and Spain. See below.

ever, shall have the right to establish and, subject only to the rules of the organisation concerned, to join organisations of their own choosing without previous authorisation'. The Convention contains a number of other important guarantees, including the right of workers' organizations to draw up their constitutions and rules, to elect their representatives in full freedom, to organize their administration and activities, and to formulate their activities.[3] Important guarantees are also to be found in the Council of Europe's 1961 Social Charter and in the European Communities' Charter of Fundamental Social Rights adopted by eleven of the twelve member states in December 1989,[4] with the latter providing that 'Employers and workers of the European Community shall have the right of association in order to constitute professional organisations or trade unions of their choice for the defence of their economic and social interests.'

Freedom of association is, perhaps uniquely, a two-dimensional freedom. On the one hand is the individual dimension, addressed most recently by the Supreme Court of Canada where it was recognized that 'The essence of the freedom is the protection of the individual's interest in self-actualization and fulfilment that can be realized only through combination with others.'[5] The second dimension is collective, the liberty or autonomy of the group to act together to promote its common interests. It might be argued of course that this dimension is merely an extension of the first, for if the autonomy of the group may be constrained with impunity, freedom of association provides a poor vehicle for the expression of individual liberty. But this would be to overlook the extent to which individual liberty can be impaired by such autonomy, with individuals becoming bound by obligations voluntarily assumed to act in association in support of some causes or conduct of which they may disapprove. Yet regardless of whether the function of group autonomy is viewed as distinct from or as an extension of individual liberty, it is clear that this too embraces two fundamental aspects. The first is the autonomy of the group to determine its own membership, and to regulate its own method and manner of government. The second is the autonomy of the group to develop its own programme of action to fulfil its goals. Before turning to consider how English law deals with these and other questions, it should perhaps be pointed out that although freedom of association is rightly regarded as being of supreme importance, it does not follow from

---

[3] ILO Conventions 98 (Right to Organize and Collective Bargaining) and 151 (Right to Organize in the Public Sector) are also important in the context of freedom of association. See K. D. Ewing, *Britain and the ILO* (1989).

[4] For a full account, see Lord Wedderburn of Charlton, *The Social Charter, European Company and Employment Rights: An Outline Agenda* (1990).

[5] *Lavigne* v. *Ontario Public Service Employees' Union* (1991) 81 DLR (4th) 545, at 623.

this that it can be or should be unlimited, in either of its dimensions. So much is recognized by Article 11(2) of the ECHR, which permits restrictions to be prescribed by law if these are necessary in a democratic society in the interests of considerations such as national security, the prevention of disorder or crime, or for the protection of the rights and freedoms of others. It should perhaps also be pointed out here that, although much of the activity in this area concerns trade unions, an attempt will be made not to trespass too far into the territory occupied by labour law.

STATE SUPPORT FOR FREEDOM OF ASSOCIATION

In Britain freedom of association enjoys no general protection in the law. There is no constitutional guarantee as there is in other jurisdictions and there is no freedom of association statute. As a result, freedom of association is protected in the same way as the other classical civil liberties: people are free to associate for any purpose provided that it is not otherwise forbidden.[6] So there is no guarantee of the right to form and join political parties, despite the fact that political parties play a crucial role in the political process. Indeed, political parties are barely recognized by law. On the other hand, there are precious few restrictions on what political parties may do.[7] Unlike many other jurisdictions, there are no registration requirements; no obligations to report and disclose income and expenditures; and no direct restrictions on the sources of funding or how these funds may be used.[8] This is not to say that members of some political parties will not be disadvantaged or to deny that some political parties will have difficulties in projecting their message. Thus, for the parties on the left in particular membership still has its prejudices. An illustration of this is the purge procedure introduced in the civil service in 1948 initially to cleanse government employment mainly of communists, deemed to be unreliable on national security grounds.[9] We return to this in a later section. But we may also note at this stage that, at the other end of the political spectrum, fascist and racist organisations will have some

[6] See A. V. Dicey, *An Introduction to the Study of the Law of the Constitution*, 10th edn. (1964), esp. ch. 7.

[7] On the law relating to political parties in Britain, see K. D. Ewing, *The Funding of Political Parties in Britain* (1987), esp. ch. 1.

[8] See *Report of the Committee of Financial Aid to Political Parties*, Cmnd. 6601 (1976). Note also that political parties in Britain have never been required to identify their members, though the absence of any formal obligation has never been a problem for state agencies anxious to acquire such information. Cf. *Communist Party of America* v. *Subversive Activities Control Board*, 367 US 1 (1961).

[9] The procedure applied also to fascists, but communists were the main target. For a full account see S. Fredman and G. S. Morris, *The State as Employer: Labour Law in the Public Services* (1989), 104–5, 232–4.

difficulty in campaigning to promote a central object of the association. So although it is not unlawful to associate in a political party which has a racist programme, legislation prohibiting incitement to racial hatred may make it difficult to campaign for that particular cause.[10]

It would be wrong to assume, however, that steps have not been taken to promote freedom of association. Three different forms of support by the state can be identified. The first is the removal of obstacles to effective association. A classic example of this is the Trade Union Act 1871 which, by providing that trade unions were not to be regarded as organizations in restraint of trade, thereby reversed an earlier court ruling which effectively classed trade unions as outlaws,[11] unable to recover funds which had been misappropriated by members. Another example of this is the Trade Union Act 1913 which reversed the famous *Osborne* judgment in which the House of Lords held that trade unions could not lawfully use their funds to finance the Labour Party.[12] The reversal of this decision,[13] with qualifications, was perhaps of greater importance to the Party than it was to individual trade unions, for without it the Party might not have survived and certainly would not have survived in the form that it did. A second form of support has been the use of executive power to encourage freedom of association: for example, the requirement that broadcasting authorities make time available to the political parties for party-political and election broadcasts.[14] Other examples lie again in the economic sphere (though not without implications for the political). Thus, in the period after the First World War, the executive, through the Ministry of Labour, actively encouraged employers to set up employers' associations and to enter into collective bargaining arrangements with trade unions through Whitley Councils, following recommendations to this effect by the post-

---

[10] Public Order Act 1986, ss. 18–22, prohibiting acts intended or likely to stir up racial hatred.

[11] *Hornby* v. *Close* (1867) LR 2 QB 153. See now Trade Union and Labour Relations (Consolidation) Act 1992, s. 11.

[12] *Amalgamated Society of Railway Servants* v. *Osborne* [1910] AC 87 (HL).

[13] Trade unions could again engage in political activity but only if they first held a ballot of their members in which a majority of those voting supported the adoption of political objects. The union was then required to adopt political fund rules whereby political objects would be financed by a separate levy of the members who each had a right to contract out of the obligation to pay it. The Trade Union Act 1984 required trade unions to conduct ballots for authority to continue to promote political objects on a ten-yearly basis. See now Trade Union and Labour Relations (Consolidation) Act 1992, Part 1, Chapter 6.

[14] See H. F. Rawlings, *Law and Electoral Process* (1988). By agreement between the broadcasting authorities and the main political parties, free time is provided for Party Political Broadcasts and Party Election Broadcasts. The time is allocated by agreement, though if this is not possible an allocation will be imposed by the broadcasters. Questions of allocation and access have from time to time given rise to difficulty and controversy. See e.g. *R.* v. *Broadcasting Complaints Commission, ex parte Owen* [1985] 2 WLR 1025. See now Broadcasting Act 1990, s. 36.

war reconstruction subcommittee under the chairmanship of J. H. Whitley, a former Speaker of the House of Commons.[15] Another example of this kind of bureaucratic intervention is the Fair Wages Resolution of the House of Commons of 1946. Although this was revoked in 1983, it provided in paragraph 4 that every government contractor was required to 'recognise the freedom of his workpeople to be members of trade unions'.[16]

So the first two strategies of support involved the use of legislation to remove obstacles to association and the use of bureaucratic pressures to promote and encourage association. The third strategy has been the introduction of specific legislation to promote and encourage association and to protect people from the hostile consequences of being in association. It is true that there is not much legislation of this nature and that the statutes which do exist operate mainly in the economic sphere to regulate what is essentially a relationship in private law, that between employer and worker. The first statute dealing with this question was the Conservative government's Industrial Relations Act 1971, which provided by section 5 that every worker shall, as between himself and his employer, have the right to be a member of a trade union of his choice, and to take part in the activities of a trade union at an appropriate time. Although the 1971 Act was repealed in 1974,[17] similar provisions were promoted by the incoming Labour government, with the law being consolidated in 1978 and again in 1992.[18] Under these rules, which still operate, employees were protected from action short of dismissal and from dismissal either because they were members of trade unions or had taken part in the activities of independent trade unions at an appropriate time. This was defined to mean a time outside working hours or a time during working hours when the conduct was engaged with the consent of the employer. Major developments since 1978 include the increase in the levels of compensation for employees dismissed because of their trade union membership or activities, an initiative taken in 1982.[19] Also noteworthy is the Employment Act 1990, which addressed a major gap in the protection which until then did not apply to people who were refused employment because of their trade union membership or activities.[20] The 1990 Act

---

[15] *Interim Report on Joint Standing Industrial Councils by the Sub-Committee on Relations between Employers and Employed of the Committee of Reconstruction*, Cmnd 8606 (1917).

[16] Fair Wages Resolution 1946, para. 4. See B. Bercusson, *Fair Wages Resolutions* (1978).

[17] Trade Union and Labour Relations Act 1974.

[18] See Employment Protection (Consolidation) Act 1978, ss. 23 and 58, and now Trade Union and Labour Relations (Consolidation) Act 1992, Part III.

[19] Employment Act 1982, s. 5. See now Trade Union and Labour Relations (Consolidation) Act 1992, ss. 156–158.

[20] Employment Act 1990, s. 1. See now Trade Union and Labour Relations (Consolidation) Act 1992, s. 137.

went some of the way, though not the whole way,[21] towards closing the gap, by providing that it is unlawful for an employer to refuse to employ someone because he or she is a member of a trade union. An important qualification to these measures of protection is to be found in the Trade Union Reform and Employment Right Act 1993 (s. 13) which authorises an employer to discriminate against trade union members in such matters as pay and other terms and conditions of employment.[21a]

RESTRICTIONS ON FREEDOM OF ASSOCIATION

Although the freedom to associate has thus been reinforced by three different strategies employed by the state, it has also been restricted on a number of grounds. The first was to deal with the threat to public order and internal security caused by the rise of fascism in the 1930s. The Public Order Act 1936 did not make it an offence to be a member of the British Union of Fascists, but it did prohibit the public display of support for fascist organizations, in the sense that it was an offence for any person in a public place or at any public meeting to wear a uniform signifying association with any political organization. Shortly after the Act was passed it was used against fascists who wore uniforms at public meetings or while selling newspapers in public places.[22] It was subsequently used in later years to deal with members of the Ku Klux Klan[23] and members or supporters of the IRA.[24] In more recent years the concerns about public order and internal security have been directed not at fascists but at terrorists, and the IRA in particular. Specific legislation enacted in response to this phenomenon, the Prevention of Terrorism Act 1989 (as it now is), contains unparalleled restrictions on freedom of association to the extent that it is an offence to belong to a proscribed organization, a term defined in the Schedule to mean the Irish Republican Army and the Irish National Liberation Army.[25] This appears to be the only example of British legislation which makes it a criminal offence simply to be a member of a particular organization. The Act also makes it an offence to solicit or

---

[21] See *Beyer* v. *City of Birmingham District Council* [1977] IRLR 211, holding that the protection does not apply in respect of trade union membership or activities in a previous employment. However, this particular problem may be overcome by the recent decision of the Court of Appeal in *Fitzpatrick* v. *British Railways Board* [1991] IRLR 376.

[21a] Reversing the Court of Appeal in *Wilson* v. *Associated Newspapers Ltd.* [1993] IRLR 63.

[22] See *R.* v. *Wood* and *R.* v. *Charnley* (1931) 81 Sol. Jo. 108.

[23] S. H. Bailey, D. J. Harris, and B. L. Jones, *Civil Liberties Cases and Materials*, 3rd edn. (1991), 158.

[24] *O'Moran* v. *DPP* [1975] QB 864.

[25] Prevention of Terrorism Act 1989, s. 2. This measure was first introduced in the Prevention of Terrorism Act 1974, s. 1. See also Northern Ireland (Emergency Provisions) Act 1991, s. 28. See further, chapter 5.

invite financial or other support for a proscribed organization, or arrange (or assist in the arrangement of) meetings or addresses of three or more persons in public or in private if the meeting is designed to support the proscribed organization. Not just membership is declared unlawful; so too is participation in the activities of the organization. The list of proscribed organizations may be extended by the Secretary of State if the organization appears to him to be concerned in terrorism occurring in the United Kingdom and connected with Northern Irish affairs.

A second basis for restricting freedom of association was the perceived threat of subversion and the consequent threat to external security created by members and sympathizers of the Communist Party during the cold war. This led to the introduction of the purge and positive vetting procedures in the civil service. The purge procedure was originally introduced with communists as the main target, though it also applied to fascists. In 1985, however, it was widened to include members of any 'subversive group, acknowledged as such by the Minister, whose aims are to undermine or overthrow Parliamentary democracy in the UK and Northern Ireland by political, industrial or violent means'.[26] This is a definition of quite extraordinary scope which effectively means whatever the Government wants it to mean. As Fredman and Morris have pointed out, if an employee is found to be associated with such a group, he or she may be transferred from a sensitive post, or, where this is not possible, dismissed.[27] The effect is to exclude members or supporters of disapproved organizations from employment in 'all posts which are considered to be vital to the security of the State, and is invoked when the reliability of a public servant is thought to be in doubt on security grounds'.[28] But whereas the purge procedure operates to remove people from sensitive work, positive vetting operates to keep them out in the first place. Under procedures revised in 1990, this seeks to ensure that no one is employed 'in connection with work the nature of which is vital to the security of the state.' Those who may be affected include members or associates of organisations which have advocated espionage, terrorism, sabotage, or action intended to overthrow or undermine Parliamentary democracy by political, industrial or violent means.[29] Positive vetting involves an investigation into 'the civil servant's character and circumstances', which includes among other matters political associations.[30] The

[26] 76 HC Debs. 621 (3 Apr. 1985). See Bailey, Harris and Jones, *op. cit.* (note 23), pp. 452–3.
[27] Fredman and Morris, *State as Employer*, 233.
[28] Ibid.
[29] 177 HC Debs. 159–161 (WA) (24 July 1990). See also E. C. S. Wade and A. W. Bradley, *Constitutional and Administrative Law*, 11th edn. (1993), 539–41.
[30] Fredman and Morris, *State as Employer*, 235.

arrangements appear to have been designed initially to exclude from sensitive posts members and sympathizers of the Communist Party.

In more recent years the concerns about external security have led to restrictions being imposed on civil servants employed in intelligence-gathering work at GCHQ. Ever since 1947 civil servants employed in this work had been both permitted and encouraged to join trade unions. By the end of 1983 several thousand people were trade union members. In the early 1980s, however, the staff employed at GCHQ had been engaged in industrial action which caused some disruption to the service. Indeed the Government estimated that no less than 10,000 working days were lost between 1979 and 1981 as a result of industrial action, though it should perhaps be pointed out that this action was undertaken for traditional trade union motives—a concern about pay and working conditions—and was not motivated in any way by a desire to threaten or undermine national security. Nevertheless the Government responded by announcing on 25 January 1984 its unilateral decision to remove all trade union membership rights from staff employed at GCHQ. This was done in two ways: first by invoking powers under the royal prerogative to vary the terms and conditions of employment of the staff in question; and secondly by issuing a ministerial order under section 138(4) of the Employment Protection (Consolidation) Act 1978, removing from the designated staff the statutory protection against dismissal and action short of dismissal for inter alia trade union membership or participation in trade union activities.[31] The only legal basis by which these decisions could be challenged was by way of an application for judicial review under Order 53 of the Rules of the Supreme Court. But although this was done, the action failed, principally because the Government acted as it did for reasons of national security which the House of Lords accepted was a complete defence to conduct which might otherwise be unlawful.[32] A complaint to Strasbourg of a breach of Article 11 of the European Convention on Human Rights met a similar fate, principally because Article 11(2) contains a permitted exclusion from the freedom of association guarantees for workers engaged in the administration of the state.[33] A complaint to Geneva of a breach of ILO Convention 87, Article 3, has, however, been upheld both by the ILO Committee on Freedom of Association and by the Committee of Experts. But so far the British government has refused to take steps to implement the recommendations of either Committee.[34]

---

[31] A similar step had been taken in 1976 by the then Labour government in respect of members of the security service. (See now Trade Union and Labour Relations (Consolidation) Act 1992, s. 275).

[32] *Council of Civil Service Unions* v. *Minister for the Civil Service* [1985] AC 374 (HL). See S. Fredman (1985) 14 *ILJ* 42.

[33] *Council of Civil Service Unions* v. *UK* (1988) 10 EHRR 269.

[34] For an account of these proceedings and enforcement procedures, see Ewing, *Britain and the ILO*.

The third basis for imposing restrictions has been to promote the objective of a politically neutral public service. This has been advanced by administrative rules and practices initiated by the Government, now to be found in the Civil Service Pay and Conditions Code. Three categories of civil servants are identified, the politically restricted, an intermediate category, and the politically free.[35] Following the report and recommendations of the Armitage Committee in 1978 the rules were relaxed in 1984 so that a greater range of people are now free to engage in partisan political activity.[36] Nevertheless, about 20 per cent of civil servants remain politically restricted, which means that they cannot hold office in national political parties, express views on controversial questions, or canvass for political parties or candidates in parliamentary elections. Britain is not alone in seeking to restrict the political activities of civil servants in order to promote political neutrality. In other countries, however, constitutional protection of freedom of association and related freedoms may make it possible to challenge these arrangements in the courts. Yet although such a challenge has been successfully mounted at least as far as the Federal Court of Appeal in Canada,[37] the US Supreme Court has consistently held restrictions of this kind to be justifiable. In *United States Civil Service Commission* v. *National Association of Letter Carriers*,[38] it upheld a restriction on the rights of federal employees, prohibiting them from taking an active part in political management or in political campaigns. According to the Court, such activities 'must be limited if the Government is to operate effectively and fairly, elections are to play their proper part in representative government and employees themselves are to be sufficiently free from improper influences'.

The concern for political neutrality has also led to restrictions being imposed on the police and, most recently, local government officers. So far as the police are concerned, the Police Regulations 1987 provide that an officer 'shall at all times abstain from any activity which is likely to interfere with the impartial discharge of his duties or which is likely to give rise to the impression amongst members of the public that it may so interfere; and in particular shall not take any active part in politics'.[39] This would appear to preclude police constables from participating in the affairs of political parties and high-profile pressure groups. In *Champion* v. *Chief Constable of Gwent*,[40] it was held by the House of Lords that the

---

[35] S. A. de Smith and R. Brazier, *Constitutional and Administrative Law*, 6th edn. by R. Brazier (1989), 194–5.

[36] *Report of the Committee on Political Activities of Civil Servants*, Cmnd. 7057 (1978).

[37] *Osborne* v. *Treasury Board* (1989) 52 DLR (4th) 241.

[38] 413 US 548 (1973).

[39] Police Regulations 1987, SI 1987/851, Sch. 2. Note also that, in 1919, concerns about public disorder led to the introduction of a statutory restriction on the right of police officers to join trade unions. See now Police Act 1964, s. 47(1).

[40] [1990] 1 All ER 116.

purpose of these provisions 'is to prevent a police officer doing anything which affects his impartiality or his appearance of impartiality. Impartiality means favouring neither one side nor the other but dealing with people fairly and even-handedly'. In contrast to the rules governing police officers, the new restrictions on politically restricted local government officers (introduced because 'it is not proper that local government officers are treated entirely differently from civil servants')[41] would appear to allow membership of a political party, but not much more.[42] Politically restricted posts include those where the rate of remuneration exceeds £19,500, unless the member of staff in question is exempt under the Act.[43] By virtue of regulations introduced in 1990, people who are in this position are prohibited from acting as an election agent; or as an officer of a political party or of any branch of such a party. A person in a politically sensitive post is also prohibited from canvassing on behalf of a political party or candidate for election. The restrictions go wider still. Thus, an affected employee 'shall not speak to the public at large or to a section of the public with the apparent intention of affecting public support for a political party'.[44] So although such people are expressly permitted to display posters (presumably also political posters) in their homes and cars, for the most part they are free only to be in political association with others but not to act in association with others by displaying that support in public or to solicit support for the association.[45]

THE RIGHT OF NON-ASSOCIATION

The discussion so far has been concerned exclusively with the position of individuals who wish to belong to associations of one kind or another. It is quite clear that if freedom of association is to mean anything, it must protect such people. But a rather different question which arises is whether freedom of association protects the position of people who do not wish to join; in other words, does it extend to protect people against what Emerson refers to as 'forced association'?[46] In those jurisdictions which have developed the idea of freedom of association through the process of constitutional adjudication, this question has met with a mixed response. The difficulty arises mainly because, in protecting the freedom to associate,

[41] Official Report, Standing Committee G, 7 Mar. 1989, col. 186 (Mr Nicholas Ridley).

[42] Local Government and Housing Act 1989, ss. 1, 2.

[43] Ibid., s. 2. Provision is made for raising the sum of £19,500. See SI 1990/1447.

[44] Local Government Officers (Political Restrictions) Regulations, SI 1990/851. See *NALGO* v. *Secretary of State for the Environment, The Times*, 2 Dec. 1990.

[45] See further, K. D. Ewing (1990) 19 *ILJ* 111; 192.

[46] T. I. Emerson, 'Freedom of Association and Freedom of Expression' (1964) 74 *Yale LJ* 1.

by no means all constitutional documents expressly protect the freedom not to associate, though there are some which do, as in Spain. In the Irish Republic, however, such a right was implied into the constitutional guarantee of 'liberty for the exercise of the right of the citizens to form associations and unions'. In the view of Budd J. in *Educational Company of Ireland* v. *Fitzpatrick (No. 2)*

If it is a 'liberty' that is guaranteed, that means that the citizen is 'free' to form, and I think that must include join, such associations and unions, and, if he is free to do so, that obviously does not mean that he *must* form or join associations and unions, but that he *may* if he so wills.[47]

A rather different approach has been taken by the US Supreme Court which, in a number of cases, including *Railway Employees Union* v. *Hanson*[48] and *Abood* v. *Detroit School Board*,[49] has declined to read into the implied constitutional right of freedom of association an absolute right of non-association which correlates with the right to associate. So it has been held that legislation authorizing union shops and agency shops, whereby workers can be compelled to become members of a union or pay a fee to a union for the services it extends to all workers, does not violate the right to freedom of association. This is provided that the union only requires financial support from unwilling employees, and does not also seek to force ideological conformity or other action in contravention of the First Amendment.[50]

The absence of any constitutional guarantee of freedom of association in Britain means that the courts here have been unable to contribute to this debate. The common law, however, would appear to encourage the view that there is a right not to associate, with compulsory union membership, for example, likely to be treated as a restraint of trade. More recently, public policy appears to be moving in the direction of a right not to associate, being most marked in the area of trade union membership. Attitudes have been fortified by the requirements of the European Convention on Human Rights. Although it is true that this does not expressly protect the right to freedom of non-association, it has been held nevertheless that the negative freedom is not completely excluded. In *Young, James and Webster* v. *United Kingdom*[51] the European Court of Human Rights found that workers should not be required to join a particular trade union but that they should have a choice of union, and also that workers should not be compelled to join any union where

---

[47] [1961] IR 345.
[48] 351 US 225 (1956).
[49] 431 US 209 (1977). See also *Communications Workers of America* v. *Beck*, 487 US 735 (1988).
[50] *Abood*, ibid. See below.
[51] [1981] IRLR 408.

membership would conflict with other Convention-protected freedoms such as freedom of conscience and religion in Article 9 and freedom of expression in Article 10. But this civil libertarian concern of the international human rights community is not the only factor at work in directing public policy away from permitting closed shop arrangements at the expense of individual freedom. This outcome coincides conveniently with the views of Hayek who, for some, was the major intellectual inspiration of the Thatcher economic revolution of the 1980s. Hayek has harsh words for compulsory union membership arrangements, which he condemned not only as undermining individual freedom but also as a means of reinforcing trade union power by coercive means. For Hayek it was crucial that trade union influence, which he saw as distorting the proper functioning of the labour market, should be rolled back. An important feature of any such policy would be to treat closed shop contracts as contracts in restraint of trade and deny them the protection of the law.[52]

Hayek's particular concern was that 'unions should not be permitted to keep non-members out of employment'.[53] Legislation to promote this goal has been introduced progressively since 1980, culminating in rules introduced in 1988 and 1990 which effectively create a right of non-association in the industrial relations sphere which parallels the right to associate. This is despite the fact that in 1968 the Donovan Royal Commission on Trade Unions and Employers' Associations[54] had argued convincingly that the two are not necessarily symmetrical:

It might be argued that the closed shop should be prohibited. As part of the argument for prohibition it might be said that since we suggest elsewhere that any condition in a contract of employment that the employee shall not join a union is to be void in law, it would be right to treat in the same way a condition that a worker shall join a union. However, the two are not truly comparable. The former condition is designed to frustrate the development of collective bargaining, which it is public policy to promote, whereas no such objection applies to the latter.[55]

It may also be noted that the right of non-association which has been secured by legislation goes much further than what is demanded by North American constitutional jurisprudence, though it does bring us closer into line with the constitutional obligations of a number of EC member states (such as Germany, Italy, and Spain) where the right of non-association is either expressly or impliedly accorded equal status with the right to

---

[52] See esp. F. A. Hayek, *The Constitution of Liberty* (1960), 264–70.

[53] Ibid. 278.

[54] Cmnd. 3623.

[55] Para. 599. It is of course no longer the case that collective bargaining is to be promoted as a matter of public policy. Indeed the reverse is probably true.

associate, notwithstanding the criticisms of this position made by Donovan. The Government's first initiative in what has been a gradual assault on the closed shop was to require ballots before a legally effective union membership agreement could be introduced. The ballot had to be conducted every five years and had to show 80 per cent of those eligible to vote or 85% of those actually voting to be in favour of the agreement.[56] But even then a large number of exemptions were created, particularly for those with conscientious or deeply held personal objections to trade union membership.[57] In 1988 the Government took steps to legalize what had become the practical reality, namely that it would be unfair to dismiss someone on account of non-union membership.[58] It would also be unlawful to take action short of dismissal against someone for this reason. In 1990 the law was extended to make it unlawful to refuse to employ someone because he or she was not a trade union member.[59]

The question of 'forced association' has also arisen in the rather different context of the compulsory payment of dues to the Bar Council by practising barristers. May barristers operate a closed shop when other workers may not? At the annual general meeting of the Bar in July 1986, approval was given to a resolution to amend the Code of Conduct for the Bar of England and Wales. The effect of the resolution was to introduce a new rule whereby 'a practising barrister . . . must pay to the General Council of the Bar at such time or times as it should become due the subscription currently payable by a barrister of his seniority'.[60] Similar arrangements were challenged in the USA, some thirty years ago in *Lathrop* v. *Donohue*,[61] where a member of the Wisconsin Bar argued that the rules and by-laws of the state bar which required him to pay dues to the treasurer violated his constitutional right to freedom of association. The plaintiff objected particularly to the fact that he was 'coerced to support an organization which is authorised and directed to engage in political and propaganda activities'. The claim was, however, rejected, with the Supreme Court closely following a previous decision in *Railway Employees Union* v. *Hanson*.[62] There it was held that the Railway Labor Act did not abridge constitutionally protected rights of association when it required employees to pay union dues as a condition of employment. Indeed, the decision in that case had been justified at the time on the ground that the

---

[56] Employment Act 1980, s. 7(3); Employment Act 1982, s. 3.
[57] 1980 Act, ibid. s. 7(2).
[58] Employment Act 1988, s. 11. See Trade Union and Labour Relations (Consolidation) Act 1992, s. 152.
[59] Employment Act 1990, s. 1. See Trade Union and Labour Relations (Consolidation) Act 1992, s. 137.
[60] *Re S. (a barrister)*, *Guardian*, 9 Oct. 1990.
[61] 367 US 820 (1961).
[62] 351 US 225 (1956).

practice being challenged 'is no more an infringement or impairment of First Amendment rights than there would be in the case of a lawyer who by state law is required to be a member of an integrated bar'.[63] Following this reasoning, the Court in *Lathrop* held that the compulsory dues served 'a legitimate end of state policy', namely the elevation of 'the educational and ethical standards of the Bar to the end of improving the quality of the legal service available to the people of the State'.[64]

The absence of any constitutional guarantee of freedom of association means that the arrangements adopted in Britain would have to be challenged, if at all, on other grounds, even if the constitutional route appeared to be a fruitful line of attack. Indeed, although the defendant in *Re S. (a barrister)*[65] objected to the new rules on traditional freedom of association grounds (he was opposed to 'compulsory subscriptions because it would turn the Bar into a "closed shop"') the absence of any applicable public law rights meant that these important questions of principle, which were at the heart of the matter, had to be relegated to a minor role in legal proceedings which focused more on matters of private rather than public law. The matter arose not by way of a direct challenge to the new rules, but collaterally, with S. being charged with professional misconduct following his refusal to pay the mandatory levy. In defending himself against the charge, S. argued successfully first that the new rules were *ultra vires* the powers of the Bar Council, and secondly that the subscription was void as being an unreasonable restraint of trade. The Bar Council had argued in terms not unlike those accepted in *Lathrop* v. *Donohue*,[66] that the subscription was lawful because it was in the public interest 'that there should be a properly regulated and remunerated Bar, maintaining a very high standard of ethical conduct which could properly be policed and enforced. That required a properly financed secretariat.' But although the tribunal hearing the case sympathized with these arguments, it was unable to sustain them, taking the view that

However desirable the goal of compulsory subscriptions might be, it could not properly be achieved by treating failure to pay as professional misconduct. The validity of the rules of professional conduct must be measured against their essential purpose, which was to ensure that the right of audience was confined to fit and proper persons. That purpose was not served by a rule which imposed, on pain of disbarment, compulsory subscriptions to a professional body whose functions included a substantial element of trade union activity.

As a result S. was acquitted.

[63] 351 US, 238.
[64] 367 US 820 (1961).
[65] *Guardian*, 9 Oct. 1990.
[66] 367 US 820 (1961).

THE RIGHT TO FREEDOM OF THE ASSOCIATION

So far we have been concerned with the individual dimension of freedom of association. If we turn now to consider the collective dimension—the freedom of the association—we may begin by examining the right of the association to regulate its own government and administration, before proceeding to consider questions relating to the powers and purposes of the association. So far as questions of government and administration are concerned, two questions of autonomy arise: the first is the autonomy of the association to control whom it accepts into membership, and retains in membership, and the second is the autonomy of the association to control and regulate its procedures for internal management. So far as the autonomy over membership is concerned, English common law appears to recognize this as an important feature of freedom of association, the point having arisen forcefully in *Cheall* v. *APEX*.[67] In that case Mr Cheall was expelled from a trade union, APEX, to comply with a ruling of the TUC Disputes Committee that he had been wrongly recruited by APEX and should transfer to his original union ACTSS. The Court of Appeal held the expulsion to be invalid, with Lord Denning relying on Article 11 of the European Convention on Human Rights to construct a fundamental principle of common law that a man has the right to join a trade union of his choice for the protection of his own interests, and that he cannot be expelled from such an association by a rule which is unreasonable. This was reversed by the House of Lords, however, which held that 'freedom of association can only be mutual; there can be no right of an individual to associate with other individuals who are not willing to associate with him.' The House of Lords thus restored the autonomy of the association to control its membership, though it should be noted that this autonomy is not unqualified, with Lord Diplock noting that different considerations might apply if the effects of Cheall's expulsion from APEX were to have put his job in jeopardy, either because of the existence of a closed shop or for some other reason.[68]

Cheall thus confirms the traditional common law rule that voluntary associations may determine who they will or will not admit into membership. This is not to say that the courts have no supervisory role: they may

---

[67] [1983] AC 180 (HL) ([1983] CLJ 207).
[68] Ibid. at p 405. See now *Industrial Relations in the 1990s: Proposals for Further Reform of Industrial Relations and Trade Union Law*, Cmnd. 1602 (1991). There the Government has signalled an intention to introduce a statutory right not to be excluded from a trade union solely on the ground that the individual was previously a member of another trade union or because of an inter-union arrangement over recruiting rights. By s. 14 the Trade Union Reform and Employment Rights Act 1993 implements this proposal by a device which significantly impairs trade union autonomy over membership and recruitment.

intervene if someone has been expelled in breach of the rules of the association or in breach of the rules of natural justice. It is true that, as *Cheall* suggests, the courts have for some time been concerned about the closed shop and the possible abuse of power by trade unions in particular; exclusion or expulsion from a trade union could have obvious implications for the employment position of the individual in question. But the courts have had difficulty in fashioning a cause of action, in the absence of tort, contract, or restraint of trade, with which to arm the disappointed applicant or expelled member who has been properly excluded or expelled under rules of which the courts disapprove. Restrictions on sex and race discrimination,[69] and the gradual attack on the legality of the closed shop by legislation since 1980 have, however, made this problem much less pressing, and the attention of the courts much less urgent.[70] The restrictions on the closed shop in particular remove much of the justification for restricting trade union autonomy over the control of admissions and expulsions. It is perhaps strange then that the Employment Act 1988, which finally provided that it would be unfair for an employer to dismiss for non-membership of a trade union, should also contain measures to override trade union membership rules to provide that a trade union could not unjustifiably expel or discipline members, this to apply in particular to protect people who refused to take part in a strike or other industrial action, even where this has been lawfully called under the rules of the union and the relevant legislation introduced since 1980.[71] It has been argued that these measures violate Article 11 of the European Convention on Human Rights,[72] and it has been held by the ILO Committee of Experts that they breach Article 3 of the ILO Convention 87 and that the 1988 Act should be amended as a result.[73]

If freedom of association means that the group must have the right to regulate its own membership, it also means that the group must have the right to regulate its own procedures for government and administration. These principles appear to be acknowledged in Britain to the extent that there is little, if any, direct statutory regulation of the internal affairs of political parties or pressure groups. The principle appears also to be recognized by common law where there is no evidence of any desire on

[69] Sex Discrimination Act 1980, s. 12; and Race Relations Act 1976, s. 11.

[70] See Employment Act 1980, s. 4, giving workers employed in a closed shop the right not to be unreasonably excluded or expelled from membership of a trade union. This was repealed in 1992, but more restrictive measures were introduced by the 1993 Act. See note 68 above.

[71] Employment Act 1988, s. 3. See now Trade Union and Labour Relations (Consolidation) Act 1992, s. 64. See *Bradley* v. *NALGO* [1991] IRLR 159 (EAT). See further Trade Union Reform and Employment Rights Act 1993, s. 16.

[72] See Leader, 'The European Convention on Human Rights, the Employment Act of 1988 and the Right to Refuse to Strike' (1989) 20 *ILJ* 39.

[73] See Ewing, *Britain and the ILO*.

the part of the courts to create structures for the proper government of voluntary associations. The role of the courts is confined to enforcing the rules by which the members have themselves agreed to be bound.[74] To this extent, the principles developed by the US Supreme Court in *Eu* v. *San Francisco County Democratic Central Committee*[75] appear to be adequately protected in British law. In that case legislation introduced in California prohibited the official governing bodies of political parties from endorsing any of the candidates who were standing in party primary elections. The legislation also dictated the organization and composition of the governing bodies, limited the term of office of a party chair, and required that the chair rotate between residents of northern and southern California. Although it was struck down, the endorsement ban does not translate directly into British experience. But so far as the regulation of internal party structures is concerned, the Court held the regulations also to be unlawful, on the ground that they 'directly implicate the associational rights of political parties and their members',[76] taking the view that a political party must be allowed to determine the structure which best allows it to pursue its political goals. In the view of the Court, freedom of association 'encompasses a political party's decisions about the identity of, and the process for electing, its leaders'. In striking down the legislation (which limited 'a political party's discretion in how to organize itself, conduct its affairs, and select its leaders')[77] the Court was unable to find any compelling state interest to sustain it.

There are, however, exceptions to this recognition of autonomy in the government and administration of associations. The most prominent exceptions both in the USA and in Britain are trade unions. Constitutional guarantees of freedom of association and principles such as those in *Eu* v. *San Francisco County Democratic Central Committee*[78] have not been effective to prevent detailed state regulation of trade union internal affairs in the USA. Nor has a commitment to Diceyesque notions of liberty prevented the introduction of equally detailed regulation here, though it is perhaps paradoxical that the USA, with its (admittedly implied) constitutional commitment to freedom of association regulated before the United Kingdom, without any apparent impediment from the courts.[79] Detailed statutory regulation of trade union government in Britain was first introduced by the Industrial Relations Act 1971. This provided

---

[74] And also requiring that the rules of natural justice be observed in disciplinary proceedings. See *John* v. *Rees* [1970] Ch. 345; *Lewis* v. *Heffer* [1978] 1 WLR 1061.
[75] 489 US 214 (1989).
[76] Ibid. 229.
[77] Ibid. 230. See also *Tashjian* v. *Republican Party of Connecticut*, 479 US 208 (1986).
[78] 489 US 214 (1989).
[79] See B. Aaron, 'The Labor-Management Reporting and Disclosure Act of 1959' (1960) 73 *Harvard LR* 851.

that the rules of registered trade unions should comply with a number of requirements specified in Schedule 4, though these left a lot of autonomy as to content to each organization. More detailed regulation is now to be found in the Trade Union and Labour Relations (Consolidation) Act 1992. This lays down mandatory arrangements for the election of trade union general secretaries and national executive committees. These people must be elected at five- yearly intervals by a secret postal ballot at which entitlement to vote is accorded equally to all members of the union other than those who belong to classes of members specified in the Act.[80] No member of the union is to be unreasonably excluded from standing as a candidate, and in particular no candidate is to be required, whether directly or indirectly, to be a member of a political party.[81] This is presumably designed to prevent unions from requiring office-holders to be members of the Labour Party. Of some interest is the fact that the statute does not prohibit unions from requiring that candidates should not be members of all political parties or any specified political party. This must be a calculated omission which allowed some trade unions to continue to exclude communists from office, the former electricians' union being historically the best-known, but not the only example of this.[82]

Although these measures were highly controversial at the time, it has been held by the ILO Committee of Experts that they do not constitute a breach of Article 3 of Convention 87. This is despite the fact that this purports to guarantee that public authorities shall refrain from any interference which would restrict or impede the right of workers' organizations to draw up their constitutions and rules, to elect their representatives in full freedom, to organize their administration, and to formulate their programmes. In truth, the freedom of association is never likely to allow or to require complete autonomy in the government and administration of organizations which wield power for and over people. (Although the US Supreme Court in *Eu*[83] took a fairly robust line on state interference in the internal affairs of a political party, in doing so it made clear that political parties have a favoured status and that in any event restrictions can be justified where there are compelling reasons.) The bottom line is that freedom of association, like all liberal values, is rooted in individual rather than group autonomy. Those who see freedom of association as an aspect of individual freedom will have no difficulty with a regime of state regulation of the internal affairs of trade unions and other associations if

---

[80] Trade Union Act 1984, pts. 1 (election of executive committees) and 2 (strike ballots) amended by Employment Act 1988. See now Trade Union and Labour Relations (Consolidation) Act 1992, Part I, Chapters 4, 5, and 6.

[81] Trade Union Act 1984, s. 3 (10). See now 1992 Act, *op. cit.* s. 47 (2).

[82] For an account of this see P. Elias and K. D. Ewing, *Trade Union Democracy, Members' Rights and the Law* (1988), 145–7.

[83] 489 US 214 (1989).

this is designed to ensure that they are organized in such a way as to maximize the potential for individual participation in their affairs. On this basis, what is the purpose of associating for any purpose—social, economic or political—if the individual is denied any opportunity to direct the affairs of the group? In other words, if the purpose of freedom of association is to provide a vehicle for individual fulfilment, little premium will be placed on the freedom of the group to regulate its own affairs. In dismissing the complaint from Britain and an earlier complaint from Australia,[84] the ILO authorities appear to have adopted this liberal political perspective and applied it to the economic sphere. In doing so, it has left itself open to the criticism of failing to appreciate that the promotion of liberal values can seriously undermine the efficiency and integrity of the group and consequently the totality of individual members whose interests it exists to serve.

THE RIGHT TO FREEDOM IN THE ASSOCIATION

So far as group autonomy is concerned, we have looked at autonomy over membership and autonomy over administration and government. A further question relates to the autonomy of the group as to the methods which it employs to promote its goals. To what extent is freedom of association recognized in this sense? So far, the problem has arisen most sharply in the area of political activity, with the courts having intervened in a number of ways to regulate the freedom of associations to promote their ends by direct involvement in the political process. The absence of any unifying theme (such as a constitutional guarantee of freedom of association) has meant that these questions have tended to arise and be answered in a rather spasmodic and *ad hoc* manner. Indeed, although questions relating to the freedom of a variety of associations to engage in political activities have arisen in the courts, there are as many answers to these questions as there have been questions to be answered. In fact, the response of the courts has varied according to whether the association has been a company, a trade union, or a charity. The courts have taken the most liberal approach with commercial organizations, insisting only that the company has sufficient powers in its objects' clause, either express or implied, to engage in political activity.[85] The matter is thus one largely within the control of the organization itself, though admittedly it is far from clear that all companies which pursue political activity in the form of

---

[84] *ILO Official Bulletin*, 60, Series B, No. 2. Report of the Governing Body Committee on Freedom of Association. Case No. 846.

[85] Otherwise companies are required only to disclose to shareholders political donations in excess of £200: Companies Act 1985, sch. 7, paras. 3–5.

donations to political parties have the necessary power to do so. The point is illustrated in unlikely circumstances by *Simmonds* v. *Heffer*[86] where the League Against Cruel Sports was restrained by Mervyn Davies J. from making donations to the Labour Party. In the absence of an express power to make such donations, the payments were held to be *ultra vires*, the court being unwilling to conclude that they could be permitted under a power which permitted the League 'to affiliate to combine or co-operate with subscribe to and or support any institution having objects similar to the main objects of the League'. Nor was the court prepared to permit the payment as being reasonably incidental to one of the other powers of the League which was 'To oppose and prevent cruelty to animals.'[87]

The position taken with regard to corporations contrasts to that which governs trade unions. Here the courts insisted that it was not enough that a trade union should have power in its objects clause, as they did in the case of companies. In *Amalgamated Society of Railway Servants* v. *Osborne*,[88] two other impediments were created. The first was the doctrine of statutory *ultra vires*, the House of Lords taking the view that trade unions could only engage in those activities which related to the statutory definition of the purposes of trade unions appearing in legislation of 1871 and 1876, which extended the cloak of legality to workers' organizations. So legislation which was designed to recognize unions as legal entities was used as a basis for restricting what they could lawfully do. Secondly, Lord Shaw of Dunfermline in particular took the view that trade union support of the Labour Party was unlawful as being contrary to public policy. In his judgment it was unlawful for members of Parliament to agree to abide by the decisions of the Parliamentary Labour Party, this being incompatible with 'that independence and freedom which have hitherto been held to lie at the basis of representative government in the United Kingdom'.[89] The decision was overturned by the Trade Union Act 1913, but only partially. Although trade unions were again permitted to levy money from their members for political purposes, unlike companies, it was not enough that the union had power in its objects clause. Additional requirements were imposed in the sense that trade unions were required to ballot their members for approval to set up a political fund which would be used for the party-political purposes specified in the Act. A majority of members voting had to vote in favour of the adoption of political objects and the creation of the fund which was the only source to be used for political purposes. But even if the union thus secured

---

[86] [1983] BCLC 298.
[87] See also *Lafferty* v. *Barrhead Co-operative Society Ltd.* 1919 SLT 257.
[88] [1910] AC 87 (HL).
[89] Ibid. 111.

authority to set up a fund, the political fund rules which it was obliged to adopt required the union to permit members to claim exemption from the obligation to contribute (to contract out) and to provide further that such members were not to suffer any disability or disadvantage as a result of their exemption. Any member aggrieved by an alleged breach of the political fund rules could complain to the Chief Registrar of Friendly Societies, a jurisdiction subsequently transferred to the Certification Officer.

One possible justification for treating trade unions and the Labour Party differently from other organizations is the closed shop. Thus, it is argued that if workers are required on pain of losing their jobs to join a trade union, they should not also be required to make a financial contribution which may be used eventually to support the Labour Party or any other political cause to which they may be strongly opposed. Indeed, it was this consideration which moved the US Supreme Court to introduce into US law a measure which was based on the British Act of 1913. In *Abood* v. *Detroit School Board*[90] the union had concluded agency shop agreements with state employers which required employees, if they were not members of the union, to pay an agency fee to the organization for the services which it provided on behalf of its members and non-members alike. The union then used its general treasury income for political causes, a practice to which the Court objected, on the ground that the First Amendment requires that people should not be required to make such contributions, on the ground in turn that in a free society one's beliefs should be shaped by one's mind and conscience rather than coerced by the state. The solution, however, was not to declare the agency shop principle unlawful but to require the unions to introduce an administrative plan whereby workers could be exempt from the obligation to finance political activity and to pay only their portion of collective bargaining-related expenditure. But although the closed shop—or similar arrangements—has thus been accepted as justifying special regulation of trade union political activity, the closed shop and similar practices have for all practical purposes been made unlawful and unenforceable in Britain by measures progressively introduced since 1980. So while there may be a case for the special regulation of trade union political activity and the operation of the principles of the 1913 Act in the USA, where union shops and agency shops are permitted, the case for such regulation in the UK is considerably weakened. The new restrictions on the closed shop make it particularly difficult to justify in principle the new restrictions on trade union political activity which were introduced in 1984. These not only required trade unions to ballot their members every ten years for authority to continue to promote political objects, but also

---

[90] 431 US 209 (1977).

widen the definition of political objects, thereby denying trade unions without political funds the opportunity to engage in an even wider range of political activity.[91]

The position with regard to charities contrasts again with that which governs companies and trade unions. Here the courts have also intervened to prohibit much political activity by such associations, though on this occasion Parliament has not stepped in to provide the necessary relief from the common law. In order to enjoy charitable status and the tax and other benefits that go with it, the purposes of a voluntary body must fall within the legal definition, which has its origins in the Statute of Elizabeth I but which is now to be found in Lord MacNaghten's dictum in *Income Tax Special Purposes Commissioners* v. *Pemsel*,[92] where he said that 'Charity in its legal sense comprises four principal divisions—trusts for the relief of poverty, trusts for the advancement of education, trusts for the advancement of religion, and trusts for other purposes beneficial to the community, not falling under any of the previous heads.' In the case of political objects the courts have consistently taken the view that these cannot be charitable. To qualify as a charity, an association must promote goals which are for the public benefit but the courts have no way of knowing whether a trust for the attainment of political objects would meet this requirement.[93] So it has been held that associations to procure the abolition of torture[94] or secure vivisection reform[95] or temperance reform[96] are not charities, the same conclusion being reached in the case of educational trusts designed to promote Conservative principles[97] or Labour Party propaganda.[98] But although an organization for a wholly or exclusively political purpose can never be charitable, it seems that the courts will allow a bona fide charity to promote its interests by some active involvement in the political arena, provided that this is merely ancillary to its main charitable purpose.[99] On the other hand, however, the trustees must be careful not to stray too far into politics, for if political activity should become a main object of the association then it would lose its right to be regarded as a charity. Moreover, although a charity might in this way be permitted to spend its funds promoting public

---

[91] See *Paul* v. *NALGO* [1987] IRLR 413. The law is now to be found in the Trade Union and Labour Relations (Consolidation) Act 1992, Part I, Chapter 6.

[92] [1891] AC 531.

[93] *Bowman* v. *Secular Society* [1917] AC 406, per Lord Parker of Waddington at 442.

[94] *McGovern* v. *Attorney-General* [1982] Ch 321.

[95] *National Anti Vivisection Society* v. *Inland Revenue Commissioners* [1948] AC 31.

[96] *Commissioners of Inland Revenue* v. *Temperance Council of the Christian Churches of England and Wales* [1927] 136 LT 27.

[97] *Bonar Law Memorial Trust* v. *Inland Revenue Commissioners* (1933) 49 TLR 220.

[98] *Re Hopkinson* [1949] 1 All ER 346.

[99] *National Anti-Vivisection Society* v. *Inland Revenue* (see n. 95). See also *Inland Revenue Commissioners* v. *Yorkshire Agricultural Society* [1928] 1 KB 611.

general legislation, charities are regarded as being precluded 'from direct or indirect financial or other support of, or opposition to, any political party or individual group which seeks elective office or any organization which has a political object'.[100]

Perhaps unsurprisingly, a number of charities have encountered some difficulty in recent years as a result of the legal restrictions, many of these being documented in the annual reports of the Charity Commissioners who exercise a supervisory role in this area. One such case related to the 1979 general election during which the Royal Society for the Prevention of Cruelty to Animals (RSPCA) sponsored an advertisement in several national newspapers by a body called the General Election Co-ordinating Committee for Animal Protection.[101] The advertisement urged electors to write to MPs and parliamentary candidates seeking their views on animal welfare, to attend political meetings to put forward their views, and to make these views known through the ballot-box. When the matter was brought to the attention of the Charity Commissioners they concluded that, while it was open to the RSPCA to press for legislation to prevent cruelty to animals, 'it was improper for them to support a direct attempt to influence voters'. The Commissioners raised the question with the Society, which after discussions agreed with their views and undertook to consult with them should any difficulties arise in the future. Yet, although the matter thus appears to have been settled amicably, it remains the case that the legal position as demonstrated by this affair seems irrational, if not bizarre, given that the election of a particular group of candidates or a particular political party could go a long way towards satisfying the objects of a particular charity. If a particular object is deemed to be charitable, why should the association not be free to promote that object by the best means available? There is, however, little prospect of the law being liberalized to allow charities more flexibility in the political arena. Indeed, if anything, the present rules are more likely to be tightened up. In a White Paper published in 1989, the Government claimed that it was vital that 'political and charitable purposes should remain distinct' and that it would be wrong if taxpayers, through the government, were to find themselves unwittingly distorting the democratic process by subsidizing bodies whose true purpose was to campaign not so much for their beneficiaries as for some political end. But although the Government detected some public anxiety about those charities which strayed beyond the boundaries of what is permissible and desirable, it did not propose any change to the law for the time being.[102]

---

[100] *Charities: A Framework for the Future*, Cm. 694 (1989), para. 240.
[101] *Report of the Charity Commissioners for 1979* (HC 608, 1979–80), para. 20.
[102] Cm. 694 (1989). Cf. Charities Act 1992. Now Charities Act 1993.

CONCLUSION

The law and practice relating to freedom of association in Britain has evolved very gradually throughout the twentieth century. There is no general right to freedom of association in English law, but governments have intervened from time to time to remove obstacles to effective association and also to encourage the growth and promote the development of associations where it was necessary or expedient to do so. This is not to deny that governments have also intervened to impose restrictions. But this was generally done only where it seemed necessary in the interests of some pressing social or political goal, as in the case of the restrictions on fascists in the 1930s and the restrictions on communists in the 1940s. It is not possible to see 1950 in any sense as a watershed, as it may be in some of the other areas covered by this book, though it is clearly the case that the 1970s and the 1980s in particular have seen an increase in the level of activity, with a number of new or extended restraints being introduced. These include the criminal liabilities for membership of proscribed terrorist organizations, the GCHQ ban, and the political restrictions on local government officers. No single factor explains these changes, which appear to be reactions to a number of unconnected concerns relating to levels of terrorist activity, the need to protect national security from perceived danger, and a desire to promote political neutrality in the higher levels of local government service. Most activity, however, has occurred in relation to trade unions, a number of measures being introduced since 1980 to regulate the closed shop and the internal affairs of trade unions. These reflect a concern in government to reduce the economic power of organized labour and at the same time to make trade unions even more accountable to their members than was already the case.

In view of the large number of issues which are covered by the rubric of freedom of association, it is difficult to draw any meaningful general comparisons with the position elsewhere in Europe. One important feature which distinguishes Britain from countries such as Germany, Italy, and Spain, however, is the absence of a constitutionally guaranteed right. This omission has made it very difficult for vulnerable groups to defend themselves in the courts, particularly in view of the fact that so many of the restraints have been introduced by way of primary legislation. It is difficult to believe, however, that a Bill of Rights or the incorporation of the European Convention would have made much practical difference. The case-law in a number of common law jurisdictions is characterized in the first place by a reluctance on the part of the courts to remove state restrictions, as reflected by the US Supreme Court's upholding of the control on civil servant's political activities. At the same time, the European Convention has proved to be of little value, with the wide exceptions to the right of freedom of association leading to the dismissal of the GCHQ complaint and with the court also prepared to accept wide restrictions on civil servants' political freedom.[103] Moreover, the constitutional right to freedom of association has been used to undermine group freedom and to undermine the power of people acting in association to promote common ends and interests. The most vivid example of this is in relation to union membership arrangements which, with the exception of Canada,[104] have been the subject of constitutional attack in a number of common-law jurisdictions as well as by the European Court of Human Rights.[105] The inevitable conclusion of all this is that, while there is a need for greater political respect to be paid to freedom of association, this is not likely to be secured or enhanced by a constitutional guarantee which relies on the courts for its enforcement.

[103] *Glasenapp* v. *Germany* (1987) 9 EHRR 25.

[104] *Lavigne* v. *Ontario Public Service Employees' Union* (1991) 81 DLR (4th) 545.

[105] See further K. D. Ewing, 'Freedom of Association in Canada' (1987) 25 *Alberta LR* 437, and Forde, 'The Closed Shop Case' (1982) 11 *ILJ* 1.

9

Privacy

JAMES MICHAEL

THEORY

Privacy is notoriously difficult to define, and its recognition as a right in various legal systems has been relatively recent. It is not mentioned as such in the French Declaration of the Rights of Man or the Bill of Rights in the United States, and problems of defining a right to privacy have hindered its recognition in the United Kingdom. The Younger Committee on Privacy decided in 1972[1] against recommending a general legal right to privacy, not least because of the difficulties of defining what was to be protected.[2] In 1990 the Calcutt Committee also recommended against the introduction of a general tort of infringement of privacy, but thought that definition would not be an insuperable problem. 'We consider . . . that it would be possible to define a satisfactory tort of infringement of privacy. . . . Our grounds for recommending against a new tort do not, therefore, include difficulties of definition.'[3] The reluctance of both committees and successive governments to propose general legislation is partly because of difficulties of definition, but also a reflection of the common law preference for specific remedies rather than abstract principles. (Governments may also have reflected on the difficulties that they have had with the general right to privacy in Article 8 of the European Convention on Human Rights.)

Theoretical arguments over the definition of privacy continue, along with more specific disputes over the details of new legal rules. In the international law of human rights, at least, 'privacy' now is clearly and unambiguously established as a fundamental right to be protected. As early as 1948, Article 12 of the Universal Declaration of Human Rights asserted that 'no one shall be subjected to arbitrary interference with his privacy, family, home, or correspondence, nor to attacks upon his honour and reputation'. However, quite apart from arguments over the definition of privacy, there are also differences of view over the categorization of privacy among human rights. Privacy is usually classified as a civil or

---

[1] Cmnd. 5092.

[2] Ibid. paras. 57–73; 665.

[3] *Report of the Committee on Privacy and Related Matters*, Cmnd. 1102 (1990), para. 12.12.

political right, principally because it is included in the United Nations Covenant on such rights; but it can be argued that, in certain respects, it also involves questions of economic, cultural, and social rights. In *Privacy and Freedom*[4] Professor Alan Westin wrote that 'privacy is at the heart of freedom in the modern state'. But Professor David Flaherty has written that 'although privacy is an important instrumental value, it is not identical to such fundamental values as liberty, freedom, and democracy'.[5] It is not necessary to decide between these two emphases here, except to suggest that, although privacy is important in an instrumental sense in that it is closely linked to the exercise of other rights such as the right to receive and impart information, it also reflects a near-universal need, although one of varying content, that makes it a fundamental value on its own.

Professor Westin's definition[6] of privacy as the desire of individuals for solitude, intimacy, anonymity, and reserve is useful, though not conclusive. So is the formulation by Dean Prosser of privacy into four torts of intrusion, disclosure of embarrassing private facts, presenting an individual in a 'false light', and appropriation of a name or likeness.[7] In considering the connotations of 'privacy', it is of some importance to recall whether one is discussing a state or condition, a desire, a claim, or a right.[8] The state or condition is familiar enough: everyone knows what it is to be withdrawn from the society of others, to find seclusion, avoid publicity, be solitary, or retire from the world's activities. Desiring to achieve such a state or condition presents no problems by itself, but to claim to be entitled to achieve it at will is quite another matter; and to be given a right to it, even against adverse claimants, is yet another.

Perceptions of all these things differ. It is a commonplace that privacy is culture-specific: the matters which a particular society regards as 'private' can vary widely.[9] Although anthropological evidence cannot be conclusive, it does seem that most societies regard some areas of human activity as not really suitable for general observation and knowledge. There is some evidence that even in small agricultural communities the boundaries of physical space which is 'private' are quite sharply defined, and that some kinds of personal information are also closely guarded.[10] This widespread, if not universal, desire for privacy is not limited to activities or information about them which would necessarily lead to

---

[4] A. F. Westin, *Privacy and Freedom* (1968 and 1970), 350.
[5] D. H. Flaherty, T. J. Donohue, and P. J. Harte (eds.), *Privacy and Data Protection: An International Bibliography* (1984).
[6] In *Privacy and Freedom*.
[7] 48 *Calif. LR* 383.
[8] For a discussion of these distinctions, see P. Sieghart, *Privacy and Computers* (1976).
[9] 'Crow men and women sometimes made love in public, in broad daylight. The Cheyenne were famous for their chastity.' I. Frazier, *Great Plains* (1990), 52.
[10] See S. Tefft (ed.), *Secrecy: A Crosscultural Perspective* (1980).

unpleasant consequences, or produce guilt or shame in the person concerned.

One explanation of the human urge for privacy, advanced by Sisella Bok, is derived from anthropology and developmental psychology. The process by which an infant becomes aware of its existence as an entity apart from others seems to be connected to the realization 'that one has the power to remain silent [which is] linked to the understanding that one can exert some control over events—that one need not be entirely transparent, entirely predictable, or . . . at the mercy of parents who have seemed all-seeing and all-powerful'.[11]

This explanation of the desire for privacy is closely linked with the development of the individual human personality. The Younger Committee did not linger over questions of definition, but it did assert that the need for privacy is nearly universal, and that it is not limited to human beings:

The quest and need for privacy is a natural one, not restricted to man alone, but arising in the biological and social processes of all the higher forms of life. All animals have a need for temporary individual seclusion or the intimacy of small units, quite as much as for the stimulus of social encounters among their own species. Indeed the struggle of all animals, whether naturally gregarious or not, to achieve a balance between privacy and participation is one of the basic features of animal life.[12]

For this chapter it is particularly important to draw another distinction: between privacy in all its aspects, and information privacy as one particular aspect of it. The general claim to privacy includes such things as not to have our territory invaded by others, even if this causes us no loss and tells those others nothing about us. But the area of 'information privacy' is much narrower; here, what we claim is that others should not obtain knowledge about us without our consent. In Professor Westin's words,[13] this is 'the claim of individuals, groups or institutions to determine for themselves when, how, and to what extent information about them is communicated to others'. Professor Arthur Miller put it even more briefly, defining it[14] as, 'the individual's ability to control the circulation of information relating to him'.

This is the aspect of the general right of privacy with which this chapter is principally concerned. This emphasis raises one fundamental question, made more important by the development of both surreptitious surveillance devices and information-processing technology. If the choice of the

---

[11] S. Bok, *Secrets* (1983).
[12] *Report on Privacy*, Cmnd. 5012 para. 109.
[13] *Privacy and Freedom*, 7.
[14] *Assault on Privacy*, 40.

individual is central to personal privacy, both in the sense of allowing physical intrusions and of sharing information, how can privacy be said to be invaded by the obtaining of information about an individual and its processing by automatic means if the individual has no knowledge that this is occurring? An easy answer lies in the 'chilling effect' doctrine first articulated by the US Supreme Court. As the West German Constitutional Court said, in its 1983 decision holding a new Census Act unconstitutional:

If someone cannot predict with sufficient certainty which information about himself in certain areas is known to his social milieu, and cannot estimate sufficiently the knowledge of parties to whom communication may possibly be made, he is crucially inhibited in his freedom to plan or decide freely and without being subject to any pressure/influence (i.e. self-determined). The right to self-determination in relation to information precludes a social order and a legal order enabling it, in which the citizens no longer can know who knows what, when and on what occasion about them. If someone is uncertain whether deviant behaviour is noted down and stored permanently as information, or is applied or passed on, he will try not to attract attention by such behaviour. If he reckons that participation in an assembly or a citizens' initiative will be registered officially and that personal risks might result from it, he may possibly renounce the exercise of his respective rights. This would not only impair his chances of development but would also impair the common good because self-determination is an elementary functional condition of a free democratic community based on its citizens' capacity to act and to cooperate.[15]

This doctrine focuses on the inhibitory effect on the exercise of other rights of knowing that one is, or may be, subject to surveillance. It is just the condition of uneasiness that was expressed in George Orwell's *1984*. As the Younger Committee reported: 'In such cases, we were told, the result would be an increase in the incidence of tension-induced mental illness or at least a decrease in the imaginativeness and creativity of the society as a whole.'[16]

Most of the writing on privacy assumes or asserts that invasions of privacy will become known to, or suspected by, those who are subject to such invasions. But a perfect system of surveillance and data processing need not even provoke suspicion. A 1972 report by the International Commission of Jurists quoted Professor Westin's description of privacy as 'the voluntary and temporary withdrawal of a person from the general society', and said:

Clearly, the 'withdrawal' of which A. Westin speaks is not possible when the individual is not conscious of the threat to or violation of his privacy, in other

---

[15] English trans. in 5/1 *Human Rights LJ* 94, at 100–101.
[16] Para. 111.

words when information is obtained by improper means, or when, having been obtained with the knowledge of the individual, it is then used for a purpose other than that contemplated or stipulated at the time of communication.

The question is far from theoretical. In the United Kingdom it is a cardinal administrative rule of police practice (*not* one of admissibility of evidence)[17] that tapes and transcripts of intercepted telephone conversations are never used as direct evidence in court. Other evidence can usually be found and if it is not, then it seems that an acquittal is an acceptable price to pay for the maintenance of secrecy. This rule of practice was explained by the British Attorney-General to the European Court of Human Rights in the *Malone* case,[18] and has since been incorporated in the Interception of Communications Act 1985. In *Malone* the Court held, unanimously, that the British system of telephone-tapping by administrative warrant, and of providing the police with 'metering' information about telephone calls with no warrant of any kind, were both violations of the right to privacy under Article 8 of the European Convention. As a matter of regional international law, then, such 'unperceived invasions' do violate the right to privacy, but the theoretical justification remains difficult.

Although it may not be easy to ascribe any psychological harm to an individual who is under thorough surveillance while mistakenly believing that he is not, it is possible to assert that such a system, particularly when imposed in secret by public authorities, is at the very least an unacceptable breach of the social contract between citizens and their governments which democracy entails. In this way, it is possible to explain one of the more significant developments in privacy-protection law since 1973 (the year of the first national data-protection statute, in Sweden). Before that time the legal protection of privacy in most countries was, apart from constitutional statements of principle, a matter of providing legal remedies which aggrieved individuals could pursue through the courts, or at times through administrative channels. However, a central feature of data-protection legislation since then has been the creation of supervisory commissions, commissioners, or boards, with the duties and powers of overseeing data users in the public and private sectors, in order to protect the privacy of what are now called data subjects. There may be a danger in such a development if it is at the expense of individual remedies, since it is always difficult to create bodies which are sufficiently independent of those to be regulated, and particularly difficult when government agencies are to be regulated. Nevertheless, this development is essential to protect

---

[17] *R. v. Keeton*, [1970] Crim. LR 402 (CA).
[18] Memorial of the Government of the UK in *Malone* (4/1983/60/94), Cour (83) 94, para. 2.19.

individuals from improper surveillance of which they may not be aware, as well as providing a practical solution to the question of how to resolve the difficult cases in which surveillance may be justifiable without defeating its possible purpose by disclosure to the person concerned.

One writer on privacy has such reservations about the definition of 'privacy' that he is reluctant to use the word at all. Professor Raymond Wacks believes that 'except to describe the underlying value, the term "privacy" ought to be resisted—especially as a legal term of art. It adds little to our understanding either of the interest that it is sought to protect or of the conduct that it is designed to regulate.'[19] As the title of his second book indicates, he prefers to describe the subject as the use (and misuse) of personal information about an individual. 'The essence of my argument . . . that at the heart of the concern about "privacy" is the use, and especially the misuse, of "personal information" about an individual.'

The argument is appealing, but it means that the US and European Convention on Human Rights cases on what might be called physical, rather than informational, privacy are defined out of consideration. In the context of US constitutional jurisprudence this is rather important, as the first case in which the Supreme Court declared or discovered a constitutional right to privacy concerned a state law forbidding the use of contraceptives, held to be unconstitutional in violating a constitutional right of marital privacy.[20] US constitutional law is particularly relevant to the United Kingdom because the US Supreme Court derived the right to privacy from common law rights, just as Warren and Brandeis based 'The Right to Privacy'[21] on common law and equitable principles such as defamation and the law of confidence. Some of the most important cases interpreting the right to 'private and family life' guaranteed by Article 8 of the European Convention on Human Rights have involved abortion[22] and homosexuality;[23] although there are also many Strasbourg cases involving personal information under the same Article.[24] Perhaps the problem is simply a characteristic of the English language: notoriously rich in synonyms, when used by lawyers the language seems to have a

---

[19] R. Wacks, *Personal Information: Privacy and the Law* (1989), 21. The word had been demoted from its place in the title of his earlier work, *The Protection of Privacy* (1980).

[20] *Griswold* v. *Connecticut*, 318 US 479 (1965). In his second book Dr Wacks considers the case and its successors in a footnote, albeit a lengthy one in the US manner, leaving space for only four lines of text on p. 32.

[21] (1890) 4 *Harv. LR* 193.

[22] *Bruggemann and Scheuten* v. *Federal Republic of Germany*, (1978) 3 EHRR 244, (1978) 10 Dec. & Rep. 100.

[23] *Dudgeon* v. *UK* (1983) 4 EHRR 149, *Norris* v. *Ireland*, European Court of Human Rights, 26 Oct. 1988.

[24] *Klass* v. *Federal Republic of Germany* (1978) 2 EHRR 214, *Malone* v. *UK* (1983) 5 EHRR 385, (Commission) (1984) 7 EHRR 14.

confusing tendency for a single word to be used to mean several quite different things (as with 'condition' in contract law).

One attempt at a sociological definition of 'the heart of the concern' was the research commissioned for the report of the Younger Committee on Privacy in 1972. This was followed by a survey of lawyers conducted by Dr Wacks. The Wacks survey measured the extent to which legal advice had been sought concerning six aspects of personal 'privacy': telephone-tapping, bugging, spying, unwanted publicity, appropriation of name or image, and misuse of confidential information. The highest number of reported complaints concerned the alleged misuse of confidential information. This is particularly significant when one considers, as Dr Wacks comments, that many attacks on an individual's 'privacy' will not be known to that individual, and that this is more characteristic of misuse of confidential personal information than of some of the other aspects selected for the survey. But by selecting aspects of personal privacy that are largely concerned with the use of personal information, the survey does not measure concern about 'physical' aspects of 'privacy' such as legal regulation of sexual conduct. A survey of solicitors asking how often they were consulted about the law regulating abortion and sexual activity would also be interesting. Laws regulating sexual activity require intrusion for their enforcement and amount to an interference by their existence even if they are not enforced.[25] Laws regulating abortion require a different analysis, but are even more difficult to classify as involving the use of personal information. Whether they should be described as concerning 'privacy' or something else, it is likely that public opinion is more intense on laws regulating sexual behaviour than on a central computer with personal information (even in the tendentious version of the question on that subject used in the Younger survey).

In the United States the separation of the action for breach of confidence (which was one of the bases for Warren and Brandeis's argument for a 'Right to Privacy' in 1890) from emerging torts of invasions of privacy had progressed so far by 1939 that the *Restatement of the Law of Torts* could say that the action for breach of confidence 'does not deal with cases in which the information relates to matters in one's life outside of his business'. Dr Wack's scholarship contributes the intriguing information that this comment was omitted from the *Restatement, Second* in 1977.[26]

The argument that 'privacy' is not a useful legal term is compelling, but leads to a question: why do jurists and legislators use the term increasingly in national and international jurisprudence and legislation,

---

[25] So the European Court of Human Rights found in *Dudgeon*. See chapter 15.
[26] Wacks, *Personal Information*, 132.

often to describe circumstances other than the use or misuse of personal information? Judge Bork indicated his impatience with the US Supreme Court's difficulties in defining 'privacy' in his testimony at the Senate hearings considering his unsuccessful nomination for the US Supreme Court. (A look at the lower court opinions of Judge Souter, whose nomination was successful, involving privacy showed random breath tests of drivers, telephone-tapping, and restrictions on sexual-history testimony in rape cases being considered by the judge in his opinions.) Perhaps what is needed now is an attempt at defining what is meant by 'privacy' when judges do not mean the use of personal information. However, the purpose of this chapter is not a theoretical analysis of privacy in the law of the United Kingdom, but a discussion of its development since 1950. The most significant developments, particularly in relation to those in other European countries, have been in 'informational privacy', especially the aspect now generally known as 'data protection'. For that reason, for reasons of space, and in order to avoid overlapping other chapters, this chapter concentrates on the law relating to personal information.

OVERVIEW OF NATIONAL DEVELOPMENTS

Privacy and the press

Developments in the law relating to privacy since 1950 can be divided roughly into privacy and the press (and broadcasting) and other developments regarding privacy. Invasions of personal privacy by the press and broadcasting are naturally far more obvious than breaches of personal privacy by the use of personal information for administrative or commercial purposes. The latter are, except for direct mail advertising, rarely known even to the person concerned. Breaches of privacy by the mass media, at least invasions by publication rather than by the means of obtaining information, are obviously exposed to the general public.

Some complaints about invasions of privacy by the press are about the techniques used in attempting to get information, others are about the publication of personal information, and many are about both. Concern about invasions of privacy by the press are often mixed with concern about defamation, and it is perhaps worthwhile to recall the difference. Invasion of privacy, in the sense of informational privacy, by the press is, in Prosser's terms, the disclosure of embarrassing private facts. Defamation is the publication of damaging information which is false. Without going into details about the burden of proving truth or not, privacy is about true information, defamation about false.

The history of privacy and the press since 1950 has largely been one of recommendations from official Commissions and Committees about self-regulation as an alternative to statutory regulation, followed by changes in the self-regulatory system. Private members' bills have also played an important part in the process, with at least five private members' bills introduced to establish a right of privacy affecting the press.[27] The first Royal Commission on the Press recommended in its report[28] the establishment of a voluntary General Council of the Press, and this was done in 1953, after a private member's bill was introduced to establish a statutory council. The second Royal Commission[29] proposed changes in the council membership, with a statutory council if nothing was done. The council changed its name (to the Press Council) and membership. The Younger Committee on Privacy[30] made recommendations, some of which were adopted, including a Declaration of Principle on Privacy issued in 1976. The third Royal Commission on the Press reported in 1977,[31] criticizing the press and proposing a legal right to privacy and possibly a statutory Press Council if self-regulation did not improve. The Calcutt Committee on Privacy and Related Matters[32] in June 1990 was the latest, although not necessarily the last, official body to urge changes in self-regulation with legislation as the alternative.

The appointment of the Calcutt Committee was a reaction to two private members' bills rather than a government initiative. Even if the Government had intended to do something about privacy and the press, under Prime Minister Margaret Thatcher Departmental Committees and Royal Commissions were not favoured. The parliamentary support for the Protection of Privacy Bill introduced by John Browne and the Right of Reply Bill introduced by Tony Worthington was sufficient, especially as they were based on similar bills in earlier sessions, to persuade the Government to appoint a Home Office Committee chaired by David Calcutt QC.

The terms of reference deserve quotation in full, if only to illustrate the difficulties in describing the law relating to privacy.

In the light of the recent public concern about intrusions into the private lives of individuals by certain sections of the press, to consider what measures (whether legislative or otherwise) are needed to give further protection to individual privacy from the activities of the press and improve recourse against the press for

[27] Lord Mancroft (1961), Alexander Lyon (1967), Brian Walden (1969), William Cash (1987), and John Browne (1989).
[28] Cmnd. 7700.
[29] Cmnd. 1811.
[30] Cmnd. 5012.
[31] Cmnd. 6810.
[32] Cmnd. 1102.

the individual citizen, taking account of existing remedies, including the law on defamation and breach of confidence.

Although the Calcutt Committee could not consider broadcasting, a Broadcasting Complaints Commission had been established by the Broadcasting Act 1981 to decide on complaints about invasions of privacy in the making and broadcasting of programmes. The statistics compiled by the Broadcasting Complaints Commission and the Press Council indicated 'less than a handful of complaints about infringements of privacy each year with no evidence of a rising trend'.[33] While the Committee was sitting, however, there was an incident that illustrated both 'intrusions into the private lives of individuals by certain sections of the press' and the limits of existing remedies.

In February 1990 a journalist and photographer from the *Sunday Sport* intruded into a hospital room where an actor was recovering from brain surgery after injuries suffered in a storm. The only part of an inter-locutory injunction against publication that survived in the Court of Appeal was a prohibition, based on malicious falsehood, on reporting that the actor had agreed to being photographed and interviewed. Glidewell LJ said:

It is well-known that in English law there is no right to privacy, and accordingly there is no right of action for breach of a person's privacy. The facts of the present case are a graphic illustration of the desirability of Parliament considering whether and in what circumstances statutory provision can be made to protect the privacy of individuals.

He then canvassed the four grounds relied on for an injunction (libel, malicious falsehood, trespass to the person, and passing off) and concluded that only malicious falsehood could support a limited injunction against reporting that the actor had agreed. Bingham LJ agreed, saying:

If ever a person has a right to be let alone by strangers with no public interest to pursue, it must surely be when he lies in hospital recovering from brain surgery and in no more than partial command of his faculties. It is this invasion of his privacy which underlies the Plaintiff's complaint. Yet it alone, however gross, does not entitle him to relief in English law.[34]

The Calcutt Committee did not, however, recommend a general tort of invasion of privacy or a statutory right of reply.[35] Instead, the recommendations were specific, divided roughly into institutional pro-posals for self-regulation, with fairly specific threats of a statutory tribunal if self-regulation seemed to fail, and legislative proposals, mostly the

---

[33] Cmnd. 1102, para 4.4.
[34] *Kaye* v. *Robertson and Sport Newspapers Ltd.*, *The Times*, 20 Mar. 1990.
[35] Recommendations 10, 9.

creation of new offences connected with the seeking of personal information for publication. The major institutional proposal, that the Press Council should be disbanded and replaced by a Press Complaints Commission, has been carried out almost exactly as described. It remains to be seen whether 'maverick publications persistently decline to respect the authority of the Press Complaints Commission', which Calcutt proposed should lead to the establishment of the Commission on a statutory basis.[36]

Calcutt also recommended that two separate 'triggers' would require the replacement of a voluntary Press Complaints Commission with a statutory Press Complaints Tribunal that would have even greater powers than a statutory Press Complaints Commission.[37] One trigger would be the failure to implement all the recommendations for a voluntary Press Complaints Commission by June 1991, with commitments to the Commission by all major newspaper and magazine publishers as the minimum acceptable.[38] The other trigger would be a serious breakdown in the system of self-regulation, as in 'a less than overwhelming rate of compliance with the Commission's adjudications'.[39]

Of the specific proposals, by far the most controversial were those to create three new crimes of acts committed to obtain information 'with a view to its publication'.

The following acts should be criminal offences in England and Wales:

(*a*) entering private property, without the consent of the lawful occupant, with intent to obtain personal information with a view to its publication;

(*b*) placing a surveillance device on private property, without the consent of the lawful occupant, with intent to obtain personal information with a view to its publication; and

(*c*) taking a photograph, or recording the voice, of an individual who is on private property, without his consent, with a view to its publication and with intent that the individual shall be identifiable.[40]

Thus far there has been little enthusiasm from the Government and Opposition political parties for the recommendations to make acts criminal when done with a view to publication, but not when committed by private investigators or government officials without such a view to publication. The system of self-regulation is now being tested, however, and there may yet be a statutory Commission or Tribunal. One indication of the new Commission's approach was an adjudication in May 1991 that the *News of the World* had violated the code of conduct by invading the personal privacy of Clare Short MP.

---

[36] Recommendation 25.
[37] Recommendations 27, 28.
[38] Para. 16.10.
[39] Para. 16.11.
[40] Recommendation 1, para. 6.33.

The year ending in December of 1992 was rich in incidents involving the privacy of public figures and the press. Just before the General Election there was the revelation in a Scottish paper that the Leader of the Liberal Party, Mr Paddy Ashdown, had had an extra-marital affair several years before. The information had been obtained from documents stolen from the office of Mr Ashdown's solicitor. In addition to provoking debate about privacy, public figures, and the press, the incident illustrated the difference in the law on interlocutory injunctions in England and Wales, where injunctions based on breach of confidence had at least temporary success, and in Scotland, where the requirement of specificity meant that they were less successful.

The Secretary of State for the newly-created National Heritage Department, Mr David Mellor, appointed Sir David Calcutt to review the effectiveness of the Press Complaints Commission. Shortly after the appointment, Mr Mellor himself was the subject of an invasion of privacy by the publication of information obtained by recording telephone conversations between him and an actress with whom he was having an extra-marital affair. The recording was almost certainly not an offence under the Interception of Communications Act 1985 because it was recorded on another extension from the subscriber's telephone. Mr Mellor resigned, although not until there was evidence in a libel case that he had accepted a free holiday from one of the litigants.

Publication of allegedly confidential information from a book about the private life of the Princess of Wales prompted the Press Complaints Commission to deplore such journalism, although later developments suggested that the publication might actually have been with her approval. Publication of surreptitious photographs taken of the Duchess of York at a swimming pool in France with her financial adviser illustrated the difference between the absence of any general civil privacy law in the United Kingdom and the effect of such a law in France, where they both recovered damages.

The publication of transcripts (and the public availability by telephone of recordings) of telephone conversations alleged to be between members of the Royal Family and others also concentrated public attention on the issues of privacy and public figures. The interceptions of cellular telephone conversations almost certainly breached the Interception of Communications Act 1985, the Wireless Telegraphy Act 1949, and possibly the Copyright, Patents and Designs Act 1988, but no action was taken. A little-noticed change in copyright law now gives a speaker copyright in his or her words once they have been recorded (although the recorder also has a copyright). The difficulty is that in order for a civil action to be taken, the party to the intercepted conversation must identify himself or herself, and such identification may also be inevitable in any criminal proceedings. Information about the status of the credit card account of the Chancellor of the Exchequer was published, British Airways admitted to gaining access to customer information on the database of Virgin Airways, and both were investigated by the Data Protection Registrar.

Parliamentary attention concentrated on two developments. One was a private member's bill introduced by Mr Clive Soley. Although not directed specifically at the protection of privacy, the bill would establish a mechanism requiring newspapers to publish corrections, which would at least provide a remedy of sorts for the publication of 'false light' information. The other Parliamentary development was the decision of the Select Committee on the Heritage Department to hold hearings on Privacy and the Press. Their report seemed likely to support several measures, including legislation on the use of surveillance devices and a general tort of invasion of privacy, but there was also argument that they should not propose any further measures to protect privacy against the press until a bill, such as Mr Mark Fisher's private member's bill, was enacted to provide for greater press freedom in the form of a general public right of access to government records.

At the end of 1992, it seemed likely that Sir David Calcutt would propose even more strongly than before that the three criminal offences which were recommended by the Committee that he chaired should be made law, and that he would also recommend a statutory system of press regulation. Government reaction to Sir David's report would also be influenced by the later report of the Select Committee on the Heritage Department.

Other developments

The Data Protection Act 1984 and the Interception of Communication Act 1985 have been major changes in legislation on privacy not involving the press since 1950. The first was adopted almost entirely for economic reasons, so that data-protection laws in other countries could not be used as non-tariff trade barriers. Its content was largely dictated by the Council of Europe's Convention for the Protection of Individuals with regard to Automatic Processing of Personal Data (to be discussed below), which was opened for signature and ratification in 1981. The second statute was adopted to comply with the decision of the European Court of Human Rights in the *Malone* case.

There had been political controversy about personal privacy at the end of the 1960s, marked by a series of private members' bills. The Younger Committee on Privacy was appointed in 1970 in exchange for the withdrawal of one such bill. The Committee reported in 1972, and did not recommend a general right to privacy. Few of the specific proposals made by the Younger Committee have become law. Neither the courts nor Parliament have been particularly active in finding or making law to protect privacy, apart from the two statutes mentioned. In terms of a general right to privacy almost nothing has changed, at least in terms of domestic law. (It should not be forgotten that Scotland's separate, and largely civil, legal system has gone further towards a general right to privacy through the principle of *actio injuriarum*, which provides a remedy for injuries to honour.)[41]

Data protection

On the subject of data protection, the concern which began with private members' bills in the late 1960s continued, heavily influenced by efforts at the Council of Europe, the EC, and the OECD. The Lindop Committee was appointed, and reported in 1978. By 1979 there had been sufficient progress for one observer to write that 'Given the inevitable international delays and pressure on the British government's legislative timetable, a reasonable projection would be for a bill to be introduced and passed during the 1982–3 session.' He concluded that 'It would perhaps be appropriate for a British Data Protection Act to take effect in 1984.' Events went more or less as predicted, despite the fact that the predictions were made before the Conservative government took power in 1979.

The chronology of events which culminated in the Data Protection Act 1984 has been set forth in detail in other publications, and this will be an outline. One of the various bills to protect aspects of privacy during the 1960s was introduced by Brian Walden MP, who withdrew it in exchange for an undertaking from the Labour Home Secretary to appoint a committee to inquire into the protection of privacy. The committee was hampered in its enquiries by terms of reference restricting it to invasions of privacy in the private sector. Requests to both Labour and Conservative ministers to permit examination of threats to privacy from government were refused. That committee, chaired by Kenneth Younger, reported in 1972. Although the Committee did not favour a general right to privacy, a number of specific proposals were made, one of which was a set of principles to be observed in the automatic processing of personal data. Those principles deserve repetition not only for their influence on British privacy law, but also for their resemblance to the principles established in 1981 by the Council of Europe Data Protection Convention.

1. Information should be regarded as held for a specific purpose and not be used, without appropriate authorisation, for other purposes; and
2. Access to information should be confined to those authorised to have it for the purpose for which it was supplied.
3. The amount of information collected and held should be the minimum necessary for the achievement of the specified purpose.

[41] Kilbrandon, 'The Law of Privacy in Scotland' (1971) 2 *Cambrian LR* 35 (1971).

4. In computerised systems handling information for statistical purposes, adequate provision should be made in their design and programs for separating identities from the rest of the data.

5. There should be arrangements whereby the subject could be told about the information held concerning him.

6. The level of security to be achieved by a system should be specified in advance by the user and should include precautions against the deliberate abuse or misuse of information.

7. A monitoring system should be provided to facilitate the detection of any violation of the security system.

8. In the design of information systems, periods should be specified beyond which the information should not be retained.

9. Data held should be accurate. There should be machinery for the correction of inaccuracy and the updating of information.

10. Care should be taken in coding value judgments.

The next specific development in legal protection of privacy was in the Consumer Credit Act 1974. The Younger Committee had recommended that 'an individual should have a legally enforceable right of access to the information held about him by a credit rating agency'.[42] (A similar recommendation had been made by the Molony Committee on consumer protection in 1970.) Just such a right was included in the 1974 Act, which had been introduced by a Conservative government and then reintroduced by the new Labour government.[43] The right of access in the Consumer Credit Act is to a transcript reduced into plain English, and so would almost certainly include computer-stored information, and require a print-out of it.[44] Appeals over rights of access, correction, and other questions (such as whether the 200-word statement to be included is 'scandalous' or not) are initially to the Director General of Fair Trading.

One of the first acts of the new Labour Home Secretary after the second general election in 1974 was to travel to the United States to inquire into the Freedom of Information Act, which had just been amended, and the Privacy Act of 1974. He returned more convinced of the need for privacy legislation than for an open government law such as the Freedom of Information Act. In 1975 the Labour government published two White Papers on computers and privacy. One of them[45] proposed a non-statutory body to prepare the way for permanent data-protection machinery. That Data Protection Committee was announced

---

[42] Cmnd. 5012, para. 298.

[43] The subject-access provision does not distinguish between manual and automated records, and it is limited to records kept on natural persons. It does, however, distinguish between those seeking credit for personal use (s. 159) and those seeking credit for business purposes (s. 160). The right of those seeking business credit is more restricted.

[44] s. 158(5).

[45] *Computers and Privacy*, Cmnd. 6353, para. 31.

early in 1976. There was a delay when the first chairman of the committee, Sir Kenneth Younger, died suddenly, and a new chairman, Sir Norman Lindop, was appointed. The terms of reference for the committee were to advise on legislation to protect personal data automatically processed. This meant that two issues on which other countries have differed in legislating on data protection, whether to include legal persons and manual records, were already decided. The committee was also directed to consider data protection in both the public and the private sectors.

The period between the publication of the committee's report in 1978 and legislation in 1984 was marked by a series of government announcements of policy short of what the committee had recommended, followed by criticism from Sir Norman Lindop and other committee members; after a period of silence the Government usually then announced a change of mind to something closer to what had been recommended. In the end, most of the recommendations in the Lindop Report became law.

The structure of the proposed statute was to include an independent Data Protection Authority with broad regulatory powers. The Government moved from an initial proposal that the Home Office carry out these functions to the final Act's establishment of a nearly independent registrar. The committee's proposal for the registration of data users was largely accepted, although the recommendation that any exemption for reasons of national security should be 'precisely limited' was not. The Act allows for a complete exemption of any data processing if it is certified to be in the interests of national security by a Cabinet Minister. The committee envisaged a major role for codes of practice in data protection, to be approved by Parliament and have the force of law. But the Government resisted any recognition of such codes until the final legislative stages, when an amendment was allowed for the registrar to encourage the development of voluntary codes.

The rules relating to medical and social-welfare files were, however, effectively deferred beyond the Act and left to ministerial rule-making. The rules[46] restrict access which 'would be likely to cause serious harm to the physical or mental health of the data subject' or 'disclose to the data subject the identity of another individual (who has not consented to the disclosure of the information)'. The Access to Personal Files Act 1987 gives rights of subject access to particular kinds of social-service records. The Access to Medical Reports Act 1988 gives individuals access to medical reports about themselves made by their general practitioners to insurance companies or employers. The Access to Health Records Act 1990, which only applies prospectively, gives individuals general rights of

---

[46] Data Protection (Subject Access Modification) (Health) Order 1987, SI 1987, No. 1903.

access to their medical records. The statutory rights of subject access were the products of adroit and expert campaigning by Maurice Frankel, director of the Freedom of Information Campaign, which was started in 1984.

One particular aspect of data protection for employees seems not to have been considered specifically by Lindop. The report recommended that there should be no contracting-out from the provisions of the law. When the government bill was being considered in Parliament several amendments were put forward to stop the possibility that employers might require all applicants, and perhaps even all employees, to exercise their rights of subject access to records of any criminal convictions and give the results to employers as a condition of employment. All of the proposals were rejected, and the use of the subject-access right in this way is now increasing. This is despite the Rehabilitation of Offenders Act 1974, which otherwise gives individuals the right to withhold information about 'spent' convictions. 'What is happening now, because individuals are being asked to obtain their records and to present them to would-be employers, is that [Rehabilitation of Offenders] Act is perhaps being flouted.'[47]

The most obvious effect of the Data Protection Act 1984 is the requirement that all data users (except for those certified to be in the interests of national security) must register and indicate what kind of personal data they process and for what purposes. Such universal registration is being abandoned in other countries with data-protection legislation, in favour of concentration on closer scrutiny of the processing of 'sensitive' personal data. Behind the registration system is the power of the registrar to issue enforcement notices of different kinds which may be appealed by a data user (although not by a data subject) to the Data Protection Tribunal. Two groups of cases decided by the Tribunal illustrate the effect of the Act and some areas of public concern.

When the Younger Committee commissioned research on privacy they found that 35 per cent of those surveyed thought that the availability of their names and address on the electoral register was an invasion of privacy.[48] Although there have been objections and the Data Protection Registrar has expressed concern about some uses of the information, electoral registers are available for public inspection and copies must be available for sale. One aspect of the community charge that provoked controversy was the collection of information on individuals. The Community Charge Register was available for inspection, but the statute

---

[47] Mr Eric Howe, Data Protection Registrar, Minutes of Evidence, Home Affairs Committee Report on the Annual Report of the Data Protection Commissioner, Session 1990–1, HC 115, p. 16, para. 56.

[48] Cmnd. 5012, app. E, table H.

effectively prevented the large-scale use of the information in the way that credit reference agencies use electoral register information.

The National Council for Civil Liberties collected forms that Community Charge Registration Officers (CCROs) were using and sent them to the Data Protection Registrar, arguing that some were requiring personal information that was not required by law. The registrar issued enforcement notices prohibiting the CCROs from requiring individuals to give their ages and the type of property they occupied. In November 1990 the Data Protection Tribunal upheld the enforcement notices against four local authorities.[49] Credit reference agencies use electoral registers and rely on unsatisfied county court judgments and other information about anyone who has lived at an address to determine credit ratings, even if the information is about someone who is unknown to the credit applicant. In 1990 the registrar issued an enforcement notice prohibiting the use of such 'third-party' information. The credit reference agencies appealed, and in February 1991 the Tribunal largely upheld the prohibition, allowing the use of such information only when the third parties had the same surname or were reasonably believed to be living as members of the same family as the credit applicant.

In a little-known case, the registrar issued a 'transfer prohibition notice' forbidding the transfer of personal information by a particular company to other companies in the United States for use in fraudulent mail solicitations.[50] The reason was that, although the United States has a federal Privacy Act, it only applies to the records of the federal government. There is no law affecting the private sector that is equivalent to the Data Protection Act.

The Data Protection Act 1984 almost certainly would not have been enacted had it not been for apprehensions that the United Kingdom needed legislation in order to ratify the Council of Europe Convention, and that other national data-protection laws might be used as non-tariff trade barriers if the UK did not ratify. The bill was usually presented as having the two objectives of protecting privacy and promoting trade in data processing, but the responsible minister conceded, in a speech to a Canadian audience, that he was only persuaded of the need for legislation by the economic argument.

Interception of communications

It was thus international law in the form of the Data Protection Convention, which derives in part from the right to privacy under Article 8 of

---

[49] *Rhondda Borough Council CCRO (DA/90 25/49/2), Runnymede Borough Council CCRO (DA/90 24/49/3), South Northamptonshire District Council CCRO (DA/90 24/49/4), London Borough of Harrow CCRO (DA/90 24/49/5)* v. *Data Protection Registrar*, 21 Nov. 1990.

[50] Transfer Prohibition Notice against Winsor International Ltd., 3 Dec. 1990.

the European Convention on Human Rights, that caused the United Kingdom to adopt its data-protection legislation. Similarly, it was a decision of the European Court of Human Rights that persuaded the UK to adopt legislation to regulate the interception of communication. In the *Malone* case the Court held unanimously that both the system of telephone interception and that of providing 'metering' information to the police violated the right to privacy.

The application was brought by a man who had been tried and acquitted on charges of handling stolen property. During his trial he discovered that his telephone had been tapped. This in itself was unusual, as direct evidence of intercepted conversations is never used in court as a matter of administrative practice. After his acquittal he brought a civil action against the Metropolitan Police Commissioner which failed because there is no general right to privacy and because none of his particular claims provided a remedy. In particular, the law of confidence, the articulation of which in *Prince Albert* v. *Strange*[51] is nearly as familiar in privacy literature as the Warren and Brandeis article, was held not to apply to intercepted communications. The Vice Chancellor did consider Article 8 of the European Convention and its interpretation in the *Klass* case, but could not apply it because the Convention had not yet been incorporated into British law. And he said that the subject was one which 'cried out' for legislation.[52]

Having exhausted his domestic remedies, Malone exercised his right of individual petition to the European Commission of Human Rights. In addition to his claim that tapping his telephone, which the police had conceded having done at least once, violated his Article 8 right, Malone also claimed that his right to privacy had been violated by police use of 'metering' information. He claimed that police had used information from this source about the people he had telephoned, and that several people he had talked to had been visited by the police. The Government denied that any metering had taken place, and argued further that no privacy issue was presented if it had, because the information would only be a record of signals sent to the exchange, and not of any conversation.

The Commission found in Malone's favour on the question of whether the British system of telephone-tapping had violated his right to privacy, but they could not find any violation of his rights by metering because there was inadequate evidence from which they could find that his telephone had been metered or that such information was ever furnished to the police. The case was then referred to the Court, which had before it not only the arguments of Malone and the British government, but also written observations from the Post Office Engineering Union, which

---

[51] 1 Mac. & G. 25 (1849).
[52] *Malone* v. *Metropolitan Police Commissioner* [1979] Ch. 344 at 380.

had been given permission under the Court's new rules. The POEU submission included evidence that metering information was given to the police without any judicial or administrative warrant.

In August 1984 the Court's judgment was published.[53] The Court held, unanimously, that there had been a breach of Article 8 both regarding the interception of communication and the release of metering records to the police. In keeping with its usual approach the Court first considered whether there had been an interference with the right to privacy, then considered whether it had been authorized by law. They found that it had not been so authorized, and so did not inquire further into whether the interference was justified in a democratic society for one of the reasons listed in paragraph 2 of Article 8. They also felt it unnecessary to consider whether there had been a violation of Article 13, which guarantees a right to an effective domestic remedy for violations of rights guaranteed by the Convention. Two partially dissenting opinions and one concurring opinion urged that the Court go further and indicate what the Convention required as an effective domestic remedy for invasions of privacy.

The British government acted quickly, introducing a bill to regulate the interception of communication. That bill became the Interception of Communications Act 1985. In outline, the law gives legislative approval to the existing system of ministerial warrants. However, a quasi-judicial system of review was introduced in the form of a tribunal. Anyone who suspects an interference with communication may appeal to the tribunal, which has wide powers to investigate. If they find that there has been an interception which was not properly authorized, they have the power to order that it be stopped, that records be destroyed, and that damages be paid. But if they find that the interception was properly authorized they will report to the applicant only that they have found no violation. One case brought after the passing of the Act challenged telephone-tapping before it was passed. In *R.* v. *Secretary of State for the Home Department, ex parte Ruddock*,[54] judicial review of the Home Secretary's authorization failed on the ground that it was not sufficiently unreasonable for the court to set aside.

The 1985 Act effectively places the previous administrative system on a statutory basis with the addition of a tribunal. It creates a criminal offence of unlawful interception of post or telecommunications, and provides that an interception under authority of a warrant from the Secretary of State under section 2 is not an offence. Warrants can be issued in the interests of national security, for the purpose of preventing or detecting serious crime, or to safeguard the economic well-being of the

---

[53] *Malone* v. *UK* (1984) 7 EHRR 14, Series A, vol. 82.
[54] [1987] 1 WLR 1482.

United Kingdom. A tribunal of five legally qualified persons is established to hear complaints and decide whether communications have been improperly intercepted. The tribunal investigates and decides whether an interception has in fact occurred, and if it was properly authorized. If the tribunal finds that the authorization for an interception was made contrary to the Act, it can quash the authorization and order compensation to be paid. If it finds that an interception was made which was properly authorized it will report that there was no violation of the Act, which could mean that there was no interception or that an interception had been properly authorized. No violations have yet been found by the tribunal. The Act also continues on a statutory basis the system of having a senior judge review the system of warrants and report annually to the Prime Minister, with edited versions of the reports published. These reports have indicated that occasionally the wrong telephone lines have been tapped.[55]

In at least one sense the Act extends the scope of permissible intercepts. The existing administrative system was said to have been used only for the prevention and detection of serious crime, with definitions of serious crime varying somewhat over the years. The Act incorporates this standard for domestic intercepts. However, the Act also authorizes what may be called 'class' intercepts for international communications. One ground for such intercepts, which are wider in scope than domestic ones, is the prevention of terrorism. Another allows international interceptions in the interests of the 'economic well-being of the country'. This does not necessarily require any evidence of violation of the criminal law. The words are taken directly from the second paragraph of Article 8 of the European Convention on Human Rights. Although it is likely that the words were adopted originally in the interests of countries with exchange controls being able to intercept communications for evidence of violations of those controls, the phrase does not necessarily require such violations. The United Kingdom does not now have exchange controls, and it is at least possible, under the new law, for the British government to inform its economic strategy by the interception of international financial communications. And the Act contains an absolute ban on the introduction of any evidence in any judicial proceeding other than before the tribunal to indicate that an interception has taken place.

Breach of confidence

The law of confidence, important in the historical development of privacy, thought by the Younger Committee to be an important potential doctrine for the protection of privacy, but insufficient to provide Mr Malone with

---

[55] Report of the Commissioner for 1987, paras. 16–21.

a remedy, may yet be given new statutory form. One of the few recommendations of the Younger Committee to be acted upon was the proposal that the Law Commissions should consider the doctrine and its possible reform. The Lord Chancellor referred the matter to both Commissions, who considered it over the years. The Law Commission for England and Wales reported in 1981[56] and the Scottish Law Commission reported in 1984.[57]

Both Law Commission proposals are a mixture of greater protection for individual privacy and measures to allow greater freedom of the press under the doctrine, which would be made a statutory tort. The proposals of the two Law Commissions are very similar, although the Scottish proposals differ from the English in two ways. The Scottish report would impose a duty of confidentiality on information obtained by illegal means or means regarded as improper by a reasonable person,[58] rather than the more detailed circumstances outlined in the English report.[59] The Scottish Commission would also impose an obligation of confidentiality on any information acquired by such means 'however trivial it may seem to an outsider'.[60] The greater protection for privacy would be in the extension of the obligation of confidentiality to information obtained by surveillance. This was in part to reverse the statement by Megarry VC in *Malone* that a person who uses a telephone must accept the risk of being overheard by tapping. The English Commission commented that they did 'not think that in a civilised society a law abiding citizen using the telephone should have to expect that it may be tapped'.[61] The English Commission's draft bill would impose such an obligation for information obtained without authority from any computer or data-retrieval system. It would also impose the obligation for information obtained by a device made primarily for the purpose of surreptitiously carrying out surveillance, or for any other device capable of being used for such purpose. The 'ordinary' surveillance device, such as a camera with a telephoto lens, would be made subject to a 'reasonable expectation' test: acquisition of information by such a device imposes an obligation of confidentiality only if 'a reasonable man in the position of the person from whom the information is acquired would have appreciated the risk'.

Thus, on the one hand, it may be thought that two people, who meet secretly in a secluded corner of a large railway station throughout which clear notices are displayed that television cameras are being used to deter criminal activities (such as malicious damage), cannot reasonably expect the fact of their meeting to be

[56] *Breach of Confidence*, Cmnd. 8388 (1981).
[57] *Breach of Confidence*, Cmnd. 9385.
[58] Paras. 4.36–41.
[59] Para. 6.46.
[60] Para. 4.38.
[61] Para. 6.35.

treated as confidential. On the other hand, it may well be that the use of an ordinary camera with a telephoto lens to obtain from the street a picture of a confidential document lying on a desk in a private house would go far beyond the reasonable expectations of the person who left it there, and that the taker of the picture should be subject to an obligation of confidence in respect of the information so obtained.[62]

The provision for the freedom to receive and impart information lay in the changes made to the existing doctrine that breaches of confidentiality may be justified in the public interest. In its present form this is a defence to an action for breach of confidence on the basis that the breach was justified to disclose 'iniquity', which until recently was largely restricted to justifying disclosure of crime. The English Commission expanded this, recommending that the 'public interest may arise in the disclosure or use of confidential information whether or not the information relates to iniquity or other forms of misconduct'.[63] The other important change would be that, if the person accused of a breach of confidence satisfies the court that the public interest is involved, then 'it should be for the plaintiff to establish that this interest is outweighed by the public interest in the protection of the confidentiality of the information'.[64]

If the English Law Commission bill were to be adopted it would go some way towards increasing the legal protection for some privacy interests, but it is not, as the Commission pointed out, a protection of privacy measure: 'to give a remedy merely because information is acquired by one of these means [by surveillance device] would amount to the creation of a right of privacy—a right, for example, not to be photographed even if the photographs were later never published . . .'.[65] To create such a right was almost certainly beyond the Commission's terms of reference, but their proposals and analysis raise fundamental questions about the legal protection of privacy. Is the initial obtaining of information, which one person prefers to keep private, by another person, an invasion of privacy? The answer may be an easy 'yes' if it is assumed that the person observed becomes aware of it. It is more difficult if the obtaining of information is by the surreptitious methods contemplated by the Commission, which are designed not to reveal the fact of observation to the person observed. Imposing an obligation of confidentiality on information so obtained is important, but it limits the public use of that information, rather than the activity of surveillance itself. A prohibition on surveillance itself would depend on the interpretation

---

[62] Para. 6.37.

[63] Para. 7.2(24)(iii).

[64] Para. 7.2(24)(v).

[65] Para. 6.35, noting that aerial photographs had been allowed in *Bernstein of Leigh (Baron)* v. *Skyviews and General* [1978] QB 479.

given to the principle in Article 5 of the Data Protection Convention, which says that 'personal data undergoing automatic processing shall be: obtained and processed fairly and lawfully'.

In one case the law of confidence has advanced from *Malone* in being used to prohibit publication of information obtained by an unlawful interception of telephone communications.[66] The case was, however, an interlocutory injunction, and one against a newspaper, whose intrusions are perhaps considered to be more objectionable than those less public. Although the Home Secretary, then Leon Brittan, gave an undertaking to the House of Commons on 12 March 1985 to introduce legislation along the lines of the English Law Commission's report, that has not yet been done. The Calcutt Committee did not endorse the Commission's proposals, partly because the doctrine was thought to be 'most effective for the protection of commercial information rather than individual privacy'.[67] The Committee may also have been influenced by a parliamentary written answer in 1989 that 'the Law Commission's report . . . amounts for the most part to a recommendation for a restatement of the common law of breach of confidence',[68] a statement that the authors of the report might dispute.

INTERNATIONAL AND EUROPEAN LAW DEVELOPMENTS

Effect of the European Convention on Human Rights

The most effective provision of a general right to privacy in the United Kingdom was anticipated in a 1972 report on privacy by the International Commission of Jurists, which noted that the United Kingdom was a party to the European Convention on Human Rights and had accepted the right of individual petition. 'It is conceivable, therefore, that a complaint could be brought before the European Commission in a proper case to enforce a right of privacy . . .'.[69] That has happened, and in the cases of *Golder*, *Silver*,[70] *Malone* (already referred to), *Dudgeon*, and *Gaskin* the Court has enforced the right of privacy under Article 8 against the United Kingdom. Although incorporation of the European Convention into domestic law could finally provide a general right to privacy enforceable in British courts, it might also deprive the institutions of the European Convention of a prime source of test cases. Although decisions

---

[66] *Francome* v. *Mirror Group Newspapers*, [1984] 1 WLR 892 (CA).
[67] Para. 8.6.
[68] Para. 8.4.
[69] International Commission of Jurists (1972), 458.
[70] Both discussed in more detail in Ch. 6.

by the Commission and Court on applications against other countries are not binding on the United Kingdom, they are a clear indication of how similar complaints against the United Kingdom would be decided.

Article 8 of the European Convention differs substantially in its formulation of the right of privacy from the Universal Declaration of Human Rights and the International Covenant on Civil and Political Rights, both as to content and as to method of interpretation. As in many of the Convention rights, it is formulated in two paragraphs. The first is a statement of the right very like that of the Declaration and the Covenant; the second sets out the qualifications:

1. Everyone has the right to respect for his private and family life, his home and his correspondence.

2. There shall be no interference by a public authority with the exercise of this right except such as is in accordance with the law and is necessary in a democratic society in the interests of national security, public safety or the economic well-being of the country, for the prevention of disorder or crime, for the protection of health or morals, or for the protection of the rights and freedoms of others.

The Commission and Court have rarely attempted a precise definition of 'private life' or the reasons for the provision, but in one case the Commission used language similar to that of the German constitution's provisions, in its Article 2(1), for the 'free determination of personality'. Although no violation was found, the Commission said:

The scope of the right to respect for private life is such that it secures to the individual a sphere within which he can freely pursue the development of his personality. In principle, whenever the state enacts rules for the behaviour of the individual within this sphere, it interferes with the respect for private life.[71]

The Commission and the Court take the elements of the Article in turn. The first question is whether the facts complained of present any question of the rights in the first sentence at all. If so, the next question is whether the facts present any 'interference' with the right. If that is established, then two basic questions are presented by the second paragraph of Article 8: was the interference in accordance with the law, and are the aims of the interference permissible for one of the interests described? The rights in Article 8 are subject to derogation in time of war and public emergency under Article 15, as well as permissible reservations under Article 64. The United Kingdom has made no such reservations or derogations regarding Article 8, although it has filed a derogation regarding personal liberty under Article 5. If a country has no law protecting particular aspects of privacy, then there may be a violation of Article 13 as well.

[71] *André Deklerck* v. *Belgium*, Application No. 8307/78 (1981) 21 Dec. & Rep. 116.

The acceptance of the right of individual petition has been very nearly the sole source of the case-law interpreting Article 8.[72] Much of this has been concerned with prisoners' rights and family life, sometimes combining the two. Although the subject of prisoners' rights is considered in detail in Chapter 6 by Genevra Richardson, it is particularly relevant to privacy because the very act of imprisonment removes one layer of privacy by subjecting the prisoner to a degree of surveillance.[73]

In the relatively early 'confidence-building' period, the Commission tended to accept interferences with the privacy of prisoners as 'inherent' in imprisonment.[74] Now, however, the Commission follows the approach outlined above, requiring each intrusion to be justified. The tendency is to find that the activity complained of is an interference with the prisoner's private life, but that it is justified by reasons such as the prevention of disorder or crime, the interests of public safety, or the protection of the rights and freedoms of others.

Some of the most important cases involving interference with correspondence have involved prisoners. Initially, there was a tendency to uphold interferences with prisoners' correspondence as inherent in imprisonment.[75] Now, however, largely as a result of two major decisions by the Court, both concerning the United Kingdom, interferences with the right of prisoners to correspond with the outside world must be specifically justified under paragraph 2 of Article 8. *Golder* was the first of these, in which the Court found an unjustifiable interference in the Home Secretary's refusal to permit a prisoner to write to a solicitor to seek legal advice.[76] *Silver* was less condemnatory, but at least required some form of discernible administrative rules about censorship of prisoners' post.[77]

The requirement that interferences with correspondence must be both provided for by law and justifiable for one of the reasons specified in the Article has also been considered in cases involving telephone-tapping. Germany's system was found to be in compliance with the Convention; those of the United Kingdom and France were not. In *Klass*[78] the Court

---

[72] The right of individual petition has now been accepted, although not always for the same period of years, by all the state parties.

[73] Jeremy Bentham's planned prison, the Panopticon, was designed for constant surveillance on inmates. It now seems standard in writing on surveillance to quote from Bentham's plans, or from Foucault's description of them in *Discipline and Punish*.

[74] Case-Law Topics No. 1, 'Human Rights in Prison', pp. 23–4 (1970), Council of Europe.

[75] As in No. 2749/66, *DeCourcy* v. *UK* 10 *YBEC* 412.

[76] *Golder* v. *UK* 4451/70, judgment of 21 Feb. 1975, Series A, No. 18.

[77] Judgment of 28 June 1984, Series A, No. 80.

[78] *Klass* v. *Federal Republic of Germany*, judgement of 6 Sept. 1978, Series A, no. 28, para. 30–38.

considered complaints that the German system of intercepting telephone communications was an unjustifiable interference with privacy. One seemingly minor point in the judgment is important in considering the justiciability of what may be called 'unperceived, but suspected' violations of privacy: although the applicants did not have clear proof that their conversations had been intercepted, they were entitled to claim the status of 'victims' of the system of interference. The German system of authorization for interceptions in the interests of national security involved scrutiny by a parliamentary panel, and relatively short time periods for authorizations.

In contrast, the British system was unanimously held by the Court to violate Article 8, despite the observation in a 1972 report by the International Commission of Jurists that there was 'unusually strict administrative control of authorized interceptions and there is no reason to believe that this practice is carried out to any greater extent than in other countries'.[79] The British Attorney-General's description of the system of administrative control did not persuade the Court that it met the basic requirement of Article 8 that interferences must be provided for by law before their justification on one of the enumerated grounds could be considered. As mentioned earlier, the Court held that the system of providing the police with 'metering' information about telephone numbers dialled was also an interference which violated Article 8. In this particular ruling, the Court went beyond the case-law of the US Supreme Court, which has provided privacy theorists with many particular illustrations of where the assertion of a general right to privacy can lead.[80]

Other claims that government surveillance violates the right to privacy under the Convention have been less successful. In one case the applicant had taken part in a non-violent but disruptive demonstration against the South African government during a rugby match. She was arrested and photographed during the demonstration and again at the police station, where she was told that her photograph would be kept so that if she caused trouble at future matches she could be identified and charged. The Commission found that there was no interference with her private life because the photographs were of a public incident in which she participated voluntarily, and that there was no indication that the photographs would be made public or used for any other purpose than identification in the future.[81] It is not entirely clear whether the case means that photographing someone in public without his or her consent is itself not an interference with privacy, or whether it was the further circumstances of a

[79] *International Social Science Journal* 24/3 (1972), 507.

[80] *Smith* v. *Maryland*, 99 S.Ct. 2577, in which obtaining information by use of a 'pen-register' (metering) was held not to violate the right to privacy.

[81] Application No. 5877/72, *X* v. *UK* 45 Coll. Dec. 90.

public demonstration followed by the limited use of photographs by the police which made it not an interference. In a much earlier case the Commission had found that police photography which was not connected with a public event was an interference with privacy, but that it was justified.[82]

Those applications involved surveillance which was known to the subjects at the time. There have been few cases in which 'unperceived' surveillance was involved (apart from the telephone-tapping in *Klass*). In one case the applicant was an Austrian communist who was charged with wilful bodily harm and destruction of property during a demonstration. He was acquitted, but learnt during the trial that a report on him had been submitted to the court by the Vienna Federal Police Directorate. The report included the information that he had participated in political children's holiday camps in 1961, 1963, and 1972, when he was 7, 9, and 18 years old, and that he had been responsible for publication of a student communist newspaper. None of these activities violated any law, and he had no criminal convictions. The Commission found that the use of such information in criminal proceedings was clearly necessary for the prevention of crime.[83] It is not obvious whether the actual surveillance which produced the information involved an interference with privacy, although the Commission found no evidence that the applicant had been individually subjected to secret surveillance (the federal police had said that they had collected the information from records kept in accordance with laws on registration with the police and press law). The question of whether this sort of 'data surveillance' interferes with privacy was considered in the case of *Leander* v. *Sweden*, in which no violation was found. The applicant was refused employment as a carpenter at a naval museum on security grounds, and he complained that the refusal to disclose the basis for the finding violated his right to privacy. The Court of Human Rights ruled that the state had a very wide margin of appreciation in the interests of national security, and that the various Swedish safeguards were sufficient.[84]

The complaints of Patricia Hewitt and Harriet Harman that they had been under unlawful surveillance when general secretary and legal officer of the National Council for Civil Liberties were more successful. The surveillance was revealed when a former officer of MI5 disclosed it in a television interview in 1985. The Commission found that there had been a violation because the surveillance was carried out by a body which had no legal authority, and thus was not authorized by law. After the experience of the *Malone* case, the British government anticipated the result, and

---

[82] Application No. 1307/61, *X* v. *Federal Republic of Germany*, 9 Coll. Dec. 53.
[83] Application No. 8170/78, *X* v. *Austria*, 16 Dec. & Rep. 145.
[84] Judgment of 26 Mar. 1987, Series A No. 116.

introduced the Security Service Act 1989. The Act is almost a copy of the Interception of Communications Act 1985, except that the actions authorized by ministerial warrant are acts of surveillance that may include trespass rather than interception of communications. The Hewitt and Harman cases were not referred to the European Court of Human Rights, and the Committee of Ministers of the Council of Europe decided to take no further action after the British adoption of the Security Service Act.

In the case of an application challenging the British census as a violation of Article 8, the Commission found in favour of the United Kingdom.[85] The Commission considered that the census was a prima-facie interference with the right to privacy, but that it was in accordance with the law and necessary for the economic well-being of the country. In so concluding, the Commission was particularly influenced by assurances that census returns were treated in complete confidence, that names and addresses were not to be included in the computer processing, and that the forms would not be passed to the Public Record Office for 100 years.

The European Commission and Court of Human Rights have now produced a considerable body of reasoned interpretations of the right to privacy in particular instances. Many of these involve privacy rights other than strictly 'informational privacy', such as cases involving family rights, and sexual identity and activity. Those subjects are considered in Chapters 10 and 15.

Effect of other international instruments

The Universal Declaration of Human Rights and the Covenant on Civil and Political Rights

International statements of the right to privacy, like all such international formulations, are cast in general terms. But the right to privacy articulated in Article 12 of the Universal Declaration of Human Rights 1948 included some reasonably specific provisions in saying that: 'No one shall be subjected to arbitrary interference with his privacy, family, home or correspondence, nor to attacks upon his honour and reputation. Everyone has the right to the protection of the law against such interference or attacks.' The concept of individual privacy was thus extended to include the kinship 'zone' of the family.[86] The physical zone of protection includes the home, and correspondence with others, which may go very far from the physical home.

The main objective of the Covenants was to reinforce the Universal

---

[85] Application No. 9702/82, *X* v. *UK*, decision of 6 Oct. 1982, 30 Dec. & Rep. 239.

[86] Samuel Warren might have approved, as his objection was at least in part on behalf of his family.

Declaration by specific treaty law. There is as yet little material interpreting the right to privacy guaranteed by Article 17 of the International Covenant on Civil and Political Rights, which approaches the jurisprudence interpreting the corresponding guarantee in Article 8 of the European Convention on Human Rights. The Human Rights Committee's examinations of national reports referred to it under Article 40 have given some indication of the areas of concern under Article 17. Questions have been asked, for example, about measures to protect individual privacy from automated information systems, and about safeguards to protect privacy from national intelligence services.[87]

International data protection law

In international circles, concern about the potential effect of automatic data processing upon the right to privacy began to grow during the late 1960s and early 1970s. But the eventual drafting of international measures specifically for the purpose of protecting privacy interests against the possible misuse of information by automatic means was carried out entirely by regional institutions, mostly European ones. Although the relatively rapid growth of this activity in data protection was deliberately directed at protecting one aspect of personal privacy, it is very unlikely that it would have grown so rapidly had it not been for a quite unexpected economic aspect. Some countries, especially those committed by treaty to reducing tariff barriers, began to fear that others might use their national data-protection laws as non-tariff trade barriers. The development of international standards would not only establish minimum requirements for national legislation, but it could also be used to create a community of countries which met those requirements and which agreed on a free market of information among themselves, to the potential exclusion of others.

These efforts were made largely by three co-operating institutions: the Council of Europe, the European Economic Community (EEC), and the Organization for Economic Co-operation and Development (OECD). Of the three, the Council of Europe took the lead in developing a legal instrument. This may have been because the Council already had responsibility for development of the right to privacy guaranteed by Article 8 of the European Convention on Human Rights. In 1971 the Legal Affairs Committee of the Council of Europe proposed two resolutions, one for the private sector and another for the public sector, which were adopted by the Committee of Ministers in 1973 and 1974.[88] The Conven-

---

[87] 33 GAOR Supp. 40, UN Doc. A/33/40 para. 348, 239 (1978).

[88] Resolution (73)22, 'On the Protection of the Privacy of Individuals Vis-à-Vis Electronic Data Banks in the Private Sector', adopted 26 Sept. 1973; and Resolution (74)29, adopted 24 Sept. 1974.

tion for the Protection of Individuals with regard to Automatic Processing of Personal Data covering the public and private sectors was opened for signature and ratification in 1981. The European Economic Community first considered data protection in a 1973 report, which was followed by debates in the European Parliament in 1974 and 1975. Thereafter, Community institutions deferred to the Council of Europe and its Convention until the summer of 1991, when the Community published a draft directive on data protection that would, among other things, extend the data-protection principles to systems of manual records. The OECD continued to co-operate with the Council of Europe and the European Community, and published its own guidelines in 1980.

The Convention came into effect in 1985, and is now legally binding on its adhering parties, which the Guidelines are not. The Convention has three major functions: it establishes basic rules for data-protection measures to be adopted by adhering states; it sets out special rules about trans-border data flows; and it establishes mechanisms for consultation, if not enforcement. The Convention is not a 'European' Convention; the absence of 'European' from its formal title emphasizes that it is open for adherence and ratification by non-European countries.

The core of principles gives considerable scope for variation to suit different constitutional and legal systems in their domestic implementation. These are found in chapter II of the Convention, beginning with the obligation on parties to take the necessary measures in domestic law to give effect to the principles. 'Measures' is not necessarily limited to laws, but it is unlikely that a domestic approach limited to voluntary codes alone would comply with the Convention. The measures must be in force at the time when the state concerned becomes bound by the Convention.

Article 5 sets out the basic principles of data protection, as to both the contents of the data and their processing. If they undergo automatic processing they are to be:

(a) obtained fairly and lawfully;
(b) stored for specified and legitimate purposes and not used in a way incompatible with those purposes;
(c) adequate, relevant and not excessive in relation to the purposes for which they are stored;
(d) accurate and, where necessary, kept up to date;
(e) preserved in a form which permits identification of the data subjects for no longer than is required for the purpose for which those data are stored.

This does not attempt to describe for what purposes personal data may be stored, but at least bars the collection of such data for no declared purpose. The last principle probably does not require an irrevocable 'anonymization' of name-linked data, but only adequate security measures.

The 'special categories' of personal data in Article 6 are not to be processed automatically 'unless domestic law provides appropriate safeguards'. The list represents a broad consensus on the sort of personal information regarded as particularly sensitive by representatives of the European countries in the late twentieth century. They are 'personal data revealing racial origin, political opinions or religious or other beliefs, as well as data concerning health or sexual life', and 'the same shall apply to personal data relating to criminal convictions'. This is a minimum list, and it is open to parties under Article 11 to give effect to cultural values by giving similar special protection to other kinds of personal data. The inclusion of criminal convictions in the list will pose problems for newspapers which use or disseminate data from automated data bases; it may be an incentive for them to adopt the Swedish practice of not identifying defendants in criminal trials (although the exception to that rule for 'public figures' would seem to violate the principle as well). Article 7 requires 'appropriate security measures'.

Article 8 is about 'additional safeguards for the data subject', but might be better called 'rights of data subjects', except that the Convention does not necessarily require the creation of rights. There is a division in data-protection circles between those who see a right such as subject access as one of several means to the end of seeing that personal data are handled according to Article 5 principles, and those who see subject access as an independent right of its own. The Article requires domestic legislation to give data subjects the right to know whether there are automated data files on them; to find out the contents of such files; to have errors corrected or deleted; and to have a remedy if these rights are violated. Although the right to know the existence of files does not require a public register of their controllers, other measures, such as notification to a data subject when the files are opened, would be required to provide equivalent protection. It is not clear from the text whether the right to rectification includes retrospective notification to everyone who has received the erroneous information, but it would seem at least arguable that it does.

None of the data-protection principles is absolute. Article 9 attempts to define the circumstances in which states are justified in departing from the principles. The resemblance of the qualifying second paragraph to several of the rights under the European Convention on Human Rights is deliberate. The easiest exception is for personal data files 'used for statistics or for scientific research where there is obviously no risk of an infringement of the privacy of the data subjects'. Less easy is the allowance of derogation from Articles 5, 6, and 8 (the core of the core principles) when such derogation is provided for by the law of the Party and constitutes a necessary measure in a democratic society in the interests of:

(*a*) protecting State security, public safety, the monetary interests of the State or the suppression of criminal offences;

(*b*) protecting the data subject or the rights and freedoms of others.

'Monetary interests' would seem to justify laws to detect tax evasion and violations of exchange controls, and perhaps also violations of social-welfare benefits. 'Suppression of criminal offences' may go beyond detection to include at least some elements of prevention. 'Protecting the data subject' is probably intended to justify the reluctance of many medical authorities (and some others) to allow uncontrolled subject access to health data, while one of the major 'rights and freedoms of others' which may conflict with data protection is the right to receive and impart information.

Article 10 says that the domestic methods of implementing the principles must include remedies, while Article 11 makes it clear that the Convention sets basic standards, and that parties can give greater protection to data subjects, such as rights of access to manual files. That is the core of requirements to protect the privacy of data subjects. The rest of the Convention is concerned with trans-border data flows and machinery for implementation.

Chapters IV and V establish the machinery and rules for implementation. Article 14 requires the designation of a data-protection authority or authorities for the purposes of co-operation with other parties. The Council of Europe's explanatory note to this section says: 'It should be underlined, however, that while the Convention requires the designation of an authority by each Contracting State, this does not mean that the Convention requires each state to have a data protection authority. A Contracting State may designate an authority for the purposes of the Convention only.' Articles 18–20 establish a Consultative Committee, which is not given any enforcement powers. Each party has a representative on the committee, which is to meet at least once every two years. Its function is not to adjudicate, but to 'make proposals' for the application or amendment of the Convention, and to express an 'opinion' on any proposed amendment or any question concerning the Convention, when requested to do so by a party. The committee is clearly expected to assist in the resolution of disputes between parties, but just as clearly does not have any authority to resolve such disputes formally. In practice, it is composed largely of representatives of the data-protection institutions of the parties. Article 23 is potentially one of the most important parts of the Convention: it authorizes accession by countries outside the Council of Europe on the invitation of the Committee of Ministers with the approval of all parties to the Convention. This was mainly intended for the benefit of the non-European members of the OECD (Canada, USA, Australia, and Japan).

The OECD Guidelines Governing the Protection of Privacy and Transborder Data Flows of Personal Data are in the language of recommendation rather than obligation, and have special provisions recognizing the special problems of federations (such as Canada, the USA, and Australia) and perhaps approving a sectoral approach more than the Convention does. The basic principles are similar to those of the Convention, but written in less specific terms. The other basic principles are roughly equivalent to those of the Convention in urging the principles of collection limitation, data quality, purpose specification, use limitation, security safeguards, and accountability. Although the Guidelines do not discuss the question of legal and natural persons, the definition of 'personal data' as 'any information relating to an identified or identifiable individual [data subject]' displays an intention to refer only to natural persons. It would even seem that the Guidelines do not contemplate supplementary measures to cover legal persons, although inclusion of manual records (as in three of the non-European member states) would seem to be within the statement (Guideline 6) that: 'These Guidelines should be regarded as minimum standards which are capable of being supplemented by additional measures for the protection of privacy and individual liberties.'

CONCLUSION

The legal protection of personal privacy since 1950 has been marked by public concern (at least as represented by the Younger, Lindop, and Calcutt Committees, and the various commissions on the press and broadcasting), reluctance to legislate in general terms, and specific legislation, such as that on data protection and interception of communications, mostly forced on the United Kingdom by its membership of the Council of Europe. The introduction of a general tort of invasion of privacy is often resisted on the ground that it would be an unacceptable infringement of the freedom of the press. However, this would seem to be a less than complete argument when weighed against the experience of the United States of America, France, and the Federal Republic of Germany, all of which have both varying types of civil actions for invasion of privacy and freedom of the press that is probably equivalent to that in the United Kingdom.

The form of privacy legislation that seems most likely in the near future is to regulate the press, probably like the regulation of broadcasting in the Broadcasting Complaints Commission.

The spate of incidents involving privacy and public figures, particularly those in the Royal Family, make it very likely that there will be some legislation during the 1993–94 session of Parliament. At one extreme would be a form of systematic regulation by a statutory authority as envisaged by the report of the Calcutt Committee and likely to be proposed even more firmly by Sir David himself. Another possibility, although perhaps less likely, would be the establishment of a general civil tort, perhaps including the four types of torts described by Prosser in the United States. Most likely will be some sort of legislation to control the use of surveillance devices, perhaps along the lines recommended by the Younger Committee in 1972.

Whatever changes are made, there is a fundamental problem. The difficulty of using the existing or future legal rules and remedies is that someone must assert them, and in the case of the civil remedies that is the person whose privacy right is being breached. It is a myth that members of the Royal Family cannot or will not go to law: they have done in the past, using the law of confidence and contract successfully. What is desired by most, but not all, of those subject to intense scrutiny by the press is a system by which the law will restrain disclosures of personal information without any disclosure of the fact that they would like the law to impose such restraint. If the judgment of third party authorities is relied upon to restrain invasions of privacy, they are very likely to get it exactly wrong, as in the condemnation by the Press Complaints Commission of disclosure of information about the Princess of Wales on the assumption that she could not possibly want people to read about such things.

If the subject is entitled to a degree of concealment, a price of exercising that right in a liberal democracy is to assert it in a public court. There is already an unfortunate tendency of the judiciary to secret hearings under the Contempt of Court Act. Another price in a liberal democracy is that there should be a fundamental bias against prior restraint, with the emphasis on remedies such as accounting for profits.

Invasions of privacy by the press are in the nature of things well-publicized, and thus more likely to provoke concern for regulation than invasions of privacy by the use of personal information by government and some industries in the private sector. Such non-press invasions of privacy are now regulated to some extent by statutes such as the Data Protection Act 1984, the Interception of Communications Act 1985, and the Security Service Act 1989 (the last of which is less concerned with informational privacy). It is almost certain that none of these would have become law without the influence of the institutions of the Council of Europe. All of them have near-absolute immunity from legal redress for invasions of personal privacy by government said to be in the interests of national security. Striking the balance between personal privacy and national security for the foreseeable future in the United Kingdom seems likely to depend on the interpretations of the European Convention on Human Rights by the Commission and Court in Strasbourg.

10

Families and Children: From Welfarism to Rights

JOHN EEKELAAR

THREE TYPES OF STATUS RELATIONSHIP

For the purposes of this chapter, a distinction is drawn between three paradigmatic forms of status relationship. In an *instrumentalist* relationship, the legal relationship is structured on the basis that A controls B in the furtherance of A's interest. In a *welfarist* relationship, A has legal powers over B which are expected to be employed for the benefit (as defined by A) of B (or of a group of people which includes B). In a *rights-based* relationship, the starting-point of A's relationship to B lies in claims which B makes of A. The theoretical basis for the distinction between welfarist and rights-based status relationships can be stated here only in summary form.[1] The argument is that the two relationships are incompatible one with the other. The duty to promote the welfare of another implies the power to determine where that other's welfare lies. The possession of such a power is inconsistent with the recognition of that other's rights. Although it might logically be held that B has the right that A should promote B's welfare in accordance with A's perception of that welfare, such a right is really no right at all. A person who surrenders to another the power to determine where his own welfare lies has in a real sense abdicated his personal autonomy.

In contrast, rights-based thinking takes as its starting-point the recognition that all people make claims. Individuals have rights to the extent that others are placed under a duty to act in accordance with those claims. Since the claims of different individuals may conflict, it would be incoherent to say that everyone has a right that his or her claim should be realized. It therefore follows that the way conflicting claims are reconciled, and the extent to which they create duties for others, reflects the image which a community holds of itself. But this image is not *simply* a perception by a dominant group of the welfare of the whole community: it is not a version of the welfarist model. This is because it is made up of, and built around, a corpus of claims which individuals in the community are

---

[1] I have developed the arguments at length in J. Eekelaar, 'The Importance of Thinking that Children have Rights' (1992) 6 *Int. J. Law and the Family* 221 and 'Parenthood, Social Engineering and Rights' (1994) *Zeitschrift für Rechts und Sozial-philosophie* (forthcoming).

making. Rights-based thinking therefore *compels* (1) recognizing that people make claims, (2) paying serious attention to those claims, and (3) justifying decisions in terms of those claims. In particular, it means that the mere assertion that action which affects an individual is in that individual's interests is never sufficient justification in itself for such action. Actions must be justified by relating the action specifically to the particular claim which the individual is making (or which the individual would make if he or she could make it). A person's claim may be said to have become a right when the disposition of powers and duties of others and of the community's resources are such as to make the realization of the claim a realistic option.

It is argued in this chapter that, while early family relationships were instrumentalist in character, this instrumentalism gave way to welfarism from the late nineteenth century until around the middle of this century. From that time, welfarism has gradually been replaced by a rights-based approach to family relationships. I have elsewhere elaborated the wider significance of rights-based thinking, especially with regard to children.[2] The instrumentalist nature of the early law will not, however, be pursued here.[3]

LEGAL RELATIONSHIPS BETWEEN ADULTS

The rights of married women

In the absence of a code, English law contained no stipulation like the welfarist provision found in the French Civil Code until 1970 that 'le mari est le chef de famille. Il exerce cette fonction dans l'intérêt commun du ménage et des enfants' (CC. Art. 213). But in the mid-twentieth century the common law position was not very different. On marrying, each partner came under a duty to live with the other (expressed as the right to consortium). The husband had to concern himself with his wife's well-being, but was given much freedom in determining its content. This is

---

[2] See n. 1.

[3] It is described in the case of the parent–child relationship in J. Eekelaar, 'The Emergence of Children's Rights' (1986) 6 *Oxf. J. Legal Studies* 161; for the husband's rights to 'consortium' (i.e. 'comfort and services') of his wife, see C. A. Morrison in R. H. Graveson and F. R. Crane (eds.), *A Century of Family Law* (1957), 101–3; and D. Mendes da Costa, ibid. at 179–80. Stephen Parker, 'Rights and Utility in Anglo-Australian Family Law' (1992) 55 *MLR* 311 prefers to characterize the progression as a movement from a 'rights' to a 'utility' model. Parker's narrow 'rights' model corresponds to the 'instrumentalism' in my analysis. Insofar as Parker's 'utility' model is welfarist, my analysis sees it as a transitional model, leading to a greater receptivity to accepting what individual family members *themselves* claim to be in their interests and decreased emphasis on general utility as perceived by the policy maker. This attitude is, I think, characteristic of thinking of people as actual or potential right holders.

apparent in a series of nineteenth-century cases, but a leading textbook writer in 1957[4] considered that they probably still represented the law at that time as being as follows.

1. Where they were living together, the husband needed to provide his wife with necessary expenses for running the house. But he controlled the extent to which she could incur him in liability for her purchases and in this way controlled the family's standard of living.[5]

2. The magistrates' courts had the power to order a husband to make periodical payments to his wife if he wilfully neglected to provide reasonable maintenance for her or a child of the family. But liability did not accrue while the parties were living together.[6]

3. Nor could a wife improve her position by separating from her husband. If she was not content with what he gave her and left him, she would have to accept whatever he agreed to give her, even if it was insufficient to feed her and the children.[7] If she was not willing to live with him on his terms, she could expect no better. In 1967 the Court of Appeal confirmed that if a husband had agreed to give his separated wife nothing, she could claim nothing.[8]

4. In his 1957 textbook, Bromley wrote that a husband's duty was primarily 'to provide his wife with the necessities of life. As Hodson LJ has repeatedly pointed out, the duty is prima facie complied with if he provides a home for her, and the wife has no right to separate maintenance in a separate home unless she can justify living apart from her husband'.[9] This meant that, unless the husband acted unreasonably, the wife was bound to accept his decision. In 1940, Henn Collins J. had said:[10]

[The husband] put forward, as he had put forward before 28 November, accommodation at a boarding house kept by a Mrs. Smith. It is common ground that that accommodation was in itself perfectly suitable, and, I think, exceptionally good of its class. The wife seems to have taken up the position that she was entitled to dictate to her husband where he should live. The rights of a husband as they used to be have been considerably circumscribed in favour of the wife, without very much, if any, curtailment of his obligations, but we have not yet got

---

[4] P. M. Bromley, *Family Law*, 1st edn. (1957), 204.

[5] *Jolly* v. *Rees* (1864) 15 CB (NS) 628.

[6] Matrimonial Proceedings (Magistrates' Courts) Act 1960, s. 7(1); it was otherwise for order in the High Court: *Caras* v. *Caras* [1955] 1 All ER 624.

[7] *Eastland* v. *Burchell* (1878) 3 QBD 432 (no power to pledge credit beyond husband's authority). *Tulip* v. *Tulip* [1951] P. 378 allowed a wife to claim for wilful neglect to provide reasonable maintenance when, through inflation, the amount the husband had agreed to provide became woefully insufficient. This case indicates a change in attitude and was regarded as problematic at the time: cf. Bromley, *Family Law* (1957), 205.

[8] *Northrop* v. *Northrop* [1968] P. 74, [1967] 2 All ER 961.

[9] *Family Law* (1957), 201.

[10] In *Mansey* v. *Mansey* [1940] P. 139, 140.

to the point where the wife can decide where the matrimonial home is to be, and if the husband says he wants to live in such and such a place then, assuming always that he is not doing it to spite his wife and the accommodation is of a kind that you would expect a man in his position to occupy, the wife is under the necessity of sharing that house with him.

Two years later this statement was approved by the President of the Divorce Division.[11]

It might be claimed that these pronouncements distort the true position and that a wife could assert rights independent from her husband through the divorce law by withdrawing from the home and seeking the remedies available in the divorce court. But that would not be correct. In a divorce case in 1949 Denning LJ said that, while he disagreed with the views of Henn Collins J. cited above,[12] and thought that each spouse had an equal voice in these decisions, nevertheless each spouse must be reasonable. The result was that, if they could not agree, but neither was unreasonable from their own point of view and they lived apart as a consequence, neither was in desertion of the other, and no divorce could be obtained. So, provided that the husband was not being 'unreasonable', the wife had to accept his terms; if she disagreed, even reasonably, she had either to continue to live with him or separate, without the protection of the divorce law. As we have seen, it was unlikely that he was even under a duty to support her in such circumstances.

But the wife's position was weaker even than this. Suppose the husband was ill treating her. Was his support to be bought at such a price? Ostensibly not: if she could establish one of the grounds of a fault-based divorce law (adultery, cruelty, desertion) she could protect her position through the divorce courts. But divorce was out of the reach of most women. It was available only in the High Court, with its deterrent costs and procedures. The most they could hope for were the pitiful remedies of the magistrates' courts: low and poorly enforced maintenance orders, and no freedom to remarry. Even if a wife was in a position to seek divorce, and the stringencies of the matrimonial offence doctrine were satisfied (she must not have pardoned him, or attempted reconciliation, or be 'guilty' of like offences), divorce was a hazardous step, especially if she wished to keep young children. She probably owned no property, and her earning potential was likely to be limited. Yet, as Willmer LJ put it in 1953, as the day of divorce approached, 'the moment is inevitably and inexorably approaching when she will in any case have to leave the matrimonial home'.[13] The courts had no power to order the husband to transfer any of his property to her. At most she might hope to do a deal

---

[11] *King* v. *King* [1942] P. 1, 8.
[12] *Dunn* v. *Dunn* [1949] P. 98.
[13] [1953] P. 266.

over maintenance. The husband was under a duty to provide her with reasonable maintenance, at least if she was 'innocent' of any offence.[14] She could therefore bargain away her rights to maintenance in return for the house.[15] But why should a husband agree to such a deal? If it was more advantageous to him to keep his property, he could do so.

Of course, wives could divorce their husbands and the statistics indicate that they did so at a slightly higher rate than husbands divorced their wives. In 1938, the number of divorce petitions by wives exceeded those by husbands by 14 per cent, in 1949 by 6 per cent, and in 1959 by 23 per cent. But in 1979 petitions by wives were 157 per cent of those by husbands and in 1989 173 per cent.[16] This is a vast change. It illustrates the extent to which the effectively 'rightless' position of a wife in mid-century has been reversed. Now she has a right to full participation in determining how her life should unfold, and the power to enforce it with the sanction of withdrawing from the partnership. How did this reversal occur?

The introduction of the reformed divorce law in 1971 was an essential element in the process that led to this change. But its immediate practical effects may have been less than its substance would suggest. Although the reform replaced the matrimonial offence doctrine with a single ground for divorce (that the marriage had irretrievably broken down), the break-down could be proved only by alleging certain facts which looked very like the former offences.[17] It may have been more important that various procedural and technical difficulties were removed: for example, it no longer mattered that the parties had agreed that the petition would not be defended.[18] Furthermore, since 1967,[19] divorce cases were no longer confined to the High Court; they could be heard in the county courts. After 1977, the so-called 'special procedure' dispensed with a court hearing altogether in undefended cases. However, perhaps even more important was the power given to the courts in 1971 to order the husband to transfer the house to the wife.[20] Although it is true that wives now frequently

---

[14] During the first half of the century, this duty fell far short of a duty to keep the wife at the standard of living she enjoyed during marriage. However, in the 1950s there is evidence that the courts began to adopt the more generous principle: see J. Eekelaar and M. Maclean, *Maintenance after Divorce* (1986), 10–12.

[15] See *Vaughan* v. *Vaughan* [1953] 1 QB 762 (per Denning LJ).

[16] See *Civil Judicial Statistics* for the relevant years.

[17] Namely, adultery, 'unreasonable behaviour', and desertion. But the additional 'facts', two years' separation if the parties consented, and five years' separation if there was no such consent, broke new ground. But these last two remained relatively little used: see G. Davis and M. Murch, *Grounds for Divorce* (1988), ch. 5.

[18] The 'bar' of collusion was gradually mitigated after the enactment of the Matrimonial Causes Act 1963.

[19] Matrimonial Causes Act 1967.

[20] Now contained in Matrimonial Causes Act 1973, s. 23.

forego maintenance in favour of the house,[21] the fact that the husband can be made to part with it has strengthened the wives' bargaining position: if he won't agree to her terms, the court will be favourably disposed to securing her occupation.[22] But these procedural and substantive changes would have had little effect were it not for the increased availability of legal aid and advice. Indeed, the cost to the legal aid fund of matrimonial causes began to cause governments growing concern during the 1970s.[23]

The legal and procedural changes which have facilitated the ability of women to withdraw from marriage have been instrumental in releasing women from the welfarist domination of their husbands. But it is unlikely that they would have sufficed were it not for the social changes which accompanied them. In 1931, 10 per cent of married women were economically active; in 1951 this had risen to 30 per cent and in 1987 to 60 per cent.[24] Although women's employment patterns are dictated primarily by their fertility pattern[25] and they are much more likely to work part-time than men, this massive entry into the work-force has clearly opened out options to most married women which were unavailable to them previously.

Other legal developments buttressed those mentioned above. Already in the 1950s the courts began to notice that, although income from women's earnings was contributing to the family's well-being, the fact that the major assets (particularly the house) were usually legally owned by the husband meant that, under the system of separation of property, if the marriage broke down (or if the husband became bankrupt), the wife had accumulated no share in the capital value of these assets. The courts then began to use the law of trusts as a means of giving such wives a share in that value commensurate to their financial contribution.[26] When in 1971 the divorce courts acquired their powers to order the transfer of property on divorce, these cases lost most of their significance for married people (but continued for unmarried cohabitees, as will be seen). For a number of years arguments were still made that a married woman should be given an automatic half share in property which was legally owned by her husband,[27] but, apart from a half-hearted and unworkable attempt to give wives a share in property acquired from a housekeeping allowance

---

[21] J. Eekelaar, *Regulating Divorce* (1991), 69.
[22] Ibid. 73.
[23] Ibid. 27–33.
[24] See K. Kiernan and M. Wicks, *Family Change and Future Policy* (1990), 26.
[25] J. Martin, *Women and Employment* (1984).
[26] The cases stretch from *Re Roger's Question* [1948] 1 All ER 328 (CA) to their culmination in *Gissing* v. *Gissing* [1971] AC 886 (HL).
[27] See Law Commission, *Family Law: Third Report on Family Property. The Matrimonial Home (Co-Ownership and Occupation Rights) and Household Goods* (1978); Law Commission, *Family Law: Matrimonial Property* (1988).

given them by their husbands,[28] these have come to nothing. Like anyone else, a wife acquires a legal or beneficial interest in property either by conveyance into her name, alone or jointly with her husband, by contributing directly to its purchase, or by legally enforceable agreement with the owner.

It seems unlikely that the fact that marriage in itself gives neither spouse a right in the property of the other has detrimentally affected the position of married women. But it might have been otherwise were it not for an unexpected development in property law. Even before the divorce courts acquired their powers to transfer property from husbands to wives it was recognized that a husband might put the home beyond the scope of any divorce settlement by selling it without the wife's knowledge. In the early 1950s Denning LJ attempted to protect the position of a deserted wife from such an event by holding that anyone who acquired the property from the husband took it subject to the wife's 'right' to remain in occupation.[29] In 1965 the House of Lords rejected this judicial creation[30] and Parliament responded in 1967 by passing the Matrimonial Homes Act which allowed wives to use the land-registration machinery to secure their rights of occupation of the matrimonial home against third parties who later acquired an interest in it.[31] This legislation had very little impact because it would be unusual for a wife to take the necessary legal measures.[32] Of much greater significance was the decision in 1981 in *Williams & Glyn's Bank* v. *Boland*[33] in which the House of Lords held that if a person with a beneficial interest in a house, which was registered in the name of another, was in 'actual occupation' at the time the third party acquired an interest in the house, the third party took subject to that interest (which included the right to continue in occupation). Since it is often difficult to know whether a wife, or any other adult occupier, has a beneficial interest, it has become standard conveyancing practice to obtain from the occupier a waiver of any rights he or she may have in favour of the third party.

The effect of this on social behaviour has not been empirically measured but it is probable that it has played a part in enhancing the wife's 'voice' in transactions which could significantly affect her interests. In 1992 the Court of Appeal dramatically underlined the importance of this development by holding that, where a married woman (and possibly other people

---

[28] Married Women's Property Act 1964.

[29] *Bendall* v. *McWhirter* [1952] 2 QB 466.

[30] *National Provincial Bank* v. *Ainsworth* [1965] AC 1175 (HL).

[31] Matrimonial Homes Act 1967; consolidated, with amendments, in the Matrimonial Homes Act 1983.

[32] Even if she did, and registered a land charge, the court has a discretion to disregard it in favour of the interests of the third party: *Kaur* v. *Gill* [1988] Fam 110 (CA).

[33] [1981] AC 487 (HL).

in analogous situations) agrees to provide security for her husband's debts, the security will be unenforceable by a creditor who knows of the relationship if the surety's consent was given under undue influence or without adequate appreciation of the transaction and the creditor failed to take reasonable steps to ensure the surety's consent was genuine and informed.[34] The Matrimonial Homes Act also has a provision which contributes usefully towards enhancing the security of a wife's occupation independently of her husband's financial position. If he simply stops paying the rent, rates, or the mortgage, she can require the landlord or the mortgagee to accept payment from her in discharge of his obligation.[35] More important, no doubt, than all these provisions has been the general increase in the practice of purchasing the matrimonial home in the joint names of the spouses.[36] The law has faltered, however, when it has come to choosing between the competing claims of a spouse and the creditors of his or her bankrupt partner. In *Re Citro (a bankrupt)*[37] a majority of the Court of Appeal held that the matrimonial home should ordinarily be sold where the trustee-in-bankruptcy seeks this, even if this could cause hardship to the non-bankrupt spouse and the children. The case was decided before new provisions in the Insolvency Act 1986 came into force. However, it is generally thought that where a trustee applies for sale under that Act, the court's powers to shield the family from the hardship this will cause are very limited.[38]

The recognition given to the claims of married women brought other reforms in its wake. These included the removal in 1962[39] of the prohibition on either spouse from suing the other in tort; the abolition in 1970 of certain anachronistic actions, such as the right of a husband (who had not 'wronged' his wife) to sue a person who refused to return her to him and to claim damages from a person who had committed adultery with her;[40] and the removal, as late as 1991, of the common law immunity of a husband from prosecution for rape on his cohabiting wife.[41]

None of these developments, it seems, have been directly influenced by steps taken under international human rights machinery. Council of Europe Resolution (78)37, On Equality of Spouses in Civil Law, and the

---

[34] *Barclays Bank plc* v. *O'Brien* [1992] 3 WLR 593.

[35] Matrimonial Homes Act 1983, s. 1(5).

[36] In 1988 the Law Commission observed that 'it is understood that virtually all matrimonial homes are purchased in joint names': *Family Law: Matrimonial Property*, para. 4.3.

[37] [1990] 3 WLR 880 (CA).

[38] See Cretney, (1991) 107 *LQR* 177. However, it is arguable that if the non-bankrupt spouse makes an application under the Matrimonial Homes Act 1983 *within a year from the bankruptcy*, the court may have power to postpone sale for a period going beyond one year from the bankruptcy: see Insolvency Act 1986, ss. 336(3) and (4).

[39] Law Reform (Husband and Wife) Act 1962, s. 1.

[40] Law Reform (Miscellaneous Provisions) Act 1970, ss. 4 and 5.

[41] *R.* v. *R. (Rape: Marital Exemption)* [1991] 2 All ER 257.

Recommendation R (81)15, On the Rights of Spouses relating to the Occupation of the Family Home and the Use of Household Contents, express sentiments which are broadly in line with them. But their legal and social impetus appear to be derived wholly from domestic sources, and the Council of Europe measures to be simply formulations of values which have already found expression in the laws of most member states.

The rights of married men

The replacement of the husband's welfarist domination with wives' rights was accompanied by the establishment of equal entitlements for husbands against their wives. So, whereas previously wives could be required to support their husbands only if disability kept them from employment, the reforming legislation commencing in 1971 treated the spouses on the basis of equality.[42] In fact, of course, the structure of employment means that husbands were seldom under *economic* restraints against bringing their marriage to an end. Nevertheless, the major cost of the legal measures described above fell on them and the reaction was not slow in coming. It was strongly argued in 1977[43] that the persistence of a maintenance obligation on men after divorce amounted to an impediment on their 'right' to remarry. A public image of divorced wives as parasites, living indolently on the spoils of the marriage, was cultivated.[44] The resulting Matrimonial and Family Proceedings Act 1984 contained a series of measures designed to encourage the limitation of a husband's post-divorce liabilities towards his former wife.[45] Yet the Act left a good deal of discretion in the hands of the registrars and judges. Nor did it affect the husband's obligations towards his children, although in *Delaney* v. *Delaney*[46] the Court of Appeal was prepared to reduce the husband's child-support payments to a nominal amount where he needed the money to buy a house for himself and his new partner and the mother's low earnings meant that his payments would merely have reduced her entitlement to family credit. Ward J. said:

This court has proclaimed and will proclaim that it looks to the realities of the real world in which we now live, and that among the realities of life is that there is life after divorce. The respondent husband is entitled to order his life in such a way as will hold in reasonable balance the responsibilities of his existing family which he carries into his new life, as well as his proper aspirations for that new future.

[42] Matrimonial Proceedings and Property Act 1970.

[43] K. Gray, *The Re-allocation of Property on Divorce* (1977).

[44] See Eekelaar, *Regulating Divorce*, 34–6, referring to the role played by the pressure group, Campaign for Justice in Divorce.

[45] The Family Law (Scotland) Act 1986 was even more explicit in pursuing this objective.

[46] [1990] 2 FLR 457 (CA).

As we shall see, this assertion of a divorced man's rights collides with the interests of his children, and of the state. These have been addressed in the Child Support Act 1991. But in 1992 the law had failed to achieve a coherent reconciliation between the claims of divorced men and the interests of their former wives whose longer-term financial standing may have been injured by bringing up their common children. This remains a task for the 1990s.[47]

The 'right' to divorce and other interests

Of course, it may be argued that women (and men) would be better off if divorce were restricted, or even prohibited. This argument prevailed in the divorce referendum in the Republic of Ireland.[48] It is essentially a welfarist argument, which refuses to accept the claims of those who desire to withdraw legally from a marriage. But the adults are not the only people whose rights are at stake. Is the unrestricted 'right' to divorce, or to dispose of one's resources to one's own best advantage after divorce, consistent with a proper recognition of the claims of children? It is a matter of great social interest that, in the debates about the 'problem' of divorce during the larger part of this century, virtually no attention has been paid to the views the children might hold of these events. We will take up this theme at a later stage.

Another competing interest is that of the state. By 1990 concerns about the costs of single parenthood to the state had assumed political significance. Over the period 1961–85 the number of lone-parent families doubled; about three-fifths were divorced or separated people. The Department of Social Security estimated in 1990 that about two-thirds were receiving income support, compared to 40 per cent in 1979 and under 20 per cent in 1961; they were receiving £3.6 bn in social-security benefits in 1988/9, compared to £1.75 bn (at 1990 prices) in 1981/2.[49] A woman might choose, or be forced, to withdraw from her marriage; but, if she had a child with her, the cost was falling on the public. In 1990 the Government decided to try to minimize that cost by recovering it from the fathers. The Child Support Act 1991 ensures that a father such as the man in *Delaney* v. *Delaney* (above) would need first to meet in full his obligation to his children (to be set by a formula) before devoting resources to a new partner. The husband's largely self-defining duty to take care of the well-being of his wife will have given way to a sharply defined

---

[47] The issue is extensively discussed in Eekelaar, *Regulating Divorce*, 76–89.

[48] See W. Duncan, 'The Divorce Referendum in the Republic of Ireland: Resisting the Tide' (1988) 2 *Int. J. Law and the Family* 62.

[49] *Children Come Frist: The Government's Proposals on the Maintenance of Children*, Cmnd. 1263 (1990).

obligation initiated by the act of procreation, whether inside or outside marriage. Put another way, his child and its carer have been given genuine rights to his economic resources. The goal is that fatherhood should become for all men, as motherhood always was for all women, a life-changing event.[50]

Spouses and cohabitees: Protection against violence

Although the women's movements which arose towards the end of the last century publicized the degree to which women were subjected to brutality within the home, concern over this issue faded in the period between the world wars. However, it re-emerged after Erin Pizzey established the first women's refuge in Chiswick in 1971.[51] The rightlessness of women described above reinforced their vulnerability to these assertions of male power. From the 1960s the courts had begun to grope towards redressing this imbalance. In 1969, in *Gurasz* v. *Gurasz*,[52] Denning LJ expressed the view that, if a husband abused his duty to provide accommodation for his wife to such an extent that she could not reasonably be expected to live with him, the court might turn the husband out of the house. This statement, though logically sound, found no further judicial response and the courts preferred to confine their jurisdiction to expel a spouse from the matrimonial home to the narrow circumstances where this was necessary to protect a spouse who had initiated matrimonial proceedings against the other. Neither approach was of much assistance to wives before 1971 because, as we have seen, at that time the husband was likely to recover the house when she ceased to be his wife.

However, after 1971, the device of seeking an injunction either to restrain a violent husband from molesting his wife or expelling him from the house after filing a divorce petition became more frequent. It was now possible that the wife might be allowed to stay in the house after the divorce. But as a mode of emergency relief the procedure was deficient. It required the initiation of matrimonial proceedings, usually by filing a divorce petition; the jurisdiction could be exercised only by the High Court; even if an order was made, the applicant needed to return to court for a committal order if it was breached. The Domestic Violence and Matrimonial Proceedings Act 1976 sought to overcome these restrictions. Henceforth applications could be made to county courts without seeking 'principal relief'; if the respondent had inflicted actual bodily harm and

---

[50] For a general analysis of the Child Support Act 1991, see M. Maclean and J. Eekelaar, 'Child Support: the British Solution' (1993) 7 *Int. J. Law and the Family* 205. For a 'rights' analysis of the Australian scheme, see Parker, n. 3 above.

[51] E. Pizzey, *Scream Quietly or the Neighbours will Hear* (1974).

[52] [1969] 3 All ER 892 (CA).

was likely to repeat this, a power of arrest could be attached to the order. More circumscribed powers were given to magistrates' courts in the Domestic Proceedings and Magistrates' Courts Act 1978.

Despite the attempt to free this procedure from the divorce context, it remains most useful as a preliminary to divorce. It is doubtful whether jurisdiction under the Act survives after divorce and the circumstances in which a power of arrest may be granted are too narrow.[53] For this reason many women must still rely on the criminal law for immediate protection against violence. It is a measure of the reluctance to confront these ills that a Minister of State in the Home Office was obliged to announce in 1990 that serious efforts were being made to improve the police reaction to domestic violence and to extend the number of special units in police forces which had training to deal with it.[54] In 1992 the Law Commission recommended legislation which would widen the powers of the courts to great appropriate remedies.[55]

The Domestic Violence and Matrimonial Proceedings Act 1976 contained a significant provision extending its protection to 'a man and a woman who are living with each other in the same household as husband and wife'. In such circumstances the order cannot, of course, be preliminary to a longer-term settlement on divorce and must necessarily operate as an emergency measure. But the fact that it can result in the expulsion of a property owner from (his) premises at the behest of a person with whom (he) has no legal relationship[56] raises important questions about what rights are at stake. In the case of married people, these are clear. The jurisdiction promotes equality of rights between persons who have taken on the legal commitments or marriage and undermines the earlier welfarist domination of the husband. But what rights are at issue where the parties have not married?

Cohabitation outside marriage

A feature of housing policy this century has been to allow members of a tenant's family to succeed to certain types of controlled tenancy when the tenant dies. The tenant's spouse could view this right as one of the benefits of marriage. But could an unmarried partner, who enjoyed a long-standing, exclusive, sexual relationship with the tenant, count as a member of the tenant's 'family'? In 1950 the Court of Appeal thought not.[57] Asquith LJ said:

[53] *Kendrick v. Kendrick* [1990] 2 FLR 107.
[54] See *The Independent*, 1 Aug. 1990.
[55] Law Commission, *Family Law: Domestic Violence and Occupation of the Family Home*, Law Com. no. 207 (1992).
[56] See *Davis v. Johnson* [1979] AC 264 (HL).
[57] *Gammans v. Ekins* [1950] 2 KB 328 (CA).

if their relationship was platonic, I can see no principle on which it could be said that the two were members of the same family. . . . If, on the other hand, the relationship involves sexual relations, it seems to me anomolous that a person can acquire a 'status of irremoveability' by living or having lived in sin, even if the liaison has not been a mere casual encounter but protracted in time and conclusive in character.

The judge's pejorative reference to 'living in sin' (and, later, to the parties as 'masquerading as husband and wife') indicates that the refusal to countenance claims which such people may make rests upon a perception that so to do would presumably threaten the moral welfare of the community. Later a different argument was found against conferring legal rights on cohabitees. It was now said that to do this would be to deny unmarried couples the option of avoiding the legal regime of marriage.[58] Remedies are to be withheld because it is thought better for potential claimants to be left without them.

Neither approach pays regard to the claims cohabitees may, and do, make and therefore to what rights they should have. To a limited degree the law has responded to these claims. The Domestic Violence and Matrimonial Proceedings Act 1976 permitted cohabitees to use its machinery for their physical protection, and cohabitees have recourse to it frequently. In the preceding year the Court of Appeal decisively rejected its 1950 ruling as to whether a long-term cohabitee could be considered a member of a tenant's family.[59] Also in 1975, legislation allowed a person who had been dependent on another at the time of that other's death to claim against the estate for provision for his or her maintenance.[60] In 1982 the Administration of Justice Act allowed a cohabitee who had been living as the husband or wife of a partner for two or more years to claim under the Fatal Accidents Acts for loss of dependency on the death of the partner.[61] Legal policy seemed to be becoming slowly more willing to give effect to certain expectations a person might hold simply on the basis that he or she had been involved in a domestic relationship with another person.

But the 'rights' of cohabitees are still very limited. There is no 'right' to financial support, either during or after the cohabitation. If a long-standing partnership breaks down, there is no jurisdiction equivalent to the divorce jurisdiction under which property adjustments may be made. A partner may be held to have acquired a beneficial interest in property under the 'ordinary' rules referred to above. But these expressly *prohibit* courts from taking into account 'domestic' behaviour (such as

---

[58] R. Deech, 'The Case against Legal Recognition of Cohabitation', in J. Eekelaar and S. N. Katz (eds.), *Marriage and Cohabitation in Contemporary Societies* (1980).

[59] *Dyson Holdings* v. *Fox* [1976] QB 503.

[60] Inheritance (Provision for Family and Dependents) Act 1975, s. 1(1)(*e*).

[61] Administration of Justice Act 1982, s. 3.

running a house and bringing up children) as evidence from which to infer an intention that beneficial ownership should be shared, which is devastating to a female cohabitee who has brought up her partner's children.[62] It may be, however, that these expectations could be protected within the doctrine of proprietary estoppel, but there is as yet no clear decision to this effect.[63]

Where, however, a cohabitee is currently looking after the child of a former partner, the cohabitee's position may be little different from that of a spouse or former spouse. The Family Law Reform Act 1987 allows substantial support to be ordered in favour of the child, and this could include an element for its carer,[64] a feature retained under the Child Support Act 1991. The 1987 Act also permits lump sum and property transfer orders to be made for the child's benefit, and this power could be used to keep the mother and child in occupation of a house.[65] But once the child leaves his or her care, the former cohabitee becomes vulnerable, especially if (she) has reduced her career prospects by bringing up the child.[66] Also, the rights given to cohabitees to claim against the estate of a deceased partner are unsatisfactory because they do not apply if the claimant has been providing more benefits than were received from the deceased[67] and the Law Commission has recommended that the 1975 Act[68] should be amended.[69] However, despite the growing popularity of extra-marital cohabitation,[70] conferring rights on cohabitees is unlikely to be politically attractive in a generally conservative political climate.

The right to marry

Unless within the prohibited degrees of kindred,[71] cohabitees of the opposite sex usually have the possibility to marry one another if they wish. But people of the same sex cannot do so. If they form a domestic relationship, some aspects of the law relating to unmarried cohabitees may apply, although not those which apply to people living together 'as

---

[62] See *Burns* v. *Burns* [1984] Ch. 317 (CA).

[63] See *Grant* v. *Edwards* [1986] Ch. 638 (CA).

[64] *Haroutanian* v. *Jennings* (1977) 121 Sol. Jo. 663.

[65] *K.* v. *K.* [1992] 2 All ER 727.

[66] This was the situation in *Burns* v. *Burns* (above, n. 62).

[67] See *Jelley* v. *Iliffe* [1981] Fam 128 (CA).

[68] Above, n. 60.

[69] Law Commission, *Distribution and Intestacy*, Law Com. no. 187 (1989).

[70] The proportion of women cohabiting outside marriage rose from 8 to 17% between 1979 and 1987: J. Haskey and K. Kiernan, 'Cohabitation in Great Britain: Characteristics and Estimated Numbers of Cohabiting Partners' (1989) 58 *Population Studies* 23.

[71] These were slightly expanded by the Marriage (Prohibited Degrees of Relationship) Act 1986 to permit a man to marry his step-daughter, and other step-relations, in certain circumstances.

husband and wife'.[72] It is also possible that such a couple will not be considered to be a 'family'.[73] Where one of the parties is a post-operative transsexual, it seems that English law will hold that person to his or her pre-operative sexual identity on the basis that an individual's sex is permanently determined by chromosomal structure at conception.[74] A post-operative transsexual cannot therefore marry within his or her new sexual identity (except, ironically, to someone who does not share his or her chromosomal sexuality, even though they may share the outward and psychic aspects of sexuality) and the Government has refused to amend birth certificates to record 'sex changes'. Although it would seem to be strongly arguable that this position contravenes Articles 8 ('right to respect for private and family life') and 12 ('right to marry') of the European Convention on Human Rights, British transsexuals have found little comfort from the European Court of Human Rights, which has held twice[75] that the UK was not in breach of the Convention. It is difficult to see what wider public interests (other than administrative convenience) are served by refusing to recognize the claims of this class of citizens.

The position of transsexuals is different from that of homosexuals because the essence of the claim of the former is that they should be recognized in their new sexual identity and may wish to form domestic relationships with people holding the opposite identity. Homosexuals, however, may seek domestic relationships with people of the same sexual identity. English law knows no provision such as that introduced in Denmark in 1989 whereby people of the opposite sex are permitted to enter into a 'registered partnership' whose legal consequences are almost identical to those of marriage.[76] It has been argued that the failure to respect the domestic relationships of these groups and the inconsistent attitude towards cohabitation in general puts the United Kingdom in breach of the International Covenant on Civil and Political Rights which it has both signed and ratified.[77] There is little evidence that the rights-

---

[72] *Harrogate Borough Council* v. *Simpson* [1986] 2 FLR 91 (CA).

[73] *Sefton Holdings* v. *Cairns* [1988] 2 FLR 109 (CA) (excluding a woman who had lived most of her life with another as if she were her sister).

[74] *Corbett* v. *Corbett* [1971] P. 83. It has been argued that the statutory codification of the common law in the Matrimonial Causes Act 1973, which prohibits marriage between a 'male' and a 'female', rather than a 'man' and a 'woman', may have changed the test from a genetic to a psychic one: S. M. Cretney and J. M. Masson, *Principles of Family Law* (1990) 48. But it is unlikely that Parliament intended such a change.

[75] *Rees* v. *UK* (1987) 9 EHRR 56; see J. Taitz, 'A Transsexual's Nightmare: The Determination of Sexual Identity in English Law' (1988) 2 *Int. J. Law and the Family* 139–54; *Cossey* v. *UK*, Series A, No. 184 (1990).

[76] See L. Nielsen, 'Family Rights and the "Registered Partnership" in Denmark' (1990) 4 *Int. J. Law and the Family* 297–307.

[77] P. R. Ghandhi and E. McNamee, 'The Family in UK Law and the International Covenant on Civil and Political Rights 1966' (1990) 5 *Int. J. Law and the Family* 104.

based thinking which has developed in other areas of family law has been applied to these groups of people; if it is to be, their claims should be taken considerably more seriously than they have hitherto.

PARENTS AND CHILDREN

The rights of parents regarding their children

(i) *The 'right' to be a parent.* A novel question has arisen in recent years: does a person have a right to become a parent? While there are no controls over private initiatives to obtain insemination,[78] questions of rights arise in the context of the provision of infertility services. However, whatever rights adults may wish to have in these matters, closer attention may need to be given to the rights of children born to them.

The Human Fertilisation and Embryology Act 1990 requires that where artificial insemination of a woman with the sperm of someone other than her husband or partner is provided as part of a public medical service, the activity must be licensed. There is no absolute prohibition against providing this treatment to single women, or women living in lesbian relationships, but section 13(5) of the Act states that the agency must take into account the welfare of any child who may be born, including the need of that child for a father. On the other hand, the Act expressly contemplates that treatment may be provided for an unmarried couple because the man, if he consents, is to be treated as the child's father.[79] The Act also gives limited blessing to parenthood acquired through surrogate birth. Although the carrying woman will be treated, initially, as the child's mother (even if the gametes are entirely those of the commissioning parents),[80] and surrogacy agreements are legally unenforceable,[81] if she hands the child over to the commissioning parents they may obtain an order whereby the child will be 'treated in law' as their child. But the surrogate mother must agree unconditionally to this, and only married people can apply for an order.[82]

The *Code of Practice* contemplates that agencies might provide infertility treatment to single women provided they pay careful attention to the child's welfare.[83] This overtly welfarist stance towards the child is

---

[78] See Gillian Douglas and Nigel Lowe, 'Becoming a Parent in English Law' (1992) 108 *LQR* 414.

[79] Human Fertilisation and Embryology Act 1990, s. 28(3).

[80] Ibid., s. 27.

[81] Ibid., s. 36.

[82] Ibid., s. 30.

[83] *Guardian*, 22 Mar. 1991. See also *R.* v. *Ethical Committee of St. Mary's Hospital (Manchester), ex parte H.* [1988] 1 FLR 299 (woman refused treatment because rejected by social services department as potential adopter).

difficult to reconcile with the statutory reference to the 'need' of the child for a father, especially if the 'need' is viewed as a potential claim. It is, of course, hard to know what rights people, as yet unborn, should have. But a right merely that others should take into account their welfare seems too weak. If a person's conception is to be consciously engineered by an agency providing a public service, must we not address that person's human rights? In doing this, should we not ask: if they had the choice, would people born in this way choose to be born into a single-parent or a two-parent family?

The problem is deeper even than that. Does an individual who is conceived as a result of donated sperm have the right to knowledge about his or her genetic origin (perhaps also the identity of the donor)? Although the Human Fertilisation and Embryology Act 1990 requires agencies to provide non-identifying information *if asked* by such an individual, it provides no mechanism for ensuring that people will know that they have been conceived by this method. A viewpoint which maintains that such information should be withheld from children 'because it is in their best interests to do this' is a stark form of welfarism which could be used to justify many forms of state manipulation deemed to be for the benefit of citizens. Rights-based thinking requires the question to be asked: would anyone choose to live his or her entire life on the basis that he or she had been deliberately deceived about their genetic origin?[84] Questions like these are prompted by Article 7 of the UN Convention on the Rights of the Child which gives children 'as far as possible the right to know and be cared for by (their) parents' and by Article 8 which requires states to 'respect the right of the child to preserve his or her identity'.

(ii) *Rights between parents.* At common law, the rights which a parent had with respect to a child were described as 'rights of custody'. They included the right to possess the child, and to make important decisions on the child's behalf, such as choice of religion. If the child was legitimate, these rights vested in the father; if illegitimate, in the mother. The struggle of the women's movement in the late nineteenth century for equality between husbands and wives in this respect was resisted by the male-dominated legislature,[85] which found a compromise by enacting, in 1886, that in any question in which the upbringing of a child was in issue the court might 'make such order as it may think fit regarding the custody of such infant and the right of access thereto of either parent, having regard to the welfare of the infant, and to the conduct of the parents, and to the

---

[84] For a perceptive discussion, see K. O'Donovan, 'A Right to Know One's Parentage?' (1988) 2 *Int. J. Law and the Family* 27–45.

[85] S. Maidment, *Child Custody and Divorce* (1984).

wishes as well of the mother as of the father'.[86] The Guardianship of Infants Act 1925 pushed the welfare of the child into the foreground by stating that the court

shall regard the welfare of the infant as the first and paramount consideration and shall not take into consideration whether from any other point of view the claim of the father, or any right at common law possessed by the father, in respect of such custody . . . is superior to that of the mother, or the claim of the mother is superior to that of the father.[87]

This provision was substantially re-enacted in 1971.[88] By subordinating parents' rights to the courts' appreciation of the children's welfare Parliament avoided giving married women equal *rights* with respect to their children. Formal equality of rights between married parents with respect to their children was achieved only in 1973.[89]

The position after divorce raised other problems. The standard practice was to grant 'custody' to one parent (often the wife) with 'reasonable access' to the other. But during the 1970s it became very unclear whether, if the wife alone had been granted 'custody' (and therefore acquired rights to make decisions respecting the child), the father retained any of his previous 'rights' with respect to the child.[90] Some courts sought to retain an absent father's 'rights' to participate in major decisions regarding the child by making 'joint' custody orders, but practice was very uneven. Some commentators objected to the practice on the ground that it permitted absent fathers, who no longer had day-to-day responsibility for the child, to meddle in the mother's life.[91] The Children Act 1989 sought a solution through the concept of 'parental responsibility'. Although each parent retains this responsibility after divorce, the Act allows either 'to act alone and without the other' in meeting his or her responsibility for the child.[92] Since children stay with their mother more frequently than with their father, it has been alleged that the lack of a requirement of consultation has significantly prejudiced absent fathers who, previously, might have obtained a joint custody order.[93] But although the absent parent would need to seek a court order[94] if (he) disagreed with a decision taken by the other, that was also true with respect to a parent

---

[86] Guardianship of Infants Act 1886, s. 3.
[87] Guardianship of Infants Act 1925, s. 1.
[88] Guardianship of Minors Act 1971, s. 1.
[89] Guardianship Act 1973, s. 1.
[90] *Dipper* v. *Dipper* [1981] Fam 31 (CA); see Cretney and Masson, *Principles of Family Law*, 540.
[91] See J. Brophy, 'Custody Law, Child Care and Inequality in Britain', in C. Smart and S. Sevenhuijsen (eds.), *Child Custody and the Politics of Gender* (1989).
[92] Children Act 1989, s. 2(7).
[93] A. Bainham, 'The Privatisation of the Public Interest in Children' (1990) 53 *MLR* 206.
[94] Under Childen Act 1989, s. 8.

with a joint custody order, so the practical effect of the change seems slight.

(iii) *The special position of unmarried fathers.* Unmarried fathers have inherited centuries of rightlessness and until recently were not even considered to fall within the definition of a 'parent' in most important legislation.[95] The prevailing attitude was vividly expressed in a case in 1969[96] in which a mother had refused to marry her child's father. Nevertheless, the man had been visiting the child for two years every Sunday under a court order. The mother now wished to marry another man, adopt the child with him, and terminate the natural father's contact. In a characteristically 'welfarist' judgment, the Court of Appeal approved of this. Harman LJ said that the advantage to the child of removing the 'stigma of illegitimacy' should 'not be thrown away in order that the father may have the pleasure of seeing the child and dandling him on his knee for two hours a week on Sunday morning before he goes off to play football'.

The unmarried father's position has now much improved. The Family Law Reform Act 1987 allowed a father to apply to a court for an order giving him full parental rights. Now, under the Children Act 1989, an unmarried father may acquire 'parental responsibility', either by court order or simply by recording in court an agreement to this effect with the mother.[97] In fact, even without parental responsibility, his position is little different from that of a married father. If his child is being looked after by a local authority, the presumption of contact and duties of consultation, mentioned above, apply to him too. If he objects to the way the mother exercises her parental responsibility, he may apply for an appropriate order[98] without first obtaining leave of the court, just as a married father would have to do. In *Re K. (a minor) (custody)*[99] the Court of Appeal effectively treated an unmarried father as having an equivalent 'right' as against third parties to possess his child as a married father. But in *Re F. (a minor) (blood tests)*[100] the court refused to allow a man to attempt to establish his paternity of a child born to a woman who was married to another man when the woman and her husband accepted the child as theirs. The justification was a welfarist concern for the child. Countries in which the European Convention on Human Rights

---

[95] See A. Bainham, *Children, Parents and the State* (1988), 36.

[96] *Re E(P) (Infant)* [1969] 1 All ER 323 (CA).

[97] Children Act 1989, s. 4(1). The Scottish Law Commission has recommended that unmarried fathers should have such 'responsibility and rights' without needing to take any step to acquire it: *Report on Family Law*, Scot. Law Com. no. 135 (1992).

[98] Under Children Act 1989, s. 8.

[99] [1990] 2 FLR 64; see above, n. 93.

[100] [1992] 2 FCR 725.

plays a larger role would resolve such problems within an analysis of the competing rights of all persons involved.[101]

(iv) *The right not to be a parent.* Despite the promotion of parental *responsibility*, the *exercise* of parenthood is closer to a right than a duty. Parents are not *obliged* to bring up their child (though if they do, they must do so consistently with the child's rights, discussed below); they may place the child out for adoption or arrange for another to exercise their 'responsibility' on their behalf.[102] They cannot, however, easily escape *financial* responsibility under the Child Support Act 1991. So an unmarried father, who does not have 'parental responsibility', may nevertheless be liable to support the child, whether or not he is entitled to, or able to, have contact with the child. Nor can he rely on the mother's refusing to disclose his identity, for if she receives various state benefits, she may be required to reveal it. Although the mother may be relieved of the duty if the social security officer thinks compliance risks her or her children undue harm or distress, she cannot challenge the officer's decision until an order reducing her benefit has been made. Whether these provisions sufficiently safeguard her rights under the European Convention on Human Rights to respect for her private and family life and to challenge discretionary decisions is open to argument.

A pregnant woman might seem to have the right to carry her child to full term (so any provision in a surrogacy agreement giving commissioning 'parents' the 'right' to require its abortion will almost certainly be void[103] and it is surely inconceivable that abortion can be performed on any woman against her will). But might this be a duty rather than a right? If so, it is not owed to the father, who has no control over her decision whether to abort,[104] thus, as Meulders-Klein has remarked, reversing the ancient domination of men over women in matters of fertility.[105] Does a mother, however, owe such a duty towards the unborn child, or, possibly, to society generally? To some feminists, any such duty reflects the persistence of male control over women's autonomy. Whatever view is taken on that point, a woman has traditionally been treated as being obliged to bear the child she is carrying, and this duty was given legal sanction in

---

[101] For such an analysis, see Caroline Forder, 'Constitutional Principle and the Establishment of the Legal Relationship between the Child and the Non-Marital Father: a Study of Germany, the Netherlands and England' (1993) 7 *Int. J. Law and the Family* 40.

[102] Children Act 1989, s. 2(9).

[103] See M. A. Field, *Surrogate Motherhood: The Legal and Human Issues* (1990), 79; Human Fertilisation and Embryology Act 1990, s. 36.

[104] *Paton v. Trustees of the United Kingdom Pregnancy Advisory Services* [1978] 2 All ER 987; see *Paton v. UK* (1981) 3 EHRR 408.

[105] M.-T. Meulders-Klein, 'The Position of the Father in European Legislation' (1990) 4 *Int. J. Law and the Family* 131, 133.

England in 1803.[106] The fact that the duty was mitigated where necessary to preserve the mother's life or her physical or mental health[107] gave legal power over abortion decisions to the (male-dominated) medical profession. The Abortion Act 1967 confirmed this medicalization in England and Scotland[108] by conditioning the legality of abortion on certification by two medical practitioners that (1) 'the continuance of the pregnancy would involve the risk' (*a*) to 'the life of the pregnant woman' or (*b*) of 'injury to the physical or mental health of the pregnant woman or of any existing children of the family greater than if the pregnancy were terminated' or (2) that 'there is a substantial risk that if the child were born it would suffer from such physical or mental abnormalities as to be seriously handicapped'. This was, however, limited by prohibition on abortion where the foetus was 'capable of being born alive', which was presumed to be the case after 28 weeks' pregnancy.[109] The Human Fertilisation and Embryology Act 1990 confined abortion on ground (1)(*b*) mentioned above to within 24 weeks of pregnancy, but allowed it *without time limit* where (i) termination 'is necessary to prevent grave or permanent injury to the physical or mental health of the pregnant woman or any children of the family', (ii) where 'continuance of the pregnancy would involve risk to the life of the pregnant woman greater than if the pregnancy were terminated', or (iii) where 'there is a substantial risk that if the child were born alive it would suffer from such physical or mental abnormalities as to be seriously handicapped'.

Glendon[110] placed British law as expressed in the 1967 Act among the eight European countries with 'soft' grounds for abortion in early pregnancy. The abortion issue in the United Kingdom has been framed less in terms of 'rights' than in the United States. This has had the advantage of avoiding irreconcilable conflicts between advocates of women's 'rights' to autonomy or privacy and of foetal 'rights' to life. The price has been the formal location of decision-making within the welfarist judgments of the medical profession, leading to variation in availability, particularly according to whether it is sought in the public or private health sector,[111] which has satisfied the 'rights' protagonists of neither side. Britain displays the same ambiguity towards the status of the foetus

---

[106] See K. O'Donovan, *Sexual Divisions in Law* (1985), 88; J. Keown, *Abortion, Doctors and the Law* (1988), 11–21.

[107] See R. J. Cooke and B. M. Dickens, *Abortion Law in Commonwealth Countries* (1979), 8, 13.

[108] See K. McNorrie, 'Abortion in Great Britain: One Act, Two Laws' [1985] *Crim. LR* 475. The Act does not apply in Northern Ireland.

[109] Infant Life (Preservation) Act 1929, s. 1.

[110] M. A. Glendon, *Abortion and Divorce in Western Law* (1987), 14.

[111] See Keown, *Abortion*, ch. 7 n. 119. There has been a strong trend towards more liberal medical practice: *Trends in Abortion*, Population Trends 64 (HMSO, Summer 1991).

as other Western societies, reflected in such contradictions as demands for the dignified disposal of miscarried foetuses alongside massive foetal destruction through abortion.[112] But it seems that the power of individuals to control their fertility through contraception[113] and early abortion is such that, while there may not yet be firm 'rights' to either which impose duties on others at the behest of claimants, it can be plausibly stated that individuals, particularly women, will rarely be in a position when they are under a duty to *become* a parent.

(v) *The rights of parents against others to care for their children.* The price paid by the strategy of subordinating rights to the welfare principle was the near-extinction of parental rights. In 1970, in *J.* v. *C.*,[114] the House of Lords decided that the statutory enshrinement of the welfare principle applied also to disputes between parents and strangers, so that where parents who had allowed their child to be brought up by another couple later used wardship proceedings to recover the child, the claims of the parents carried no more weight than any other factor which made up the calculus of what was in the child's best interests. This decision reduced any parental right to possess a child to a mere shadow. It might have been the same for parental rights of access. However, in 1988, under pressure from a series of decisions by the European Court of Human Rights,[115] which viewed access rights as 'civil rights' which could not be infringed without a court hearing,[116] the House of Lords attempted to reconcile the dominance of the welfare principle with parental rights of access.[117] Those rights, said Lord Oliver, were not extinguished by the principle; they were merely subservient to it. The Court of Appeal, however, in *F.* v. *Wirral BC*,[118] cast doubt on that conclusion. The court held that there was no tort of interference in parental rights. The basis for the decision seems to have been that the only parental 'right' protected by the common law was the right to a child's 'services'. These 'rights' were progressively abolished in 1970 and 1982.[119] The only other remedies available to a parent lie in the wardship jurisdiction which, as has been seen, gives little or no independent weight to parental rights.

It is astonishing that the Court of Appeal felt unable to assert that the

---

[112] See C. Wells, D. Morgan, and D. Leat, 'Fetuses and Burials' (1991) 141 *New LJ* 1046.

[113] In 1948 the House of Lords held that conception was not a fundamental purpose of marriage: *Baxter* v. *Baxter* [1948] AC 274 (HL).

[114] *J.* v. *C.* [1970] AC 668 (HL).

[115] *R.* v. *UK*, Series A, No. 121 (July 1987).

[116] European Convention on Human Rights and Fundamental Freedoms, Art. 6.

[117] *Re K.D. (a minor)* [1988] AC 806 (HL).

[118] [1991] 2 All ER 648 (CA).

[119] Law Reform (Miscellaneous Provisions) Act 1970, s. 5; Administration of Justice Act 1982, s. 2.

abolition of the 'instrumentalist' character of the parent–child relationship (the right to a child's 'services') could be succeeded by any form of parental right to the company of a child. (It is, of course, another matter whether, and in what circumstances, interferences in such a right might be actionable in damages.) The underlying assumptions of two other Court of Appeal decisions suggest that there must be such a right. In 1990 the court overturned the decisions of two judges who refused to return children to parents who had informally arranged for them to be cared for by others on the ground that the children would do better in their new homes.[120] Butler-Sloss LJ said that the mother 'must be shown to be entirely unsuitable before another family can be considered'. The question was not 'With whom would the child be better off?' but 'Did the welfare of the child "positively demand" the displacement of the parental right?' Although this position was articulated in terms of parental 'responsibility' rather than 'rights', it reflects a foundational social rule that, special circumstances apart, 'natural' parents have a (defeasible) right to *initiate* the process of bringing up their children. The right is weakened to the extent that its exercise is delayed or sporadic and the child develops relationships with others.

(vi) *The Children Act 1989.* The origins of state welfarism as it operated with regard to families can be traced to nineteenth-century concerns over the potential social threat contained in the circumstances of mass social deprivation and escalating criminality. The goals of the poor-law authorities, enshrined in legislation until after the Second World War, were to endeavour to save these children for service to the national economy. But an emerging welfarist ideal found expression in the Children Act 1948, which transferred the power of those authorities to children's departments within local authorities, and charged them with the duty to 'exercise their powers with respect to (the child) so as to further his best interests, and to afford him opportunity for the proper development of his character and abilities'.[121] By the 1970s, however, unease was spreading that child-care policies were failing. Many children drifted into care, losing contact with their parents, and seldom integrated into new families. It was felt that social workers were inhibited in planning for these children's future by anxieties over their residual ties with their natural parents. The Children Act 1975 tried to remedy these failures by increasing the power of local authorities. They were empowered to require parents who had left their children in care for over six months to give twenty-eight days

---

[120] *Re K. (a minor) (custody)* [1990] 2 FLR 64; *Re K. (a minor) (Wardship: Adoption)* [1991] 1 FLR 57 (CA).

[121] Children Act 1948, s. 12(1); later, Child Care Act 1980, s. 18.

notice before removing them; they could acquire parental rights by administrative resolution simply because a child had been in their care (or in the care of a voluntary society) for three years or more; new procedures were introduced for facilitating adoption.

In retrospect, the 1975 Act may prove to be the high-water mark of welfarism in family law. Expectations of improvements in child-care services were disappointed by the economic crisis of the 1970s. Inadequately resourced bureaucratic power could be represented by the political left as a form of class oppression. The 'new right', whose fortunes rose with the Conservative victory of 1979, were ideologically hostile to all forms of state direction. These forces drew strength from a series of inquiries exposing deficiencies in the management of child abuse cases, culminating in a judicial inquiry into the handling of child sexual abuse allegations in Cleveland in 1987.[122] Many of these concerns were articulated by the House of Commons Social Services Select Committee in 1984[123] and examined in the light of extensive research, and representations by pressure groups like Justice for Children, the Family Rights Group, and the Children's Legal Centre, by an interdepartmental committee. The resulting report[124] provided the groundwork for the Children Act 1989, which represents a significant dismantling of the welfarist powers accumulated by local authorities since the war.[125]

When that Act came into effect in October 1991, local authorities lost their former power to acquire parental rights by administrative resolution; nor could they impose a notice requirement for the recovery of children for whom they were providing care on a voluntary basis. Married parents, and unmarried mothers, are fixed with 'parental responsibility' for their children from the time of their birth, and can lose it only when the child is adopted. In the context of the Act, the expression 'responsibility' represents the perception that it is the *function* ('responsibility') of parents and not of public agencies to care for their children.[126] The Act is therefore constructed upon the basis that parental claims are to be respected unless defeated for clear reasons and by clear procedures. So, while a parent may arrange for an authority to look after a child, section 20(8)

---

[122] See N. Parton, *Governing the Family: Child Care, Child Protection and the State* (1991), ch. 4.

[123] HC 360 (1984).

[124] *Review of Child Care Law* (1985).

[125] See H. Geach and E. Szwed, *Providing Civil Justice for Children* (1983); R. W. J. Dingwall and J. M. Eekelaar, 'Families and the State: An Historical Perspective on the Public Regulation of Private Conduct' (1988) 10 *Law and Policy* 341; S. M. Cretney, 'Defining the Limits of State Intervention: The Child and the Courts', in H. K. Bevan and D. Freestone (ed.), *Children and the Law* (1990); N. Parton, *Governing the Family* (1991), ch. 2.

[126] See J. M. Eekelaar, 'Parental Responsibility: State of Nature or Nature of the State?' (1991) *J. Soc. Welfare and Family Law* 37.

states starkly that: 'Any person who has parental responsibility for a child may at any time remove the child from accommodation provided by or on behalf of the local authority under this section.'

This right can be defeated only by proving before a court that grounds for coercive intervention exist. If these are proved, the court may order the child into the care of a local authority. Under earlier law, such an order gave the authority a complete discretion, based on its assessment of the child's welfare, whether or not to allow the parents to continue to see the child.[127] However, prompted by the European Court of Human Rights,[128] parents were in 1983 given limited opportunities to challenge these decisions before a court.[129] The 1989 Act goes much further than the previous law and creates a presumption that the parents are to be allowed to maintain contact with their children in care unless this is expressly ruled out by the court. Local authorities are required to consult with parents over their decisions concerning children they are looking after[130] and 'endeavour to promote contact' between the child and its parents and relatives.[131] The courts, too, have indicated that they expect authorities to share information with parents when they are contemplating legal proceedings against them with respect to the care they have provided their children.[132] Government guidelines issued in 1988 differed sharply from earlier guidance by urging that parents be invited to attend part or the whole of case conferences unless the chairman believed this would preclude proper consideration of the children's interests.[133]

But despite the reiteration of parental rights and responsibility in the Act, the position asserted in an influential American book published in 1979 that a child has a basic 'right' to 'family integrity', defined as the 'privacy of family life under the guardianship of parents who are autonomous'[134] did not prevail. That would simply have reasserted the welfarist domination of parents. The Act needed to reconcile the parents' rights with those of children, to which we now turn.

---

[127] A. v. *Liverpool City Council* [1982] AC 363 (HL).

[128] *H. and O.* v. *UK*, Series A, No. 120 (23 Oct. 1986); *B.R. and W.* v. *UK*, Series A, No. 121 (July 1987).

[129] Health and Social Services and Social Security Adjudications Act 1983.

[130] Children Act 1989, s. 22(4).

[131] Children Act 1989, sch. 2, para. 13(1).

[132] *R.* v. *Hampshire County Council, ex parte K.* [1990] 2 QB 129, [1990] 2 All ER 129. Watkins LJ, however, expressed the duty to share information with the parents as part of the authority's duty to promote the child's welfare. Despite this 'welfarist' justification, it seems clear that parents have rights in the matter: see *Re M. (a minor) (disclosure of material)* [1990] 2 FLR 36; *R.* v. *Norfolk County Council, ex parte M.* [1989] QB 619, [1989] 2 FLR 120.

[133] See Parton, n. 122 above, pp. 130–1.

[134] J. Goldstein, A. Freud, and A Solnit, *Before the Best Interests of the Child* (1979). See the critique by M. Freeman, 'Freedom and the Welfare State; Child-Rearing, Parental Autonomy and State Intervention' (1983) *J. Soc. Welfare Law* 70.

Children's rights

It is arguable that it was not until towards the end of the nineteenth century that the interests of children began to be perceived as being worthy of legal protection *for the sake of the children themselves* rather than for the sake of others, such as parents or the wider community, who may have derived a benefit from their protection.[135] For children the transition from a position of rightlessness to the holders of rights has been slow, and remains incomplete. Nineteenth-century reformers were without doubt motivated by a wish to improve children's welfare, but employed the conceptual framework of moral condemnation of exploiters they shared with animal-welfare protagonists rather than a rights-based ideology founded on the actual or perceived viewpoint of children.[136] This was replaced by a welfarist ideal, in which first the courts, and then public authorities, were enjoined to make decisions respecting children according to their perceptions of the children's welfare.

No doubt the requirement to place children's welfare above competing claims of parents or the state is a significant humanitarian achievement. It has found its place in Article 3(1) of the UN Convention on the Rights of the Child of 1990, where it is stated that 'in all actions concerning children, whether undertaken by public or private social welfare institutions, courts of law, administrative authorities or legislative bodies, the best interests of the child shall be a primary consideration'. But on the view taken in this chapter, one cannot conclude from the mere fact that A has a duty to act in accordance with A's perceptions of B's interests that B has rights in anything other than a formalistic sense. B will acquire rights only to the extent that A's duties coincide with claims which B actually makes. The difficulty in the case of children is, of course, that they may be too young to make any claims; or their claims may be based on limited intellectual or emotional capacity. This problem does not absolve rights-based thinking from the requirement of engaging in a process which attempts to adopt a child's viewpoint and, where necessary, to endeavour to answer the hypothetical question: what decision would this child, when it acquires sufficient capability, have wished to have been made?[137] The consequence of this approach is that genuine efforts should be made to ensure that the child's voice should be heard; that his or her capacity to express an opinion be fully evaluated and that the decision taken should be articulated and related with specificity to the child's

---

[135] See J. M. Eekelaar, 'The Emergence of Children's Rights' (1986) 6 *Oxf. J. Legal Studies* 161.

[136] See J. M. Eekelaar, R. W. J. Dingwall, and T. Murray, 'Victims or Threats? Children in Care Proceedings' (1982) *J. Soc. Welfare Law* 68.

[137] See M. D. A. Freeman, *The Rights and Wrongs of Children* (1983), ch. 2; J. Eekelaar, 'The Interests of the Child and the Child's Wishes: The Role of Dynamic Self-Determination', (1994) 8 *Int. J. Law and the Family* (forthcoming).

actual or hypothesized viewpoint. The ultimate objective is that the child should be placed in a social context which permits it scope to influence outcomes in accordance with its own personal development. This objective should be overridden only on the ground that another (articulated) right, either of others or of the child itself, should prevail instead.

(i) *Self-determination (autonomy)*. Can children be said to have any rights, in this sense, under present English law? The power of parents to make decisions regarding their children has traditionally been untrammelled, apart from the potential of the wardship court to superimpose *its own view* of the children's welfare if the decision was challenged.[138] This position, which negates rights for children in the sense understood here, was held so strongly that in the *Gillick* case in the mid-1980s the Court of Appeal effectively held that a doctor who treated a child under 16 without first obtaining the consent of a parent or of the wardship court was acting unlawfully, and might be committing an assault on the child, even if the child had requested the treatment and the doctor thought it was in her interests. The House of Lords reversed this astonishing decision.[139] The majority seemed to hold that, once a child had reached sufficient maturity to make a decision on a matter (including contraception), the parental right to control the outcome was terminated and the child was *pro tanto* emancipated. One interpretation of the case holds that the child's decision has effect only if the court, or perhaps (in the case of medical treatment) the medical profession, believes this to be in its best interests.[140] Of course, no one, whether child or adult, is entitled to be treated in a manner contrary to a doctor's medical judgment (and, in the case of a minor seeking contraceptive medication, the doctor's judgment might sometimes include advice that the child should consult her parents). But it is doubtful whether doctors could properly withhold the medication for non-therapeutic 'welfarist' reasons and the *Gillick* case confirms that a doctor who wishes to treat a 'competent' child is not *required* to obtain parental permission. In 1991 statute confirmed this to be the position also in Scotland.[141]

The judiciary has not, however, been willing to renounce its own power, under the 'inherent' jurisdiction, to impose measures on a competent and mature child where the court deems them to be in the child's best interests, even though the child objects.[142] Lord Donaldson has even been prepared to hold that such measures could be imposed on an

---

[138] See J. Eekelaar, 'What are Parental Rights?' (1973) 89 *LQR* 210.

[139] *Gillick* v. *West Norfolk and Wisbech Area Health Authority* [1986] AC 112 (HL).

[140] See Bainham, *Children, Parents, State*. For a contrary view, see J. M. Eekelaar, 'The Eclipse of Parental Rights' (1986) 102 *LQR* 4.

[141] Age of Legal Capacity (Scotland) Act 1991, s. 2(4).

[142] *Re R. (a minor) (wardship: medical treatment)* [1992] Fam. 11.

objecting child if the parents consented, confining the *Gillick* principle to situations where a child consents but the parents disagree.[143] His declared purpose is to protect doctors from legal action should they *mistakenly* assess a child *not* to be 'competent' and intervene with the parents' consent. But such immunity is not given if a doctor misjudges the validity of the consent or refusal of an *adult*. Then liability might lie in negligence.[144]

It is in the context of divorce that children's wishes are most often likely to conflict with adults' ideas about their welfare. Ann Mitchell's research in Scotland revealed that parents consistently underestimated the unhappiness which their divorce caused the children and that children often felt that their views went unheard.[145] Yet children are very rarely represented in divorce proceedings. If the adults dispute the living arrangements, a welfare officer will usually report to the court, and children old enough to express a view will usually be given a chance to speak to the officer, but the weight placed on this and whether to pass it on to the court is very much a matter for the officer. If there is no dispute, the court will normally accept the parents' proposals without consulting the children. Even if the child's views are known, the judge may discount them in favour of his own views of what is best. In *Re DW (a minor) (custody)* in 1984,[146] for example, the judge forcibly transferred a 'mature' boy of 10 from the stepmother he loved to his natural mother, with whom his connections were 'tenuous', simply because he thought he would be better off in a two-parent family.[147]

It is less likely that such a decision will occur after the commencement of the Children Act 1989, for the Act sets out a 'check-list' of factors which must be taken into account in decisions regarding children. The first of these is: 'the ascertainable wishes and feelings of the child concerned (considered in the light of his age and understanding)'.[148] But this falls short of recognizing a prima-facie *entitlement* in the child to a disposition in his or her favour, even if he or she is mature enough to hold a decided view. The child's wishes are merely one among a number of factors to be taken into account in determining where his or her welfare lies. Furthermore, if the parents have agreed the outcome between themselves, the check-list does not apply. But the Act can be credited

---

[143] *Re W. (a minor) (medical treatment: court's jurisdiction)* [1992] 3 WLR 758; see (1993) 109 *LQR* 182.

[144] *Re T. (an adult: refusal of medical treatment)* [1992] 3 WLR 782; *Sidaway* v. *Bethlem Royal Hospital Governors* [1985] AC 871.

[145] A. Mitchell, *Children in the Middle: Living through Divorce* (1985).

[146] (1984) 14 *Family Law* 17.

[147] In *M.* v. *M. (Custody Appeal)* [1987] 1 WLR 404 (CA) the Court of Appeal granted a 12-year-old girl's wish to stay with her father, but only because her objections to going to her mother were so extreme that her health would have been endangered, not out of recognition for her claims.

[148] Children Act 1989, s. 1(3)(a).

with having made a considerable attempt to sensitize decision-makers to the views of the children who will be affected by their decisions. A child over 16 who is being looked after by a local authority cannot be compelled to return to his or her parents just because they wish it;[149] the authority must ascertain and give due consideration to a child's wishes before providing him with accommodation[150] and before making any decision regarding any child it is looking after.[151]

In late 1992 a number of cases were reported in the press[152] that, for the first time, children were seeking to 'divorce' their parents. This inaccurate expression referred to instances where children, whose relationships with their parents had already fractured, applied in their own name for an order granting 'residence' to a third party. The decision is still made according to the welfare principle. The appearance of children instructing lawyers on their own behalf may simply be a consequence of decreased eligibility of adults for legal aid. But it does thrust the children's claims to the fore, and there is evidence that these views are given increasing weight.[153] But sensitive evaluation of these views is complex and poses a challenge to present procedures.

(ii) *Rights to protection.* A child's rights include far more than the right, where appropriate, to determine what will happen to her or him. They must include a right to protection from neglect and ill-treatment. Hence the provision in the Prevention of Cruelty to and Protection of Children Act 1889, which created an offence of wilfully ill-treating, neglecting, or abandoning a child in a manner likely to cause unnecessary suffering or injury to health, can be characterized as a recognition of a right in children to be safeguarded from such treatment. During the twentieth century, however, official perception of child abuse declined and children who were victims were treated as being in need of care and protection in the same way as delinquent children. Although the Committee on Children and Young Persons which reported in 1960 was charged with inquiring into 'the prevention of cruelty to, and exposure to moral and physical danger of juveniles' as well as juvenile delinquency, its report said almost nothing about the former.[154] The result was that the Children and Young Persons Act 1969 applied the same procedures to both categories of children[155] and in neither case was it necessary to show that the child's

---

[149] Children Act 1989, s. 20(11). If the *Gillick* principle applies as interpreted here, the child may be able to refuse to return at an even younger age.

[150] Ibid., s. 20(6).

[151] Ibid., ss. 22(4) and (5).

[152] See *The Independent*, 12 Nov. 1992; *The Times*, 6 Nov. 1992.

[153] Sally Jones, 'The Ascertainable wishes and feelings of the Child', (1992) 4 *Jo. Child Law* 181.

[154] *Report of the Committee on Children and Young Persons*, Cmnd. 1191 (1960).

[155] The history is recounted in Eekelaar, Dingwall, and Murray, 'Victims', n. 136 above.

parents were responsible for the circumstances occasioning the intervention. Indeed, the parents were not even permitted to be parties to the relevant court proceedings.

The growing awareness of child abuse during the 1970s coincided with the disillusion over state intervention referred to above. The Children Act 1989 sought a resolution by locating the *responsibility* for bringing up children primarily on parents, but seeking to ensure that parents exercised this power *responsibly*.[156] But it also needed to avoid replacing the welfarism of parents with the welfarism of the state. Its broad strategy was as follows:

1. State authorities may acquire parental responsibility only on proof, before a court, that the child is suffering, or is likely to suffer, significant harm.[157] 'Harm' means 'ill-treatment or the impairment of health or development'; 'development' means physical, intellectual, emotional, social, and behavioural development and 'ill-treatment' includes 'sexual abuse and forms of ill-treatment which are not physical'.[158] These conditions are more precise than the previous condition which allowed intervention if the child's 'proper development' was being 'avoidably impaired or neglected'.[159] Furthermore, local authorities will no longer be able to by-pass these statutory provisions by using the wardship court which, under the earlier law, could have committed a child into care simply because the court thought this was in the child's best interests.[160]

2. The authority will have to demonstrate that the harm is attributable to 'the care being given to the child, or likely to be given to him if the order were not made, not being what it would be reasonable to expect a parent to give him' or that the child is beyond parental control.[161] It follows that parents are given full party status in court proceedings.

3. The authority will need to specify in some detail to the court how its plans for the child will advance the child's interests.

This approach is consistent with seeing children as rightholders. Intervention in their lives is not justified by the general assertion that the action is deemed to be in their best interests. Each intervention must be expressly related to a specific justificatory ground: for example, that it is designed to put the child in an environment which will be an improvement over one in which, as a consequence of lack of reasonable parental care, he or she is suffering (or is likely to suffer) significant impairment to his or her emotional development. Intervention in these circumstances can

---

[156] See n. 126 above.
[157] Children Act 1989, s. 31(2).
[158] Ibid., s. 31(9).
[159] Children and Young Persons Act 1969, s. 1.
[160] Children Act 1989, s. 100(2)(*a*).
[161] Children Act 1989, s. 31(2).

be defended as a vindication of the rights of the child only on the basis that this is what the child wishes or, if he or she is not in a position to express such a wish, what the child is likely, retrospectively, to have wished.

Although the dilution of welfarism in favour of more sharply defined 'rights' enhances a 'rights-based' evaluation of children's interests, it may also in other respects have weakened the overall protection given to children. Hence the strongly welfarist innovation of the Matrimonial Causes Act 1958, that no divorce decree should be made absolute unless the court was 'satisfied' that the arrangements made for the children were satisfactory or the best that could be devised in the circumstances,[162] was replaced in the Children Act 1989 with a duty merely to consider whether the court should make an order. The policy visible throughout the Act of shifting the responsibility for promoting children's interests from the state to their parents implies a diminution of the state's supervision over parental behaviour, and hence a weaker policing of children's rights against the welfarist domination by their parents.

(iii) *Other rights children may have.* The strategy of the Children Act 1989 stops short of proclaiming that children have the right to be protected from these harms if they originate from sources *other than* lack of care from their parents or immediate caregivers. We should not conclude however that children should not, or do not, have such rights. But these would be social rights, to be found (if at all) in general community provision for children in areas such as health, education, and environmental policies. These fall outside the scope of this chapter, and lie beyond the function of family law. Nevertheless, reference should be made to claims children may make to reasonable financial support to protect them from the economic deprivations they suffer when one of their parents exercises a 'right' to terminate a relationship. Family law and, in particular, the Child Support Act 1991, establishes that a child has a right against its parents' resources for such support. But where this support fails, British children must rely on the residual, poverty-line provisions of the social-security system. The Government has refused to follow the Council of Europe recommendations[163] that the state should underwrite the parents' obligations, even in cases of temporary lapses.

This chapter does not explicitly consider the possible impact in the United Kingdom of the UN Convention of the Rights of the Child (1989).[164] The Government ratified the Convention in December 1991,

---

[162] Subsequently re-enacted in Matrimonial Causes Act 1973, s. 41.

[163] Recommendation No. R (82) 2.

[164] B. Walsh, 'The United Nations Convention on the Rights of the Child: A British View' (1991) 5 *Int. J. Law and the Family* 170.

with reservations in the substantial areas of immigration, nationality, juvenile custody, conditions for employment of children between 16 and 18, children's hearings in Scotland, and abortion. The Convention could prove to be an important measure in the furtherance of the personal and social rights of children[165] and in 1992 the Children's Rights Development Unit[166] was established to monitor the United Kingdom's compliance with the Convention.

(iv) *Illegitimate children.* The European Convention on the Status of Children born out of Wedlock (1975) both reflected and gave further impetus to the general movement in Western countries towards abolishing the legal disadvantages to children born out of wedlock. Opportunities for their legitimation had already been expanded,[167] and the Family Law Reform Act 1969 almost completely assimilated the position of children born within and outside wedlock as far as succession rights *vis-à-vis* children and parents were concerned. However, it was not until the Family Law Reform Act 1987 that the legal status of a child became almost entirely[168] independent of the marital status of its parents.

CONCLUSIONS

If the assumptions upon which this analysis is based are accepted, it seems that greater attention is now given to the claims (and hence, actual or, potential rights) of people living in familial relationships than hitherto. This perception should be seen as a corrective to the widespread academic belief that the presence, or even the absence, of law is always a manifestation of control over or diminution of the subject. People can rightly refer to enjoying 'the benefit of the law'. So the rights of wives *vis-à-vis* husbands, between heterosexual cohabitees and of children towards their parents are better recognized. The child protection system pays closer heed to the views of parents and children, and the courts are more attentive to children. But there remain desolate areas. Transsexuals and homosexuals are largely ignored (or worse); and this chapter specifically excludes consideration of the claims individuals might make against the community (e.g. welfare rights). Perhaps the relative strengthening of

---

[165] See D. McGoldrick, 'The United Nations Convention on the Rights of the Child' (1991) 5 *Int. J. Law and the Family* 132; P. Alston, S. Parker and J. Seymour (eds.), *Children, Rights and the Law* (1992).

[166] At 235 Shaftesbury Avenue, London WC2H 8EL.

[167] See Matrimonial Causes Act 1950, s. 9; Legitimacy Act 1959.

[168] But only a marital child acquires British nationality through his father. Nor will the father's name necessarily appear on the birth certificate: see S. M. Cretney and J. M. Masson, *Principles of Family Law* (1990), 479.

rights of family members *inter se* is a logical corollary of a diminution of rights against the community. Finally, the adult world, quite understandably, is struggling with the implications of extending the rights of young people. We need to find a way which removes the potentially oppressive aspects of welfarism and which reconciles the proper role of adults and institutions as guardians entrusted to guide young people with respect both for the individuality of those people and for their potential themselves to influence the world that is waiting for them when they emerge from the status of minority.

11

Immigration and Nationality

ANN DUMMETT

INTRODUCTION

Immigration laws are aimed in the first place at persons outside the jurisdiction of the state who wish to enter the state. The need to gain admission to a new state is most acute for the asylum-seeker, who fears persecution, imprisonment, torture, or even death in his or her own country of nationality. But a need arises too where people are fleeing from danger of death in civil war, from imminent starvation, or from a country where civil administration and the economy have utterly broken down. Western European governments have sought to distinguish between 'genuine refugees' and 'economic migrants'. But it is often far from easy to distinguish between the moral claims of individuals fleeing from countries which may have oppressive governments, on the one hand, and those suffering from civil war, economic breakdown, or famine, on the other. The term 'economic migrant' is very vague: it could be applied to the Albanians who tried to enter Italy in August 1991, to Ethiopians at immediate risk of starvation, or to doctors who are overworked and underpaid in one country and who seek a better salary and shorter hours in another.

In the second place, immigration laws establish a regime for migrants *after* their admission, defining their position within the state. Even when admitted to a new state, migrants often lack security of residence there and are dependent upon the rules made by state authorities for the opportunity to seek employment or set up a business or service. The overwhelming majority of people need a family life: migrants may be forbidden to have close relatives with them. Breaches of immigration law, such as taking a job without permission, may be criminal offences. The worst problems arise for poor migrants, who may constitute one of the most vulnerable groups in a society, and commonly have worse housing, a lower quality of health care, and less desirable forms of work than non-immigrants. They frequently experience discrimination on grounds of race or nationality, and feel menaced by officialdom and the police.

Third, even persons whose entry and stay are not controlled may be affected by immigration laws. For example, some countries penalize an employer who takes on illegal migrant workers. In Britain anyone who harbours an illegal entrant or overstayer is liable to prosecution. A British citizen has no right to be joined in the country by a non-citizen spouse. Immigration laws, then, can touch the rights and interests of anyone in the population.

This chapter will consider the political history of United Kingdom immigration and nationality law, before examining some provisions of that regime in some detail. It then considers rights of appeal within the immigration system, before assessing the position in international and European law. In the light of these considerations, some basic reform proposals are then put forward.

DOMESTIC IMMIGRATION LAW: ITS POLITICAL HISTORY

The Home Office, and various ministers echoing its views, have repeatedly said for the last sixty years and more that Britain is not a country of immigration. This is to state a modern policy, not a historical fact. For about a thousand years, a small but steady stream of immigrants has settled in English towns, and sometimes in the countryside, often with official encouragement for the sake of the economy. These immigrants have played an important part in the cultural and political, as well as the economic, life of the country as have many of their descendants, as familiar names like Handel, Hallé, Barbirolli, van Dyck, Disraeli, Conrad, and Eliot as well as Brunel, Ferranti, and Courtauld testify. Periodically there have been limitations and expulsions aimed at particular groups, including Jews and Catholics. In the Victorian era, immigration was legally completely uncontrolled; settlers included Germans, Russians, French, Italians, a small number of Africans and Indians, and, towards the end of the century, Chinese. However, this legal *laissez-faire* approach was challenged at the end of the nineteenth century. A large influx of East European Jews from the 1880s onwards produced a hostile reaction from a small number of politicians, whose campaign for restrictions to be introduced resulted in the Aliens Act 1905.

The Act did not impose rules on all aliens but only on those travelling in the cheapest class on certain types of ship arriving at certain ports. The Act created powers under which variations could be made, and these were used to prevent the entry of Chinese on one occasion and gypsies on another.[1] However, most aliens remained free from control.

The lasting importance of this measure lay, first, in the establishment of an aliens inspectorate, later to develop into the immigration service,

[1] Order of 19 Apr. 1911, PRO HO 45/11843.

which from the beginning exerted considerable powers over applicants for entry on the decision of comparatively minor officials, and, second, in the allotting of immigration control to the Home Office. As the Department of State responsible for national security, the Home Office saw its role from the beginning as the defence of the country against dangerous outsiders: a view strengthened by the politics of the early twentieth century, when anarchy and revolution were feared throughout Europe and when Germans were suspected of infiltrating Britain in order to destroy her. There was close co-operation from then on between the immigration authorities and the intelligence services.[2]

The First World War marked an important stage in immigration control. The 1905 Act had come up against strong liberal unease; the original bill had to be much amended, and an appeal system for immigrants had been established which implicitly recognized that alien applicants for entry had the right to a hearing before an independent board. But the Aliens Restriction Act 1914, together with the Defence of the Realm Acts, suspended this right of appeal and gave enormous discretionary powers to the Home Secretary, not only over enemy aliens but over all aliens, within the country as well as at the point of entry. The immigration appeal boards were formally abolished in 1919. Apart from a short-lived advisory panel to review deportation decisions, there was no independent review of aliens' claims to enter or remain from 1919 until 1955, when Britain ratified the European Convention on Establishment, which required that deportees be given the right to make representations to an independent authority. From then until 1969, alien deportees could make representations to the Chief Metropolitan Magistrate at Bow Street, who could recommend revocation of a deportation order.

Also from 1919 onwards, *all* aliens were subjected to controls on and after entry. The shock of the Russian revolution made British governments throughout the inter-war period go to great lengths to exclude foreigners suspected of political subversion, particularly anyone of Russian origin or with suspected communist sympathies. Work-permits for aliens, first introduced during the 1914–18 war, were continued as political rather than economic controls. The Aliens Restriction (Amendment) Act 1919 confirmed and strengthened restrictions originally intended as wartime emergency measures and empowered the Secretary of State to make Orders in Council. The new Act had to be reapproved annually under the Expiring Laws Continuance Act, and remained in force until 1973.

This history has been of lasting importance because the pattern of restrictions on British *subjects* gradually took its shape from this model originally designed for the regulation of *aliens*. In most countries, the concept of citizens' rights is fundamental to the national constitution and

---

[2] C. Andrew, *Secret Service* (1986).

is given written definition there. No ordinary legislation can take away the citizen's right to live in the national territory and to re-enter it after absence: the only recourse for a government is to take a person's citizenship away, as the USSR used to do with dissidents and as Nazi Germany did in the Nuremberg laws. Outside the United Kingdom, it is taken for granted that citizens have the right to live in the national territory but aliens do not: immigration laws exist to regulate *aliens'* entry and stay. The British pattern is radically different. Because there has been no clear-cut underlying theory of citizenship, Britain's restrictions on immigration have dealt indiscriminately with whichever group of people the government of the day wanted to exclude, whether subjects or aliens.

In 1905, the main group in question consisted of East European Jews, aliens, fleeing from persecution. In the same period, however, the self-governing Dominions (Canada, South Africa, Australia and New Zealand) had the consent of the British government at Westminster when they passed legislation to exclude British subjects of Indian and Chinese origin from free entry to their territories, though at the time British subjecthood was in theory the nationality of both the excluders and the excluded, and when war came in 1914 the King declared war on Germany on behalf of all of them, and all of them sent soldiers and sailors into the fray. After the First World War, under the provisions of the Special Restriction (Coloured Alien Seamen) Order 1925, made under the powers conferred by the Aliens Restriction (Amendment) Act 1919, the burden of proof lay on a coloured seaman to show he was not an alien and so did not need to register with the police, and the authorities repeatedly refused to accept that evidence offered by coloured subject seamen proved their status. The policy, according to Haldane Porter, Chief Aliens Inspector, was 'calculated to prevent the arrival of other coloured seamen' in British ports, and Sir John Pedder of the Home Office wrote that, 'the whole point of the Order is to deal with the trouble caused by the accumulation of these coloured seamen (Aliens and British Subjects mixed) especially at the South Wales ports'.[3]

Thus, citizenship has never had the same meaning in Britain as it has in other countries' laws and in international law. The theoretical basis on which certain persons are subjected to entry controls is not that they are aliens but that they are people Parliament has decided to control. The immigration service, in concert with the Home Office, developed highly restrictive habits of thought and work practices; the notion of rights for immigrants was lost to sight, racial discrimination became entangled in the system, some British subjects came under its controls, and a large discretion for the executive became taken for granted. In effect, statute law was used to legalize administrative licence in dealing with immigration.

[3] 24 Apr. 1925, PRO HO 45/12314.

This pattern of restrictions in turn influenced policy on immigration after the Second World War, with the influx of 'coloured' immigration. There was no need to specify race or colour in new legislation in order to discriminate in practice. Controls over 'coloured' subjects were already being discussed by Colonial Office officials under the Labour government of 1945–50,[4] and although the country was suffering from a severe labour shortage, such discussions continued, involving also the Home Office and the Ministry of Labour in the 1950s. Only the fear that public opinion would not stand for a colour bar held ministers back from introducing controls. According to Cabinet papers released under the 30-year rule, a memo by Sir Alec Douglas-Home in 1955 said:

On the one hand it would presumably be politically impossible to legislate for a colour bar, and any legislation would have to be non-discriminatory in form. On the other hand, we do not wish to keep out immigrants of good type from the old Dominions. I understand that, in the view of the Home Office, immigration officers could, without giving rise to trouble or publicity, exercise such a measure of discrimination as we think desirable.[5]

The Home Secretary, Gwilym Lloyd George, then suggested that immigration officers should be empowered to refuse entry to anyone with a British passport who did not have suitable housing and employment available, and the Cabinet secretary, Sir Norman Brook, pointed out reassuringly that in practice these conditions would not be applied to immigrants from the old Dominions, that is, from Canada, Australia, New Zealand and South Africa. The 'new' Dominions pointedly excluded were India and Pakistan. 'New' signified postwar acquisition of self-government.

From the late 1950s onwards, a small number of MPs and others on the right of politics began to campaign vociferously against non-white immigrants. Some used blatantly racial language; others protested that they had no racial motives but were worried about pressure on housing and hospitals. White alien immigrants, however, who were continuing to arrive on work-permits, were not mentioned. The numbers cited were always numbers from colonies and 'new' Commonwealth countries.

Champions of strict immigration control ever since then have habitually used arguments about numbers to justify demands for more controls. But the numbers game has always been disingenuous, and has conveyed a seriously false impression of the scale of immigration. Emigration has exceeded immigration in almost every year since 1945. The United Kingdom's population was then about 48 million; the number of civilian aliens then

---

[4] S. Joshi and B. Carter, 'The Role of Labour in the Creation of a Racist Britain' (1984) 25 *Race and Class* 53.
[5] See report in the *Guardian* on Cabinet papers then just released, 2 Jan. 1986, p. 4.

registered with the police (including many long-standing residents) was 287,000. By 1948 a few hundred West Indian British subjects had arrived in search of work; their numbers grew rapidly to about 30,000 a year in 1956. At the same time, however, many British people were emigrating: in the peak years of 1952 and 1957 over 200,000 people a year left Britain for other Commonwealth countries. By far the largest group of entrants came from Ireland, but many Irish immigrants were temporary workers: out of around 700,000 a year in the 1950s only around 40,000 would settle permanently. Immigration has done nothing to add to population size since 1945; population growth has resulted from natural increase—excess of births over deaths—which in turn is due less to a high birth rate than to much longer life-expectancy than before.[6]

From the late 1950s onwards, the entry of people of Indian descent began to grow and became eventually the target of most anti-immigrant campaigning. Ironically, the growth arose in the early 1960s as a direct result of the introduction of a work-voucher scheme in the Commonwealth Immigrants Act 1962. Travel agents in the Indian sub-continent, showing an entrepreneurial flair not yet fashionable in Britain, energetically advertised their services in helping applicants get vouchers, and applications rocketed to 300,000. The 1962 Act, introduced by a Conservative government partly in response to pressure from constituency associations and opposed by some back-benchers and by the Labour front bench, did nothing to reduce 'coloured' immigration. In 1965, a Labour government used the large powers in the 1962 Act to slash the number of work-vouchers and restrict entry of relatives. Labour had become nervous of losing some of its racially prejudiced voters. In 1968, a Labour government rushed a second Commonwealth Immigrants Act through all its parliamentary stages in less than a week, in order to stop United-Kingdom-and-Colonies citizens from East Africa from coming to Britain if they were of Indian origin. Denounced by *The Times*, traditionalist Conservatives, Liberals, some but not all Labour left-wingers, many

[6] Office of Population, Censuses, and Surveys (OPCS), *International Migration*, Series MN; Home Office, *Control of Immigration Statistics*, published annually; H. Booth, 'Immigration in Perspective: Population Development in the United Kingdom' in A. Dummett (ed.), *Towards a Just Immigration Policy* (1986). The statistics which support these statements are available but difficult to use. Statistical information on immigration is mainly derived from the Home Office annual figures on entry, the quarterly returns of the OPCS, and the census. All these sources have to be used with care. The Home Office shows some entrants twice over—first on conditional entry and later on acceptance for settlement—and others (Commonwealth-country citizens with right of abode) not at all. It does not record emigration. OPCS records immigration and emigration on the basis of a 10% sample, prepared by asking people at ports whether they are intending to remain, or to be away, for over 12 months. The census, before 1991, recorded place of birth but not citizenship, and includes among many people born overseas persons of British citizenship. These and other sources are not readily comparable with each other.

lawyers, and civil rights groups, the 1968 Act left the British East African Asians effectively stateless. The Government claimed it was delaying, not denying, their entry to Britain, by means of a voucher (not work-voucher) scheme for gradual admissions, but the annual quota of vouchers was far too small to meet the need. Many of the people concerned, who had no legal right to work in East Africa, became destitute.

THE RELATIONSHIP BETWEEN IMMIGRATION AND NATIONALITY LAW

Since 1962, immigration law and nationality law have been so closely and confusingly entangled that it is difficult to describe them separately. The name of 'citizenship' had been introduced into British nationality law by the British Nationality Act 1948, which created a citizenship of the United Kingdom and Colonies to stand alongside the citizenships of independent Commonwealth countries, like Canada, within the larger framework of British subjecthood. This British subjecthood was supposed to be a nationality for the entire Commonwealth, and the historic rights of the subject were still attached to it. No rights were attached to citizenship of the UK-and-Colonies. So a person born in the United Kingdom and a person born in Canada or India were still entitled to vote in the United Kingdom and to move freely in and out of it by virtue of being subjects: the former had no extra rights derived from the new citizenship. Parliament, being sovereign, could alter the rights both of the larger group (subjects) and of the sub-group (UK-and-Colonies citizens), and did so in subsequent Immigration Acts, making some in each group subject to immigration control in the United Kingdom while leaving others free of it. In Britain, the clear-cut line between citizens and aliens which is found elsewhere does not exist: the line between first-class and second-class rights cuts across nationality.

From the late 1960s, the weight of immigration control began to shift on to the dependent relatives of non-white immigrants already settled in Britain. Because wives and children were entitled to come under the legislation, further reductions in entry figures could be achieved only by administrative means, which have been used from the 1960s to the present day to delay or deny entry. From 1969, a relative seeking admission was required to apply for entry clearance overseas before travelling, and from then on it was impossible to seek the help of friends, relatives, and legal advisers on arrival at a British port to argue a case if entry clearance had not been obtained. Long waiting times soon developed in the Indian sub-continent for entry certificates, with the authorities taking several years to process each application.

In 1971, the Conservatives returned to power, having promised in their election manifesto to bring the 'primary' immigration of workers to an end. The Immigration Act 1971 brought alien and Commonwealth immigration ('Commonwealth' including some British) under a single regime. All entrants for work were to be admitted in the first place on permits valid for only 12 months. The new Act, which was not subject to annual review, included numerous new restrictions. Its most important innovation was the concept of patriality, or right of abode, a sort of quasi-nationality denoting freedom from immigration control but not corresponding to any existing category in nationality law. Patrials included most, but not all, UK-and-Colonies citizens in the United Kingdom, certain (mostly white) ones outside it, and several million (mostly white) Commonwealth-country citizens. Everyone else—British, Commonwealth, or alien—was non-patrial and subject to control. Ten years later, the British Nationality Act 1981 (introduced by another Conservative government in fulfilment of an election pledge to reduce *future* sources of immigration) renamed the 1971 categories in nationality terms.[7]

The 1971 Act empowered the Secretary of State from time to time to make immigration rules, which would come into effect immediately but could be subsequently disapproved by a simple majority in either House of Parliament. The rules are not strict rules of law, but the courts have to have regard to them and they must be taken into account on an application for judicial review or habeas corpus where there is a question whether those administering the Act have done so fairly. The rules are published, but they are supplemented by unpublished instructions to immigration officers and entry clearance officers. The character of these instructions remained unknown outside official circles until, in 1985, the Commission for Racial Equality published excerpts from them in its report, *Immigration Control Procedures*.

The British Nationality Act 1981 divided the existing holders of United-Kingdom-and-Colonies citizenship into three groups: British citizens, British Dependent Territories' citizens (BDTC), and British Overseas citizens. British citizens were given one right: the right of abode in the United Kingdom. This signifies freedom from immigration control. The right is not exclusive to British citizens; it is also held by those citizens of independent Commonwealth countries who had been given it by the Immigration Act 1971. British Dependent Territories' citizens had no right of abode in the United Kingdom under the Act, and were dependent for the right to live anywhere on the separate laws of the dependencies: in

---

[7] It is impossible to summarize the provisions of the 1971 Act without falsifying them, so complicated are they. Readers are referred to J. Evans, *Immigration Law* (1983) and I. Macdonald and N. Blake, *Immigration Law and Practice in the United Kingdom* (1991).

most but by no means all cases a BDTC would be free from control in one dependency, with which he or she was connected. British Overseas citizens had no right of abode anywhere in the United Kingdom or colonies.

All three groups remained in law 'Commonwealth citizens'; this term has now replaced 'British subject' in its old usage for a person with nationality of any territory in the Commonwealth. The 1981 Act gave the name of British subject from 1 January 1983 onwards only to a small group known since 1948 as British subjects without citizenship of any Commonwealth country. Most of these were people who had been left out of citizenship in the Indian sub-continent after independence. They have no right of abode in the United Kingdom.

Another group of people existed in 1981 whose status was regarded as British on the international plane but who had never been subjects or citizens. These were British-protected persons. They remained British-protected persons, but unlike the other four categories listed above they do not have the rights of the subject in the United Kingdom, such as voting. They have no right of abode in the United Kingdom, and as there are now no protected or trust territories under British jurisdiction left in the world they have to rely for their residence rights on some other state.

A sixth category of Britishness, that of British National (Overseas), has been defined for use in Hong Kong after 1997, when Hong Kong reverts to China. Most of the inhabitants of Hong Kong, being ethnic Chinese, are Chinese citizens in Chinese law, but some thousands of inhabitants (mostly of Indian or mixed descent) have British Dependent Territories' citizenship as their only nationality. They are to be allowed to apply for British Overseas citizenship—a dubious benefit—after 30 June 1997. If expelled from Hong Kong territory, they will not be readily admitted by Britain if present policy continues. The status of British National (Overseas) confers no right of abode in the United Kingdom. It is intended for use by Hong Kong business people in third countries: they will be able to use it to request consular assistance from British authorities. Hong Kong BDTCs must apply before 1997 to obtain it.

There is one further complication. When the United Kingdom joined the European Community, the British government appended a unilateral Declaration to the Treaty of Accession in 1972 defining who would be British nationals for Community purposes. This definition was slightly amended, again unilaterally, in 1982 just before the British Nationality Act 1981 came into force. The definition includes British citizens, those British Dependent Territories' citizens who are connected with Gibraltar by reason of their or their fathers' birth, registration, or naturalization there, and a small number of British subjects (as redefined in 1981), namely any who already had right of abode in the United Kingdom. Neither Declaration was submitted to Parliament for approval.[8]

The numbers in all these categories are not known with certainty, and official estimates have fluctuated considerably. Roughly speaking, there are about 57 million British citizens, some two or three million of whom live permanently abroad. Most existing Dependent Territories' citizens live in Hong Kong and will lose that citizenship in 1997. There are perhaps three million of them there. BDTCs in the other dependencies probably now number about 60,000. (Falklanders were made full British citizens by the British Nationality (Falkland Islands) Act in 1983.) British Overseas citizens lacking any other nationality number perhaps 200,000; the remaining two categories are impossible to estimate.[9]

Virtually all the people in the categories lacking right of abode are of non-European descent. Within British citizenship, there is a substantial minority—around two million—who are not white. The British government has argued that our nationality law is not racially discriminatory because this minority is included among those with right of abode. It is harder to claim that those who lack right of abode have not been discriminated against on racial grounds: most are of Indian or Chinese ethnic origin. By contrast, Portugal has a single citizenship, which includes inhabitants of Macao born before 1980, and France has a single nationality which includes inhabitants of the French Overseas Departments in the Caribbean. French and Portuguese nationals have all the rights of their respective citizenships, including right of abode in the metropolitan territory.

In recent years, the unprecedented imposition of visa requirements on certain Commonwealth countries, and a growing emphasis in the rules on requirements that some applicants must have enough means not to need recourse to public funds after admission, have worked together with the very tight control over work-permits exercised by the Department of Employment to make settlement in Britain virtually impossible for all but three groups of people. These are: Commonwealth citizens with right of abode under domestic law, European Community nationals with rights under Community law (see below), and a few rich or exceptionally highly qualified persons of other origins who are admitted as a matter of practice under domestic law and remain at discretion rather than as of right. Several thousand asylum-seekers are admitted annually in addition.

---

[8] On the British Nationality Act 1981 see L. Fransman, *Fransman's British Nationality Law* (1989).

[9] Reports by the all-party Select Committee on Home Affairs have demonstrated the unreliability of government estimates: see *Numbers and Legal Status of Future British Overseas Citizens Without Other Citizenship* (HC 158, 1980–1) and *Immigration from the Indian Sub-continent* (HC 90–1, 1981–2).

DENIAL OF RIGHTS UNDER BRITISH
IMMIGRATION LAW

Denial of family unity has been the cruellest work of our immigration law. Section 2(2) of the Commonwealth Immigrants Act 1962 provided that the power to refuse admission to the United Kingdom or to impose conditions of stay under section 2 'shall not be exercised, except as provided by subsection (5) in the case of any person who satisfies an immigration officer that he or she . . . is the wife, or a child under sixteen years of age, of a Commonwealth citizen' who fulfils certain conditions. This subsection has often been loosely referred to as one giving right of admission to certain wives and children, but it does not: the immigration officer must be satisfied of the relationship, and this requirement of satisfaction is frequently repeated throughout the Act. Section 1(5) of the Immigration Act 1971 provides that the immigration rules 'shall be so framed that Commonwealth citizens and their wives and children are not, by virtue of anything in the rules, any less free to come into and go from the United Kingdom than if this Act had not been passed'. Their position continued to be governed by the 1962 provisions until 1988; the Immigration Act of that year removed the entitlement. Meanwhile, however, a large proportion of the wives and children in the Indian sub-continent who sought to join men settled here was refused. An investigation, published by the Runnymede Trust in 1977, showed that 55 wives and children out of 58 who had been refused and whose cases were independently investigated were genuinely related as claimed to men here; two were fraudulent and one case was inconclusive. More recently, the introduction of DNA testing showed that the vast majority of child applicants is genuine[10] though a majority is refused.

Husbands have never had a right to join wives in the United Kingdom, though the immigration rules have stated circumstances in which they *may* be admitted. These have varied considerably over time. Throughout, the authorities have had to be satisfied of the relationship, but this has not been the main hurdle. After 1962, instructions were issued to immigration officers saying husbands and fiancés were 'normally' to be admitted.[11] On 30 January 1969, the Home Secretary, James Callaghan, told the Commons in a written parliamentary answer that this 'concession' had been 'abused' providing young men with the opportunity to come and take employment without a work-voucher.[12] He cited as evidence of abuse the strikingly small

[10] UK Immigrants' Advisory Service, *Annual Report* (1987–8) and the Home Office's own Quarterly Statistical Bulletins.

[11] Instructions issued under the 1962 and 1968 Acts were published, but like the rules instituted in 1971 were supplemented by unpublished instructions.

[12] HC Debs., 776, cols. 366–7.

number of 1,676 Commonwealth-citizen husbands admitted altogether in 1968. Admission of husbands and fiancés was henceforth to be 'restricted to cases presenting special features', and an entry certificate was to be obtained before travelling to the United Kingdom. Men already admitted on a temporary basis who married wives settled in the United Kingdom would not be permitted to settle 'save in exceptional circumstances'. The woolliness of the language used here is characteristic of the immigration rules and instructions over the last 30 years, and highly important. What were 'special features' and 'exceptional circumstances'? It was for the authorities to define them. A truer and simpler statement of the position would have been, 'No husband of a woman settled in Britain shall have the right to join her, though the Secretary of State may decide he can'. But the style of the law has been, throughout, to state elaborate conditions in vague verbiage, so that much time is spent in claim and counter-claim, producing administrative delays of many years before a final decision.

The rules on husbands were liberalized in 1974, but in 1977 were changed to say that a husband would be admitted provisionally for 12 months before being granted settlement, and would be allowed to remain only if the marriage was still subsisting. Both on entry and after 12 months, permission depended upon the Secretary of State's being satisfied that the marriage was genuine, and genuineness did not arise from the marriage's validity in civil or religious law but from the Secretary of State's opinion that it had not been entered upon in order to circumvent immigration control. In subsequent changes to the rules over the next decade, this proviso was elaborated and produced *inter alia* the infamous 'primary purpose rule' still in operation. The husband has to satisfy the authorities that marriage was not entered into with the primary purpose of entering the United Kingdom. Originally, the argument for restricting husbands was that marriages of convenience were being used to circum-vent controls, but the real reason has always been to exclude men of working age. The rules have been administered so as to exclude many husbands who have been married for some years and have children by their wives; in most cases of refusal there is no evidence at all of a marriage being one of convenience only. The complexity of administrative practices used cannot be adequately described here but has been docu-mented in the annual reports of the Joint Council for the Welfare of Immigrants (JCWI) and the United Kingdom Immigrants' Advisory Service (UKIAS).

There are numerous other powers which fail to meet human rights standards. One of the most objectionable in principle, although it affects very few people, is the Secretary of State's power to deport the wife and children under 18 of a male deportee, and the children under

18 of a female deportee, when these family members have committed no offence.[13]

APPEAL RIGHTS

Since 1969 there has been an immigration appeals system. It was redefined in the Immigration Act 1971. A statutory right of appeal exists against refusal of entry clearance, leave to enter at a port, or a certificate of entitlement (documenting a person's right of abode), against the making of a deportation order, against refusal to revoke a deportation order, and against a decision concerning country of destination for a deportee. A would-be visitor has no appeal before removal, however; nor does an allegedly illegal entrant within the country. There is no appeal against refusal of a work-permit or of a voucher under the non-statutory scheme for admitting British Overseas citizens of East African origin. Appeal is in the first instance to a single adjudicator. The Asylum Bill published in 1992 would, if passed, remove appeal rights from visitors, short-term and prospective students and their dependents, against refusal of an entry clearance or leave to enter.

The appeals system has been frequently and heavily criticized, notably in the Commission for Racial Equality's formal investigation of immigration control procedures published in 1985. As discussed in chapter 13, the CRE has powers, under the Race Relations Act 1976, to conduct formal investigations for any purpose connected with its statutory duties. The CRE could not, however, investigate racially discriminatory immigration legislation itself, because section 41(1) of the Race Relations Act 1976 excepts from its scope any racial discrimination done in pursuance of any statute, subordinate instrument, or Order in Council. Over the objections of the Home Office, the High Court determined in 1980 that an investigation could, however, be carried out as part of the CRE's statutory duty to promote good race relations. The immediate cause of the investigation was a scandal over 'virginity tests' carried out on Indian women seeking entry as fiancées of men settled in Britain: publicity for one case, in 1978, brought others to light.

The CRE was able to interview immigration staff and inspect unpublished instructions. The report which resulted is rare in that it is the only empirical study of British immigration law which has been able to base its findings on current, internal communications from the Home Office and Foreign and Commonwealth Office. Some features of immigration practice are of particular interest in limiting the effectiveness of the appeal system.

[13] Immigration Act 1971, ss. 3(5)(*c*), 5(4)(*a*) and (*b*).

These Departments' instructions to their officers, who present the case against the appellant in immigration cases, for example, caution officers against making any assertion which might shift the burden of proof from the appellant:

if our statement says that the Secretary of State 'concluded that the appellant intended to remain here permanently' then the presenting officer may be asked to prove this conclusion. Thus a positive assertion of this kind should be made only where it is supported by firm evidence: otherwise it is safer to stick to the negative form of words ('the Secretary of State was not satisfied . . .').[14]

Presenting officers are also told that they should 'normally resist any request' by the appellate authorities for 'documents to be submitted, other than those which are already attached as appendices to the explanatory statement'.[15] The explanatory statement is a written statement of the facts relating to a refusal, and is frequently prepared by an officer other than the one who conducted interviews, studied papers, and issued the refusal. It is, moreover, often written many months after a refusal took place at an entry clearance post overseas, and can include the only available account of the appellant's responses at the interview: in entry clearance cases the appellant is normally absent from the hearing, which takes place in the United Kingdom. Charles Blake comments on the explanatory statements as follows: 'Hardly ever tested by cross-examination (often because representatives do not ask for it) it contains a mixture of facts, assertion, law, hearsay, conclusion and inference. Paradoxically the evidence is often of a higher quality in port appeals because there has been insufficient time to prepare an explanatory statement'.[16]

There is an appeal from the adjudicator to a three-person tribunal on a point of law and in certain other cases; leave must be given, in most instances, to appeal to the tribunal. Its chairman must be legally qualified. There is no appeal from the tribunal to the courts. Judicial review and informal representations on an appellant's behalf by an MP to the Home Secretary have therefore been the only recourse for failed appellants, and over a long period the work of the Divisional Court has been overloaded with immigration cases. Judicial review is not an adequate mechanism to deal with all the wrongs the present system can produce. Moreover, since 1988 the scope for MPs' representations has been reduced by the Home Office. Legal aid is not available in immigration appeal cases. Lawyers have a right of audience before the adjudicator or tribunal, but many cases

[14] Commission for Racial Equality, *Immigration Control Procedures: Report of a Formal Investigation* (1985), 116.
[15] Ibid. 119.
[16] C. Blake, 'Immigration Appeals: The Need for Reform' in Dummett (ed.), *Towards a Just Immigration Policy*.

are conducted by JCWI, UKIAS, or other voluntary helpers. There is little incentive for lawyers to specialize in immigration law, except at its more lucrative end dealing with applications from business people, persons of independent means and top earners seeking work-permits. Meanwhile, most appeals, which are made on behalf of relatively poor people from the Indian sub-continent, Ghana, Nigeria, and, in the case of asylum-seekers, Sri Lanka, East African countries, and the Middle East, are unsuccessful.

It has often been said that the shortcomings of the appeal system arise largely from the immigration regime itself. 'The system', says the CRE report, 'gives every appearance of being biased against appellants', but the authors go on to point out, 'it would be a mistake to suppose that if an administrative system is seriously flawed, a system of appeals against it will correct those defects. It will not'.[17] The appeals structure fails the test of upholding human rights not only because the appellant is often absent and the rules of evidence are not followed but because of the vast powers that statute law has conferred on the immigration authorities.

INTERNATIONAL LAW

It is a well-established principle of international law that the admission of aliens is at the discretion of each state. There are certain limitations on the powers of exclusion[18] but these are exceptional. Migrants' rights for the most part depend on the municipal laws of each state. International law, therefore, takes for granted that each state may determine its own immigration laws. The basis on which states make their selection may be morally questionable, but it is seldom legally so, except where a state has voluntarily bound itself by an international agreement. If a state enters into an international agreement to admit certain aliens, it is bound by that agreement.

Since 1945, it has been accepted, however, that the protection of individual human rights is a matter for international concern. There are now numerous international instruments governing the protection of the rights of aliens within a state's jurisdiction[19] but protection of the individual seeking entry to a state remains very weak. The United Kingdom has ratified the UN Convention on Civil and Political Rights 1966, and the European Convention on Human Rights and Fundamental Freedoms

---

[17] CRE report, n. 14, p. 123.
[18] See R. Plender, *International Migration Law* (1972), 94–126.
[19] R. Plender, 'Human Rights of Aliens in Europe', in Council of Europe Directorate of Human Rights, *Human Rights of Aliens in Europe* (1985), 34.

1950, but since our immigration law excludes many British nationals from entry, has withheld ratification from the article common to both the UN Convention and the Fourth Protocol (1963) to the European Convention which says, 'No one shall be deprived of the right to enter the territory of the State of which he is a national'. The United Kingdom has also ratified the UN Convention on the Status of Refugees 1951 (the Geneva Convention), but this does not give an unqualified right of asylum. It protects a refugee only against *refoulement* to the country where he or she fears persecution, and so Britain (and other states party to the Convention) can refuse entry to a person even if it is admitted that he has a well-founded fear of persecution, and send him to a third country where there is no guarantee of admission either.

It was expected when the Geneva Convention was drawn up that most refugees to Western countries would be from Eastern Europe. Since 1980, the number of asylum-seekers world-wide has risen sharply, and most now come from the Middle East, Asia, and Africa. One method of resisting their entry is to deny that a well-founded fear of persecution exists, a practice British officials have used even in cases where there is clear evidence a person has been tortured.[20] A second is to impose a visa requirement on refugee-producing countries. Another is to prevent the arrival at a port of entry of persons who may claim asylum by imposing sanctions on air and shipping lines. Belgium, Denmark, Germany, Italy, France, and Britain have all introduced legislation since 1987 imposing fines on carriers who bring in passengers without valid visas and travel documents or whose papers are forged. The effect is to oblige airline staff 'to act as *international* immigration officers . . . in order to be able to respect fully the legislation on fines, airline staff are expected to be well informed of immigration requirements of practically every country served at large international airports'.[21]

Since Britain's Immigration (Carriers' Liability) Act came into force in 1987, 'just about every regular air and sea carrier has been found to be in breach of the Act with fines totalling £26,604 million' up to 31 January 1991.[22] Some foreign airlines refused to pay and were threatened with loss of landing rights. British Airways has however zealously tried to observe the controls. On 9 April 1990, BA staff at Heathrow allegedly seized and removed three Tamils before they had a chance to present their claims. Many refusals naturally take place in other countries, where applicants who are refused tickets lack any legal redress and dare not publicize their claims for obvious reasons. For those who reach a British

---

[20] e.g. the Medical Foundation for Victims of Torture gave evidence in the cases of five Tamils (see below n. 36).

[21] A. Cruz, 'Carrier Sanctions in Five Community Countries', pamphlet issued to subscribers to *Migration News Sheet* (Brussels, 1991).

[22] Ibid.

port and are ordered to be removed, there has been no appeal before removal. However, the Asylum Bill published in 1992 proposes a right of appeal before removal for *all* applicants refused asylum. Unfortunately, the bill's other provisions have the effect, despite this apparent improvement, of reducing the chance of acceptance to an unprecedently low level. In some cases, an asylum-seeker will be allowed only 48 hours in which to lodge an appeal. An asylum-seeker is required to be finger-printed, and can be arrested without warrant for failure to comply. A person admitted already under some other category, for example as a student, who applies for asylum will, if refused it, lose leave to remain in the original category. The opportunity for judicial review is lost: an appeal from the final determination of the Immigration Appeals Tribunal may be made to the Court of Appeal but only on a point of law. New detention centres for asylum-seekers are being built: the effect of all these moves is to criminalise the asylum-seeker rather than to observe the spirit and intention of the Geneva Convention.

As well as the blatant injustice of imprisoning persons whose only offence is to have applied for asylum, one must consider the mental anguish inflicted on those applicants who have fled from a country where they have been in jail and suffered torture and who, on arrival in a country where they hoped to find sanctuary, are jailed again, not knowing what will then happen to them. Doctors from the Medical Foundation for the Care of Victims of Torture have said that imprisonment here imposes 'unnecessary cruelty and degradation on people already distressed by torture and loss'.[23]

Perhaps the most serious of Britain's violations of international standards occurs under the immigration legislation's provisions on detention and administrative removal. These came suddenly under the glare of publicity during the Gulf war early in 1991 but had been in force ever since the Immigration Act 1971 came into effect on 1 January 1973. Despite the efforts of JCWI and others to publicize them, most people in Britain had remained unaware that immigration law authorized virtually indefinite detention at the will of the executive with no right for the individual to a hearing or to be told what the allegations were against him or her. Under no other modern legislation, not even the Prevention of Terrorism Acts, have such sweeping powers over the liberty of the individual been conferred on the executive.

A person may be held immediately on arrival and detained until a decision is made whether to allow entry or not. If the decision is negative, the person may be held until removed from the United Kingdom.[24] Anyone authorized by the immigration officer may detain the passenger:

[23] M. Ashford, *Detention without Trial* (1993).
[24] Immigration Act 1971, Sch. 2, para. 16.

usually this is a police officer, a prison officer, or an employee of Group 4, a private security firm with which the Home Office has a contract for these duties. Anyone liable to be detained can be arrested without warrant by an immigration officer or police officer.[25] A person being detained is in lawful custody, and an application for habeas corpus is almost certain to fail because the onus of proof is on the detainee to show he is being held illegally.[26]

A person who has already entered the country and who is suspected of being an illegal entrant, having been allowed to enter because of fraud or misrepresentation, is also liable to arrest without warrant and subsequent detention and removal. There is no appeal until after removal. The courts have been very reluctant to grant habeas corpus in illegal entry cases. Since 1987 many applicants for asylum have been detained on arrival in prison cells or a prison ship, and since 1989 some have been placed in widely scattered places of detention outside London, where solicitors and voluntary agencies have difficulty in discovering their whereabouts and reaching them before removal. There have been since 1987 three suicides of asylum-seekers while in detention. These practices clearly violate those international instruments, to which Britain is a party, which require judicial control of detention and which provide that a person arrested, or a detainee, must be informed of the reasons for arrest or detention.[27]

Deportation is a different process from administrative removal. The Secretary of State must make a deportation order, and a deportee must be served with notice to deport and informed of rights of appeal. Regulations require that service must be made to the last known address of the deportee; if the person does not then receive the notice it still counts as good service. An appeal may be made before deportation takes place, except where an order is made on the ground that a person's presence in the country is not conducive to the public good. In cases concerning 'the interests of national security or of the relations between the United Kingdom and any other country or for other reasons of a political nature'[28] representations on the deportee's behalf may be heard by a panel of three wise men (colloquially so called) appointed by the Secretary of State, but the deportee has no right to hear or cross-examine Home Office or security services' witnesses nor to subpoena witnesses on his own behalf. The appellant is not entitled to legal representation, nor may

---

[25] Immigration Act 1971, para. 17.

[26] *R.* v. *Secretary of State for the Home Dept., ex parte Phansopkar* [1976] QB 606, [1975] 3 All ER 497 (CA); also *Re Wajid Hassan* [1976] 2 All ER 123.

[27] International Covenant on Civil and Political Rights, Arts. 9(2) and 9(4); ECHR 5(2) and 5(4).

[28] Immigration Act 1971, s. 15(3).

he hear the evidence against him. If, despite all these difficulties, the panel advises against deportation, the Secretary of State is not bound to follow their advice. In short, his discretion to deport is almost unlimited.[29]

In practice, national-security cases appear to be influenced by political considerations that have little to do with national security. The panel was first invoked in the case of Franco Caprino, an Italian immigrant worker with Marxist views who had been an active trade unionist; the order was revoked shortly before the panel met. Much more worrying were the cases of people of Kuwaiti, Bahraini, Iraqi, and Palestinian origin which arose in 1990 and 1991 and were described in a report by Amnesty International on United Kingdom Human Rights concerns in June 1991.[30] Two Kuwaitis were deported in May 1990 on non-conducive, national-security grounds, which were never tested in a court. Amnesty considered the real reason lay in the men's non-violent political activities; one of them, Anwar al-Harby, was well known for human rights work in Kuwait. After the Gulf war began in January 1991, about 90 Arabs—some Iraqi, some Palestinian—were arrested and detained on national-security grounds. Some were told they had 'known links' with organizations which might take terrorist action; others were given no explanation. 'Although the detainees had the right to apply for a *habeas corpus* writ, the courts stated that they were not in a position to question the specific reasons for the detention, once the government cites national security.'[31] One case, that of Abbas Shiblak, a Palestinian who had lived since 1976 in Britain and was an academic and writer well known for his human rights work, was widely publicized in the newspapers, and strong political pressure built up for his release. The panel of three heard his case on 1 February and he was released on 7 February; as the hearing was secret one cannot know whether the evidence or the publicity persuaded the Home Office to let him go.

But the most obvious breach of human rights to be found in British immigration law is that it is, in intention and effect if not in terms, racially discriminatory. Richard Plender remarks, 'There is some evidence to show that racial discrimination in the selection of immigrants may be considered as contravening a generally accepted norm or standard, if not a rule of customary international law.'[32] If such a norm, standard, or customary rule exists, it has been flouted by Britain throughout the twentieth century.

---

[29] Except concerning European Community nationals: see below.
[30] Amnesty International, *United Kingdom Human Rights Concerns* (1991).
[31] Ibid. 31.
[32] Plender, 'Human Rights of Aliens'.

EUROPEAN CONVENTION ON HUMAN RIGHTS

Individuals who believe they have been treated contrary to international law have little recourse in Britain to a hearing by an international body. The great exception is the right of individual petition to the European Commission of Human Rights. On three occasions, cases alleging breaches of human rights by the United Kingdom in immigration matters have been submitted under the European Convention on Human Rights. In the first case, a group of East African citizens of the United Kingdom and Colonies complained to the European Commission on Human Rights that the manner in which they had been forced to become destitute in East Africa, because of Britain's refusal to admit them, constituted degrading treatment, and that the Commonwealth Immigrants Act 1968 had been racially motivated. Many of the British 'East African Asians' who tried to arrive without an entry voucher after 1968 were shuttled back and forth on planes or ships several times before being admitted only to be imprisoned for a period; the policy was intended to deter others from coming.[33] Pressure was brought on airlines to stop them carrying these citizens to Britain, and the Indian government was persuaded to accept temporarily several thousand of them pending admission to the United Kingdom (not all have yet, even now, been admitted). The Commission ruled their applications admissible, but the cases did not proceed to the European Court of Human Rights. The British government admitted the individual complainants, and increased the number of vouchers available to East Africa, but did not repeal the 1968 Act.[34]

In the second case, in 1985, three women of non-European origin legally resident in Britain, who had been refused permission to have their husbands join them, claimed that the British rules were racially discriminatory, infringed the right to family life, and were sexually discriminatory (because at the time husbands were entitled to bring in wives but wives could not bring in husbands). This case[35] did reach the Court, which rejected the first two grounds for complaint but upheld the claim that the immigration laws discriminated on grounds of sex. The British government's response was to pass new legislation, the Immigration Act 1988, which took away the apparent right of husbands to bring in wives and thus established sex equality in lack of rights.

The third case involved five Tamils who were removed from Britain in 1987 to Sri Lanka and subsequently suffered persecution there. The

---

[33] See a letter from the Foreign and Commonwealth Office to JCWI, quoted in A. Dummett and A. Nicol, *Subjects, Citizens, Aliens and Others: Nationality and Immigration Law* (1990), 204.

[34] *East African Asians* v. *UK* (1981) 3 EHRR 76.

[35] *Abdulaziz, Cabales and Balkandali* v. *UK* (1985) 7 EHRR 471.

Commission on Human Rights ruled that Article 3 of the Convention, prohibiting inhuman and degrading treatment, had not been breached by the British government, but that Article 13, which requires an effective remedy before a national authority, had been. The Court subsequently held that the British government's deportations had violated neither Article 3 nor Article 13 (the vote being 8 to 1 on the first point and 7 to 2 on the second). The decision marks a regression in legal protection for asylum-seekers.[36]

The European Convention is not incorporated into domestic law. In many other countries international agreements become part of domestic law once they have been ratified by the government concerned. In Britain, implementation is left within the power of the executive, and only if specific provisions from an international agreement are embodied in statute law do they have direct effect. One result is that lawyers and the public in Britain pay less attention to the precepts of international law than they do in certain other countries, notably the Netherlands. This is probably one reason why political discussion of migrants' rights in Britain is theoretically jejune, insular in tone, and overshadowed by considerations concerning immigration policy generally which have nothing to do with human rights safeguards.

THE EUROPEAN COMMUNITY

This insularity appears in a different and ambiguous form in Britain's role in the European Community. In theory, the countries of the EC are developing a common immigration policy towards everyone outside the EC's external frontier. In theory again, internal frontiers are to disappear, with movement unchecked between the states concerned, not only for the nationals of EC countries but for everyone within EC territory: an American visitor arriving from New York is to be checked on arrival in Paris but not checked again on flying from there to London; a Turkish resident in Germany is to be able to drive to the Netherlands without any immigration checks at the border. In other words, for immigration purposes, the EC is to be a single territory like a nation-state. Clearly this plan requires harmonisation, if not complete identity, between the immigration and asylum laws of the countries concerned. Article 8A of the Single European Act, ratified by all Member States in 1986, required internal frontier controls to be suppressed by 1 January 1993, but this has not happened. An essential step, agreement on the Convention on the crossing of external frontiers, has been held up since 1991 by a dispute

---

[36] *Vilvarajah and Others* v. *UK* (1992) 14 EHRR 248.

between Spain and the United Kingdom about the policing of the border at Gibraltar. Moreover the Home Office has declared its determination to maintain port controls for all entrants, despite the threat of being taken to the European Court of Justice. Ireland has difficulties because its immigration control structure is closely linked to that of the United Kingdom. And Denmark has a problem because it is a member of the Nordic group of countries which have their own internal free movement rules. At the time of writing, it is uncertain at what date the plan for one external frontier around an internal free-movement area in the Community will come into effect.

The original plan, envisaged when the Treaty of Rome was amended by the Single European Act in 1986, was part of an intended, continual progression towards closer political and economic unity in the Community, of which the next stage was marked by the Maastricht Treaty on Political Union, agreed by heads of government in December 1991 and signed in February 1992. However, this progression has been halted by delays in ratifying the Maastricht Treaty and thrown into confusion for a number of reasons. The intention to develop a *Community* immigration policy, worked out by the EC's institutions and subject to the scrutiny of the European Commission and the European Court of Justice, has been thwarted because several states, with Britain in the lead, insisted at Maastricht on an inter-governmental framework instead. As a result, negotiations have been conducted by Ministers and officials of national governments, working in secret and not effectively controlled either by their own national Parliaments or by the European Parliament and Commission. Any agreements they reach will take the form either of an ordinary international agreement outside the Community framework (and so not justiciable by the ECJ) or, in effect, and even more importantly from the point of view of human rights, of concerted action at the administrative level. A common immigration policy thus formed, even if given the nod of approval by Ministers sitting as the Council of Ministers of the EC, does not make a Community policy proper.

British policy is, consistently, to resist any growth in competence for the Community and to insist on national sovereignty even where (as on the lifting of internal border controls) Britain is already legally bound by its own agreement to yield to Community competence. British citizens and residents from ethnic minorities would benefit from the lifting of border controls; at present, both on entering another EC country and on returning to Britain, they risk harassment from officials. The British government has argued, however, that the lifting of border controls would require the introduction of identity cards here as a check on undesirables, and that this would be bad for human rights. It must be emphasised that the EC does not require ID cards: this is a matter for each state's internal jurisdiction. Arguably, the large powers which already exist in the Immigration Act 1971 are more than enough for the internal controls necessary.

Another complicating factor in the development of a common entry policy is the Schengen Agreement, an inter-governmental agreement between nine states of the Community (who may be joined by some non-EC states). This provides for an area without internal border controls, comprising all the EC except Britain, Ireland and Denmark. Although Britain is outside Schengen, its scheme is certain to influence the common policy of the 12. For example, there is to be a Schengen Information System (SIS), incorporating a central, computerised databank of information on persons considered undesirable or already refused entry to any one Schengen country. There is also to be, for the 12, a European Information System of the same kind. The two are likely to be merged. Worries about the implications of the SIS for justice and human rights have already delayed the coming into force of the Schengen Agreement, which was due in July 1992.

Unfortunately, a common policy is being discussed against the background of increasingly restrictive attitudes on the part of all West European governments. Worried by recession and unemployment at home, an influx (smaller than had been expected but still significant) from Eastern Europe, and an increase in asylum-seekers, governments which formerly tolerated illegal entry have begun to clamp down on it; respect for family reunion has weakened; the claims of asylum-seekers are being denied. In this context, the prospects for liberalisation of British immigration law look dim. However, if the EC recovers from the problems that have beset it since 1989, and its institutions recover the initiative they have lost recently, there is still some hope that the Community's approach to human rights will make a positive difference to applicants for entry in future.

Meanwhile, the Community's rules on migration between Member States have already driven a wedge into British immigration law. Britain declined to take part in the formation of the Community in the 1950s, despite pleas from other states to do so. Only in 1973 did Britain become a member (on 1 January, by coincidence the same day the Immigration Act 1971 came into force). Both before and after joining, British attitudes, outside a limited circle of Euro-enthusiasts, have been suspicious of foreign domination, uninterested in the character and ideals of the EC as a whole, and yet determined at the same time to get advantages for Britain out of it. Reflecting these attitudes, the British media have never reported Community affairs fully, have played up the notion of a dictatorial Brussels bureaucracy, and have concentrated on a few economic and financial issues to the exclusion of political and legal affairs. Many British people, even now, are unaware of their right to work and reside anywhere in EC territory, and of the corresponding rights of other EC nationals to live and work in the United Kingdom.

It was characteristic that, although British entry to the EC was being negotiated at the same time as the 1971 Immigration Bill was drafted and then debated in Parliament, the drafters and most of the debaters ignored the implications of entry for domestic immigration law. The 1971 Act nowhere mentions the EC, whose nationals' special position was described only in the immigration rules until 1988. But the Immigration Act of that year does not indicate all the ways in which Britain is now bound by EC law on free movement.

Community law overrides domestic law where the two conflict. The Treaty of Rome and subsequent EC legislation, notably Regulation 1612/68, lay down rules on free movement for workers and their families throughout the territory of member states. An EC national is to be admitted to an EC country other than his or her own without need of a visa and may remain in search of employment, access to which is to be allowed on the same terms as for nationals of the receiving state. There is also a right of free movement for EC nationals seeking to set up a business, provide a service, or establish themselves in professions. An EC national who has found an occupation in Britain under any of the above heads acquires right of residence and the right to be joined by a spouse of either sex, dependent children under 21, and dependent grandchildren, parents, and grandparents. Favourable treatment is to be given to any relative who has habitually lived under the same roof as the sponsor. In none of these cases is it required that the relatives be EC nationals themselves.[37] Time limits are imposed so that administrative delays cannot frustrate admission. Thus EC law acknowledges that the right to family unity is a very powerful claim, overriding member states' powers over immigration of non-EC nationals.

An EC national is a person with citizenship of a member state. This definition needs qualifying in the British case, since the British government has twice made unilateral Declarations restricting the definition of British nationals for Community purposes to British passport-holders with right of abode under domestic law, together with Gibraltarians, whose free movement is required by Article 227(4) of the Treaty of Rome. Excluded from EC nationality are British Dependent Territories' citizens other than Gibraltarians; British Overseas citizens, British-Protected Persons, British Nationals (Overseas), and those British subjects under Part IV of the British Nationality Act 1981 who lack right of abode. The rights of

---

[37] This summary needs qualification and addition, impossible for reasons for space here. For a fuller account, see I. Macdonald and N. Blake, *Immigration Law and Practice*, 3rd edn. (1991).

all these British nationals have thus been curtailed, in contrast to the approach taken by some other European states. The inhabitants of Martinique and Guadeloupe, for example, are full French citizens and therefore enjoy EC freedom of movement, together with their families.

Within Britain, there are some respects in which an EC national from another state has rights superior to those of a British citizen. An Irish worker in Britain with an Indian spouse has a right to bring in that spouse, though a British worker with an Indian spouse has not, for example. However, the British worker, by moving to work in another EC state, can be joined by the spouse. It was assumed over a long period that a worker in the latter situation could not, having been joined by a spouse in another EC country, use EC law to claim the right to bring the spouse back to the United Kingdom. In other words, if a man from Pakistan who had spent ten years trying to bring his wife to Britain went to the Netherlands, got a job and was able to send for her within three months, he would then have to stay there, or in any EC country other than Britain, or be parted again from his wife. In July 1992, however, the European Court of Justice determined in the case of *Surinder Singh* that Article 52 of the Treaty and Council Directive 73/148/EEC of 21 May 1973, properly construed, 'require a Member State to grant leave to enter and reside in its territory to the spouse, of whatever nationality, of a national of that State who has gone, with that spouse, to another Member State in order to work there as an employed person . . . and returns to establish himself or herself . . . in the State of which he or she is a national'.[38] This decision throws into relief the restrictiveness of United Kingdom law on families.

EC nationals may be refused entry or stay on the grounds of public policy, public health, or public security, but these grounds are strictly defined in Community law and are not open to varying interpretation by different states' immigration authorities. A final decision rests with the European Court of Justice. In *Van Duyn* v. *Home Office* in 1975, for example, the Court considered the case of Mrs Van Duyn, a Dutch national and member of the Church of Scientology, who had been re-fused permission to remain in the United Kingdom on the ground that her presence was not conducive to the public good. The Court upheld the refusal on public-policy grounds, but made clear that the Home Office could not interpret 'public policy' in a manner derived from national law. The large discretion the Home Secretary can use to deport outside the scope of EC law is here limited. Similarly, exclusion on public-health grounds is limited; an annex to Council Directive 64/221/EEC of 25 February 1964 lists the disabilities which justify refusal of entry. Public

---

[38] Case C-370/90, judgment of 7 July 1992.

security can be invoked only on the basis of the personal conduct of the migrant, not as part of a blanket ban nor as a deterrent to others. Anyone refused entry or stay on any of the three named grounds must be given reasons for the refusal and has the right to some form of judicial review.[39] Comparison of these provisions with domestic law shows that the rights of the migrant under EC law are markedly superior.

This too-brief summary of the rules in place at the end of 1992 has taken a narrow view, singling out specific effects of EC membership on Britain. But EC law needs to be understood in the context of the Community as a whole: its effects in all member states, its debt to different countries' habits of legal thought and attitudes towards rights (as enshrined, for instance, in their own written, national constitutions), and its regard for the standards of international law. Moreover, the Community itself must be seen as a body whose aims have never been solely economic but which from its inception looked to eventual political union as a goal. In this light, it is no surprise that it was agreed at Maastricht that a European citizenship be established, to be held by all EC nationals alongside their existing citizenships, and that common policies on immigration and asylum were to include common policy measures on the rights of third-country nationals (i.e. non-EC nationals) already legally settled in member states. The European Commission had argued for some time that it was essential to bring the rights of these migrants more closely into line with the rights of EC nationals;[40] the same policy has been urged, not so much from a human rights point of view as from an economic one, by some who hold that the single market cannot function effectively if it includes two classes of person on the labour market, one with full freedom of movement in EC territory and the other confined within a particular national territory.

One can distinguish more than two classes of rights if one takes into account the various Association Agreements with third countries already in force, guaranteeing certain limited rights to these countries' nationals after admission to EC territory. Such agreements exist with Turkey, Algeria, Tunisia and Morocco, and are being concluded with a growing number of East European countries. Agreement in principle has already been reached, for the future, with the seven countries of the European Free Trade Area (EFTA) to establish at a date not yet fixed a measure of free movement of labour between EC and EFTA countries (Switzerland excluded). All these matters, like Schengen, have implications for British immigration law and policy, but are hardly known and rarely discussed in Britain at the time of writing.

Minority groups in Britain have expressed fears that the Community

---

[39] See Macdonald and Blake, *Immigration Law*.

[40] See e.g. 'Commission Communication to the Council and the European Parliament on Immigration', SEC (91) 1855 final (Brussels, 23 Oct. 1991).

represents a 'Fortress Europe', hostile to non-European immigrants and to minority rights. The situation is far more complicated than the picture they paint. It is true that a common immigration policy will require a common visa policy, but not necessarily an increase in visa requirements: the pressure for tighter control is coming from national governments and not from the Community institutions. The pressure for a more liberal approach is coming from the European Commission, while the case-law of the Court of Justice, as examples quoted above demonstrate, upholds the rights of migrants. In Article K, the Treaty on Political Union is concerned with the rights of third-country nationals.

Overall, human rights would probably be better served by granting more competence to the Community institutions concerning immigration and asylum than by reserving these matters to national and inter-governmental decision. The Court and Commission, which are independent of national, political pressures, and the Parliament, which is democratically elected, are more open and efficient than the work of national officials and their ministers.

CONCLUSION: REFORM OF THE LAW

Unhappily, the attention given to individual cases over many years, although it has sometimes solved individuals' problems (as with Anwar Ditta whose children had consistently been refused entry over years) has never yet been successfully transferred to the Immigration Act powers themselves. Campaigns to repeal the Act as a whole attract little support, and there has been no organized campaign, despite the anxieties of some lawyers, to undertake reform of those parts of it which allow abuse of human rights.

Despite the broad effects of immigration law, and the importance of the interests they touch, the claims of migrants and those closely affected by migrants are also given little attention in either political or legal theory. Among political theorists, Joseph H. Carens[41] is rare in arguing that poor people should be free to enhance their life chances by moving to countries which offer them better opportunities than do their own. Alan Dowty,[42] on the other hand, argues fiercely against control of exit but believes that control of entry is essential to the idea of sovereignty. Michael Walzer[43] agrees with Dowty that 'immigration and emigration

---

[41] J. H. Carens, 'Aliens and Citizens: The Case for Open Borders' (1987) 49 *Rev. of Politics* 251.

[42] A. Dowty, *Closed Borders: The Contemporary Assault on Freedom of Movement* (1987).

[43] M. Walzer, *Spheres of Justice: A Defence of Pluralism and Equality* (1983).

are morally asymmetrical' and holds that communities, as communities, have a moral right to determine who shall join them. Few political theorists have tackled the question of the rights of migrants at all,[44] and this hiatus is clearly visible now that migration to the United Kingdom and to Western Europe has become a major political question.

Nevertheless, British immigration and nationality laws are clearly in need of radical reform if they are to comply with even minimum standards of human rights. The first step should be to establish a single British nationality whose holders would all have defined rights, including the right of abode in the national territory. The second should be to curb the large discretion which immigration legislation confers on the Secretary of State over entry, stay, removal, detention, and deportation. Third, EC rules on family reunions should be extended to all legally settled residents. Some of the faults apparent in the present appeals system would be corrected by such legislation, but other faults will still need correction; in particular, legal aid should be available and an appeal should exist to the courts.

These changes would be enormously important in both principle and effect, yet they would still not touch upon the fundamentals of immigration policy: namely, who besides British nationals, EC nationals, and the relatives of settled immigrants should be admitted. On such primary policy, the upholding of human rights requires two things: that the rights of asylum-seekers should be guaranteed by examination procedures which fulfil both the letter and spirit of the Geneva Convention, and that there should be no racial discimination in immigration law and control procedures. The latter goal might be achieved by an amendment of the Race Relations Act 1976, subjecting statute law and ministerial actions to the possibility of investigation.

Such proposals sound, perhaps, impossibly radical when one looks at our constitution and our current political debates; yet looked at from an international standpoint, or indeed from a point of view that is sincere rather than hypocritical about individual freedom, they are minimal.

---

[44] Dummett and Nicol, *Subjects, Citizens* (1990), has a full discussion of this question in the final chapter.

12

Women's Rights in Employment and Related Areas

EVELYN COLLINS and ELIZABETH MEEHAN

INTRODUCTION

The question of women's rights is a broad one—arising in the private, civil, and political spheres. Several aspects of women's rights are discussed in other Chapters in this book. For example, sexual expression is one theme in Chapter 7 by Gardner; abortion and aspects of women's rights in family life are taken up by Eekelaar in Chapter 10; legal status in nationality and citizenship is discussed by Dummett in Chapter 11. The main focus of this chapter is the issue of women's rights at work and in related areas. It outlines the main legislative reforms since 1945 and discusses the concepts underpinning these provisions, with some of the critiques of these concepts. This is followed by a brief account of the political pressures which contributed to the legislative reforms, with particular reference to the participation of women. The chapter then deals more extensively with the scope of rights, through a discussion of domestic case-law and the significance of Community law and other international instruments. Some of the discrimination law aspects discussed in this chapter are relevant to Chapter 13 on race discrimination and readers are referred to it for discussion of common points. This chapter also attempts to assess the impact of the law by referring to empirical evidence of employment changes and the success— or otherwise—of law enforcement in the United Kingdom, with some reference to how the United Kingdom compares to other countries.

MAIN LEGAL DEVELOPMENTS

The most significant legal developments in respect of women's rights in employment since 1945 have been the Sex Discrimination Act 1975 and its amendments[1] and the Equal Pay Act 1970 and its amendments.[2] The former applies the principle of equal treatment to non-contractual employment matters, such as recruitment, promotion, training, and working conditions, and, in addition, to education and to the provision of housing, goods, facilities, and services. Direct and indirect discrimination on grounds of sex are prohibited and, in respect of the employment aspects of the legislation only, direct and indirect discrimination on grounds of married status. Victimization is also unlawful. The Act applies equally to women and men, although it is primarily aimed at providing rights for women who were considered 'more likely to be the victims of unfair treatment on grounds of sex'.[3] The Equal Pay Act applies to contractual terms and conditions including pay. As well as these laws, there have been some positive changes to the social security and taxation systems, excluded from the scope of the anti-discrimination legislation, and these are discussed below.

In administrative terms, two Equal Opportunities Commissions—one in Great Britain and one in Northern Ireland—were set up with the statutory duty to work towards the elimination of discrimination, to promote equality of opportunity generally between men and women, to keep under review the working of the Sex Discrimination Act and the Equal Pay Act, and, when so required by the Secretary of State or when the Commissions otherwise think it necessary, draw up and submit to the Secretary of State proposals for amendment to the legislation.[4]

In addition to domestic legislation, the European Community, both through its substantive law and the jurisprudence of the European Court of Justice, has played a critical role in shaping the development of the nature and standard of rights available in the United Kingdom. Article 119 of the Treaty of Rome 1957 required member states to implement the principle of equal pay for equal work and the Equal Pay Directive of 1975 further developed this to place member states under an obligation to ensure that women and men received the same pay for work of equal value,[5] the standard of the International Labour Organiza-

---

[1] Amendments include the Sex Discrimination Act 1986, the Sex Discrimination (Amendment) Order 1988 (SI 1988, No. 249), and the Employment Act 1989. In Northern Ireland the comparable legislation is the Sex Discrimination (Northern Ireland) Order 1976, the Sex Discrimination (Amendment) Order 1987, the Sex Discrimination (Amendment) (Northern Ireland) Order 1988 and the Employment Act 1989. The chapter will refer to British legislation, except where necessary in relation to cases in Northern Ireland.

[2] Amended by the Equal Pay (Amendment) Regulations 1983; Industrial Tribunal Regulations 1983, SI 1983/1794; and in Northern Ireland by the Equal Pay (Amendment) Regulations (Northern Ireland) 1984.

[3] *Equality for Women*, Cmnd. 5724. (1974), para. 1.

[4] SDA 1975, s. 53; SD(NI)O 1976, Art. 54.

[5] Council Directive 75/117/EEC of 10 Feb. 1975 on the approximation of laws of the member states relating to the application of the principle of equal pay for men and women, *OJ* L45/19, 19 Feb. 1975.

tion.[6] In 1976 the Equal Treatment Directive was adopted, guaranteeing the principle of equal treatment in access to employment, vocational training, promotion, and working conditions.[7] For the purposes of this Directive, the principle of equal treatment means that there shall be no discrimination whatsoever on grounds of sex, either directly or indirectly, by reference in particular to marital or family status.[8]

Three other equality directives have also been adopted by the Council: in 1979, one concerning the progressive implementation of the principle of equal treatment in statutory social security schemes;[9] in 1986, one concerning the implementation of the principle of equal treatment in occupational social security schemes;[10] and, also in 1986, one concerning equal treatment for women and men in self-employed occupations including agriculture.[11] Thus, a platform of rights has been created by the Community which has had a major impact on the scope of rights available in the United Kingdom, as will be demonstrated in the substantive discussion which follows.[12] Indeed, Community law has become increasingly important over the last decade in the United Kingdom, and recent caselaw from the European Court indicates that the influence of European-derived rights will have a greater impact in the coming years.[13]

CONCEPTS

There is considerable legal and political controversy over the meanings of discrimination and the proper scope of action against it. These disputes are discussed extensively elsewhere by McCrudden[14] and they are also dealt with in chapter 13 of this book. In respect of the aims of anti-discrimination law, one dominant approach promotes the view that the proper aim of the law is the establishment of fair processes or the

---

[6] ILO Convention 100 of 1951 concerning Equal Remuneration for Men and Women Workers for Work of Equal Value, which entered into force on 23 May 1953.

[7] Council Directive of 9 Feb. 1976 on the implementation of the principle of equal treatment for men and women as regards access to employment, vocational training and promotion and working conditions, *OJ* L39/40, 14 Feb. 1976.

[8] Ibid., Art. 2(1).

[9] Council Directive of 19 Dec. 1978 on the progressive implementation of the principle of equal treatment for men and women in matters of social security, *OJ* L6/24, 10 Jan. 1979.

[10] Council Directive of 24 July 1986 on the implementation of the principle of equal treatment in occupational social security schemes, *OJ* L225/40, 12 Aug. 1986.

[11] Council Directive of 11 Dec. 1986 on the application of the principle of equal treatment between men and women engaged in an activity, including agriculture, in a self-employed capacity, and on the protection of self-employed women during pregnancy and motherhood, *OJ* L359/56, 19 Dec. 1986.

[12] See also C. McCrudden (ed.), *Women, Employment and European Equality Law* (1987).

[13] *Marleasing SA* v. *La Commercial Internacional de Alimentacion SA*, Case C-106/89, [1990] ECR 4125 (ECJ).

[14] C. McCrudden (ed.), *Anti-Discrimination Law* (1991).

elimination of the harmful consequences of decisions based on prejudice.[15] It is essentially an individualistic approach, concentrating on securing fairness for individuals, by removing specific arbitrary obstacles which lead to less favourable treatment of one individual compared to another.[16] Law based on this approach is expressed in universalistic terms and requires equal treatment for women and men in similar situations, to bring about the same chance of success for similarly situated women and men. Such an approach has its roots in classical liberal theory, which presupposes that people start out as equals with the freedom to make rational choices about their destinies, or that they would do so if formal juridical barriers were removed.[17] The approach is evidenced in the campaigns for equality for women over the last two centuries, which aimed to remove the formal legal barriers which denied women the same access as men to education, employment, and suffrage.[18]

A second approach moves away from the process of decision-making and its impact on individuals to a concern with the outcome of decisions and practices, with the ways in which it impacts on groups. The proper aim of anti-discrimination law is to ensure an improvement in the social and economic position of disadvantaged groups or to ensure a redistribution of benefits and opportunities from advantaged groups to disadvantaged ones.[19] The approach recognizes the importance of differences between groups and it requires an understanding of the characteristics of particular groups and of their social and historical experience.[20] The view that a legitimate reason for legislative intervention is that of distributive justice also has liberal origins and this approach requires, like the former one, comparisons either between a woman and a man or between women and men as groups.

The sex discrimination legislation of the 1970s combines elements of both these broad liberal approaches within its scope, which is unsurprising given the influence of the United States on the development of anti-discrimination law generally[21] and the political links made between the

---

[15] e.g. P. Brest, 'In Defence of the Antidiscrimination Principle' (1976) 90 *Harvard LR* 1; and see K. O'Donovan and E. Szyszczak, *Equality and Sex Discrimination Law* (1988) for discussion of M. Weber, *Law, Economy and Society* (1954).

[16] See R. H. Fallon and P. C. Weiler, 'Firefighters v. Stott: Conflicting Models of Racial Justice' (1984) *The Supreme Court Rev.* 1; J. Gardner, 'Liberals and Unlawful Discrimination' (1989) 9 *Oxf. J. Legal Studies* 1; R. S. Wasserstrom, 'Racism, Sexism and Preferential Treatment: An Approach to the Topics' (1977) 24 *UCLA LR* 581.

[17] See K. O'Donovan, *Sexual Divisions in Law* (1985), ch. 7.

[18] See A. Carter, *The Politics of Women's Rights* (1988), for general account.

[19] O. M. Fiss, 'Groups and the Equal Protection Clause' (1976) 5 *Philosophy and Public Affairs* 107; Gardner, 'Liberals'.

[20] C. McCrudden, 'Institutional Discrimination' (1981) 2 *Oxf. J. Legal Studies* 303; L. Lustgarten, 'The New Meaning of Discrimination' (1978) *PL* 178.

[21] See E. Meehan, *Women's Rights at Work: Campaigns and Policy in Britain and the United States* (1985); J. Gregory, *Sex, Race and the Law: Legislating for Equality* (1987), generally.

legal nature of race and sex discrimination legislation at the time.[22] The individual approach may be seen in the provisions relating to direct discrimination and equal pay for like work, which are premised in the legislation on direct comparison between the treatment of similarly situated persons; the distributive justice approach may be seen in the provisions relating to indirect discrimination, positive action, and, to an extent, equal pay for work of equal value, which allow for taking group membership into account.

McCrudden discusses some of the challenges to these liberal approaches in chapter 13, but it is worth mentioning here some of the challenges from feminist perspectives. The diversity of feminist thought is considerable. Liberal feminism and welfare-socialist feminism adhere to the view that women's rights can be advanced within a broadly democratic framework and thus that improvements are required to the operation of existing legislation. Marxist-socialist feminist thought, which locates the primary source of women's oppression in the economic system, promotes the view that equality will not be attained unless capitalism is overthrown. Radical feminist thought, which locates women's inequality in the sexual and political power of men over women, requires a different kind of transformation.[23] Each of the various strands of feminism implies different views of the role of anti-discrimination legislation.

One of the most significant feminist challenges to the anti-discrimination legislation is to its comparative approach which, while facilitating the removal of formal barriers to equality, is considered limited because women are not in a position to compete with men on an equal footing due to their reproductive capacity, domestic responsibilities, and historical disadvantage.[24] Particularly problematic has been the minimization of the significance of biological differences between women and men, which has created real problems for the way in which discrimination on grounds of pregnancy has been dealt with.[25]

The lack of attention given to the private sphere of social organization and family life and women's subordination within it has also been the subject of criticism, as too has the fact that the legislation contains little to address the problem that the law and various institutions, for example

---

[22] K. Karst, 'Women's Constitution' (1984) *Duke LJ* 447; McCrudden, *Anti-Discrimination*, p. xxiv.

[23] These are, of course, overly simplistic descriptions. The complexities of feminist politics and the difficulty of labelling them are treated elsewhere, e.g. V. Randall, *Women and Politics*, 2nd edn. (1987); E. Meehan and S. Sevenhuijsen (eds.), *Equality Politics and Gender* (1991); Carter, *Politics*.

[24] O'Donovan and Szyszczak, *Equality*, ch. 3; Gregory, *Sex, Race*, ch. 1.

[25] N. Lacey, 'Note: Dismissal by Reason of Pregnancy' (1986) 15 *Industrial LJ* 43; C. Bacchi, 'Pregnancy, the Law and the Meaning of Equality', in Meehan and Sevenhuijsen, *Equality Politics*; see also S. Law, 'Rethinking Sex and the Constitution' (1984) *Univ. Penn. LR* 953.

the labour market, have been designed by men to suit male needs and maintain male privilege.[26] The limited number of tools in the legislation which had the potential to work towards dismantling institutionalized discrimination have, as will be seen later, been hampered in their effectiveness by an unsympathetic judiciary and an inability, or reluctance, on the part of the statutory agencies to capitalize fully on their powers.

POLITICAL PRESSURES

Despite the passage of the Sex Disqualification (Removal) Act in 1921, women have, throughout the twentieth century, been poorly represented in the House of Commons and in government. Bodies such as the Council of Europe, the United Nations, and the Inter-Parliamentary Union are concerned to encourage greater participation by women everywhere but the proportion of women in the United Kingdom Parliament remains, at 9 per cent, amongst the lowest in the European Community countries and in the bottom half of the league table for parliaments across the world, despite Britain's relatively early democratization.[27] During the 1960s and 1970s, the period of the major sex equality reforms, the numbers of women in the House of Commons fluctuated between 19 and 29 and very few held ministerial office in or out of the Cabinet.[28] At critical points, however, there were key women ministers with responsibility for the laws discussed in this chapter. There were also redoubtable women backbenchers and members of the House of Lords who contributed to the advent and content of legislation. The former were Barbara Castle, Secretary of State for Employment between 1968 and 1970 and Shirley Summerskill, Under-Secretary in the Home Office between 1974 and 1979; the latter included Lady Seear as a particularly significant promoter of change.[29] In addition, women's groups generally have played a part in the lobby for, and the content of, domestic legislation.

Prior to the major reforms of the 1960s and 1970s, liberal feminists had

---

[26] O'Donovan, *Sexual Divisions*; K. Mackinnon, *Feminism Unmodified: Discourses on Life and Law* (1987).

[27] Council of Europe, *The Situation of Women in the Political Process in Europe*, part ii. *Women in the Political World in Europe* (1984); Inter-Parliamentary Union, *1989 Participation of Women in Political Life and in the Decision-Making Process* (1988); UNESCO, *Meeting of Experts (Cat IV) for the Europe Region to Examine Ways in Which Women May Exert a More Effective Influence on the Action of Public Authorities and Decision-Making Processes: Oslo, Norway, 5–9 February 1990 Final Report* (1990); Woman of Europe Newsletter No. 32, March 1993.

[28] House of Commons, *Factsheet no. 5: Women in the House of Commons* (1990).

[29] See Meehan, *Women's Rights*, chs. 2 and 3.

been active at the end of the Second World War in advocating equal compensation for women and men who had been wounded.[30] This campaign flowed into another for equal pay in civilian life for teachers and public servants. The law was reformed in this respect between the late 1940s and the early 1950s.[31] Liberal feminists of this period also recognized what was at the forefront of welfare-socialist feminist thinking in the 1960s and 1970s—that equal treatment in pay and recruitment would be meaningless without other policies that took account of the different situations of men and women, for example, in taxation, family, and income maintenance policies. Nevertheless, the main political pressure during the late 1960s and early 1970s from both kinds of feminists and the labour movement was for equal opportunities in employment and education.[32] As one would expect, the content of the legislation reflects diverse influences: an interplay of ideas about equality, economic rationality, and justice, the interests of ministers and producer groups, and the various strands of feminist thinking that informed the public debates at that time.

The Equal Pay Act did not originate solely from feminist pressure but the women's movement played a part. At this time, Marxist and radical strands of feminism had not yet emerged into the wider public domain. Liberals and welfare socialists supported equal pay strikes in the 1960s and early 1970s and when Barbara Castle became Secretary of State for Employment she accelerated governmental action because she saw the law as appealing to the interests of working-class women. The Confederation of British Industry (CBI) tried to persuade the Government to restrict the bill to the right to equal pay for equal work, the European Community standard contained in Article 119 of the 1957 Treaty of Rome. The Trades Union Congress (TUC) wanted the International Labour Organization's standard of equal pay for work of equal value, as did those feminists whose analysis primarily highlighted the different occupational situations of women and men and the need for distributive justice.[33] The result was a compromise that made some acknowledgement of the different recruitment patterns of women and men, a compromise which lasted until 1983 when the United Kingdom had to amend its legislation to introduce the right to equal pay for work of equal value to comply with its obligations under the European Community's Equal Pay Directive of 1975.[34]

---

[30] Carter, *Politics*, ch. 1.
[31] *Royal Commission on Equal Pay, 1944–6 Report*, Cmnd. 6937 (1946).
[32] Meehan, *Women's Rights*, ch. 2.
[33] Ibid., ch. 3.
[34] *Commission* v. *UK*, Case No. 61/81, [1982] ECR 2601 (ECJ).

Implementation of the 1970 Equal Pay Act was voluntary for five years and women's groups tried, unsuccessfully, to speed up the process of compliance. But between 1970 and 1972 their main aim was to draw attention to wider anti-discrimination bills that were being proposed by back bench members of Parliament and to influence their substantive content by giving evidence to the House of Lords Select Committee, which was considering a bill proposed by Lady Seear.[35] Lady Seear's approach appealed to an economic rationality that encouraged, instead of inhibited, the use of labour, combined with notions of justice. The pressure was intensified when the Government announced that it would bring in its own legislation.[36] Three hundred groups and individuals responded to its consultative document. In 1974, when Labour replaced Conservatives in office, feminists, trade unionists, and the National Council for Civil Liberties co-operated, often under the auspices of the Fawcett Society, to strengthen the new Government's proposals, covering employment, education, and the provision of goods, facilities, and services. At least two key improvements can be attributed partly to this activity and partly to the persuasiveness to the Home Secretary, Roy Jenkins, of American ideas: the inclusion of indirect discrimination and a limited version of positive action.[37] Although the conception of the Sex Discrimination Act owes a great deal to Roy Jenkins, it was Shirley Summerskill who did much to secure it a safe passage through the House of Commons.[38]

There was one major disagreement among those campaigning for legislation, concerning the future of protective legislation. The most classically liberal feminists shared, for different reasons, the CBI view that the same treatment should be achieved by removing restrictions on women's conditions of employment, as they believed that protective legislation was being used as an arbitrary barrier to circumvent equality. The TUC and other feminists wanted the same treatment by having protection extended to men on the grounds that working-class employees, and working-class women in particular, would be rendered even more vulnerable to occupational hazards if protection were to be removed.[39] While the Sex Discrimination Act left the existing protective legislation intact, the Equal Opportunities Commissions were asked to make recom-

[35] Meehan, *Women's Rights*, 66.

[36] *Equality for Women*.

[37] See Meehan, *Women's Rights*, ch. 3; Gregory, *Sex, Race*; also C. Callendar, 'The Development of the Sex Discrimination Act, 1971–75' (unpublished dissertation presented at Bristol Univ. 1978).

[38] Meehan, *Women's Rights*, 55.

[39] Carter, *Politics*, 122.

mendations on how the matter should be dealt with,[40] in line with the obligations which would be forthcoming under the European Community's Equal Treatment Directive. Subsequent changes in this area are discussed below.

Pressure groups continue to lobby government for changes to the sex discrimination and equal pay legislation; they have also been active around issues not covered by this legislation, particularly for example in respect of care of dependent relatives, including children, social security benefits, and taxation.[41] Success has been limited under successive Conservative governments since 1979[42] although, for the first time, a Cabinet Minister was designated with responsibility for women's issues following the 1992 general election and a Cabinet Subcommittee, chaired by this minister, was set up to review and develop policy on women's issues and oversee its implementation.[43] Increasingly, women's groups look to international platforms on which to organize their demands—such as during the United Nations Decade for Women, 1976–85—and, since 1987, women's groups in the United Kingdom have been active in the setting up of a European Women's Lobby, which aims to lobby the Community's institutions for improvements to standards of rights available at Community level.[44]

SCOPE OF RIGHTS

This section examines the meaning of discrimination contained in the legislation and its coverage; the scope and availability of the right to equal pay; changes to provide equal treatment in the social security system; and a number of specific issues which have proved particularly significant since the introduction of the legislation. It also briefly discusses the exceptions to the principle of equal treatment permitted by the legislation.

[40] SDA 1975, s. 51; s. 55.
[41] E. Meehan, 'British Feminism from the 1960s to the 1990s', in H. Smith, *British Feminism in the Twentieth Century* (1991).
[42] D. Bouchier, *The Feminist Challenge: The Movement for Women's Liberation in Britain and the United States* (1983); J. Lovenduski and V. Randall, *Contemporary Feminist Politics* (1993).
[43] (1992) 44 *Equal Opportunities Rev.* 2; and see 'Shepherd to press for woman's opportunities' *Independent*, 21 May 1992.
[44] C. Hoskyns, 'Women, European Law and Transnational Politics', 1986 *Int. J. Sociol. of Law* (special issue); S. P. Mazey and J. J. Richardson, 'British Pressure Groups in the European Community: The Challenge of Brussels', (1992) 45 *Parliamentary Affairs* 92–107; C. Hoskyns, 'The European Women's Lobby' (1991) 38 *Feminist Rev.* 67–9.

The meaning of discrimination in the Sex Discrimination Act 1975 and its coverage

The Sex Discrimination Act provides that 'A person discriminates against a woman in any circumstances relevant for the purposes of any provision of this Act if . . . on the ground of her sex he treats her less favourably than he treats or would treat a man'.[45] This has come to be known as 'direct discrimination' and, although it has been criticized as having limited usefulness, a number of important issues have been raised concerning its scope.[46] In the early days of interpretation, for example, there was some doubt as to whether a discriminatory motive constituted an essential element in determining whether particular treatment was unlawful within the meaning of the Act. In *Peake* v. *Automotive Products Ltd.*,[47] the Court of Appeal held that male workers were not discriminated against on grounds of their sex where the employer allowed women workers to leave work five minutes earlier than the male workers, for reasons of safety. It considered that the Sex Discrimination Act could admit instances where women and men were treated differently but for a good reason. However, in *Grieg* v. *(1) Community Industry and (2) Ahern*,[48] the Employment Appeal Tribunal held that the motive for the behaviour leading to less favourable treatment was irrelevant, distinguishing *Peake* on its facts, and subsequently the Court of Appeal itself, in *Jeremiah* v. *Ministry of Defence*,[49] held that it would not follow the reasoning in *Peake*. In *R.* v. *Birmingham City Council*,[50] the House of Lords held that the correct test for interpreting 'on ground of sex' is an objective one— less favourable treatment is not saved from being unlawful by the fact that there was a 'benign' motive for the treatment.

Questions concerning motive and intention continue to be handled with some difficulty by the courts and the House of Lords dealt with it again in 1990, following *R.* v. *Birmingham City Council*, in *James* v. *Eastleigh Borough Council*.[51] The Court of Appeal had construed the phrase 'on ground of sex' as implying a subjective examination of the reasons for the discriminatory behaviour, but the House of Lords held that the appropriate test was whether the complainant would have received the same treatment from the respondent 'but for' his or her sex. According to Lord Goff, the 'but for' test is advantageous because it covers decisions overtly

---

[45] SDA 1975, s. 1(1)(*a*).
[46] O'Donovan and Szyszczak, *Equality*, ch. 3.
[47] [1978] QB 233; [1977] ICR 968 (CA).
[48] [1979] IRLR 158 (EAT).
[49] [1980] QB 87; [1979] IRLR 436 (CA).
[50] *R.* v. *Birmingham City Council, ex parte Equal Opportunities Commission* [1989] AC 1155; [1989] IRLR 173 (HL).
[51] [1990] 2 AC 751; [1990] IRLR 288 (HL).

or covertly based on sex as well as those founded on gender-based criteria. Also, according to Lord Goff, 'it avoids, in most cases at least, complicated arguments relating to concepts such as intention, motive, reason or purpose, and the danger of confusion arising from the misuse of those elusive terms'.[52] The dissenting opinions in this case are strong, however, and further discussion on this point seems likely.[53] What is clear is that less favourable treatment 'on ground of sex' can include stereotypical assumptions about the characteristics of one sex, and this is of value in tackling some of the outdated assumptions which can limit women's employment opportunities.[54]

A second element of the direct discrimination provision concerns 'less favourable treatment'. Despite some concerns in the early years that a claimant had to prove that she was subjected to different treatment than a person of the opposite sex had been or would have been accorded *and* that the different treatment caused them detriment,[55] the House of Lords held in 1989 that, in order to establish less favourable treatment, it is enough to show that the claimant was deprived of a choice which was valued by her and is one valued, on reasonable grounds, by many others. It is not necessary to prove that what was lost was 'better'.[56]

Finally, in relation to direct sex discrimination, there are similar difficulties relating to the burden of proof, as those discussed in chapter 13 relating to direct race discrimination. Shifting the burden of proof to the employer to produce non-discriminatory reasons for the actions has the support of the Equal Opportunities Commissions,[57] and indeed the European Commission, which produced a Proposal for a Directive on this issue in 1988.[58] This has not yet been adopted by the Council.

What has come to be known as 'indirect discrimination' is set out in section 1(1)(*b*) of the Sex Discrimination Act:

A person discriminates against a woman in any circumstances relevant for the purposes of any provision of this Act if he applies to her a requirement or condition which he applies equally to a man but (i) which is such that the

[52] Ibid., at 774; 295.

[53] See comment on case in (1990) 33 *Equal Opportunities Rev*. 42–4, which refers to the decision as one of 'transcedent importance'; and also the dissenting judgment of Lord Lowry at [1990] IRLR 295–8 for a discussion of the issues. See also the criticism of the EAT's analysis of the implications of the *James* decision in *Bullock* v. *Alice Ottley School* [1991] 1 CR 838 (EAT) in (1991) 39 *Equal Opportunities Rev* 44–45. The decision was reversed by the AC [1992] IRLR 564.

[54] e.g. *Horsey* v. *Dyfed County Council* [1982] IRLR 395 (EAT); *Coleman* v. *Skyrail Oceanic Ltd*. [1981] IRLR 398 (CA).

[55] O'Donovan and Szyzsczak, *Equality*, 65.

[56] *R*. v. *Birmingham City Council*.

[57] See Equal Opportunities Commission, *Equal Treatment for Men and Women: Strengthening the Acts* (1988), 21–2.

[58] COM (88) 269 final, 24 May 1988.

proportion of women who can comply is considerably smaller than the proportion of men who can comply with it, and (ii) which he cannot show to be justifiable irrespective of the sex of the person to whom it is applied, and (iii) it is to her detriment that she cannot comply with it.

Importantly, as already noted, this provision provides room for arguments which take into account differences between the sexes, but how has it been translated into practice? There are three main elements in the indirect discrimination provision.

First, there is a 'requirement or condition' to be complied with. The courts over the years have adopted a fairly broad approach to the range of practices and factors which might be considered a 'requirement or condition': age bars, promotion procedures, the obligation to work full-time, and implicit conditions that candidates should not have young or dependent children.[59] The courts have held practices not laid down formally as requirements or conditions to be such within the meaning of the Act.[60] In *Clarke* v. *Eley (IMI) Kynoch Ltd.*[61] the Employment Appeal Tribunal held that it was important to construe the words broadly in order to further the statutory purpose of rendering unlawful any discriminatory practices which could not be justified on grounds other than sex. The purpose of the legislature in introducing the concept of indirect discrimination was to seek to eliminate those practices which had a disproportionate impact on women and were not justifiable for other reasons.

However, in later years a stricter approach has been adopted. In a race discrimination case, *Perera* v. *Civil Service Commission*,[62] the Court of Appeal held that a requirement or condition means something which has to be complied with in the sense that non-compliance would mean an absolute bar. It is a 'must'. The Court reiterated this in 1988 in *Meer* v. *London Borough of Tower Hamlets*.[63] Byre has criticized this line of thinking as potentially undermining the entire indirect discrimination concept, since it makes it difficult to challenge practices or criteria which are not absolute bars, such as preferences for certain qualifications, but which may be decisive in an employment decision.[64] However, not all recent decisions have been so restrictive. The decision, in 1990, of the Northern Ireland Court of Appeal in *Briggs* v. *North Eastern Education and Library Board*[65] is to be welcomed as an important step in removing

[59] e.g. *Price* v. *Civil Service Commission (No. 2)* [1978] IRLR 3; *Home Office* v. *Holmes* [1985] 1 WLR 71; [1984] IRLR 299 (EAT); *Hurley* v. *Mustoe* [1981] IRLR 208 (EAT).

[60] *Steel* v. *Union of Post Office Workers* [1977] IRLR 288 (EAT).

[61] [1982] IRLR 482 (EAT).

[62] *Perera* v. *Civil Service Commission and Dept. of Customs and Excise (No. 2)* [1983] IRLR 166 (CA).

[63] [1988] IRLR 399 (CA).

[64] A. Byre, *Indirect Discrimination* (1987), 25.

[65] [1990] IRLR 181 (NICA).

some of the other difficulties faced by complainants on this issue. The Court held that the fact that the nature of a particular job requires full-time attendance does not prevent that being a requirement which is applied by the employer—this goes to the employer's justification for imposing the requirement rather than whether the requirement exists.

In relation to proving the second necessary element—that a smaller proportion of women than men can comply with the requirement or condition—a number of difficulties has arisen in the way this has been interpreted. The need for a claimant to show that she can or cannot comply with a requirement or condition has been held to mean that she can *in practice* comply with it, not that she can, in the sense of it being a theoretical possibility. This principle was clarified early on, in *Price* v. *Civil Service Commission (No. 2)*,[66] a case concerning upper age limits for recruitment into the Civil Service. In determining whether women can comply with the requirement or condition, it was held to be relevant to take into account the current usual behaviour of women in this respect. However, in *Turner* v. *Labour Party and the Labour Party Superannuation Scheme*,[67] the Court of Appeal adopted a more restrictive interpretation. The case concerned benefits which were to accrue at a future date—the pension scheme at issue provided pensions automatically for surviving spouses on the death of a member but applied conditions in respect of an unmarried member. This was held not to be to the detriment of the claimant (a divorcee) because the conditions for benefit from the pension fund were to be satisfied at a future date and it could not be said that the claimant could not marry in the future.

For years, there were no clear judicial guidelines on how to establish that the proportion of one sex who could comply with the condition or requirement was considerably smaller than that of the other sex. The High Court did hold, however, in a 1987 case concerning the allocation of education grants, that in defining the pool for comparison, the risk of incorporating an act of discrimination into the definition must be avoided.[68] In *Jones* v. *Chief Adjudication Officer*,[69] the Court of Appeal has also provided guidelines on the process of reasoning necessary to establish adverse impact. However, unlike the approach to 'disparate impact' cases in the United States, courts in the United Kingdom have been somewhat reluctant to rely heavily on statistical evidence, so it remains to be seen how the *Jones* formulation will develop in practice.[70]

---

[66] [1978] IRLR 3.

[67] [1987] IRLR 101 (CA).

[68] *R.* v. *Secretary of State for Education, ex parte Schaffter* [1987] IRLR 53.

[69] [1990] IRLR 533 (CA).

[70] Compare *Price* v. *Civil Service Commission (No. 2)* [1978] IRLR 3, where statistical evidence was accepted, with *Perera* v. *Civil Service Commission* [1983] ICR 428, where the EAT opined that Industrial Tribunals should not be over concerned with elaborate statistical evidence.

Further problems relating to the establishment of adverse impact concern the lack of judicial determination on what may constitute a 'considerably smaller' proportion[71] and also the lack of data on the workforce (and the potential workforce) generally.[72]

The third and final element of the indirect discrimination provision which needs to be proved in order for a prima-facie case to be established is that of personal detriment.[73] There is, however, some confusion about what the applicant must show to prove detriment—for example, over the separateness of the tests of 'can comply' and proof of detriment and over whether a claimant has to show that she would have gained something if she had been able to comply.[74] For example, in *Steel* v. *Union of Post Office Workers and the General Post Office*,[75] the fact that in the fullness of time the complainant might have been able to comply with the requirement or condition was irrelevant; it was the date at which the requirement or condition must be fulfilled which was relevant for establishing detriment. In both *Turner*[76] and *Clymo*,[77] however, it was held that no detriment had been established, as in both cases the complainants were deemed as capable of complying—in the former, possibly at some stage in the future, as noted above, and in the latter in circumstances in which she had been offered child-minding by the employers which it was considered that she could afford. It is clear that much clarification is required here and, indeed, in the *Briggs* decision, the Northern Ireland Court of Appeal criticized the decision in *Clymo* and ruled that the requirement in question was to the complainant's detriment because she could not comply with it in practice, a construction must to be preferred.[78]

Once a claimant has established the three necessary elements of indirect discrimination, the burden of proof moves to the employer and effectively the issue becomes one of whether the discriminatory policy or practice is justifiable. This is a question of fact for the courts, which originally employed a strict interpretation, that it had to be justified by the needs of the business or enterprise, not just its convenience, for example.[79] If there were other ways of achieving a particular aim by the employer,

---

[71] In the USA, a 'four-fifths' rule is used, whereby adverse impact is shown if there is a 20% difference in impact between the groups involved. See O'Donovan and Szyzsczak, *Equality*, 104–8 generally.

[72] J. Carr, *New Roads to Equality: Contract Compliance for the UK?* (1987), 21.

[73] It was clearly the intention of Parliament that a complainant should have a personal *locus standi* to bring a complaint: Official Reports, HC, 18 June 1975, 1491–2.

[74] See O'Donovan and Szyzsczak, *Equality*, 108–10 for fuller discussion.

[75] See n. 60.

[76] See n. 67.

[77] *Clymo* v. *Wandsworth London Borough Council* [1989] IRLR 241 (EAT).

[78] See n. 65.

[79] See n. 60.

which did not involve indirect discrimination, then the employer could not argue that the policy was justifiable.[80] However, following a number of race discrimination cases, the justifiability test was watered down somewhat—a test of 'reasonableness' crept in, thus undermining both the objectivity and the strictness of the test.[81]

The European Court of Justice clarified the scope of the justifiability defence in Community law in 1987. In a landmark decision concerning unequal access for part-time workers—predominantly women—to a company's occupational pension scheme, the Court in *Bilka-Kaufhaus*[82] held that for such a policy to be justified, it must be explained by objectively justified factors unrelated to any discrimination on grounds of sex. To satisfy this, the national courts must ask whether the measures chosen correspond to a *real need* on the part of the undertaking, whether the measures are an *appropriate means* of achieving the objective pursued, and whether they are *necessary* to achieve the objective. Since then United Kingdom courts have followed the guidance provided and reverted to a stricter approach. The House of Lords in *Rainey* v. *Greater Glasgow Health Board*[83] adopted the test laid down in the *Bilka* case as the correct standard for interpreting the justifiability test in the Sex Discrimination Act.

The scope of the right to equal pay

Under the Equal Pay Act 1970, employers are placed under an obligation to pay women the same as men when they are doing the same or broadly similar work, in the same establishment or in different establishments of the same employer where these are covered by common terms and conditions.[84] Equal pay is also required where women and men are doing work rated as equivalent if, but only if, their jobs have been given an equal value, in terms of the demands made on the workers under various headings (for instance, effort, skill, decision), in a study intended by the employer to evaluate in those terms the job to be done by all or any of the employees in an undertaking or group of undertakings.[85] To these provisions was added in 1983, the right to equal pay for work of

---

[80] *Price* v. *Civil Service Commission* [1977] 1 WLR 1417; [1977] IRLR 291 (EAT).

[81] See e.g. *Singh* v. *Rowntree Mackintosh Ltd.* [1979] IRLR 199; *Ojutiku and Oburoni* v. *Manpower Services Commission* [1982] IRLR 418; *Orphanos* v. *Queen Mary College* [1985] AC 761; [1985] IRLR 349.

[82] *Bilka-Kaufhaus GmbH* v. *Weber von Hartz*, Case No. 170/84, [1986] IRLR 317 (ECJ).

[83] [1987] AC 224; [1987] IRLR 26 (HL).

[84] Equal Pay Act 1970 (as amended), s. 1(4).

[85] Ibid., s. 1(5).

equal value, even where the employer had not conducted such a study.[86] The Regulations enacting these provisions were reluctantly introduced following successful infringement proceedings against the United Kingdom by the European Commission for its failure, contrary to its obligations under the Equal Pay Directive, to provide a right for a woman to claim equal pay for work of equal value to a male comparator unless a job evaluation scheme had been carried out.[87]

Unfortunately, progress towards the attainment of equal pay remains slow and it is clear that there will continue to be a need for national courts to rely on Community law in this area to interpret national legislation and, where necessary, to refer questions of interpretation of the scope of European obligations to the European Court of Justice.

This has occurred with respect to several issues. In relation to the meaning of 'pay', for example, European Community law specifies that pay is 'the ordinary basic or minimum wage or salary and any other consideration, whether in cash or in kind, which the worker receives, directly or indirectly, in respect of his employment from his employer'.[88] Thus its scope is broad, covering all aspects of the remuneration package, and this issue has been subject to clarification by the European Court on numerous occasions, including in a number of references from courts in the United Kingdom.[89] Yet in the first case to proceed through the national courts under the equal value provisions, there was significant cause for concern about the interpretation given to 'pay' by the lower courts. In *Hayward* v. *Cammell Laird Shipbuilders Ltd.*,[90] the House of Lords had to overturn the decisions of the three lower courts, stating clearly that they had wrongly concluded that an employer does not have to pay a woman employed on work of equal value the same basic hourly wage or overtime rates as her male comparator if, considered as a whole, her remuneration is not less favourable. The House of Lords held that each term of the contract could be compared, rather than the totality of the contract.

The basic approach of the equal value legislation is one of job content, the examination of one employee's job compared to another's. There is some guidance in the legislation as to what may be considered relevant—factors such as effort, skill, and decision—but beyond that there is no guidance on how to go about evaluating job content. There are both

---

[86] Ibid., s. 1(2)(c).
[87] See n. 34.
[88] Art. 119, Treaty of Rome 1957.
[89] e.g. *Garland* v. *British Rail Engineering Ltd.*, Case No. 12/81 [1982] ECR 359; *Worringham* v. *Lloyds Bank Ltd.*, Case No. 69/80 [1981] ECR 797; *Newstead* v. *Dept. of Transport.* Case No. 192/85 [1988] 1 All ER 129; *Barber* v. *Guardian Royal Exchange Assurance Group*, Case No. 262/88, [1990] IRLR 240 (ECJ).
[90] [1988] IRLR 257 (HL).

conceptual problems and practical problems arising from this approach. Conceptually, job evaluation is a system which essentially developed with an emphasis on securing the agreement of those involved in the scheme on what may be considered acceptable. Despite its apparent objectivity, therefore, it involves a number of value judgements, each of which may result in discriminatory evaluation of women's work.

In practical terms, the technical issues involved in evaluating jobs under a job evaluation scheme have given rise to lengthy arguments about how a job may be evaluated. The Court of Appeal in *Bromley* v. *Quick*[91] has provided some guidance on what may constitute an adequate job evaluation scheme for the purposes of the Act: it must be analytical, non-discriminatory, and apply to the jobs of both the claimant and the comparators. Some guidance has also been provided by the European Court of Justice,[92] but there is a real need for clarification of the type of factors which should be taken into account and the weight to be given to them, as well as training for tribunal members in the meaning of job evaluation.[93]

The scope of the comparison available to a claimant is limited to a man in the 'same employment', which can include employment by an associated employer or at another establishment at which common terms and conditions are observed.[94] A claimant has to establish that she is engaged on work of equal value to that done by an actual (not hypothetical) man, although the comparison can be made with a male predecessor.[95] The House of Lords has held that the existence of a man employed on like work to a woman does not preclude her from making a claim for equal value with another man. It recognized that an employer could otherwise avoid the obligations of the equal value provisions by employing a token man on the same work as a group of female applicants who were paid less than a group of men employed on work of equal value to the women.[96]

The limits imposed in the legislation in respect of comparisons are problematic, because of existing labour market segregation and the increased use of forms of atypical employment such as sub-contracted services, where women are unable to claim equal pay for work of equal

---

[91] [1988] IRLR 249 (CA).

[92] e.g. *Rummler* v. *Dato-Druck Gmbh*, Case No. 237/85 [1987] IRLR 32 (ECJ); *Handels-og Kontorfunktionaerernes Forbund i Danmark* v. *Dansk Arbejdgiverforening (acting for Danfoss)*, Case No. 109/88, [1989] IRLR 532 (ECJ).

[93] The European Commission announced, in its *Third Action Programme on Equal Opportunities, 1991–1995*, that it intended to produce a Memorandum to define the scope and content of equal pay for work of equal value and to give guidance on the criteria to be taken into account in job evaluation and job classification. COM (90) 449 final, 6 Nov. 1990, p. 8.

[94] Equal Pay Act 1970 (as amended), s. 1(6)(*c*).

[95] *Macarthys Ltd.* v. *Smith*, Case No. 129/79 [1980] IRLR 210 (ECJ).

[96] *Pickstone* v. *Freemans PLC* [1988] 3 WLR 265; [1988] IRLR 357 (HL).

value with the men they work alongside as they are not technically employed by the same employer.[97] There is also a question, as yet untested, as to whether the limits to the scope of comparison are lawful under Community law.[98]

The defences available to employers under the equal pay provisions are also problematic. Under the original 1970 Act, an employer could argue that there was a genuine material *difference* which explained any pay differential and this was interpreted as meaning factors personal to the employees concerned, for example merit, qualifications, or length of service.[99] The Government extended the scope of the defence under the equal value provisions, to allow employers to argue that there was a genuine material *factor* explaining the difference and it foresaw that this would include issues going beyond any personal factors between the woman and the comparator, such as skill shortages or other market forces.[100] In the event, the difference in the defences available is probably no longer significant as the House of Lords, in *Rainey* v. *Greater Glasgow Health Authority*,[101] made no distinction between the two forms of defence but allowed for variations other than personal factors to be used in 'like work' and 'work rated as equivalent' claims. It accepted a market forces defence in this case, a 'like work' claim.

The extent to which the genuine material factor defence is allowed has been clarified at European level, in *Bilka-Kaufhaus*, as discussed above, which laid down a test for the national courts to follow in adjudicating the justification of a particular practice or policy which has an adverse impact on women. While this test should lend uniformity to the approach to such cases by national courts, it is clear also that it has opened the way for economic justifications to override the discriminatory impact of employment practices.[102] What remains uncertain is the range of issues which may be regarded as constituting a material factor defence. The Employment Appeal Tribunal, in *Davies* v. *McCartneys*,[103] held that there was no limitation on the factors which may be regarded as 'material'—a decision seen by some as having the potential to undermine the effectiveness of the equal pay law.[104]

However, subsequently, the Court of Appeal referred a number of

---

[97] C. McCrudden, 'Options for Amending the Equal Pay Legislation', in *Equal Pay for Work of Equal Value: Conference Report* (1990).
[98] See O'Donovan and Szyzsczak, *Equality*, 143 for discussion.
[99] *Clay Cross (Quarry Service) Ltd.* v. *Fletcher* [1978] IRLR 288 (CA).
[100] Official Reports, HC, 20 July 1983, col. 486.
[101] See n. 83.
[102] O'Donovan and Szyzsczak, *Equality*, 149.
[103] [1989] IRLR 439 (EAT).
[104] 26 *Equal Opportunities Rev*. 13.

questions to the European Court of Justice concerning the scope of the material factor defence, in the *Enderby* case.[105] These relate to the extent to which an employer has objectively to justify a pay difference between jobs done primarily by women and jobs done primarily by men, to whether an employee can rely on the fact that the pay levels were arrived at by a different collective bargaining process for the different groups, and to whether proven labour shortages in particular areas can be relied upon objectively to justify all or part of the pay difference.

Of particular concern in relation to defences is the extent to which job evaluation schemes can be relied upon as a defence. It is a defence to an equal value claim if a job had already been rated as not equivalent, as long as the evaluation scheme used follows the guidelines laid down in *Bromley* v. *Quick*.[106] A job evaluation scheme was cited as an employer's 'most effective safeguard against oppressive equal value claims' in *Leverton*, one which would afford 'complete protection'.[107] Employers, unsurprisingly, are implementing job evaluation schemes either before cases are taken against them or, indeed, after claims have been lodged.[108]

However, the Northern Ireland Court of Appeal has held that a job evaluation scheme can only be relied upon to defend an equal value claim by employees in the undertaking or group of undertakings in which the study was undertaken.[109] In this case, the employers had sought to rely upon a job evaluation scheme carried out in Great Britain, not Northern Ireland, in relation to jobs similar to those of the complainants.

As regards the procedures involved in claiming equal pay, it is not proposed to enter a lengthy discussion. The introduction of the equal value amendment, however, was accompanied by Rules of Procedure which set out preliminary stages unique to equal value claims.[110] Litigation at each of these stages proliferates[111] and, while the procedures under the original 1970 Act were criticized as complex, the procedures for equal value cases were described as 'scandalous' and amounting to 'a denial of justice' by the President of the Employment Appeal Tribunal in 1990.[112]

---

[105] *Enderby* v. *Frenchasy Health Authority and the Secretary of State for Health* [1992] IRLR 15 CA.

[106] See n. 91.

[107] *Leverton* v. *Clwyd County Council* [1989] 2 WLR 47; [1989] IRLR 28 (HL).

[108] *Dibro Ltd.* v. *Hore and others* [1990] IRLR 129 (EAT).

[109] *McAuley and others* v. *Eastern Health and Social Services Board* [1991] IRLR 467 (NICA).

[110] The Industrial Tribunal (Rules of Procedure) (Equal Value Amendment) Regulations, SI, 1983/1807.

[111] See 'Equal Value Update' (1991) 38 *Equal Opportunities Rev.* 12–29 for a review of equal value cases and the procedures involved.

[112] *Aldridge* v. *British Telecommunications plc* [1989] 1 CR 790, [1990] IRLR 10 (EAT).

Equal treatment and social security

One of the major implications of women's inequality in the labour market is that women rely more heavily than men on social security and yet they may be entitled to lower benefits. While a discussion of the complexities of the social security system is beyond the scope of this chapter, it is worth noting that one of the enduring aspects of discrimination against women, particularly married women, has been the concept underpinning the social security system which defines women as economically dependent on men.[113]

The social security system was not affected by the introduction of the equal pay and sex discrimination legislation in 1975, although changes were introduced in both the Social Security Act and the Social Security Pensions Act of 1975 to provide for equal treatment in some areas, such as the right to unemployment benefit and sickness pay for married women on the same basis as men.[114] The same legislation, however, also introduced new discriminatory elements into the system, such as invalid care allowance[115] and the non-contributory invalidity pension,[116] for which married and cohabiting women were ineligible. A 'housewives' version' of the non-contributory invalidity pension was introduced in 1977, giving these women an entitlement to the pension if they could prove that they were incapable of carrying out normal household duties, a test which did not apply to the general invalidity pension.[117]

In general terms, however, the main changes which have been introduced since the 1970s have been as a result of Community directives applying the principle of equal treatment to both statutory and occupational social security schemes. The former, adopted in 1979, aimed to ensure the application of the principle of equal treatment to statutory social security schemes, and member states were obliged to implement any necessary changes by 23 December 1984.[118] The Directive is designed primarily to cover the main employment-related risks, such as sickness, invalidity, old age, accidents at work, and unemployment,[119] and it applies to the 'working population' which is defined to cover those in work and seeking work, as well as those whose employment is interrupted by accident, sickness, or involuntary unemployment.[120] There are specific

---

[113] See Carter, *Politics*, generally, and S. Atkins and B. Hoggett, *Women and the Law* (1984), ch. 9.
[114] See S. Atkins and L. Luckhaus, 'The Social Security Directive and UK Law', in McCrudden (ed.), *Women, Employment*, for detailed discussion.
[115] Social Security Act 1975, s. 37.
[116] Ibid., s. 36.
[117] SI 1977/1312 made under the Social Security Act 1975, s. 36(7) then in force.
[118] See n. 9.
[119] Ibid., Art. 3.
[120] Ibid., Art. 2.

exclusions from the scope of the Directive, such as the determination of pensionable age for the purposes of granting old-age and retirement pensions;[121] and derogations from the principle of equal treatment, such as survivors' benefits and some family benefits, and maternity benefits.[122]

A series of changes was implemented from the end of 1983 to bring the United Kingdom into compliance with the provisions of the Directive. This included changes to the discriminatory rules relating to adult and child dependency additions, for which a variety of approaches was adopted, involving both levelling up (extending benefit to previously excluded groups) and levelling down (taking benefits away from previously included groups).[123] The rules regarding eligibility for supplementary benefit were amended at this time, to allow married and cohabiting women to claim for the first time—but only if they could establish themselves as the 'breadwinner' rather than a 'dependant', a concept which has remained problematic for women under social security rules.[124] Changes were also made to family income supplement and the non-contributory invalidity pension and its housewives' equivalent were repealed.[125] Invalid care allowance was not repealed at this time, however, and it was not until the judgment of the European Court in the *Drake*[126] case, which challenged the allowance, was pending that the Government took steps to provide the entitlement on the same basis as men and single women to married and cohabiting women.[127]

Despite these changes, there has been a significant amount of litigation about the scope of European obligations since the Directive came into force. For example, the European Court ruled in *Johnson* v. *Chief Adjudication Officer*,[128] a case concerning eligibility for severe disablement allowance, the successor to non-contributory invalidity pension, that the personal scope of Article 2 of Directive 79/7 does not apply to a person who had interrupted her occupational activity for child-caring purposes and who is prevented from returning to work because of illness *unless* the person was seeking employment and her search was interrupted by the onset of one of the risks set out in Article 3 of the Directive, the reason for previously leaving employment being irrelevant. It also held that the Directive could be relied on to set aside national legisla-

---

[121] Ibid., Art. 7.
[122] Ibid. Art. 4(2).
[123] See Atkins and Luckhaus, 'Social Security Directive', 106–9.
[124] See Supplementary Benefit Act 1976, Sch. 1, para. 3(1) as amended by the Social Security Act 1980, Sch. 2, Part 1, and SI 1983/1004.
[125] Health and Social Security Act 1984, s. 11, amending the Social Security Act 1975, s. 36 supplemented by SI 1984/1303.
[126] *Drake* v. *Chief Adjudication Officer*, Case No. 150/85 [1987] QB 166; [1986] 3 CMLR 43.
[127] Social Security Act 1986, s. 37, amending Social Security Act 1975, s. 37(3).
[128] Case No. 31/90 [1992] 2 All ER 705.

tion which makes entitlement to a benefit subject to rules for eligibility to a preceding benefit which contained discriminatory conditions. In the absence of appropriate measures implementing Article 4 of the Directive, women placed at a disadvantage by the maintenance of discriminatory conditions are entitled to be treated in the same manner as men.

Questions concerning the requirement on a man to pay national insurance contributions for five years longer than women in order to be entitled to the same basic pension and the requirement that, should a man continue to work between the ages of 60 and 65, he still has to pay national insurance contributions when women over 60 years of age do not, whether they are working or not, were also considered by the European Court in a case in the name of the Equal Opportunities Commission.[129] The Court ruled in 1992, that the United Kingdom is not in breach of Directive 797 by imposing this requirement. Cases concerning eligibility for supplementary benefit and income support were also decided by the European Court in 1992.[130] It held that Directive 797 does not apply to such benefits.

National courts themselves of course have to apply the provisions of the Directive and, in respect of social security, a 1990 judgment of the Court of Appeal augurs well. In *Thomas* v. *Adjudication Officer and Secretary of State for Social Security*,[131] it held that the female claimants, who were over 60 years of age, were entitled to rely on the 1979 Directive to claim severe disablement allowance or invalid care allowance, despite the fact that these benefits were restricted to those under the state retirement age. The Court held that the Secretary of State had failed to prove that the restriction properly fell within the scope of the exclusion in the Directive relating to the determination of pensionable age for the purposes of old-age and retirement pensions and the possible consequences thereof for other benefits. The House of Lords referred questions arising from this case to the European Court in November 1991.

The other social security Directive, concerning occupational schemes, was due to come into effect on 1 January 1993, though some aspects will not be effective until 1999.[132] Its aim is to extend the principle of equal treatment to schemes not covered by the 1979 Directive whose purpose is to provide workers with benefits intended to supplement or replace benefits provided by statutory schemes, whether membership of such schemes is optional or compulsory.[133] The Directive allows member states

---

[129] Case 69/91 *R.* v. *Secretary of State for Social Security, ex parte Equal Opportunities Commission*, [1992] 2 All ER 577.

[130] Case 63/91 *Cresswell* v. *Chief Adjudication Officer* and Case 64/91, *Jackson* v. *Chief Adjudication Officer*, Decision 16 July 1992, unreported to date.

[131] [1990] IRLR 436 (CA). HL referred case to ECJ on 27 Nov. 1991—Case 328/91.

[132] See n. 10.

[133] Ibid., Art. 2.

to defer the compulsory application of a number of its provisions. These include the determination of pensionable age for the purposes of granting old-age or retirement pensions until a directive requires such application or until the date on which such equality is achieved in statutory schemes, and the application of the Directive to survivors' benefits until a directive requires such application. The Directive also provides that different actuarial calculation factors may be taken into account until 1999.[134] The decision in the *Barber*[135] case, considered below, led to uncertainty over the precise scope of these wide exceptions and the scope for further Community legislation is unclear, despite being in the pipeline for some time.[136] In national terms, however, the Government has sought to implement the Directive with the Social Security Act 1989.

Specific issues

Progress has been made towards equality in a number of specific areas since the introduction of the legislation, particularly because of membership of the European Community. These are considered below. The first few relate to what may be termed the economic circumstances of women; the second group involve considerations of reproduction, sexuality, and assumptions about appropriate female labour.

Retirement and pensions

In the Sex Discrimination Act 1975, there was an exception for provisions in relation to death and retirement[137] and in cases such as *Roberts* v. *Cleveland Area Health Authority*,[138] and *Duke* v. *Reliance Systems Ltd.*,[139] differences in normal retirement age for women and men were held to be excluded from the scope of the Act. In *Marshall* v. *Southampton and South West Hampshire Health Authority*,[140] however, the European Court of Justice held that the provisions of the Equal Treatment Directive 1976 prohibited such an exception, and that it rendered unlawful the maintenance of different age limits for women and men for compulsory retirement in the public sector. The Government thus had to amend the Sex Discrimination Act and in doing so it extended the requirement to the

---

[134] Ibid., Art. 9(*a*), (*b*), and (*c*).

[135] *Barber* v. *Guardian Royal Exchange Assurance Group*, Case No. 262/88 [1990] IRLR 240 (ECJ).

[136] Proposal for a Directive completing the implementation of the principle of equal treatment of men and women in matters of social security schemes, COM/87/484 final, *OJ* C309, 19 Nov. 1987.

[137] SDA 1975, s. 6(4).

[138] [1979] 1 WLR 754 (HL).

[139] [1988] AC 618; [1988] IRLR 118 (HL).

[140] Case No. 152/84 [1986] QB 401; [1986] IRLR 140 (ECJ).

private sector also and, as regards dismissal, to voluntary redundancy schemes and early retirement schemes.[141]

In relation to pensions, there has been heavy reliance on the jurisprudence of the European Court of Justice, which has held that occupational pensions based on negotiations between an employer and employees and which were not covered by state legislation were 'pay' within the meaning of Article 119.[142] In *Barber* v. *Guardian Royal Exchange Assurance Group*, the Court held that a pension paid under a contracted-out private occupational scheme also constitutes pay.[143] This was followed in October 1990 by the Employment Appeal Tribunal, in *Roberts* v. *Birds Eye Walls Ltd.*,[144] which held that Article 119 prohibits an employer from reducing a woman's occupational pension scheme from age 60, on grounds that she will be in receipt of state pension, but not reducing a man's pension until age 65. This issue is now being considered by the European Court of Justice, on a reference from the Court of Appeal.[145]

Problems remain about the precise nature of entitlements under both Community and national law, and it is likely that the Community will seek to clarify the position.[146] In respect of national law, several other issues arising from *Barber* are being referred to the European Court for guidance. In *Neath* v. *Hugh Steeper Ltd.*,[147] for example, an industrial tribunal has decided to refer questions relating to the retrospective effects of the *Barber* decision. The European Court in *Barber* had held that its decision could not be relied upon to make claims retrospectively—that is, to claim entitlement to a pension with effect from a date prior to its decision. The main point on which guidance is sought is whether equal treatment in relation to pensions required by Article 119 applies to the proportion of Mr Neath's pension benefits which accrued before the date of the *Barber* decision.

In *Coloroll Pension Trustees Ltd* v. *Russell and Others*, the High Court has also asked the European Court of Justice to clarify the retrospection issues raised by *Barber* as well as a series of other questions relating to such issues as: the enforceability of rights under Article 119 against both trustees and the assets of pension schemes as well as employers, whether Article 119 outlaws discrimination in contracted-in pension schemes, and the permissibility of using actuarial assumptions.[148] The Department of Social Security was providing financial backing for the pension fund in

[141] SDA 1986, ss. 2 and 3.
[142] *Worringham* v. *Lloyds Bank Ltd.*, Case No. 69/80 [1981] ECR 797.
[143] See n. 135.
[144] [1991] IRLR 19 (EAT).
[145] [1992] IRLR 23 (CA).
[146] *Third Action Programme*, 8.
[147] See 'European Court to clarify Barber', 38 *Equal Opportunities Rev.* 2.
[148] Coloroll questions to European Court (1991) 39 *Equal Opportunities Rev.* 35.

this instance, in light of the importance of the issues. The case was likely to be heard early in 1993, together with a number of references on similar points from courts in other Member States.[149]

Also referred to the European Court were several questions concerning the lawfulness of an employer's action in raising female employees pension age (previously 60) to that of male employees (65), in order to comply with the *Barber* decision. Assisted by the Equal Opportunities Commission, the women were challenging this, arguing that equality is achieved only by improving the terms of workers not reducing them.[150]

Problems concerning pensions are, of course, compounded by the continuing discrimination in the age at which state pensions are provided—60 for women and 65 for men—although, following *James* v. *Eastleigh Borough Council*,[151] the position regarding discriminatory treatment between men and women of the same age, where one is in receipt of a state pension and the other is not, in respect of payment for use of public facilities, has been held by the House of Lords to constitute direct discrimination. The Government announced in June 1991 that it was committed to equalizing the state pension age after the general election[152] and, in December 1991, it published a consultation document canvassing opinion on various options for doing so.[153] The Equal Opportunities Commissions in Great Britain and Northern Ireland have urged Government to allow women and men alike to receive a full state pension at 60, as the fairest option available,[154] but, although no decision has yet been taken, it appears from press reports that Government intends to raise the pension age for women to 65 as its preferred option.[155]

Income taxation

The personal taxation system was also excluded from the scope of the 1975 Act and yet the system in place at the time and until recently contained blatantly discriminatory features, which undermined the spirit of the legislation and impeded advances made with regard to pay and employment opportunities of women, particularly married women, in respect of economic independence. Among the features of the tax system subject to the most vociferous criticism was the lack of privacy for married women in respect of their financial affairs, the husband being

---

[149] (1993) 47 *Equal Opportunities Rev.* 32.
[150] *Smith and other* v. *Award Systems Ltd.*, 30 November 1992; Case No. 6684/92.
[151] See n. 51.
[152] 'Equal Retirement Ages Promised'. *Independent* 27 June 1991.
[153] Options for Equality in State Pension Age, Cmnd 1723, Dec. 1991.
[154] A Question of Fairness: Response by EOC to Department of Social Security's Discussion Paper Options for Equality in State Pension Age, EOC, June 1992; EOCNI, press release, 18 June 1992.
[155] 'State may lift pension age limit for women to 65', *The Times*, 12 Dec. 1992.

responsible for declaring his wife's income. A wife's earned income was automatically aggregated with that of her husband unless she elected to be taxed separately and, in the case of her investment income, it had to be aggregated. The Married Man's Allowance, payable to the husband except on very rare occasions when he was not working and after lengthy negotiations with the Inland Revenue, was also considered an anachronism.[156]

The European Commission, while respecting the competence of the member states over the question of personal taxation, has taken an interest in this issue and in 1984 presented a Memorandum to the Council on Taxation and Equal Treatment for Women and Men.[157] This recommended that there should be a system of totally independent taxation for all and that at least the option of separate taxation should be available to couples. Whether this led to the changes announced by Government in 1988 or not, the changes were broadly welcomed as long overdue but regarded as not extensive enough.[158] These changes made provision for women to manage their own financial affairs from 1990. Specifically, married women, like all others, will have their own personal allowance and the capacity to make separate returns to the Inland Revenue in respect of their earned and unearned income. The Married Man's Allowance has been replaced with a Married Couples Allowance, although this remains payable to the husband and the wife does not have the same entitlement except in limited circumstances.[159] While this falls short of a completely independent taxation system and it remains to be seen how it operates in practice, the changes are at least partially positive.

Pregnancy discrimination and maternity rights

One of the issues which has caused most difficulty under the sex discrimination legislation relates to discrimination against pregnant women. While the state has made special provision to provide pregnant women with some statutory employment rights—such as the right to protection against unfair dismissal, the right to maternity leave and maternity pay, the right to time off for antenatal care, and the right to return to work—these are limited in their scope, dependent on service with the same employer for a minimum period, and extremely complicated.[160] The Sex Discrimination Act allows for such special provision as an exception to the principle of

[156] See Equal Opportunities Commision for Northern Ireland, *Twelfth Annual Report* (1988), 24–5, for discussion.
[157] COM (84) 695 final, 14 Dec. 1984.
[158] See Fawcett Society, *Annual Report for 1987–1988* (1988).
[159] Income and Corporation Taxes Act 1988, Taxation of Income of Spouses, ss. 279–89.
[160] e.g. Employment Protection Act 1975; Employment Protection (Consolidation) Act 1978, s. 31A, s. 33, ss. 45–8, s. 56 inserted by the Employment Act 1980, s. 13, ss. 11–12; Social Security Act 1986, s. 24.

equal treatment[161] and the Equal Treatment Directive also contains a provision which allows derogation from the principle of equal treatment for provisions concerning the protection of women, particularly as regards pregnancy and maternity.[162]

However, the form of the legislation has given rise to problems for the accommodation of women's reproductive role, with its 'assimilationist' approach, which minimizes biological differences and defines equality as the similar treatment of those who are similarly situated.[163] It has led tribunals to find initially that, since a man could not become pregnant, it was impossible to compare the treatment—a dismissal—of a pregnant applicant with the treatment of a man and therefore a claim of direct discrimination could not be sustained.[164] Although this approach was subsequently rejected, the courts then held that the 'proper approach' was to ask whether pregnancy with its associated consequences was capable of being matched with 'analogous circumstances' such as sickness applying to a man;[165] and that a woman's absence from work due to pregnancy and a man's medical condition are not 'materially different' and are therefore comparable.[166] The dangers inherent in these approaches are apparent: either there is no remedy for discrimination on grounds of pregnancy as there is no 'comparator', or a comparison must be made with sickness. This is inappropriate as pregnancy is not a sickness and it is unhelpful, in the context of improving women's access to the labour market, for it to be considered as such. The tribunals in Northern Ireland have taken a more purposive approach to this question, holding for example, in one case, that the correct comparison was between the treatment accorded to a mother-to-be and that which would be accorded to a father-to-be;[167] and, in another case, that discrimination against motherhood is sex discrimination.[168]

While this issue has created difficulties for national courts—and for pregnant women—the matter was clarified in 1990 at European level, with the European Court of Justice holding in *Dekker* v. *Stichting Vormingscentrum voor Jonge Volwassen (VJV-Centrum) Plus*[169] and *Hertz* v. *Aldi*

---

[161] SDA 1975, s. 2(2).

[162] Art. 2(3).

[163] Wasserstrom, 'Racism, Sexism'; Lacey, 'Note: Dismissal'.

[164] *Reaney* v. *Kanda Jean Products* [1978] IRLR 427; *Turley* v. *Allders Dept. Stores Ltd.* [1980] ICR 66 (EAT).

[165] *Hayes* v. *Malleable Working Men's Club* [1985] ICR 703. The case was heard by the EAT with *Maughan* v. *NE London Magistrates Court Committee* which, when reheard by the IT, resulted in a finding that dismissal for pregnancy was contrary to the SDA 1975.

[166] *Webb* v. *EMO Cargo Ltd.* [1990] IRLR 124 (EAT). See further discussion on reference to ECJ on this case below.

[167] *Donley* v. *Gallaher*, Case 66/86 SD. Decision 6 Nov. 1987.

[168] *McQuade* v. *Dabernig*, DCLD 1, Case No. 427/89 SD, 31 August 1989.

[169] Case No. 177/88 [1991] IRLR 27 (ECJ).

Marked K/S[170] that discrimination on grounds of pregnancy is direct sex discrimination. In the former case, the Court held that the employer was in breach of Articles 2(1) and 3(1) of the Equal Treatment Directive if he refused to employ a suitable female candidate on grounds of possible adverse consequences arising from employing a woman pregnant at the time of her application. The Court was clear that as employment can only be refused because of pregnancy to women, such a refusal is direct discrimination on grounds of sex and that the fact that there was no male candidate for the post to be filled did not alter this. It also held that such discrimination cannot be justified by the financial detriment in the case of recruitment of a pregnant woman suffered by the employer during her maternity leave.

The *Hertz* case concerned a woman who had been dismissed following an extended period of sick leave, commencing some months after her return from maternity leave. The grounds for her dismissal were repeated absences due to illness. The Court was clear that Article 2(3) of the Equal Treatment Directive allows for national provisions which ensure specific rights for women in respect of pregnancy and maternity as an exception to the principle of equal treatment. Consequently the dismissal of a female worker because of her absence due to maternity leave from which she benefits under national law would constitute unlawful discrimination in the same way as refusal to recruit a pregnant woman. It went on to rule, however, that in respect of an illness arising after maternity leave, there was no reason to distinguish an illness associated with pregnancy or maternity from any other illness, and that, if comparable sick leave would lead to dismissal of a male worker, there was no direct discrimination on grounds of sex.

The implications of these cases are of course significant in the United Kingdom. First, it is clear that discrimination on grounds of pregnancy constitutes direct sex discrimination, contrary to the Equal Treatment Directive. Second, the decisions reduce the importance of the comparative approach to such cases, evident from the national rulings mentioned above. The European Court treats pregnancy, for the purposes of anti-discrimination law, as a condition unique to women. Thus it is sufficient to show unfavourable treatment as a result of pregnancy in order to prove a breach of the principle of equal treatment, rather than have to establish a comparison with treatment of a man or a hypothetical man. It is also unlikely that any defence based on the costs of employing a pregnant woman will be accepted in future, as the Court was most clear on this point. In addition, as the Equal Opportunities Review points out, the principles laid down in *Dekker* particularly have implications for the ex-

---

[170] Case No. 179/88 [1991] IRLR 31 (ECJ).

clusion of pregnancy and maternity from sick pay and disability benefits or from other forms of insurance.[171]

Despite such guidance from the European Court, national courts remain in some difficulties and a number of important questions have been referred for further clarification to the European Court. In *Webb* v. *Emo Cargo Ltd*,[172] the House of Lords overturned the Court of Appeal's decision which sought to distinguish the case on its facts from *Dekker*. The Court of Appeal had ruled that if a women is dismissed for a reason arising from her pregnancy—that is, unavailability for work at a particular time—and claims sex discrimination, it is necessary for the Court to decide whether a man with a condition as nearly comparable as possible which had the same effect upon his ability to do the job would or would not have been dismissed. The House of Lords was clear that dismissal on grounds of pregnancy is sex discrimination but sought to make a further distinction between pregnancy and the consequences of pregnancy, re-introducing a comparative approach. It asked the European Court to consider questions relating to the particular facts of the *Webb* case, where a woman recruited to replace a pregnant employee became pregnant herself and was unavailable to work at the relevant period. The Northern Ireland Court of Appeal is also seeking guidance from the European Court of Justice in *Gillespie and Others* v. *various Health Boards and the Department of Health and Social Services*, on questions concerning the level of pay to which women are entitled during maternity leave. The Equal Opportunities Commission for Northern Ireland, which is supporting the case argues that, in light of European equality law, there should be no reduction in pay during maternity leave.[173]

In addition to the European Court's jurisprudence expanding rights in this area, the Council adopted a Directive in October 1992 on the protection of pregnant women and women who have recently given birth.[174] The Commission had proposed the Directive as part its actions to implement the 'Social Charter',[175] primarily as a health and safety measure under Article 118A of the Treaty which allows for qualified majority voting. Several difficulties were encountered during the Directive's legislative process, with disagreements between the various Community institutions and between a number of the Member States about the

---

[171] See (1991) 35 *Equal Opportunities Rev.* 40–4.

[172] 1992 2 All ER 942.

[173] *Gillespie and Others* v. *various Health Boards and the Department of Health and Social Services*, Case No. 308/89 EP; 310/89 SD, Date of decision 10 June 1991; see also *Todd* v. *Eastern Health and Social Services Board and Department of Health and Social Services*, Case No. 1149/88 EP; 1150/88 SD, Date of decision 16 Oct. 1989. EOC Press Release, 'Court of Appeal refers maternity case to Europe', 20 Dec. 1991.

[174] OJ L 348/1, 28 Nov. 1992.

[175] COM (89) 471 final, Oct. 1989.

precise scope of the Directive, particularly in relation to its provisions on maternity pay.[176] In the event, the Directive provides for 14 weeks continuous maternity leave[177] and the maintenance of a worker's contractual rights, except pay, during this period.[178] However, she must receive a payment of at least equivalent to that which she would receive in the event of requiring sick leave, although Member States may make this dependent on employment in the previous 12 months.[179] Also, importantly, the Directive proves protection against dismissal[180] and a number of health and safety requirements are included.[181] The adoption of the Directive, in such a watered-down form, received a lukewarm reaction from national equality agencies.[182]

The Government is seeking to introduce a number of its obligations under this Directive, which should be implemented by all Member States by October 1994, in the Trade Union Reform and Employment Rights Bill which received its second reading in the House of Commons on 17 November 1992.[183] Its provisions relating to maternity have been criticized as likely to create further confusion for employers and employees alike, in what is already a very complex area.[184]

Sexual harassment

'Sexual harassment' as a phrase was employed by the women's movement, first in the United States and then in the United Kingdom, to describe unwanted sexual attention of a variety of forms—physical, verbal, and non-verbal—which has an adverse impact on either the employment opportunities of a particular woman or on her working conditions.[185] Despite recognition of the problem and some attempts to deal with it through employment law generally, such as the laws on unfair dismissal, it was not until the early 1980s that the problem began to be addressed as one of sex discrimination. The TUC produced guidelines in 1983[186] which called attention to the problem as one of unequal power relations between

---

[176] 'Pregnant Workers Directive Adopted' (1993) 47 *Equal Opportunities Rev.* 2.
[177] Pregnant Workers Directive Article 8.
[178] Article 11.
[179] Ibid.
[180] Article 10.
[181] Articles 3–8; Annex I and II.
[182] See EOCNI, press release, 28 October 1992, 'Qualified Welcome for EC Pregnancy Directive'.
[183] Bill 78.
[184] 'EOC NI Comments on the Trade Union Reform and England Rights Bill, January 1993; EOC GB, Formal Response to the Trade Union Reform and England Rights Bill, January 1993.
[185] C. MacKinnon, *Sexual Harassment of Working Women: A Case of Sex Discrimination* (1979), has been particularly influential.
[186] Trades Union Congress, *TUC Guide: Sexual Harassment at Work* (1983). An updated version was published in 1991.

men and women in the workplace and a number of complaints began to surface. The first case in the United Kingdom in which sexual harassment was considered to constitute unlawful sex discrimination came from a Northern Ireland industrial tribunal, where a young female apprentice motor mechanic successfully complained of sex discrimination after being subjected to physical ill-treatment and denial of opportunities to train.[187] The leading national authority to date is *Strathclyde Regional Council* v. *Porcelli*.[188] It was made clear in this case that 'sexual harassment is a particularly degrading and unacceptable form of treatment which it must be taken to have been the intention of Parliament to restrain', and that a woman who was subjected to a campaign of unpleasant treatment, including an element of sexual harassment, by two male colleagues was treated less favourably on the ground of sex, despite the fact that adverse treatment would have been meted out to an equally disliked man. As part of the treatment was concerned with sex, to which a man would not have been subjected, a case of sex discrimination had been established. The Court thus accepted that the uncomfortable work environment caused by the sexual harassment could constitute a detriment within the meaning of the Sex Discrimination Act, as well as harassment leading to adverse employment consequences.

An increased number of complaints of sexual harassment have been forthcoming as a result of litigation in this area[189] and it is an issue which has been taken up extensively by the trade union movement. At European Community level, there has also been significant progress, following a European-wide study and lobbying for a Directive specifically to outlaw sexual harassment.[190] Both the Council and the Commission have stated that sexual harassment may, in certain circumstances, be contrary to the principle of equal treatment contained in the Equal Treatment Directive.[191] The Council, in a Resolution of May 1990, stated that conduct of a sexual nature or other conduct based on sex is 'an intolerable violation of the dignity of workers' and that it is unacceptable if it is unwanted, unreasonable, and offensive to the recipient; if a person's submission to or rejection of, such conduct leads to adverse

---

[187] *Mortiboys* v. *Crescent Garage Ltd.*, Case 24/83 SD, Decision 15 Feb. 1984.

[188] [1986] IRLR 134 CS.

[189] See Equal Opportunities Commission, *Annual Report* (1990), 16; Equal Opportunities Commission for Northern Ireland, *Fourteenth Annual Report* (1990), 27. See also 'Profile' (1991) 37 *Equal Opportunities Rev.* 24–5, for discussion of the work of a voluntary organization established in 1985 to deal with complaints of sexual harassment (Women Against Sexual Harassment).

[190] M. Rubenstein, *The Dignity of Women and Men in the Workplace: The Problem of Sexual Harassment in the Member States of the European Community*, V/412/1/87.

[191] Council Resolution of 29 May 1990 on the protection of the dignity of women and men at work, *OJ* C157/3, 27 June 1990, preamble and para. 2.2; and Commission Recommendment on the protection of the dignity of women and men at work of 27 Nov. 1991. OJ L49/1, 24 Feb. 1992.

decisions affecting a person's access to employment, vocational training, or dismissal; and/or that it creates a hostile, intimidating, or uncomfortable work environment—thus providing a basis for a legal definition at Community level.[192] The Commission, following consultation with the social partners, ECOSOC and the European Parliament, adopted a Recommendation on the protection of the dignity of women and men at work and a Code of Practice on measures to combat sexual harassment on 27 November 1991.[193] The Recommendation urges Member States to take action to promote awareness that sexual harassment is unacceptable and may be contrary to the Equal Treatment Directive, and to promote the implementation of the Code of Practice in the public and private sectors. The Code itself comprises a series of guidelines on means of ensuring that sexual harassment does not occur and, if it does occur, that adequate procedures are readily available to deal with it and prevent its recurrence. There is some evidence that employers are increasingly adopting sexual harassment policies in recent years and that such policies are becoming more sophisticated.[194] There would also appear to be some reliance on the Recommendation in the industrial tribunals[195] although the view that the European Commission Code has a legal status which means it ought to be taken into account in industrial tribunals, based on a European Court of Justice decision concerning the status of a health and safety Recommendation, has yet to be endorsed in national courts.[196]

Protective legislation

The controversy surrounding discussion of protective legislation has already been referred to. Protective legislation, placing restrictions on the employment of women, had been in place since the nineteenth century[197] and section 51 of the 1975 Sex Discrimination Act left intact the provisions of such legislation as the Health and Safety at Work Act 1974, as an exception to the principle of equal treatment, thus providing for the exclusion of women from certain types of work or certain patterns of work. Under the Equal Treatment Directive 1976, member states were

[192] Ibid., para. 1.
[193] OJ L 49/1, 24 Feb. 1992.
[194] 'Harassment policies show growing sophistication' (1992) 46 *Equal Opportunities Rev.* 32.
[195] *Tofield* v. *Pollicino t/a Donnabella Hair Design* Case 10862/92, Date of decision 1 Oct. 1992; *Donnelly* v. *Watson Grange Ltd.*, Case No. 5/119/92, Date of decision, 16 Nov. 1992.
[196] See Rubenstin M. 'Sexual Harassment Recommendation and Code', (1992) 41 *Equal Opportunities Rev.* 27.
[197] J. Jarman, 'Equality or Marginalisation: The Repeal of Protective Legislation', in Meehan and Sevenhuijsen (eds.), *Equality Politics*.

obliged to revise all protective laws, regulations, and administrative procedures contrary to the principle of equal treatment, 'when the concern for protection which originally inspired them is no longer well founded', and it gave member states four years within which to review such legislation and make any necessary changes.[198]

In 1979 the Equal Opportunities Commission in Great Britain published its review of the provisions of the Health and Safety Act 1974 and recommended the abolition of restrictions on the hours during which women might work, accepting the argument that the existence of regulation was a pretext for the denial of equality, and that women, would want the opportunity of working at night and access to higher earnings.[199] A bitter controversy resulted, with criticism of the Commission that the survey on which it had based its recommendations had been limited in that it had not probed whether women wanted both paid work and a regular family life, facilitated by good child-care arrangements, and that its findings did not seriously challenge the interests of employers as articulated by the Confederation of British Industry.[200]

The European Community's influence continued to be significant for domestic legislation. The Commission had issued a reasoned opinion to the Government, stating that it considered the exception provided by section 51 concerning prior legislation to be too broad and contrary to the Equal Treatment Directive.[201] In 1987 it published a Communication on protective legislation which focussed attention on the fact that many of the provisions still in place across the Community had a negative influence on women's employment prospects.[202]

Many of the restrictions on women's employment have now been removed. The Sex Discrimination Act 1986 repealed restrictions on women's working hours,[203] for example, and the Employment Act 1989 dealt with what was left of protective legislation concerning women. It did so in three main ways. First, it repealed some restrictions outright, such as that on the employment of women in mines and on cleaning machinery in factories and offices. Second, it amended section 51 of the Sex Discrimination Act so that it now requires the principle of equal treatment to prevail over prior protective legislation except where it relates to pregnancy and maternity and risks specially affecting women.

---

[198] Art. 3(2)(c); Art. 5(2)(c); Art. 9(1).

[199] Equal Opportunities Commission, *Health and Safety Legislation: Should we Distinguish between Men and Women?* (1979).

[200] Jarman, n. 197.

[201] C. Docksey, 'The Promotion of Equality', in McCrudden (ed.), *Women, Employment*, 14.

[202] *Communication on Protective Legislation for Women in the Member States of the European Community*, COM (87) 105 final, 20 Mar. 1987.

[203] SDA 1986, ss. 7 and 8.

This is done by means of a new statutory 'override' contained in section 1 of the Act, which nullifies any provision predating the 1975 Act where it imposes a requirement to do an act of sex discrimination. This is matter for judicial interpretation. In addition, the Secretary of State is given the power to repeal by order any legislation passed before the 1989 Act which he considers requires the commission of such discriminatory acts. However, he also has the power to save particular legislative provision by 'disapplying' the statutory override of section 1. Finally, certain legislation is specifically exempted from the 'override' by the Act itself, including legislation aimed at the protection of women during pregnancy.[204]

Thus the 1989 Act's provisions are complex and provide for different categories of protective legislation. Particularly problematic are the wide powers given to the Secretary of State to vary or modify the scope of the statutory override. In addition, there is 'serious doubt' about whether it goes far enough to comply with the requirements of the Equal Treatment Directive and whether it strikes a good balance between promoting equal opportunities and protecting women against risks.[205] At the time of passage of the 1989 legislation, the Government apparently took the view that differential protection was permitted under Community law by virtue of Article 2(3) of the Equal Treatment Directive on the protection of women as regards pregnancy and maternity.[206]

In practice, this is likely to involve cases where there is a reproductive or foetal risk. Yet such risks do not affect only women. The European Commission pointed out in its Communication that the reproductive capacity of men may be more at risk from exposure to lead than women for example and it recommended, in respect of ionizing radiation, that provisions relating to the protection of women of reproductive capacity should be extended to men who are able or intending to have children.[207] Moreover, a general policy of excluding women of child-bearing age from exposure to substances which are dangerous to the foetus rests on assumptions which exclude women's autonomy—such as that child-bearing is a 'natural' function of all fertile women, or that women cannot plan for pregnancies. Under national law now, the validity of a foetal or reproductive protection policy will be assessed under section 51 of the Sex Discrimination Act as amended by the 1989 Act. It is open to the employer to show that the job entailed a reproductive or foetal hazard, so

---

[204] See S. Deakin, 'Equality under a Market Order: The Employment Act 1989' (1990) 19 *Industrial LJ* 1, for full discussion.

[205] 'The Employment Bill and Equal Opportunities' (1989) 23 *Equal Opportunities Review* 28–34, 30.

[206] 'US Court Bars Foetal Protection Policy' (1991) 37 *Equal Opportunities Rev.* 26–7.

[207] *Protective Legislation*, 22.

that it was necessary to exclude women in order to comply with the provisions of the Health and Safety Act.

In this context it is worth noting a recent decision of the Supreme Court in the United States. In *International Union, United Auto Workers v. Johnson Controls Inc.*,[208] it held that discrimination based on a woman's child-bearing capacity constitutes sex discrimination and is unlawful unless it can be proved that her reproductive potential prevents her from performing the duties of her job. The company's foetal protection policy, under which women of child-bearing age were excluded from certain high-paying jobs involving exposure to lead unless they could prove they were infertile, was held to be unlawful. The Equal Opportunities Review points out that

as a pure matter of statutory construction, ... protection of a foetus is not protection of a woman. Nor is exposure to lead a risk specially affecting women. Therefore, it is at least arguable that a woman of reproductive capacity forced out of her job when a man would be permitted to remain would not be precluded from claiming that she had been discriminated against by s. 51.[209]

The precise impact of the 1989 Act remains to be seen but it is clear that, once again, the role of the Community has been significant. In the particular case of protective legislation, however, it must also be acknowledged that removal of protections for workers falls within the general labour market deregulation policies of the Government in power since 1979.[210]

Exceptions

One of the major criticisms of the Sex Discrimination Act has been that the exceptions from its scope are too broad. Some of these exceptions have already been discussed—such as those relating to social security, retirement and pensions, and protective legislation—each of which has been amended in the light of European Community obligations. Other matters which were excluded from the scope of the legislation, such as employment in small firms[211] and acts done in the interests of national security,[212] have also been subject to challenge at the Community level and changes have resulted. In addition, a number of the exceptions for employment in which the sex of the worker is considered a 'genuine occupational qualification',[213] such as for reasons of physiology, privacy,

---

[208] [1991] 55 FEP Cases 365.

[209] 'US Court Bars Policy', 27.

[210] 'Equality under Market', 1.

[211] *Commission* v. *UK*, Case No. 165/82 [1984] 1 All ER 353; [1984] 1 CMLR 44.

[212] *Johnston* v. *Chief Constable of the Royal Ulster Constabulary*, Case No. 222/84 [1987] QB 129; [1986] 3 CMLR 240.

[213] SDA 1975, s. 7.

and decency, welfare or education considerations, has been successfully challenged at European Community level.[214] Thus the scope of the exceptions is gradually being eroded.

One of the more significant exceptions remaining in the legislation relates to 'positive action'. The legislature recognized the limits of formal equality to address the actual and practical inequalities between the sexes and makes some allowance for the tackling of past and present discrimination. Training bodies, employers, and trade unions may make special provision for women, relating mainly to training or special encouragement, under certain defined circumstances, which would otherwise be unlawful sex discrimination.[215] Further detail is provided in chapter 13 as there are similar provisions in the race discrimination legislation, but it is worth emphasizing that these provisions are permissive only. They provide no incentive for those employers or training bodies who are not already committed to taking action to effect positive changes in the labour market or in their organization. The provisions are in line with Article 2(4) of the Equal Treatment Directive.[216] We return to the issue of positive action below.

ENFORCEMENT OF RIGHTS AND REMEDIES

Individuals with complaints under the sex discrimination and equal pay legislation have the right of access to the industrial tribunal system—a specialized tribunal dealing with employment matters generally. Set up in the 1960s, the tribunal system was designed to dispense justice informally, quickly, and cheaply, without the need for legal representatives. The membership (normally three on each tribunal panel) is drawn from both sides of industry, who sit with a legally qualified Chair presiding over each case. There is no right to legal aid for applicants before the industrial tribunal but assistance may be available from the Equal Opportunities Commissions.[217]

Both Commissions receive a significant and growing number of applications for assistance and have provided a considerable source of support and assistance to complainants over the years.[218] Both experienced difficulties in their early years of existence with their approach to law

[214] See n. 211; SDA 1986, s. 1.

[215] SDA 1975, ss. 47–9.

[216] See n. 7. Art. 2(4) states that 'This Directive shall be without prejudice to measures to promote equal opportunity for men and women, in particular by removing existing inequalities which affect women's opportunities in the areas referred to in Article 1(1)'.

[217] SDA 1975, s. 75.

[218] See successive Annual Reports from both Commissions.

enforcement, which the Equal Opportunities Commission for NI resolved more quickly than its counterpart in Great Britain, gaining a reputation for its robust and strategic application of the law through the support of individual complaints.[219] Assistance with individual complaints, however, forms only part of the work envisaged by Parliament for the Commissions. They were given powers of formal investigation which have not been used to the fullest extent possible, for a variety of reasons. In general terms, the 1975 Act placed the main responsibility on individuals for raising complaints and pursuing them through the industrial tribunal system, with a discretion rather than a duty given to the Commissions whether to provide assistance, leaving them 'free to concentrate on the vital strategic role of seeking out patterns of discrimination and enforcing the law in the public interest'.[220] A full discussion of the powers of formal investigation, which were mirrored in the 1976 Race Relations Act, can be found in Chapter 13.

The formal investigation approach to tackling discrimination has many advantages, in principle, over the individual litigation approach. For example, resources can be strategically targeted at particular sectors or types of discrimination, perhaps those areas in which many people are affected or in which there are no identifiable victims, rather than being responsive to areas from which complaints arise; and the impact is likely to be wider in circumstances where a non-discrimination notice is issued than in the resolution of an individual case.[221] There are two main reasons for the lack of progress achieved through formal investigations. First, it appears that the Equal Opportunities Commission in Great Britain, in its early years, preferred to try a persuasive approach to tackling discrimination. It did not develop a strategy for the use of formal investigations and consequently did not address many of the problems it could have through this approach. Second, the powers of formal investigation of the Commission began to be challenged by respondents through the courts with the result that procedural difficulties began to outweigh other considerations.[222]

The use of judicial review was increasingly considered by both Commissions in the late 1980s as a means of tackling discrimination by a public body. Judicial review has been used successfully to challenge discriminatory arrangements relating to education, both in Great Britain

---

[219] See V. Sachs, 'The Equal Opportunities Commission: Ten Years On' (1986) 49 *MLR* 560–92; 'Ulster team wins credibility by applying the law' *Independent*, 13 April 1987.

[220] G. Appleby and E. Ellis, 'Formal Investigations: The Commission for Racial Equality and the Equal Opportunities Commission as Law Enforcement Agencies' (1984) *PL* 236–76, 239.

[221] Ibid.

[222] Ibid., and see chapter 13 for discussion of legal difficulties.

and in Northern Ireland.[223] However, its use suffered a setback in the Court of Appeal in November 1992 in a case seeking to challenge the statutory qualifying thresholds under the Employment Protection (Consolidation) Act 1978 to claim unfair dismissal or statutory redundancy pay.[224] These rights are conditional upon the claimant working 16 hours or more per week for at least 2 years (or 5 years if the claimant works between 8 hours and 16 hours per week). Given that the majority of part-time workers are women, this would appear to constitute indirect discrimination against women. The European Court of Justice had already held, in a German reference, that to treat part-time workers less favourably in relation to statutory sickness schemes amounted to indirect discrimination.[225] The Court of Appeal in this case held that the Equal Opportunities Commission had no locus standi to bring proceedings against the Secretary of State for Employment on this question. As of December 1992, leave to appeal this decision to the House of Lords had not yet been granted.

Remedies

The remedies available to a successful claimant in a sex discrimination case include a discretionary declaration of the rights of the parties, compensation, and a discretionary recommendation relating to the discrimination to which the complaint relates.[226] In practice, compensation is the most common form of remedy and it is available for both financial loss and injury to feelings, to a maximum of £10,000.[227] This upper limit is currently the subject of a reference to the European Court of Justice, in *Marshall* v. *Southampton and South-West Hampshire Area Health Authority (No. 2)*.[228] The House of Lords has asked for guidance on whether the limits are in breach of Community law, in particular of Article 6 of the Equal Treatment Directive which requires the introduction into national legal systems of 'such measures as are necessary to enable all persons who consider themselves wronged by the failure to apply to them the principle of equal treatment . . . to pursue their claims by judicial process after possible recourse to other competent authorities'. The Court of Appeal has already held in this case that Article 6 does not

---

[223] *R.* v. *Birmingham City Council ex parte Equal Opportunities Commission* (n. 50); also *In re the Equal Opportunities Commission and others, No. 1 and No. 2*, NICA, [1988] NILR, 10 *Bulletin of Judgements* 44–87, 88–105.

[224] *R.* v. *Secretary of State for Employment ex parte EOC*, [1993] 1 All ER 1022; Also EOC GB press release 'EOC continues to fight for part-time workers rights', 24 Nov. 1992.

[225] *Rinner-Kuhn* v. *FWW Spezial-Gebaudereinigung GmbH and Co.*, Case No. 171/88, [1989] IRLR 493 (ECJ).

[226] SDA 1975, s. 65. See further chapter 13.

[227] (1991) 37 *Equal Opportunities Rev.* 4.

[228] Case C-271/91, [IRLR] 1990 481-CA.

have 'direct effect' as its meaning is not sufficiently 'clear, precise and unconditional' to be relied upon in order to award compensation beyond the limit laid down by the Sex Discrimination Act.[229]

A further important element of the provisions relating to remedies concerns indirect discrimination; while the intention of the employer is irrelevant in proving unlawful indirect discrimination, it is relevant with regard to compensation. The Act precludes an award of compensation if it is proved by the employer that the requirement or condition was not applied with the intention of treating the applicant unfavourably.[230] It is open to question whether this provision complies with the need to ensure real and effective judicial protection against discrimination, as the European Court has held is required by Article 6 of the Equal Treatment Directive.[231]

In equal pay cases, there is no limit on the level of the award, which can include damages and backpay for up to two years prior to the date on which proceedings were implemented and the claimant's contract is deemed to be modified. The tribunal may also make a declaration of the rights of the parties.[232]

Voluntary compliance

Voluntary compliance through the use of 'positive action' is another means of addressing the problem of discrimination. 'Positive action' here means any action taken generally to make positive, conscious efforts to overcome the effects of discrimination, both past and present. Taking a broad view of the possibilities of positive action, it is clearly not confined to the situations outlined discussed above to which the exceptions to the Sex Discrimination Act would apply. It also covers actions to identify and replace discriminatory practices, a common form of positive action in the United Kingdom; indeed, as McCrudden points out, it is a legal duty.[233] It is clear that a number of employers have voluntarily taken steps to eradicate discrimination in employment policies and practices. There has been a growth in the number of organizations developing 'equal opportunities programmes' for example and in the 'professionalism' of equal opportunities officers.[234]

---

[229] [1990] IRLR 481 (CA).

[230] SDA 1975, s. 66(3).

[231] M. Rubenstein, 'The Equal Treatment Directive and UK Law', in McCrudden (ed.), *Women, Employment*, 102.

[232] Equal Pay Act 1970 (as amended), s. 2.

[233] See C. McCrudden, 'Rethinking Positive Action' (1986) 15 *Industrial LJ* 219, generally for an analysis of positive action in the UK.

[234] See C. Cockburn, *In the Way of Women: Men's Resistance to Sex Equality in Organisations* (1991).

Attempts have been made to go further, particularly in local government in the 1980s. The now defunct Greater London Council attempted to use local government purchasing power to promote equal opportunities practices amongst the contractors and suppliers from whom it bought goods and services, with some evidence of success in promoting change to employment practices amongst potential contractors.[235] Its use, however, was restricted by the Local Government Act 1988 which outlawed contract compliance in respect of sex discrimination issues altogether, and allows a limited form only to be continued by councils in respect of race discrimination issues, as discussed in chapter 13.[236] This is a matter of great regret, as the use of measures such as contract compliance and mandatory forms of positive action could act as powerful incentives for change and could complement existing legislation greatly.

IMPACT OF THE LEGISLATION

It is difficult to establish causal connections between legal reform and material change. There is a perennial debate about whether the law promotes or follows change which has taken place because of wider socio-economic factors. Perhaps it is possible, at most, only to establish correlations (or their absence) between the existence of legal rights and the scope of opportunities in the labour market.

As regards equal pay, the 1970 legislation appears to have had some impact on the average pay differentials between women and men. There was an initial improvement in the ratio of female and male pay until about 1978, from between 50 or 60 per cent, depending on whether hourly or weekly figures are considered, to a plateau of about 75 per cent, but this has remained unchanged since then.[237] Criticisms were levelled at the limited scope and content of the legislation, at procedural difficulties and judicial interpretations, and, by the early 1980s, many felt that it had outlived its usefulness. Gregory points out that the number of equal pay applications to the industrial tribunals dropped from 1,742 in 1976 to 26 in 1983, for example.[238] There was a growing acceptance, too, that the

---

[235] See Inner London Education Authority Equal Opportunities Contract Compliance Unit, *Contract Compliance: A Brief History*, (1990) for analysis.

[236] Attempts to introduce amendments to the bill in the House of Lords, to allow for the use of contract compliance in respect of sex equality, failed: (1988) 18 *Equal Opportunities Rev.* 31.

[237] See P. Z. Tzannatos and A. Zabalza, 'The Anatomy of the Rise of British Female Relative Wages in the 1970s: Evidence from the New Earnings Survey', (1984) *Brit. J. Industrial Relations*, 177–94.

[238] Gregory, *Sex, Race*, 23.

falling off in improvements in pay and the drop in the number of cases reflected the fact that most women do different work from men and therefore could not claim under the 'like work' provision of the Act. Nor could many women claim that their jobs had been 'rated as equivalent', as only a small proportion of employers had introduced sufficiently all encompassing job evaluation schemes.[239] Central government has remained hostile to the implementation of the equal value provision and, indeed, is the respondent in a number of major cases. Even when some employers, such as local government and service providers, undertook and implemented reviews of their pay structures, in anticipation of equal value claims, pay differentials remained, because of differences in productivity schemes and the like.[240]

One test of the effectiveness of the Sex Discrimination Act might be a reduction in occupational and industrial segregation but on the basis of official statistics there has been little change since the mid-1970s.[241] We have shown that the development of the right to equal treatment under the Sex Discrimination Act since 1975 has not always been straightforward. The indirect discrimination provisions particularly should have enabled the courts and tribunals to take a purposive approach to tackling the structural discrimination which is embedded deeply in the labour market. Instead, they have focused on technicalities and procedures and there is evidence of 'a judicial reluctance to widen the ambit of the debate over equality'.[242]

The industrial tribunal system has also been found to be a disappointment as a mechanism for dealing with individual complaints and ensuring the more widespread adoption of non-discriminatory employment practices. Leonard concluded, for example, from her analysis of successful claimants in the period 1980–4, that compensation awarded to individual applicants was inadequate; that often claimants were out of pocket themselves, despite winning their cases; that there were often real difficulties in collecting the compensation; and that the costs for complainants in terms of emotional stress and damaged future employment prospects were extremely high. Also, the fact of winning a case seemed

---

[239] C. McCrudden, 'Between Legality and Reality: The Implementation of Equal Pay for Work of Equal Value in Great Britain' (1991) 3 *Int. Rev. of Comparative Public Policy*, 177–217, 180.

[240] London Equal Value Steering Group (LEVEL), *A Question of Earnings: A Study of the Earnings of Blue Collar Employees in London Local Authorities* (1987); *Job Evaluation and Equal Value: A Study of White Collar Job Evaluation in London Local Authorities* (1987).

[241] See 'The Position of Women in the Labour Market: Trends and Developments in the Twelve Member States of the European Community, 1983–1990', *Women of Europe* Supplement No. 36, March 1992.

[242] O'Donovan and Szyzsczak, *Equality*, 115.

to have little impact on the conditions of co-workers, in that it led to little change in the organization.[243] Gregory, in her analysis of unsuccessful applicants in the 1985–6 period painted a similar, dismal picture.[244] Gregory and Leonard both recorded problems with lack of specialized knowledge of discrimination law in the tribunal system, often due to problems such as the allocation of cases widely across membership and the absence of compulsory formal training. Chambers and Horton looked at the impact of tribunal decisions on employers and their conclusions are also an indictment of a mechanism which had the potential to effect change in the workplace.[245] It is clear that changes are necessary not only to the legislation but also to the systems of advice, assistance and representation; to the tribunal procedure itself; and to the remedies and sanctions which can be applied, to ensure that they impact on employers both as a deterrent and as an incentive to introduce changes.

Comparison between the UK and other countries regarding the scope of rights available and the effectiveness of any equal treatment laws is difficult without a full discussion of the different legal systems, the different laws, and the different social and economic contexts within which the countries operate. Formally, all Community countries have, for example, fulfilled their obligation to implement the equality directives in their jurisdiction, but as directives leave it open to member states to decide the means to do this, the approaches have been different and it is difficult to say which is 'better' or 'worse'. While a significant number of references to the European Court have come from the United Kingdom, for example, it cannot be said that it is the worst in the Community. The references are as likely to be as a result of the fact there are larger independent equality Commissions here than elsewhere, which have strategically used Community law as a means of forcing change nationally.[246] On some counts, however, particularly in respect of child-care provision and parental leave, it is clear that the United Kingdom ranks among the lowest (with Ireland) in the Community.[247]

In respect of labour market statistics, comparative analyses reveal hazards in using simple indices. Norris, for example, draws attention to the fact that, in comparing OECD countries, overall measures of

---

[243] A. Leonard, *Pyrrhic Victories: Winning Sex Discrimination and Equal Pay Cases in The Industrial Tribunals, 1980–1984* (1987).

[244] J. Gregory, *Trial by Ordeal: A Study of People who Lost Equal Pay and Sex Discrimination Cases in the Industrial Tribunals during 1985 and 1986* (1989).

[245] G. Chambers and C. Horton, *Promoting Sex Equality: The Role of Industrial Tribunals* (1990).

[246] E. Szyszczak, 'L'Espace sociale européenne: Reality, Dreams or Nightmares?' (1990) 33 *German Yearbook of International Law* 284–307, 304.

[247] See P. Moss, *Childcare and Equality of Opportunity* (1988).

segregation produce different orderings when broken down by social class.[248] Norris also argues that success in desegregating occupations is not conclusively related to improvements in earnings. The entry by some women into middle-class occupations does not affect overall earnings' ratios if most women continue to enter poorly paid occupations. If, as in Italy, there are relatively fewer women in prestigious jobs than in the United States but working-class women are better paid than elsewhere, women as a group are better off.[249] This would lend support to the view in the United Kingdom that it is too difficult to improve women's economic status, on average, through changing shifting recruitment and promotion patterns and that the promotion of 'equal value' is a promising alternative, despite its problems.

EUROPEAN AND INTERNATIONAL INFLUENCES

It has been demonstrated in the discussion of the development of the legislation and of case-law that the European Community has had an enormous impact on women's employment rights in the United Kingdom. In addition to ensuring improvements to the substantive scope of rights available in the United Kingdom, the influence of the Community has also had symbolic value, keeping alive and even generating 'a legal discourse on equality and individual rights in a political climate which utilizes the principle of "de-regulation" to mask hostility towards individual and collective rights in the labour market'.[250]

In relation to other international obligations relevant for women's rights, the United Kingdom has signed the main human rights instruments of the United Nations, although it has not incorporated any of these into domestic law.[251] It is a party to two of the Conventions which deal specifically with women's rights: the Convention on the Political Rights of Women, 1953, which entered into force on 7 July 1954; and the International Convention on the Elimination of All Forms of Discrimination against Women, 1979, which entered into force on 3 September 1981; neither is incorporated into domestic law. Indeed, the influence of the United Nations generally has been largely symbolic rather than of practical significance in the furtherance of women's rights, although the United Kingdom is required, under Article 18 of the 1979 Convention, to

[248] P. Norris, *Politics and Sex Equality* (1987), 62–9.
[249] Ibid. 69.
[250] Szyszczak, 'L'Espace', 297.
[251] Universal Declaration of Human Rights 1948; International Covenant on Economic, Social and Cultural Rights, 1966; International Covenant on Civil and Political Rights, 1966.

produce regular reports on the position of women for the consideration of the Committee for the Status of Women. This ensures that some attention is paid, if only at four-yearly intervals, to the range of issues addressed in the Convention.[252] The United Kingdom also participated in the activities surrounding the UN Decade for Women, 1976–85, which stimulated an internationalization of the women's movement and focused further attention on the need to improve the condition of women.[253]

The more specific focus of the International Labour Organization (ILO) which, as a specialized body under the auspices of the United Nations, works towards the establishment of international labour standards, could be said to have had more significance. While it has not incorporated it into domestic law, the United Kingdom is a party to the Convention concerning Equal Remuneration for Men and Women Workers for Work of Equal Value, 1951, and the standard of equal pay contained therein—equal pay for work of equal value—did feature in discussions surrounding the implementation of domestic equal pay legislation. The ILO has adopted a number of other conventions concerning women's employment[254] and, of course, on other labour issues, but the United Kingdom in later years has denounced several of these—such as that on minimum-wage-fixing machinery.[255] Indeed, in some respects, its recent labour legislation has been condemned by ILO Committees as being incompatible with ILO standards.[256]

The European Convention on Human Rights has had limited practical impact on the development of the right to non-discrimination in employment for women in the United Kingdom.[257] It is a party to the European Social Charter of 1961, which specifies in its preamble that the enjoyment of social rights should be secured without discrimination on grounds of sex and other factors. It has been said that the Charter has had some impact on domestic labour legislation, for example the Employment Protection (Consolidation) Act 1978 was influenced by Article 8(1) of the Charter on paid leave for women employees before and after child-

---

[252] The scope of the Convention extends well beyond employment issues, to cover health, education, welfare, public and political life, traffic in women and exploitation of prostitution, nationality, equality before the law, etc.

[253] See Home Office, *The Nairobi Forward Looking Strategies for the Advancement of Women: A Review* (1987), for the UK response to the UN Decade for Women.

[254] e.g. Convention on Night Work for Women, revised 1990.

[255] See C. Brewster and P. Teague, *European Community Social Policy: Its Impact on the UK* (1989), 6.

[256] See chapter 8.

[257] Art. 14, which provides that the rights and freedoms contained in the Convention generally shall be available without discrimination on a number of grounds, including sex, has had minimal effect in this context.

birth.[258] Despite the limited impact of these conventions through the domestic courts, however, the European Court of Justice has stated that it will take the aims of these instruments into account as far as possible in its deliberations.[259]

CONCLUSION

There has been a significant improvement in the scope of women's rights in the last forty-five years, particularly in the field of employment but also in other areas, as other chapters have demonstrated. The combination of the existence of European Community sex equality law and the role of the independent equality Commissions in promoting change has been decisive, particularly in comparison to some of the other rights discussed in this book. However, it cannot be said that equality for women has yet been achieved. Many obstacles remain, some of which this chapter has sought to highlight: for example, the conceptual underpinnings of the sex discrimination and equal pay legislation, and procedural and interpretational difficulties. Further research is necessary on the impact of these problems, in order to give support to the need for legislative change. In particular, consideration needs to be given to the appropriate conceptual basis for any future legislation.

The debate about women's rights that is taking place all over the world does not focus solely upon specific legislation or proposals for reform. It is also about the meaning of citizenship and what is necessary for women to be full participants in society. In political theory, as in legal theory, there are complex arguments about whether women can become autonomous citizens through being treated as if they were the same as men or whether it is necessary to acknowledge differences in their practical situations through the provision of special treatment. The choice of approach has a profound effect upon the capacity of women to participate on an equal footing but the 'correct' one is not always self-evident. What is important is for women to have the power and opportunity to define their own needs and corresponding rights and for the political will to exist to expand the boundaries of what may be considered 'possible' and to contribute to new agenda for those who have the power to bring about legal change.

---

[258] Brewster and Teague, *Social Policy*, 12.

[259] The Court has invoked the provisions of the European Convention on Human Rights in proceedings before it on many occasions, e.g. *R.* v. *Kirk*, Case No. 63/83, [1984] ECR 2689; *Johnston* v. *Chief Constable of the Royal Ulster Constabulary*, Case No. 222/84 [1986] IRLR 263 (ECJ). In 1977, the Community institutions issued a joint declaration expressing their commitment to the principles of the European Convention on Human Rights, OJ C103, 27 April 1977.

13

Racial Discrimination

CHRISTOPHER McCRUDDEN

INTRODUCTION

This chapter considers the development and use of legislation against racial discrimination in Britain, in particular the three major pieces of legislation against racial discrimination: the Race Relations Acts of 1965, 1968, and 1976. It considers the influence of foreign and international developments on the growth and interpretation of this legislation, and discusses the limited empirical research so far conducted into its operation in practice.

IMMIGRATION AND THE POLITICS OF ANTI-DISCRIMINATION LEGISLATION

Legislation against racial discrimination has been primarily concerned with discrimination against those of Afro-Carribean, African, and Asian origin. This is not to say that other minority groups are not discriminated against (such as those of Jewish or Irish origin, for example) and members of such groups have taken advantage of this legislation. The primary focus of this chapter, however, is largely on the development and use of the legislation in the context of discrimination against those of Afro-Carribean, African, and Asian origin. The 1991 British Census included, for the first time, a question on ethnic group.[1] Those who described themselves as White formed 94.5 per cent of the population. The remainder consisted of Black Carribean (0.9 per cent), Black African (0.4 per cent), other Black (0.3 per cent), Indian (1.5 per cent), Pakistani (0.9 per cent), Bangladeshi (0.3 per cent), and Chinese (0.3 per cent). Other groups amounted to 0.9 per cent (including 0.4 per cent Asian). These figures, however, hide the wide differences between regions; the geographical distribution of groups other than White was uneven. Black groups (taken together) comprised 13.5 per cent of Inner London, for example, but 0.1

---

[1] Office of Population Censuses and Surveys, 1991 Census: Great Britain, CEN 91 CM 56 (OPCS, London, 1992). See also J. Haskey, 'The Ethnic Minority-Populations Resident in Private Households: Estimates by County and Metropolitan District of England and Wales' (Spring 1991) 63 *Population Trends* 22–35.

per cent of Northern England. So too those describing themselves as Indian, Pakistani, or Bangladeshi comprised 9.7 per cent of the West Midlands but 0.4 per cent of South-West England.

The presence of these groups is largely (though not entirely, as Dummett shows in Chapter 11) a post-Second-World-War development. Over nine-tenths of black and brown adults in Britain in the mid-1970s were people born abroad who had come to Britain within the previous twenty years; half had come within the previous twelve years.[2] The main sources of non-white immigration to Britain were the West Indies and the Indian sub-continent. Immigration from the former took place in the 1950s and early 1960s, with more than half from Jamaica. Many came with official encouragement to take up jobs which could not otherwise be filled. Restrictions on immigration into the United States also encouraged West Indian emigrants to look elsewhere and Britain was an obvious alter-native, given the former colonial links. Immigration from the Indian sub-continent took place mainly in the 1960s and to a lesser extent in the 1970s. In addition, however, Asians from East Africa came from Uganda and Kenya during both decades, in part because of policies of African-ization adopted in both countries. The diversity of the origins of the minority-group population is thus considerable.

The relative recency of this immigration meant that Britain had for the first time to find solutions to the new problems which have arisen from its transformation to a multi-racial society. The perception within Britain of the newness of the issues arising from these developments is one of the most striking features of the initial political reactions to this post-Second-World-War immigration.[3] Katznelson, writing in 1973 captured the point well: 'For the mid-twentieth-century politician, class issues presented few unknowns. In discussing issues of class, he knows, broadly, what is expected of him; his rhetoric and behaviour conform on the whole to relatively clearcut norms . . . [O]n the other hand, domestic racial issues, when first raised, were worrying, confusing, incoherent, anomic.'[4] Katznelson argued that mainstream British political reaction to racial minorities attempted 'to structure the politics of race out of conventional politics'.[5] He identified three distinct, yet chronologically overlapping stages in this process. In the first stage, the pre-political consensus of 1948–61, 'both [major political parties] were in substantial agreement, if not on what to do, at least on not what to do'. Second, in the period of fundamental debate (1958–65) the parties, 'largely because of internal imperatives, reacted differently to

² D. J. Smith, *Racial Disadvantage in Britain* (1977), 14.
³ An overview of the history of race relations in Britain can be found in D. Hiro, *Black British, White British*, 2nd edn. (1991).
⁴ I. Katznelson, *Black Men, White Cities* (1973), 125.
⁵ Katznelson, *Black Men*, 125–6, from which the following quotations are also taken.

the politicisation of race. The pre-political consensus broke down as they differed over immigration control legislation. In potential and in fact, race became a potent partisan electoral issue.'

Third, there was the period of political consensus (from 1965 until the mid-1970s), 'when the front benches of the two major parties developed a new consensus, politically arrived at, to depoliticise race once again'. Non-discrimination within Britain was the policy of all three major political parties, a stand which was reinforced by the growing internationalization of the issue of race relations. This position was also assisted by low unemployment and considerable economic growth during the 1950s and 1960s, but also by the gradual acceptance by both Labour and Conservative Parties of the desirability of immigration controls at the point of entry, beginning in 1962. Anti-discrimination legislation also emerged as part of the political concensus on racial issues.

THE FIRST RACE RELATIONS ACT

Prior to the first race relations legislation in 1965, there was no general rule, policy, or principle in common law directly relevant to combating racial discrimination, though in specific situations the common law, as a by-product of its regulation of other areas, did provide a limited remedy indirectly to one who could prove such discrimination.[6] 'English law', wrote Griffith prior to the 1965 legislation, 'has very little to say about discrimination.'[7]

There were several reasons why this was so. Most important was the entrenched concept of freedom of contract. 'When this freedom came into conflict with the post-Second World War concept of "freedom from discrimination"', wrote Hepple, 'it is hardly surprising that the older and better appreciated freedom has prevailed.'[8] In the context of employment discrimination, freedom of contract was supplemented by a tradition of legal abstention from industrial-relations issues, which we shall examine subsequently. The virtual abstention by the courts from effective substantive or procedural review of the acts of public bodies by way of administrative law from the First World War until around the beginning of the 1960s also contributed to the absence of a non-discrimination principle becoming embedded in the common law.[9]

---

[6] See e.g. *Constantine* v. *Imperial Hotels, Ltd.* [1944] KB 693; *In re Dominion Students' Hall Trust* [1947] Ch 183; *Re Meres' Will Trusts, The Times*, 4 May 1957; *Clayton* v. *Ramsden* [1943] AC 320; but see *Re Lysaght, Hall* v. *Royal College of Surgeons* [1966] 1 Ch. 191, [1965] 2 All ER 888; *Blathwayth* v. *Baron Cawley* [1975] 3 All ER 625 (HL).

[7] J. A. G. Griffith *et al.*, *Coloured Immigrants in Britain* (1960), 171.

[8] B. Hepple, *Race, Jobs and the Law in Britain*, 2nd edn. (1970), 155.

[9] See e.g. *Weinberger* v. *Inglis* [1919] AC 606; *Short* v. *Poole Corporation* [1926] Ch. 66.

The inadequacies of the common law and an emerging perception of a growth in racial discrimination against the new black and brown immigrants stimulated several unsuccessful initiatives by back-bench MPs during the 1950s and early 1960s to secure legislation designed to combat racial discrimination directly. Initially this legislation was opposed by the Conservative government. The Labour Party in opposition was neutral. By 1964, however, the Labour Party had moved from a position of neutrality to one of support, a decision which Katznelson attributes in part to the effect of the Notting Hill and Nottingham 'race riots' in the summer of 1958. The resurgence of neo-Nazism in the early 1960s also contributed. In part, perhaps, support for anti-discrimination legislation was also thought to counterbalance support for the increasingly restrictive immigration legislation, which began in 1962 with the Commonwealth Immigrants Act, discussed by Dummett in Chapter 11.

The first Race Relations Bill to be proposed by government was formally introduced into the House of Commons in 1965 by the recently elected Labour government. In addition to provisions concerning incitement to racial hatred (discussed by Gardner in Chapter 7), the bill proposed to make discrimination in some limited respects a criminal offence punishable with a fine. Discrimination in hotels, public houses, restaurants, theatres, cinemas, public transport, and any place maintained by a public authority would have been made punishable by fines of up to £100. Prosecutions would only have been undertaken, however, with the authority of the Director of Public Prosecutions. Though racial restrictions by property sellers on the disposal of tenancies were made unenforceable, discrimination in housing was not to be made unlawful, nor was discrimination in employment.

Considerable pressure was put on the Government to change this proposed criminal-law method of enforcement. Drawing on North American experience of the enforcement of anti-discrimination legislation (at that time the American Civil Rights Movement was at the height of its international reputation), a number of groups constituting an incipient British civil rights movement argued for a system of enforcement through a specially constituted administrative body as an alternative to the use of the criminal or the ordinary civil legal process.[10] Enforcement by such a body would emphasize the elimination of discrimination in the public interest, rather than the punishment of the individual discriminator. Such a body, it was thought, could also be given powers which would make it more effective than the ordinary civil or criminal processes.[11]

Partly as a result of the pressure inside and outside Parliament, significant

---

[10] B. Heinemann, *The Politics of the Powerless* (1972), *passim*.
[11] J. Jowell, 'Administrative Enforcement of Laws Against Discrimination' (1965) PL 119–86.

changes were made in the bill's anti-discrimination provisions before it became law.[12] Criminal sanctions were retained only for incitement to racial hatred. A specialized agency, the Race Relations Board (RRB), was established to investigate breaches of the legislation and secure compliance. Local conciliation committees were set up by the Board to investigate complaints from those who considered that they had been discriminated against. If their attempts to settle complaints by conciliation failed, the local committees reported to the Board. If, in turn, the Board found that there had been discrimination and considered it likely that the discrimination would continue, it could refer the case to the Attorney-General who would then be empowered to bring proceedings in court seeking an injunction which would require the discrimination to cease.

Neither the Board nor the local committees, however, had the power to summon witnesses, subpoena documents, require answers to questions, or issue orders. Rather than being adopted as a more effective method of enforcement, conciliation was included, according to a government spokesman, 'to avoid bringing the flavour of criminality into the delicate question of race relations'.[13] It was a continuing theme throughout the parliamentary debates that the Government hoped that litigation would not arise under the Act and actively wanted to prevent it.

THE SECOND RACE RELATIONS ACT

Between 1965 and 1968 there was a well-organized campaign by a number of pressure groups (and by the Race Relations Board itself), for an extension and a strengthening of the provisions of the 1965 Act.[14] A number of the influences apparent before the 1965 Act remained in evidence. The influential Street Report[15] proposed a model of enforcement based on the American experience and this influenced the approach subsequently adopted in the 1968 Act, in particular the inclusion of a proactive power of investigation by the Race Relations Board which did not depend on a complaint being made by an individual. The enactment of the Commonwealth Immigrants Act in 1968 may again have persuaded some to support fresh anti-discrimination legislation as another counterbalance. There was also a substantial amount of cross-party support for new legislation.

To these influences were added some new elements. Substantial empirical

[12] Race Relations Act 1965.
[13] Official Report, HL 286, 26 July 1965, col. 1006.
[14] See A. Lester and G. Bindman, *Race and the Law* (1972).
[15] H. Street, G. Howe, and G. Bindman, *Street Report on Anti-Discrimination Legislation* (1967).

research demonstrated authoritatively the extent of racial discrimination in areas not covered by the 1965 Act. The campaign for a new Act stressed that it should for the first time prohibit discrimination in housing and employment, and drew support from an influential and well-publicized report by Political and Economic Planning (PEP), the forerunner of the Policy Studies Institute, which found that racial discrimination in these and other areas varied in extent but was generally substantial.[16] Additional influences not present during the campaign for the 1965 Act were the arguments made by the Race Relations Board itself for an extension of its powers, and the presence of a notably receptive minister at the Home Office in the shape of Mr Roy Jenkins.

A new Race Relations Act was passed in 1968, and this prohibited discrimination in both public and private employment and housing, subject to certain exceptions. The 1968 Act retained the two-tier enforcement mechanism of Race Relations Board plus local concilation committees. Once the Board had received a complaint, it first determined whether it had jurisdiction and whether there had in fact been discrimination, and then tried to conciliate the dispute. If this was unsuccessful the Board was empowered to bring a case against the discriminator in one of a number of specially designated county courts, in which 'race relations assessors' sat with the judge. Individuals could not take discrimination cases directly to the county courts; that was solely the responsibility of the Board. The Board was also given an additional power in section 17 of the Act to initiate investigations without an individual complaint; this was limited, however, by the need for the Board to suspect that a particular person had been discriminated against.

There was, however, a different procedure for the settling of complaints concerning employment discrimination. Prior to 1967, the Confederation of British Industry (CBI) and the Trades Union Congress (TUC) opposed legislation which would have prohibited discrimination in employment on the ground that it conflicted with the British tradition of 'voluntarism' or self-regulation in industrial relations. In this context, voluntarism involved a preference for collective bargaining over state regulation as a method of settling wages and other terms and conditions of employment; a preference for keeping industrial disputes out of court by preserving a non-legalistic type of collective bargaining; and a preference for retaining the autonomy of the bargaining parties.[17]

When the CBI and the TUC finally agreed to the inclusion of employment in the Act, it was only on condition that industry dispute procedures

---

[16] Political and Economic Planning, *Racial Discrimination* (1967).

[17] A. Flanders, 'The Tradition of Voluntarism' (1974) 12 *Brit. J. Industrial Relations* 352, at p. 354.

should be exhausted first. The procedure that was adopted in the 1968 Act required that any complaints of discrimination in employment should be dealt with, initially, not by the Race Relations Board but by the Department of Employment. If there was suitable voluntary machinery to deal with the complaint within the industry concerned, the Department would send it back to that 'industry machinery'. Only if none existed, or if the complainant was appealing against a decision of this industry machinery, did the Race Relations Board have any jurisdiction to hear the complaint. Only after *that* might the courts be involved. The use of industry machinery was thus a compromise between the value of self-regulation in industrial relations and the need for government intervention where the parties themselves were unable to solve the problem.

The 1968 Act also established another institution parallel to the Race Relations Board: the Community Relations Commission (CRC), which was given the tasks of encouraging 'harmonious community relations',[18] co-ordinating on a national basis the measures adopted for that purpose by others, advising the Home Secretary on any matter referred to the Commission by him, and making recommendations to him on any matter which the Commission considered should be brought to his attention. 'Community relations' was defined as 'relations within the community between people of different colour, race, or ethnic or national origin'.[19]

ENFORCEMENT PROBLEMS OF THE ACT BETWEEN 1968 AND 1975

In practice, there were substantial deficiencies in the coverage and enforcement provisions of the 1968 Act. Discrimination, as defined by the Act, was difficult to prove.[20] The Act did not apply to the present effects of past discrimination or to unintentional discrimination. The operation of the Act demonstrated many of the limitations of a largely complaints-based approach. There appears to have been a lack of perception of discrimination by the ethnic groups themselves. Even where individuals understood that discrimination was widespread, it was exceptional for it to be perceived in their particular circumstances. There appears to have been a lack of knowledge of the Board's existence and of what its functions were. There was a reluctance to complain. This appears to have been greatest, understandably, where discrimination in terms and conditions of employment, training, or promotion were involved.[21] Growing

[18] Race Relations Act 1968, s. 25.
[19] RRA 1968, s. 25(3).
[20] See Lester and Bindman, *Race and Law*, 100.
[21] See RRB Report for 1969–70 (HC 309, 1970), para. 46; *RRB Report for 1970–1* (HC 448, 1971), para. 65.

unemployment was identified by the Board as a significant deterrent to making complaints. These factors contributed to the low number of complaints.

The limited powers given to the Race Relations Board to investigate when no complaint had been received were small compensation. In most cases the Board waited for a specific informant to give the Board information rather than itself seek instances where this section 17 power might be used. In part, this may have been due to the discontinuance by the Board of its research function and the lack of available information.[22] The Board was, moreover, required to find a specific instance of discrimination against an identifiable person before a finding of discrimination could be made, even in a section 17 case. This had the consequence that, even where it seemed that equality of opportunity was not being afforded to members of a minority group, the Board was often unwilling to intervene without a complaint by an individual.[23] The Board did not necessarily depend on complaints from the victims themselves but in the main it did depend on fairly specific information. In most cases, then, the Race Relations Board could do little until it received a complaint. It was unable to conduct a systematic planned campaign using legal enforcement against discriminatory practices.

The Board's investigations were further handicapped because it had no power to require relevant evidence to be given to it prior to litigation. On occasion investigations were abandoned for lack of information without forming an opinion.[24] The lack of such powers also increased the number of decisions in which an opinion of no discrimination was formed. Although it had a wide discretion to form an opinion that there had been unlawful discrimination, it was 'inevitably reluctant to do so in the absence of convincing evidence because of the risk that their opinion [would] subsequently be rejected in the courts'.[25] Thus the frequency with which complaints were rejected was thought 'not [to] reflect the incidence of discrimination or the weakness of the complaints but rather the inability of the conciliation machinery to obtain information on its own initiative to plan an effective strategy against the most important areas of racial discrimination'.[26] The lack of such powers had the further effect of extensively delaying investigation of complaints.[27]

---

[22] See Nuffield Foundation, *Problems and Prospects of Socio-Legal Research: Proceedings of a Seminar, Nuffield College, June–July, 1971* (Mar. 1972), passim.

[23] See RRB evidence to House of Commons Select Committee on Race Relations, *The Organisation of Race Relations Administration* (HC 448, 1975), Q. 252; *RRB Report for 1971–2* (HC 296, 1972), paras. 77–80. But see contra *RRB Report for 1973* (HC 144, 1974), para. 42.

[24] See e.g. *RRB Report for 1973* (HC 144, 1974), 40 (case 3) and para. 42.

[25] Lester and Bindman, *Race and Law*, 305.

[26] Ibid. 305–6.

[27] House of Commons Select Committee on Race Relations, Employment, *Report* (HC 448, 1975), para. 19.

A number of other features of the 1968 Act proved disappointing. The two-tier structure of conciliation committee decisions being reviewed by the Board increased the time spent on investigation, and, particularly in the early years of the Board's existence, provoked some bitter controversies between Board and committees. The hope that the special enforcement provisions for dealing with employment complaints would stimulate the growth of voluntary procedures was not borne out; there were few cases in which industry machinery found discrimination; and on the whole it proved cumbersome and of questionable value to both industry and race relations groups. Finally, even when discrimination was proven the remedies available to the Board, and more particularly to the courts, were extremely limited.

Relatively few cases under the 1968 Act reached the courts and even fewer involved substantial questions of interpretation by the appellate courts. In those which did, the courts adopted an approach to the interpretation of the statute which was narrow and unhelpful, in the main. This approach is best illustrated by three cases which reached the House of Lords. In *Ealing London Borough Council* v. *Race Relations Board*[28] the House of Lords held that to discriminate on grounds of nationality did not contravene the 1968 Act's prohibition on discrimination 'on grounds of national origin'. In *Charter* v. *Race Relations Board*[29] the House of Lords held that the decision of a Conservative club to reject an application for membership was not unlawful under the Act. The Act provided that it was unlawful for any person concerned with the provision to the public or a section of the public of any services to discriminate in the provision of those services. The majority of the House of Lords held that the club members were not a 'section of the public'. Nor, in a second case[30] involving discrimination by a Labour club against the members of another Labour club with which they were associated, was the House of Lords prepared to hold that the associates constituted a 'section of the public', even though there were some 4,000 clubs in the association. Since the Act was a restriction on the common law liberty to discriminate, it should, the House of Lords held, be read restrictively.[31] These decisions were subsequently overturned or modified when the third Race Relations Act was passed in 1976, to which we now turn.

---

[28] [1972] AC 342.

[29] [1973] AC 868 (HL).

[30] *Dockers' Labour Club* v. *RRB* [1976] AC 285, [1974] 3 WLR 533 (HL).

[31] Rather strangely, however, in light of these decisions, in *RRB* v. *Applin* [1975] AC 259 (HL) the House of Lords held that foster parents were concerned with the provision of services or facilities to a section of the public.

THE CURRENT STATUTORY POSITION

The meaning of discrimination

The Race Relations Act 1976, which replaced the 1968 Act, is the main Act currently in force dealing with race discrimination. It makes discrimination on grounds of nationality unlawful, in addition to the discrimination on grounds of race, colour, and ethnic and national origins prohibited in the 1968 Act, thus overturning the *Ealing* case. The interpretation of discrimination adopted in the 1968 Act was limited for two additional reasons. First, it was difficult to establish that discrimination had occurred because proof was required of a person's discriminatory intention. Second, legislation against discrimination as defined in these terms did little to prevent the use of criteria which had the effect of excluding disproportionate numbers of minority groups, irrespective of intention. For example, the 1968 Act made it unlawful for an employer to consider race when a black worker sought a job in a factory. However, the 1968 Act ignored the fact that this person was less likely to be hired for reasons other than those intentionally connected with race. That is, there would be a greater chance that a black worker would be deficient in those attributes which make an applicant successful. This could be due to the present effect of past intentional discrimination, immigrant disadvantage, or 'institutional discrimination' (the unjustified use of criteria having an exclusionary effect even though exclusion may not have been the intended effect of adopting those criteria).

The 1976 Act took account of these limitations of the 1968 Act by expanding the meaning of discrimination to include not only 'direct' but also 'indirect discrimination'.[32] A person alleging indirect discrimination has to prove three elements. Consider an employment situation. First, does the employer have a 'requirement or condition' which he applies to both his Pakistani-origin and non-Pakistani-origin workers? (For example, does he require passing a language proficiency test before any worker can be considered for promotion?) Second, if so, is this requirement or condition such that the proportion of Pakistanis who are able to comply with it is 'considerably smaller' than the proportion of non-Pakistani-origin workers who are able to comply with it? (In our example, do considerably fewer Pakistani workers pass the language proficiency test than the others?) Third, is the person alleging discrimination 'unable to comply' with the requirement? (In our example, is the person who is complaining of discrimination unable to pass the test himself?) If the person alleging discrimination has been able to establish the three elements, then the employer must show that the requirement is 'justifiable'. If the

---

[32] RRA 1976, s. 1(1)(*b*).

employer is not able to do so, then indirect discrimination has been established.

Positive action

The 1976 Act also brought to British law a considerably broadened concept of equality of opportunity between the races.[33] The 1976 Act explicitly permits, though it does not require, a limited measure of 'positive action' in favour of racial minority groups, in the form of exceptions to the general prohibition of discrimination. Section 35 of the Act contains a general exception for conduct intended to meet the special needs of particular racial groups as regards education, training, welfare, or ancillary benefits. In addition, employers, training bodies, trade unions, and employers' organizations may, though they need not, operate a system of 'positive action'.[34] For example, if a racial group has been under-represented in an occupation within twelve months prior to commencement of a training programme, then the training body may lawfully discriminate by limiting access to training for such work to that racial group, or take steps to encourage members of that racial group to take advantage of opportunities for doing that work. Employers may take similar action under similar conditions, taking into account the population of the area from which the employer normally recruits. In addition, if appropriate conditions are fulfilled, trade unions and employers' organizations may lawfully encourage members of a particular racial group to take advantage of opportunities to hold posts in the organization or afford members of that racial group access to training facilities which will prepare them to hold such posts.

The Act, however, does not generally make it lawful for the employer to discriminate at the point of selection for such work or for organizations to discriminate in admitting people to membership or in appointing to posts in the organization. In only one circumstance is this permitted: where being of a particular racial group is a 'genuine occupational qualification'.[35] This includes the situation where part of the job is to provide persons of a particular racial group with personal services promoting their welfare, and those services can most effectively be provided by a person of that racial group.

---

[33] We shall see below that the provisions were amended by the Employment Act 1989. The discussion in this paragraph reflects the post-1989 position.

[34] RRA 1976, ss. 37–8.

[35] Ibid., s. 5. Several cases have considered this provision in this context, particularly *London Borough of Lambeth* v. *CRE* [1990] IRLR 230 (CA), *Totterham Green Under-fives' Centre* v. *Marshall* [1989] IRLR 147 (EAT), and *Totterham Green Under-fives' Centre* v. *Marshall (No. 2)* [1991] IRLR 162 (EAT).

Scope of the anti-discrimination provisions

The 1976 Act prohibits direct and indirect discrimination in many circumstances. It is unlawful for an employer to discriminate in the recruitment of new employees or in the treatment of existing employees (for example, in such matters as promotion, training, transfer, and dismissal).[36] It is unlawful for a trade union to discriminate against its members.[37] Discrimination in recruitment to the police force is unlawful, as is discrimination in terms and conditions afforded to officers once they have joined.[38] Discrimination is unlawful as regards admissions to, and the provision of facilities in, educational establishments in both the public and private sectors.[39] Bodies in the public sector of education are also placed under a duty to ensure that the facilities for education are provided without racial discrimination.[40] Discrimination in the provision to the public or a section of the public of goods, facilities, and services (including, for example, loans, finance, mortgages, and facilities for entertainment),[41] and of housing accommodation[42] is unlawful. In an attempt to deal with the 'clubs cases' discussed above, it is unlawful for clubs and associations with 25 or more members to discriminate in the selection of new members.[43] In general, the provisions of the Act apply to acts done by Ministers of the Crown and Government Departments.[44] Racial discrimination by prison officers against prisoners is unlawful.[45] It is also unlawful to victimize any person for alleging discrimination under the Act.[46]

Section 71 of the Act imposes an additional general duty on local authorities to take account of the racial dimension in the exercise of their functions.[47] It requires local authorities to make appropriate arrangements with a view to securing that their functions are carried out with due regard to the need to eliminate unlawful discrimination and to promote equality of opportunity and good relations between persons of different racial groups.

The scope of the 1976 Act is, however, subject to several important restrictions. There is an exception for all clubs whose main object is to benefit members of a particular race or ethnic group or national origin, though not members of a particular colour.[48] With regard to employment, there is a limited exception where being of a particular racial group is a 'genuine occupational qualification' for a particular job.[49] The Act does not apply to employment within a private household.[50] The Act is also

---

[36] The Employment Act 1989, s. 4.   [37] Ibid., s. 11.   [38] Ibid., s. 16.
[39] Ibid., ss. 17–18.   [40] Ibid., s. 19.   [41] Ibid., s. 20.
[42] Ibid., s. 21.   [43] Ibid., s. 25.   [44] Ibid., ss. 75 and 76.
[45] *Alexander* v. *Home Office* [1988] 1 WLR 968 (CA).
[46] RRA 1976, s. 2.
[47] Ibid., s. 71.   [48] Ibid., s. 26.
[49] Ibid., s. 5.   [50] Ibid., s. 4(3).

limited in its geographical scope. It applies to England, Scotland, and Wales, but not to Northern Ireland. An employer may also discriminate in, or in connection with, employment on a ship, if the person applied or was engaged for that employment outside Great Britain.[51] With respect to housing, there is a limited exception for owner-occupiers, and the leasing of small premises is excluded.[52] Differential educational arrangements relating to people who are non-residents in Britain are excepted.[53] Fostering and the boarding of people in need of special care is excluded from the scope of the Act.[54] Employment in the service of the Crown or public bodies, although generally covered by the Act, is allowed to be restricted to persons of particular birth, nationality, descent, or residence.[55] There are a number of general exceptions in the Act: for discrimination (save on the ground of colour) in order to comply with charitable instruments;[56] for nationality, place of birth, or residence requirements relating to participation in sports and games.[57] More generally still, exceptions exist for acts taken to safeguard national security.[58] Lastly, there is an important exception for acts done under statutory authority.[59] This last exception effectively rules out challenges under the Act to any discriminatory actions taken under immigration legislation (as discussed by Dummett in Chapter 11).[60]

Legal enforcement and the Commission for Racial Equality

Two principal methods were incorporated in the 1976 Act to enable the legal process to be used to enforce the anti-discrimination requirements.

[51] Ibid., ss. 6 and 9. However, the Race Relations (Offshore Employment) Order 1987, SI 1987/929, brings employment relating to exploration of the seabed, etc. within the scope of the Act.

[52] Ibid., s. 22.

[53] Ibid., s. 36.

[54] Ibid., s. 23(2). This provision overturned the *Applin* decision, see above n. 31. This issue has remained highly controversial. Several cases have considered issues of inter-racial adoption and child-care outside the context of the RRA 1976: see e.g. *Re P. (a minor) (transracial placement)* [1990] FCR 260, [1990] 1 FLR 96 (CA); *Re N. (a minor) (adoption)* [1990] 1 FCR 58. The Children Act 1989 provides that local authorities are under the general duty, in relation to children looked after by them, to have regard to the child's religious persuasion, cultural and linguistic background, and racial origin (section 22(5)(*c*)). Furthermore, in making arrangements either for the provision of day care within their area or designed to encourage persons to act as local authority foster parents, the local authority is required to have regard to the different racial groups to which children in need in their area belong (Sch. 2, para. 11).

[55] RRA 1976, s. 75(5).

[56] Ibid., s. 34.   [57] Ibid., s. 39.   [58] Ibid., s. 42.   [59] Ibid., s. 41.

[60] *Amin* v. *Entry Clearance Officer, Bombay* [1983] 2 AC 818 (HL). The CRE succeeded, despite opposition from the Home Office (see *Home Office* v. *CRE* [1982] QB 385), in holding a general formal investigation into the immigration service, but such investigations may not allege unlawful discrimination, see pp. 422–3 below.

One method was investigation by a new regulatory agency. The second was the initiation of complaints by individuals aggrieved by actions which they consider to be discriminatory. Both methods differ in significant respects from the previous equivalent methods in the 1968 Act.

The new regulatory agency established under the Act (the Commission for Racial Equality) replaced both the Race Relations Board and the Community Relations Commission. In addition to its duty to work for the elimination of discrimination, the CRE has a duty to promote equality of opportunity, and good relations, between persons of different racial groups generally.[61] It is empowered to issue Codes of Practice for promoting equality of opportunity in employment,[62] and since 1988 in housing.[63] These Codes may be taken into account in litigation, though they are not binding. The Commission is also authorized[64] to conduct formal investigations for any purpose connected with the carrying out of its statutory duties. Given its importance in the scheme of the Act, it is worth discussing this power in somewhat greater detail.

Formal investigations

It appears to have been envisaged that there would be three types of formal investigation available to the Commission. The first type was an investigation into the activities of a 'named person' alleging that those activities unlawfully discriminated. The second type was an investigation into the activities of a 'named person' but in which there was no, or no sufficient, evidence known to the CRE that the person had unlawfully discriminated. The third type was an investigation not into a named person, but into a particular area of activity, for example banking, insurance, etc. Again, this third type would not make allegations of unlawful discrimination. (We shall see subsequently that the courts have held that the second type was not, in fact, envisaged in the Act.)[65]

Before a formal investigation begins, where the terms of reference of the proposed investigation confine it to the activities of named persons and the Commission proposes to investigate any unlawful discrimination which it believes a person may have done, it is required to inform that person of its belief and its proposal to investigate.[66] In addition it is required to offer him an opportunity of making oral and written representations, and of being represented by counsel or a solicitor or by some other person of his choice, subject to an objection by the Commission on the ground that the choice is 'unsuitable'.[67] Investigations into specific

---

[61] RRA 1976, s. 43.   [62] Ibid., s. 47.   [63] Housing Act 1988, s. 137.
[64] RRA 1976, s. 48(1).   [65] See text at nn. 149, 150 below.
[66] RRA 1976, s. 49.   [67] Ibid., s. 49(4).

alleged acts of discrimination automatically attract powers of subpoena; such powers are conferred on the CRE only at the discretion of the Secretary of State in the case of an investigation other than into the action of a named person.

Where a formal investigation by the CRE has disclosed unlawful discrimination, it is empowered to issue a non-discrimination notice requiring the respondent not to commit any such discrimination and, where compliance with that requirement involves changes in any of his practices or other arrangements, to inform the CRE that he has effected those changes and what those changes are.[68] The CRE cannot technically prescribe what changes in practice are to occur. It is for the respondent to tell the CRE what changes he will make, though the CRE can tell the respondent to take such steps as may reasonably be required for the purpose of communicating that information to other persons concerned. A non-discrimination notice may also require the person to provide the CRE with information 'reasonably required' in order to verify that the terms of the notice have been complied with. A period of five years after the notice has become final is the statutory maximum time for which the CRE may impose these requirements. Within this constraint the notice may specify the time at which and the manner and form in which the information is to be furnished to the CRE. The CRE is empowered to make a further investigation at any time within five years to ascertain whether the recipient has complied with its terms.[69] The remedies available after a formal investigation are solely general in character, allowing a wide range of practices to be commented on, but allow no remedy to any individuals whom the CRE has discovered have been discriminated against individually.

The Commission may also, on the basis of the formal investigation, recommend to any person changes in policies or procedures with a view to promoting equality of opportunity.[70] These powers of recommendation are not restricted to circumstances where the CRE has made findings of unlawful discrimination. Though the CRE may not require it, it is empowered to recommend a degree of positive action. No legal sanctions are available, however, for failure to accept these recommendations.

A person served with a non-discrimination notice may appeal against any requirement of the notice.[71] If it is an employment issue, the appeal is to an industrial tribunal which may quash the requirement if it is unreasonable, if it is based on an incorrect finding of fact, or for any other reason; and may substitute other requirements. An appeal against a non-discrimination notice must be made to an industrial tribunal not later

[68] Ibid., s. 58. [69] Ibid., s. 60.
[70] Ibid., s. 51. [71] Ibid., s. 59.

than six weeks after a non-discrimination notice is served on any person. Where the court or tribunal considers a requirement in respect of which an appeal has been made to be unreasonable, because it is based on an incorrect finding of fact or for any other reason, the court or tribunal is required to quash the requirement. On quashing a requirement the court or tribunal may direct that the non-discrimination notice shall be treated as if, in place of the requirement quashed, it had contained a requirement in terms specified in the direction.

Where the CRE has reasonable cause to believe that the person intends not to comply with a requirement in a non-discrimination notice, it may apply to a county court for an order requiring him to comply with it. This is not an injunction, however, and non-compliance with it is not contempt. A small fine is the only remaining sanction. An injunction from the county court is only available where, after a non-discrimination notice (or a finding by a tribunal) has become final, it appears that 'unless restrained he is likely to do one or more' unlawful discriminatory acts in the future (so called 'persistent discrimination').[72]

Litigation

Unlike under the 1965 and 1968 Acts, individuals complaining of discrimination have considerable say over the conduct of their case.[73] Industrial tribunals (in most employment-related complaints) and designated county courts (in other complaints) are empowered to adjudicate directly on complaints of racial discrimination, replacing the complaints investigation and adjudication role of the Race Relations Board.[74] Individuals are permitted, within a limited period of time after the act complained of,[74a] to take their cases directly to the county courts or industrial tribunals; they do not have to process them first through a Race Relations Board or an equivalent type of body.

Though litigation at the initiative of those who perceive themselves to be the victims of discrimination is given a much greater role in the enforcement of the legislation than under the 1968 Act, it is a limited role. An individual is not permitted to bring proceedings with regard to discriminatory practices, discriminatory advertisements, instructions to discriminate, pressure to discriminate, or persistent discrimination. En-

---

[72] RRA 1976, s. 62. The same procedure is available for following up persistent discrimination after tribunal and court decisions.

[73] Ibid., ss. 54–7.

[74] A different procedure for enforcing the general duty of non-discrimination in the public sector of education was introduced. Here the only sanction is action by the minister responsible. Most allegations of discrimination against specific pupils of a school or applicants for admission, however, can be contested in the 'designated' county courts.

forcement of these provisions may only be initiated by the Commission.[75] The logic behind restrictions such as these on the role of individual plaintiffs drew on (and in turn reinforced) the procedural attributes of the traditional model of the private plaintiff, namely one taking an individualized complaint and not acting 'in the public interest'. It was envisaged that the latter role would be carried out by the Commission, and (this is the important point) *only* by the Commission.

Where an industrial tribunal upholds a complaint by an individual of unlawful discrimination, it is empowered ('as it considers just and equitable') to make one or more of the following orders:[76] first, an order declaring the rights of the respondent and complainant 'in relation to the act to which the complaint relates'; secondly, an order requiring the respondent to pay damages; thirdly, the tribunal may make a recommendation that the respondent take, within a specified period, 'action appearing to the tribunal to be practicable for the purpose of obviating or reducing the adverse effect on the complainant of any act of discrimination to which the complaint relates'.[77] The 1976 Act stipulates that damages may include compensation for injury to feelings, whether or not they include compensation under any other head.[78] Damages have another function as well. Recommendations by the industrial tribunal are not specifically enforceable. If not complied with, the tribunal may only make an order of damages or increase an already existing order of damages subject to a maximum.[79]

It is important to distinguish the remedy available for an individual proving a breach of the direct discrimination provisions from that of an individual proving indirect discrimination. In the case of the latter no damages may be awarded 'if the respondent proves that the requirement or condition in question was not applied with the intention of treating the claimant unfavourably on racial grounds'.[80] Where a recommendation has not been complied with in an indirect discrimination case the sanction of increasing the damages (or providing them in the first place) as a substitute does not, therefore, apply.

No injunctive remedy is available from an industrial tribunal under any circumstances in race discrimination cases. Industrial tribunals have no contempt powers. In employment discrimination cases, only county courts

---

[74a] In the case of industrial tribunal the time limit is three months, RRA, 1976, s. 68(1). A discriminatory regime maintained over a period of time which adversely affects an individual is regarded as an act which continues for so long as the regime continues, RRA, 1976, s. 6817(6). See further *Barclays Bank plc.* v. *Kapur* [1991] IRLR 136 (HC).

[75] RRA 1976, ss. 28–31, 62. See, for example, *CRE* v. *Imperial Society of Teachers of Dancing* [1983] IRLR 315 (EAT).

[76] Ibid., s. 56.   [77] Ibid., s. 57(3).
[78] Ibid., s. 57(4).   [79] Ibid., s. 56(4).
[80] Ibid., s. 57(3).   [81] Ibid., s. 62(1).

are empowered to enjoin unlawful acts. Individuals are not permitted to apply for an injunction under the 1976 Act. Injunctions may only be issued by the county courts, at the behest of the Commission for Racial Equality, in a certain limited range of such cases where persistent discrimination is suspected. Even in such limited circumstances as injunctions are available, negative orders only appear to be permitted. Where the Act permits injunctions to be issued it is in terms of 'restraining the respondent from doing the acts complained of'.[81]

With a view to helping a person who considers he may have been discriminated against in contravention of the Act to decide whether to institute proceedings and, if he does so, to formulate and present his case in the most effective manner, the Act provides that the person may question the respondent on his reasons for doing any relevant act, or on any other matter which is or may be relevant, and the respondent may if he so wishes reply to any questions.[82] The questions the person aggrieved put to the respondent, and any reply by the respondent are admissible as evidence in the proceedings. If it appears to the court or tribunal that the respondent deliberately, and without reasonable excuse, omitted to reply within a reasonable period or that his reply is evasive or equivocal, the court or tribunal may draw any inference from that fact that it considers it is just and equitable to draw, including an inference that he committed an unlawful act.

Full legal aid is not available before the industrial tribunals, only coming into effect at the stage of appeal before the Employment Appeal Tribunal and beyond. However, where, in relation to proceedings or prospective proceedings under the 1976 Act, an individual who is an actual or prospective complainant applies to the CRE for assistance, the CRE is required to consider the application and may grant it if it thinks fit to do so, on the ground that the case raises a question of principle, or that it is unreasonable, having regard to the complexity of the case or the applicant's position in relation to the respondent or another person involved or any other matter, to expect the applicant to deal with the case unaided, or by reason of any other special consideration.[83] Assistance by the CRE may include giving advice; procuring or attempting to procure the settlement of any matter in dispute; arranging for the giving of advice or assistance by a solicitor or counsel; arranging for representation by any person, including all such assistance as is usually given by a solicitor or counsel in the steps preliminary or incidental to any proceedings, or in arriving at or giving effect to a compromise to avoid or bring to an end any proceedings, or any other form of assistance which the CRE may consider appropriate.

[82] RRA 1976, s. 64.
[83] Ibid., s. 65.

INFLUENCES ON THE DEVELOPMENT OF
THE 1976 ACT

Several of the influences noted above as playing an important role in the development of the 1965 and 1968 Acts were also present in the development of this 1976 Act. There was strong support within sections of the Labour Party for new legislation and Labour came to power in 1974. Mr Roy Jenkins was again Home Secretary and, though less enthusiastic than before, was successfully pressed by his advisers to accept that new legislation was necessary.[84] The Conservative opposition was in general not vociferously opposed to such legislation. There was pressure from both the RRB and the CRC for new legislation because of the deficiencies of the 1968 Act; important empirical evidence of continuing discrimination was again produced by PEP.[85]

The American influence was again potent. Americans who were knowledgeable of and involved in anti-discrimination work in the United States were consulted and occasionally brought to Britain to give their assessments of the American experience with a view to suggesting which elements in it might be of relevance to the United Kingdom.[86] British civil servants, parliamentarians, and others also sought direct information about the situation in the United States. The Department of Employment commissioned a report on overseas practice on equal pay and equal opportunities, including the legal position in the United States.[87] The House of Commons Select Committee on Race Relations and Immigration visited the United States in 1975 in order to study employment discrimination issues there. Mr Roy Jenkins visited the United States in January 1975 with his Permanent Under-Secretary, Sir Arthur Peterson, and his special adviser, Mr Anthony Lester QC, partly in order to examine the operation of equal opportunity legislation.[88] This more recent American influence was particularly important in the development of the broadened notion of discrimination, and the incorporation of positive action provisions in the legislation. Acceptance of the American approach was not wholesale, however. In two important respects the American experience was rejected: monitoring of the racial composition of employers' work-forces was not

---

[84] R. Jenkins, *Life at the Centre* (1991).

[85] Smith, *Racial Disadvantage*.

[86] See e.g. L. H. Pollock, *Discrimination in Employment: The American Response* (1974).

[87] Dept. of Employment, *Women and Work: Overseas Practice* (1975).

[88] Home Office Evidence to the Select Committee on Race Relations and Immigration, *The Organization of Race Relations Administration, Oral Evidence*, 13 Feb. 1975, qq. 22 and 23.

made compulsory; and proposals to set up a system of effective 'contract compliance'[89] were stillborn.

An important new influence leading to the form of new legislation in 1976 was the campaign for legislative intervention to help secure greater equality between the sexes, which became prominent in the late 1960s, as discussed in detail by Collins and Meehan in Chapter 12.[90] In 1970 an Equal Pay Act had been enacted, and was to come fully into effect at the end of 1975. By 1973 it was becoming clear that opposition to the idea of some kind of further legislative action against sex discrimination was diminishing. The Government decided that there were strong arguments in favour of adopting, almost in their entirety, the coverage and enforcement details of what was to become the Sex Discrimination Act 1975 in the new race relations legislation. Not only might this have the practical advantages of increasing public understanding of how the two Acts operated and of enabling both enforcement agencies to work on similar lines; it might also ease the passage of race relations legislation, since Parliament would already have approved virtually identical enforcement provisions in the less controversial Sex Discrimination Act. Harmonization would not, however, go so far as to include a single enforcement agency at that time. In September 1975 the White Paper, *Racial Discrimination*,[91] was issued proposing that almost exactly similar coverage and enforcement provisions as in the Sex Discrimination Act should be enacted to deal with racial discrimination. This led to the Race Relations Act 1976 and the establishment of the Commission for Racial Equality.

One influence vital to the development of the 1968 Act's approach had significantly declined, however, namely the British industrial relations ideology of voluntarism. Industrial tribunals were originally established in the mid-1960s to deal with individual claims under statutes relating to redundancy. The decade 1969 to 1979 saw the importance of industrial tribunals grow considerably. The Donovan Report had recommended in 1969 that industrial tribunals become the major institution in the administration of individual labour law. The Ministry of Labour had invited the Royal Commission to consider whether the jurisdiction of the existing industrial tribunals should be enlarged so as to comprise 'all disputes between the individual worker and his employer'.[92] Having considered the matter, the Commission came to the conclusion that, subject to

---

[89] 'Contract compliance' in this context means the use of contractual terms inserted in government contracts to enforce government policy ends, see IPM, *Contract Compliance: The UK Experience* (1987).

[90] See also A. Byrne and J. Lovenduski, 'The Equal Opportunities Commission' (1978) 1 *Women's Studies Int. Q.* 131.

[91] Cmnd. 6234 (1975).

[92] Donovan Report, para. 572.

certain limitations, this should be done. The recommendation that jurisdiction over individual labour law issues thenceforth be given to industrial
tribunals was clearly influential in persuading the Government to allocate
individual disputes under the Equal Pay Act 1970, the Sex Discrimination
Act 1975, and the Race Relations Act 1976 to these tribunals. Industrial
tribunals were used because, in the words of Mrs Castle, the Secretary of
State for Employment at the time of the Equal Pay Act, they provided a
means of redress 'which is speedy, informal and accessible'. They were
'ideal because the tribunals are experienced in dealing with employment
matters, they include representatives of workers and they sit at various
centres scattered throughout the country'.[93] Though the influence of
labour law developments was thus considerable in so far as the 1976 Act
adopted industrial tribunals, the influence of voluntarism had declined
markedly in comparison with its impact on the 1968 Act.

THE OBJECTIVES OF LEGISLATION

It will be useful at this point to step back a little from the mass of political
and legal detail, and consider some more general issues. One of the
puzzling elements in considering the theory of anti-discrimination law is
that there is really relatively little consensus on why racial discrimination
is considered to be wrong, and why legal intervention is a desirable way
of addressing the phenomenon.[94] This is not to say that there is not a
consensus in Britain that such discrimination *is* generally wrong (opinion
polls tend to show a consensus against racial discrimination in Britain and
in favour of such legislation).[95] But there is, perhaps surprisingly, little
agreement on why such discrimination is objectionable, and why governments should intervene to stop it.

Some have argued that a reduction in racial discrimination is necessary
if public order is to be maintained: discrimination breeds resentment and
this is likely to spill over on to the streets unless adequate alternative
means of redress are provided. Or discriminatory actions are said to be
unfair to the individual discriminated against, because that person is
being judged on the basis of personal characteristics which are immutable
(the colour of his or her skin) and therefore ones over which the person

---

[93] Official Report, 795 HC, 9 Feb. 1970, col. 918.

[94] See generally, C. McCrudden (ed), *Anti-Discrimination Law* (1991), The editor's
introduction has been drawn on in the following discussion.

[95] See e.g. C. Airey, 'Social and Moral Values', in R. Jowell and C. Airey (eds.), *British
Social Attitudes* (1984), which quotes 69% in support of such race legislation, 28% opposed
and 3% 'other' in the poll conducted in 1983, an apparent increase over an equivalent poll
conducted in the late 1960s.

discriminated against has no choice. Preventing discrimination, others argue, is justified because discriminatory activity is economically inefficient, so that for example, in the labour market, legal intervention helps to achieve a more economically efficient system by widening the pool of available labour and thus allows for a more effective use of the talents and resources of the society. Yet others have justified anti-discrimination legislation on grounds similar to the ideal of 'cultural pluralism' discussed by Poulter in Chapter 14.

The objectives of the 1965 and 1968 Acts

In its first annual report, published in 1967, the Race Relations Board summarized the role of legislation in what became a classic statement of the functions of law in this field:

1. A law is an unequivocal declaration of public policy.
2. A law gives support to those who do not wish to discriminate, but who feel compelled to do so by social pressure.
3. A law gives protection and redress to minority groups.
4. A law thus provides for the peaceful and orderly adjustment of grievances and the release of tensions.
5. A law reduces prejudice by discouraging the behaviour in which prejudice finds expression.[96]

The objectives of the 1976 Act

Whereas the objectives of the 1968 Act were relatively clear, if limited, the same cannot be said of the 1976 Act quite so categorically. In some respects the aims of the 1976 Act *were* clear, particularly with regard to the institutional arrangements established: there was clearly an intention to free individual litigants from dependence on the regulatory agency; and to permit the regulatory agency to adopt a strategic approach to enforcement. When one attempts to set out the deeper aims of the legislation, however, there is considerable difficulty. To explain why, it will be useful to sketch out two alternative models of the objectives of anti-discrimination law popular in the United States at around the time the 1976 Act was passed: the individual justice model and the group justice model.[97]

---

[96] Quoted in C. McCrudden, D. Smith, and C. Brown (with the assistance of J. Knox), *Racial Justice at Work: The Enforcement of the Race Relations Act 1976 in Employment* (1991), 3.

[97] This discussion follows closely that in McCrudden *et al.*, *Racial Justice*, 5–7.

Individual justice model

The individual justice model is generally consistent with a view that the aim of anti-discrimination law is to secure the reduction of discrimination by eliminating from decisions illegitimate considerations based on race which have harmful consequences for individuals. This approach concentrates on cleansing the process of decision-making, and is not concerned with the result, except as an indicator of a flawed process. It is markedly individualistic in its orientation: concentrating on securing fairness for the individual. It is generally expressed in universal and symmetrical terms: blacks and whites are equally protected. It reflects respect for efficiency, merit, and achievement and, given the limited degree of intervention permitted, it preserves and possibly enhances the operation of the market. It is 'manageable' in that its aims can be stated with some degree of certainty, and its application does not depend on extensive factual enquiries and judgements.

Despite its obvious attractions, this model has been criticized as deeply flawed. Various arguments tend to recur. The model is said to misconceive the deep structure of racial discrimination. Race discrimination is as often institutional as individual, it is argued, and therefore there is little likelihood that a highly individualistic model will work. The problem, it is said, is misconceived as being one of intention rather than effect. The model is said not to take adequately into account the surrounding and reinforcing nature of racial disadvantage, and concentrates only on particular actions in assessing whether there should be legal intervention, ignoring the wider picture.

Group justice model

Out of these criticisms come various alternative approaches. Some have argued for an approach which is concerned with the results of decision-making and which seeks to redistribute resources from the group of advantaged whites to the group of disadvantaged blacks. Common to such approaches is a view that the aim of anti-discrimination law should be to fix on the outcomes of the decision-making processes. The basic aim is the improvement of the relative position of racial minority groups, whether to redress past subordination and discrimination, or out of a concern for distributive justice at the present time. The approaches tend to be redistributive and to be concerned with the relative position of groups or classes, rather than individuals. These approaches depend on a recognition of social classes or groups. The principle is often expressed in asymmetrical terms as focusing on the betterment of minority groups and is less concerned with symmetrical protection for non-minority groups.

These approaches too have been subject to criticism. Some criticisms relate to the open-endedness and unmanageability of the principles espoused. They have been criticized for elevating concern for groups above concern for individuals, for their asymmetrical tendency, and for the lack of emphasis given to (if not actual hostility towards) efficiency, merit, and individual achievement. Other criticisms have focused on the implications these approaches have for remedies. In particular, these approaches have been said to take insufficient account of the extent to which the burden of helping racial minority groups falls on third parties who may be 'innocent' of past wrong-doing, who may have gained no benefit from discrimination against these groups in the past, and who comprise some of the least advantaged sections of the community in terms of their economic circumstances.

The alternative models and the 1976 Act

The difficulty with specifying the aims of the 1976 Act is that it would seem most consistent with the evidence to view the legislators as intending to adopt modified versions of both the individual justice *and* the group justice models.[98]

The intention to further the individual justice model is reflected in the following attributes of the legislation: the individualistic nature of the county courts and industrial tribunals; the concentration on the eradication of discrimination as the prime target of legal enforcement; the restriction of remedies to individuals after litigation; the general prohibition of reverse discrimination and the generally symmetrical nature of the protection accorded. However, the individual rights model as adopted in the legislation is modified in two particular respects. First, within the framework of the Act, an individual's claim is addressed to only a limited extent: in particular, the remedies available for harm done to the individual are very limited. Second, the individual rights model as adopted in the Act is modified by the adoption of aspects of the group justice model.

The intention to further a modified version of the group justice model is reflected in the following attributes of the legislation: the prohibition of indirect discrimination; the requirement on the CRE to further equality of opportunity; the statutory duty on local authorities to further equality of opportunity; the powers of the CRE to conduct formal investigations without individual complaints; the recognition of the existence of 'racial groups'; the inclusion of exceptions permitting limited positive action which would otherwise have amounted to reverse discrimination. However,

---

[98] For an alternative perspective, see J. Gardner, 'Liberals and Unlawful Discrimination' (1989) 9 *Oxf. J. Legal Studies* 1.

the intention to adopt only limited aspects of the group justice model is reflected in the following: the refusal to require employers to engage in racial monitoring; the rejection of widespread contract compliance; the absence of group remedies; the absence of a more general legal requirement to secure equality of opportunity; and the absence of any requirement to engage in positive action to remedy imbalances. We have therefore a complex picture of mixed intentions in the sphere of legal enforcement.

Divining the intentions behind the Act is further complicated by the fact that the White Paper *Racial Discrimination*[99] suggested that the Act was but one part of what would become a comprehensive strategy for tackling racial disadvantage. The strategy which subsequently was to have been put into effect would have complemented the Act. This largely failed to materialize, however, and the Act was left as the main plank in the Labour government's anti-discrimination strategy.

Challenges from the left, the right, and from minorities

Ideological positions seldom remain unchallenged, and the objectives which characterized the 1976 Act have been challenged from both the left and the right of the political spectrum, adding to the complexity of the ideological debates surrounding race discrimination legislation in Britain. For reasons which we shall discuss subsequently, the most significant intellectual challenges have originated in the United States, but these have also percolated through into British debates.

From the right have come concerns which gained considerable political strength in the United States and Britain since the late 1970s. These include the need to protect the operation of the 'free market' system, a concern to advance efficiency and individual choice, and the desirability of recognizing merit and individual achievement. This has been, of course, a general political development, but anti-discrimination law has also been affected by its influence. Reliance on the 'free market' is a central theme in Abram's work on the proper role of anti-discrimination legislation.[100] He distinguishes between two approaches to anti-discrimination law. On the one hand are 'fair shakers' who are preoccupied with 'equality of opportunity'. This is the view he subscribes to. On the other hand are 'social engineers' who are preoccupied with equality of results. For the latter, the 'only way to measure equality is in terms of . . . representation, and . . . it is the government's role to bring about proportional representation in short order'. This is the approach he opposes. He presents a considerable range of arguments in favour of the former and against the

---

[99] See above, n. 91.

[100] M. B. Abram, 'Affirmative Action: Fair Shakers and Social Engineers' (1986) 99 *Harvard LR* 1312–26.

latter, but a conception of the limited role of law in the free market seems central to his concerns. Intervention is justified to eliminate discriminatory barriers that deny individuals the opportunity to compete, that is to participate in the market. Such a role for law is 'consistent with our meritocratic view of the relevant differences between individuals—a view through which our society rewards the individual for attainment'. Occupational and professional standards should enable the 'best available' candidates for jobs to be hired. Not only is the social engineering approach unjustified in seeking to go beyond these goals, it is also dangerous because it threatens the operation of this system by politicizing the process of selection and distribution. In the absence of the 'neutral decision-making' of the free market, the process 'must inevitably degenerate into a crude political struggle between groups seeking favored status'. Societies that have departed from such neutral decision-making have suffered a history 'not of liberation but of crippling oppression'. Nor is a 'social engineering' approach likely to deliver the higher standard of living desired for disadvantaged groups because of the overall decrease in the efficiency of the economy.

From the left, and from some who identify themselves as coming from a racial minority, have come arguments questioning whether an emphasis on anti-discrimination law is necessary, effective, or desirable. For some, there is the obvious concern as to whether such reform efforts concentrating on legal enforcement are worth the effort.[101] Others, however, challenge what they perceive as the basic justifications for intervention of this type. In Britain, some writers from the Asian community, for example, have challenged what they perceive as an assumption behind the legislation that all the ethnic-minority groups share a similar set of beliefs and needs.[102] They object in particular to the assumption that 'blacks' should be treated as a homogenous group, and argue that anti-discrimination legislation mistakes the nature of the problem which the Asian community faces in Britain, which they define as lack of recognition of their cultural distinctiveness. To the extent that anti-discrimination law encourages a goal of integration, rather than a recognition of difference, it is at best irrelevant, at worst harmful.

For others, talk of 'rights' is either a chimera or a diversion, and 'legalism' is a term of abuse. Some have argued that the purpose of the legislation was merely to give the impression that something was being done, or was an attempt to deprive the ethnic-minority community of autonomous leadership by establishing buffer institutions such as the CRE which would head off a direct assault on the Establishment by the

---

[101] See L. Lustgarten, 'Racial Inequality and the Limits of Law' (1986) 49 *MLR* 68.

[102] See e.g. T. Modood, 'The Indian Economic Success: A Challenge to Some Race Relations Assumptions' (1991) 19 *Policy and Politics* 177–89.

black community.[103] Commentators hostile to 'rights discourse' in this context sometimes go beyond viewing legal rights as merely ineffective in providing protection, and view such discourse as positively undesirable, because of its perceived tendency to emphasize atomistic individualism. In particular, some argue that the good which should be promoted is one which does not necessarily address the well-being of the individual, but rather the well-being of the 'community', conceived as encompassing, but going beyond the interests of the individuals which make it up. Anti-discrimination law serves, in Freeman's view,[104] largely 'to legitimize the existing social structure and, especially, class relations'. Anti-discrimination laws were adopted, it is said, in part because it was in the interests of the ruling classes to do so, in part for internal economic reasons, and in part to justify the state to the world, to avoid embarrassment and to stabilize its position. Anti-discrimination law, it is said, fails to challenge the most important determinant of inequality: class. But without challenging the class structure, a credible measure of tangible progress for minorities will not be possible to any great extent. The movement for greater justice for minorities 'has been compromised by the implicit limitations associated with reliance on rights and the legalism of formal equality'.[105]

More recently, there has been a heated debate which has sought to challenge aspects of these approaches. While Ben-Tovim and Gabriel regard these arguments as partially convincing in terms of certain effects of the legislation, they challenge the view that the effects identified were those intended:

If the Acts . . . were simply drawn up as a diversionary exercise, one designed to legitimate racist practices elsewhere (e.g. immigration controls), then it becomes necessary to explain the role played by progressive elements . . . in the development of those initiatives. If capitalism somehow requires this legislation, then one might expect to find its representatives and not their opponents instrumental in drawing it up.[106]

Delgado argues that analyses such as those discussed above contain features 'that repel and in fact threaten minorities'.[107] He considers, for

[103] See e.g. A. Sivanandan, 'Race, Class and the State: The Black Experience in Britain' (1976) 17 *Race and Class* 347; P. Fitzpatrick, 'Racism and the Innocence of Law', in P. Fitzpatrick and A. Hunt, *Critical Legal Studies* (1987), 119–32.

[104] A. Freeman, 'Legitimating Racial Discrimination through Anti-Discrimination Law' (1982) 62 *Minn. LR* 96.

[105] A. Freeman, 'Racism, Rights and the Quest for Equality of Opportunity: A Critical Legal Essay' (1988) 23 *Harvard Civil Rights-Civil Liberties LR* 295, 335.

[106] G. Ben-Tovim and J. G. Gabriel, 'The Politics of Race in Britain, 1962–79: A Review of the Major Trends and of Recent Debates', in C. Husband, *'Race' in Britain* (1982), 155.

[107] R. Delgado, 'The Ethereal Scholar: Does Critical Legal Studies have What Minorities Want' (1987) 22 *Harvard Civil Rights-Civil Liberties LR* 301–22, 303.

example, the disparagement of legal rules and rights to be particularly worrying. He argues that much of the misfit, between those advocating approaches such as those discussed above and minority aspirations, is due to the former opting for 'consciously informal processes that rely on good will, inter-subjective understanding and community'. Delgado argues that rights (and the formality that goes with them) are indispensable for the empowerment they provide to otherwise unprotected persons and groups. Whites, who have little to lose, may feel free to gamble by exchanging relative security for the promises of power-sharing and the benefits of community. For minorities to embrace the informality espoused in the rhetoric of anti-legalism is to risk losing the very rights that constitute the only means of securing, and protecting their hard-won gains. Crenshaw[108] argues that such authors as those discussed above disregard the 'transformative potential' that liberalism and rights consciousness in fact offer to minorities. Where she advances the discussion is in her sensitive treatment of the dilemma that the oppressed face when engaging in the process of reform by using legal methods: the problem of how to take advantage of the transformative potential which anti-discrimination law offers, without at the same time becoming powerless to go beyond the limited vision from which it has emerged.

LEGISLATIVE AND GOVERNMENTAL DEVELOPMENTS AFTER 1979

Given the ambiguity of the objectives of the 1976 Act, and the increasing polarization of intellectual views challenging any significant role for such legislation from both the right and the left, the 1976 legislation could well have become an ideological battleground but, perhaps surprisingly, this has not happened to any great extent in Britain.

It is true that the arguments from the right of the political spectrum have been influential generally in government since the election in 1979 of a Conservative administration dedicated to a free-market approach, pledged to reduce government spending, and sceptical of regulatory mechanisms which restrict the operation of the market. It is not surprising therefore that there has been no substantial legislative expansion of race discrimination legislation. Indeed the most controversial legislation affecting racial discrimination law enacted since 1976 fits quite closely the free-market, anti-regulation model, namely the Local Government Act

---

[108] K. W. Crenshaw, 'Race, Reform and Retrenchment: Transformation and Legitimation in Anti-Discrimination Law' (1988) 101 *Harvard LR* 1357.

1988.[109] The increased scrutiny of local government activities, which was such a notable feature of government since the 1980s, provided on additional ideological explanation for these provisions.

Deregulatory influences

From 1969 there has been a clause in all central-government contracts requiring contractors to conform to the employment provisions of the 1976 Act. No attempt has been made by Government to monitor compliance with this provision. Particularly during the 1980s, however, some local authorities, particularly those which were Labour-controlled, adopted equivalent clauses in their contracts, so-called 'contract compliance' policies. Such policies often sought to use the local authorities' contracting powers to impose broadly defined equal opportunity requirements on contractors, particularly with regard to monitoring the work-force by race. However, such policies were regarded with disfavour by central government and this resulted in the passage of a number of provisions in the Local Government Act 1988 restricting the power of local government in Britain to use contract compliance. In general, local authorities are required to exercise a number of functions in relation to proposed or subsisting supply-of-works contracts without reference to matters which are noncommercial. The Act defines eight matters as 'non-commercial'. Most important, for the purposes of restricting contract compliance in the area of race equality, the Act specifies that the terms and conditions of employment of the contractor's work-force or the composition of or arrangements for promotion, transfer, or training are all regarded as 'non-commercial' considerations. There is a limited exception to these restrictions for the purpose of complying with section 71 of the Act, which places responsibility on local authorities to promote equality of opportunity, but the CRE has argued that the exception is too limited to be of substantial use, and the use of contract compliance policies by local authorities appeared to have declined significantly after the Act came into force.

Benign neglect?

However, what is perhaps more surprising, given its ideological preferences, is the extent to which the 1976 Act remained relatively untouched under the Conservative government elected in 1979. Indeed, on a number of occasions, minor amendments were made to expand its coverage, largely due to initiatives by the CRE to persuade Government or sympathetic back-benchers to introduce amendments during the course of largely unrelated bills. The Education Acts of 1980 and 1981 extended the duty on local education authorities not to discriminate to cover their functions under those Acts. The Housing and Planning Act 1986 amended the 1976 Act to make it unlawful for a planning authority to discriminate against a person in carrying out its planning functions. The Housing Act 1988 gave power to the CRE, subject to the approval of Parliament, to issue codes of practice in the field of *rented* housing, which the Local Government and Housing Act 1989 extended to *all* forms of housing,[110] and imposed a duty on the Housing Corporation and Housing Action Trusts to make appropriate arrangements with a view to ensuring that their various functions are carried out with due regard to eliminating unlawful discrimination and to promoting equality of opportunity and good relations between persons of different racial groups. The Employment Act 1989 amended the positive action provisions to remove certain designation requirements, thus making it easier for various bodies to engage in such positive action. The Courts and Legal Services Act 1990 extended the protection of the 1976 Act to cover discrimination in professional relationships between barristers, and between barristers and solicitors. The Broadcasting Act 1990 required applicants for licences to run independent television stations to make arrangements to promote equal opportunities in employment.[111] The Criminal Justice Act 1991 introduced a non-discrimination duty throughout the criminal justice system.[112] In 1992, the Further and Higher Education Funding Councils were prohibited from racially discriminating,[113] as were employees of members of the staff of the House of Commons.[114]

In non-legislative activity the Government approved measures which were broadly in line with the spirit of the 1976 Act. The Police (Discipline) Regulations introduced in April 1985 under the Police and Criminal Evidence Act 1984 included the specific offence of racially discriminatory behaviour. Guidelines were introduced which required future legislation and future policy initiatives to be assessed for any racial discrimination. Codes of Practice in employment and housing were approved by the relevant Secretaries of State and Parliament. Monitoring of the composition of the civil service by race was introduced progressively during the 1980s and in 1990 a programme of action was put in place for

[109] Another deregulatory-inspired amendment to the 1976 Act was the removal of the designation requirement relating to positive action training in s. 37 in the Employment Act 1989, and the removal of the designation requirement for the inclusion of work experience trainees within the coverage of the 1976 Act.

[110] Housing Act 1988, s. 137; Local Government and Housing Act 1989, s. 180.

[111] Broadcasting Act 1990, s. 38.

[112] Criminal Justice Act 1991, s. 95.

[113] Further and Higher Education Act 1992, Sch. 8, para. 84.

[114] Trade Union and Labour Relations (Consolidation) Act 1992, Sch. 2, para. 7.

[115] Office of the Minister for the Civil Service, *Programme for Action to Achieve Equality of Opportunity in the Civil Service for People of Ethnic Minorities* (1990).

furthering equality of opportunity in the civil service for people of ethnic minorities.[115] For the first time, the 1991 Census introduced a 'race' question.[116] The Home Office issued instructions to prison officers on avoiding discrimination in prisons in 1991.[117] In 1987 the Ministry of Defence had started monitoring the ethnic origins of applicants and entrants of the Armed Forces and in 1992 it was announced that this would be extended to serving personnel.[118]

Stability rather than change has, however, been the dominant policy. There has been little tampering with the CRE; nor have controversial decisions by the courts been overturned by the legislature. To echo a phrase used to describe an earlier American administration, the Government's policy on race discrimination legislation has, perhaps, been largely one of 'benign neglect'. It has not sought to extend the legislation substantially, but it has not sought to repeal or substantially weaken it either. How to avoid controversy, rather than how to create it, has been the dominant motive. Why?

No single explanation is wholly convincing, but taken together several factors probably contributed to the degree of stability manifested after 1979. Political attention on race issues from the right of the political spectrum largely focused on further tightening of immigration rules (see Dummett, Chapter 11), and attacks on Labour local-government policies such as contract compliance (see above) and anti-racism in schools, rather than on anti-discrimination legislation as such.[119] There was no strong body of opinion within the Conservative Party or right-wing 'think-tanks' which urged substantial change in the 1976 Act.[120] From the mid-1960s, anti-discrimination issues were at the margins of British political thinking and therefore they were unlikely to be a major focus of attention for a radical reforming government (whether of left or right). An additional factor was that only two years into the first Conservative government several riots broke out which were popularly perceived as having a racial origin and focus. The report to the Government (by Lord Scarman, a respected liberal on race issues) concerning the most violent of these riots in Brixton, tended to confirm rather than challenge acceptance of the need for anti-discrimination policies.[121] It is noteworthy, for example,

[116] M. NiBhrolchain, 'The Ethnicity Question for the 1991 Census: Background and Issues' (1990) 13 *Ethnic and Racial Studies* 542.

[117] Directorate of Inmate Administration, HM Prison Service, Home Office, *Race Relations Manual: A Manual for Staff Working in Prison Service Establishments* (Apr. 1991).

[118] Twelfth Periodic Report of the United Kingdom to CERD, CERD/C/226/Add. 4, 17 December 1992, paras. 90 and 91

[119] W. Ball and J. Solomos, *Race and Local Politics* (1990).

[120] There have been occasional exceptions, e.g. R. Lewis, *Anti-Racism: A Mania Exposed* (1988).

[121] *The Brixton Disorders, 10–12 April 1981, Report of an Inquiry by the Rt. Hon. the Lord Scarman, OBE*, Cmnd. 8427 (1981).

that the acceptance by the Government of the CRE's Code of Practice in employment was somewhat easier after these riots than before,[122] and that experimental ethnic monitoring in the civil service was first introduced only months after the Scarman Report.[123]

Tackling racial disadvantage

One further additional element must be considered. Government has sought, to some extent, to develop policies which tackle the disadvantages of ethnic minorities in Britain in ways other than by way of anti-discrimination law.[124]

One of the earliest attempts to target government spending on ethnic minorities was under Section 11 of the Local Government Act 1966. This provision, which continued in force from its enactment throughout our period of study, enabled local authorities (in the main) to employ additional staff to help members of communities of Commonwealth origin to overcome linguistic or cultural barriers and thereby to gain access to services and facilities. The rate of grant was at 75 per cent of approved expenditure. Provision was made for £129 million to be paid in grant in 1992/93, of which around 90% was allocated to education, mainly to the provision of teachers of English as a second language. In 1991, an additional Ethnic Minority Grant was introduced to fund projects run by voluntary organizations to help members of ethnic minorities to gain employment, to take up vocational training, or to set up their own business. Provision was made for £4 million to be paid in grant in 1992/93.

Several other programmes were developed, targeted at inner city problems in particular, which were considered by government to be of especial relevance to ethnic minorities. First, City Action Teams were established to co-ordinate government action in eight inner city areas. Equipped with small budgets, the teams funded projects targeted on employment, environmental improvement and enterprise promotion problems. Second, the Urban Programme aided some 9,000 projects within 57 inner city local authorities, encouraging investment in defined target areas; spending on projects directly benefiting ethnic minority groups was at about 10 per cent of total Urban Programme spending of £226 million in 1990/91 and ethnic minorities are a priority for Urban Programme support. Third, the City Challenge initiative was established to enable

---

[122] C. McCrudden, 'Codes in a Cold Climate: Administrative Rule-Making by the Commission for Racial Equality' (1988) 51 *MLR* 409.

[123] U. Prashar, 'The Need for Positive Action', in J. Benyon, *Scarman and After* (1984), 207.

[124] See Twelfth Periodic Report to CERD, *supra*, for a further discussion of these initiatives.

local authorities to tackle some of the worst economic, social and environmental conditions through creating a partnership with local businesses; the guidance issued to authorities emphasised the importance of considering the special needs of ethnic minorities.

All these programmes and initiatives are, however, heavily dependant on the vicissitudes of government expenditure reviews. During 1992, promised reductions in government spending were thought likely to reduce significantly the budgets of programmes funded under Section 11, the Urban Programme, and the City Challenge initiative.[125]

INTERNATIONAL INFLUENCES

International influences played a limited role in affecting the general political and legal climate in favour of anti-discrimination measures and thus provided a stabilizing influence on the domestic legislation. We have already identified the important American influence on the 1968 and 1976 Acts. However, changes in the United States in both the political and judical spheres have led to that country being much less the recognized leader in using the law to tackle racial discrimination than it was up till the end of the 1970s. Other international influences have also been apparent (though not of the same force).

Since the Second World War there has been a proliferation of international treaties prohibiting racial discrimination generally, and in specific situations: the Convention on the Prevention and Punishment of the Crime of Genocide, 1948; the European Convention for the Protection of Human Rights and Fundamental Freedoms, 1950; the ILO Convention No. 111 on Discrimination in Employment and Occupation, 1958; the Unesco Convention against Discrimination in Education, 1960; the European Social Charter, 1961; the UN Convention on the Elimination of All Forms of Racial Discrimination, 1965; the Covenant on Civil and Political Rights, 1966; and the Covenant on Economic, Social and Cultural Rights, 1966. Only in the case of the 1958 ILO Convention has the United Kingdom failed to ratify the international provision. The United Kingdom considers itself unable to adhere to this Convention because 'of difficulties in a number of areas'.[126] These difficulties concern exclusions and qualifications in the Race Relations Act relating to members of the armed forces, restrictions in certain cases on the employ-

---

[125] P. Gosling, 'Poor Relations in Race for Government Cash', *The Independent* (10 Dec. 1992), 20.

[126] International Labour Conference, 75th Session, 1988, Report III (part 4B), *Equality in Employment and Occupation: General Survey by the Committee of Experts on the Application of Conventions and Recommendations* (1988).

ment of foreigners in the public service, and distinctions in the wages paid to seafarers employed on board vessels flying the flag of the United Kingdom, according to whether they are domiciled in the United Kingdom or not. The United Kingdom has, however, signed and ratified the other treaties. But neither in the detail of the drafting of this legislation, nor in its subsequent interpretation and implementation, have these international standards played a critical role.

European Convention on Human Rights

Despite its apparently widely phrased wording, Article 14, the principal anti-discrimination provision of the European Convention, has been little used as a weapon against racial discrimination in Europe generally, or in the United Kingdom specifically. One major limitation is that Article 14 does not provide for any independent right not to be discriminated against. There must be some connection established between the alleged discrimination and another substantive right or freedom. What the exact connection must be is not entirely clear since the case-law of the European Commission on Human Rights and the European Court of Human Rights has not always been consistent.[127] In general, however, it would seem that Article 14 'has no independent existence, since it has effect solely in relation to the "enjoyment of the rights and freedoms" safeguarded by those provisions.'[128] Nevertheless, it is not necessary to establish actual violation of the other Article before Article 14 can come into play. If the facts in issue 'fall within the ambit' of one or more of the substantive rights, then Article 14 may be relied upon. Article 14 forms 'an integral part of each of the Articles laying down rights and freedoms'.[129]

However, even with this limited role, Article 14 has relatively consistently been played down by the Commission and the Court. When a violation of one of the substantive rights and freedoms has been found, it has generally been held that it is not necessary to inquire further into the possible violation of Article 14. In such cases Article 14 'is not treated as an autonomous and complementary, but only as a subsidiary guarantee'.[130] So, too, the content of the non-discrimination standard has been interpreted restrictively by the Court. The concept of discrimination adopted appears to be largely 'direct' discrimination and has not yet been clearly stated to include 'indirect' discrimination. In addition, discrimina-

---

[127] See O. van Dijk and G. J. H. van Hoof, *Theory and Practice of the European Convention on Human Rights*, 2nd edn. (1990), 533–7.

[128] *Abdulaziz, Cabales and Balkandali* v. *UK* (1985) 7 EHRR 471.

[129] *Belgian Linguistics Case* 1 EHRR 252.

[130] Van Dijk and van Hoof, *Theory*, 537.

tion may be justified by the state on considerations derived from the public interest, and a wide margin of discretion is given to the national authorities in appreciating the weight to be accorded to the public interest.

In light of these interpretations, it is perhaps not surprising that the Court has not held the United Kingdom to be in breach of the prohibition of racial discrimination in Article 14. In Article 3, however, the Commission has found an interesting substitute for the lack of a general prohibition against racial discrimination. Article 3 prohibits 'torture, and inhuman or degrading treatment or punishment'. Discriminatory treatment on the basis of race has on occasion been interpreted by the Commission as constituting 'degrading treatment'. In the *East African Asians* case,[131] the Commission held that the United Kingdom was in breach of Article 3 on these grounds. Article 3 is likely to be interpreted in this way, however, only in extreme cases. In general, therefore the Convention is likely to add little to current domestic protection against racial discrimination. Attempts to reform the Convention to include an autonomous right to non-discrimination have thus far been unsuccessful.[132]

International Convention on the Elimination of All Forms of Racial Discrimination

Unlike the European Convention, the Convention on the Elimination of All Forms of Racial Discrimination contains detailed provisions for eliminating racial discrimination.[133] The Convention was the first legally binding international agreement in the field of human rights concluded under the auspices of the UN which embodied international measures of implementation, including the establishment of international machinery for overseeing implementation (the Committee on the Elimination of Racial Discrimination). The Committee, which began its operations in 1970, is composed of eighteen experts serving in a personal capacity, chosen by the states parties to the Convention from among nationals nominated by the states parties.

The Convention provides for three procedures of scrutiny: a reporting procedure, obligatory upon all states parties; a procedure for state-to-state complaints, which is open to all state parties but recourse to which is discretionary; and an optional procedure for complaints by individuals or groups of individuals. The Committee has responsibilities under each of

---

[131] *East African Asians* v. *UK* (1981) 3 EHRR 76.
[132] See Council of Europe, *Explanatory Reports on the Second to Fifth Protocols to the European Convention for the Protection of Human Rights and Fundamental Freedoms* H(71)11, pp. 52–3.
[133] See generally, N. Lerner, *Group Rights and Discrimination in International Law* (1991).

these procedures. In addition it has advisory responsibilities relating to the attainment of the principles and objectives of the Convention in territories for which state parties have responsibilities. There have, however, been no inter-state cases involving the United Kingdom; and the United Kingdom does not recognize the competence of the Committee to receive communications from individuals or groups claiming to be victims of a violation of the Convention. In practice, the primary method of scrutiny has proven to be the compulsory reporting procedure. Every two years each state party is required to submit for consideration by the Committee a report on the legislative, administrative, judicial, or other measures that it has adopted to give effect to the provisions of the Convention.[134] In addition to these 'periodic reports' the Committee may request 'supplementary reports'. The Committee considers and examines these reports, makes suggestions and general recommendations, and reports to the General Assembly of the United Nations.

The practice has been to examine the reports at meetings at which representatives of the relevant states party are present, and questions are put by the members of the Committee to that representative. However, in exercising its powers, the Committee has, on its own admission, 'felt constrained to tread cautiously'.[135] Over the years several issues recur in questioning by the Committee of the UK representative: relations with South Africa, immigration policies, incitement to hatred legislation, the activities of the police force, and the absence of racial discrimination legislation applying to Northern Ireland. Apart from these issues there is often questioning about specific exceptions provided in the domestic legislation. In general, however, few members of the Committee appear particularly well briefed, and the questioning is somewhat desultory and lacking in depth. Even more seriously, the Committee has been starved of adequate funds. During the latter part of the 1980s and the early 1990s many sessions were curtailed or cancelled due to the failure of some states to contribute toward the expenses of the Committee.

European Community and race

Unlike with regard to sex discrimination, there is no provision in the Treaty of Rome, or in subsequent amendments, which provides specifically that racial discrimination is contrary to European Community law. Nor has there been any subsidiary legislation in the form of directives

---

[134] See e.g. the *Tenth Periodic Report of the United Kingdom*, CERD/C/172/Add. 11.

[135] CERD, *Study of the Work of the Committee on the Elimination of Racial Discrimination and Progress Towards the Achievement of the Objectives of the International Convention on the Elimination of all Forms of Racial Discrimination*, A/Conf. 92/8, 19 Apr. 1978, para. 194, p. 43.

or regulations to that effect. The nearest that the Treaty comes to in addressing the issue is in Articles 48–51, requiring the free movement of workers. This has been interpreted extensively to prohibit discriminatory provisions restricting the ability of citizens from one member state from moving to other member states for work. This prohibition of discrimination on grounds of nationality of a member state clearly falls short of the much wider prohibition of racial discrimination in Britain.

There have, however, been limited developments since the mid-1980s which may lead to the EC taking a more active role.[136] The European Parliament has taken some interest in the question of racism and xenophobia, beginning in 1984, after an increase in the number of elected representatives of extreme right-wing groups in the European Parliament. A committee of inquiry was established to study the rise of fascism and racism in Europe and to submit a report to the Parliament. The report was submitted in December 1985.[137] Following this, a joint declaration against racism and xenophobia was signed in June 1986 by the Presidents of the European Parliament and the Council, the representatives of the member states, and the Commission.[138]

The European Commission was pressed to go further but considered that it had no legal competence under the Treaty to do so. The Parliament disagreed. In 1988 the Commission proposed a limited formal resolution on racism and xenophobia, together with an action programme. However the resolution finally passed by the Council in 1990 was so changed from even the limited draft resolution proposed by the Commission that the Commissioner responsible disassociated herself from the final product. The resolution called upon member states to adopt 'such measures as they consider appropriate' to counter racism and xenophobia.[139] A further European Parliament committee of inquiry reported in July 1990, with further recommendations for action.[140]

More specifically related to employment, the preamble to the Social Charter, agreed to by all member states except the United Kingdom in December 1989, provides that 'in order to ensure equal treatment, it is important to combat every form of discrimination, including discrimination on grounds of sex, *colour, race*, opinions and beliefs, and . . . in a spirit of solidarity, it is important to combat social exclusion [*emphasis added*]'.[141] The general introduction to the Action Programme accompanying the Social Charter stressed the need for member states to

---

[136] For a general discussion see M. Spencer, *1992 and All That: Civil Liberties in the Balance* (1990).

[137] Evrigenis Report, PE 97.547.

[138] *OJ* C158, 25 June 1986.

[139] *OJ* C157, 27 June 1990, Art. 3.

[140] PE 141. 205/fin, Document A3-195/90.

[141] Community Charter of the Fundamental Social Rights of Workers, Dec. 1989.

eradicate discrimination on the grounds of race, colour, or religion, 'particularly in the workplace and in access to employment',[142] but no plans were announced for carrying these declarations into legislation. However in 1991 the Directorate-General for Employment, Industrial Relations, and Social Affairs commissioned a study seeking a 'comparative assessment of the legal instruments implemented in the various Member States to combat all forms of discrimination, racism and xenophobia and incitement to hatred and racial violence'.[143] The report of this study recommended that each state 'review existing legislation for gaps in coverage and consider the adoption of comprehensive anti-racism and anti-discrimination legislation'.[144] Achieving a limited degree of consensus on the content of any future legislation by the Community may be facilitated by the adherence of all but one EC member state (Ireland) to the Convention on the Elimination of All Forms of Racial Discrimination, though, according to a comparative analysis by Forbes and Mead of measures to combat racial discrimination in the member states of the Community, few of the other EC states have anything approaching the conceptual sophistication and detailed enforcement apparatus of the British legislation.[145] In December 1991, the European Council asked Ministers of the Member States *and the Commission* 'to increase their efforts to combat discrimination and xenophobia, and to strengthen the legal protection for third country nationals in the territories of the Member States'.[146]

INTERPRETATION AND IMPLEMENTATION OF THE 1976 ACT

Given the general neglect by government of race discrimination law since 1979, and the absence of significant external influences of the kind which has been so important in the development of sex discrimination law (see Chapter 12), much of the focus of development has therefore shifted to the domestic courts, tribunals, and agencies. The ideological conflicts,

[142] Action Programme, COM (89) 568 final, 29 Nov. 1989.

[143] *OJ* C84/16, 28 Mar. 91.

[144] Commission of the European Communities, *Legal Instruments to Combat Racism and Xenophobia* (1993), p. 75.

[145] I. Forbes and G. Mead, *Measure for Measure* (1992). This report was commissioned by the Department of Employment. A second report, commissioned by the Home Office in 1992, comprised a literature survey of the anti discrimination laws operating in the U.S.A., Canada, Australia and New Zealand, see Scottish Ethnic Minorities Research Unit, *Report on Anti-discrimination Law on the Grounds of Race* (1992).

[146] Conclusions of the Maastricht European Council (9–10 December 1991), *Declaration on Racism and Xenophobia*.

between right and left, and between the two models defining the mixed objectives of the Act have tended to be played out in the courts and in the regulatory body, rather than in more public party-political debate. How have these bodies coped? It is to the approach adopted by the superior courts that we turn first.

Judicial decisions

In general, there is some evidence in judicial decisions of the acceptance of a philosophy hostile to regulatory agencies (in the context of formal investigations) and favouring a free-market approach (in the context of the interpretation of indirect discrimination, for which see Chapter 12). Nevertheless, the approach taken by the higher courts has also recognized the concern that such legislation may become a dead letter, and the courts have in various ways sought to prevent the 'transformative possibilities' of the legislation from becoming atrophied through too restrictive an interpretation of the legislation. Within the constraints of the legislation the higher courts have, in general, been relatively sympathetic: in the context of the definition of the racial groups protected (for which see the discussion in Chapter 14), in decisions relating to the problems of tackling direct discrimination through litigation (demonstrated in several decisions dealing with discovery of documents, the burden of proof, and remedies), and in the scope of judicial review. It is to these issues that we now turn.

One preliminary point should be borne in mind. The dominant approach adopted by the House of Lords to the interpretation of the 1968 Act (that the legislation should be restrictively interpreted because it interfered with common law liberties) has been quietly jettisoned. In part, this relatively favourable response was conditioned by the growing involvement of the European Communities in gender discrimination issues.[147] As between the two models defining the mixed objectives of the Act, however, no clear preference is as yet apparent.

Formal investigations

In both *R. v. CRE, ex parte Hillingdon London Borough Council*,[148] and in *In re Prestige*,[149] the Commission's initiation of two formal investigations was successfully challenged. In *Hillingdon* a 'belief' investigation

---

[147] e.g. in *Hampson v. Dept. of Education and Science* [1988] IRLR 87, the Court of Appeal reinterpreted *Ojutiku and Oburoni v. Manpower Services Commission* [1982] IRLR 418 holding that, so far as race discrimination cases were concerned, there was no significant difference between it and the test laid down in *Rainey*.

[148] *R. v. CRE, ex parte LBC Hillingdon* [1982] AC 779.

[149] *In re Prestige* [1984] 1 WLR 335, [1984] ICR 483.

was in issue. The Commission's decision to conduct a formal investigation was held by the House of Lords to be *ultra vires* on the grounds that the terms of reference of the investigation were too wide. In *Prestige* the House of Lords considered the powers of the Commission to embark on an investigation into a named person in which unlawful discrimination was not alleged, and held that the Act permitted only two types of investigation, not three as had been supposed. One type was against a named person where an unlawful act of discrimination is suspected. The second type was a general investigation not alleging discrimination and without named persons.[150]

Proving direct discrimination

The superior courts have been relatively more sympathetic to the difficulties of proving direct discrimination. In *Science Research Council* v. *Nasse*[151] an applicant alleging discrimination in promotion sought discovery of confidential reports on himself and two other people who had been promoted, and discovery of notes made by a local review board. The House of Lords held that there was no public-interest immunity for disclosure in respect of private confidential documents. The test for deciding whether discovery should be ordered was whether it was necessary for the fair disposal of the proceedings. In *West Midlands Passenger Transport Executive* v. *Singh*,[152] the Court of Appeal ordered discovery of details of the ethnic origins of applicants for and appointees to posts within a band of grades comparable to that for which he had applied. Such evidence, it was held, may have a probative effect as indicating intention in the particular case in issue.

With regard to the burden of proof of direct discrimination, there has been considerable difference of emphasis between different courts and, indeed, different judges. The EAT set out its preferred approach in *Khanna* v. *Ministry of Defence*.[153] The proper approach was to look at all the evidence and

> to take into account the fact that direct evidence of discrimination is seldom going to be available and that, accordingly, in these cases the affirmative evidence of discrimination will normally consist of inferences to be drawn from the primary facts. If the primary facts indicate that there has been discrimination of some kind, the employer is called on to give an explanation and, failing clear and specific explanation being given by the employer to the satisfaction of the

---

[150] See further, Munroe, 'The Prestige Case: Putting the Lid on the Commission for Racial Equality' (1985) 14 *Anglo-American LR* 187.

[151] [1980] AC 1028, [1979] 3 WLR 762, [1979] ICR 921, [1979] 3 All ER 673, [1979] IRLR 465 (HL).

[152] [1988] 1 WLR 730, [1988] ICR 614, [1988] 2 All ER 873, [1988] IRLR 186 (CA).

[153] [1981] ICR 653 (EAT).

industrial tribunal, an inference of unlawful discrimination from the primary facts will mean the complaint succeeds.

Although this approach was questioned by a later EAT in *Barking and Dagenham London Borough Council* v. *Camara*,[154] the *Khanna* approach was subsequently been given strong support by the English Court of Appeal in *Baker* v. *Cornwall County Council*,[155] holding that the tribunal 'should be prepared to draw the inference that the discrimination was on such grounds unless the alleged discrimination can satisfy the tribunal that there was some other innocent explanation'. In *Dornan* v. *Belfast City Council*,[156] the Northern Ireland Court of Appeal also endorsed *Khanna*. Subsequently, in *British Gas plc* v. *Sharma*[157] the EAT again interpreted *Khanna* and *Noone* as requiring tribunals, having looked at the whole of the evidence, to decide whether there are primary facts which in the absence of an explanation point to discrimination. If there was no acceptable or adequate explanation then it was open to the tribunal to find by inference that there was discrimination (though they are not bound to do so).

In *King* v. *The Great Britain-China Centre*,[158] the English Court of Appeal attempted a comprehensive restatement of the approach which should be taken. Neill L. J. set out five principles which he considered could be extracted from the authorities, of which the fourth and fifth were of particular importance:

Though there will be some cases where, for example, the non-selection of the applicant for a post or for promotion is clearly not on racial grounds, a finding of discrimination and a finding of a difference in race will often point to the possibility of racial discrimination. In such circumstances the Tribunal will look to the employer for an explanation. If no explanation is then put forward or if the Tribunal considers the explanation to be inadequate or unsatisfactory it will be legitimate for the Tribunal to infer that the discrimination was on racial grounds. This is not a matter of law but, as May L. J. put it in *Noone*, 'almost common sense'.

It is unnecessary and unhelpful to introduce the concept of a shifting evidential burden of proof. At the conclusion of all the evidence the Tribunal should make findings as to the primary facts and draw such inferences as they consider proper from those facts. They should then reach a conclusion on the balance of probabilities, bearing in mind both the difficulties which face a person who complains of unlawful discrimination and the fact that it is for the complainant to prove his or her case.[159]

---

[154] [1988] ICR 865, [1988] IRLR 14 (EAT).
[155] [1990] IRLR 194 (CA).
[156] [1990] IRLR 179 (NICA).
[157] [1991] IRLR 101 (EAT).
[158] [1991] IRLR 513 (CA).
[159] [1991] IRLR at 518.

Remedies

Within the constraints of the Act, the courts have proven flexible, recognizing the importance of remedies. In *Noone* v. *NW Thames Regional Health Authority*,[160] the Court of Appeal held that an award of £3,000 was appropriate in this case as an 'injury to feelings' award. In *Alexander* v. *Home Office*,[161] the Court of Appeal laid down the principles upon which awards for injuries to feelings should be based, including that, though such awards should be restrained, they should not be minimal. In subsequent cases, an award of £500 for injury to feelings was regarded as at or near the minimum appropriate level of award.[162] More generally, damages should reflect compensation for the consequences of the discrimination. Aggravated damages might also be awarded where the circumstances of the discrimination were such that the resulting sense of injury was heightened by the manner in which, or the motive for which, the discrimination was carried out. In *Noone* v. *NW Thames Regional Health Authority*[163] the Court of Appeal also saw no reason why exemplary damages, which are designed to punish the discrimination, should not be awarded. In *Bradford Metropolitan Council* v. *Arora*,[164] however, the Court of Appeal held that cases in which exemplary damages are appropriate will be rare in employment discrimination cases though they may be awarded where compensatory damages are inadequate to punish the discriminator for outrageous conduct. More recently, however, the EAT has held that exemplary damages cannot be awarded in a discrimination case.[165]

Judicial review

Lastly, the courts have been willing to open up the process of judicial review, in which the applicant argues that a public body is acting *ultra vires*, unreasonably, or contrary to procedural fairness, as a method of enforcing anti-discrimination standards. This technique was first developed in the context of sex discrimination, but the courts have also opened up this possibility in the context of race discrimination as well.[166]

---

[160] [1988] ICR 813, [1988] IRLR 195 (CA).

[161] [1988] 1 WLR 968, [1988] 2 All ER 118, [1988] ICR 685, [1988] IRLR 190 (CA).

[162] *Sharifi* v. *Strathclyde Regional Council* [1992] IRLR 259 (EAT); *Deane* v. *London Borough of Ealing* [1993] IRLR 209 (EAT).

[163] See n. 147.

[164] [1991] IRLR 165 (CA).

[165] *Deane* v. *London Borough of Ealing* [1993] IRLR 209 (EAT).

[166] *R.* v. *London Borough of Hammersmith and Fulham, ex parte NALGO* [1991] IRLR 249; *R.* v. *Army Board of the Defence Council, ex parte Anderson* [1991] IRLR 423 (DC); *R.* v. *Department of Health, ex parte Ghandhi* [1991] IRLR 431 (DC).

The effects of the 1976 Act

Despite the legislation, however, and despite relatively favourable interpretation by the higher courts, research has shown that there is a
widespread view, as gleaned from opinion polls, that there remain high
levels of racial discrimination in contemporary Britain. This view seems
to be more the case amongst black people than amongst Asian and white
people.[167] There is also empirical research which tends to show that these
perceptions are accurate. Racial discrimination and inequality of opportunity between racial groups continue at substantial levels, for example,
in employment and housing.[168] Racial minority groups differ substantially
from the white population in economic conditions generally, in housing
conditions, in being unemployed, and in employment opportunities.[169]
In 1992, for example, the total of black people unemployed was more
than twice the national average. Some 25 per cent of black adults were
jobless.[170] These are, however, substantial differences between the ethnic
minority groups in economic conditions, with Indian. African-Asian and
Chinese groups being closer to the white population than the Afri-
Carribean, Pakistani, and Bangladeshi ethnic groups.[171]

Policy Studies Institute research of the legal enforcement of the 1976
Act in employment, covering the period up to 1989, and published in
1991, argued that the 1976 Act was not working as intended.[172] An
important aim of the 1976 Act was, as we have seen, to tackle institutional
discrimination: the whole range of policies and practices that are unfair to
racial minorities in effect, even though that may not have been the
intention. The PSI research concluded, however, that the new concept of
'indirect discrimination', introduced as part of the attempt to tackle
institutional discrimination, had been of limited use. (As discussed by
Poulter in Chapter 14, the most prominent cases in which applicants have
successfully used the indirect discrimination provision in a race context
have involved the more limited areas which may best be classified as
involving questions of cultural pluralism.)

The 1976 Act gave the CRE power to carry out formal investigations.

---

[167] NOP opinion survey reported in the *Independent on Sunday*, 7 July 1991 and discussed
in *Race and Immigration*, No. 247, July/Aug. 1991.
[168] See e.g. C. Brown and P. Gay, *Racial Discrimination: 17 Years after the Act* (1986);
CRE, *Sorry its Gone: Testing for Racial Discrimination in the Private Rented Housing Sector*
(1990). For a survey of the evidence, see CRE, *Second Review of the Race Relations Act
1976* (1992), 11–15.
[169] 'Ethnic Origins and the Labour Market', *Employment Gazette*, Feb. 1991 (unemployment among ethnic minorities was 14% compared with 9% among white people).
[170] Official Report, House of Commons, 14 Jan. 1993, Weekly Hansard, Issue No. 1605,
col. 777 (written answer).
[171] T. Jones, *Britain's Ethnic Minorities* (1993).
[172] McCrudden *et al.*, *Racial Justice*.

Between 1977 and the end of 1991, the Commission conducted some 54 formal investigations.[173] However, there were, according to the PSI Report, major difficulties in using this power to tackle institutional discrimination. Between 1977 and 1982, for example, the CRE mounted twenty-four formal investigations in employment, many of them large and complex. The investigations often ran into difficulties because of the complexity of the procedures imposed by the law and because of legal challenges. The combined effect of the *Hillingdon* and *Prestige* cases was to call into question a substantial number of investigations under way at that time and to derail the formal investigation strategy adopted by the Commission.[174] The CRE, in managing employment investigations, was not able to find a way through these obstacles. From 1983 to 1989, the CRE mounted many fewer employment investigations: four into specific employers and five general investigations. The CRE issued no non-discrimination notices in employment investigations started from 1983.[175]

As regards individual complaints, PSI studied the process of litigation in the industrial tribunals. The intention behind the Act, as we have seen, was to open up the complaints system to everyone with a grievance. The industrial tribunals were to be a more informal and friendlier alternative to the ordinary courts. The PSI research found that to some extent this new strategy had worked. The number of complaints of racial discrimination made to the tribunals had increased, and so had the number of complaints upheld. But in many ways the system for dealing with individual complaints had not developed as expected. The industrial tribunal process was quicker and simpler than civil actions in the courts, but race cases were longer and more complex than others litigated in tribunals. Success depended on applicants making formal legal moves and presenting specific kinds of evidence, so the unrepresented applicant was at a severe disadvantage. There had been little development of campaigning organizations to sponsor individual complaints. Consequently, the CRE had been much the most important source of help for individual applicants.

The CRE provided legal representation to one-quarter of applicants and these applicants were found to have a much better chance of success than the remaining three-quarters. This was partly because the CRE tended to select the stronger cases; but it was also because the CRE

[173] CRE, Annual Report 1991, 16 (1992).

[174] At least three investigations were regarded by the CRE as *ultra vires* as a result of *Hillingdon*, and at least six as a result of *Prestige*. See CRE *Annual Reports* for 1983 and 1984.

[175] There have, however, been important formal investigations in other areas, particularly housing, in which non-discrimination notices have been issued, though the effect of these notices is not entirely clear, see J. Solomos, 'The Politics of Race and Housing' (1991) 19 *Policy and Politics* 147–57, on the effects of the Hackney formal investigation.

provided more effective help to individuals than any other body, and also provided moral support at the earlier stages when many unsupported applicants withdrew. What mostly determined the outcome of a case was whether the CRE decided to assist. So a twin-track system had developed, with the CRE in the dominant position, deciding which track an applicant was to follow. Thus the objective of freeing individuals from dependency on the regulatory agency had not been achieved.

It is hard to disagree with Solomos's assessment in 1989 that: 'The translation of policies into practice has been hampered by a weak legal framework, organisational marginality and a lack of political legitimacy. This lack of political legitimacy has become increasingly evident during the past decade.'[176]

REFORM PROPOSALS AND ISSUES FOR THE FUTURE

The PSI research proposed several ways in which the aims of the Race Relations Act could be better achieved in the future. Some improvements could be made within the present law; others would require new legislation. Reform proposals were also made by the CRE in its 1985 and 1992 reform recommendations.[177] These sets of proposals cannot adequately be summarized here. Some of the proposals, however, are particularly relevent for several issues discussed above. The PSI report and the CRE proposals, for example, argued for the following changes to be considered: legal aid should be extended to applicants in industrial tribunals race cases; more specialist knowledge of race issues should be developed among chairmen and members of industrial tribunals; and the scale of damages awarded in race cases should be increased. In addition, PSI recommended that the CRE should use individual complaints as a strategy for changing organizations, should increase the use of formal investigations for legal enforcement, should improve its management of formal investigations, and should improve the skills of its staff through training in research methods and law. Similar proposals were also made in 1991 by the Labour Party.[178]

More generally, however, the PSI report argued that government must decide whether the first priority is justice for aggrieved individuals, or change in the position of racial minorities as a group. If group justice was to be the first priority, then a new approach was needed both in the drafting of legislation and in its enforcement. A possible model, it sug-

---

[176] J. Solomos, 'Equal Opportunities Policies and Racial Inequality: The Role of Public Policy' (1989) 67 *Public Administration* 79, at 90.

[177] CRE, *Second Review of the Race Relations Act 1976* (1992).

[178] Labour Party, *The Charter of Rights* (1991).

gested, would be the Fair Employment (Northern Ireland) Act 1989.[179] This model also strongly influenced the reform proposals of the CRE, including the recommendation that the provision of equality of opportunity should become the objective over and above the elimination of discrimination and the proposal to require employers to engage in monitoring.

The Commission for Racial Equality in its review of the 1976 Act also argued strongly for a more secure legal basis to be provided in European Community law. Two main reasons were set out in a 1991 Consultative Paper. First, those protected by racial discrimination laws in Britain 'lose the protection of our laws and are not so well protected elsewhere', when travelling to the rest of Europe. Second, 'there is the real danger that one day a process of harmonisation of laws might lead not to an improvement in the protection for racial discrimination across Europe, but to a reduction to, as it were, the lowest common denominator'.[180] The Labour Party has also argued that the Treaty of Rome should be amended 'to make it unlawful to discriminate on grounds of race in the same way that the Treaty currently outlaws sex discrimination'.[181] In December 1992, a group of independent experts from several member states published a proposal for a draft Council directive concerning the elimination of racial discrimination which would require all member states to pass appropriate legislation.[182]

Lastly, we have seen that the 1976 Act does not apply to Northern Ireland. This has been the cause of some concern within CERD and there have been various calls to introduce equivalent legislation in Northern Ireland from the Standing Advisory Commission on Human Rights,[183] the Committee on the Administration of Justice,[184] and the Commission for Racial Equality.[185] In December 1992 the government published a

---

[179] For the details of the Act, see C. McCrudden (ed.), *Fair Employment Handbook*, 2nd edn. (1991). For discussion of its implications for race discrimination law, see B. Hepple, 'Discrimination and Equality of Opportunity: Northern Ireland Lessons' (1990) 10 *Oxf. J. Legal Studies* 408.

[180] CRE, *Second Review* of the Race Relations Act 1976: A Consultative Paper (1991), para. 5.3.

[181] Labour Party, *Charter of Rights*, 11. The Law Society of England and Wales has also argued in favour of an EC Directive to harmonize race discrimination legislation, see *New Law Journal*, 13 Sept. 1991, p. 1214.

[182] See CRE Press Release No. 452, 10 Dec. 1992, 'European Summit Must Act on Racism, Major Urged.'

[183] Standing Advisory Commission on Human Rights, *Religious and Political Discrimination and Equality of Opportunity in Northern Ireland: Second Report* (Cn. 1107, 1990, Belfast).

[184] Committee on the Administration of Justice, *Racism in Northern Ireland* (CAJ Pamphlet No. 20, CAJ, Belfast, 1992).

[185] Commission for Racial Equality, *Second Review of the Race Relations Act 1976* (CRE, London, 1992).

Consultative Document which considers such legislation along with several other possible initiatives.[186]

CONCLUSION

The way British law views racial discrimination has changed dramatically since the Second World War. From a position of abstention in the 1940s and 1950s, there is now a detailed apparatus of legal control. The legal techniques have, however, changed considerably over time. There have been significant shifts in strategy: from voluntarism to intervention in employment situations, from regulatory agency control to a heavier emphasis on individual litigation, from a limited notion of discrimination to a broader notion of equality of opportunity. But which are the most appropriate legal techniques to enable the issues to be addressed effectively is the subject of continuing dispute, as indeed is the question of what the aims of such legislation are, or should be.

[186] Central Community Relations Unit, *Race Relations in Northern Ireland* (CCRU, Belfast, 1992).

14

Minority Rights

SEBASTIAN POULTER

THEORETICAL PERSPECTIVES

The minority rights addressed in this chapter comprise those broadly 'cultural' rights granted to minority groups by Article 27 of the International Covenant on Civil and Political Rights when it declares: 'In those States in which ethnic, religious or linguistic minorities exist, persons belonging to such minorities shall not be denied the right, in community with the other members of their group, to enjoy their own culture, to profess and practise their own religion, or to use their own language.' This narrow concept of what constitutes a minority for the purposes of this chapter means that while Jews, Muslims, Sikhs, Hindus, gypsies, and Welsh-language-speakers are included, other groups such as the gay community, the physically handicapped, and the mentally ill are not. Although the true ambit of Article 27 is far from clear and it possesses no counterpart in the European Convention on Human Rights, it is set out here as a beacon to guide the overall direction of the discussion which follows, for it admirably summarizes the field of inquiry.

The location of such a wide-ranging provision relating to cultural rights within the International Covenant on Civil and Political Rights rather than as part of the complementary Covenant on Economic, Social, and Cultural Rights, demonstrates the considerable degree of overlap between the two areas. As we shall see, cultural rights can affect employment, family life, and education,[1] just as vitally as they do freedom of expression and freedom of religion.[2] Indeed, both Covenants expressly endorse the rights of peoples freely to pursue their own cultural development by virtue of their right of self-determination[3] and both prohibit discrimination on the basis of, *inter alia*, race, language, and religion.[4]

Most theoretical analysis of the appropriate response to minority cultures in modern Britain tends to revolve around the dichotomy between 'assimilation', on the one hand, and 'cultural pluralism' on the other.[5]

---

[1] See Arts. 6, 10, and 13 of the Covenant on Economic, Social, and Cultural Rights.
[2] See Arts. 18 and 19 of the Covenant on Civil and Political Rights.
[3] Art. 1(1) of each Covenant.
[4] Art. 2 of each Covenant.
[5] For illuminating analyses of the concepts and their implications, see generally B. Parekh, *Colour, Class and Consciousness* (1974), ch. 15; B. Parekh, 'Britain and the Social Logic of Pluralism', in *Britain: A Plural Society* (1990), 58–76.

These concepts are seen as representing profoundly different options for policy-makers to choose between, across a broad spectrum of issues. A policy of assimilation is generally understood as entailing the absorption of the minorities into the mainstream culture of the majority community. Minority groups are required to surrender the distinctive characteristics of their separate identities and blend into the wider society. That many white people in Britain feel this to be a moral obligation on the part of migrants and their descendants is apparent from the frequently heard utterance of the old adage, 'When in Rome do as the Romans do'. A variety of ulterior motives may obviously be ascribed to some assimilationists, including feelings of prejudice and xenophobia. However, there are at least four rational arguments which merit more serious consideration. First, there may be genuine concern that failure by some minorities to assimilate with respect to certain important English values may involve a departure from minimum standards of acceptable behaviour. On this basis it might be felt justifiable to outlaw, for example, female circumcision, polygamy, and child-marriage. Secondly, it may be argued that the vital principle of 'equality before the law' will be jeopardized if minorities are accorded special exemptions and privileges by the English legal system on the basis of their adherence to traditional practices. Such equality has long been regarded as a cardinal virtue of our legal system and more than a century ago Dicey made particular reference to it as part of the fundamental concept of the 'rule of law'.[6] Thirdly, anxiety may be expressed about the degree to which members of minority groups can attain economic integration and advancement and break free from a cycle of disadvantage and discrimination if they insist on retaining their separate customs and identities. The suggestion here is that white employers, for example, may be less willing to recruit minority workers whom they fear may make cultural or religious demands upon them or who may not 'fit in' and work well with other employees. Fourthly, there may be a feeling that the active promotion of separate ethnic identities may foster divisiveness and be incompatible with the degree of general social cohesion required for the maintenance of a sense of national unity.

Those theorists who favour the alternative policy of 'cultural pluralism' believe in according proper respect to the distinctive cultures and identities of members of the various ethnic communities. They regard this as the appropriate response to diversity in a liberal democracy, in which individual choice, personal freedom, and religious toleration are cherished as important values. Advocates of cultural pluralism usually accept some limits to the ambit of the doctrine and are willing to restrict the freedom

---

[6] See now A. V. Dicey, *Introduction to the Study of the Law of the Constitution*, 10th edn. (1959).

of minorities occasionally, in particular when their cultural practices appear to violate the minimum standards laid down in widely ratified international treaties for the protection of fundamental human rights.[7] However, leaving aside these comparatively rare instances, while cultural pluralists are naturally concerned to uphold the principle of equality before the law in a formal (Diceyan) sense, they also seek genuine equality for minorities in terms of social justice. Often this may require the law to affford minority cultures, traditions, and values equal respect and recognition with those of the majority, if justice is to be done in individual cases. Identical treatment, irrespective of cultural or religious differences, may be an inappropriate legal response and international human rights law has long recognized the permissibility of according minorities special differential treatment in certain circumstances.[8]

Advocates of cultural pluralism not only seek to affirm the value of greater diversity in enriching the overall life of the nation; they also reject the notion that minorities should be encouraged or induced to surrender their identities in order to gain economic advancement. It would, in their view, be unwise to assume that prejudice and discrimination on the part of white employers are caused by cultural differences rather than skin colour. Such differences may be put forward merely as a smokescreen to disguise blatant racist attitudes.

In social and political terms there is no field more significant for making the broad choice between 'cultural pluralism' and 'assimilation' than the sphere of education and in recent years no official report has been so authoritative and influential as *Education for All*, the Report of the Committee of Inquiry into the Education of Children from Ethnic Minority Groups (the *Swann Report*).[9] The report decisively rejected a deliberate social policy of assimilation on the ground that it would amount to 'a denial of the fundamental freedom of all individuals to differ on aspects of their lives where no single way can justifiably be presented as universally appropriate'.[10] Instead, the members of the Committee favoured a policy of pluralism, while making it very clear that the minority communities could not, in practice, preserve all the elements of their cultures and life-styles unchanged, for this would prevent them from taking on the shared values of the wider pluralist society.[11]

The *Swann Report*'s vision of creating 'a genuinely pluralist society . . . both socially cohesive and culturally diverse'[12] is, as its authors openly

---

[7] See e.g. S. Poulter, 'Ethnic Minority Customs, English Law and Human Rights' (1987) 36 *ICLQ* 589.

[8] See W. B. McKean, *Equality and Discrimination under International Law* (1983).

[9] Cmnd. 9453 (1985).

[10] At p. 4.

[11] At p. 5.

[12] At p. 6.

acknowledged, by no means shared by everyone in Britain.[13] There are many who continue to favour the assimilationist approach and in recent years there have been a number of utterances by senior Conservative politicians along these lines.[14] However, statements made by Home Office ministers reveal a surprising continuity of government policy, stretching from the Labour administration of the mid-1960s right up to the present time. Whereas in the early years of New Commonwealth immigration between 1948 and 1965 the general assumption among policy-makers was that a process of assimilation was desirable, a major change occurred once it was appreciated that the largest number of immigrants would be from the Indian sub-continent rather than the Caribbean. The change was heralded in official circles in a notable speech in 1966 by the Labour Home Secretary, Roy (now Lord) Jenkins, in which he commented on the meaning of 'integration':

Integration is perhaps a rather loose word. I do not regard it as meaning the loss, by immigrants, of their own national characteristics and culture. I do not think we need in this country a melting pot, which will turn everybody out in a common mould, as one of a series of carbon copies of someone's misplaced vision of the stereotyped Englishman . . . I define integration, therefore, not as a flattening process of assimilation but as equal opportunity, coupled with cultural diversity, in an atmosphere of mutual tolerance.[15]

Eleven years later, when the United Kingdom submitted its first periodic report to the UN Human Rights Committee under Article 40 of the International Covenant on Civil and Political Rights, it echoed exactly these sentiments in describing the manner in which it was giving effect to 'minority rights' pursuant to Article 27 of the Covenant.[16]

Public support for this policy of cultural pluralism has since been given by three Conservative ministers at the Home Office, Timothy Raison[17] in 1980, and Douglas Hurd[18] and John Patten[19] in 1989, as well as by John Major as Prime Minister in 1991.[20] It thus seems probable that the three objectives of equal opportunity, cultural diversity, and mutual tolerance will constitute part of official government policy towards the minority

---

[13] Ibid., see also C. Brown, *Black and White Britain: The Third PSI Survey* (1984), 273, 289 (40% of white respondents disagreed with the idea that 'people of Asian and West Indian origin should try to preserve as much of their own culture as possible').

[14] See e.g. *Independent*, 5 Oct. 1987 (John Biffen) and 30 May 1989 (Sir John Stokes); HC Debs. 159, cols. 1105–6 (John Townend, 8 Nov. 1989); *The Field*, May 1990 (Norman Tebbit).

[15] R. H. Jenkins, *Essays and Speeches* (1967), 267.

[16] See *Yearbook of the Human Rights Committee 1977–8* (1986), ii. 111.

[17] 'Cultural Diversity, Adaptation and Participation' (1980) *New Community* 96.

[18] Speech at Birmingham Central Mosque, repr. in *New Life*, 3 Mar. 1989.

[19] Letter to leading British Muslims, repr. in *The Times*, 5 July 1989.

[20] Speech reported in (1991) 250 *Race and Immigration* 4–5.

communities for the foreseeable future, regardless of which party holds the reins of power. However, significant differences of emphasis can be anticipated, together with varying degrees of commitment to the implementation of such policies, as has been the position during the past twenty-five years. In looking back at legal developments since 1950 we can also expect to find some 'assimilationist' tendencies. However, it needs to be borne in mind that these might either reflect a general assimilationist policy or else merely represent minor departures from the norm of cultural pluralism in those rare instances where shared core values, minimum international standards, and fundamental human rights issues were felt to be at stake.

Concern has sometimes been expressed that a policy of cultural pluralism may involve a person's individuality being submerged within his or her membership of a particular ethnic group.[21] For example, if 'group rights'[22] are to be championed at the expense of individual rights this may unduly constrain personal freedom and confer too much power on the collectivity, for a person's ethnicity is only one part of his or her overall identity. Hence if English law were to confer upon British Muslims, as a group, the right to regulate their family affairs exclusively by the rules of Islamic personal law this would impose considerable restrictions on, for instance, any Muslim women who might feel that Islamic principles of family law were more oppressive than those operating under English law. Similarly, it might be argued that to extend the present blasphemy laws to protect non-Christian faiths might stifle legitimate doctrinal dissent within those religious communities.

However, a more liberal version of group rights can take the form of a claim by the members of a particular minority that their institutions and values be accorded legal recognition so long as individuals retain the freedom to exercise their general rights under English law.[23] On this basis there can be no objection in principle to the existence of separate religious institutions, community organizations, and independent educational establishments catering for the special needs of minority groups. Nor can there be any legitimate complaint if these enjoy public funding, provided this is available on a non-discriminatory basis, for example in the form of rate relief, tax concessions, or voluntary-aided status for schools. Even so, as we shall see, the most common way in which the law will accord recognition to cultural diversity will tend to be through allowing *individual*

[21] See e.g. H. Goulbourne, 'Varieties of Pluralism: The Notion of a Pluralist, Post-Imperial Britain' (1991) *New Community* 211.

[22] For helpful discussion of the variety of meanings that can be accorded to the concept of 'group rights', see J. Montgomery, 'Legislating for a Multi-Faith Society: Some Problems of Special Treatment' in B. Hepple and E. Szyszczak (eds.), *Discrimination: The Limits of Law* (1992), 195–9.

[23] See e.g. *Swann Report*, 178–9.

members of minority communities to assert claims and rights that are based on their ethnic, religious, or linguistic identity, where this is a relevant factor in the proceedings. The individual will thus be relying upon a particular status conferred by membership of a wider class or group of people, just as citizens, spouses, and minors are entitled to do in appropriate circumstances. This corresponds with the approach adopted in Article 27 of the International Covenant on Civil and Political Rights, which confers rights upon 'persons belonging to' minority groups rather than the minorities themselves. In this way the concept of group rights can be rendered wholly compatible with the notion of individual rights, and the antipathy generally felt towards group rights by modern liberalism can be demonstrated to be without foundation.[24] The goal of equality of individual opportunity may sometimes only be achievable, for instance in the fields of employment and education, through legal recognition of cultural diversity. This fact needs to be clearly understood, for some writers have sought to confine cultural diversity to a 'private' domain removed from the realm of legal regulation.[25] On this basis it is argued that while family and community matters can safely be left to control by all the separate groups themselves, the public domain of law, politics, and the economy must function on the basis of uniformity. This leads Rex, for example, to the following erroneous conclusion: 'Any suggestion that individuals or groups should receive differential treatment in the public domain is a move away from the multi-cultural ideal towards the plural society of colonialism.'[26] As we shall see, the multi-cultural ideal can only be made to work effectively in some spheres through such differential treatment on the part of the law. While principles of race equality may well require that the law almost invariably be 'colour-blind',[27] the imperatives of ethnic pluralism dictate that it cannot ignore important cultural differences if justice is to be attained.

OVERVIEW OF NATIONAL DEVELOPMENTS

Legislation

The legislative changes which have been introduced since 1950 can be seen as falling into two distinct categories. On the one hand, there are several instances where Parliament has enacted provisions designed to

---

[24] See generally, W. Kymlicka, *Liberalism, Community and Culture* (1989).

[25] See e.g. J. Rex, 'The Concept of a Multi-cultural Society' (1987) *New Community* 218, and, from the other end of the political spectrum, R. Honeyford, *Integration or Disintegration* (1988), ch. 2.

[26] Ibid. at 222.

[27] See above, ch. 13.

recognize and support cultural diversity. On the other hand, there are a number of examples of 'assimilationist' provisions aimed at outlawing or denying legal significance to ethnic and religious traditions. Sometimes both tendencies are apparent in the same Act and, particularly in the case of the Education Reform Act 1989, a serious attempt has been made to achieve a fair balance between them. In other cases, while endorsing pluralism, Parliament has been careful to circumscribe it within reasonable bounds by imposing various restrictions and limitations, so as to ensure that minorities were not being afforded unwarrantable privileges.

Enactments endorsing pluralism

In 1967 the Welsh Language Act provided that, in any legal proceedings in Wales (or the former county of Monmouthshire), Welsh could be spoken by any party, witness, or other person who wished to use it. This marked a significant advance on the earlier provision for the language under the Welsh Courts Act 1942, which only authorized parties and witnesses to use Welsh if they felt they would otherwise be at a disadvantage by reason of this being their natural language of communication. This had meant that if such a person could understand and speak English perfectly adequately but merely preferred to employ Welsh, no right to use it was recognized.

In 1968 Parliament acknowledged the desire of many gypsies to maintain their nomadic lifestyle by imposing upon all local authorities a duty, under the Caravan Sites Act of that year, to provide adequate sites for gypsies residing in or resorting to their areas.[28] Hitherto the official assumption had been that gypsies would, over time, settle down as members of the house-dwelling community and that they should be encouraged to do so.

In 1972 the notoriously unjust rule in *Hyde* v. *Hyde*[29] was abolished by the Matrimonial Proceedings (Polygamous Marriages) Act which allowed the courts for the first time to grant matrimonial relief to parties to actually or potentially polygamous marriages contracted abroad.[30] Since polygamy had long been perceived by the judiciary as an odious, alien custom this was a major advance.

Four years later Parliament enacted two further Acts with profound significance, one in a practical sense and the other symbolically. The first of these, the Race Relations Act 1976, incorporated into English law the

---

[28] See Part II of the Act, esp. s. 6(1), as amended by Local Government Act 1972, sch. 30. Government plans to relieve local authorities of this duty were announced in March 1993, following the issue of a consultation paper—see DOE, 'Gypsy Sites Policy and Illegal Camping: Reform of the Caravan Sites Act 1968' (1992).

[29] (1866) LR 1 P&D 139.

[30] s. 1, now embodied in Matrimonial Causes Act 1973, s. 47.

novel concept of 'indirect' discrimination[31] which renders unlawful certain
neutral acts by employers and others which are not designed to discriminate against ethnic minorities but which nevertheless have a disproportionately adverse impact upon them because of their cultural and religious
backgrounds. In particular, the Act's provisions have the effect of making
it unlawful for employers and educational establishments to impose stan-
dardized rules about uniforms, dress, and appearance with which members
of minority groups cannot conscientiously comply, unless such rules can
be demonstrated to be 'justifiable'.[32]

The other piece of legislation in 1976, the Motor-Cycle Crash Helmets
(Religious Exemption) Act was passed in order to relieve turbaned Sikhs
from the duty being imposed on all motor-cyclists to wear helmets complying with high safety standards.[33] Parliament took the view that the
health and safety considerations underlying the obligation to wear a
helmet were outweighed by the religious concern felt by many orthodox
Sikhs about having to discard their turbans.[34] More recently, the Sikh
community has built upon the precedent established by the 1976 Act and
persuaded Parliament to pay further attention to its needs. One of the
five distinctive symbols of Sikhism ('the five Ks') is the *kirpan* (a small
sword or dagger), which should be carried as a matter or religious obligation by all orthodox Sikhs. The Criminal Justice Act 1988, which contains
provisions designed to penalize those who carry knives and other sharply
pointed instruments in public places, deals with the *kirpan* by means of a
specific exemption for those carrying sharply pointed articles for 'religious
reasons'.[35] Similarly, section 11 of the Employment Act 1989 contains an
exemption for turbaned Sikhs from the duty recently imposed upon all
those working on construction sites to wear safety helmets.[36] Furthermore, any employer who refuses to employ a Sikh on a construction site
simply because he is unwilling to wear a safety helmet in place of his
turban will be barred from even being able to argue that such a policy is
'justifiable' on grounds of safety under the indirect discrimination provisions of the Race Relations Act.[37]

In 1989 the Children Act also imposed a new obligation upon all local
authorities to give due consideration to the religious persuasion, racial
origin and cultural and linguistic background of any child whom they

---

[31] See s. 1(1)(*b*). For analysis of the concept, see above, ch. 13.
[32] See below, pp. 470–1.
[33] See now Motor-Cycles (Protective Helmets) Regulations 1980, enacted pursuant to
Road Traffic Act 1972, s. 32.
[34] The exemption is now contained in Road Traffic Act 1988, s. 16(2).
[35] s. 139(5)(*b*).
[36] Construction (Head Protection) Regulations 1989.
[37] Employment Act 1989, s. 12.

were looking after, or preparing to look after, in making any decision about that child's future.[38]

While each of these legislative enactments is clearly motivated by a philosophy of cultural pluralism, it is worth noting their limitations. By no means all Sikh men in Britain are either orthodox or observant enough to wear a turban and comply with 'the five Ks'.[39] Only if a Sikh is actually wearing a turban can he rely on the exemptions from the duty to wear a safety helmet. Moreover, a turbaned Sikh who is injured on a construction site as a result of the negligent act of another person can only claim damages for any injuries that would have been suffered even if he had been wearing a helmet.[40] In this sense the Sikh is deemed to take the extra risk of injury upon himself. Furthermore, any Sikh who carries a *kirpan* in a public place with a view to using it as an offensive weapon will naturally be guilty of an offence.[41] Again, the fact that a general defence of justifiability is afforded to employers under the Race Relations Act in cases of indirect discrimination means that in some instances minority cultural requirements can be overridden by wider commercial considerations.[42]

Enactments promoting assimilation

While the main aim of the Matrimonial Proceedings (Polygamous Marriages) Act 1972 was to allow the courts to afford matrimonial relief to parties to polygamous marriages, section 4 of the Act struck a sharp blow against cultural pluralism. It incorporated a new situation into the statutory list of grounds upon which a marriage is void, namely where either party to an actually or potentially polygamous marriage contracted abroad was at the time of the marriage domiciled in England and Wales.[43] Although section 4 purported to be merely a restatement of the pre-existing common law position, its enactment seems to have occurred without sufficient regard to the extent of its application. The provision is framed so widely that it is liable to bar, for example, Muslims from the Indian sub-continent who have acquired a domicile of choice here, from returning to their countries of origin to enter into a first marriage through an Islamic wedding.[44] This restricts them in their choice of ceremony since an Islamic form of marriage cannot validly be contracted in England.

---

[38] s. 22(5)(*c*).

[39] The five Ks are *kesh* (long hair), *kanga* (a comb), *kara* (a steel bangle), *kaacha* (long underpants), and *kirpan*.

[40] Employment Act 1989, s. 11(5).

[41] See Prevention of Crime Act 1953, s. 1.

[42] See below, pp. 470–1.

[43] See now Matrimonial Causes Act, s. 11(*d*).

[44] For its interpretation, see *Hussain* v. *Hussain* [1986] Fam 134 (CA), discussed below, pp. 469–70.

At around the same time, section 16 of the Domicile and Matrimonial Proceedings Act 1973 introduced another rule into English family law which removed the rights of Muslims and other minorities to follow their own norms and procedures. This provision reversed the previous common law rule that English law would generally recognize the validity of extra-judicial divorces obtained here, provided they were acceptable to the law of the domicile.[45] Henceforth, no proceeding would be regarded as validly dissolving a marriage unless it was instituted in a court of law. This meant a ban both on the Muslim *talaq* and on consensual forms of divorce used by many societies in different parts of the world.[46]

Twelve years later the Prohibition of Female Circumcision Act 1985 was enacted in the wake of reports that a small number of such operations had been occurring within this country. Female circumcision was probably a common law crime or an offence under the Offences Against the Person Act 1861 prior to 1985, but many felt it was desirable, in the absence of any judicial precedent, to put the matter beyond doubt by means of a well-publicized and unequivocal enactment. During the course of the parliamentary debates on the subject a Government spokesman described female circumcision as 'not compatible with the culture of this country' and 'thoroughly repugnant to our way of life'.[47] Although the Act creates special exemptions for operations needed for a person's physical or mental health, no account is to be taken in assessing mental health of anyone's belief that the operation is required 'as a matter of custom or ritual'.[48]

From a practical point of view, by far the most significant assimilationist tendencies have come in the field of immigration law. The Immigration Act 1971 empowers the Secretary of State to make rules[49] and since 1971 a variety of changes to the immigration rules have restricted entry to the UK in such a way as to strike at the family customs and traditions of several of the minority communities, with the result that the unity of many families has been destroyed. The most notorious of these provisions has been the 'primary purpose' rule, in terms of which spouses will be refused entry-clearance certificates to join partners in this country unless the entry-clearance officer is satisfied that 'the marriage was not entered into primarily to obtain admission to the UK'.[50] A similar rule applies to the entry of fiancés and fiancées who seek to come to the UK to marry and reside with a partner here.[51] The details of these rules are discussed

---

[45] See e.g. *Qureshi* v. *Qureshi* [1972] Fam 173.
[46] The ban is now contained in Family Law Act 1986, s. 44(1).
[47] HL Debs. 447, col. 86 (Lord Glenarthur).
[48] s. 2(2).
[49] s. 3(2).
[50] For the current version of this rule, see 'Statement of Changes in Immigration Rules', HC 251 of 1989–90, para. 50.
[51] Ibid., para. 47.

elsewhere in this book,[52] but the point to be made here is that they clearly have an adverse impact upon the traditional pattern of arranged marriages favoured by several ethnic minority communities.

Two further changes in immigration law are worthy of note because they demonstrate the determination of the authorities to exclude from entry parties to certain unacceptable types of marriage. From 1986 persons under the age of 16 have been barred from entering the UK in reliance upon their status as a spouse of someone living here.[53] This followed the discovery early in that year of a 12-year-old Iranian bride and a 13-year-old Omani bride, each living in England with her student husband.[54] Since 16 is the 'age of consent' for sexual intercourse[55] and the minimum age for marriage in England,[56] it was felt to be intolerable to have these child brides performing their wifely duties here.[57] Further, in 1988 steps were taken to prevent second and subsequent wives of polygamous men from joining their husbands here for settlement purposes.[58] The policy underlying this change was to prevent a husband simultaneously having two or more wives living with him in the UK on the grounds that actual polygamy is not 'an acceptable social custom in this country'.[59]

A balancing exercise: The religious education provisions of the Education Reform Act 1988[60]

(i) *Collective worship*. The 1988 Act retains the requirement laid down in the Education Act 1944 that all pupils at state schools must take part in an act of collective worship on each school day, subject to a right of parental withdrawal.[61] As before, in LEA schools the act of worship must not be distinctive of any particular Christian denomination, but under the 1988 Act it is required to be 'wholly or mainly of a broadly Christian character'.[62] This does not mean, however, that every single day's worship has to be Christian in character. It is sufficient if, 'taking any school term as a whole', most acts of worship are wholly or mainly of a broadly Christian character.[63] Furthermore, despite the express reference to Chris-

---

[52] See above, ch. 11.

[53] HC 306 of 1985–6; see now HC 251 of 1989–90, para. 2.

[54] See e.g. *The Times*, 6 and 20 Mar. 1986.

[55] Sexual Offences Act 1956, s. 6(1).

[56] Matrimonial Causes Act 1973, s. 11(*a*)(ii).

[57] Cf. *Alhaji Mohamed* v. *Knott* [1969] 1 QB 1 (DC) upholding the validity of a Nigerian marriage, involving a 13-year-old girl, in the eyes of English law.

[58] Immigration Act 1988, together with amendments to Immigration Rules, HC 555 of 1988; see now HC 251 of 1989–90, paras. 3–5.

[59] See HC Debs. 122, col. 785 (Douglas Hurd MP).

[60] For a detailed examination and critique, see generally S. Poulter, 'The Religious Education Provisions of the Education Reform Act 1988' (1990) 2 *Education and the Law* 1.

[61] Education Reform Act 1988, ss. 6(1), 9(3).

[62] s. 7(1).

[63] s. 7(3).

tianity (of which there was no explicit mention in the 1944 Act), there are two provisions specifically designed to take account of the diversity of faiths represented in the pupils of many schools. First, in working out the precise forms of collective worship to be adopted, individual LEA schools may themselves decide what is appropriate and in the process have due regard to the family background of their pupils.[64] Secondly, a LEA school may be exempted from the 'broadly Christian' requirement if the LEA's Standing Advisory Council on Religious Education (SACRE) decides this would be appropriate.[65] Then the worship can either be distinctive of a non-Christian religion or else give no particular emphasis to any one faith. Moreover, the exemption may either relate to the school as a whole or to 'any class or description of pupils' there.[66]

(ii) *Religious education classes.* Under the 1988 Act religious education classes remain compulsory, forming part of the new 'basic curriculum', subject to the right of parental withdrawal.[67] In LEA schools the classes must, as before, follow an 'agreed syllabus' formulated by a locally convened conference of four committees.[68] These committees are appointed to represent the LEA, the teachers' associations, the Church of England, and such other Christian and other denominations as reflect the principal religious traditions of the area.[69] The greatest change made by the 1988 Act is that any new agreed syllabus produced after 29 September 1988 must 'reflect the fact that the religious traditions in Great Britain are in the main Christian whilst taking account of the teaching and practices of the other principal religions represented in Great Britain'.[70] It is to be hoped that this rather opaque provision will be construed in a sensible manner as a genuine attempt to achieve a balance between the primacy to be accorded to Christianity, on the one hand, and the importance of recognizing religious diversity, on the other.[71]

Judicial decisions

Family law

(i) *Consent to marriage.* There have been conflicting Court of Appeal decisions upon what constitutes duress sufficient to have a marriage annulled

---

[64] s. 7(4), (5).
[65] s. 12(1), (9). By mid-1990 one quarter of Bradford's schools had been granted an exemption: *TES*, 29 June 1990.
[66] s. 12(1).
[67] ss. 2(1)(*a*), 9(3).
[68] Education Act 1944, s. 26(1), as amended by Sch. 1 of the 1988 Act.
[69] Education Act 1944, Sch. 5, as amended by Sch. 1 of the 1988 Act.
[70] s. 8(3).
[71] This had, in fact, been the practice in many LEAs for several years prior to the 1988 Act.

for lack of consent.[72] The basic problem lies in identifying the stage at which the entirely acceptable tradition of arranged marriages becomes distorted to produce a forced marriage. The right of an individual not to be coerced into a marriage may conflict with the claims made by the family (and perhaps the wider minority community) that by religion or tradition their concerns should be paramount in the choice of spouse.

In *Singh* v. *Singh*[73] in 1971 the Court of Appeal endorsed the view expressed by Simon P. in an earlier case that for duress to vitiate a marriage the will of one of the parties must have been overborne by genuine and reasonably held fear caused by threat of immediate danger to life, limb, or liberty.[74] The petitioner, a 17-year-old Sikh girl who had only entered into the marriage because of the pressures imposed upon her by her parents and her feeling that she was bound to obey their wishes as a matter of traditional custom, was held not to have satisfied this burdensome test. However, eleven years later in *Hirani* v. *Hirani*[75] the Court of Appeal laid down far less onerous conditions in granting an annulment to a 19-year-old Hindu girl who had been faced with very similar family and community pressures. Ormrod LJ declared simply: 'The crucial question in these cases . . . is whether the threats, pressure, or whatever it is, is such as to destroy the reality of consent and overbears the will of the individual.'[76]

(ii) *Polygamy*. In *Hussain* v. *Hussain*[77] the Court of Appeal managed, with ingenious (if rather dubious) reasoning, to limit the damage done by section 4 of the Matrimonial Proceedings (Polygamous Marriages) Act 1972, later re-enacted as section 11(*d*) of the Matrimonial Causes Act 1973. Instead of ruling, as everyone expected, that a Muslim marriage in Pakistan contracted by a Pakistani man domiciled in England was void as being potentially polygamous, the Court declared it to be monogamous and thus valid. The Court held that the question whether the marriage was polygamous had to be examined by investigating not its nature under the *lex loci contractus* but the capacities of the individual parties. Since the husband had no capacity to take a second wife by reason of his English domicile and the wife no capacity to take a second husband by reason of Islamic law the marriage was deemed monogamous. In reaching this conclusion, Ormrod LJ drew attention to our 'increasingly pluralistic society'[78] and to the widespread and profound repercussions on the Muslim

---

[72] See Matrimonial Causes Act 1973, s. 12(*c*).
[73] [1971] P. 226.
[74] *Szechter* v. *Szechter* [1971] P. 286 at 297–8.
[75] (1982) 3 FLR 232.
[76] At 234.
[77] [1983] Fam 1.
[78] At 32.

community here if the decision had gone the other way.[79] However, the decision has created an anomaly because if the couple's domiciles had been reversed the husband would have had the capacity to take a second wife and the marriage would therefore have been polygamous and void.

Discrimination law

In *Mandla* v. *Dowell Lee*[80] the House of Lords ruled that Sikhs are an 'ethnic' as well as a religious group and are thus protected against discrimination by the provisions of the Race Relations Act 1976. Hence a headmaster who had denied a Sikh boy admission to his school on the ground that the boy wished to wear a turban in contravention of the school's rules about pupils' dress was held to have acted unlawfully. Lord Fraser identified two characteristics which had to be possessed by any community if it was to qualify as an ethnic group under the Act: first, a long shared history, of which the group is conscious as distinguishing it from other groups and the memory of which keeps it alive, and secondly, a cultural tradition of its own, including family and social customs and manners, often but not necessarily associated with religious observance.[81] Jews clearly qualify as an ethnic group, using these criteria, as do gypsies,[82] but Muslims[83] and Rastafarians[84] apparently do not.

In several employment cases, Sikh men against whom indirect discrimination had been practised, because they wore beards or turbans, have been unsuccessful in their claims. Their employers have been able to demonstrate that company rules requiring employees in confectionery factories to be clean-shaven[85] or those in engineering workshops to wear a hard hat[86] were 'justifiable' on grounds of health or safety. However, the decision of the Court of Appeal in *Hampson* v. *Department of Education*[87] has recently stiffened the test of justifiability, bringing the interpretation of the Race Relations Act into line with comparable provisions and decisions on sexual equality.[88] As a result, employers will now only be able to prove an act of discrimination is justifiable if they can show that

---

[79] At 33.

[80] [1983] 2 AC 548.

[81] At 562.

[82] *CRE* v. *Dutton* [1989] QB 783, [1989] 1 All ER 306 (CA).

[83] *Nyazi* v. *Rymans Ltd.* (E.A.T., 10 May 1988, unreported); *CRE* v. *Precision Manufacturing Services* (Sheffield Industrial Tribunal, 26 July 1991, unreported).

[84] *Dawkins* v. *Department of the Environment* [1993] IRLR 284 (CA).

[85] See *Singh* v. *Rowntree Mackintosh Ltd.* [1979] ICR 554; *Panesar* v. *Nestle Co Ltd.* [1980] ICR 144 (CA).

[86] *Kuldip Singh* v. *British Rail Engineering Ltd.* [1986] ICR 22.

[87] [1990] 2 All ER 25 (CA), discussed above, ch. 13.

[88] See e.g. *Rainey* v. *Greater Glasgow Health Board* [1987] AC 224 (HL) discussed above, chs. 12 and 13; the test has been endorsed by the House of Lords in *Webb* v. *EMO Air Cargo (UK) Ltd.* [1992] 4 All ER 929 at 936.

there was no viable alternative method of achieving the outcome they desired.[89] Beards, for instance, can easily be netted in 'snoods' and lightweight turbans can sometimes be worn under specially constructed helmets.

Criminal law

(i) *Blasphemy.* In *R.* v. *Chief Metropolitan Stipendiary Magistrate, ex parte* Choudhury[90] the Divisional Court ruled that the common law offence of blasphemy is confined to the protection of the Christian religion and cannot be extended judicially to cover other faiths, such as Islam. Hence the Chief Metropolitan Stipendiary Magistrate had acted correctly in refusing to issue a summons for the private prosecution of Salman Rushdie and his publishers for allegedly vilifying the Muslim faith and the Prophet Muhammad in the novel *The Satanic Verses*.[91] More than a decade earlier Lord Scarman had criticized this limitation upon the ambit of the blasphemy law in *Whitehouse* v. *Lemon*[92] when he commented:

The offence belongs to a group of criminal offences designed to safeguard the internal tranquillity of the Kingdom. In an increasingly plural society such as that of modern Britain, it is necessary not only to respect the differing religious beliefs, feelings and practices of all but to protect them from scurrility, vilification, ridicule and contempt.[93]

Subsequently the possibility of reforming the law was referred to the Law Commission, but while the minority of the Commissioners (including the chairman) was sympathetic to the line taken by Lord Scarman, the majority favoured outright abolition of the offence with nothing being enacted in its place.[94]

(ii) *General principles.* In recent years several cases have affirmed the long-established principle[95] that, in determining the question of guilt or innocence, a uniform and consistent standard must be applied to all those who are accused of criminal offences, regardless of their national origins, their ignorance of English law and their different social values, traditions, and beliefs.[96] However, when the court comes to the matter of sentencing

---

[89] See *Bilka-Kaufhaus* v. *Weber von Hartz* [1986] ECR 1607, discussed above, chs. 12 and 13.

[90] [1990] 3 WLR 986, [1991] 1 All ER 306.

[91] Viking/Penguin, 1988.

[92] [1979] AC 617.

[93] At 658.

[94] Law Commission, *Offences against Religion and Public Worship* (1985).

[95] See e.g. *R.* v. *Esop* (1836) 7 C&P 456; *R.* v. *Barronet and Allain* (1852) Dears CC 51.

[96] See e.g. *R.* v. *Dad and Shafi* [1968] Crim. LR 46 (CA); *R.* v. *Moied* (1986) 8 Crim App R (S) 44 (CA).

a person who has been convicted, these types of factor can certainly be borne in mind by way of mitigation.[97] On this basis a Muslim father has been given a conditional discharge following his conviction of failing to ensure his 15-year-old daughter's attendance at the coeducational school to which she had been allocated by the LEA.[98] The father's religious objections to coeducational schools, the only ones available in his area, were held to afford him no defence. Similarly a Nigerian mother was given an absolute discharge following her conviction of the offence of assault for scarifying the faces of her two young sons.[99] Her claim that she was merely following the tribal custom of her people, the Yoruba, was equally no defence to the charge.

National political controversies

Of the legal developments outlined above only three can be said to have generated substantial political controversy at the national level. The 'Rushdie affair' has been by far the most divisive, with many deep differences of opinion being expressed even between members of the same political party and between those of the same particular religious faith or of none.[100] In the absence of any emerging consensus as to what direction any reform of the current blasphemy law should take, the Government has elected, perhaps not surprisingly, to do nothing. In a letter written to leading British Muslims in 1989 John Patten MP, Minister of State at the Home Office, stated: 'We have considered [the] arguments carefully and reached the conclusion that it would be unwise for a variety of reasons to amend the law of blasphemy, not least the clear lack of agreement over whether the law should be reformed or repealed.'[101] Those who favour repeal have chiefly relied upon the importance our society attaches to freedom of expression and the need to keep exceptions to this principle to a bare minimum.[102] Blasphemy operates as a form of censorship and although it appears only to strike at those who express themselves in an outrageous manner it can actually restrict the content of what can be published with impunity. In the words of the majority report of the Law Commission, which were cited with approval

[97] See e.g. *R.* v. *Bibi* [1980] 1 WLR 1193 (CA).
[98] *Bradford Corporation* v. *Patel* (1974, unreported).
[99] *R.* v. *Adesanya* (1974, unreported).
[100] For a variety of perspectives on the affair, see the three seminar reports published by the Commission for Racial Equality in 1990, *Law, Blasphemy and the Multi-Faith Society*, *Free Speech*, and *Britain: A Plural Society*. See also L. Appignanesi and S. Maitland (eds.), *The Rushdie File* (1989); S. Akhtar, *Be Careful with Muhammad* (1989).
[101] The letter is repr. in *Law, Blasphemy and the Multi-Faith Society*, 84–7.
[102] See e.g. S. Lee, *The Cost of Free Speech* (1990).

by the International committee for the Defence of Salman Rushdie and his Publishers:[103]

Such restrictions would in particular have adverse consequences for what many would consider to be proper criticism of matters pertaining to religion and religious belief. Ridicule has for long been an acceptable means of focusing attention upon a particular aspect of religious practice or dogma which its opponents regard as offending against the wider interests of society, and in that context use of abuse or insults may well be a legitimate means of expressing a point of view upon the matter at issue. The imposition of criminal penalties upon such abuse or insults becomes, in our view, peculiarly difficult to defend in the context of a 'plural' or multi-racial, multi-religious society. Here one person's incisive comment (or indeed seemingly innocent comment) may be another's 'blasphemy', and to forbid use of the strongest language in relation, for example, to practices which some may rightly regard as not in the best interests of society as a whole would, it seems to us, be altogether unacceptable.[104]

Furthermore, even if the strongest justification for retaining an offence of blasphemy is to protect religious believers from suffering hurt to their feelings (rather than to protect religions as such), it is questionable whether the law should accord special reverence to what devout people regard as sacred while ignoring the sensitivities of those who attach great spiritual significance to other objects or persons.[105] This is especially so in the predominantly secular society which Britain has now become.

On the other hand, those who favour reform of the blasphemy law feel that it is worthwhile attempting to preserve intact a sense of the sacred in our multi-faith society.[106] As the minority report of the two dissenting Law Commissioners aptly put it: 'we believe that, in this country, people generally share that respect for feelings of reverence even when they themselves do not adhere to any religion. If that is right, they would regard it as morally wrong and offensive to cause outrage—if done deliberately and without reason—to that feeling of reverence.'[107]

Reformers have proposed two main alterations to the existing offence.[108] First, religions other than Christianity should be protected. Secondly, in order to meet criticism that this would unduly restrict freedom of expression, the offence would change from being one of strict liability, as it is at present, to one in which there existed a specific intent on the part of the defendant to cause outrage to religious feelings. A key argument made on

---

[103] *The Crime of Blasphemy: Why it should be Abolished* (1989), 7.

[104] *Offences against Religion*, 18.

[105] Ibid. 20–2.

[106] See e.g. *Report of the Bishop of London's Group on Blasphemy* (GS Misc. 286, 1988), 19.

[107] *Offences against Religion*, 41–2.

[108] See e.g. S. Poulter, 'Towards Legislative Reform of the Blasphemy and Racial Hatred Laws' (1991) *PL* 371.

behalf of Muslims for an extension of the law has been that they identify themselves primarily as a religious rather than an 'ethnic' community and feel that, having been an object of Christian oppression in the past, they now form a vulnerable minority in Britain today.[109] Vilification of their faith can thus profoundly harm their interests here and protection should be accorded to them by analogy with the law banning the incitement of racial hatred which protects racial and ethnic, but not religious, groups.[110] It is not, of course, easy to frame a satisfactory statutory definition of religion with a view to clarifying exactly which groups would gain protection under a reformed blasphemy law, but the matter could simply be left for the courts to decide.[111]

During 1990 two private members' bills were introduced and each given a first reading, one designed to extend the crime of blasphemy to encompass Islam, Hinduism, Sikhism, Judaism, and Buddhism,[112] and the other proposing to abolish the offence altogether,[113] but neither bill made any further progress.

A further political controversy was created by the Education Reform Act 1988 in general, and its religious education provisions in particular, both of which provoked substantial debate about the nature of British society today. By establishing a new, uniform, national curriculum for all pupils in the state sector, central government was seen to be taking greater control over what is taught in schools and this generated concern about whether there is, in fact, any public agreement on, for example, what makes up British culture, history, and values. Is there an unproblematic 'British heritage' which can properly be prescribed for all pupils? Anxiety was also expressed as to whether the content of the history syllabus would adequately reflect the perspectives of minority groups.[114]

The religious education debate was sparked off when Baroness Cox and others moved amendments in the House of Lords to provide that such education should be 'predominantly Christian' on the grounds, *inter alia*, that the subject had become trivialized through the consumption by pupils of 'a fruit cocktail of world faiths'.[115] Both the Archbishop of Canterbury and the Bishop of London expressed reservations about this proposal and the latter subsequently conducted an extensive round of consultations with representatives of the churches, other faiths, LEAs, and others, before returning to the House of Lords with a package of

---

[109] See e.g. T. Modood, 'British Asian Muslims and the Rushdie Affair' (1990) 61 *Political Quarterly* 143.

[110] Public Order Act 1986, s. 17.

[111] For discussion of this problem, see Law Comm., *Offences against Religion*, 26–7, 43.

[112] Blasphemy Bill, 89 of 1990.

[113] Blasphemy (No. 2) Bill, 79 of 1990.

[114] The National Curriculum History Working Group was determined that it should do so—see *Final Report* (1990), 1, 15–17.

[115] See *HL Debs*. 496, col. 502; 498, cols. 640–1.

amendments which (together with some government amendments) form the balanced compromise which is reflected in the provisions mentioned earlier. Dissatisfied with this compromise, Baroness Cox has since turned to campaigning for more religious denominations to be entitled to gain voluntary aided status for their schools so that they can teach their own separate faiths in their own way. For many years the Department of Education and Science tended to reject applications for such status (and the substantial state funding which goes with it) on the ground that falling school rolls have left many surplus places at existing schools and that further public expenditure on separate religious schools cannot therefore be justified. However, while there are thousands of Anglican and Roman Catholic voluntary-aided schools and a small number of Jewish schools, no such approval has yet been accorded to a Muslim school. Some accuse the government of racial or religious prejudice, while others fear that separate Muslim schools would be divisive and create misunderstandings between faiths. Baroness Cox and her supporters see government funding for separate religious schools as a logical extension of the Conservative commitment to parental choice and in 1991 she promoted (but later withdrew) a private member's bill which would have required the DES to disregard the question of surplus places in deciding upon applications for voluntary-aided status.[116] The view taken by the Commission for Racial Equality has been that, while no less favourable treatment should be accorded to minority faiths, the public debate about the future of religious schools should be broadened to consider the role of all such schools in a multi-racial society, not just those owned by minority faiths.[117] In so far as the demand for minority faith schools reflects dissatisfaction with racism and discrimination in other schools, policies and practices operating there concerning multicultural and anti-racist education clearly need to be developed and strengthened.

Finally, there has been a long-standing controversy over the 'primary purpose rule' in immigration law. It was originally introduced with a view to protecting the domestic labour market from primary male immigration from the Indian sub-continent, but opponents rightly claim that it serves to keep families apart and discriminates against the system of arranged marriages.[118]

Foreign comparisons

In the limited space available here, it is not possible to make any detailed comparisons with legal developments abroad. However, it is clear that, in

---

[116] Education Bill, 8 of 1990; see *HL Debs*. 526, cols. 1247–1308.

[117] CRE, *Schools of Faith: Religious Schools in a Multicultural Society* (1990).

[118] See e.g. R. Sondhi, *Divided Families: British Immigration Control in the Indian Subcontinent* (1987).

general terms, recent years have witnessed a world-wide upsurge in the
identification of many peoples with their particular ethnic, religious, or
linguistic communities. An appreciation of their cultural heritage, a concern for the preservation of their distinctive customs and traditions, and
a commitment to maintain their deeply held values and beliefs, have led
them to stake out claims upon the wider polity. This 'ethnic revival'[119]
obviously takes different forms and requires different legal responses
according to the circumstances prevailing in a particular country. However, within the states of the European Community it is apparent that the
claims of some of the newer migrants (notably North African Muslims in
France) are being met by a growth in hostility and racism on the part of
the majority communities.[120] The tension between the rival policies of
cultural pluralism and assimilation is just as strong there as it is in Britain.
A huge political furore arose in France in 1989 over whether Muslim girls
should be allowed to wear religious headscarves at school.[121] Eventually a
ruling by the Conseil d'État upheld their right to do so as a matter
of freedom of religious expression, subject to a number of conditions,
despite the strongly secular nature of state education in France.[122] A
similar dispute occurred in a girls' grammar school in Cheshire in 1990
and was quickly resolved with the same outcome.[123]

Explanations for the changes

Virtually all the changes described above have been achieved by means of
legislation, supplemented by judicial interpretation. The explanations for
them lie in a mixture of factors including economic and social developments, the activities of pressure groups, and the recommendations of
various commissions. All of these have been operating against the general
background of the pluralist and assimilationist philosophies outlined earlier.
None of the issues has been central to mainstream party political debate
or been adopted by any of the main political parties as a major plank in
its policies. Very little interest has been shown in developments abroad.

Economic considerations have clearly influenced legislation outlawing
racial discrimination in the field of employment. Recruitment of workers
on merit rather than on the basis of 'race' or ethnicity makes for greater
efficiency.[124] Moreover, the fact that there were thought to be as many as
40,000 Sikh building workers employed on construction sites in 1989 must

---

[119] See e.g. A. Smith, *The Ethnic Revival* (1981).
[120] See *Report of Committee of Inquiry into Racism and Xenophobia* (1990), 56–60.
[121] Ibid. 58.
[122] See (1990) *PL* 434–5.
[123] See S. Poulter, *Asian Traditions and English Law* (1990), 91.
[124] See further above, ch. 13.

have helped to persuade Parliament that they needed to be exempted from the requirement to wear a safety helmet if the industry was not to be severely damaged.[125]

The most important social development has naturally been the growth of a multicultural society in which significant numbers of people follow different religious faiths and practise different customs and traditions. In addition to older Jewish and gypsy communities, there are now thought to be approaching a million Muslims living in this country, as well as about 300,000 Sikhs and around 400,000 Hindus. This helps to explain the balancing exercise accomplished by the religious education provisions of the Education Reform Act 1988, the indirect discrimination provisions of the Race Relations Act 1976, and the changes in family law outlined above.

A variety of pressure groups have been involved in pushing for reforms, but so far with rather limited success. The Sikh community as a whole has, of course, achieved three notable triumphs in securing parliamentary action, making much play of the esteem in which they were held by the British in India during the days of Empire and arguing cogently that if they could be allowed to wear turbans in the British Army there could surely be no reason why they should not do so on motor-cycles and construction sites. However, among those groups which have nothing concrete to show for their pains brief mention may be made of the following. The Union of Muslim Organizations of the UK and Eire has failed to persuade successive governments to introduce a separate system of Islamic personal law for all British Muslims.[126] The RSPCA has failed to obtain the repeal of the special statutory exemptions accorded to the Jewish and Muslim methods of slaughtering animals for food.[127] Muslim organizations campaigning for an extension of the blasphemy laws[128] and secular groups seeking abolition of the offence[129] have both failed to make any progress. Nor have organizations opposed to the 'primary purpose rule'[130] had any success in achieving its abolition, although its application was modified in 1992.[131]

On several of these and other issues, recommendations have been made by official bodies, but their views have also largely been ignored.

---

[125] See *HL Debs*. 511, col. 744 (Lord Strathclyde, 16 Oct. 1989).

[126] For an analysis of the claim and the reasons for its rejection, see S. Poulter, 'The Claim to a Separate Islamic System of Personal Law for British Muslims', in C. Mallat and J. Connors (eds.), *Islamic Family Law* (1990), 147–66.

[127] See e.g. RSPCA, *Humane Slaughter* (1981); *Ritual Slaughter* (1984).

[128] See e.g. UK Action Committee on Islamic Affairs, *The British Muslim Response to Mr Patten* (1989).

[129] See e.g. Int. Comm. for the Defence of Salman Rushdie, *Crime of Blasphemy*.

[130] See above, pp. 466–7.

[131] *HC Debs*. 210, cols. 523–4 (written answers).

The Farm Animal Welfare Council recommended the abolition of the animal slaughter exemptions for Jews and Muslims in 1985.[132] The same year the Law Commission recommended (by a slender majority) the demise of the offence of blasphemy.[133] It also proposed the rectification of the anomaly produced by the Court of Appeal in *Hussain* v. *Hussain*[134] on potentially polygamous marriages contracted abroad.[135] On the other hand, its earlier recommendation[136] that the rule in *Hyde* v. *Hyde* be abolished was implemented.[137] In 1985 the Swann Committee Report on the Education of Children from Ethnic Minority Groups recommended the abolition of the compulsory daily act of collective worship in schools on the basis that it could no longer be justified 'with the multiplicity of beliefs and non-beliefs now present in our society'.[138] However, as we have seen, the Education Reform Act 1988 retained such daily worship.

Methods of conflict resolution

In recent years probably the greatest legal threat to the maintenance of Jewish culture and religious practice has come from campaigns to remove the statutory exemption for *shechita*, their religious method of slaughter. In addition to the activities of the RSPCA and the recommendations of the Farm Animal Welfare Council, private members' bills to abolish or limit the exemption have been introduced in Parliament on no fewer than six occasions since 1955.[139] The responsibility for defending *shechita* has principally been borne by the relevant committee of the Board of Deputies of British Jews which has successfully co-ordinated the discreet lobbying of MPs to ensure the defeat of all the bills.[140] In this they have been helped both by the Council of Christians and Jews and by the fact that in recent years the Jewish community has been well represented in the House of Commons.[141]

The Sikhs, despite having no representation in Parliament, have similarly been able to employ orthodox political methods to ensure that their needs are met. As we have seen, on no fewer than three occasions since 1976 legislation has been enacted specifically to accommodate their reli-

---

[132] *Report on the Welfare of Livestock when Slaughtered by Religious Methods* (1985), 25.
[133] Law Comm., *Offences against Religion*.
[134] [1983] Fam 26.
[135] Law Com. Report No. 146, 'Polygamous Marriages'.
[136] Law Com. Report No. 42, 'Polygamous Marriages'.
[137] By Matrimonial Proceedings (Polygamous Marriages) Act 1972.
[138] Cmnd. 9453, p. 497.
[139] See R. Charlton and R. Kaye, 'The Politics of Religious Slaughter: An Ethno-religious Case Study' (1985–6) *New Community* 490.
[140] Ibid.
[141] See G. Alderman, *The Jewish Community in British Politics* (1983), 174–5. Since 1930 there have never been fewer than 16 Jewish MPs; in 1974 there were 46.

gious traditions.[142] However, in *Mandla* v. *Dowell Lee*[143] they did have to resort to litigation and take a test case all the way to the House of Lords, to establish that they were protected as an 'ethnic group' by the Race Relations Act 1976, despite the fact that this had clearly been Parliament's intention.[144]

Muslims have experienced only very limited success. They have failed to persuade Parliament either to introduce a system of Islamic personal law or to reform the law of blasphemy. They have tried to employ judicial review to establish that Islam is protected by the current blasphemy law, but to no avail.[145] They were, however, granted judicial review to quash the refusal by the Department of Education and Science to grant voluntary-aided status to an independent Muslim school.[146] Judicial review has also been used on several occasions by gypsies in order to try to force speedier action on the part of local authorities in implementing their duties to provide enough sites for encampments under the Caravan Sites Act 1968.[147] However, while such court actions have occasionally been effective in persuading ministers to give directions to dilatory local councils[148] and in preventing the unreasonable eviction of gypsies from unauthorized sites,[149] they cannot actually force the hand of ministers[150] or guarantee that all such evictions will be halted until enough authorized sites are available.[151]

It seems highly likely that a large number of family and community disputes among members of minority groups (perhaps especially in the various Asian communities) are being settled within those communities themselves, without recourse to legal proceedings or indeed access to detailed knowledge of English law through contact with a solicitor. Traditional and religious values will no doubt contribute substantially to such

---

[142] Motor-Cycle Crash Helmets (Religious Exemption) Act 1976, Criminal Justice Act 1988, Employment Act 1989.

[143] [1983] 2 AC 548 (HL).

[144] See 'Racial Discrimination', Cmnd. 6234 (1975), para. 55; *Official Report*, Standing Committee 'A', 4 May 1976, cols. 102–3.

[145] *R.* v. *Chief Metropolitan Stipendiary Magistrate, ex parte Choudhury* [1991] 1 All ER 306. A similar attempt to use judicial review to have Viking Penguin prosecuted under the Public Order Act 1986 for distributing copies of *The Satanic Verses* failed: see *R.* v. *Horseferry Road Metropolitan Stipendiary Magistrate, ex parte Siadatan* [1991] 1 QB 280, [1991] 1 All ER 324.

[146] *R.* v. *Secretary of State for Education, ex parte Islam*, *The Times*, 22 May 1992. The Court ordered a reconsideration of the application because of procedural impropriety and manifest unfairness.

[147] See e.g. *R.* v. *Secretary of State for the Environment, ex parte Ward* [1984] 1 WLR 834, [1984] 2 All ER 556, and the cases cited below.

[148] *R.* v. *Secretary of State for the Environment, ex parte Lee* (1987) 54 P & CR 311; *West Glamorgan CC* v. *Rafferty* [1987] 1 WLR 457, 1 All ER 1005.

[149] *West Glamorgan CC* v. *Rafferty*, n. 148.

[150] *R.* v. *Secretary of State for Wales, ex parte Price* [1984] JPL 87.

[151] *West Glamorgan CC* v. *Rafferty* 1 All ER at 1023.

settlements. Within certain limits defined by public policy, 'private order-ing' of family disputes is regarded as an eminently sensible and satis-factory method of resolving domestic difficulties in modern English family law.[152] This is, of course, always subject to the proviso that the interests of wives and children are properly safeguarded and there is clearly some risk of wives being oppressed where informal means of dispute resolution are employed.

EUROPEAN AND INTERNATIONAL DIMENSIONS

European Convention on Human Rights

The principal effect of Britain's ratification of the European Convention has been to assist in providing a broader legal framework within which policy decisions with respect to ethnic, religious, and linguistic minority rights can be taken and evaluated. In the absence of a written constitution, appeals to arguments about fundamental human rights and values tend to focus on the Convention. In particular, since freedom of religion is not expressly enshrined in English law, the protection afforded to this freedom by the Convention[153] has become a potent source of reference.[154] Other provisions of direct relevance are those guaranteeing freedom from dis-crimination on the basis of, *inter alia*, sex, race, colour, or religion;[155] the right to the free assistance of an interpreter for anyone who cannot understand or speak the language used in court proceedings;[156] and the right of parents to ensure education of their children in conformity with their religious convictions.[157]

The most significant use of the European Convention by the English courts occurred in the case of *Ahmad* v. *ILEA*[158] in 1978. Ahmad, a devout Muslim, who had been employed by ILEA as a full-time teacher at several of its schools, complained of unfair dismissal following his resignation (under protest) after ILEA sought to make him change his status to that of a part-time teacher, working only four-and-a-half days a week instead of five. The reason why ILEA had pressed this alteration on

---

[152] See e.g. S. Cretney and J. Masson, *Principles of Family Law*, 5th edn. (1990), ch. 21 and pp. 562–3.

[153] Art. 9.

[154] See e.g. the reliance placed upon it, albeit unsuccessfully, in *R.* v. *Crown Court at Aylesbury, ex parte Chahal* [1976] RTR 489 by a turbaned Sikh motor-cyclist convicted of not wearing a helmet prior to the enactment of the Motor-Cycle Crash Helmets (Religious Exemption) Act 1976.

[155] Art. 14.

[156] Art. 6(3)(*e*).

[157] Art. 2 of the First Protocol.

[158] [1978] QB 36 (CA).

him, with a resulting reduction in salary, was that he was regularly absenting himself from his school classes early on Friday afternoons in order to visit a nearby mosque for Friday prayers. Section 30 of the Education Act 1944 seeks to guarantee that a teacher in the state system will not 'receive any less emolument' by reason only of his 'attending . . . religious worship' and this provision is automatically incorporated in the contracts of all such teachers. However, the majority of the Court of Appeal felt unable to apply section 30 at its face value in affording protection to the applicant and the Court ruled that his dismissal had not been unfair. In a vigorous dissenting judgment Scarman LJ (as he then was) expressed alarm that the result of the case would render it impossible for devout Muslims to become or remain full-time teachers. He argued that education authorities should make suitable administrative arrangements to ensure that Muslims could attend Friday prayers, that pupils would be taught, and that other staff were not unduly burdened, even if this involved extra public expenditure in employing a few more teachers.

A central issue in the case for all the judges was the impact of Article 9 of the European Convention guaranteeing freedom of religion. Scarman LJ took the view that the majority decision would be 'almost certainly a breach of our international obligations'.[159] As to the relevance of the European Convention in English law he declared:

it is no longer possible to assume that because the international treaty obligations of the United Kingdom do not become law unless enacted by Parliament our courts pay no regard to our international obligations. They pay very serious regard to them; in particular, they will interpret statutory language and apply common law principles, wherever possible, so as to reach a conclusion consistent with our international obligations . . .[160]

While Scarman LJ emphasized the importance of the guarantee of freedom of worship in Article 9(1), Lord Denning MR and Orr LJ stressed the significance of the restrictions authorized by Article 9(2), which states: 'Freedom to manifest one's religion or beliefs shall be subject only to such limitations as are prescribed by law and are necessary in a democratic society in the interests of public safety, for the protection of public order, health or morals, or for the protection of the rights and freedoms of others.' In their view the rights of ILEA under the contract and the rights of the pupils to be taught prevailed over Ahmad's freedom of religion.

When Ahmad ultimately took his complaint to the European Commission, alleging a violation of Article 9(1), it was the stance taken by the

---

[159] At 50.   [160] At 48.

majority of the Court of Appeal, not Scarman LJ, which was finally vindicated.[161] Ahmed's petition was dismissed as manifestly ill-founded, on the basis that ILEA was entitled to rely on its contract being fulfilled and had given due consideration to Ahmed's religious freedom by offering him a part-time post. The upshot seemed to be that Ahmed's exercise of his freedom of worship had to be subordinated to the contractual rights of the authority which were entitled to protection under the provisions of Article 9(2).

A similar concentration on the restrictions upon freedom of religion authorized by Article 9(2) is to be found in the decision of the Divisional Court in *R*. v. *Chief Metropolitan Stipendiary Magistrate, ex parte Choudhury*[162] in refusing to extend the blasphemy law to protect Islam. The Court took the position that any extension to non-Christian faiths would be likely to encourage intolerance, divisiveness, and unreasonable interference with the rights of others to freedom of expression under Article 10.[163]

Dissatisfied litigants have had recourse to the European Commission of Human Rights on eight occasions. As we have seen, Ahmad's allegation of a violation of his religious freedom was declared inadmissible. So, too, was an application by a turbaned Sikh who had been prosecuted, convicted, and fined twenty times between 1973 and 1976 for failing to wear a crash helmet before the Motor-Cycle Crash Helmets (Religious Exemption) Act 1976 was brought into force.[164] The Commission decided that there had been no violation of Article 9 because the compulsory wearing of helmets was considered a necessary safety measure for the protection of health within one of the exceptions in Article 9(2). The third application was made by a Muslim man who complained that his religious freedom had been violated because he was not permitted by English law to marry a 14-four-old Muslim girl, as authorized by Islamic doctrine.[165] His application was rejected as manifestly ill-founded by the Commission, on the ground that marriage could not be considered simply as a form of religious practice. The right to marry was regulated specifically by Article 12 of the Convention which leaves the requirements for a marriage to national law. The fourth application, by Choudhury, alleging that the blasphemy law violated both his freedom of religion under Article 9 and the prohibition of religious discrimination under Article 14, was also rejected.[166] The Commission ruled that there had

---

[161] *Ahmad* v. *UK* (1982) 4 EHRR 126.
[162] [1991] 1 All ER 306.
[163] At 321–2.
[164] *X*. v. *UK* (1978) 14 Dec. & Rep. 234.
[165] *Khan* v. *UK* (1986) 48 Dec. & Rep. 253.
[166] Application 17439/90, (1991) 12 *Human Rights LJ* 172.

been no violation of Choudhury's freedom of religion because such freedom did not extend to guaranteeing a right to bring any specific form of
proceeding (such as a private prosecution for blasphemy) against those
who, by authorship or publication, offended the sensitivities of an individual or a group of individuals. Further, the Commission decided that
since the complaint under Article 9 had been rejected as outside the
scope of the Convention, so too must the complaint under Article 14 be
rejected for the same reason.[167] The fifth application was made by a
gypsy who alleged that section 10 of the Caravan Sites Act 1968, which
makes it a criminal offence for gypsies to camp on unauthorized sites in
certain designated areas, constituted both a violation of her right to
respect for her private and family life under Article 8 of the Convention
and a breach of Article 14 since it only relates to gypsies and not to
other campers or caravaners.[168] The case is still pending, as is the sixth
application, made by the International Society for Krishna Consciousness
(ISKCON).[169] The Society alleges a violation of Article 9 through the
imposition of restrictions upon freedom of worship by Hindus at a temple
in Hertfordshire, pursuant to an enforcement notice issued under the
planning laws.[170] There have also been two unsuccessful applications by
Jewish prisoners, who claimed that they were denied the opportunity of
practising their religion in gaol, particularly in relation to dietary norms.[171]

Hardly any explicit reference has been made to the provisions of the
European Convention either in political discussion or in parliamentary
debates on minority rights. The Council of Europe's Social Charter has
had no direct impact on the cultural rights of minorities, being concerned
rather with social and economic rights.

European Community law

Britain's membership of the European Community has had virtually no
impact on the development of the cultural rights of minorities, as opposed
to their economic rights, save in one small particular. Many parents in the
minority communities are eager for their children to acquire or develop a
proficiency in their 'mother tongue' or the language of their family's
country or community of origin.[172] While some believe this can be achieved
satisfactorily through use of this language at home, coupled with evening
or weekend tuition at the many 'supplementary schools' organized locally
by the various communities themselves, others believe the task is far

---

[167] For criticism of the decision, see Poulter, 'Towards Legislative Reform'.
[168] *Smith* v. *UK*, application 18401/91.
[169] *ISKCON* v. *UK*, application 20490/92.
[170] See *ISKCON* v. *Secretary of State for the Environment* (1992) 64 P & CR 85.
[171] *X* v. *UK* (1976) 5 Dec. & Rep. 8; *X and Y* v. *UK*, application 13669/88.
[172] See e.g. *Swann Committee Report*, 404–5, 663, 682–3.

better accomplished as part of the ordinary curriculum during normal school hours.[173] With no legal obligation in this regard resting on local education authorities under English law,[174] protagonists of the latter approach have turned their attention towards the 1977 EEC Directive on the Education of Children of Migrant Workers,[175] Article 3 of which declares: 'Member States shall, in accordance with their national circumstances and legal systems, and in co-operation with States of origin, take appropriate measures to promote, in co-ordination with normal education, teaching of the mother tongue and culture of the country of origin for the children referred to in Article 1.'

While Article 1 only specifically covers dependants of migrant workers from other states within the European Community (and hence fulfills the obvious purpose of facilitating the free movement of workers by easing their anticipated return to and reintegration within their countries of origin) and certainly does not apply to British citizens, in 1977 the UK concurred in a political agreement reached at the Council of Ministers that the benefits of the directive should be extended to the nationals of non-member states.[176] However, in view of the exclusion of the children of British citizens (and hence the majority of ethnic-minority children resident here today) and the very restricted interpretation attributed by the Government to the phrase 'take appropriate measures to promote' in Article 3 (mainly sponsoring research),[177] the impact of the directive has been negligible. Indeed the EC Commission's own 'Report on the Implementation of the Directive'[178] in 1984 revealed that the UK was doing far less in this regard than other member states and in some cases clearly breaching the directive through the practice of certain local authorities in exacting a charge for the use of school premises by supplementary schools. However, no proceedings have been taken, either in the English courts or in Luxembourg on the subject. By way of contrast, mother-tongue teaching and bilingual education are flourishing in Wales, where the Welsh language is used as the medium of instruction in many schools and is given statutory recognition as a 'core subject' in Welsh-speaking schools and a 'foundation subject' in non-Welsh-speaking schools in Wales by the Education Reform Act 1988.[179]

---

[173] See e.g. CRE, *Ethnic Minority Community Languages: A Statement* (1982).

[174] A number of non-European languages have been designated as 'foundation subjects' within the national curriculum pursuant to s. 3(2) of the Education Reform Act 1988, but there is no duty placed on any school to teach any particular language at the request of parents.

[175] 77/486/EEC.

[176] See *Swann Committee Report*, 401–2.

[177] See DES Circular 5/81, paras. 7, 8; DES, 'Memorandum on Compliance with Directive 77/486/EEC' (1983).

[178] See Com. (84) 54.

[179] s. 3(1)(a), (2)(c).

Specialized international instruments

The provisions of a small number of specialized international instruments ratified by the UK affect minority rights in two very different ways. The first group reinforces the rights of minorities to enjoy their own cultures and practise their own religions pursuant to Article 27 of the International Covenant on Civil and Political Rights. For example, the UN Declaration on the Elimination of Intolerance and Discrimination Based on Religion or Belief[180] elaborates in greater detail the freedom of religion accorded in the European Convention and the International Covenant. Similarly, the International Convention on the Elimination of All Forms of Racial Discrimination[181] reinforces the prohibition of discrimination based on 'ethnic origin' which is contained in both those treaties and the Convention against Discrimination in Education outlaws discrimination in this field on grounds such as race, colour, language, and religion.[182]

The second group of provisions, by contrast, severely restricts cultural diversity and outlaws certain customary practices, whether they are derived from religion or tradition, usually in the interests of affording protection to women and children. For example, the Convention on Consent to Marriage, Minimum Age for Marriage and Registration of Marriages insists that a marriage have the 'free and full consent' of the intended spouses.[183] The Convention on the Elimination of All Forms of Discrimination against Women reiterates this demand and also requires that contracting states take all appropriate measures to eliminate discrimination against women in all matters relating to marriage and family relations. It also proscribes child-marriages.[184] The Convention on the Rights of the Child,[185] while reaffirming minority rights in general[186] and a child's right to recognition of its ethnic identity in particular,[187] requires states to take all effective and appropriate measures with a view to abolishing traditional practices prejudicial to the health of children,[188] apposite terminology to justify a legislative ban on the mutilation of girls through circumcision.[189]

This latter group of provisions serves to promote and defend human

---

[180] Resolution 36/55 of 25 Nov. 1981 adopted unanimously. No agreement has yet proved possible on a UN Convention on the subject.
[181] Cm. 4108 of 1969.
[182] Cmnd. 1760 of 1962.
[183] Cmnd. 4538 of 1970, Art. 1(1).
[184] Ibid., Art 16(2).
[185] Cm. 1976 of 1992.
[186] Art. 30.
[187] Arts. 8, 20.
[188] Art. 24(3); see also Art. 19(1).
[189] See Prohibition of Female Circumcision Act 1985. The ban might also have been justified by reference to Art. 7 of the International Covenant on Civil and Political Rights which outlaws 'cruel, inhuman or degrading treatment'.

rights just as vitally as the former category and demonstrates the necessity of placing certain restraints upon the maintenance of cultural diversity on the part of minority groups in order to preserve individual liberties. Even so, none of the provisions mentioned above has played any part in any decisions of the English courts, nor have any of them featured in political debates or legislative discussions in relation to minority rights. No use has yet been made of any of the institutional mechanisms established by these international instruments.

EMPIRICAL ASSESSMENTS

A preliminary issue here is whether accurate information is available about the size of Britain's ethnic, religious, and linguistic minorities. The 1991 national census was the first to include a question about ethnic origin and when the full, detailed results are known there will be much more reliable statistics than hitherto as to the ethnic composition of the population and its geographical distribution. This should greatly assist both central and local government in policy-planning and the allocation of resources, especially in the fields of education, housing, and personal social services. However, while the census question about ethnic origin will provide figures for the numbers of people of, for example, Indian, African, Chinese, or 'white' origin, it will not furnish information as to a person's religious affiliation or mother tongue, nor will gypsies,[190] for instance, be differentiated as a separate ethnic group. No official government statistics are kept concerning religious affiliation and the figures regularly published by the Central Statistical Office in *Social Trends*[191] are derived from estimates originally provided by the various denominations themselves[192] and can thus hardly be regarded as at all reliable. Some (if not all) religions will naturally tend to exaggerate their membership and in recent years many Muslim 'leaders' have claimed in comments to the media that there are up to two million adherents to Islam living in Britain, whereas a scholarly and impartial analysis undertaken in 1990 suggested that a figure of between 550,000 and 750,000 was then far more likely.[193] In any event, religious statistics must always be viewed with caution for they tell us nothing of the strength of religious beliefs, nor do they reveal the variety of degrees of religious observance, for instance

---

[190] An authoritative study prepared for the European Commission in 1989 concluded that there were 31,000 gypsies living in caravans and a further 29,000 living in houses: see D. Kenrick and S. Bakewell, *On the Verge: The Gypsies of England* (1990), 8, 58.

[191] See e.g. (1990) 20 *Social Trends* 166.

[192] See *UK Christian Handbook 1989/90* (1990).

[193] C. Peach, 'The Muslim Population of Great Britain' (1990) 13 *Ethnic and Racial Studies* 414.

in terms of public worship or compliance with traditional rituals and customs. It is possible to count the number of mosques in England and Wales,[194] but it is far less easy to estimate the proportion of Muslims who regularly attend Friday prayers there or those who comply with dietary rules and it is virtually impossible to calculate the degree to which Muslims in general have been 'offended' by the publication of *The Satanic Verses*. Similar problems arise in relation to linguistic minorities, for it is hard to assess the amount of usage of mother tongues or the degree of commitment to their preservation in this country. Hence, in seeking to answer the question whether, on the basis of empirical evidence, minority rights have been enhanced or undermined in Britain during the past half-century it needs to be borne in mind that such assessments are intrinsically hazardous. Furthermore, any decrease in the adherence of minorities to their ethnic traditions, religious practices, and mother tongues (assuming any such estimate could be accurately made) might simply reflect their voluntary assimilation into the majority culture, rather than any decline in their civil liberties.

Despite these limitations it is possible to draw attention to a number of valuable surveys. In 1985 a major assessment was undertaken of the provision made by local authorities for the special needs of Muslims in relation to a variety of issues of particular concern, notably burial facilities, provision of meat slaughtered according to Muslim requirements, policies concerning the granting of planning permission for mosques, the opportunities available to Muslim council employees to take time off work for Friday prayers and Islamic festivals, and the policies and practices of schools in accommodating the religious needs of Muslim pupils.[195] Another survey has examined the attitudes of the various regional river authorities to requests by Hindus and Sikhs to scatter the cremated ashes of deceased relatives in English rivers.[196]

Several surveys have demonstrated the keen desire of Asian parents that their children should acquire and maintain a fluency in their mother tongue[197] and the large number of supplementary schools operating in the evenings and at weekends is further testimony to this.[198] Research[199] also

---

[194] The number of mosques registered in 1989 was 452; see OPCS, *Series FM2, No. 18: Marriages* (1992), 59. However, since registration is not compulsory this figure is unlikely to reflect the total number accurately.

[195] J. Nielsen, 'A Survey of British Local Authority Response to Muslim Needs', summarized in T. Gerholm and Y. G. Lithman (eds.), *The New Islamic Presence in Western Europe* (1988), ch. 3.

[196] S. Poulter, 'The Scattering of Cremated Ashes, River Pollution and the Law' (1989) 1 *Land Management and Evironmental Law Report* 82.

[197] See e.g. D. Brown, *Black and White Britain* (1984), 131, 143; *Swann Committee Report*, 404–5.

[198] See e.g. J. Nagra, 'Asian Supplementary Schools' (1981–2) *New Community* 431.

[199] See K. Polack and A. Corsellis, 'Non-English Speakers and the Criminal Justice System' (1990) 140 *NLJ* 1634, 1676.

suggests that linguistic minorities are almost certainly not receiving in practice the proper standard of arrangements for interpretation in criminal proceedings which they are entitled to expect both under English law[200] and in terms of the European Convention on Human Rights.[201] In 1986 a committee of Justice described these arrangements as 'casual and haphazard in operation'.[202] A striking illustration of this inadequacy was revealed in the case of *R. v. Iqbal Begum*[203] in 1985, in which the Court of Appeal eventually had to quash the conviction of a Muslim woman of the murder of her husband because the defendant and the interpreter did not share the same Asian language and dialect.

In considering how best to assess empirically whether minority rights are being adequately safeguarded in future years it is thought that attention should primarily be focused on two areas, employment and education. So far as employment is concerned, attempts should be made to monitor the extent to which members of minority groups are being unjustifiably denied equal opportunities on the basis of their adherence to religious beliefs or traditional cultural practices. Rather than relying on isolated test cases in the courts, surveys need to be undertaken to see if barriers are in practice being imposed by employers against Sikhs wearing beards and turbans, Asian women wearing *shalwar-qemiz*, and Muslims demanding time off for Friday prayers or religious festivals. It might also be worth keeping records of accidents involving turbaned Sikh building workers on construction sites to see whether there is any evidence of their safety being unduly endangered by their exemption from the need to wear helmets.

In relation to education, it is important to attempt to discover the extent to which the religious and cultural freedoms of both parents and children from non-Christian faiths are being denied in the state system. This can occur through, *inter alia*, the application of insensitive school rules about dress, inadequate treatment of minority faiths in religious education classes and collective worship, an insufficient number of single-sex schools, and a disregard of dietary taboos in the provision of school meals. A number of safeguards already exist to promote minority rights, such as the existence of an independent sector for those who can afford private schooling, the right of withdrawal from religious education classes and collective worship, and the power of schools to obtain exemptions from the requirement that such worship be 'wholly or mainly of a broadly Christian character'.[204] It would also, therefore, be desirable to monitor the use of such facilities to assess their practical significance.

---

[200] See *R. v. Lee Kun* [1916] 1 KB 337; Prosecution of Offences Act 1985, s. 19(3)(*b*).
[201] Art. 6(3)(*e*).
[202] 'Witnesses in the Criminal Court', 7.
[203] (1991) 93 Crim App R 96.
[204] Education Reform Act, s. 7(1).

CONCLUSION

The concept of members of ethnic, religious and linguistic minorities in Britain claiming distinctive 'cultural' rights, in addition to the ordinary rights they are entitled to as individuals, is a comparatively novel one. Neither the Universal Declaration of Human Rights (1948) nor the European Convention on Human Rights (1950) contained a specific provision on minority rights and modern recognition of such rights in international law derives its inspiration from Article 27 of the International Covenant on Civil and Political Rights (1966). Since the mid-1960s successive British governments have publicly endorsed a broad policy of cultural pluralism, while accepting the need for assimilationist measures in certain circumstances, for instance where minimum international standards would be breached if certain ethnic traditions and religious customs were to be practised in the United Kingdom. A growing appreciation of the importance of ethnic identity, as well as of the diversity of religious faiths in British society, has led both Parliament and the courts to give increasing legal recognition to cultural differences. Although this has occurred on an ad hoc basis rather than being grounded in any statement of general legal principle, it can be seen as a positive development in the advancement of human rights. However, insofar as legal recognition of ethnicity is strongly tied to racial differences, as in the Race Relations Act 1976, there is a danger that the needs of some religious minorities may be overlooked. Muslims, in particular, are now pressing for greater legal protection against religious discrimination and this question is likely to dominate the field of minority rights over the next few years.[205] Several applications have been made to the European Commission of Human Rights alleging violations of religious freedom or discrimination on the basis of religion and although none has been upheld so far, the prospect of a successful challenge being mounted in the near future should not be discounted.

---

[205] See e.g. Commission for Racial Equality, *Second Review of the Race Relations Act 1976* (1992), 58–61.

15

Sexual Orientation Discrimination

ROBERT WINTEMUTE

INTRODUCTION

This chapter will deal with discrimination against gay, lesbian, and bisexual men and women and same-sex emotional-sexual conduct, whether by the law or a public or private policy or decision. In the next section, I will define what I mean by 'sexual orientation', 'same-sex emotional-sexual conduct', and 'sexual orientation discrimination', and discuss ways in which such discrimination can be viewed as an issue of human rights. I will then consider the many instances of sexual orientation discrimination contained in, or permitted by, English law in 1993, changes in the law since 1950, and possible explanations both for those changes and for the general lack of changes. Finally, I will look at the influence of the European Convention on Human Rights and European Community law in this area and discuss proposals for law reform.

CONCEPTS

Sexual orientation

Before turning to the different ways in which sexual orientation discrimination can be conceived of as an issue of human rights, it is necessary first to define what I mean by 'sexual orientation' and 'sexual orientation discrimination'. By the 'sexual orientation' of a person of a given sex, I mean whether that person, in deciding with whom to engage in 'emotional-sexual conduct', is attracted to, or actually chooses to engage in such conduct with, persons of the opposite sex (i.e. they are 'heterosexual'), persons of both sexes (i.e. they are 'bisexual'), or persons of the same sex (i.e. they are 'gay', if they are male, or 'lesbian', if they are female). I will use 'gay or lesbian' and 'same-sex' interchangeably, but will not use 'homosexual', which (in my opinion) is irrevocably tainted by years of use in a pejorative sense and means nothing more than 'same-sex' (as 'heterosexual' means 'other-sex').

I use the term 'emotional-sexual conduct' to describe any kind of activity or relationship involving two (or more) persons where the activity or relationship may be treated differently according to the sexes of the participants, including private sexual activity, public displays of affection, and the formation of partnerships. Its use is intended to stress the fact that sexual orientation is not just a matter of private sexual activity, but also involves the expression of emotions and the formation of emotional relationships, i.e. it is about love as well as sex. Thus, a person's sexual orientation is the *direction* (to the opposite sex, the same sex, both, or neither) of their attraction to emotional-sexual conduct, or of their actual choice of emotional-sexual conduct. In addition, each specific instance of emotional-sexual conduct (e.g. a sexual act or a partnership) can also be said to have a sexual orientation, in that (if it involves two persons) it will necessarily be either heterosexual or same-sex.

Although it is fairly easy to state a definition of sexual orientation, it is much more difficult to determine the sexual orientation of a person (as opposed to a specific instance of emotional-sexual conduct). A person's sexual orientation may be (*a*) the direction of their attraction to emotional-sexual conduct, (*b*) the direction of their actual choice of emotional-sexual conduct (taken as a whole), (*c*) the direction of their actual choice of a specific instance of emotional-sexual conduct, or (*d*) the sexual orientation with which they identify (i.e. whether they consider themselves heterosexual, bisexual, gay, or lesbian). There is no necessary consistency in these four criteria and, in the case of an individual who has engaged in a specific instance of emotional-sexual conduct, their application could lead to several different answers. For example, a married man who has just engaged in sexual activity with another man, does so frequently, and is primarily attracted to men, but considers himself heterosexual and frequently engages in sexual activity with his wife, might be gay under (*a*) or (*c*), bisexual under (*b*) and heterosexual under (*d*).

The law, however, can usually ignore these conflicting answers. This is because a discriminator will often have no knowledge of (*a*), (*b*), (*c*), or (*d*), and will base their decision whether or not to discriminate on (*e*) their perception of that person's sexual orientation (e.g. in a job interview). Thus, it makes no difference how a discriminator defines a person's sexual orientation, or whether their definition or perception is consistent with that person's self-identified sexual orientation. All that matters is that the discriminator has treated the person less favourably, because they consider the person to be of one sexual orientation, than if they had considered the person to be of another sexual orientation.

Sexual orientation discrimination

One person may discriminate directly against another person either because of the sexual orientation of the other person, or because of the sexual orientation of a specific instance of emotional-sexual conduct in which the other person has engaged. This will involve treating a person less favourably than persons of another sexual orientation, or than persons who have engaged in a specific instance of emotional-sexual conduct of another sexual orientation. One person may also discriminate indirectly against another person by applying an unjustifiable requirement (other than being of a particular sexual orientation) with which a disproportionate number of persons of the other person's sexual orientation are unable to comply.

The significance of defining sexual orientation as whether or not a person, or a specific instance of emotional-sexual conduct, is heterosexual, bisexual, gay, or lesbian, is that many general issues of 'sexuality' (i.e. a person's capacity to engage in emotional-sexual conduct and every aspect of their exercise of that capacity) or 'sexual freedom' may not involve any discrimination on the ground of sexual orientation. Thus, the law's treatment of such issues as paedophilia, incest, sado-masochism, prostitution, pornography, or polygamy may interfere with sexual freedom or discriminate on other grounds. But it does not raise an issue of sexual orientation discrimination, provided that all such conduct (or expression), whether it is heterosexual, bisexual, or same-sex, is treated in the same way. It is important to remember that sexual orientation (as I have defined it) is just one aspect (the direction, as between the sexes) of sexual freedom. As a result, sexual orientation discrimination may be prohibited without having to address every controversial aspect of sexual freedom, and without necessarily precluding legal regulation of aspects other than sexual orientation (e.g. the parties' ages or relatedness, consensual use of force, commercial sexual activity), each of which can be considered on its own merits. Similarly, sexual orientation discrimination is not the same as discrimination against transvestite or transsexual persons, because transvestism and transsexualism are not sexual orientations (as I have defined them). They involve choosing to adopt the appearance associated with a particular sex, or permanently to change one's physical sex characteristics, and not the choice of the sex of the persons with whom one engages in emotional-sexual conduct.

Sexual orientation discrimination as a human rights issue

In asking how sexual orientation discrimination can be conceived of as an issue of human rights, one is asking what the right to be free from such discrimination has in common with other human rights, and particularly with the rights to be free from other kinds of discrimination. Is there any general principle that can explain why certain kinds of discrimination are considered prima facie wrongful and such that they are permitted only in limited circumstances in which they can be justified? Such a question arises in a judicial context in countries with constitutional guarantees of 'equality', which require courts to decide which kinds of discrimination are constitutionally permissible and which are not.[1] It does not arise in the same way in Britain, given the absence of a domestic, constitutionally entrenched Bill of Rights, and the non-incorporation of the European Convention on Human Rights into UK law. However, a British legislator or academic, when faced with a novel kind of discrimination, could pose the same question in attempting to decide whether that kind of discrimination should be prohibited, like discrimination on the grounds of race, sex, being married, trade union membership, or (in Northern Ireland) religious belief or political opinion.[2] Many general principles could be suggested, but only three possibilities will be considered here: (i) prohibited grounds of discrimination are initially unchosen and currently immutable statuses over which a person has no control; (ii) prohibited grounds of discrimination are fundamental rights or freedoms or choices which should not be interfered with; and (iii) prohibited grounds of discrimination protect politically, economically, or socially disadvantaged groups that have historically been subjected to prejudice, stereotyping, or discrimination. Does the phenomenon of sexual orientation, or do gay, lesbian, and bisexual persons as a group, come within one of these principles?

The argument that sexual orientation is an unchosen and immutable status assumes that immutability explains the prohibitions of race and sex discrimination.[3] If this is the case, one must ask what aspects of sexual orientation are immutable and whether a finding of immutability would support a comprehensive prohibition of sexual orientation discrimination. Of the four criteria for identifying sexual orientation mentioned above, the only one that is arguably immutable is the direction of a person's attraction to emotional-sexual conduct. Their actual choices of the direction of such conduct (as a whole), or of specific instances of such conduct, and the sexual orientation with which they identify,[4] are all clearly voluntary and are not either initially unchosen or currently immutable (even though they may see their attraction as constraining

---

[1] See e.g. Canadian Charter of Rights and Freedoms, Section 15(1); US Constitution, Fifth and Fourteenth Amendments ('equal protection').

[2] Race Relations Act 1976, s. 1 (race); Sex Discrimination Act 1975, ss. 1–2 (sex), 3 (being married); Employment Act 1990, ss. 1–2 (trade union membership); Fair Employment (Northern Ireland) Act 1989, s. 49(1) (religious belief and political opinion).

[3] See e.g. H. Miller, 'An Argument for the Application of Equal Protection Heightened Scrutiny to Classifications Based on Homosexuality' (1984) 57 *So. Calif. LR* 797, 812–13, 817–21.

[4] On 'importance to identity' as an alternative to 'immutability', see e.g. Note, 'The Constitutional Status of Sexual Orientation: Homosexuality as a Suspect Classification' (1985) 98 *Harv. LR* 1285 at 1300–5; Note, 'Custody Denials to Parents in Same-Sex Relationships: An Equal Protection Analysis' (1989) 102 *Harv. LR* 617 at 621–3.

their choice of conduct). It is certainly true that many gay men and lesbian women feel that their attraction to same-sex emotional-sexual conduct is an innate characteristic, and some recent scientific studies suggest the possibility of a genetic cause.[5] However, there is as yet no conclusive scientific proof, and some persons (especially lesbian women) assert that they rationally chose the direction of their attraction.[6]

Even if the direction of attraction of some (if not all) persons could be proved to be immutable, this would only support an argument for prohibiting sexual orientation discrimination that is based on a person's direction of attraction, as opposed to the direction of their actual choice of conduct. This might be true of certain cases of discrimination in employment, housing, or services, where the discriminator had no knowledge of an individual's actual conduct and the decision to exclude them was based on their attraction. This would, however, effectively preclude the individual from admitting that he or she had a same-sex partner or had ever engaged in same-sex sexual activity.[7] Moreover, as will be seen in the next section, much sexual orientation discrimination involves direct interference with the choice of same-sex emotional-sexual conduct, through the criminal law's discriminatory treatment of sexual activity between men and public displays of same-sex affection, and the civil law's non-recognition of same-sex partnerships. In such cases, the immutability of an attraction would provide no basis for protection against discrimination based on an actual choice of conduct.

If immutability is not the appropriate principle, could the actual choice of the direction of a person's emotional-sexual conduct be a fundamental right or freedom or choice? One approach would be to argue that the choice of sexual orientation comes within a widely recognized right or freedom,[8] such as freedom of expression or freedom of association.[9] However, this would tend to collapse the distinctions between expression and conduct (most emotional-sexual conduct is probably not intended to

---

[5] See e.g. S. Levay, 'A Difference in Hypothalamic Structure between Heterosexual and Homosexual Men', *Science* (30 Aug. 1991), 1034; M. Bailey and R. Pillard, 'A Genetic Study of Male Sexual Orientation', *Archives of General Psychiatry* (Dec. 1991), 1089 (study of identical and fraternal male twins and pairs of adoptive brothers).

[6] See e.g. P. Crane, *Gays and the Law* (1982) at p. 4.

[7] The US Supreme Court's refusal to find constitutional protection for same-sex sexual activity, in *Bowers* v. *Hardwick*, 478 US 186 (1986), has forced gay, lesbian, and bisexual plaintiffs to resort to a strained 'status vs. conduct' distinction. See e.g. *Watkins* v. *US Army*, 875 F. 2d 699, 712–17, 725 (9th Cir. 1989).

[8] For examples of statutory protection against employment discrimination based on a person's exercise of a fundamental freedom, see n. 2 (being married, trade union membership, religious belief, political opinion).

[9] See e.g. *Gay Law Students Assoc.* v. *Pacific Tel. & Tel. Co.*, 595 P. 2d 592, 609–11 (Calif. 1979) (expression); K. Karst, 'The Freedom of Intimate Association' (1980) 89 *Yale LJ* 624, 682–6.

be expressive, in the sense of communicating a message to third parties),[10] and between the right to associate with others and the right of the association to engage in particular activities. An alternative approach is to argue that the choice of sexual orientation comes within some residual guarantee of 'liberty', 'personal autonomy', or 'privacy'. This argument claims, not that sexual orientation comes within an existing fundamental freedom, such as religion, expression, or association, but that it is a deeply personal choice of such importance to the individual that it is deserving of the same respect and protection as their religious or political beliefs. In a constitutional context, the main difficulty facing this argument is not necessarily establishing the importance or 'fundamentality' of the choice (so long as the choice is defined generally as that of sexual orientation rather than specifically as that of same-sex or bisexual sexual orientation).[11] It is, rather, finding a textual basis for a residual guarantee and defining criteria for identifying which freedoms or choices are 'fundamental' and which are not.[12] Where such a textual basis exists, as in the reference to 'private life' in Article 8 of the European Convention on Human Rights, this approach has had some success. It may also be used in a legislative context, even if it only has persuasive value, in that a legislature (particularly in the absence of a constitutional Bill of Rights) may have no duty to treat analogous freedoms or grounds of discrimination in the same way.

To invoke the third possible general principle, gay, lesbian, and bisexual persons must be characterized as a 'disadvantaged group'. In applying this principle, the focus is not on sexual orientation as a neutral, universal characteristic, but on the persons who have in fact been affected by its use as a ground of discrimination in the past. In the case of gay, lesbian, and bisexual persons, it is easy to establish the history of discrimination against them, the prejudice and stereotyping that has often motivated that discrimination, their lack of political power, and their resulting current status as a 'disadvantaged group'. There are, however, at least two difficulties with this principle. It may need to be supplemented, in that it provides no basis for distinguishing between groups that are justifiably disadvantaged (e.g. persons convicted of assault or theft) and those that are not.[13] And it is often said to be an asymmetrical prin-

---

[10] See 'Constitutional Status' at 1292–7.

[11] See *Hardwick* (n. 7) at 190–2 (insistence on defining right in issue as engaging in 'homosexual sodomy'). See also J. Rubenfeld, 'The Right of Privacy' (1989) 102 *Harv. L. Rev.* 737.

[12] In *Hardwick* (n. 7) at 195–6, the Court refused to extend its controversial 'right of privacy' case-law to include private, consensual adult sexual activity, partly because it did not want to have to consider in which cases such activity could justifiably be regulated.

[13] See L. Tribe, 'The Puzzling Persistence of Process-Based Constitutional Theories' (1980) 89 *Yale LJ* 1063.

ciple, in that it protects only members of disadvantaged groups against discrimination, and not members of 'advantaged groups'.[14] This asymmetry presents a problem in Britain, in that anti-discrimination law has so far been symmetrical in most cases, protecting all groups defined by a particular ground (e.g. ethnic minorities and the ethnic majority, women and men), especially by not permitting positive discrimination.[15]

In addition to the three general principles discussed above, a fourth approach is possible. It involves arguing that sexual orientation discrimination is nothing more than a kind of sex discrimination and that, whatever the principle that explains prohibitions of sex discrimination, every such prohibition applies to cases of sexual orientation discrimination. The argument has been developed by David Pannick in Britain, and by several authors in the USA.[16] Essentially, it maintains that, in many cases, whether or not a law (or policy or decision) restricts a person's choice of the direction of their emotional-sexual conduct depends entirely on that person's sex.[17] In the UK, a male member of the armed forces may not engage in sexual activity with a man (a female member may do so); a woman may not marry a woman (a man may do so). In spite of the argument's strength, almost every court in Canada and the USA has rejected it. The most common reasons have been the illusory argument that denying a different option to each sex means that both are treated equally, and a perceived absence of legislative intent that a prohibition of sex discrimination would apply in such cases.[18] However, in May 1993, the Supreme Court of Hawaii became perhaps the first court in the world to accept the argument, in holding that the denial of marriage licenses to same-sex couples is a prima facie violation of the Hawaii Constitution's prohibition of sex discrimination, requiring the state to show that a 'compelling state interest' justifies the denial.[19]

Finally, it should be noted that this chapter uses 'sexual orientation discrimination' as the relevant concept rather than 'gay, lesbian and bisexual rights'. This is intended to suggest that prohibitions of such discrimination should be symmetrical, protecting not only gay, lesbian,

---

[14] See e.g. O. Fiss, 'Groups and the Equal Protection Clause' (1976) 5 *Phil. and Pub. Aff.* 107.

[15] See *Lambeth London Borough Council v. CRE* [1990] IRLR 231 (CA); grounds in note 2, (only 'being married' is asymmetrical).

[16] See D. Pannick, *Sex Discrimination Law* (1985), ch. 8; B. Capers, 'Sex(ual Orientation) and Title VII' (1991) 91 *Colum. LR* 1158; 'Custody' Note, note 4; A. Koppelman, 'The Miscegenation Analogy: Sodomy Laws as Sex Discrimination' (1988) 98 *Yale LJ* 145.

[17] This is also true of discrimination against transvestite and (in many respects) transsexual persons.

[18] See e.g. *Singer v. Hara*, 522 P. 2d 1187 (Wash. Ct. App. 1974); *DeSantis v. Pacific Tel. & Tel. Co.* 608 F. 2d 327 (9th Cir. 1979); *Vogel v. Manitoba* (1983) 4 CHRR D/1654.

[19] See *Baehr v. Lewin* [1993] Haw. LEXIS 26 (5 May), 30 (27 May).

and bisexual persons (against whom almost all such discrimination is directed), but also heterosexual persons. It is also intended to emphasize the importance of thinking of sexual orientation as a neutral, universal characteristic, with several different manifestations, rather than a phenomenon unique to gay, lesbian, and bisexual persons. Heterosexual persons need to be reminded that they too have a sexual orientation, and that theirs is not the only possibility. Just as there is more than one religion, political opinion, race, or sex in Britain, there is more than one sexual orientation.

THE STATE OF ENGLISH LAW IN 1993

To what extent does English law actively discriminate on the ground of sexual orientation, or fail to prohibit such discrimination by public or private actors? As will be seen below, it does so to such an extent that any person who considers themselves gay, lesbian, or bisexual, or who ever engages in any aspect of same-sex emotional-sexual conduct, is clearly a second-class citizen. It would require an entire book[20] to discuss in detail the variety and extent of this discrimination, so only a brief summary will be attempted here. I have divided the many kinds of sexual orientation discrimination into four categories: (*a*) discrimination in the criminal law; and all other discrimination against gay, lesbian, and bisexual persons (*b*) as individuals, (*c*) as members of same-sex couples, and (*d*) as existing or prospective parents. The division between (*a*) ('criminal law discrimination') and (*b*), (*c*), and (*d*) (sub-categories of 'all other discrimination') corresponds to the division between discrimination that actively seeks to prohibit (altogether or in certain circumstances) the choice of same-sex emotional-sexual conduct (or its depiction or discussion) through fines or imprisonment, and discrimination that seeks to discourage that choice through other sanctions (inadequate protection against violence or incitement to hatred, or denial of jobs, housing, local-authority funding, benefits for a partner, custody of children).

Discrimination in the criminal law

Same-sex sexual activity

The most basic discrimination against sexual activity between men is that sections 12 and 13 of the Sexual Offences Act 1956 prohibit so-called 'buggery' (i.e. anal intercourse) and 'gross indecency' (i.e. any other

---

[20] The best treatment to date is Paul Crane's excellent *Gays and the Law.*

sexual activity between men). Consequently, all sexual activity between men is illegal unless an exception in another act makes it legal in defined circumstances. Although section 1 of the Sexual Offences Act 1967 creates an exception covering sexual activity between two consenting men over 21 in private, it remains the case that illegality is the general rule and legality the exception. No such blanket prohibition applies to heterosexual sexual activity[21] (e.g. 'all vaginal intercourse or other sexual activity between a man and a woman is illegal, except if it is in private and both parties consent and are over 16'), or sexual activity between women. The law starts from a presumption of legality and deals only with the exceptional cases of illegality (e.g. lack of actual consent or capacity to consent).

In two cases, the merchant navy and armed forces, the 1967 Act provided no exception to the general rule that sexual activity between men is illegal.[22] The vague prohibitions of 'disgraceful conduct of a[n] . . . indecent or unnatural kind' were used to prosecute members of the armed forces for same-sex sexual activity, between men or between women, on or off duty, and with civilians, and provided for imprisonment of up to two years. In 1992, however, the Government announced that such prosecutions would end and that the discriminatory treatment of armed forces members would be repealed.[23] It also began consultations regarding a repeal of the merchant navy provision.[24] Where the 1967 Act's exception applies, it discriminates by establishing a narrower zone of legal sexual activity between men than is the case between a man and a woman or between women. The most seriously discriminatory feature is the difference in the age of consent, which is 21 rather than 16,[25] and is estimated to have caused the imprisonment of 33 men in 1989 for consensual (and, at least in some cases, private) sexual activity with males aged 16 to 20.[26] In addition, no more than two men may be present (which is not the case for heterosexual or lesbian sexual activity).[27] Even where a particular sexual act between men would be legal, steps taken by one of the participants or by a third party to arrange or facilitate the act may be illegal as constituting the 'procuring' of the act,[28] 'solicit[ing] or

---

[21] There is one bizarre exception: heterosexual anal intercourse is illegal and punishable by life imprisonment! See 1956 Act, s. 12; Criminal Law Revision Committee, *Fifteenth Report: Sexual Offences* Cmnd. 9213 (1984), 52 (such intercourse should be legal at 16).

[22] 1967 Act, ss. 1(5), 2; Army Act 1955, s. 66; Air Force Act 1955, s. 66; Naval Discipline Act 1957, s. 37.

[23] H.C. Deb., Vol. 209, 989–93 (17 June 1992).

[24] See *Capital Gay* (13 Nov. 1992), 1.

[25] 1967 Act, s. 1; Sexual Offences Act 1956, ss. 14, 15. See C. Hindley, 'The Age of Consent for Male Homosexuals' (1986) *Crim. LR* 595.

[26] See letter from P. Tatchell, *Independent* (11 June 1991), 16.

[27] 1967 Act, s. 1.

[28] 1956 Act s. 13; 1967 Act, s. 4.

importun[ing] [by a man] in a public place for immoral purposes',[29] or a 'conspiracy to corrupt public morals'.[30] These offences do not apply to, or are rarely enforced against, comparable steps taken in relation to heterosexual or lesbian sexual activity.

A more controversial kind of sexual activity between men is that which takes place in 'semi-public' places, such as public toilets (particularly inside cubicles) or parks, beaches, or other outdoor areas, and accounts for the vast majority of prosecutions of men for sexual activity with men. Such activity is 'semi-public' in that, although there may be a theoretical possibility that a member of the public might come along and see it, this is often extremely unlikely because of the infrequent use of the location, or its isolation or darkness. Indeed, in most cases, the only person who witnesses the activity is the police officer who searches for it.[31] Because comparable heterosexual or lesbian sexual activity is not entirely legal, the underlying question of a right to engage in sexual activity in semi-public places is a general one of interference with sexual freedom, and not one of sexual orientation discrimination.[32]

However, semi-public sexual activity between men does raise issues of sexual orientation discrimination in that prosecution is far more likely and results in far more severe penalties. Persons found engaging in heterosexual sexual activity in parks, parked cars, and other public places would rarely be prosecuted, and any prosecution would probably be under public-order legislation or for 'outraging public decency'. Men engaging in comparable activity are actively hunted by police, who go to extraordinary lengths (including using *agents provocateurs* or video surveillance) to catch them in the act, and face maximum penalties of 2 years' imprisonment (5 if with a male aged 16 to 20) under sections 12 and 13 of the 1956 Act. The availability of these offences is a consequence of the general rule of illegality: any sexual activity between men that is not strictly 'in private' falls outside of section 1 of the 1967 Act and into the 1956 Act. Treatment of semi-public sexual activity between men also raises serious issues as to the use of police resources (while sexual and other assaults go undetected), the abuse of police powers (entrapment, coercive 'chain' interrogations,[33] fabrication of evidence), and extra-legal

---

[29] 1956 Act, s. 32. See *R.* v. *Kirkup* (1992) 96 Cr. App. R. 352; H. Power, 'Entrapment and Gay Rights' (1993) 143 *New L.J.* 47; R. Cottrell, 'Soliciting for Immoral Purposes' (1993) 137 *Sol. J.* 303.

[30] See e.g. *Knuller* v. *DPP* [1973] AC 435 (HL).

[31] See e.g. *Chief Constable of Hampshire* v. *Mace* (1986) 84 Cr. App. R. 40; *Parkin* v. *Norman* [1983] Q.B. 92; *R.* v. *Preece* (1976) 63 Cr. App. R. 28.

[32] The same is true of private sado-masochistic sexual activity, although the unprecedented prosecutions and harsh prison sentences for such activity in *R.* v. *Brown* [1993] 2 All ER 75 (HL) may have been influenced by the fact that the activity was between men (see the references to 'homosexual sado-masochism' at 91–2, 100).

[33] See Crane, *Gays and Law*, at 35–9, 51–5.

penalties wholly out of proportion to any public nuisance involved in the offence. Publicity often leads to loss of a job, harassment by the local community, and, in some cases, suicide.[34] The discriminatory treatment of semi-public sexual activity between men was graphically illustrated by a 1992 decision regarding a very public instance of heterosexual sexual activity. A man and a woman were fined £50 and £25 costs each for engaging in vaginal intercourse 'in a crowded . . . second-class compartment' of a British Rail train.[35] Two men committing a similar act would face a heavy fine, if not prison.[36]

Public displays of same-sex affection

Few people would attempt to engage in sexual activity on the Underground, in a busy shopping street, or in a crowded park on a sunny afternoon, but the same is not true of displays of affection or foreplay, ranging from relatively chaste holding hands, kissing, or hugging to passionate embraces. Heterosexual couples do this constantly without even thinking about it, in almost any public place and regardless of how close they may be to other people. But for same-sex couples, a seemingly trivial, yet fundamental and particularly galling, aspect of sexual orientation discrimination is that they are effectively denied the right to express their love for each other in public in the same way as heterosexual couples. The most compelling reason for refraining from doing so is the fear of being physically attacked by onlookers.[37] And a closeted gay, lesbian, or bisexual person may fear that their employer, landlord, tutor, parent, or neighbour might see them. But the criminal law provides an additional deterrent. It is astounding but true that two men or two women can be fined simply for kissing in the street.[38] The statutory or common law offences used to prosecute same-sex affection in public do not expressly discriminate on the ground of sexual orientation, but their

---

[34] See e.g. *Independent* (15 Feb. 1992), 3.

[35] See *Independent* (7 August 1992), 2.

[36] See *R. v. Kirkup* (1992) 96 Cr. App. R. 352 (costs of £590 against man who masturbated in front of plain clothes police officer in lavatory); P. Tatchell, *Europe in the Pink: Lesbian and Gay Equality in the New Europe* (1992), 87 (37-year-old man jailed for 18 months for 'fondling and kissing' 17-year-old male in deserted church courtyard in middle of night).

[37] See e.g. *Capital Gay* (5 Aug. 1988), 1 (man 'brutally beaten over the head' after being seen kissing another man); 'MUMS FIGHT GAYS IN PLANE BRAWL AT 30,000 FT', *The Sun* (16 May 1990), 1 ('[f]our homosexuals kissing and caressing in front of young children were attacked by angry mums on a holiday jet').

[38] See e.g. *Masterson v. Holden* [1986] 3 All ER 39 (two men, seen kissing and cuddling at a bus stop at 1.55 a.m., convicted of 'insulting behaviour'); *Guardian* (26 Apr. 1988), 7 (two men fined £40 each for kissing in the street under the Public Order Act 1986); *Gay Times* (July 1988) 19 (two men walking hand in hand in a park threatened with prosecution for 'disorderly behaviour').

enforcement is clearly discriminatory in that similar heterosexual conduct would never be prosecuted.

Gay, lesbian, and bisexual publications and other expression

In addition to prohibiting same-sex sexual activity in certain circumstances, and potentially all public displays of same-sex affection, the criminal law may prohibit depictions or discussions of any aspect of same-sex sexual activity or other emotional-sexual conduct, or of the lives and concerns of gay, lesbian, and bisexual persons. When, if ever, the criminal law should be used to prohibit a publication solely because its content is deemed 'obscene', 'indecent', or 'blasphemous' is a general question of freedom of expression. However, the apparently neutral laws on 'obscene', 'indecent', or 'blasphemous' publications[39] are enforced in such a way that they have a much more severe effect on publications dealing with same-sex emotional-sexual conduct or gay, lesbian, and bisexual persons. The most blatant examples of this disparity are the harassment of gay, lesbian, and bisexual bookshops, or any person who imports, sends by mail, or displays a gay, lesbian, or bisexual book, newspaper, or newsmagazine.[40] The notorious Customs and Excise campaign against the 'Gay's the Word' bookshop in London began with a raid in 1984 in which over 800 books were seized and ended with the dropping of all charges of importing 'indecent or obscene' books in 1986.[41] Similarly, the national gay, lesbian, and bisexual newspaper of the Netherlands, *De Gay Krant*, has been seized *en route* to British subscribers.[42] And while *The Joy of Gay Sex* and *The Joy of Lesbian Sex* were once banned by Customs and Excise, *The Joy of [Heterosexual] Sex* was not.[43]

Another example of discriminatory treatment of gay, lesbian, and bisexual expression was the private prosecution of the publisher and editor of *Gay News* for 'blasphemous libel' in publishing a poem describing the sexual fantasies of a male Roman soldier about Jesus Christ.[44] It is doubtful whether the resurrecting of this obsolete offence would have succeeded had the fantasizing party been a woman. Finally, the offences of 'solicitation', 'procuring', and 'conspiracy to corrupt public morals' can all be applied to cases of pure expression, where no same-sex sexual

---

[39] Obscene Publications Act 1959; Indecent Displays (Control) Act 1981; Local Government (Miscellaneous Provisions) Act 1982, s. 2; Customs Consolidation Act 1876, s. 42; Post Office Act 1953, s. 11.

[40] See Crane, *Gays and Law*, at 87–98.

[41] See S. Jeffery-Poulter, *Peers, Queers and Commons: The Struggle for Gay Law Reform from 1950 to the Present* (1991), 168–9, 200–1, 248. The seizures continued. See *R. v. Bow Street Magistrate, ex parte Noncyp Ltd.* [1988] 3 WLR 827.

[42] See *Gay Times* (July 1989), 25.

[43] Telephone conversation with 'Gay's the Word', 12 June 1993.

[44] See *Whitehouse* v. *Lemon* [1979] A.C. 617 (HL); Crane, *Gays and Law*, 93.

activity has actually taken place and there have only been statements preliminary to, or a publication facilitating, such activity.

Other discrimination against gay, lesbian, and bisexual individuals

Protection against violence or incitement to hatred

The criminal law of assault and murder does not expressly discriminate on the ground of sexual orientation. It is an offence to assault or kill *any* person, whether they are heterosexual, bisexual, gay, or lesbian. However, because of prejudice and hostility towards gay, lesbian, and bisexual persons, or anyone engaging in same-sex emotional-sexual conduct, on the part of members of the police and the judiciary, the many incidents of violence against such persons tend not to be treated as seriously by the police or the courts as similar incidents against heterosexual persons.[45] The enforcement of the criminal law of assault or murder thus has the effect of discriminating on the ground of sexual orientation. In the case of the police, prejudice and hostility may lead not only to a lack of interest in investigating and prosecuting assaults and murders, but also to verbal and physical abuse of gay, lesbian, and bisexual persons and the zealous enforcement of laws prohibiting semi-public sexual activity.[46] In the case of judges, hostility to gay, lesbian, and bisexual crime victims may be reflected in lighter sentences, particularly where the accused alleges that the victim made a sexual advance.[47] A sign of change was the agreement of the Metropolitan Police to begin monitoring attacks against gay, lesbian, or bisexual persons in the same way as racially motivated attacks.[48]

The absence of protection against incitement to hatred is a discriminatory omission in the criminal law, rather than a problem of its discriminatory enforcement. Prohibitions of incitement to hatred raise a general question as to whether they are justifiable restrictions of freedom of expression. But the enactment of Part III of the Public Order Act 1986 (prohibiting incitement to racial hatred in Great Britain), and Part III of the Public Order (Northern Ireland) Order 1987[49] (prohibiting incitement to religious or racial hatred in Northern Ireland) means that the decision

---

[45] See e.g. *Independent* (18 Dec. 1990), 12.

[46] See Gay London Policing Group, *7th Annual Report 1992*; Crane, *Gays and Law*, 40–55; *Independent* (26 Sept. 1992), 6 (Metropolitan Police pay gay man £17,000 in settlement of false imprisonment and malicious prosecution action).

[47] See e.g. *Capital Gay* (25 Mar. 1988) 1, (judge describes gay victim of beating and robbery as 'the little sodomite from Glasgow'); *Pink Paper* (15 June 1991), 2 (male accused sentenced to 6 years 3 months for manslaughter after killing man who had 'made an advance'); J. Meldrum, *Attacks on Gay People* (1980), 10–14.

[48] See *Independent* (2 Aug. 1991), 2.

[49] N. I. Statutes, S.I. 1987/463 (N.I. 7).

that such prohibitions are justified has already been taken. The only issue would seem to be whether incitement to hatred based on sexual orientation is as serious a problem as that of incitement to racial or religious hatred. There can be no doubt about the quantity of statements inciting hatred of gay, lesbian, and bisexual persons. One need only peruse tabloid newspapers to see a torrent of abusive language.[50] And, as with incitement to racial or religious hatred, there may be a causal connection between such statements and violence or discrimination against gay, lesbian, and bisexual persons. Yet decisions of the press's self-regulatory bodies seem to have had little effect,[51] and the Government has 'no plans' to legislate against incitement to this kind of hatred, because such a prohibition would 'be open to objection on grounds of eroding freedom of speech and attracting malicious litigation'.[52]

Employment

Because there is no legislation, equivalent to the Sex Discrimination Act 1975 and the Race Relations Act 1976, that expressly prohibits sexual orientation discrimination in employment, a gay, lesbian, or bisexual applicant who is refused a job or a promotion seems to have no legal recourse.[53] Although there is a very strong argument that sexual orientation discrimination is merely a kind of sex discrimination, and therefore prohibited by the 1975 Act, there do not appear to be any reported cases in which a gay, lesbian, or bisexual plaintiff has relied on the 1975 Act. However, in what may be the first such case (involving the dismissal of Marie O'Rourke and Simone Wallace, two lesbian women who were partners and recognized as such in their common workplace), a Scottish industrial tribunal ruled at an initial hearing that the 1975 Act applies, and will determine at a full hearing whether they were dismissed because of their relationship or because of misconduct. Their solicitor had asked:

[50] Terry Sanderson's 'Media Watch' column in *Gay Times* provides numerous examples of such abusive language every month.

[51] See *Independent* (14 May 1990), 5 (ruling of Press Council that 'use of the words "poof" and "poofter" by *The Sun* were an "unnecessary crude abuse"'); *Gay Times* (Nov. 1991) 6 (decision of Press Complaints Commission that *Daily Star* title 'Poofters on Parade' and reference to 'strident, mincing preachers of filth' breached clause 14 of the Commission's code of practice prohibiting 'prejudicial or pejorative reference to a person's . . . sexual orientation'); *Gay Times* (March 1993), 6 (Commission rejects complaint about 'No man is safe from gay rape' head line in *The Sun*).

[52] See *Gay Times* (Sept. 1990), 7 (citing letter of Home Office minister John Patten in report on British National Party stickers saying 'Protect us from Aids—outlaw homosexuality').

[53] For examples of employment discrimination, see Crane, *Gays and Law*, 99–118; P. Greasley and M. Williams, *Gay Men at Work: A Report on Discrimination against Gay Men in Employment in London* (1986); C. Beer, R. Jeffery, and T. Munyard, *Gay Workers: Trade Unions and the Law*, 2nd edn. (1983), 13–19; *What about the Gay Workers?* (1981).

'If Simone Wallace was Simon Wallace, would she have been dismissed?'[54] If a gay, lesbian or bisexual plaintiff is not an applicant but an existing employee who has been dismissed, they may be able to take their case to an industrial tribunal under the unfair dismissal provisions of the Employment Protection (Consolidation) Act 1978, if they have been in the job for two years. However, their chances of success will not be great in view of the reported decisions. Most of these have upheld dismissals of gay, lesbian, or bisexual employees as for a 'substantial reason', often relying on the prejudices of co-workers or customers, or stereotypes about the 'threat' of such an employee to adolescents or children.[55] But in a few cases tribunals have declined to uphold an employer's decision.[56]

In the absence of statutory protection, a gay, lesbian, or bisexual applicant or employee could look to any equal opportunities policy their employer may have. Over 50 per cent of London local authorities now specifically mention 'sexual orientation' or 'lesbians and gay men' in their equal opportunities policies,[57] and often include the list of grounds or groups covered by these policies in job advertisements. At least twenty trade unions have non-discrimination policies, which may affect their relations with their own employees, and cause them to support members experiencing sexual orientation discrimination and to negotiate the inclusion of sexual orientation in the policies of their members' employers.[58] And the Bar Council has agreed to add sexual orientation to the anti-discrimination clause (paragraph 204) of its Code of Conduct (whereas the Law Society has so far decided to delete sexual orientation from a proposed Solicitors' Anti-Discrimination Rule that would include race, sex and disability).[59] However, such non-discrimination (and anti-harassment)[60] policies are relatively uncommon among private-sector

---

[54] *O'Rourke and Wallace* v. *BG Turnkey Services (Scotland) Ltd.*, Case Nos. S/457/93, S/458/93. See *Independent* (9 June 1993), 7; *Capital Gay* (11 June 1993), 1.

[55] See *Saunders* v. *Scottish National Camps Assoc. Ltd.* [1980] IRLR 174 (EAT), aff'd, [1981] IRLR 277 (Ct. Session); *Wiseman* v. *Salford City Council* [1981] IRLR 202 (EAT); *Nottinghamshire County Council* v. *Bowly* [1978] IRLR 252 (EAT); *Boychuk* v. *H. J. Symons Holdings Ltd.* [1977] IRLR 395 (EAT); *Gardiner* v. *Newport County Borough Council* [1974] IRLR 262.

[56] See *Bell* v. *Devon and Cornwall Police Authority* [1978] IRLR 283; unreported decisions discussed in Crane, *Gays and Law*, and Beer, Jeffery, and Munyard, *Gay Workers*, 34–6, 43–5; *Gay Times* (Jan. 1989) 6 (cash settlement where tribunal said it was likely to find in favour of gay factory worker dismissed because of 'gross indecency' conviction).

[57] See *Gay Times* (Feb. 1990), 6; Greasley and Williams, *Gay Men*, ch. 5.

[58] See *LAGER Trade Union Survey: Equal Opportunities* (May 1989); Greasley and Williams, *Gay Men*, ch. 6.

[59] See (18 Nov. 1992) *Law Soc. Gaz.* 9; *Stonewall News* (Summer 1993), 1.

[60] See *Capital Gay* (27 March 1992), 3 (NHS makes harassment because of sexual orientation a disciplinary offence); *Cambridge Evening News* (10 Nov. 1992), 5 (gay postman suffers verbal abuse, taunting and graffiti).

employers.[61] And although they may provide access to an internal grievance procedure, it may not be clear how such a policy could legally be enforced if the employer decides not to abide by it.

Certain public-sector employers have traditionally been extremely hostile to gay, lesbian, and bisexual employees. The worst of these has been the armed forces, whose pre-1992 policies prohibited the recruiting of gay, lesbian, and bisexual persons and required at least that they be 'administratively discharged', if not court-martialled, imprisoned, and 'dismissed with disgrace'.[62] In 1992, the Government announced an end to criminal prosecutions, but emphasized 'that homosexual activity remains incompatible with military service, and that those who engage in it must [expect] to be discharged'.[63] Similarly, civil service jobs involving access to classified information and requiring 'positive vetting', particularly in the security and intelligence services and in the Foreign Office, have traditionally been denied to gay, lesbian, and bisexual persons.[64] This ban would appear to have been lifted by John Major, who said in 1991 that 'in future there should be no posts . . . for which homosexuality represents an automatic bar to security clearance, except in . . . the armed forces'.[65] But it is not yet clear whether the change in policy will protect previously closeted gay, lesbian, and bisexual civil servants who become open with their employers.[66] If the change is effective, it will extend to these jobs the policies of non-discrimination that have been encouraged in the case of non-sensitive civil service jobs since 1986.[67] At least 20 departments have adopted such policies.[68] Another hostile employer, the police, has slowly begun to change. Non-discrimination policies have been adopted by a small number of police forces in England, including the Metropolitan Police.[69] As for the judiciary, from which at least two judges have resigned since 1988 following allegations that they were gay,[70] the Lord Chancellor stated in 1991 that he 'did not have a policy not to appoint

---

[61] See e.g. 'Barclays Bank Faces Boycott Call', *Gay Times* (Oct. 1988), 7 (refusal of union's request for inclusion of sexual orientation in equal opportunities policy).

[62] See *Independent* (12 June 1991), 15; Tatchell, *Europe*, 90–2.

[63] H.C. Deb., Vol. 209, 991 (17 June 1992).

[64] See e.g. *R. v. Director of GCHQ, ex parte Hodges*, *Independent* (21 July 1988), 32; L. Moran, 'The Uses of Homosexuality: Homosexuality for National Security' (1991) 19 *Int. J. Sociol. of Law* 149.

[65] H.C. Deb., Vol. 195, 474 (Written Answers) (23 July 1991).

[66] See 'Homosexual Diplomats Told to "Come Out" Face Sack', *Independent* (15 Aug. 1991), 1. Surely, such a policy change must, in this respect, be retroactive. Would the ending of a ban on Jewish civil servants not apply to existing Jewish civil servants who had been forced to hide the fact?

[67] See *Guardian* (27 Apr. 1988), 20; *Gay Times* (June 1988), 15.

[68] See *LAGER Gay Men's Newsletter* (Spring 1990).

[69] See *Independent* (12 Nov. 1991), 6.

[70] See *Independent* (18 Jan. 1990), 1; *Independent* (26 Jan. 1988), 1.

homosexual judges' and that an applicant's 'sexual orientation and behaviour is a matter of the individual circumstances'.[71]

Housing, education, and services

As in the case of employment, the absence of any protection like the Sex Discrimination Act 1975 and the Race Relations Act 1976 means that any provider of a product or service (including housing or education) is free to refuse it to a gay, lesbian, or bisexual person (or any association of such persons), if they have no right to it under a lease or other contract.[72] A particularly common example of discrimination in services is that practised by life-insurance companies against gay and bisexual men since the HIV epidemic began. Such companies may require gay or bisexual men applying for life insurance to complete supplementary questionnaires and to submit to medical examinations, including HIV tests, and may refuse cover or charge inflated premiums, even if these men are HIV negative.[73] Inability to obtain life insurance can make it more difficult to purchase a home in Britain. And the companies' additional requirements and coverage policies are not automatically imposed on women or heterosexual men (unless they report a negative HIV test result).

HIV-motivated direct sexual orientation discrimination may be contrasted with direct discrimination against persons with HIV that constitutes indirect sexual orientation discrimination because of the disproportionate number of gay and bisexual men with HIV. Discrimination on the basis of HIV status could be dealt with as direct disability discrimination (rather than indirect sexual orientation discrimination), if disability discrimination were prohibited in Britain, as it is in most jurisdictions in Canada and the USA.

Section 28 of the Local Government Act 1988

While employers and providers of services are generally free to decide whether or not to discriminate on the ground of sexual orientation (in

[71] See *Independent* (30 Oct. 1991), 8; C. Richardson, 'Homosexuality and the Judiciary' (1992) 142 *New L.J.* 130.

[72] See e.g. *Capital Gay* (31 Jan. 1992), 3 (gay-themed play scheduled for theatre within pub cancelled to prevent 'influxes of the gay community' into a 'community pub'); *Gay Times* (Aug. 1991), 6 (bank's initial refusal to open account for lesbian and gay youth group); *Capital Gay* (18 Oct. 1991), 12 (sign in window of barber shop saying '[i]n the interests of health and hygiene, no homosexuals please'); ' "No Gays" ' *Gay Times* (Mar. 1988), 9 (sign on door of motorway cafe); *Gay Times* (Dec. 1988), 13 (newspapers' refusals of advertisement for gay switchboard); *Gay Times* (Oct. 1991), 17 (council's refusal to permit conference of Campaign for Homosexual Equality).

[73] See W. Gryk, 'AIDS and Insurance', in D. Harris and R. Haigh (eds), *AIDS: A Guide to the Law* (1990), 145–61; 'Gay Men Still Face Bias from Insurers', *Independent* (28 Dec. 1991), 3; I. Kennedy and A. Grubb, 'HIV and AIDS: Discrimination and the Challenge for Human Rights' in A. Grubb (ed.), *Challenges in Medical Care* (1992), 11–14.

practice against gay, lesbian, and bisexual persons, but in theory also against heterosexual persons), local authorities are in the unique position of having to consider whether the law requires them to discriminate against gay, lesbian, and bisexual persons (or associations of such persons) in certain circumstances. This is because the notorious section 28 of the Local Government Act 1988 ('Section 28') prohibits local authorities from 'intentionally promot[ing] homosexuality', 'publish[ing] material with the intention of promoting homosexuality', or 'promot[ing] the teaching . . . of the acceptability of homosexuality as a pretended family relationship'. The scope of these prohibitions has yet to be interpreted by the courts, because no proceedings have been brought alleging that a local-authority decision (e.g. to grant funding to a gay, lesbian, or bisexual group) violated Section 28, or challenging the validity of a decision (e.g. not to grant such funding) that relied on Section 28.[74]

Without any judicial interpretation, and given the vague wording of the prohibitions, it remains unclear what local-authority activities, if any, are prohibited by Section 28. The broadest possible interpretation of 'intentional promotion of homosexuality' would encompass all non-discrimination on the ground of sexual orientation, and consequent equal treatment of gay, lesbian, and bisexual persons, on the assumption that equal treatment would encourage more people to become gay, lesbian, or bisexual. Thus, to avoid 'promotion of homosexuality', local authorities would need actively to discriminate against gay, lesbian, and bisexual persons and organizations, by denying them access to employment or services (including funding or licenses), and censoring any mention of their existence in local-authority-funded publications, plays, exhibitions, schools, and libraries. This interpretation was urged by opponents of Section 28 and rejected by the Government.[75] The narrowest possible interpretation would treat 'promotion of homosexuality' as a contradiction in terms, in that, as some opponents of Section 28 argued, a person's sexual orientation is an immutable status and cannot be changed by any amount of 'promotion'.[76]

A court would probably adopt an intermediate interpretation, whereby 'promotion of homosexuality' is theoretically possible (in that actually engaging in same-sex emotional-sexual conduct, rather than being attracted to it, involves a choice that any person could make), but requires active encouragement of same-sex emotional-sexual conduct (or preferential treatment of persons who engage in such conduct with a view to

[74] See M. Colvin and J. Hawksley, *Section 28: A Practical Guide to the Law and its Implications* (1989), 10–11, 17, 57–9 (arguing that a decision not to make a grant could be challenged by seeking judicial review); *Gay Times* (Dec. 1989), 17 (fear of adverse decision halts judicial review of denial of grant for lesbian and gay festival).

[75] Ibid. at pp. 17–23, 67–8; K. Norrie, 'Symbolic and Meaningless Legislation' (Sept. 1988) *J. of Law Soc. of Scotland* 310, 311.

[76] See Norrie, ibid. at p. 312; Jeffery-Poulter, *Peers, Queers*, 238–9, 263–4.

encouraging others to do so).[77] If this is the meaning of Section 28, it would seem that Parliament has prohibited something which a local authority could theoretically do, but which in fact no local authority appears to have done, either before or after the passage of Section 28.[78] If Section 28 is aimed at policies or practices that do not exist (which may explain, in part, the absence of enforcement to date), what effects does it have?

The most important effect is symbolic, in that it expresses the view of Parliament that 'homosexuality' is something negative which must not be 'promoted' (even if it need not be actively discouraged).[79] And it adds insult to injury by asserting that the relationship between same-sex partners is a 'pretended' (i.e. 'falsely avowed') family relationship. (Imagine a law prohibiting local authorities from 'intentionally promoting Judaism or Islam' or 'the teaching of the acceptability of Judaism or Islam as pretended religions'.) But, even though it may not actually prohibit any local authority activities, it has had, and will probably continue to have, a 'chilling effect' on the funding of gay, lesbian, and bisexual organizations or events, the employment of gay, lesbian, and bisexual teachers, and the discussion of gay, lesbian, and bisexual persons and same-sex emotional-sexual conduct in local authority-funded schools. Local authorities concerned about violating Section 28 may be over-cautious in their decisions, or may use Section 28 as a justification for prejudiced decisions to discriminate.[80] A study of the effects of Section 28 found only 5 to 7 refusals by local authorities to fund gay, lesbian, or bisexual organizations.[81] Those refusals (and other reported refusals) have related to funding, providing premises for or permitting outdoor festivals, films, and plays, art gallery and library exhibitions, youth groups and student societies, community centres, 'International Women's Day' events, and various publications. But the fact that these refusals are relatively isolated suggests that the local authorities that were funding gay, lesbian, and bisexual organizations before Section 28 are continuing to do so.[82]

The most specific of the three prohibitions in Section 28 is the third,

---

[77] One legal opinion defined 'promot[ing] homosexuality' as 'active advocacy directed by local authorities towards individuals in order to persuade them to become homosexual, or to experiment with homosexual relationships'. Colvin and Hawksley, *Section 28*, 12.

[78] Colvin and Hawksley's study, ibid. at p. 57, concludes that 'it would appear that there are few, if any, existing local authority activities which are outlawed by Section 28'.

[79] See Norrie, 'Symbolic', 314 (Section 28 says 'we cannot tolerate tolerance of homosexuals').

[80] See P. Thomas and R. Costigan, *Promoting Homosexuality: Section 28 of the Local Government Act 1988* (1990), 31–2; Norrie (Section 28 'is legally meaningless, but . . . will encourage, indeed promote, intolerance of minorities').

[81] Thomas and Costigan, ibid. at pp. 28–30. See also Colvin and Hawksley, *Section 28*, 5–6; 'Tragic or Toothless? Section 28—One Year On' *Gay Times* (June 1989), 10.

[82] See Colvin and Hawksley, *Section 28*, 39.

dealing with teaching in maintained schools. It expressly prohibits the teaching of a particular idea, i.e. 'the acceptability of homosexuality as a pretended family relationship'.[83] This, too, has probably had a limited impact, because the primary responsibility for sex education in schools in England and Wales lies with school governing bodies rather than local authorities.[84] However, it may have had a 'chilling effect' on discussion of gay, lesbian, and bisexual issues in classrooms, and seems to be related to a few cases of teachers being dismissed or suspended or resigning.[85]

Other discrimination against same-sex couples

The rights of married heterosexual couples

The most basic discrimination against same-sex couples is their exclusion from the right to marry each other (i.e. entering into a civil, rather than religious, marriage). The Court of Appeal rejected the possibility of same-sex marriage in *Corbett* v. *Corbett*, where a transsexual woman attempted to marry a man,[86] and section 11(*c*) of the Matrimonial Causes Act 1973 provides that a marriage is void if 'the parties are not respectively male and female'. There do not appear to be any reported decisions in Britain where two men or two women have challenged a refusal to grant them a marriage license.[87] Because same-sex couples are denied the right to marry, they are excluded not only from any symbolic or intangible benefit that may accompany legal recognition of their partnership by the state (as opposed to a religious institution), but also from the tangible benefits (in the form of special rights) that married heterosexual couples enjoy. Although these benefits often constitute direct marital-status discrimination, in that unmarried heterosexual couples may also be excluded from them, they also discriminate directly on the ground of sexual orientation by incorporating a discriminatory criterion (being married) with which same-sex couples are not legally permitted to comply.[88]

For same-sex couples of mixed nationality (where one partner is a UK national and one is a non-EC national), a vital benefit of marriage

---

[83] A DfE circular (No. 11/87) prohibits sex education that 'advocates homosexual behaviour, . . . presents it as the 'norm', or . . . encourages homosexual experimentation by pupils'. See Tatchell, *Europe*, 98; *Capital Gay* (30 April 1993), 3 (draft revision of circular retains prohibition).

[84] Colvin and Hawksley, *Section 28*, 44–50; Norrie, 'Symbolic', 313.

[85] See e.g. *Gay Times* (May 1988), 18 (teacher suspended for talking to pupils about being gay); *Gay Times* (Aug. 1990), 5 (heterosexual teacher dismissed for discussion of same-sex sexual activity in sex education class).

[86] [1970] 2 All ER 33. For a pioneering discussion of same-sex marriage, see I. Kennedy, 'Transsexualism and Single Sex Marriage' (1973) 2 *Anglo-Am. LR* 112, 133–6.

[87] See *Capital Gay* (27 March 1992), 3 (Westminster registry office refuses to marry four same-sex couples).

[88] Cf. *James* v. *Eastleigh Borough Council* [1990] 2 AC 751 (HL).

is the right (subject to the 'primary purpose' test) to bring one's spouse into Britain to live and work. Without such a right, a same-sex couple may find, not only that they cannot choose to live in Britain (as a married heterosexual couple could), but that there is no country in which they have a legal right to live together. Immigration laws could thus force their relationship to end.[89] Other benefits of marriage include national-insurance pensions for spouses, favourable tax treatment, inheritance rights, hospital or prison visiting rights, and rights to maintenance or division of property in the event of divorce. In addition, many public and private employment benefits (such as pensions or private medical insurance) may be granted only to the married heterosexual partner of an employee (or, in some cases, any heterosexual partner, married or unmarried). In fact this kind of employment discrimination is far more common than dismissals or refusals to hire or promote, in that it affects every person with a same-sex partner, whether or not they are openly gay, lesbian, or bisexual.[90]

The rights of unmarried heterosexual couples

In addition to the right to marry, and the benefits that flow from marriage, same-sex couples are also denied the rights of unmarried heterosexual couples.[91] It is easier to extend such rights to same-sex couples, in that the issue of treating a same-sex couple as 'equivalent' to a married heterosexual couple does not need to be addressed. However, compared with some countries, Britain maintains a fairly rigid distinction between married and unmarried heterosexual couples, granting the latter relatively few rights. Examples include sections 1(2) and 2(2) of the Domestic Violence and Matrimonial Proceedings Act 1976, section 1(3)(*b*) of the Fatal Accidents Act 1976, the discretionary recognition of unmarried heterosexual couples for immigration purposes,[92] and the right of an unmarried heterosexual partner to succeed to a local authority tenancy upon the death of their partner, now found in sections 87, 89 and 113(1)(*a*) of the Housing Act 1985. In *Harrogate Borough Council* v.

---

[89] See e.g. *Wirdestedt* v. *Secretary of State for the Home Dept.* [1990] 2 Imm. AR 20 (CA); note 172 below.

[90] See e.g. *Guardian* (3 June 1991), 6 ('gay partners of civil servants at the British Council and British Library have won the right to benefit from a relocation package for married [heterosexual] couples'; survey shows unions beginning to campaign for same benefits for same-sex partners as opposite-sex partners); *Pink Paper* (7 Sept. 1991), 2 (union threat of legal action if British Rail does not extend free travel passes to same-sex partners of employees).

[91] See P. Ghandi and E. Macnamee, 'The Family in UK Law and the International Covenant on Civil and Political Rights 1966' (1991) 5 *Int. J. Law and Fam.* 104, 116–22.

[92] See I. Macdonald and N. Blake, *Immigration Law and Practice in the United Kingdom*, 3d ed. (1991), 256–7.

Simpson,[93] the Court of Appeal refused to extend the latter right to the partner of a lesbian woman. A recent attempt to amend the 1985 Act so as to permit succession by a same-sex partner was defeated.[94] But the Government later announced that it would issue guidelines encouraging local authorities to grant joint tenancies to same-sex couples and confirming its view that 'discrimination on the grounds of . . . sexual orientation . . . is [not] acceptable in deciding whether or not to grant a tenancy'.[95] At least 69 local authorities in England and Wales already grant joint tenancies.[96] As for other public or private services, same-sex couples may find occasionally that they are refused a service that a married or unmarried heterosexual couple would be permitted.[97]

Other discrimination against gay, lesbian, and bisexual parents

Existing parents

Where a gay, lesbian, or bisexual person is already a biological and legal parent of a child, usually because of a dissolved heterosexual marriage, the courts frequently discriminate against them on the ground of their sexual oxientation in deciding whether they, their former husband or wife, a relative, or a local authority should have custody or care and control of, or access to, the child.[98] The general attitude of the courts to lesbian and bisexual mothers is that their being lesbian or bisexual is not an absolute bar to their obtaining custody of their children, because the paramount consideration of the child's welfare[99] may require it in some cases. However, it is virtually always treated as a relevant and substantial negative factor, in that the courts effectively raise a presumption in favour of a heterosexual former spouse seeking custody that a 'normal' heterosexual home environment is better for a child.[100] Many lesbian and bisexual mothers are unable to rebut this presumption, especially if

---

[93] [1986] 2 Fam LR 91. An attempt to amend the Housing Act 1980 to include same-sex partners had been defeated. See Crane, *Gays and Law*, 142.

[94] See *Capital Gay* (22 Jan. 1993), 1.

[95] See *Capital Gay* (9 April 1993), 1.

[96] See Stonewall Group, *Parliamentary Briefing: Housing and Urban Development Bill* (Jan. 1993).

[97] See e.g. *Pink Paper* (20 Oct. 1988), 11 (refusals by some hotels to accept telephone reservation of double room by openly gay or lesbian couple).

[98] The concepts of 'custody', 'care and control' and 'access' have been replaced under ss. 2–5, 8 of the Children Act 1989 by concepts of 'parental responsibility', 'residence orders', and 'contact orders'.

[99] Children Act 1989, s. 1(1).

[100] See e.g. *C.* v. *C.* [1991] 1 F.L.R. 223, 228 (CA) (inadequate weight given to mother's lesbian relationship in awarding her care and control: it was 'axiomatic that the ideal environment for the upbringing of a child is the home of . . . her father and her mother'), [1992] 1 F.C.R. 206 (award upheld on rehearing); S. Boyd, 'What Is a "Normal" Family?' (1992) 55 *MLR* 269.

they are seen as 'militant' (i.e. open about their sexual orientation and actively challenging discrimination).[101] Where they succeed in doing so, the court often makes it clear that a custody order is made with reluctance because a heterosexual father or a local authority cannot provide a better alternative.[102] Even more difficult is the position of a gay or bisexual father who seeks custody. In a competition with a heterosexual mother, he is extremely unlikely to succeed. Indeed, so hostile are the courts to the idea of a gay or bisexual man raising children (particularly boys) that they have gone as far as denying all access to a child, and even extinguishing the father's parental rights altogether by permitting an adoption without his consent.[103]

Prospective parents

Where a same-sex couple (or a gay, lesbian, or bisexual individual) would like to become parents (or a parent) by adopting or fostering a child, or by using donor insemination or a surrogate motherhood arrangement to have a child with genetic input from one of the partners (or from the individual), they may face discriminatory legal or other obstacles. In the case of adoption, a same-sex couple will face the currently insurmountable legal barrier of section 14(1) of the Adoption Act 1976, which permits only a married heterosexual couple to adopt jointly.[104] If one partner (or an individual) applies to adopt as a single person, under section 15 of the Act, an adoption agency or local authority is free to hold their sexual orientation against them, although the equal opportunities policies of some local authorities seem to prohibit sexual orientation discrimination in adoption decisions.[105] Even if an application is successful, it will probably be for a 'hard to place' child that is older or has a disability.[106] A Department of Health consultation document cited such cases in proposing no changes in 'the law relating to single applicants, including lesbians and gay men'.[107] But a minister stressed that councils should

[101] See *B.* v. *B.* [1991] 1 F.L.R. 402, 410 (mother and partner did not 'believe in advertising their lesbianism and acting in the public field in favour of promoting lesbianism').

[102] See e.g. *Re P.* [1983] F.L.R. 401, 405 (CA) (only if 'no other acceptable alternative form of custody'). See also D. Bradley, 'Homosexuality and Child Custody in English Law' (1987) 1 *Int. J. Law and Fam.* 155; Crane, *Gays and Law*, 121–38.

[103] See *Re D.* [1977] AC 617 (HL); Bradley, 'Homosexuality', 173–7, 196, 200 (discussing successful application by gay father against lesbian mother).

[104] This barrier is required by Article 6(1) of the European Convention on the Adoption of Children (E.T.S. No. 58).

[105] See e.g. Bradley, 'Homosexuality', 162–3; Ghandi and Macnamee, 'Family', 112–13.

[106] See 'Adoption by Lesbians Attacked', *Independent* (12 Oct. 1990), 3 (2-year-old disabled boy); *Gay Times* (Feb. 1991), 7 (judge ordered return of the boy to foster mother 'not because . . . placement with a lesbian couple was inherently unsuitable or . . . they were unable to look after him properly').

[107] Department of Health, *Review of Adoption Law: Report to Ministers of an Inter-departmental Working Group* (Oct. 1992), para. 26.13.

make 'strenuous efforts' to find married heterosexual couples.[108] The chances of a same-sex couple or a gay, lesbian, or bisexual individual succeeding in adopting a newborn baby with no disabilities must be extremely small (compared with those of a married heterosexual couple).[109]

The issues are similar in the case of fostering, except that there is no legal barrier to same-sex couples acting jointly as foster parents[110] and, perhaps because fostering is often temporary and may be easier to undo than adoption, there are probably more gay, lesbian, and bisexual foster parents than adoptive parents. A number of local authorities have equal opportunities policies that include sexual orientation and apply to fostering, and some have encouraged gay, lesbian, and bisexual individuals and same-sex couples to apply, in one case, in spite of a threatened challenge under Section 28.[111] Fostering of a troubled gay or lesbian teenager in council care by a gay or lesbian couple may be seen as particularly appropriate.[112] Other local authorities have effectively banned such applications.[113] A proposed statement in paragraph 16 of the guidelines to the Children Act 1989 that '"[e]qual rights" and "gay rights" have no place in fostering services' was dropped after a vigorous campaign against it,[114] but the final version (paragraph 3.14) retained a statement that '[t]he chosen way of life of some adults may mean that they would not be able to provide a suitable environment for the care and nurture of a child'.[115]

If a lesbian or bisexual woman chooses to have a child (and raise it alone or with a female partner), it is relatively easy for her to become pregnant by donor insemination. Although arrangements for a donation of semen and self-insemination can be made privately (and thus, apart from any necessary advertising for a donor, would be difficult to prohibit), some women prefer to use fertility clinics, which can test semen for HIV and provide an anonymous donor.[116] During the passage of the Human Fertilisation and Embryology Act 1990 (HFEA), attempts both expressly

[108] See *Independent* (20 Oct. 1992), 6.

[109] But see *Re H*. [1993] Fam Law 205 (interim residence order granted to one member of lesbian couple where biological mother had 'given' them a baby girl at birth and local authority had denied their attempts to adopt or foster the girl); *Pink Paper* (1 Nov. 1992), 1 (order made permanent).

[110] See *Independent* (21 April 1993), 3 (lesbian couple approved after initial refusal).

[111] See Thomas and Costigan, *Promoting*, 30–1 (Waltham Forest).

[112] See *Re W*. [1992] 1 F.L.R. 99 (CA).

[113] See *Capital Gay* (11 Oct. 1991), 11 (Wandsworth).

[114] See *Independent* (17 Apr. 1991), 2.

[115] Department of Health, *The Children Act 1989: Guidance and Regulations*, Vol. 3 (Family Placements).

[116] See e.g. 'Clinics Can Cut Risks Posed by Infected Semen', *Independent* (12 Mar. 1991), 3.

to prohibit clinics from providing donor insemination to unmarried women and to require that clinics have equal opportunities policies were unsuccessful.[117] The final version of HFEA section 13(5) prohibits treatment 'unless account has been taken of the welfare of any child who may be born as a result of the treatment (including the need of that child for a father)'. It is not yet clear what effect this will have on the access of lesbian and bisexual women to clinics. Whether a woman arranges a donation privately or uses a clinic, her female partner will not be recognized as a legal co-parent, as the married or unmarried male partner of a woman receiving donor insemination could be under HFEA section 28.[118] And if she claims income support and knows the donor, she may be required to name him (making him liable for maintenance) or face a reduction in benefits.[119]

A gay or bisexual man who wishes to father a child (and raise it alone or with a male partner) faces the biological reality that the woman's contribution to the creation of a child is obviously far more important than the man's, and that finding a woman (or women) who is (or are) willing to donate an egg or become pregnant (by donor insemination or implantation of an embryo after IVF), and take no role in raising the child, is extremely difficult (as it is for a heterosexual couple where the woman is infertile). Assuming he were able to make such a surrogacy arrangement (which would be unenforceable under s. 1A of the Surrogacy Arrangements Act 1985), HFEA section 30 would not permit the recognition of his male partner as a legal co-parent (instead of the surrogate mother), as it would in the case of a married heterosexual woman.[120]

EXPLAINING THE CHANGES, AND LACK OF CHANGES, SINCE 1950

The preceding section has attempted to define the status quo in 1993 by describing the extent of sexual orientation discrimination contained in, or permitted by, English law. The achievement of legal equality for gay, lesbian, and bisexual persons and same-sex emotional-sexual conduct would require the elimination of all four of the categories of discrimination discussed above. Given the extent of that discrimination, it is

---

[117] See *Gay Times* (June 1990), 6; *Capital Gay* (16 Feb. 1990), 3 (amendment in House of Lords restricting donor insemination to married women defeated by one vote); G. Douglas, *Law, Fertility and Reproduction* (1991), 121–2 (only about 10% of clinics treated lesbian women in 1990).

[118] Several adoptions by lesbian co-parents have recently been approved by US courts. See [1992] *Lesbian/Gay Law Notes* 19.

[119] Child Support Act 1991, ss. 6, 46. See *Gay Times* (Feb. 1993), 6.

[120] See Douglas, *Reproduction*, 141–67.

clear that many changes are required before that goal is achieved. But have there been any improvements between 1950 and 1993? If so, how can they be explained and why have there not been more?

Examining the changes that have occurred since 1950, one can only conclude that they have been minimal, compared with the discrimination that remains and with the changes that have occurred with respect to most of the other rights discussed in this book. A crude summary of the net progress in legislation since 1950 might be 'two steps forward (the Sexual Offences Act 1967), one step back (Section 28), twenty or more steps to go'. While the 1967 Act substantially reduced the criminal law's interference with sexual activity between men (and therefore with the choice of gay or bisexual sexual orientation by men), Section 28 symbolically rejected the idea of government neutrality or non-discrimination as among sexual orientations. Although private sexual activity between adult men and between adult women may be legal, same-sex and bisexual sexual orientation are not to be 'promoted' or presented as 'acceptable'. And the whole range of other disincentives to choosing such an orientation (including criminalization of public displays of affection, employment discrimination, non-recognition of partnerships, and discrimination against existing or prospective parents) remains in place.

Does this assessment minimize the impact of the 1967 Act and exaggerate the importance of Section 28? It is true that the 1967 Act was a 'leap forward' in the sense that it freed gay and bisexual men over 21 from the fear of prosecution for private sexual activity with men over 21. And the greater openness it permitted may have facilitated the growth since 1970 of a vibrant and increasingly visible gay, lesbian, and bisexual community, with its parades, festivals, conferences, bars, restaurants, cafes, bookshops, community centres, newspapers, and political, professional, cultural, athletic and religious organizations.[121] Although a community may have existed in 1967, in 1950, and before, its manifestations were much less diverse and far more underground. In addition, substantial decriminalization (for men) paved the way for proposals for other legislative reforms, and for the piecemeal extra-legislative progress that has been made in the equal opportunities policies of many local authorities and trade unions, a few companies and police forces, and the civil service (now apparently including the Foreign Office and security services). The issues of employment discrimination, recognition of same-sex partners, and adoption or fostering would have been almost impossible to raise before 1967.

However, the 1967 Act provided for a very limited tolerance of sexual

---

[121] See J. Weeks, *Coming Out: Homosexual Politics in Britain from the Nineteenth Century to the Present*, revised ed. (1990), 185–230.

activity between men. How did this 'limited tolerance' arise, and why were the limits strengthened by Section 28? In the early 1950s, a rising level of prosecutions of men for same-sex sexual activity (much of it semi-public), and the convictions of a number of public figures, led to the establishment of a Home Office committee, chaired by Sir John Wolfenden, to investigate the issues of 'homosexual offences' and prostitution.[122] The committee's landmark report, published in 1957, recommended that 'homosexual behaviour between consenting adults in private should no longer be a criminal offence', with the meanings of 'consent' and 'in private' being determined according to 'the same criteria as apply to heterosexual acts'.[123] However, when it came to defining the age of 'adulthood', the committee faltered. Their finding that 'it [is] hard to believe that [a young man] needs to be protected from would-be seducers more carefully than a girl does', and the opinion of the majority of medical witnesses that 'the main sexual pattern . . . was usually fixed . . . by the age of sixteen', both supported equality with the age of consent for heterosexual and lesbian sexual activity (16).[124] But the committee pre-ferred 'the legal age of contractual responsibility' (21 at the time) because 'a boy is incapable, at the age of sixteen, of forming a mature judgment about actions of a kind which might have the effect of setting him apart from the rest of society'.[125] Thus, the social discrimination a 16-year-old male choosing same-sex sexual activity might face was effectively used to justify continued discrimination by the criminal law.

Viewed in isolation, the committee's main recommendation could be seen as providing for a substantial degree of equal treatment of sexual activity between men by the criminal law. However, many statements by the committee made it clear that they viewed such activity as some-thing negative, which could only be tolerated to a limited extent[126] and otherwise must be discouraged. For example, the committee noted that recognizing 'a realm of private morality and immorality . . . is not to condone or encourage private immorality',[127] that it deplored the 'damage to . . . the basic unit of society' (the traditional heterosexual family) caused by 'homosexual behaviour between males',[128] that 'the influence of detailed [press] reports of [homosexual offences] is considerable and

---

[122] See Jeffery-Poulter, *Peers Queers*, 11–27; Weeks, *Coming Out*, 156–67; A. Grey, *Quest for Justice: Towards Homosexual Emancipation* (1992), 19–33.

[123] *Report of the Committee on Homosexual Offences and Prostitution*, Cmnd. 247 (1957), 25 (the 'Wolfenden Report').

[124] Ibid. 26.

[125] Ibid. 27.

[126] 'It is true that a change of this sort would amount to a limited degree of . . . toleration [by the Legislature]'. Ibid. 23.

[127] Ibid. 24.

[128] Ibid. 22.

almost wholly bad',[129] and that it did 'not wish to see legalised any forms of behaviour which would swing towards a permanent habit of homosexual behaviour a young man who without such encouragement would still be capable of developing a normal habit of heterosexual adult life'.[130] It recommended 'research into the aetiology of homosexuality and the effects of various forms of treatment'.[131] In recommending against a change in the prohibition of 'solicitation', it stressed that 'the [proposed] limited modifications . . . should not be interpreted as an indication that the law can be indifferent to other forms of homosexual behaviour, or as a general licence to adult homosexuals to behave as they please'.[132] Finally, in the area of employment, it suggested that the retention of criminal penalties in the armed forces may be necessary 'for the sake of good management and the preservation of discipline and for the protection of those of subordinate rank',[133] and that vulnerability to blackmail 'may be a valid ground for excluding from certain forms of employment men who indulge in homosexual behaviour'.[134] It also approved of the Government's policy of ensuring 'that men guilty of homosexual offences are not allowed to continue in the teaching profession'.[135]

The 'limited tolerance' recommended in the Wolfenden Report took ten years to be implemented, and the 1967 Act provided it in an even more limited form.[136] The Act ignored the committee's view that punishment of sexual activity between men not 'in private' should be according to the same criteria as heterosexual sexual activity, by adopting a narrow definition of 'in private' and criminalizing all sexual activity between men falling outside the definition under a different offence than would apply to sexual activity between a man and a woman or between women in similar circumstances. The Act also increased the penalty for 'gross indecency' where one man is under 21 from 2 to 5 years, created a new offence of 'procuring' an act of 'buggery', and allowed 'gross indecency' prosecutions in magistrates' courts.[137] In Parliament, supporters of the reform expressed a mixture of disgust and patronizing compassion.[138] Lord Arran, the bill's sponsor in the Lords, stressed that sexual activity between men 'ha[d] been universally condemned' during the debates.[139] Leo Abse, sponsor in the Commons, referred to 'these wretched men' and

---

[129] *Report of the Committee*, 78.
[130] Ibid. 26.   [131] Ibid. 77.   [132] Ibid. 44.
[133] Ibid. 53.   [134] Ibid. 22.   [135] Ibid. 78.
[136] For detailed accounts of the Act's passage, see Grey, *Quest*, 34–126; Jeffery-Poulter, *Peers, Queers*, 28–89; Weeks, *Coming Out*, 168–82.
[137] 1967 Act, ss. 3(2), 4(1), 9(2).
[138] One opponent, the Earl of Dudley, expressed only hatred: 'I cannot stand homosexuals. They are the most disgusting people in the world . . . I loathe them. Prison is much too good a place for them.' H.L. Deb., Vol. 275, 158 (16 June 1966).
[139] Ibid. at 160.

the 'appalling misfortune', 'terrible fate', 'curse' and 'disability' of being gay.[140] He argued that 'the question of punishment of homosexuals . . . has . . . prompted us to avoid the real challenge of preventing little boys from growing up to be adult homosexuals', of 'reduc[ing] the number of faulty males . . . [with] men's bodies but feminine souls'.[141] Thus, 'homosexuality' would remain 'unlawful, although not criminal' and 'contrary to public policy', and 'no recognised status of homosexuality'[142] or 'new freedom'[143] would be created. Indeed, Lord Arran warned that '[a]ny form of ostentatious behaviour now or in the future, any form of public flaunting, would be utterly distasteful and would . . . make the sponsors of the Bill regret that they have done what they have done'.[144]

The limited nature of the 1967 reform, making sexual activity between men lawful in certain circumstances but otherwise 'unlawful', 'contrary to public policy', and 'immoral', was reflected in the continuing use of section 32 of the 1956 Act,[145] and in such House of Lords decisions as *Knuller* v. *DPP*[146] (treating a newspaper with advertisements by men seeking sexual activity with other men as a 'conspiracy to corrupt public morals' because, although the 1967 Act permits people to 'choose to corrupt themselves in this way', 'no licence is given to others to encourage the practice') and *Re D*.[147] (extinguishing a gay father's legal status as parent of his son because legal reform did not 'entitle the courts to relax . . . the vigilance and severity with which they should regard the risk of children . . . being exposed . . . to ways of life which . . . may lead to severance from normal society . . . psychological stresses and unhappiness and . . . physical experiences which may scar them for life'). It would also explain the absence, since 1967, of any legislation addressing the many other forms of sexual orientation discrimination, and the increase in prosecutions for semi-public sexual activity. These had dropped between 1954 and 1967, but subsequently rose, reaching pre-Wolfenden levels in

---

[140] H.C. Deb., Vol. 731, 260–2 (5 July 1966).
[141] H.C. Deb., Vol. 738, 1078 (19 Dec. 1966). His argument illustrates the link between sexual orientation discrimination and sex discrimination.
[142] Ibid. at 1120–1.
[143] H.L. Deb., Vol. 285, 523 (21 July 1967).
[144] Ibid.
[145] Ss. 7(2)(*b*) and 9(1) of the 1967 Act expressly contemplated 'solicitation' cases 'where the immoral purpose is the commission of a homosexual act'. See *R.* v. *Kirkup* (1992) 96 Cr. App. R. 352 (in view of 'the attitudes of right-thinking members of society', a properly directed jury could only conclude that private, consensual sexual activity between adult men is an 'immoral purpose'). Cf. *Stephens* v. *Avery* [1988] 2 WLR 1280, 1284 ('[t]here is no common view that sexual conduct of any kind between consenting adults is grossly immoral').
[146] [1973] AC 435, 457 (HL).
[147] [1977] AC 617, 629 (HL).

1989 before dropping again.[148] The most recent judicial view of the effect of the 1967 Act is Lord Templeman's in *R. v. Brown*. He said that '[b]y the 1967 Act Parliament recognised and accepted the practice of homosexuality,' that 'sexual activities conducted in private between not more than two consenting adults of the same sex or different sexes are now lawful', and that '[s]ubject to the respect for private life embodied in the 1967 Act, Parliament has retained criminal sanctions against the practice, dissemination and encouragement of homosexual activities'.[149]

Notwithstanding the message of 'limited tolerance' emanating from Parliament, many local authorities were willing to take steps to eliminate sexual orientation discrimination, to the extent possible within their powers. By the mid-1980s, it was becoming increasingly common for local authorities to adopt policies prohibiting sexual orientation discrimination in their own employment decisions and in the provision of their services. As a result, gay, lesbian, and bisexual organizations were able to apply for small grants, on an equal footing with other organizations with similar activities. And some local authorities sought to introduce 'positive images' of gay, lesbian, and bisexual persons into the curricula of state schools.[150] Unfortunately, these steps, aimed at non-discrimination and the teaching of tolerance rather than hatred, were hysterically misconstrued and characterized as the preferential treatment and 'promotion' or 'proselytization' of same-sex or bisexual sexual orientation, at the expense of heterosexual sexual orientation. Aggravating the controversy was the perception of the idea of non-discrimination on the ground of sexual orientation as a policy of the 'loony left', rather than as a non-party-political question of human rights. These local authority policies thus were not seen as a legitimate exercise of local autonomy, and became swept up in the ideological battle between Conservative-controlled central government and Labour-controlled local authorities. The attempt to discourage them through Section 28 could be seen as part of a package of legislation aimed at limiting the powers and altering the financing of local government.[151] Putting aside the matter of the level of government and political party introducing these policies, Section 28 clearly rejects them and the idea of non-discrimination, especially in education. Although it purports to deal with 'promotion', and implicitly with preferential treatment of one sexual orientation over another, any such treatment could

---

[148] See R. Walmsley, 'Indecency between Males and the Sexual Offences Act 1967' [1978] *Crim LR* 400 (2322 'gross indecency' offences in England and Wales in 1955 vs. 840 in 1967); P. Tatchell, *Gay Times* (Sept. 1990), 20 (2022 in 1989); *Gay Times* (April 1992), 15 (964 in 1991).

[149] [1993] 2 All ER 75, 81 (HL).

[150] See Weeks, *Coming Out*, 237–44; Jeffery-Poulter, *Peers, Queers*, 204–17.

[151] See e.g. Local Government Acts 1986, s. 2, and 1988, ss. 17–22.

have been dealt with by a neutral prohibition of sexual orientation discrimination by local authorities, or of promotion of one sexual orientation over another in schools. By singling out 'homosexuality', Section 28 made it clear that neither local nor central government should be neutral as among sexual orientations, and that the 'limited tolerance' of the 1967 Act was not to be extended.

What conceptual approaches have been adopted in arguing for changes, and by whom have the arguments been made? It is not clear that any single concept underlies the arguments that have been made for law reform. Nor could one be expected, in that these arguments (at least in the domestic context) have been directed primarily to Parliament, rather than to a court charged with interpreting a domestic Bill of Rights. The concepts of 'immutability' and 'privacy' both seem to have figured in the case for the 1967 Act. Lord Arran argued 'that these men do not choose to be homosexual; that no man would deliberately and obstinately forgo the choice of loving women—that no man would be so stupid', and compared persecution of gay men to 'persecution of Jews or Negroes— people who are born what they are, not of their own choosing'.[152] The case for the Act also emphasized the private location in which the decriminalized sexual activity takes place, but not the fundamental importance of the choice involved. Such a concept of 'privacy' is difficult to apply to activities that may be of a less spatially 'private' nature (e.g. public displays of affection, openness about one's sexual orientation in the workplace, state recognition of partnerships). 'Immutability' has also surfaced as an argument against Section 28, and in support of a claim of refugee status by a gay Cypriot man,[153] but 'sex discrimination' has rarely been used. Under the European Convention on Human Rights, the relevant concept has been that of interference with 'private life' under Article 8, because the limited scope of Article 14 (which only prohibits discrimination affecting the enjoyment of a Convention right or freedom and not, independently, an interest such as employment) restricts the effectiveness of the 'immutable status' and 'sex discrimination' concepts.[154]

The arguments for reform have primarily been made by what might be described as 'pressure groups', i.e., gay, lesbian, and bisexual campaigning organizations. These range from the Homosexual Law Reform Society

---

[152] H.L. Deb., Vol. 274, 606–7 (10 May 1966).

[153] See *R. v. Secretary of State for the Home Dept., ex p. Binbasi* [1989] Imm. A.R. 595 (claimant mentioned 'immutability' in arguing that he feared persecution because of 'membership of a particular social group'; government argued that he could refrain from sexual activity in Cyprus). Cf. n. 7 above.

[154] The European Commission of Human Rights rejected an argument that imprisonment for sexual activity interfered with the applicant's 'freedom to express his feelings of love for other men', contrary to Art. 10. See *X v. UK* (No. 7215/75) (1978) 19 Dec. & Rep. 66 at 80.

of the 1960s and the Campaign for Homosexual Equality of the 1970s to the *ad hoc* campaign against Section 28 in 1988 and the Stonewall Group, OutRage, the Lesbian Custody Project, and the Campaign for Access to Donor Insemination in the 1990s. These organizations have had some success since 1967, failing to stop Section 28 (in spite of several large demonstrations against it), but succeeding in preventing expressly discriminatory legislation or guidelines on donor insemination and foster parenting, and clarifying section 31 of the Criminal Justice Act 1991 (which was seen as increasing the likelihood of imprisonment for 'solicitation' or semi-public sexual activity).[155] As for the impact of these organizations on the policies of the major political parties, Labour has pledged to permit a free vote on a private member's bill on the male-male age of consent, to reform other aspects of sexual offences law, to repeal Section 28, and to introduce legislation prohibiting both discrimination and incitement to hatred on the ground of 'sexuality'. The Liberal Democrats' policies are similar, but are unequivocal on the age of consent. The Conservatives have no plans to lower the male-male age of consent and consider anti-discrimination legislation 'inappropriate and impracticable', because 'sexuality is a personal matter where legislation would be out of place'.[156] But the reforms dealing with non-military security clearances and with same-sex sexual activity by armed forces members, and John Major's September 1991 meeting with gay actor and Stonewall co-founder Sir Ian McKellen,[157] have provided the first signs of potential change.

What are the obstacles that stand in the way of further changes that would move Britain from 'limited tolerance' to 'equality'? The major obstacle is, of course, the opinion of a significant section of the public that sexual orientation discrimination is justified, and therefore not truly 'discrimination', because one sexual orientation (heterosexual) is 'right' while others (same-sex or bisexual) are 'wrong'. In many cases, this opinion is based on pure prejudice (i.e. an irrational dislike of gay, lesbian, and bisexual persons) or on religious doctrines, against either of which rational argument has no effect. (Nor should religious arguments for legislation have any place in a society that respects freedom of religion and conscience.) In other cases, the asserted 'wrongness' of same-sex or bisexual sexual orientation is tied to negative effects it is alleged to have. The most recent of these is AIDS, which has been cited as a reason for passing Section 28 and for refusing to lower the male-male age of consent.[158] This opportunistic argument falsely equates sexual activity

---

[155] See s. 31(3).
[156] See *Equal Opportunities Review* (March/April 1992), 20.
[157] See 'Major is Urged to Review Law on Gays', *Independent* (25 Sept. 1991), 6.
[158] See Thomas and Costigan, *Promoting*, 33–4; Jeffery-Poulter, *Peers, Queers*, 254.

between men (as opposed to unprotected sexual activity of any kind) with sexual activity likely to transmit HIV, and proposes discrimination as an effective means of disease control.

The other negative effects that are usually cited are the threat to the traditional heterosexual family (as the foundation of society) posed by the gay, lesbian, and bisexual minority (not unlike that posed to the traditional culture of Britain by immigrants of different religions or ethnic or racial origins), and the harm caused to any child or adolescent who is exposed to an openly gay, lesbian, or bisexual person and later chooses (for whatever reason) same-sex or bisexual sexual orientation.[159] At the core of this argument is the concern that if an individual (or an unacceptable number of individuals) chooses such an orientation, they will therefore be unable to have or raise children. Apart from assuming that having children is both essential to happiness and unproblematic in a crowded world, this argument ignores the fact that gay, lesbian, and bisexual individuals and same-sex couples can and do have or raise children and that not all heterosexual persons are able or willing to do so. Although a same-sex couple cannot have a child with genetic input from both partners, and gay couples (but not lesbian couples using donor insemination) are realistically confined to adoption or fostering (where there is genetic input from neither partner), most barriers to their having or raising children are legal or social and not biological. As for the biological limitations, any person choosing same-sex emotional-sexual conduct is most certainly aware of them and chooses such conduct in spite of them (and in spite of the discrimination that will accompany their choice).

What advocates of sexual orientation discrimination must be persuaded to do is to *accept* and *respect* this extremely difficult and deeply personal choice, whether or not they *understand*[160] or *approve* of it. Most would be willing to do so in the case of persons who choose minority religious beliefs, and would not see the teaching of the existence of and respect for those beliefs as 'promoting' them. A similar view of minority sexual orientations would permit progress beyond the current state of 'limited tolerance'. The rejection of such a view, given the emptiness of the arguments about 'negative effects', amounts to the same kind of fear of difference that underlies much discrimination based on sex, race, or religion.

---

[159] See e.g. F. Tasker and S. Golombok, 'Children Raised by Lesbian Mothers: the Empirical Evidence' [1991] *Fam. Law* 184.

[160] A heterosexual person's choice of heterosexual emotional-sexual conduct may be just as difficult for a gay man or lesbian woman to understand.

THE INFLUENCE OF THE
EUROPEAN CONVENTION ON
HUMAN RIGHTS AND EUROPEAN COMMUNITY LAW

European Convention on Human Rights

In the area of sexual orientation discrimination, the only international human rights treaty that has affected United Kingdom law is the European Convention on Human Rights, probably because its enforcement mechanism has made it the most attractive vehicle for challenging such discrimination in Europe. Although not insignificant, its impact on UK law has so far been confined to laws criminalizing all sexual activity between men. In declaring inadmissible at least 9 applications between 1955 and 1967 challenging such laws in Germany and Austria, the European Commission on Human Rights initially took the view that 'the Convention allows a High Contracting Party to punish homosexuality since the right to respect for private life [in Article 8] may, in a democratic society, be subject to interference . . . for the protection of health or morals'.[161] Thus, the Commission would probably have rejected a pre-1967 challenge to the blanket prohibition in England and Wales (had one been possible),[162] and seen the 1967 Act as a voluntary measure, not required by the Convention.

By 1978, the Commission's interpretation of the Convention had begun to change and it declared admissible Jeffrey Dudgeon's application challenging the blanket prohibition in Northern Ireland.[163] In 1980, it found that 'the legal prohibition of private homosexual acts between consenting males over 21 . . . breaches the applicant's [Article 8] right to respect for private life'.[164] The European Court of Human Rights agreed with the Commission's opinion in 1981, concluding that the law could not be justified as 'necessary in a democratic society' for the protection of 'morals' or 'the rights and freedoms of others'. The Court relied on the absence of such a law 'in the great majority of the member States of the Council of Europe' and the non-enforcement of the law with respect to men over 21 in Northern Ireland.[165] It reached the same conclusion in 1988 and 1993, in similar cases brought by David Norris against the Republic of Ireland and Alecos Modinos against Cyprus.[166]

---

[161] See e.g. *X* v. *Federal Republic of Germany* (No. 530/59) (1960), 3 YBEC 184 at 194.
[162] The UK did not recognize the right of individual petition under Art. 25 until 1966.
[163] *X* v. *UK* (No. 7525/76) (1978) 11 Dec. & Rep. 117.
[164] *Dudgeon* v. *UK* (No. 7525/76) (1980) Series B, No. 40, 11 at 41.
[165] *Dudgeon* v. *UK* (1981) Series A, No. 45 (judgment of 22 Oct. 1981), 23–4.
[166] *Norris* v. *Ireland* (1988) Series A, No. 142 (judgment of 22 Oct. 1988), 20–1; *Modinos* v. *Cyprus* (1993) Series A, No. 259 (judgment of 22 April 1993), 8–9.

The judgments of the European Court in *Dudgeon* and *Norris* have certainly spurred the substantial decriminalization of sexual activity between men in several parts of the UK (and in territories for which the UK is responsible). The adoption of the Homosexual Offences (Northern Ireland) Order 1982[167] was a direct result of *Dudgeon*, and the Commission's 1978 declaration of admissibility in that case may have influenced the passage of section 80 of the Criminal Justice (Scotland) Act 1980. Similar reforms followed in Guernsey in 1983, Jersey in 1990, the Isle of Man in 1992, and Gibraltar in 1993.[168] But apart from the single issue of a prohibition of all sexual activity between men, the Convention has had no effect on sexual orientation discrimination in the law or government decisions in the UK. This is because, in at least 15 cases dealing with other aspects of sexual orientation discrimination (10 against the UK, all of which originated in England and Wales), the Commission has found no violation of the Convention.

In the area of criminal law, 6 cases (3 against the UK) have upheld (under Articles 8 and 14) a higher age of consent for male-male sexual activity than male-female or female-female, with one of those (*Johnson*) also upholding the prohibition of sexual activity between men 'when more than two persons take part or are present'.[169] Another UK case upheld (under Articles 7, 9, 10, and 14) the conviction of the publisher and editor of *Gay News* for 'blasphemous libel'.[170] In the area of employment discrimination, the Commission has found justified (under Article 10) the dismissal of a lesbian Belgian teacher who said on television that she thought her sexual orientation had cost her the headship of a school mainly for girls, and (under Articles 8 and 14) the discharge of a gay British soldier over 21 who had engaged in sexual activity with a 20-year-old male soldier (junior in rank) and a male civilian over 21.[171]

In the area of discrimination against same-sex couples, 5 decisions on UK cases have found no violation (of Article 8, 12, or 14) where same-sex couples are not treated like married or unmarried heterosexual couples

---

[167] N. I. Statutes, S.I. 1982/1536 (N.I. 19).

[168] Sexual Offences (Bailiwick of Guernsey) Law, 1983; Sexual Offences (Jersey) Law, 1990; Isle of Man, Sexual Offences Act 1992; Gibraltar, Criminal Offences (Amendment) Ordinance 1993.

[169] See *Zukrigl* v. *Austria* (No. 17279/90) (13 May 1992); *Johnson* v. *UK* (No. 10389/83) (1986) 47 Dec. & Rep 72; *Desmond* v. *UK* (No. 9721/82) (1984) 7 EHRR 145; *X* v. *Belgium* (No. 9484/81) (1 March 1982); *X* v. *UK* (No. 7215/75) (1978) 19 Dec. & Rep. 66; *X* v. *Germany* (No. 5935/72) (1975), 3 Dec. & Rep. 46.

[170] See *X and Y* v. *UK* (No. 8710/79) (1982) 28 Dec. & Rep. 77 (arising from [1979] AC 617 (HL)).

[171] See *Morissens* v. Belgium (No. 11389/85) (3 May 1988); *Bruce* v. *UK* Application No. 9237/81 (1983) 34 Dec. & Rep. 68.

for the purposes of local-authority housing or immigration.[172] The Commission has concluded that the relationship of a same-sex couple (with or without children) does not fall within the scope of the right to respect for 'family life' under Article 8 or give rise to a right to 'marry and found a family' under Article 12, that the eviction of one partner from a council house after the death of the other or the deportation of one partner does not interfere with 'private life', and that differences in the treatment of same-sex and heterosexual couples are 'objectively and reasonably justified' and therefore not 'discrimination' under Article 14.[173] Finally, in its first decision dealing with discrimination against gay, lesbian and bisexual parents, the Commission has upheld (under Articles 8 and 14) the denial of a Dutch woman's request that parental authority over her female partner's child by donor insemination be vested in both partners (which is possible in the case of an unmarried heterosexual couple). The Commission found no discrimination, noting that 'as regards parental authority over a child, a homosexual couple cannot be equated to a man and a woman living together'.[174]

Why has the Commission given such a narrow interpretation to 'interference' with 'private life', to 'family life', and to ECHR Article 12, or so easily accepted the justifications for restricting Convention rights asserted by governments under Articles 8(2), 10(2), and 14? A thorough analysis is beyond the scope of this chapter. But one explanation is the absence of 'consensus' among the member states of the Council of Europe on the unjustifiability of any aspect of sexual orientation discrimination other than criminal prohibitions of all same-sex sexual activity. It has been convincingly argued that an important factor in the decisions of the Court and Commission is the degree of 'European consensus' that they can discern regarding a particular issue, and that they are more likely to defer to a government where there is little or no consensus.[175] Thus, a violation is far more likely to be found where few other member states have a law similar to the one being challenged than where the majority or a substantial minority of states have such a law.

This might explain the change in the Commission's opinion on blanket prohibitions of sexual activity between men. In December 1966 (just before last upholding of Germany's law), 6 of 14 member states (that had

---

[172] See, on housing, *Simpson* v. *UK* (No. 11716/85) (1986) 47 Dec. & Rep. 274 (arising from [1986] 2 Fam LR 91); on immigration, *Z.B.* v. *UK* (No. 16106/90) (10 Feb. 1990) (arising from [1989] Imm. AR 595), *C. and L.M.* v. *UK* (No. 14753/89) (9 Oct. 1989), *W.J. and D.P.* v. *UK* (No. 12513/86) (11 Sept. 1986), *X and Y* v. *UK* (No. 9369/81) (1983) 32 Dec. & Rep. 220.

[173] See *C. and L.M.*, ibid.

[174] See *Kerkhoven* v. *The Netherlands* (No. 15666/89) (19 May 1992).

[175] See L. Helfer, 'Finding a Consensus on Equality: The Homosexual Age of Consent and the European Convention on Human Rights' (1990) 65 *NYUL Rev.* 1044.

ratified the Convention) had such prohibitions, whereas in March 1980 (violation found in *Dudgeon*), only 3 (Cyprus, Ireland, and parts of the UK) of 20 still had them, following reforms in England and Wales, Germany, Austria, and Norway.[176] The 'European consensus' test would also justify a decision by the Commission and the Court that the discriminatory ages of consent in the UK violate Article 8 or 14.[177] Of the 29 member states in May 1993 (plus the Czech Republic and Slovakia, assuming they will be readmitted), at least 21 appear to have set an equal (or substantially equal) age of consent for heterosexual, gay, and lesbian sexual activity, and the figure may soon rise to 23.[178] The argument will be tested in an application to the Commission against the UK by Ralph Wilde, William Parry and Hugh Greenhalgh, with the support of the Stonewall Group.[179] A survey of the criminal law of other member states might also reveal that other examples of sexual orientation discrimination in UK criminal law deviate from a 'European consensus'.

Outside the area of criminal law, however, there is as yet no 'European consensus' on eliminating most other kinds of sexual orientation discrimination. Norway, France, Sweden, Denmark and the Netherlands prohibit such discrimination in employment or the provision of goods and services,[180] and the Scandinavian countries, the Netherlands and Ireland prohibit incitement to hatred based on sexual orientation.[181] Denmark and Norway have created separate (but almost equal) institutions of 'registered partnership', permitting same-sex couples to acquire most of the rights of married heterosexual couples, while Sweden grants them the same rights as unmarried heterosexual couples, as does the Netherlands in some areas.[182] In the Scandinavian countries and the Netherlands,

---

[176] See V. Berger, *Case Law of the European Court of Human Rights* (1989), Vol. I at 442; Tatchell, *Europe*, 138–9.

[177] See Helfer, 'Finding'.

[178] See Tatchell, *Europe*, 139 (15 states plus Czech Republic and Slovakia); K. Waaldijk, *Tip of an Iceberg: Anti-Lesbian and Anti-Gay Discrimination in Europe* (1991), 21 (San Marino); K. Waaldijk and A. Clapham (eds.), *Homosexuality: A European Community Issue* (1993), 85 (Luxembourg); *Capital Gay* (16 Oct. 1992), 16 (Iceland); *Gay Times* (June 1992), 22 (Estonia). Equal ages have been proposed for Germany (Tatchell, *Europe*, 112) and Ireland (*Capital Gay*, 21 May 1993, 1). The admission of other Eastern European states will affect these numbers.

[179] See *Capital Gay* (9 April 1993), 4.

[180] See Waaldijk, *Tip* 45–59; Waaldijk and Clapham, *Homosexuality* 79–81; Tatchell, *Europe* 132–3; *Gay Times* (April 1993), 7 (Ireland's Unfair Dismissal Bill includes sexual orientation). In the area of military employment, at least 10 member states do not ban gay, lesbian, and bisexual persons from any position. See Tatchell, *Europe* 81–2.

[181] See Waaldijk, ibid. at 81–2; Waaldijk and Clapham, ibid., Sweden, Criminal Code, c. 5, para. 5.

[182] See L. Nielsen, 'Family Rights and the "Registered Partnership" in Denmark' (1990) 4 *Int. J. Law and Fam.* 297; D. Bradley, 'The Development of a Legal Status for Unmarried Cohabitants in Sweden' (1989) 18 *Anglo-Am. LR* 322 at 327 nn. 29–30; Waaldijk and Clapham, ibid. at 91–100 (Netherlands); *Capital Gay* (9 April 1993), 3 (Norway).

same-sex partners of citizens may immigrate.[183] However, the majority of member states have yet to follow their example, and no member state has yet decided expressly to permit same-sex couples to adopt children jointly.[184] It therefore seems unlikely that the Convention will have a significant effect on most of these kinds of discrimination in the UK until legislative reforms in these areas take place in a greater number of member states, or the Court and Commission adopt a stricter, less deferential approach to interpreting the Convention than looking for 'European consensus'.[185] Otherwise, they will continue to order changes in UK law only where a part of the UK can be seen as one of a small group of 'recalcitrant laggards' within the Council of Europe.

As for other Council of Europe institutions, in 1981, the Parliamentary Assembly adopted Recommendation 924, which urged member states to apply the 'same minimum age of consent' to same-sex and heterosexual sexual activity, and ensure that gay, lesbian, and bisexual persons receive equal treatment in employment and child-custody decisions. The Assembly declined to recommend that 'sexual preference' be added to Article 14, but such an amendment may have had little impact, given the deferential test the Court and Commission apply to justifications for discrimination.[186]

European Community law

Although no existing 'hard law' of the European Community expressly mentions sexual orientation or gay, lesbian, and bisexual persons, the existing sex discrimination provisions of EC law could be interpreted as prohibiting sexual orientation discrimination. This is because all sexual orientation discrimination is arguably nothing more than a form of sex discrimination.[187] Thus, the Equal Treatment Directive (which provides that 'there shall be no discrimination whatsoever on grounds of sex' in access to employment or promotions) could be interpreted as requiring member states to amend their laws to prohibit sexual orientation discrimination in employment. The European Court of Justice gave a broad interpretation to sex discrimination in *Dekker* v. *Stichting*,[188] holding that pregnancy discrimination is direct sex discrimination. A similar argument could be made with respect to Article 119 of the Treaty of Rome, the Equal Pay Directive, and the two Social Security Directives,[189] which are particularly relevant to discrimination against same-sex couples in

---

[183] See Waaldijk and Clapham, ibid. at 100–1; Tatchell, *Europe*, 124, 133.

[184] A few joint adoptions have been approved in the USA, often of children who are older or have disabilities. See *Gay Times* (Feb. 1990), 20.

[185] See e.g. *Cossey* v. *UK* (1990) Series A, No. 184 (judgment of 27 Sept. 1990) (dissenting opinion of Judge Martens).

[186] See Helfer, 'Finding', 1086–91.

[187] See above text accompanying nn. 16 to 19.

[188] [1991] IRLR 27.

[189] The directives are 75/117/EEC, 79/7/EEC, and 86/378/EEC.

employment benefits and social-security schemes. Such discrimination causes women who choose partnerships with women to receive lower pay or benefits than men who do so with women, and men who choose partnerships with men to receive lower pay or benefits than women who do so with men.[190] The fact that this kind of sex discrimination may affect only a minority of women and a minority of men should make no difference. A refusal to hire Muslim women, but not men or non-Muslim women, would discriminate on the basis of sex (as well as religion), even though only a minority of women in the UK would be affected.

If current EC law cannot be interpreted as prohibiting sexual orientation discrimination, legislation expressly prohibiting such discrimination in employment or services might be difficult to achieve.[191] Although it has been argued that such legislation would promote freedom of movement within the internal market (because it currently exists only in France, Denmark and the Netherlands), it would probably not be eligible for qualified majority voting under the Treaty of Rome, Article 100A (as it relates to 'the free movement of persons' or 'the rights and interests of employed persons'), Article 49 (which seems to be confined to nationality discrimination), or Article 118A ('health and safety of workers'). It could thus be vetoed by a single member state under Article 100 or 235. In 1984, the European Parliament urged member states to equalize the age of consent and called on the EC Commission to 'submit proposals to ensure that no cases arise in the Member States of discrimination against homosexuals with regard to access to employment and working conditions'.[192] In 1989, it recommended that the 'Community Charter of Fundamental Social Rights' be amended to provide for 'the right of all workers to equal protection regardless of their . . . sexual preference'.[193] The EC Commission has yet to propose any action, but commissioned a recently published report on sexual orientation discrimination in the EC,[194] and included harassment of lesbian women and gay men in the EC 'Code of Practice on Measures to Combat Sexual Harassment'.[195]

PROPOSALS FOR LAW REFORM

The Sexual Offences Act 1967 made Britain a leader of sorts in the area of sexual orientation discrimination. Although the law prohibiting sexual

[190] This is sex discrimination even where benefits are limited to married heterosexual couples, because the criterion of being married to the chosen partner itself discriminates on the basis of sex.

[191] An EC attempt to harmonize criminal or family law is even more unlikely.

[192] (1984) 27 O.J., No. C104, 46–8.

[193] (1989) 32 O.J., No. C323, 46. See also (1986) 29 O.J., No. C176, 75 (para. 12); Tatchell, *Europe*, 16–23; Jeffery-Poulter, *Peers, Queers*, 249–50.

[194] See Waaldijk and Clapham, *Homosexuality*.

[195] (1992) 35 O.J., No. C27, 1, 4–11.

activity between men was amended after the majority of European countries had already done so (and long after such countries as France, which did so in 1791), the reform in England and Wales took place at a time when similar laws remained in place in Norway, Finland, Germany, Austria, Israel, Canada, New Zealand, and every state of Australia and the USA except Illinois. Yet, despite this initial display of leadership, there have been no improvements in English law since 1967 and a major regression in 1988. In having no anti-discrimination legislation, Britain now lags behind not only Norway, France, Sweden, Denmark, the Netherlands and (probably before the end of 1993) Ireland, but also Israel and a growing number of jurisdictions in Canada (Quebec, Ontario, the Yukon, Manitoba, Nova Scotia, New Brunswick and British Columbia), the USA (the District of Columbia, Wisconsin, Massachusetts, Hawaii, Connecticut, New Jersey, Vermont, California and Minnesota and dozens of cities and counties),[196] and Australia (New South Wales, South Australia, Queensland and the Australian Capital Territory).

Elimination of sexual orientation discrimination has tended to proceed in stages in most countries, starting with (a) the criminal law and then moving to (b) discrimination against individuals before addressing (c) discrimination against couples. In most countries, the 'final frontier' has tended to be (d) discrimination against parents. This is because fears of 'contamination' tend to make any issue of sexual orientation discrimination involving contact with children highly sensitive, especially as the duration of the contact increases.[197] Prohibiting sexual orientation discrimination in decisions about the custody, adoption, or fostering of children (as opposed to affirmatively requiring such discrimination, or leaving the decision whether or not to discriminate to the discretion of judges or local authorities) is thus perhaps the most controversial issue. It would seem that Britain is still at stage (a), having yet to eliminate discrimination in the criminal law, and has not begun to address stages (b), (c), and (d). The attempts by some local authorities to introduce balanced discussion of sexual orientation into schools (perhaps a child-related stage (d) issue) may have helped trigger the Section 28 backlash in a jurisdiction that has yet to accept the stage (b) right of individual gay, lesbian, and bisexual adults to be free from discrimination in such fields as employment.

What reforms of English law would be required to eliminate the sexual orientation discrimination described above, and what reforms are currently

---

[196] An ominous development is the 1992 adoption by referendum of Section 30 of the Colorado Constitution, which prohibits state or local legislation protecting gay, lesbian and bisexual persons against discrimination, and could be adopted by other states. On the situation in the USA, see generally 'Developments in the Law: Sexual Orientation and the Law' (1989) 102 *Harv. L. Rev.* 1508.

[197] See e.g. *B.* v. *B.* [1991] 1 F.L.R. 402, 405, 411 (son permitted to remain with lesbian mother because of 'her desire to see that he will be brought up on a heterosexual basis').

being proposed? The extremely varied nature of the discrimination means that a broad package of measures would be necessary, dealing with criminal law, anti-discrimination law, and family law. The criminal law measures would have to remove the 1956 Act's presumption of the illegality of sexual activity between men and the 1967 Act's limit of two participants, decriminalize such activity in the armed forces and merchant navy, equalize the age of consent at 16,[198] repeal the offences of 'procuring' and 'soliciting', treat semi-public sexual activity between men like similar activity between men and women, and establish a principle that no 'neutral' offence (including public-order or 'obscene or indecent' publications legislation) shall be applied to any instance or depiction of same-sex emotional-sexual conduct where it would not apply to similar heterosexual emotional-sexual conduct.

The anti-discrimination measures would have to create an offence of incitement to hatred on the ground of sexual orientation, prohibit sexual orientation discrimination in the areas covered by the Sex Discrimination Act 1975 and the Race Relations Act 1976, and repeal Section 28. Finally, the family law measures would have to grant same-sex couples the right to a civil (as opposed to religious) marriage, provide that all rights of married heterosexual couples apply to married same-sex couples and all rights of unmarried heterosexual couples apply to unmarried same-sex couples,[199] and prohibit discrimination on the ground of sexual orientation in all decisions regarding adoption, fostering, and custody of children, or in access to donor insemination, surrogate motherhood, or other reproductive technology. These legislative reforms could always be repealed (unless they were in compliance with a European Community directive), and thus ideally should be supplemented by an express constitutional prohibition of sexual orientation discrimination, either in an entrenched domestic Bill of Rights or an amended European Convention on Human Rights. Such a constitutional prohibition could also be interpreted as requiring reforms where legislation or a public-sector policy is the source of the discrimination, such as in government services or policies (including immigration) that fall outside the scope of current anti-discrimination legislation.[200]

What kinds of reforms are currently being proposed? In 1990, the Stonewall Group drafted a Homosexual Equality Bill which addresses most areas of discrimination in the criminal law, includes incitement to

---

[198] See *Report on the Age of Consent in Relation to Sexual Offences*, Cmnd. 8216 (1981), 17, 27 (8 votes for 18, 5 for 16, 2 for 16 except where older man in position of authority).

[199] The separate question of whether a new option of 'registered partnership' should be created for all couples (heterosexual and same-sex) who reject the institution of marriage is not one of sexual orientation discrimination, unlike the question of access to marriage.

[200] See *In re Amin* [1983] 2 A.C. 818, 835.

hatred and anti-discrimination provisions, repeals Section 28, and creates an institution of 'domestic partnership' for same-sex couples who wish to acquire many of the rights of married heterosexual couples (excluding the right to adopt children jointly). One difficulty that arises in drafting such a bill is the absence of a comprehensive statute prohibiting discrimination on such grounds as sex, race, religion, disability, and age, and creating a single enforcement agency. The existence of such statutes in Canada and the USA makes it simple to insert sexual orientation as an additional ground. In Britain, the separate, specialized acts for race and sex discrimination leave the drafter with the options of preparing a separate, equally detailed act for sexual orientation discrimination, or amending one of the existing acts. In some respects, the Race Relations Act 1976 is more suited to amendment, because it contains few exceptions and protects minority groups exposed to prejudice and hostility. However, the Sex Discrimination Act 1975 is conceptually better suited to amendment,[201] provided that many inappropriate exceptions are expressly made inapplicable to cases of sexual orientation discrimination. As for constitutional protection, two recently proposed domestic Bills of Rights,[202] and a proposed protocol to the European Convention on Human Rights[203] would prohibit discrimination on the ground of sexual orientation.

What are the prospects for reform? A February 1992 Harris opinion poll for Stonewall suggest that a substantial majority of people in Britain would support most of these reforms: 71 per cent agreed with equal rights under the law for gay men and lesbian women; 74 per cent with an equal age of consent;[204] 86 per cent with equal treatment of semi-public sexual activity (gay or heterosexual); 72 per cent with monitoring of anti-gay or anti-lesbian violence; 71 per cent with the right of gay men and lesbian women 'to be open about their sexuality without fear of harassment from the police or discrimination by employers or society in general'; 57 per cent with their right to serve in the armed forces;[205] and 70 per cent with

---

[201] See Equal Opportunities Commission for Northern Ireland, *The Sex Discrimination (Northern Ireland) Order 1976: Proposals for Amendment (Part 1)* (1983), 13 (recommendation that the Order 'be amended to outlaw discrimination on the grounds of sexual orientation'). Section 3 of Jo Richardson's unsuccessful Sex Equality Bill would have prohibited discrimination 'on the ground of homosexuality'. H.C. Deb., Vol. 50, 586 (9 Dec. 1983).

[202] See Institute for Public Policy Research proposal (Article 17(3)) ('homosexuality'); Liberty proposal (Articles 14, 19, 20). The German state of Brandenburg's 1992 Constitution expressly prohibits discrimination based on '*sexuelle Identität*' (Article 12(2)).

[203] See L. Helfer, 'Lesbian and Gay Rights as Human Rights: Strategies for a United Europe' (1991) 32 *Va. J. Int. Law* 157.

[204] Cf. *Capital Gay* (18 Oct. 1991), 5 (in Gallup poll, 74% disapproved of an equal age of 16).

[205] If the USA drops its ban on gay, lesbian and bisexual members of the armed forces in 1993, as Canada and Australia did in 1992, it will increase the pressure on the UK to do the same.

equal treatment of same-sex and unmarried heterosexual couples (where one partner dies). But, as might be expected, only 17 per cent agreed (72 per cent disagreed) that gay men and lesbian women should be allowed to foster or adopt children.

CONCLUSION

This chapter has attempted to show the extent of sexual orientation discrimination contained in, or permitted by, English law, and in so doing, to measure both the changes that have taken place since 1950 and the vast amount that remains to be done. For although life is certainly much easier for gay, lesbian, and bisexual persons in Britain in 1993 than in 1950, there is still a very long way to go before they can be said to enjoy the same rights as heterosexual persons. Indeed, of all the rights discussed in this book, the right to be free from sexual orientation discrimination may be the least well established. Will that still be the case in 2050? Or will it then be as difficult to imagine that same-sex couples were once forbidden to marry or adopt children jointly as it is now to imagine that people from Africa could be sold as slaves or that women, Roman Catholics, or Jews could not vote? Ultimately, Britain will have to decide whether or not it is willing to respect the right of every individual, regardless of their sexual orientation, to love the person of their choice.

16

Individual Rights and British Law:
Some Conclusions

CHRISTOPHER McCRUDDEN and
GERALD CHAMBERS

INTRODUCTION

We have chosen in this book to concentrate almost entirely on an intensive study of the legal protection of traditional civil and political rights in Britain, rather than a more extensive analysis of all issues which might be considered as human rights questions (such as the adequacy of welfare provision, health care, employment levels, or housing). We have also sought to be sensitive to changes over time by taking a forty-year period of study for each topic. In this concluding chapter we examine the mechanisms and institutions which underpin the legislative developments and judicial decisions outlined in the previous chapters. What are these mechanisms? How have they been used in the recent past? How well have they responded to historical change? What new mechanisms have been adopted or developed to meet new needs? What is the relative weight to be attached to domestic as opposed to European and international institutions and mechanisms? In short, how well have these mechanisms performed in the past forty years?

This essentially qualitative assessment is limited by the research available. Our initial task, therefore, will be to raise questions about empirical and comparative research in the area of human rights. On what basis and against which criteria of measurement is it possible to assess whether rights, liberties, and freedoms have increased over time in Britain? Do some British institutions defend and protect rights and liberties better than others and better than institutions elsewhere? How do our rights and liberties compare with those enjoyed by the people of other countries? What kind of available data would enable us to find answers to these questions? Having made this assessment, we then attempt a preliminary audit of the central legal and political institutions which have featured in the previous chapters and their contributions, or otherwise, to the advancement of human rights within Britain over the past forty years.

We conclude with a brief consideration of the underpinnings of civil liberties and human rights in Britain. How important is a rights consciousness among the general public? What part have political élites played in the significant developments that have been charted in earlier chapters? To what extent are the political and legal institutions discussed necessary or sufficient? How, if at all, might we think about alternative arrangements and reforms?

MEASURING THE EXTENT OF HUMAN RIGHTS PROTECTION

Attempts to theorize about the adequacy of human rights standards or about the reasons for such levels are limited, if we look only at the extent and methods of *legal* protection, as reflected in statutes or appellate court cases. We would improve our understanding considerably if we had consistently and accurately collected empirical information over time relating to human rights issues in Britain. We would also improve our understanding of these basic issues if our national study of each right or freedom could be compared with developments in other countries.

For these reasons, previous chapters have sought to indicate the extent to which such empirical information has been collected on a national basis and, in addition, the extent to which comparisons and developments in other countries have been seen as particularly relevant, and with what effect. We are now able to step back and assess this information in a more systematic way. Before considering the question of international *comparison*, we review first some of the *sources* of information which are available on the domestic front. Having examined developments in the law and in our legal institutions, what sources of data might we want to look at to find out if Britain had a better record on, say, freedom of expression in 1993 compared with 1950?

One commentator has recently noted that,

... the data resources for monitoring progress in the area of economic and social rights are in a much more advanced state of development. . . . Data [on civil and political rights] are fragmentary, incomplete and lacking comparability. There is little agreement on concepts and definitions. Whereas data on economic and social rights, more often that not, reflect achievements and progress, data on civil, political and personal security rights are more likely to emphasize violations and failures to meet international standards.[1]

It may also be the case that some of the topics covered in this volume, such as freedom of association, privacy, and freedom of expression lend themselves less easily to empirical evaluation than other topics, such as women's rights, racial and sexual orientation discrimination, and minority rights. In their respective chapters, Michael notes the conceptual dif-

---

[1] R. P. Claude and T. B. Jabine, 'Statistical Issues in the Field of Human Rights', editors' introduction (1986) 8 *Human Rights Quarterly* 563.

ficulties of defining privacy and Ewing argues that freedom of association does not exist as a specific right in English law. Thus some rights are more fragile than others and in this context the pressing issue may first be to define in what sense a right or freedom exists in law, rather than assessing the extent to which it is upheld in practice. Some rights may be emergent (children's rights or prisoners' rights, for example), whereas others (the rights of ethnic minorites not to be discriminated against in employment applications, for example) may be more clearly defined in statute. A further complicating factor is that data on violations of rights which are perpetrated by the state or its organs (the security services, for example) may be more difficult to come by than data on violations by corporate bodies or individuals. Nevertheless it is possible to look to some sources of data on social indicators of the extent to which civil and political rights are protected.

Sources

We can isolate four main types of sources which have been, or can be, drawn on. Where authors have considered it relevant, the previous chapters in the book have referred to one or more of these areas. The first source of empirical data on Britain is the important groundwork done by various international organizations through the reporting procedures under particular treaties.

Comparison over time is only possible if the information against which comparison can be made is collected and made available. In this context the work done by the Human Rights Committee in carrying out its functions under Article 40 of the ICCPR is relevant. While the Human Rights Committee produces no league tables of international performance, the five-yearly reporting procedure ensures that signatories of the Convention address human rights issues on a regular basis and lay themselves open to questioning on the implementation of the various articles of the Convention. The three reports of the British government have been discussed in Chapter 1. A similar reporting and question-and-answer procedure exists for signatories of the Convention on the Elimination of All Forms of Racial Discrimination. In addition, several of the European Community directives on equality between men and women require periodic reports on implementation. Our conclusion on the usefulness of this reporting procedure is that the Human Rights Committee has had little impact on the development of a domestic legal framework for the protection of human rights and civil liberties (none of the authors have referred to the Human Rights Committee as an important actor in domestic developments), but that the reporting procedure itself improves governmental accountability and has yielded relevant empirical information.

A second source of data against which progress on civil liberties and human rights can be measured is that available from official sources in Britain. It is of two types: statistics routinely collected and published by government departments, by quangos (such as the Equal Opportunities Commission, the Commission for Racial Equality, and the Broadcasting Complaints Commission), or by the Central Statistical Office; secondly, data gathered in the course of official fact-finding missions, parliamentary select committee enquiries, Royal Commissions, and official inquiries. (Enquiries relevant to the topics discussed in this volume are collected together in the second part of Appendix C.[2]) Good statistical data are not available in all areas: Poulter, for example, notes that the 1991 Census was the first to ask about ethnic origin, but it will not provide information on religious affiliation (except in Northern Ireland) or on mother tongue. However, even if such information were available, Poulter considers that assessing whether minority rights have been enhanced or undermined is intrinsically hazardous in view of the difficulty of evaluating the strength of commitment to the maintenance of religious, cultural, and linguistic traditions and the tendency towards voluntary assimilation. Dummett points out that official statistical information on immigration has to be derived from different sources and that these data are not readily comparable.

Thirdly, there is data collected and analysed by academics and independent researchers. Reiner and Leigh, McCrudden, Jackson, and Poulter have all separately indicated relevant academic or other independent research highlighting the erosion of civil liberties or demonstrating a need for further legislation to protect civil liberties and rights. For example, Reiner and Leigh indicate that research based on observations of police work has been important in shedding light on police discretion and on the difference between the 'law in the books' and 'the law in action'. McCrudden has argued that empirical research demonstrated authoritatively the extent of racial discrimination in activities not covered by the 1965 Race Relations Act. The campaign in 1967 for a new Race Relations Act drew support from research by Political and Economic Planning (PEP), the forerunner of the Policy Studies Institute.[3] Jackson cites numerous studies which have resulted in an increasing flow of empirical evidence on the day-to-day workings of many aspects of the criminal process. Such studies helped assessments to be made of the extent to which there was compliance with the rules. Poulter refers to research on the inadequacy of arrangements for foreign language interpretation in criminal proceedings to safeguard the interests of non-English-speakers.

Why has empirical research informed the debate in some areas but not

---

[2] See below, p. 610.
[3] Political and Economic Planning, *Racial Discrimination* (1967).

in others? It is widely and officially recognized that discrimination against women and ethnic and racial minorites exists in modern-day Britain. Governments of all persuasions have accepted a responsibility to reduce discrimination in certain areas, such as employment and housing, and government funding has therefore been directed to exploring the causes of discrimination and towards examining the policies and practices of employers. There has been less research on immigration and nationality policy, on the security services, and on prisons, where the Government is itself the prime agent. However, there exists a long tradition of Home-Office-sponsored research on criminal justice issues, including research on police interrogation, on black people in the criminal justice system, and on victim surveys. A large amount of criminological research carried out in Britain since 1950 has had official sponsorship and has taken place within or been directly funded by the Home Office.

A fourth source of data is that produced by non-governmental organizations and other campaigning groups. Such groups may of course fund their own research enquiries or make use of official government statistics but their resources are generally limited. A common strategy is therefore to focus on particular abuses of power or on individual cases and thereby highlight wider deficiencies and inadequacies in mechanisms for the protection of rights. The most prominent of such groups in the British context has been Liberty, formerly the National Council of Civil Liberties (NCCL), which was founded in 1934. Liberty describes itself as having a watchdog role, that is, monitoring new laws and reviewing official decisions. It also publishes guides to civil rights, takes test cases through British and European courts, publishes its own 'unofficial' reports,[4] as well as contributing to official inquiries on major issues, and engaging in joint campaigning with other groups. It has been instrumental in setting up other British civil liberties pressure groups.[5] A digest of significant events, case-law, and new legislation is published regularly in *Agenda*, the NCCL newsletter.

In addition to the work of Liberty, Amnesty International regularly reports on human rights in Northern Ireland,[6] and in the United Kingdom as a whole;[7] Helsinki Watch has published data on the treatment of young people by the security services in Northern Ireland;[8] and Article 19 has examined restrictions on media reporting on Northern Ireland affairs.[9]

---

[4] See e.g. ch. 2.

[5] See Liberty/NCCL, *Standing Up for Your Rights!* (1991).

[6] See e.g. *Amnesty International Annual Report* (1992), *Allegations of Human Rights Abuses in Northern Ireland* (Nov. 1991), and *Northern Ireland: Killings by the Security Forces and 'Supergrass' Trials* (1984).

[7] Amnesty International, *United Kingdom: Human Rights Concerns* (June 1991).

[8] Human Rights Watch, *Human Rights in Northern Ireland: A Helsinki Watch Report* (1991).

[9] Article 19, *No Comment: Censorship, Secrecy and the Irish Troubles* (1989).

Human rights issues are of course reported on and pursued by organizations such as trade unions which, as Collins and Meehan indicate, have been instrumental in getting recognition for sexual harassment as sex discrimination[10] and, as Ewing indicates, in defending a right to freedom of association. Dummett refers to investigations carried out by the Runnymede Trust and by UKIAS which have questioned the validity of immigration officers' decisions to refuse admission to wives and children of immigrants. Wintemute has drawn on material from pressure groups such as LAGER and GALOP which have been active in documenting human rights abuses against lesbians and gay men.

Comparative studies

All these sources of data are essentially generated within the UK and their analysis permits some assessment to be made of human rights conditions in Britain itself. In addition to this body of domestic work it is now becoming possible to compare the record of Britain with that of other countries. Writing in 1988 Paul Sieghart commented:

Judged by any objective measure, human rights and fundamental freedoms are better respected in the UK today than they are in the great majority of the world's other countries. Quite where at the top of the league table she stands may be debatable but it must certainly be somewhere in the top ten, if not in the top five.[11]

Comparison between countries with differing traditions, constitutions, and legal machinery can be complex and fraught with difficulty and the objective measures assumed by Sieghart may be elusive. When comparing the record of European countries commentators now have the benefit of data from the Council of Europe on cases before the European Commission of Human Rights (Table 1 in Appendix D), and judgments of the European Court of Human Rights (Table 2 in Appendix D) and there are also judgments of the European Court of Justice to which reference can be made. These legal decisions and their implications have been extensively cited in previous chapters. Table 2 in Appendix D compares the record of the UK with that of other European countries on cases heard by the European Court of Human Rights. The UK, it would seem, has a worse record of violation than other countries. However, such league tables require further interpretation and explanation. It has to be noted, for example, that the UK granted individuals the right of petition earlier than many other countries. Furthermore, the very fact that the Convention is not incorporated into UK law means that, unlike the situation in many

---

[10] See also the work of groups like LEVEL referred to in ch. 12.
[11] Paul Sieghart, *Human Rights in the UK* (1988), 4.

other European countries, individuals sometimes have no option other than recourse to the Convention institutions.

In addition to official data from European institutions, it is also necessary to mention at least two types of international comparative work which provide empirical data on the situation in Britain. First, there are studies which seek to rate the human rights record of various countries, including Britain, and thereby allow comparison of the record of the UK *vis-à-vis* other countries. The most exhaustive of these types of studies are carried out by the United States State Department,[12] by Freedom House,[13] and by Charles Humana.[14] In 1991, the United Nations issued a Human Freedom Index which rated countries on the basis of the level of freedom enjoyed by their citizens, but this was discontinued only one year later.[15] From the mid-1970s, the United States State Department was required by the United States Congress to prepare a report each year on the status of human rights across a range of countries. The entries 'relate specific incidents [and] quote conclusions of non-governmental monitoring groups'. As McNitt argues, the State Department reports are, however, 'a frustrating source of information, too good to be ignored and yet flawed enough to require caution when using them'.[16]

A guide by Charles Humana has periodically assessed the human rights performance of 120 countries. The method of assessment for most of those countries is based 'on a questionnaire concerning 40 human rights, all of which are drawn from the articles of the major United Nations treaties'.[17] Assessments are made by the editor himself and are derived, it seems, on the basis of questionnaires completed by informants in the various countries, from data gathered from human rights organizations, and from secondary statistical sources. Humana acknowledges that 'the great majority of embassies either declined to give information or offered descriptions of paradise'.[18] In the 1986 edition of the guide the United Kingdom achieved a rating of 94 per cent (compared to a world average of 55 per cent). Of forty human rights examined the UK achieved an unqualified 'yes' on thirty-two and a 'qualified yes' on eight (torture or coercion by the state, instances of telephone tapping, political and legal equality for women, social and economic equality for women, social and economic equality for minorities, legal rights to be brought promptly

---

[12] United States Dept. of State, *Country Reports on Human Rights Practices* (annual).

[13] R. D. Gastil (ed.), *Freedom in the World: Political Rights and Civil Liberties 1989–90* (1990).

[14] C. Humana (ed.), *World Human Rights Guide* (1986, 1992).

[15] *The Independent*, 24 Apr. 1992.

[16] A. D. McNitt, 'Some Thoughts on the Systematic Measurement of the Abuse of Human Rights', in D. L. Cingranelli, *Human Rights: Theory and Measurement* (1988), 97.

[17] Ibid., 1986, n. 14, p. 1.

[18] Ibid., 1986, p. vi.

before a judge or court, and equality of sexes during marriage and for divorce proceedings). In the 1992 edition the UK achieved an unqualified 'yes' on thirty-three rights examined and a qualified 'yes' on seven. Its overall rating has, however, slipped marginally from 94% to 93%.

Freedom House, an American foundation, has for several years prepared a series of country reports (the 'Survey of Freedom'), assessing states as comparatively free or unfree. Commenting on the Freedom House reports, Bollen states that 'measurement efforts rely mostly on U.S. and European information supplemented by local information where possible' and that 'there is ambiguity in the exact criteria or check lists that are employed to move from raw information to the measures'.[19] Bollen's general view of ratings studies such as these is that 'it is highly likely that every set of indicators formed by a single author or organisation contains systematic measurement error. . . . Selectivity of information and various traits of the judges fuse into a distinct form of bias that is likely to characterise all indicators from a common publication.'[20]

Related to but distinct from these scaled measurement assessments are the more qualitative studies carried out by groups such as Amnesty International and Article 19. Amnesty produces an annual review of human rights concerns for most countries in the world, concentrating on the issues of torture, capital punishment, due process, and imprisonment of prisoners of conscience. Amnesty 'does not grade countries according to their record on human rights; instead of attempting comparisons it concentrates on trying to end the specific violations of human rights in each case'.[21] Another international group, Article 19, produces comparative reports on free speech and freedom of expression issues world-wide. A recent report focused on freedom of expression in fifty countries 'representative of all regions and of different political and ideological systems'.[22] No attempt was made to rank or classify a country on its record.

Finally, there have been scholarly studies which have been comparative but adopted an issue-based approach. They have concentrated on a specific aspect of human rights such as sex or race discrimination, adopted an explanatory framework, and recognized that abuse of human rights and civil liberties are aspects of wider political processes and strategies. It is suggested that such studies provide a valuable way forward for comparative empirical work. Flanz, for example, compares women's rights

---

[19] Kenneth A Bollen, 'Political Rights and Political Liberties in Nations: An Evaluation of Human Rights Measures, 1950 to 1984' (1986) 8. *Human Rights Quarterly* 585. See article generally for a review of data sources and indicators of political freedom political rights and democracy.

[20] Ibid. 586.

[21] *Amnesty International Report* (1992), facing frontispiece.

[22] K. Boyle (ed.), *Article 19, World Report 1988: Information, Freedom and Censorship* (1988).

in four geo-political bands within Europe: the Scandinavian countries, Western European countries, Southern and Mediterranean countries, and East European countries, concluding that 'The Scandinavian countries continue to lead the Western and Southern European countries in term of legislation and political participation.'[23] A novel approach combining qualitative and explanatory work with scaling has been adopted by Forbes and Mead, who have examined measures taken by EC countries to combat racial discrimination and found three distinct groupings of countries. A scoring system based on domestic and international legal provisions, access to remedies, and implementation measures has been developed. Britain achieves the highest score among all EC countries. Such scorings systems need replication to test and improve their reliability and there is inevitably a subjective element since not all of the measures considered are evaluated similarly. Thus the authors decide to allocate a maximum possible 3 points if the country has a government-funded agency but only a maximum of 1 point for an ombudsman.[24]

Comparative assessment has found a place in the development of legal judgments and social policy. Examples of other country's experiences being drawn on *ad hoc* in Britain have been discussed in previous chapters. Several purposes of international comparison have become apparent from the previous chapters.

1. Other country experience is used as an active tool for campaigning organizations, to illustrate possibilities, and to persuade policy-makers. American influence seems to have been particularly drawn on by British pressure groups in certain areas. In the preparation of the 1968 Race Relations Act American experience was drawn on by several groups. For example, Americans who were knowledgeable about and involved in anti-discrimination work in the United States were consulted and occasionally brought to Britain by the Runnymede Trust, a campaigning organization on race relations, to give their assessments of the American experience with a view to suggesting which elements in it might be of relevance to the United Kingdom.[25]

2. Other countries' experience is used as a source of concepts and institutional mechanisms which are viewed as ripe for transplantation. In the area of race relations, women's rights, and privacy, for example, we have seen the importance of transplantation from the United States, especially the use of a Commission enforcement model.

3. Comparison with other European countries has been drawn on as a

---

[23] Gisbert H. Flanz, *Comparative Women's Rights and Political Participation in Europe* (1983), 321.

[24] I. Forbes and G. Mead, *Measure for Measure* (1992).

[25] See e.g. L. H. Pollock, *Discrimination in Employment: The American Response* (1974).

basis for assessing violations by European institutions. Wintemute has stressed that the approach taken by the European Commission and Court of Human Rights relies heavily on finding a 'consensus' among European countries before these bodies are willing to expand the interpretation of the rights found in the Convention.

4. As a basis for harmonization in the European Community, heavy use has been made of comparative material by the EC Commission in drawing up directives on women's rights.

5. International comparison has been used as a basis for assessment of the moral quality (or the 'human rights record') of the country concerned. However, the empirical basis of comparisons between countries often remains unclear, being based on subjective assertion rather than objective evaluation.

Having described the sources of data on which empirical investigation might be conducted, and its varied use it is immediately necessary to draw attention to two major problems with these sources. The first is the patchiness of the available material. There is a significant dearth of reliable empirical and comparative material in certain areas spanning an appropriate amount of time. It is clear, for example, that whereas there is extensive empirical data relating to policing, race relations, and prisons in Britain, equivalent information is significantly lacking on, for example, the operation of the security services and of immigration and customs and excise officers, on employment and other forms of discrimination and abuse against gay men and lesbians, on the extent of child abuse and maltreatment. So too, while there is extensive comparative assessment in the area of women's rights and on rights of association, it is noticeably absent on criminal justice, policing, and due-process issues. The apparent randomness of the availability is misleading, of course, and is related significantly to the availability of research funding. It may also be due to a lack of interest in comparative work in certain areas among policy makers and social scientists. In the due-process area, for example, Jackson has observed that there has been little interest in the criminal justice system of other EC community countries and often a great degree of scepticism about the fairness of such systems. This lack of interest has even extended to systems within the United Kingdom such as Scotland. Whatever the reasons, the absence of such material is a frustrating limitation.

The second problem in using the material lies in the biases of the reporting groups. Often the problems of bias are relatively clear and can be taken into account by the reader. The reporting mechanisms of the international human rights conventions depend heavily on the biases of the governments which supply the information. Freedom House 'has been severely criticised for its conservative, free market, pro-United States biases'.[26] In other circumstances the biases are less clear. Academic researchers may have certain political agendas which they wish to further, but which may remain undisclosed.

AN ASSESSMENT OF THE ROLE OF SOME MAJOR DOMESTIC INSTITUTIONS IN THE ADVANCEMENT OF HUMAN RIGHTS IN BRITAIN

Taking these considerations into account, we turn to our second task: an examination of the major institutions and mechanisms of the British state and their role in the advancement of human rights in Britain. Several traditional methods have been identified as characteristic of the British approach: legislative measures to deal with specific problems, constitutional conventions, a sense of responsibility and fair dealing in legislators and administrators, the influence of a free press and the force of public opinion, the independence of the judiciary in upholding the rule of law, and free and secret elections.[27] Previous chapters provide some useful material for attempting an assessment of these traditional methods, as well as some evidence of new mechanisms. However, the limitations of the material mean that little more than a preliminary assessment can be made at this time. We look, in the following pages, at the role of Parliament, government, several independent political actors, and the judicial process, before turning to consider European and international requirements.

Parliament

Although replete with ambiguities and uncertainties, the doctrine of parliamentary sovereignty provides at its core that a fundamental feature of current constitutional arrangements (at least in England and Wales) is that the Queen in Parliament can legally do what that institution wants. Within the institution of the 'Queen in Parliament', the House of Commons is equally clearly the lead chamber because of its greater legitimacy as the only democratically elected element. The Government must have the support of the House of Commons in order to remain in office and the vast majority of the members of the government sit as members of Parliament in the Commons. Also common to much British constitutional and political thought over the past forty years is the argument that it is in practice the Cabinet which runs Parliament, not the other way round,

[26] McNitt, 'Some Thoughts', 97.
[27] SACHR, 'Bill of Rights: A Discussion Paper', Mar. 1976, para. 11, repr. in SACHR, *The Protection of Human Rights by Law in Northern Ireland*, Cmnd. 7009.

and that the system is crucially dependent on tight party control of MPs. Also, the current electoral system ('first past the post') appears to have helped, for most of the period under discussion, to produce a clear working majority for either the Labour Party or the Conservative Party. Such an arrangement, it is said, gives rise to strong, effective, and stable government.

This set of constitutional arrangements, it could justifiably be argued, has enabled Parliament to enact some important advances in human rights protection discussed in previous chapters. Unlike some other countries where constitutional arrangements do not encourage strong central government, and where human rights protection becomes the responsibility of other branches of government, or none, Britain has suffered from no such power vacuum. Unlike other countries, too, the doctrine of parliamentary sovereignty can also have the effect of protecting human rights legislation from attack.[28] Parliamentary sovereignty permitted the imposition of mechanisms of human rights protection in Northern Ireland, over local objections. Parliamentary threats to impose legislation led to changes in the Isle of Man regarding homosexuality, following European intervention.

However, on the debit side, parliamentary sovereignty also allowed the Prevention of Terrorism Act 1974, passed through all stages in both Houses in forty-two hours, to escape effective scrutiny. A Labour government rushed the second Commonwealth Immigrants Act through all its parliamentary stages in less than a week. So too, although strong clear government may emerge under British constitutional arrangements, there is no guarantee that this will always happen in particular areas when it is needed. We have seen examples of the unwillingness or the inability of Parliament to enact defined powers for those in authority. The powers given the police to deal with 'breaches of the peace' have been left undefined by Parliament, despite the problems for the ordinary citizen this occasions, permitting the police, as Reiner and Leigh have argued, to use vague powers virtually uncontrolled by the rule of law.

Some civil liberties issues, then, are left off the Government's legislative agenda for reasons of convenience. Others are treated as issues of individual conscience for MPs and are non-party political, for example the issues of capital punishment and abortion have traditionally been regarded as subject to a 'free' vote in the House of Commons, with no party discipline being imposed.

---

[28] See e.g. *R. v. Jordan* (1967) *Crim. LR* 483 (application for habeas corpus on ground that s. 6 of the Race Relations Act 1965 was invalid as a curtailment of free speech; held unarguable due to the supremacy of Parliament).

Members of Parliament

Although highly constrained in the influence they can wield, individual MPs can clearly make a difference on specific civil liberties and human rights issues, perhaps particularly where an individual's grievance is in issue. Previous chapters have shown that on privacy issues, for example, there have been several significant private members' interventions which have stimulated the appointment of official committees of inquiry. The reforms of the law on sexual offences and on abortion in 1967 were introduced as private members' bills, though the government provided time and support. Back-bench pressure was very important too on the development of women's rights in the 1960s and 1970s. More recently, the influence of parliamentary committees has on occasion been significant. There have been influential reform proposals from the House of Commons Select Committee on Home Affairs in the area of public order, and in the context of prisons, Richardson notes several reports by select committees of the House of Commons. However, it is also true to say that the role of the House of Commons as the protector of individual rights is perceived to be neither strong nor central.[29] It should be remembered, for example, that there is little scope for private members' legislation as such Bills, to be successful, generally require at least tacit support from government business managers in the House of Commons.

House of Lords

The House of Lords has, on occasion, second-guessed the House of Commons and the Cabinet on certain civil liberties and human rights issues and thus acted as a constraint on an overzealous elected House of Commons. Brazier, indeed, considers that the House of Lords has become 'the only counter-weight in the British constitution to elective dictatorship'.[30] Ewing and Gearty consider that '[p]aradoxically . . . it is the unelected House of Lords which has been the major institutional bulwark to executive power'.[31] Though, ultimately, it is an ineffective constraint when the Government or the House of Commons wants its own way, previous chapters have given several examples where intervention by the House of Lords has been influential.

Reiner and Leigh have pointed out, for example, that a move towards a general exclusionary rule was incorporated into PACE as a result of an amendment introduced by Lord Scarman at the report stage of the bill, leading to a somewhat modified version being subsequently introduced by

[29] R. Brazier, *Constitutional Reform* (1991), 128.
[30] Ibid. 71.
[31] K. D. Ewing and C. A. Gearty, *Freedom under Thatcher: Civil Liberties in Modern Britain* (1990), 256.

the Government. Jackson has indicated that the House of Lords was instrumental in protecting access to the jury system in criminal trials in 1977. Thirdly, there is the example of the religious education debate sparked off when Baroness Cox and others moved amendments in the House of Lords to provide that such education should be 'predominantly Christian'. The issue was ultimately resolved with a somewhat more tolerant outcome, when the Bishop of London conducted an extensive round of consultations before returning to the House of Lords with a package of amendments which (together with some government amendments) formed the more balanced compromise in the Education Reform Act 1988. Lastly, Richardson has described how, in 1988, following disquiet expressed in the House of Lords concerning the structure of release from life sentences, a House of Lords select committee was set up to consider the crime of murder, the penalty it should attract, and the arrangements for determining release. Having taken evidence as to the requirements of the European Convention on Human Rights with regard to release procedures, the Committee recommended the removal of executive discretion and the creation of an independent tribunal to make all decisions concerning the release and recall of life-sentence prisoners. In 1990, during consideration of the Criminal Justice Bill, there was a revolt in the House of Lords over the retention of the mandatory life sentence for murder and the Government finally introduced provisions to enable the parole board to order the release of discretionary life-sentence prisoners. Brazier is, however, right to conclude that although the House of Lords has proved 'a significant inconvenience for Labour Governments and an irritant for Conservative governments, . . . it lacks both legal power and political authority to do much more'.[32]

Elections

One significant mechanism for the translation of citizens' values into public policy is the requirement of regular, open, and free elections of members of Parliament to the House of Commons. Indeed Griffith and Ryle regard the requirement in the Bill of Rights for the holding of free and frequent elections as being 'at the heart of the "unwritten" constitution'.[33] The need to face the electorate at least every five years is regarded, by some, as a guarantee of sorts that the House of Commons will not abuse its powers. And a regular change of government may lessen the possibility of corruption and contribute to a greater degree of tolerance between the parties. However, this depends significantly on the extent to which concern about tolerance and corruption is central to the electoral process.

[32] Brazier, *Constitutional Reform*, 10.
[33] J. A. G. Griffith and M. Ryle, *Parliament: Functions, Practice and Procedures* (1989), 523.

If there is no difference between the parties on these issues, the electorate has only a relatively limited opportunity to exercise its preferences. How central, then, were the issues raised in the previous chapters to party-political debate?

On the one hand, we have seen instances where human rights issues were central to party politics and electoral campaigning. In the 1979 general election campaign the issue of 'law and order' was a significant factor in the Conservative victory. Public-order and freedom-of-assembly, with a heavy emphasis on the former, were key issues to mainstream political debate, over the whole period of our study. Freedom-of-association issues, in so far as they involved trade union organization, dominated party-political debate over much of the past twenty years. Immigration control has been an important element in Conservative appeals to voters. In 1970 the Conservative Party was returned to power having promised in their election manifesto to bring primary immigration of workers to an end. In 1981 the British Nationality Act was introduced by another Conservative government in fulfilment of an election pledge to reduce future sources of immigration.

On the other hand, several areas of civil liberties and human rights have been relatively unaffected by party-political debate. In the minority rights area, Poulter finds a remarkable continuity of government policy, stretching from the Labour administration of the mid-1960s up to 1992. McCrudden argues that the 1976 Race Relations Act remained relatively untouched under the Conservative government elected in 1979, despite apparent ideological scepticism by that government of the Act. The issue of the appropriate amount of press freedom, and use of information by the press, has been a political issue, but not particularly a *party*-political issue. The issue of homosexuality played a role in the 1987 election, especially regarding local-government activities in the area, and contributed to the Local Government Act 1988, but it has not otherwise been an issue in mainstream party-political debate.

Some issues change from not being an issue dividing the major political parties to becoming one. How to react to terrorism was not a controversial party-political issue for the first seven years of the operation of the anti-terrorism legislation. The Labour governments in power in 1974–9 were supported in their introduction and maintenance of draconian legislation by the Conservative opposition. A change in approach occurred some time after Labour went into Opposition, with Labour finally deciding to vote against renewal of the Prevention of Terrorism legislation in 1983, and from then was consistently opposed to the legislation, but the issue has never surfaced as a serious electoral issue between the parties. In general, it has been left to the minor parties in British politics to consider wider questions about the adequacy of constitutional safeguards concerning individual human rights, while the division between the main parties has been on questions of economic and social policy. At least this was the case until the 1992 general election when the Labour Party appeared willing to go further than before in embracing a version of a Bill of Rights. However, in many important areas of civil liberties and human rights, there has been relatively little party-political debate and the influence of the electorate is to that extent limited. The question remains, however, where there is party-political dispute, whether the influence of public opinion is civil libertarian or not.

Public opinion

The argument that open regular elections, in which human rights issues are fully debated, results in increased protection of civil liberties is based on an assumption that the British electorate will ultimately decide in favour of tolerance and respect for others. In other countries, however, we have seen that populism and unbridled majoritarianism lead to gross violations of human rights.

How concerned has British public opinion been about human rights issues? The launch of a new political initiative on civil rights is often accompanied by evidence from surveys of public opinion which indicate support for the initiative being proposed. For example, evidence from public attitude surveys consistently show large majorities in favour of a Bill of Rights. In the 'State of the Nation' poll, conducted by MORI in 1991, 72 per cent agreed that 'Britain needs a Bill of Rights to protect the liberty of the individual'[34] and the 1990 British Social Attitudes Survey (BSA) reported a three-to-one majority in favour of a Bill of Rights, albeit to a question which was worded quite differently.[35] In 1986 a poll carried out by MORI for the Campaign for Freedom of Information found that 65 per cent of those polled favoured a freedom-of-information bill.[36] The 'State of the Nation' poll conducted by MORI in 1991 asked a similar question and found that 77 per cent of those polled agreed that 'there should be a freedom of information act'.[37]

However, because of the complexity of the issues being addressed in questions on rights and liberties, and especially so with regard to the need for a Bill of Rights, there are considerable problems of interpretation regarding attitudinal data. It is difficult to frame a question in simple terms which presents the many sides of what has, with respect to a Bill of

---

[34] MORI, *The State of the Nation* (Mar. 1991).
[35] L. Brook and E. Cape, 'Interim Report: Civil Liberties', in Jowell *et al.* (eds.) *British Social Attitudes, the 8th Report* (1991).
[36] MORI, *British Public Opinion* (Aug. 1986), 6.
[37] MORI, *The State of the Nation* (Mar. 1991).

Rights or a Freedom of Information Act, become a controversial matter. For example, the statement 'Britain needs a Bill of Rights to protect the liberty of the individual' (the statement with which respondents were asked to agree or disagree in the State of the Nation Poll) is arguably a leading question because it equates the bill with the protection of liberty, an equation which is rejected by some opponents of a Bill of Rights. In addition, variations in the way a question is asked can influence the response: the increase in the numbers broadly supporting a Freedom of Information Act between 1986 and 1991 is likely to be a reflection of the absence of a 'don't know' category in the 1991 question and a change in the way the question was introduced.[38] More importantly, attitude questions often fail to tell us how strongly individuals feel about the issues being examined in comparison with other topical issues or concerns.[39]

Some of these qualifications, then, suggest that care is required in making inferences from polling data about public support for civil liberties measures and in knowing how much political weight such public support should be given. It is suggested, therefore, that the significance of public-opinion data lies in its usefulness in analysing changes over time. Comparative study of trend data can provide an insight into changes in public awareness of civil liberties and in the public's reaction to changing circumstances. Unfortunately, very few good-quality trend data are available charting changes in public attitudes towards civil liberties. It is therefore not possible to make a comprehensive review of changes in public attitudes during the period covered by this volume. However, more attitudinal data has become available during the 1980s.

British Social Attitudes (BSA) is an annual survey of public opinion on a variety of issues and has been carried out since 1983. A module on civil liberties was included for the first time only in the 1990 survey, and it is therefore not possible to describe in detail changes in public attitudes to the wide variety of rights examined in that year. However, some individual questions from the civil liberties module had been asked in 1983; other questions are even more recent, having been asked only since 1985.

Analysis of comparative data from 1985 produces the following brief summary. There has been little change during the second half of the decade in public attitudes towards the right to protest publicly against government actions: the right to peaceful protest is supported by large

---

[38] From 'would you favour or oppose . . .' in 1986 to 'to what extent do you agree or disagree that . . .' in 1991.

[39] One exception to this was the research commissioned by the Younger Committee on Privacy (para. 99, p. 31): 'In the importance which people attached to it "protecting people's privacy" fell behind the dominating economic issues such as "keeping prices down", "reducing unemployment" and "stopping strikes". It was also thought less important than "raising old age pensions" and "improving the health service".'

majorities. By 1990 more thought that the law in general should always be obeyed, but there had been a softening of the large majorities favouring the death penalty for murder. By 1990, we see more censoriousness about revolutionaries and *much* greater censoriousness about racists. The proportion prepared to allow racists to hold public meetings to express their views dropped by 12 percentage points, and the proportion tolerating the publication of books expressing white supremacist views dropped by no less than 16 percentage points in five years.[40] Regarding police powers with respect to the surveillance and detention of persons suspected of planning crime the report notes that 'there has been a marked rise in the five years in the proportions unable to choose an answer category'. The authors interpret this to mean that there was 'an increase in uncertainty over the five years as to the sorts of powers the police should have'.[41]

Some questions from the BSA 1990 were asked in the first survey in 1983. A question on immigration control shows that in both 1983 and 1990 'there is much greater support for tightening up on black immigration than on white', but it is thought that there is an awareness by 1990 that controls were much stricter, since the proportion of respondents saying that there should be less black settlement had actually fallen by 1990.[42] The BSA report notes that police powers have increased during the 1980s and that the police were relatively better paid at the end of the decade. However, there had been 'a sharp drop (of 11 percentage points. . . .) in the proportion thinking that the police are "well run", and a similar fall between 1983 and 1990 in the proportion saying they are "very satisfied" with the way the police in Britain do their job'.[43]

An important survey of public opinion on racism and xenophobia in the European Community was carried out in its member states at the end of 1988.[44] The survey reported *inter alia* that 78 per cent of Europeans found democracy to be the best of regimes; 60 per cent said respect for human rights was one of the great causes 'which are worth the trouble of taking risks and making sacrifices for'; 8 out of 10 people disapproved of racist movements. Throughout the Community, human rights achieved a ranking of second among the 'great causes'. (World peace was first, protection of wildlife third equal, along with the fight against poverty, and freedom of the individual came fifth.) In the UK the fight against poverty displaced human rights which dropped to third place.

[40] Brook and Cape, 'Interim Report', 186.
[41] Ibid. 196.
[42] Ibid. 191. An alternative explanation is that there was an increased unwillingness to appear to be overtly racist.
[43] Ibid. 194.
[44] Commission of the European Communities, *Eurobarometer: Racism and Xenophobia* (1989).

Government

Checks and balances

Parliament and public opinion are not the only mechanisms for human rights protection under the British constitutional tradition. The following line of argument would have been as familiar to someone in the Britain of the early 1950s as it is to those in the Britain of the 1990s: the practical constraints on Parliament and on Cabinet government are as important as the theory of parliamentary sovereignty. Although Parliament (and thus, through it, the Cabinet) can in theory do what it wants, in practice it is said to be significantly constrained in ways other than by public opinion. There are several additional checks and balances which in practice act as mechanisms for the protection of human rights.

Dispersal of power

One method has been the dispersal of power. Several examples may be drawn from the previous chapters. In the area of police powers, we have seen that it was long regarded as a protection against the abuse of police power that, outside London, responsibility for policing was divided between the chief constable, the Home Secretary, and over 120 local police authorities. We have seen, however, that subsequent case-law, statutes, and organizational changes (notably the amalgamation of the 121 forces into the 43 large forces of 1992) have rendered the tripartite structure unbalanced in the extreme, with the power of the local police authorities atrophying to virtual insignificance. Further specific examples of centralization are the national co-ordination of police efforts in dealing with civil disturbance during the miners' strike in 1984–5 and in tackling terrorism. These centralizing developments have been seen as having worrying implications for the protection of some human rights.

Profound changes have, indeed, taken place in many areas since 1950 in the role of the state and of central government institutions. There have been two parallel developments of particular importance. First, it seems that there has been an increasing concentration of power at the centre of British government which, it has been argued, has unfortunate consequences for the protection of civil liberties. This concentration has been a particular feature of political developments during the 1980s as power bases alternative to Westminster or intermediary between the citizen and Whitehall have had their authority weakened and their functions reduced. The most notable upheaval has been in relations between central and local government, with local authorities being increasingly deprived of revenue-raising functions and independence of action through rate-capping. Some power bases in local government, most notably the Greater London Council, have been abolished altogether. We have seen how the reduction in local-authority powers has affected human rights developments in several ways, most notably in reducing the previous powers of local government to use contract compliance methods to achieve racial and sexual equality, and precipitating the introduction of legislation to prohibit them from 'promoting' homosexuality.

Evidence of centralization can also be derived from the expulsion (again during the Thatcher administration) of the trade unions from the industrial and political decision-making arena, at the industrial and political levels, and the introduction of extensive new regulatory controls on their conduct, giving rise to concerns that freedom of association was being adversely affected. There was, during the 1980s, a lack of tolerance in government for political dissent and a tendency for government to equate its own interests and survival with the interests of the state and the body politic in general, but perhaps any differences between that government and previous ones in these regards was more one of degree and stridency, than preference.

It would also be misleading to argue that centralizing tendencies were necessarily anti-civil-libertarian in their consequences. We have seen several examples in previous chapters where the extension of central controls has been carried out with results beneficial for human rights. Westminster intervention in Northern Ireland from the late 1960s on brought the belated imposition of some human rights protections on a devolved government, though it also permitted internment without trial. A voluntary approach to increased legal regulation in industrial relations overcame a significant ideological barrier to the enforcement of anti-discrimination standards in the employment relationship.

Government ministers

We have seen that one method by which human rights are protected is by Parliament enacting specific *ad hoc* legislative measures to protect civil liberties and promote human rights when this seems necessary, either because the government perceives it as important because of its ideological preferences or because it has been convinced that it is the right thing to do by lobbies and pressure groups, or because it is deemed to be electorally popular. But the government is, of course, not a monolithic body and the influence of individual ministers seems, on occasion, determinative.

Thus, in the context of anti-discrimination legislation, the personal commitment of individual ministers to the issue appears to have been significant. So far as race discrimination was concerned, the personal influence of Mr Roy Jenkins was particularly important in 1966 and 1967, when he was Home Secretary.[45] An equivalent role was played by Mrs

---

[45] R. Jenkins, *Life at the Centre* (1991).

Barbara Castle in the context of equal pay legislation when she was the Secretary of State for Employment during 1969–70.

Self-regulation

A substantial amount of discretionary power is accorded to those who exercise authority in the public arena. It is not uncommon to see this justified on the basis that such people can be trusted 'to do the right thing', with sensitive interpretation of the law and reasonable exercise of discretion by those entrusted with public powers. It has been argued that police discretion could be justified in that it permitted non-enforcement of the law when enforcement might violate common-sense notions of justice in particular cases. The law in the area of freedom of assembly, says Gearty, is designed in such a way as to enable the police to choose either alternative without fear of acting beyond their legal powers. Public bodies were to be trusted to regulate themselves; self-regulation was infinitely preferable to external regulation. This was particularly so in the context of the press and broadcasting media. The first self-regulatory mechanism of the General Council of the Press was established in 1953, later becoming the Press Council, and later still the Press Complaints Commission, all of them self-regulatory bodies. The civil service, though it had extensive access to information on individuals, could be trusted to use that information reasonably and sensibly. Considerable emphasis, then, is placed on the trustworthy nature of civil servants and others in positions of authority.

Gardner points out that the British Board of Film Classification has on occasion taken successful steps to discourage private prosecutions by mediation and powerful advocacy, with the effect of protecting freedom of expression. In the area of obscenity, he also shows that, from the mid-1970s, the DPP, to whom all prosecution decisions in this area are referred for the sake of consistency, largely gave up prosecuting other than for crude pornography. So too the Theatres Act 1968 applied obscenity law to theatre productions, but prosecutions were subject to the consent of the Attorney-General, and he has successfully resisted prosecuting despite substantial political pressure.

However, other evidence calls into question the desirability of such discretion. Though in some circumstances police discretion may, in theory, be beneficial in permitting non-enforcement of the law, the criteria on which non-enforcement is based may not have been arrived at by consensus or be open to examination. There would appear to be substantial evidence that the outcome of the exercise of police discretion is unjustly discriminatory. In the area of freedom of assembly, Gearty argues in Chapter 2 that the significant and increased amount of police discretion means that 'It is not unreasonable to conclude that in this branch of civil liberties we have freedom under the police rather than freedom under the law.' We have seen, too, that self-censorship on the part of the broadcasting and print media is a significant limitation on freedom of expression. In particular, coverage of Northern Ireland on television has become the subject of an elaborate system of regulation by senior management, with programmes being withdrawn or modified if they are judged by executives to be too sensitive.

'Independent' political actors

Thus far, we have concentrated on the central political institutions which have been said to protect civil liberties. We may turn now to some of the less public mechanisms: those institutions which have some independence from the traditional party political debate but are often essential to it. We can identify five such institutions: the civil service, the 'great and good', regulatory agencies, pressure groups, and the media.

The civil service

In the British political tradition, the civil service plays the key role of policy adviser to minister and executor of ministerial decisions. Civil servants are required to be independent in that they serve any elected government with loyalty and dedication. It is neither incompatible with this role, nor undesirable, that civil servants individually and collectively should have views on policy matters and that they should express them in their policy advice. Human rights and civil liberties are not likely to be immune from this, and thus it is more than likely that, in several areas under discussion in previous chapters, civil servants played a key role.

The role of senior civil servant in Britain carries with it a high expectation of confidentiality, however, and it is therefore not surprising that little is said directly in previous chapters about their influence. Given, too, the forty-year period of our study, many of the internal papers relevant to the issues are still subject to restrictions on disclosure. Only when they are released will a fuller picture be possible.

It will be particularly interesting in the future to discover the extent to which attempts to build human rights thinking into civil service training have been successful. On various occasions over the past forty years, civil servants have been directed to have regard to particular human rights values in devising policies or administering programmes. Apart from the guidelines to civil servants regarding the avoidance of race and sex discrimination, perhaps the most wide-ranging directives are those dealing with the need to take European human rights and Community law into account.[46] We have almost no knowledge how successful these have been, however.

---

[46] A. W. Bradley, 'Protecting Government Decisions from Legal Challenge' (1988) *PL* 1, 3–4.

The 'great and good'

For a significant part of the forty years under consideration, tensions and disputes in selected circumstances, including those in the field of civil liberties, were addressed through the mechanism of Royal Commissions and departmental inquiries conducted by respected public persons, the 'great and the good'. A list of the major such inquiries influencing issues discussed in previous chapters, which we include as the second item in Appendix C, demonstrates the frequent use of such methods in post-Second World War Britain. It also demonstrates, however, a sharp falling off of their use since 1979 (only one Royal Commission, on criminal justice, was appointed from 1979 to 1992), and this marks a significant difference between the more consensual approach to government before 1979 and that after 1979.

A general conclusion appears to be that, with some exceptions, the influence of such inquiries on human rights has been, more often than not, more benign than many of the other institutions considered so far. Perhaps the nature of such inquiries leads them to be more rational, less prone to extremes, more tolerant of dissenting views, and therefore more prone to a position of respect for human rights. In some areas, however, inquiries have been notably unsuccessful. With regard to political violence and terrorism, Gearty has noted the failure of official inquiries by the great and good to have any impact on government policy due to the strength of public opinion and to the security-forces lobby.

Regulatory agencies and commissions

A major change noticeable after the 1950s is the emphasis which is now placed on regulatory agencies and commissions as a primary mechanism in the domestic protection of civil liberties and human rights. No such bodies existed before 1950. Since then we have seen the development of the Race Relations Board and the Community Relations Commission, and their amalgamation to form the Commission for Racial Equality. An Equal Opportunities Commission has been established for Britain in the area of women's rights. The Police Complaints Board established in 1976 has been replaced by a Police Complaints Authority. A Data Protection Registrar and a Data Protection Tribunal have been established in the area of privacy. In the area of broadcasting, a Broadcasting Complaints Commission and a Broadcasting Standards Council have been established. In Northern Ireland, the Standing Advisory Commission on Human Rights, the Fair Employment Commission, and an Equal Opportunities Commission have been established, among others.

There are several useful roles which such bodies may serve. We have seen that, in the context of anti-discrimination law, enforcement by such a body, it was thought, would emphasize the elimination of discrimination in the public interest, rather than the punishment of the individual discriminator. Such a body could also be given powers which would make it more effective than the ordinary civil or criminal processes.[47] Michael argues that the legal protection of privacy used to be a matter of providing legal remedies which aggrieved individuals could pursue through the courts. Now, however, the central feature has been the creation of commissions with the duties and powers of regulating. This development was essential in his view to protect individuals from improper surveillance of which they may not be aware, as well as providing a practical solution to the question of how to resolve the difficult cases in which surveillance may be justifiable, without defeating its possible purpose by disclosure to the person concerned.

Similar agencies might be proposed in other areas. In the prisons' context, for example, if the defence of rights is left solely to enforcement by individuals, the result will be unsystematic compliance. More generally, several bodies, such as Liberty, have advocated the establishment of a Human Rights Commission, with similar powers to those of the CRE and the EOCs, but with a more wide-ranging jurisdiction over the entire field of civil and political rights.

Previous chapters have provided some evidence, then, of the relative success of such bodies and of their popularity as protectors of civil liberties and human rights. But there is evidence, too, of the problems which such bodies encounter in the British legal and regulatory contexts. British regulatory styles seem adversely to affect the enforcement policies of such Commissions. Collins and Meehan point to the reluctance or inability on the part of the Equal Opportunities Commissions to capitalize on their powers. The powers of formal investigation are not used to the fullest extent possible. For the EOC (GB) a persuasive approach to tackling discrimination is preferred. There are also major difficulties in using this power to tackle institutional racial discrimination in employment, as McCrudden suggests. In the area of police powers, the Police Complaints Board established by the Police Act 1976 was almost universally seen as a toothless body. In 1981 the Board's chairman declared that the 'existing board had kept so low a profile that it had climbed into a ditch'.[48]

For this reason amongst others there is a certain unwillingness on the part of several contributors to rely too much on the 'public interest' role of Commissions to the exclusion of individually enforceable rights. Michael argues, for example, in Chapter 9 that 'there may be a danger in

---

[47] J. Jowell, Administrative Enforcement of Laws against Discrimination, (1965) *PL* 119–86.
[48] Quotation taken from Chapter 3.

the development of regulatory agencies if it is at the expense of individual remedies, since it is always difficult to create bodies which are sufficiently independent of the regulated, and particularly difficult when government agencies are to be regulated'.

Pressure groups

Previous chapters have frequently pointed to the importance of pressure groups in the area of human rights (and a list of the main pressure groups which have been operating in the field of the human rights and civil liberties covered in this book appears as the first part of Appendix C).[49] It is useful to distinguish pressure groups from other collectivities such as broad interest groups (for example, 'the City', the security forces, the 'law and order' lobby), from social movements (for example, feminism, the women's movement, or even the human rights movement),[50] and from quasi-non-governmental organizations (quangos) such as the Commission for Racial Equality and the Equal Opportunities Commission. Grant characterizes an interest group as an *unorganized* body of opinion which shares certain common assumptions and views on what is important. Social movements may spawn pressure groups, some of which may even be antithetical to each other; quangos may engage in law reform campaigns, but they usually have some statutorily defined responsibilities which compromise their independence.[51]

A useful distinction can be made between 'sectional' and 'cause' groups, with the former looking after the common interests of some group or section of society and the latter espousing or seeking to advance a belief or set of values. Sectional groups may have only a limited set of aims or objectives or a restricted membership, while cause groups accept membership from anyone who adheres to the principles of the group.[52] Liberty (formerly the National Council for Civil Liberties) is one of the longest standing 'cause' groups while LAGER (Lesbian and Gay Employment Rights) and the Union of Muslim Organizations are examples of sectional groups.

---

[49] No attempt has been made to include various single-issue pressure groups such as the 'Free the Birmingham Six' campaign or to include British-based international groups such as the Anti-Apartheid Movement.

[50] L. Wiseberg has described the human rights movement as 'essentially a post-World War II phenomenon, and it is really in the 1970s that it took on its present character as a collection of independent national, regional, and international NGOs seeking to hold governments accountable to internationally defined standards of human rights': 'Protecting Human Rights Activists and NGOs: What More Can Be Done' (1991) 13 *Human Rights Quarterly* 528–9.

[51] W. Grant, *Pressure Groups, Politics and Democracy in Britain* (1989).

[52] Ibid. 12.

In seeking to understand what makes a pressure group successful, Grant argues that it is necessary to take into account not only the resources available to the group and its internal organizational strength but also the extent to which the group has 'insider' status and finds itself consulted by government and policy-makers. Examples of insider groups from Appendix C include Amnesty International, Justice, the Howard League, the United Kingdom Immigrants Advisory Service (UKIAS), and the National Association for the Care and Resettlement of Offenders (NACRO). It could be argued that some insider groups operate in a similar manner to quangos since they are dependent on government funding. A withdrawal of government funds from the UKIAS threatened its survival, although this withdrawal seems to have been in part precipitated by internal disagreements and management problems. Some pressure groups may seek to obtain insider status while others prefer to remain on the margins. A good example of this in recent years has been the different and sometimes opposing strategies adopted by campaigners for reform of the law on same-sex sexual activity. Following the decline during the 1970s of the Campaign for Homosexual Equality (CHE) as an effective homosexual law reform pressure group, campaigning has been taken up by groups seeking insider status like Stonewall and by more radical 'outsider' groups such as the shortlived Organisation for Lesbian and Gay Action (OLGA) and the more successful Outrage.[53]

Some organizations have achieved recognition by international bodies, such as the UN or the CSCE, and have been accredited with either roster or consultative status. These groups are normally referred to as non-governmental organizations (NGOs) and some relevant British NGOs which have had input into the activities of the UN Human Rights Committee have been referred to in Chapter 1.[54] Tolley has examined the operation and effectiveness of the International Commission of Jurists (ICJ) as an international pressure group seeking to influence governments to adopt human rights standards. He argues that the ICJ has 'contributed significantly to several notable successes: the 1977 Protocols to the Geneva Convention, the UN Convention Against Torture, the European Torture Convention, the African Charter of Human and Peoples Rights . . .'.[55] One reason for the success of the organization has been that 'the Human Rights Division (of the UN)

---

[53] One indication of Stonewall's attempts to gain insider status was the meeting arranged between Sir Ian McKellen, one of Stonewall's spokespersons, and Prime Minister John Major to discuss law reform in Sept. 1991.

[54] NGOs with consultative status participate in the Conference of Nongovernmental Organisations (CONGO) and a human rights NGO coalition has become a permanent subcommittee of CONGO.

[55] H. Tolley, jun., 'Popular Sovereignty and International Law: ICJ Strategies for Human Rights Standard Setting' (1989) 11 *Human Rights Quarterly* 563.

had limited resources for drafting UN standards and the ICJ provided legal expertise which in national law making would come from executive departments'.[56]

In the domestic context, insider pressure groups have no recognized status in the policy-making process, although it would appear that some groups and particularly those funded by government (e.g. Victim Support, NACRO) are regularly consulted and have access both to civil servants and members of Parliament. Such groups may be invited to discussions which take place under 'Chatham House' rules.[57] A network of pressure groups exists on criminal justice issues and reference is sometimes made to the existence of a 'race relations industry'. Richardson and Jordan have used the term 'policy community' to describe the network of departmental officials and client groups which consult together, discuss proposed legislation, and make policy decisions.[58] These networks are likely to include representatives of the main sectional and professional pressure groups and less representation of 'cause' groups. Thus, policy-making on police, criminal justice, and prisons issues is unlikely to proceed without consultations having taken place with the Association of Chief Police Officers (ACPO), the Police Federation, the Prison Officers Association, the Bar Council, the Law Society, and the Magistrates' Association, among others.

What evidence is there of the existence of and success of pressure-group activity in the areas covered by this book? Gearty notes that pressure groups have 'enjoyed some degree of influence'[59] in the development of public-order policies but he states that 'reform in the law in this area has generally been reactive to events and incidents'.[60] By contrast, Gearty notes the failure of pressure groups such as Amnesty International, Liberty, and the Committee on the Administration of Justice (CAJ) to make any significant impact on government policy on terrorism and political violence. Reiner and Leigh describe the contents of PACE as emerging out of a conflict between the 'law and order' lobby and penal reform groups. (Jackson makes a similar distinction between the 'crime control' and 'due process' lobbies.) Reiner and Leigh describe how the Royal Commission on Criminal Procedure provided a forum at which various pressure groups and lobbies were able to present evidence and make a case. Michael notes that 'adroit and expert campaigning' by the Freedom of Information Campaign led to private access to medical

---

[56] Ibid. 582.

[57] An agreement between the discussants that remarks will not be attributed to individuals and that the contents of the discussion remain unreported.

[58] J. J. Richardson and A. G. Jordan, *British Politics and the Policy Process: An Arena Approach* (1987).

[59] Ch. 2.

[60] Ch. 2.

and other records, and that intervention by NCCL (now Liberty) led to amendments to community charge registration forms. Eekelaar argues that input by such groups as Justice for Children and the Family Rights Group helped shape the Children Act 1989 since they were able to take advantage of the anti-welfarist and anti-state intervention ideologies of Conservative governments after 1979. McCrudden partly explains racial discrimination legislation initially by reference to Labour politicians reacting to the American experience, and, by the time of the 1976 Act, as a consequence of lobbying for better legislation by the quangos (Race Relations Board, etc.) set up under the previous legislation. Minority pressure groups have tended in the UK to adopt advisory and welfare functions for migrants and refugees rather than engage in overtly political interventions. Political interventions by Muslim organizations seeking to change the blasphemy laws in the wake of the Rushdie affair have been unsuccessful. Gardner states that the Society of Authors played a large part in the Campaign that led up to the Obscene Publications Act of 1959, which introduced a 'public good' defence against material which might be found depraving or corrupting. Collins and Meehan state that various pressure groups, often co-operating under the auspices of the Fawcett Society, were responsible for key improvements to the Sex Discrimination Act 1975. Wintemute describes how a mobilization of popular protest by various *ad hoc* pressure groups set up to campaign against Clause 28 (now Section 28) of the 1988 Local Government Act (the Stop the Clause Campaign) was unsuccessful. However Stonewall has claimed success in amending section 31 of the 1991 Criminal Justice Act.[61]

The media

In some circumstances, the perception that 'something must be done' seems uppermost in the minds of politicians and others. The role of public crises may be an important stimulus for political change of this type. In this context, the role of the news media is often crucial in providing a forum for discussion, and in uncovering abuses that should be remedied. This has been true particularly in the area of due process. We have seen in previous chapters that only the consistent exposure of celebrated miscarriages of justice produced changes which have increased safeguards for defendants in the criminal process.

On the other hand, the perception of a crisis can lead to precipitate action deleterious to civil liberties. In the area of anti-terrorism, Gearty argues, that in no other field are our laws so much the consequence of a desire to be seen to be doing something. 'They owe their existence not

---

[61] See Wintemute's discussion in ch. 15.

to any strategic policy goal or covert master plan, but rather to the "something must be done" syndrome that afflicts all politicians trying to assuage public indignation.'[62]

The judiciary and the judicial process

Residual rights

It was Dicey's proud boast that liberties were better protected in England than in continental Europe, in part because certain vital interests were safeguarded by an idea of residual common law rights. The question of whether this is a convincing argument recurs as a common theme in several of the previous chapters. The notion of residual rights is that a person may do, or not do, anything he or she wishes unless this is specially prohibited or required by law. Several examples would have been familiar in the 1950s: the presumption of innocence and of a right to silence in the context of criminal investigations, privilege against self-incrimination in the criminal law (chapter 3), freedom of contract (chapter 13), freedom of association (chapter 8) and freedom of assembly (chapter 2). In conformity with the traditional residual approach to such matters, British law also chose merely to regard the prisoner as retaining the rights possessed by non-prisoners, provided these rights were not removed by imprisonment (chapter 6). Policemen were regarded as citizens in uniform who lacked either legal powers or the coercive capacity to police other than by the consent of the populace.

Residual rights have indeed given rise to some civil libertarian effects. So, for example, the decision in *Rice* v. *Connolly*[63] in which it was held that a person had no legal duty to answer police questions, is a classic example of the negative-rights approach in operation. So, too, as Ewing points out in chapter 8, although there is no guarantee of the right to form and join political parties, there are precious few restrictions on what political parties may do either, though limitations on their authority to expel members, and the broadcasting ban have been notable exceptions to this 'hands-off' approach.

However, we have also seen the problems with relying on the residual rights approach. Reiner and Leigh have argued in chapter 3 that the myth of the police as 'citizens in uniform', whatever its attractions as an ideal, conceals the true nature of the growth of police power and powers, and the consequential problem of regulating their exercise. The decision in the *Malone*[64] case demonstrates, as Michael points out in chapter 9,

---

[62] Gearty, ch. 5.
[63] [1966] 3 WLR 17.
[64] *Malone* v. *Metropolitan Police Commissioner (No. 2)* [1979] Ch. 344.

residual rights operating against the right of privacy: the Government could intercept communications unless specifically legally prevented from doing so. As Gearty argues in chapter 2, little of Dicey's residual liberty of assembly is left intact. According to Gearty, 'The freedom is trapped by its complacent classification as a residual liberty, and without legislation or positive common law rules to act as its defender, it is being slowly squeezed into extinction.' The interpretation of the 1968 Race Relations Act by the House of Lords rested on a residual common-law right to discriminate which significantly limited the ability of that legislation to control racial discrimination. To sum up, the residual rights approach may serve human rights ends, but equally clearly may not.

Interpretation by an independent judiciary

An independent judiciary, through its interpretation of common law and statute, may take into account human rights values and thus smooth out the rougher edges of legal provisions, making them consistent, where possible, with the consensual values of the community. McWhinney has called this 'judicial braking', which he describes as occurring 'where a court . . . says in effect . . . that the legislature may or may not have the claimed legislative power, but it has not . . . employed that power'.[65] Several examples which secured a civil libertarian result by the use of such a technique may be seen in the previous chapters.

We have seen that there is a judicially sponsored rule of interpretation that penal statutes must be construed strictly and in favour of the accused.[66] The common law would appear to encourage the view that there is a limited right not to associate: compulsory union membership has, on occasion, been held to be a restraint of trade. Common law appears to recognize the autonomy of the association to control whom it accepts into membership and retains in membership of the association as an important feature of freedom of association. The interest in freedom of expression has been held to be an obstacle to putting intemperate critics of the judiciary behind bars,[67] and to local authorities maintaining an action for defamation.[68] There is the continued affirmation by the courts of the existence of a common-law discretion to exclude evidence whose probative value is outweighed by its prejudicial effect. The Court of Appeal has issued guidelines governing the way a judge should direct a jury when the evidence against an accused rests substantially on identification evidence based on personal impression. The judiciary have developed the common-law discretion to exclude unreliable confessions, aside

---

[65] *Judicial Review* (1969), 13.
[66] See Craies, *Statute Law*, 7th edn. (1971), 529–31.
[67] *R. v. Metropolitan Police Commissioner, ex parte Blackburn* [1968] 2 All ER 319 (CA).
[68] *Derbyshire CC v. Times Newspapers Ltd.* [1993] 1 All ER 1011 (HL).

from any inadmissibility test. We have seen, too, statements by judges asking for remedies to be provided by statute where they were unable to do anything under existing law.[69]

However, we have also seen the common law and statute being interpreted by the judges in ways that many of the authors of the previous chapters have seen to be anti-civil libertarian. Thus in *Shaw* v. *DPP*[70] the common-law offences of 'outraging public decency' and 'conspiracy to corrupt public morals' were held to have survived the 1959 Obscenity Act reform. In *Knuller* v. *DPP*[71] we saw the extension of the offence of conspiracy to corrupt public morals. In *Whitehouse* v. *Lemon*[72] the common-law offence of 'blasphemous libel' was resuscitated. In the *Spycatcher*[73] litigation, where the use of the law of confidence to limit disclosure of information was in issue, the public-interest defence was held to be limited, warranting disclosure to the relevant authorities only and not to the public at large. The courts have responded to the increase in terrorism by narrowing the political exception in extradition law so as to exclude many politically motivated crimes involving violence.[74] Since the start of the most recent Northern Irish disorders in 1969, six major challenges to executive action have reached the House of Lords and all have been unsuccessful. Gearty argues, in the context of political violence, that the record of the British courts has been remarkable for its total and consistent deference to executive authority.

Given the extensive human-rights-related legislation which has been enacted since the 1950s, a further question arises: to what extent are the judiciary willing to uphold specifically civil libertarian legislation? Some (most notably Griffith[75]) have argued that the judiciary not infrequently interpret such legislation in order to minimize its impact. How much evidence on this issue has been provided in the previous chapters? We have pointed out already that, under the 1968 Race Relations Act, the courts adopted an approach to the interpretation of the statute which was narrow and unhelpful, in the main. However, although under the 1976 Race Relations Act there is also some evidence in judicial decisions of the acceptance of a philosophy hostile to regulatory agencies (in the context

---

[69] By Megarry VC in *Malone* v. *Metropolitan Police Commissioner* [1979] Ch. 344 and by Glidwell LJ in *Kaye* v. *Robertson and Sport Newspapers Ltd., The Times*, 20 Mar. 1990. Procedures in equal value cases were described as 'scandalous' and amounting to a denial of justice by the President of the Employment Appeal Tribunal in *Davies* v. *McCartney* [1989] IRLR 439 (EAT).

[70] [1962] AC 220.

[71] [1973] AC 435.

[72] [1979] AC 617.

[73] *Attorney-General* v. *Guardian Newspapers Ltd.* [1987] 3 All ER 316; *Attorney-General* v. *Guardian Newspapers Ltd.* (No. 2) [1988] 3 WLR 776.

[74] *Cheng* v. *Governor of Pentonville Prison* [1973] AC 931.

[75] J. A. G. Griffith, *The Politics of the Judiciary*, 4th edn. (1991), 151–77.

of formal investigations) and favouring a free-market approach (in the context of the interpretation of indirect discrimination), nevertheless, the higher courts have, in general, been relatively sympathetic. The dominant approach adopted by the House of Lords to the interpretation of the 1968 Race Relations Act (that the legislation should be restrictively interpreted because it interfered with common-law liberties) has been quietly jettisoned. Reiner and Leigh argue that, with exceptions, the typical judicial response to PACE seems to be that, since it reflects a balanced package in which adequate powers are given to the police, the corresponding safeguards set out in the Act must be followed. The judiciary have shown greater willingness on the whole to supervise the new settlement than they did the earlier Judges' Rules, and somewhat contrary to expectations.

Judicial review of administrative action

A striking difference between 1950 and 1993 is the extent to which issues of political debate and controversy have come to be considered and adjudicated on in the ordinary courts. One of the major methods by which this has been accomplished has been the development of judicial review of administrative action, whereby a person who wishes to challenge the validity of an exercise of public power may apply to the Queen's Bench Division to bring an application for judicial review. The validity of the exercise of the power may be challenged on the grounds of illegality, procedural impropriety, or unreasonableness.

Since the 1960s a much more interventionist approach the judiciary has adopted. The grounds of judicial review have expanded considerably, the procedures have given substantially greater discretion to judges, and the impact of judicial review on government decision-making has been significant in a number of areas. Some commentators have seen in this development an important civil liberties mechanism at work, or one with that potential at least, and judicial review is being used increasingly to try to convince the courts to incorporate civil libertarian and human rights perspectives into their decision-making. Thus we have seen in the previous chapters significant use of judicial review in the context of race and sex discrimination, minority rights, immigration, privacy, prisoners' rights, due process, and police powers. The police complaints system was strengthened; judicial review was applied to scrutinize certain exercises of the prerogative; in prison life the courts have been willing to recognize and uphold some public-law rights; some wrongful criminal convictions have been quashed using judicial review; and anti-discrimination requirements have been enforced against some public bodies.

Judicial review has, on occasion, also proved of some assistance in addressing an issue which recurs in many of the chapters: lack of information. Often, as we have seen, there is a dearth of information as to how (or whether) discretion is controlled and structured, for example in the operation of the security vetting procedure in the civil service, in jury vetting in terrorist cases, and in covert action by the security services engaged in counter-terrorism. On several occasions, judicial review, or other similar legal process, has partially prised off the lid. The Public Order Manual of Tactical Options and Related Matters, prepared by the Association of Chief Police Officers, drafted in secret, was partly made public when the manual was referred to in the course of a criminal trial arising out of the miners' strike.[76] The immigration rules, supplemented by unpublished instructions to immigration officers and entry clearance officers, remained unknown outside official circles until the CRE investigation in 1985 was permitted to go ahead by the High Court despite opposition from the Home Office. In some circumstances, tactical use of discovery in judicial review will enable more effective political agitation.

While there are persuasive reasons for an aggrieved person's deciding not to go to court, the incidence of judicial review appears to be increasing. One reason may be that some decision-makers are less concerned than before about losing in court. A second may be a lack of awareness by government generally of developments in human rights law. A third is the advent of greater political dissensus in Britain, which provided a strong impetus towards seeking a legal rather than a political remedy. In addition, the successful challenging of some decisions has led to an increase in the number of challenges generally. Not surprisingly, enthusiasm for mounting challenges grows throughout the legal profession in proportion to the success of earlier challenges. We shall also see subsequently that the European Convention on Human Rights and European Community law have probably added to the perceived utility of judicial review.

However there are problems with the use of judicial review as a civil liberties mechanism. Apart from the considerable procedural difficulties, and the increasing delays, there are deeper problems. The incidence of judicial review is, of course, small in comparison with the number of decisions public bodies take. Even in areas where judicial review has been sought comparatively often, it is still infrequent and discontinuous. More problematic still, the courts' role is, at least in part, to set standards in cases where these are unclear or contested. Criticism of standard-setting has focused on three problems: uncertainty, inconsistency, and inappropriate values.

There is uncertainty as to the standards which will be drawn on: when,

---

[76] Although in other contexts it has been held to be protected on the ground of public interest immunity, *Gill* v. *Chief Constable of Lancashire, The Independent*, 11 November 1992.

for example, is freedom of expression regarded as relevant or not? There is uncertainty, also, as to the depth of intervention likely on any particular occasion. And there is uncertainty as to the breadth of intervention: the distinction between what is a decision by a public body, and therefore open to judicial review, and what is a decision by a private body, and therefore not, is baffling to many. As regards inconsistency, the judiciary is seen by some as supporting individuals who challenge the use of governmental powers in certain domains but not in others, without any coherent expression of the differences between the cases. There is, for example, a greater willingness to challenge the exercise of discretion by regulatory agencies than to challenge the exercise of discretion by the police. As regards the adoption of inappropriate values, many have argued that the almost complete unwillingness of the courts to become involved in seriously reviewing issues which the Government indicates have national-security implications is unacceptable from a human rights viewpoint.[77]

We have seen that judicial review applies to the exercise of powers by *public* bodies. Apart from the degree of uncertainty surrounding its application, the distinction may have two further consequences from the point of view of furthering human rights through the use of judicial review. First, on occasion, human rights will be violated by private bodies, not government. We have seen, for example, the inapplicability of public-law rights to the Bar Council, despite the importance of their decisions for freedom of association. The second consequence is that judicial review is also open to those who wish to stop public mechanisms for achieving human rights ends. We have seen the use of judicial review to stop action by bodies set up to achieve anti-discrimination objectives, for example, and the consequent lessening in the effectiveness of the formal investigation mechanism available to those bodies.

Lay element in the administration of justice

Reliance on a lay element in the administration of justice has also been seen historically as an important civil liberties mechanism. Lay magistrates and juries, it was argued, would not convict defendants in criminal trials when it considered that the law it was being asked to enforce was either unworkable or unduly oppressive. Several of the chapters have pointed to this mechanism working as intended. In a prosecution in 1970, arising out of the disclosure of documents concerning of the Biafran war, the jury appears to have insisted on interpreting the defences in the Official

---

[77] *Council of Civil Service Unions* v. *Minister for the Civil Service* [1984] 3 All ER 935; *R.* v. *Secretary of State for the Home Department, ex parte Hosenball* [1977] 3 All ER 452; *R.* v. *Secretary of State for the Home Dept., ex parte Ruddock* [1987] 1 WLR 1482. Also *R.* v. *Secretary of State for the Home Dept., ex parte Cheblack* [1991] 2 All ER 319.

Secrets Act very widely indeed. We have seen, too, the jury exercising its discretion not to convict in other more recent Official Secrets Act cases, such as that involving the civil servant Clive Ponting, who supplied documents relating to the Falklands war to MPs whom he considered had been misled by his minister.

However, there are also several problems with relying on the lay administration of justice as a bulwark against the abuse of human rights. Although juries may be now more socially diverse than in the 1950s, they may not yet adequately represent ethnic minorities. There has been some scepticism about the legitimacy of an unrepresentative lay magistracy, again particularly as regards under-representation of black and Asian magistrates. Yet the Court of Appeal has held that the power of the judge to stand by jurors was limited to removing jurors who were not competent to serve and any attempt to influence the composition of a particular jury, for example on basis of racial balance, detracted from the principle of random selection.[78] We have seen also that there has been extensive vetting of juries. Indeed, Gearty argues that this is standard practice in terrorist cases.

In other cases, the principle of trial by jury has been reduced in importance. Majority verdicts were introduced during the 1960s. Trial by jury has been removed for scheduled offences in Northern Ireland with the introduction of Diplock courts, partly because of the possibility of biased verdicts and intimidation. Nor are juries immune from abusing human rights themselves. We have seen excessive awards of damages being awarded by juries in defamation cases. We have seen too that over-reliance on juries has contributed to the difficulty which judges had in remedying miscarriages of justice.

Access to justice

Lastly, the effectiveness of several of the previous mechanisms is, in practice, determined not by the quality or otherwise of the judiciary, nor by the quality or otherwise of the precedents, but on the ability of those seeking to advance human rights issues to get to court or tribunal and present their case in the best possible way. The role of legal representation is, therefore, often crucial, particularly in the context of the largely adversarial British legal system. We have seen that the presumption of innocence in the criminal context is an important common-law safeguard. However, we have also seen that the presumption was significantly strengthened in practice by the advent of criminal legal aid. So too, in the civil context, it is of little use having the right to go to court or tribunal if the ability to make an adequate case there is not also safeguarded.[79]

---

[78] *R. v. Royston Ford* (1989) 89 Cr. App. R. 278 (CA).
[79] See further, Law Society, *Access to Justice: The Law Society Manifesto* (1991).

Where legal aid is not provided, several of the chapters have noted consequential problems for the effective use of rights. Failing adequate legal aid, through the Legal Aid Scheme, we have seen the vital importance of having an adequate alternative. Thus, though there is no legal aid for applicants before the industrial tribunals, assistance available from the CRE and the EOCs has been a considerable source of support and assistance to complainants in discrimination complaints.

We have discused above the significant role played by pressure groups in the advancement of human rights issues in Britain. Traditionally, activity by such groups was (and is) primarily political in its focus, that is, its advocacy is largely targeted at politicians, administrators, the media, and opinion-formers. Over the period of our study, however, a significant new development occurred which links some of the ideals behind legal aid with the tradition of pressure-group activity. This development, termed 'public interest law', has been defined as consisting in 'the use of litigation and public advocacy . . . to advance the cause of minority or disadvantaged groups, and individuals, or the public interest'.[80] Several of the pressure groups mentioned earlier have developed litigation strategies as part of their activities: for example, the Joint Council for the Welfare of Immigrants,[81] the Child Poverty Action Group,[82] Liberty,[83] and MIND.[84] Such groups have used domestic legal mechanisms,[85] the European Convention on Human Rights,[86] and European Community law. In a previous chapter of this book, Richardson has noted in the prisons' context that what success there has been before the courts, both domestic and European, owes much to the persistence of penal and mental health pressure groups and their astute case selection, and similar observations are appropriate also in the context of immigration. More recently, the Public Law Project has been active more generally in stimulating litigation.

EUROPE AND INTERNATIONAL REQUIREMENTS

The European Convention on Human Rights

Previous chapters have discussed the influence of the Council of Europe and its activities on developments in Britain, such as its influence on the

---

[80] J. Cooper and R. Dhavan (eds.), *Public Interest Law* (1986), 5.

[81] I. Martin, 'Combining Casework and Strategy: The Joint Council for the Welfare of Immigrants', in ibid. 261.

[82] R. Smith, 'How Good are Test Cases?' in ibid. 271.

[83] B. Cohen and M. Staunton, 'In Pursuit of a Legal Strategy: The National Council for Civil Liberties', in ibid. 286.

[84] L. Gostin, 'Professional Pressure Groups and Party Politics', in ibid. 427 at 428–30.

[85] C. Harlow, 'Public Interest Litigation in England: The State of the Art', in ibid. 90.

[86] S. Grosz and S. Hulton, 'Using the European Convention on Human Rights', in ibid. 138.

development of the Data Protection Act 1984, and the impact of the European Social Charter of 1961 on individual employment rights.[87] However, the major influence emanating from the Council of Europe *vis-à-vis* the United Kingdom is the role of the European Convention on Human Rights in protecting civil liberties and advancing human rights. We have sketched out the development and operation of the Convention in general terms in Chapter 1. The issue which we consider in this chapter is the extent to which the Convention is practically important, drawing on various examples of its use in previous chapters.

There are, in general, two main methods by which the Convention may play a role domestically. The first is direct, where a violation of the Convention has been found and the United Kingdom is required to come into compliance, whether by compensating an individual harmed by breach of the Convention or by changing existing legislation. Since the Convention is not incorporated into the law of the United Kingdom, the only way by which a violation can be established authoritatively is by having recourse to the ECHR institutions. With the exception of the case taken by Ireland against the United Kingdom regarding security policy in Northern Ireland, there have been no other inter-state cases and so the burden of activating the Convention machinery falls on individuals. In 1991, a total of 202 applications were registered against the United Kingdom by the European Commission of Human Rights, of which 38 were declared admissible by the Commission. In that same year, a total of 217 applications were declared admissible by the Commission relating to all countries. Were it not for the acceptance by the United Kingdom of the right of individual petition in the mid-1960s, the Convention would have the same minimal effect on British law as the ICCPR.

The record of the United Kingdom has been discussed extensively in the previous chapters and it is not proposed to repeat that analysis here. Up to 1989, the Committee of Ministers had found one or more violations of the Convention in 18 of the 30 United Kingdom cases referred to it. Of the 252 cases referred to the Court between 1959 and 1990, 41 concerned the United Kingdom, the highest number for any state. Up to the end of 1992 the Court found at least one violation or potential violation of the Convention in 30 out of 41 judgments involving the United Kingdom (not including decisions by the Court as to whether a successful applicant received just satisfaction under Article 50).[88] Table 3 in Appendix D

---

[87] The Employment Protection (Consolidation) Act 1978 was influenced by Art. 8(1) of the Charter on paid leave for women employees before and after childbirth.

[88] For discussion of these cases, see in particular, A. W. Bradley, 'The United Kingdom before the Strasbourg Court 1975–1990', in W. Finnie, C. Himsworth, and N. Walker (eds.), *Edinburgh Essays in Public Law* (1991), 185–214; F. J. Hampson, 'The United Kingdom before the European Court of Human Rights' (1989) 9 *Yearbook of European Law* 121.

shows the details of the United Kingdom cases before the Court, by outcome.

We have seen several successful cases against the United Kingdom across a broad range of issues and areas: the closed shop, contempt of court, sexual orientation, anti-terrorism legislation, prisoners' rights, immigration policy, and interrogation methods, to name but a few. We have seen, too, that significant changes have been introduced into British law as a result of such successful cases (as well as following complaints to the Commission which have not proceeded to the Court or the Council of Ministers).[89] Though there are difficulties involved in trying to answer the question of how far the United Kingdom has complied with decisions, research by Churchill and Young[90] has estimated that, in 23 out of 41 cases between 1975 and 1989, there is not much doubt that the UK has complied. In most cases compliance involved the enactment of legislation. Sexual activity between men in Northern Ireland, the Isle of Man, and the Channel Islands was decriminalized; parents were given opportunities to use the judicial process to challenge decisions regarding access by them to their children; a new Interception of Communications Act was enacted; several interrogation techniques were abandoned; a new Contempt of Court Act was brought into force. In 17 cases they consider it debatable whether there had been compliance (see Table 3 in Appendix D). In only one case, however, did they find that there clearly was no compliance by that time: judicial birching in the Isle of Man.[91]

In addition to the direct enforcement of the Convention there are, also, important *indirect* influences which the Convention may have, on public opinion and political debate, and on the exercise of the discretion of administrators and the interpretation of judges. We have seen an example of the influence of the Convention in a non-judicial context, when Article 3 of ECHR influenced the Royal Commission on Criminal Procedure's decision to recommend an admissibility test for confessions that excluded confessions obtained by torture, inhuman, or degrading treatment.[92] It

[89] See Liberty, *Standing Up*, 111–12, for a discussion.

[90] R. R. Churchill, 'Aspects of Compliance with Findings of the Committee of Ministers and Judgements of the Court with Reference to the United Kingdom', in J. P. Gardner (ed.), *The European Convention on Human Rights: Aspects of Incorporation* (1992). See also R. R. Churchill and James Young, 'Compliance with Judgements of the European Court of Human Rights and Decisions of the Committee of Ministers: The Experience of the United Kingdom, 1975–1987' (1991) 62 *British Yearbook of International Law* 283–346.

[91] *Tryer* (1980) 2 EHRR 1.

[92] For a useful discussion of other examples, see C. Symmons, 'The Effect of the European Convention on Human Rights on the Preparation and Amendment of Legislation, Delegated Legislation and Administrative Rules in the United Kingdom,' in M. P. Furmston, R. Kerridge and B. E. Sufrin (eds.), *The Effect on English Domestic Law of Membership of the European Communities and of Ratification of the European Convention on Human Rights* (1983), 387.

seems clear, however, that the most significant use of the Convention has been in the context of legal proceedings, the context to which we now turn.

Research by Bratza[93] in 1991 indicated that the Convention was first referred to in a judgment of the British courts in 1974.[94] A Lexis search found that the Convention has been referred to in the judgments of domestic courts and tribunals well over 200 times. His analysis of these cases indicated that 'in the majority of cases the reference has been a passing one. But what an analysis of the authorities does show is an increasing readiness on the part of the Courts to consider the Convention even in cases which do not at first sight appear to raise Convention issues.' The use of the Convention was apparent, first, as an aid to statutory interpretation, particularly in the context of the Immigration Act 1971.[95] After some uncertainty, it seems that the courts are now consistent in requiring that recourse may be had to the Convention for the purpose of resolving ambiguities. Where the statute is clear and unambiguous, however, there is no room for introducing the Convention.[96] Even this limited use of the Convention, in Bratza's view, 'has in practice proved of little value'. The usefulness of the principle has been undermined 'by . . . an over readiness on the part of the courts to hold that the words of a statute are plain and unambiguous'.[97] He does, however, consider that one area in which the Convention has proven of some value in statutory interpretation is in the use of Article 3 jurisprudence concerning torture, or inhuman or degrading treatment.[98]

The second use of the Convention was as guidance in the development of the common law. Despite early reticence regarding the use of the Convention, for example by Megarry J. in *Malone*, the Convention has proven 'of some limited value in resolving uncertainties in the common law in some areas', particularly in the context of contempt of court,[99] prisoners' rights,[100] and freedom of speech.[101] In 1992 the Court of

---

[93] N. Bratza, 'The Treatment and Interpretation of the European Convention on Human Rights: Aspects of Incorporation', in Gardner (ed.), *European Convention* (forthcoming).

[94] *R.* v. *Miah* [1974] 1 WLR 683.

[95] *Birdi* v. *Secretary of State for Home Affairs*, unreported 11 Feb. 1975; *R.* v. *Home Secretary, ex parte Bhajan Singh* [1976] 1 QB 198; *R.* v. *Secretary of State for the Home Dept., ex parte Phansopkar* [1976] 1 QB 606.

[96] Lord Bridge in *R.* v. *Secretary of State for the Home Dept., ex parte Brind* [1991] AC 696 (HL).

[97] e.g. *R.* v. *Secretary of State for the Home Dept., ex parte K.* [1990] 1 WLR 168.

[98] e.g. *Williams* v. *Home Office* (No. 2) [1981] All ER 1211, *R.* v. *Secretary of State for the Home Dept., ex parte Herbage* [1987] QB 872 and *Woldon* v. *Home Office* [1990] 3 WLR 465.

[99] Use of Art. 10: *AG* v. *BBC* [1981] AC 303 (HL) (but see in contrast *R.* v. *Liverpool Post plc* [1990] 2 WLR 494).

[100] Art. 6 (and *Golder* v. *UK* (1975) 1 EHRR 524) relied on in *Raymond* v. *Honey* [1983] 1 AC 1, and *R.* v. *Secretary of State for the Home Dept., ex parte Anderson* [1984] QB 778.

Appeal in the *Derbyshire County Council* case appeared to have accepted this development.[102] The Court considered that the common law was uncertain on the issue of whether a local authority had the right to maintain an action for defamation, and that in this situation it was appropriate to have regard to the Convention. Laws J., extra-judicially, has argued that this trend of decisions justifies the use of the Convention as a body of relevant jurisprudence, which judges may have regard to in the continuing process of developing the common law and may, indeed, use the Convention as a text to inform the common law.[103]

Third, in the field of judicial review, initially Lord Denning considered that ministers and officials exercising statutory powers were obliged to take account of the Convention.[104] The next year, however, he regarded this obligation as asking too much of officials.[105] This revised opinion was supported by the Court of Appeal in 1981, with Ackner LJ holding that the Secretary of State was not obliged to take the Convention into account.[106] Though there was relatively little discussion of the issue judicially during the 1980s,[107] there was a substantial debate extra-judicially as to the appropriate use of the Convention in public-law. It was argued by some[108] that the ground of judicial review which required decision-makers not to act unreasonably could and should be interpreted as requiring adherence to the European Convention requirements by that decision-maker.[109] However, in the decision in *Brind* v. *Secretary of State for the Home Department*,[110] in which this argument was put directly to the House of Lords, their Lordships appear to have largely rejected it, while leaving open the possibility that the courts would consider whether a decision-maker had taken account of the Convention where relevant,

[101] Blasphemy: *R.* v. *Lemon* [1979] AC 617; *R.* v. *Chief Metropolitan Stipendiary Magistrate, ex parte Choudhury* [1991] QB 429 (DC); and in the interlocutory appeal in *Spycatcher* [1987] 3 All ER 342.

[102] *Derbyshire CC* v. *Times Newspapers Ltd.* [1992] 3 All ER 65. The House of Lords, however, reached its conclusion upon its interpretation of the common law without finding any need to rely on the Convention, [1993] 1 All ER 1011 (HL).

[103] J. G. Laws, 'Is the High Court the Guardian of Fundamental Constitutional Rights' (1993) PL 59.

[104] *R.* v. *Secretary of State for the Home Dept., ex parte Bhajan Singh* [1978] 2 All ER 1081 at 1083.

[105] *R.* v. *Chief Immigration Officer, Heathrow Airport, ex parte Salamat Bibi* [1976] 3 All ER 843 at 847–8.

[106] *Fernandes* v. *Secretary of State for the Home Dept.* [1981] Imm. AR 1.

[107] Except for *R.* v. *Secretary of State for the Home Dept., ex parte McAvoy* [1984] 1 WLR 1408; *R.* v. *Board of Visitors of HM Prison The Maze, ex parte Hone* [1988] 2 WLR 177; *R.* v. *General Medical Council, ex parte Colman* [1990] 1 All ER 489.

[108] A. Lester and J. Jowell, 'Beyond Wednesbury: Substantive Principles of Administrative Law' (1987) PL 368.

[109] See further discussion in J. G. Laws, 'The Ghost in the Machine: Principle in Public Law' (1989) PL 27; Sir H. Woolf, *Protection of the Public: A New Challenge* (1989).

[110] [1991] 1 All ER 720.

while stopping short of examining how that decision-maker had evaluated the Convention.[111] It is most unlikely that the *Brind* decision is the final word on the subject.

Bratza concluded, and it is difficult to fault his analysis, that

in the vast majority of the 200 plus cases where the Convention has been referred to, the Convention appears to have made little or no difference to the result at which the Court has arrived. More often than not, the Convention has been invoked by the Court to reinforce the view which has already plainly been formed by the Court as to the proper outcome of the case.

We can see, therefore, that while the Convention has been highly significant directly, indirect use in the courts has proven of limited utility thus far. Could it be made more effective? Several contributors to this book have tended to be sceptical for two main reasons. First, several contributors point to the limited coverage of the Convention articles, or the breadth of the exceptions for which they provide. The potential influence of the Convention in the area of police powers is not likely to be pervasive, simply because certain cherished rights do not fall within the Convention or do so only to a limited extent, for example the right to silence, or the privilege against self-incrimination. The Convention, indeed, is sometimes seen as seldom more than of rhetorical significance outside the particular context of terrorism. On due process, Jackson points to the fact that Article 6 is designed to deal with procedural irregularities in the administration of justice and is not concerned with whether the domestic courts have correctly assessed the evidence. The Article does not therefore provide a remedy for miscarriages of justice *per se*. Given the breadth of the exceptions in Article 11, Gearty considers it is unlikely that incorporation of the Convention would presage a wholesale revision of public-order law. As regards minority rights, Poulter points out that there is no equivalent of Article 27 of the ICCPR in the ECHR. Collins and Meehan draw attention to the limited practical impact of the Convention on the development of the right to non-discrimination in employment for women in the UK because, despite its apparently widely phrased wording, Article 14 does not provide for any independent right not to be discriminated against.

A second reason for scepticism regarding the future utility of the Convention arises from what several contributors perceive as the restrictive interpretation by the European Court of Human Rights of those rights which are guaranteed. On sexual orientation, Wintemute points to the limited interpretation of 'family life' and 'private life' in Article 8(1). And

---

[111] See further A. R. Mowbray, 'Administrative Law and Human Rights' (2 Aug. 1991) *NLJ* 1079.

the use of the consensus test under Articles 8(2) and 14 has little bite in the absence of a European consensus. For freedom of association, the Convention has proven to be of little value, with the wide exceptions to the right of freedom of association leading to the dismissal of the complaint regarding de-unionization at GCHQ, with the Court being prepared to accept wide restrictions on civil servants' freedom to associate. Moreover, according to Ewing, the Convention has been used to undermine group freedom and the power of people acting in association to promote common ends and interests. In the area of race discrimination, the concept of discrimination in Article 14 adopted by the European Commission and Court of Human Rights appears to be largely 'direct' discrimination and has not yet been clearly stated to include 'indirect' discrimination. In addition, discrimination may be justified by the state on considerations derived from the public interest, and a wide margin of discretion is given to the national authorities in appreciating the weight to be accorded to the public interest.

Gardner goes even further in his criticism of the European Court of Human Rights, arguing that, in the context of freedom of expression, the Court has become increasingly deferential and increasingly compromise-based in its approaches. This, in his view, is making the Convention a safe instrument for domestic courts to invoke. Gardner draws attention to what he regards as the decreasingly rigorous approach to freedom of expression adopted by the European Court of Human Rights. Reliance on the 'margin of appreciation' has become an increasingly predictable feature of Article 10 cases. There has been a perceptible move towards the idea that the interests enunciated in Article 10(2) are to be balanced against the interests of freedom of expression, rather than treated as specific narrow exceptions to its supremacy.

Several further problems arise in the direct use of the Convention before the Convention institutions, which limit its impact. The first problem relates to the substantial delays in taking cases through the system.[112] The number of individual applications to the Commission has increased enormously, resulting in unacceptable delays. By January 1993 the backlog of cases amounted to some 2,500 applications (of which the Commission had not even looked at over 1,500). Nor is this increase likely to stabilize at even this level, particularly given the recent and forthcoming ratifications of the Convention discussed in Chapter 1. Nor is the position before the Court of Human Rights any more satisfactory. On average it takes the Court five years to reach a decision. Proposals have been put forward to simplify and streamline the machinery, concentrating on

---

[112] The information in this paragraph draws on A. Drzemczewski, 'The need for a radical overhaul', *NLJ*, January 29, 1993, p. 126.

replacing the existing two-tier mechanism with a single full-time European Court of Human Rights.

Other problems concern the relationship between the Convention machinery and national remedies. The two crucial provisions in the Convention are Article 13 and Article 26. Article 13 of the Convention, as we saw in Chapter 1, requires that states must provide an effective remedy before a national authority to anyone whose Convention rights and freedoms are violated. Article 26 permits the Convention institutions to consider an individual case only after all domestic remedies have been exhausted. Some have argued, first, that Article 13 requires a remedy by which the Convention itself can be pleaded before the domestic authorities concerned, and second, that where this is not possible Article 26 should be regarded as more easily satisfied. However, as the Secretary to the European Commission has stated, the first argument 'has not been adopted in the jurisprudence of the Convention organs. It is now established that the text of the Convention does not impose any obligation on the High Contracting Parties to incorporate the Convention rights into their domestic law.'[113] As regards the consequences of non-incorporation for the interpretation of Article 26, the same author has written that,

The basic approach adopted by the Convention organs for the interpretation of Article 26 is a substantive rather than a formal one. It is the substance of any complaint raised before the Convention organs that must previously have been put before the competent domestic authorities in order to allow them to redress the situation complained of on the national level.

It is, he continues,

the intrinsic aim of the domestic remedies rule to enable the state to redress any possible violation of the Convention on the national level. Having regard to the substantive approach applicable in this respect, this aim can be achieved both in States which have and which have not incorporated the Convention. However the former will as a rule be better equipped to avoid findings of violations of the Convention, in particular if their law provides for the Convention to take precedence over other legal norms or for the interpretation of those norms in the light of the Convention, including its interpretation of the Strasbourg Convention organs.

The approach taken in the *Abdulaziz, Cabales and Balkandali* case[114] is a good example of this view being applied by the European Court of Human Rights. In this case the applicants argued that they had been discriminated against on several grounds and that they had no effective

---

[113] H. C. Kruger, 'Does the Convention Machinery Distinguish between States which have and have not Incorporated Textually, or Statistically, or Substantively?', in Gardner (ed.), *European Convention*.

[114] *Abdulaziz, Cabales and Balkandali* v. *UK* (1985) 7 EHRR 417.

domestic remedy for their complaints of discrimination. The Court, having found that the discrimination on the ground of sex was contrary to the Convention, went on to hold that there was no effective domestic remedy. The United Kingdom had not incorporated the Convention into its domestic law and so the immigration rules could not have been challenged as contrary to the Convention. The other available channels of complaint (the immigration appeals system, representations to the Home Secretary, and application for judicial review) could have been effective only if the complaint alleged that the discrimination resulted from a misapplication of the immigration rules. Yet this was not the case. Since the discrimination did not therefore contravene domestic law, there was a violation of Article 13.

However, the United Kingdom government has consistently argued before the Convention institutions that the judicial review procedure provides an effective remedy in many other situations. In several cases, the European Court of Human Rights has accepted this argument. In the *Soering* case,[115] the Court considered judicial review proceedings to be an effective remedy. Soering had complained that his extradition by the UK government to the United States to stand trial for murder would breach Article 3, so long as there was a risk of the death sentence being imposed. The Court was satisfied that the domestic courts could review the 'reasonableness' of an extradition decision in the light of the kind of factors relied on by Soering before the Convention institutions. It accepted the argument of the United Kingdom government that a court would have jurisdiction to quash a challenged decision to send a fugitive to a country where it was established that there was a serious risk of inhuman or degrading treatment, on the ground that in all the circumstances of the case the decision was one that no reasonable Secretary of State could take.

In the *Vilvarajah* judgment,[116] in which a decision to remove five Sri Lankan asylum seekers to Sri Lanka was argued to be in breach of Article 3, the Court also held that judicial review proceedings provided an effective remedy for the purposes of Article 13. The applicants sought to distinguish their case from that of *Soering*, arguing that while judicial review might be an effective remedy where the facts were not in dispute between the parties and the issue was whether the decision was such that no reasonable Secretary of State could have made it, judicial review was not an effective remedy where the factual question of the risks to which they would be exposed if sent back to Sri Lanka was the substance of the dispute. Judicial review, in the submission of the applicants, does not

---

[115] *Soering* v. *UK* (1989) 11 EHRR 439.
[116] *Vilvarajah and others* v. *UK* (1992) 14 EHRR 248.

control the merits of the Secretary of State's refusal of asylum, only the manner in which the decision on the merits was taken. In particular, judicial review does not ascertain whether the Secretary of State was correct in his assessment of the rules to which the applicants would be subjected. The Court, however, accepted the UK government's argument that judicial review does have the effect of controlling the merits of the decision, despite a strongly worded dissent by the Irish judge, Judge Walsh.

The European Community

One of the most significant differences from the 1950s is the increasing role and importance of European Community law for Britain. There are several reasons why European Community law may be advantageous for applicants unable to rely on statute or common law. In those areas in which there are Community law norms, it has proven in turn more flexible, more principled, more dynamic, more open to argumentation, more general, wider, and deeper. Most important of all, it is superior to domestic legislation. It can be used to fill in the gaps of domestic legislation, but also to attack legislation directly.

In the context of the protection of human rights in Britain, these possibilities have obvious attractions to some, and the popularity of attempting to inject an element of European law into a dispute has increased considerably in the last ten years. Britain's membership of the Community has been of immense importance, in particular, for the legal treatment of women's rights. Collins and Meehan argue that the EC's influence has, indeed, been decisive. On other human rights issues, however, the Community plays a lesser role. In the context of immigration, as Dummett shows, there are respects in which an EC national from another state has rights superior to those of a British citizen, but it has had relatively limited impact on the general regulation of immigration in Britain. Although, in the context of minority rights, the 1977 EEC Directive on the Education of Children of Migrant Workers is relevant, Poulter describes the impact of the directive as 'negligible' and argues that the gaps revealed in its implementation have not been subject to proceedings either in the United Kingdom or in Luxembourg. We can note in passing that civil liberties issues may also arise indirectly by virtue of the economic aspects of the Treaty, for example the free movement of goods provisions can affect freedom of expression questions.[117]

The interest in the human rights role of the Community, apart from the

---

[117] In *Conegate Ltd.* v. *Customs and Excise Commissioners* [1987] QB 284 (ECJ) free movement of goods was applied to pornography.

area of women's rights, is more directed to its potential, rather than its achievement to date. Several authors indicate areas in which it could in the future play a much more significant role *vis-à-vis* civil liberties and human rights in Britain. Gardner points out that the EC will be more involved in freedom of expression issues in the future. In the area of sexual orientation, Wintemute mentions that the harassment of lesbians and gay men has been included in the recent Code of Practice on Sexual Harassment and the Commission has arranged for a study by the Stonewall group on sexual orientation discrimination in the EC. McCrudden, too, points to a study on racial discrimination recently sponsored by the Commission as indicating some possibility of further EC activity in this area. There have been limited developments since the mid-1980s which may lead to the EC taking a more active role.[118] And the impact of the Maastricht Treaty is potentially considerable, though we should note that freedom of association issues have been specifically excluded from the ambit of the Treaty. What has so far been absent, however, is any clear conception of a human rights dimension to Community policy as such. It remains to be seen whether the increasingly close links between the Community and the European Convention on Human Rights, discussed in Chapter 1, helps to encourage such a dimension. Of relevance too is the extent to which effective democratic accountability of the EC institutions themselves is improved.

Other international obligations

The importance of the other international human rights treaties, organizations, and initiatives discussed in the introduction has been considerable both for the international community as a whole, and for specific countries in particular. Without gainsaying this, it is also true, however, that their impact on British domestic policy has been largely confined to standard-setting. Domestic legal changes have undoubtedly occurred before some international conventions have been ratified, and in order to ratify them. Thereafter, however, the general requirement to report progress periodically to a supervisory committee has usually been the limit of further involvement.

The only notable exception to this general trend is with regard to Northern Ireland which has been raised as an issue more or less continuously in various human rights forums since the early 1970s, not only by the Republic of Ireland, and by pressure groups internal and external to Northern Ireland, but also by other states, much to the embarrassment

---

[118] For a general discussion see M. Spencer, *1992 and All That: Civil Liberties in the Balance* (1990).

of the United Kingdom. Most recently, as we have seen above, Watch Committees, drawing their inspiration from CSCE standard-setting, have produced various reports on aspects of government policy in Northern Ireland.

UNDERPINNING HUMAN RIGHTS

In previous pages, we have attempted to bring together an assessment of the role of various mechanisms in furthering human rights in Britain. It is appropriate at this point to attempt to systematize this discussion somewhat further and examine which kinds of mechanisms provide a firm foundation for the protection and development of human rights in Britain. This may give some assistance in deciding where to give priority for future reforms.

Barnum, Sullivan, and Sunkin[119] have considered recent theories of the underpinnings of political freedom in Britain and the United States. Drawing on this analysis[120] we may usefully distinguish between three broad competing theories of what particularly affects the protection of civil liberties and human rights of the types considered in the previous chapters. The first theory identifies the values held by ordinary citizens as constituting the key underpinning. Citizens must have 'internalised a critical array of fundamental norms'.[121] Without internalization, such freedoms and rights will not be protected; with it, such freedoms and rights may well be protected. Citizen awareness and support is regarded, acording to this approach, as a necessary, and even by some as a sufficient condition. Thus Ewing and Gearty have argued that 'liberty flows *only* from democracy'.[122]

A second theory argues that it is politicians, opinion-formers, civil servants, and other élites whose support is necessary to protect civil liberties and human rights from pressures from a public which may be unsympathetic to such protection. As the authors point out, an assumption on which this second theory is based is that 'particular political and legal institutions will be staffed by individuals who are especially supportive of the rules of the game of democratic politics'.[123] Those institutions, in turn, 'will play a key role in protecting political freedom and other

---

[119] D. G. Barnum, J. L. Sullivan, and M. Sunkin, 'Constitutional and Cultural Underpinnings of Political Freedom in Britain and the United States' (1992) 12 *Oxf. J. Legal Studies* 362–79.

[120] Though our analysis differs from theirs to the extent that we initially conflate their third and fourth categories.

[121] Barnum, Sullivan, and Sunkin, 363.

[122] Ewing and Gearty, *Freedom under Thatcher*, 275 (emphasis added).

[123] Barnum, Sullivan, and Sunkin, 365.

democratic norms'.[124] There are several variants of a third theory which may draw on both of the previous theories to some extent, but tends rather to concentrate on such institutional and organizational features of constitutional arrangements as the independence of the judiciary, the availability of specific legal remedies for violations of rights, federalism, the separation of powers, Bills of Rights, judicial review of administrative action, and other such mechanisms of constitutional checks and balances.[125]

Political culture and public opinion

A major focus of this chapter has been a discussion of the institutional and organizational aspects of Britain's changing constitution and on the degree of protection these afford to the civil liberties and human rights of its citizens. The importance of these mechanisms lies in their ability to be used as tools. The use of legal machinery and institutions, for example, creates judgments and precedents which clarify and develop the extent and context of rights and liberties, as has clearly occurred with prisoners' rights. And this is why access to the courts and to the legal machinery is such an important topic. However, in addition to describing the legal parameters of rights and liberties, it is necessary to examine what use is made of the mechanisms, and by whom and with what success, and what are the determinants of that success. This is to raise questions about the depth of and commitment to human rights concerns in British political culture. If it is correct to think of constitutional mechanisms as being tools available for use, and to characterize political and legal institutions as reactive rather than proactive, questions need to be raised about the level of commitment to a civil liberties culture. Hence our review earlier in this chapter of opinion poll evidence on public attitudes towards human rights issues, and on the role of pressure groups, political parties, and the electoral process. We may tentatively conclude from the evidence of the previous chapters that citizen's values, as expressed at the present time through existing institutions, may beneficially affect the depth and degree of protection of human rights, but not necessarily so, and not in a simple linear way.[126]

---

[124] Ibid. C. Palley combines the first and second theories when she argues: 'Unless public, officials and lawyers are imbued with human rights ideology, lip service to, rather than respect in practice for, human rights will frequently be the outcome.' *The United Kingdom and Human Rights* (1991), 3.

[125] Testing these theories rigorously is a daunting task and we do not attempt it here. In this conclusion we offer merely a preliminary attempt to draw together some evidence from the previous chapters and elsewhere which seems to us relevant for others attempting a more rigorous analysis, which we hope will occur.

[126] Comparative opinion poll evidence tends to support this conclusion. Barnum, Sullivan, and Sunkin tested levels of tolerance in Britain and the United States with results which

Élite values

However, it could be argued that concentration on public opinion, political parties, and pressure groups fails to understand the process of policy-making in Britain, since it omits an important element in the process: the pervasive role of British political élites. It has not been the purpose of this book to examine systematically the contribution of political élites (other than lawyers) to developments in the human rights field. It would, however, be a serious omission to pass up this opportunity to point out their importance in the promotion and development of human rights and civil liberties legislation.

The theory of élite values, it will be remembered, argues that it is political and other élites whose support is necessary for the protection of human rights against an actually (or potentially) unsympathetic populace. There certainly seems to be some evidence of a tradition, in the British context, that regards the élite theory as more credible than the citizen's values theory. This is at least a reasonable inference, we think, from the discussion earlier in this chapter. There is also evidence that this theory is supported by some of the experiences in Britain over the past forty years. Barnum, Sullivan, and Sunkin argue also that, on the basis of their researches, political élites in both Britain and the United States are 'substantially more tolerant than ordinary citizens',[127] finding that tolerance among national legislators exceeds tolerance among the general public by 30 to 60 points. On this they argue that their findings are consistent with the theory that 'support for democratic norms among political élites is a critical ingredient in the process of safeguarding political freedom'.[128] However, the previous chapters have also shown that reliance on such élites would be to overestimate their potential effectiveness as protector of human rights.

Constitutional mechanisms

What then of the British constitutional tradition? Can it be said to pro-vide a crucial and stable underpinning of human rights? Before addressing

---

indicated 'that levels of attitudinal tolerance are strikingly similar' in both countries (p. 368). The study also indicated, however, that 'tolerance was not very widespread in *either* Britain or the United States' (pp. 368–9). This finding they regard as consistent with earlier studies. They conclude from this that, on the assumption that both Britain and the USA have in fact a high level of tolerance in the areas polled, 'that widespread grassroots support for democratic norms is evidently not a prerequisite of democratic politics' (p. 370).

[127] Barnum, Sullivan, and Sunkin, 375.

[128] Ibid. 376.

[129] E. Burke, *Reflections on the Revolution in France* (1790), Penguin edn. (1986), 117–53.

these questions, however, we need first to try to identify more specifically of what this tradition consists.

An important theme running through British thought concentrates on history and tradition when evaluating the processes by which political and legal decisions are made and the results of such processes. The process and the results are assessed pragmatically. Problems are solved, in this empiricist tradition, on the basis of experience. Solutions are what works and what lasts. Institutions should therefore operate flexibly, learn from the past, and develop to suit the conditions of their time.[129] This is the essence of the common-law tradition. The British *constitutional and human rights* tradition is also heavily imbued with this approach. A number of the principles which are said to describe, inform, and underpin the British constitution (majoritarian democracy, parliamentary supremacy, and constitutional conventions) may be seen as the embodiment of this tradition, concentrating as they do on 'the authority of experience and the continuity of practice'[130] and ensuring the flexibility of the process by which decisions are made.[131] In this tradition, authoritative constitutional structures *evolve*, they are seldom *made*.

The previous chapters and this chapter have sought to provide some information on which to judge the successes and problems of the pragmatic empiricist model. The picture which emerges is, in our view, a complex and nuanced one. The tradition has, on the one hand, allowed intolerable breaches of human rights to continue unremedied, while, on the other hand, it has provided a general level of human rights protection which is not noticeably worse than that of many other modern democratic states with different constitutional traditions. For some, however, the reliance on pragmatism is, at best, disappointing and should be altered. Robertson argues: 'Our preference for pragmatism rather than principle has been convenient in the short term; its cost may be measured in the gaps which still remain in the rights of British citizens to obtain speedy, effective or indeed any remedy against abuses of private and public power.'[132]

Several features of the British political and legal system have altered radically during the course of the period 1950 to 1993. Perhaps more surprisingly, most features have remained relatively constant, though from the perspective of the 1990s, both the constitutional edifice and these methods of civil liberties protection look considerably more battered than they did in 1950. This is hardly surprising. As we have seen, the

---

[130] M. Loughlin, 'Tinkering with the Constitution' (1988) 51 *MLR* 531, 536.

[131] This section draws extensively on an earlier discussion by C. McCrudden, 'Northern Ireland and the British Constitution', in J. Jowell and D. Oliver, *The Changing Constitution*, 2nd edn. (1989), 297–342.

[132] G. Robertson, *Freedom, the Individual and the Law*, 6th edn. (1989), 18.

British legal and political systems have had to cope with domestic and international developments of the most enormous range, importance, and complexity in the forty-year period since the 1950s. If some features of the current British constitutional and legal approach to human rights issues would strike someone in the 1950s as very familiar, other features would seem significantly different. The British constitutional, legal, and political structure has reacted to the developments outlined in the previous sections in significant ways. The traditional approach has been both supplemented and to a degree challenged in several respects.

The pragmatic empiricist model of constitutional practice is so deeply embedded in the British legal and political tradition that it requires a leap of the imagination for us to consider any fundamentally different arrangement. An alternative model of constitutional practice may, however, be simply stated. Imagine a constitutional tradition which concentrates first and foremost on setting out a number of values which political and legal institutions are required to further. Imagine that the decisions of these institutions are assessed on the basis of their conformity to this set of values, and are to that extent constrained and confined. What counts in this alternative tradition are the values captured by the processes of decision-making and furthered by particular decisions, rather than the fact that the process has evolved and survived. We may call this the idealist approach. A major development in constitutional mechanisms of this type has been the adherence of the United Kingdom to the European Convention on Human Rights. So too, the expansion of judicial review of administrative action in the past thirty years has been interlaced with considerable idealist tendencies, despite the absence of a written constitution. The United Kingdom's membership of the European Community brings another explicitly 'idealist' element into an otherwise largely pragmatic empiricist constitutional tradition.

One of the questions which has dominated much debate over constitutional reform in Britain is whether a significantly greater idealist element should be injected into British political and legal decision-making. A second question is what those values should be. A third, separable, question is how those values should be incorporated, refined, changed, and interpreted.

There is a tendency in some of the recent British literature on human rights to assume that answers to the first and second questions automatically supply an answer to the third. Others reformulate the third question as being largely about the merits of a written constitution and an entrenched bill of rights for Britain. This would, in our view, be a mistake. It is true, of course, that, written constitutions are more likely to be idealist in their approach, but not necessarily. Thus, for example, it is clear that the Supreme Court of Israel not infrequently adopts an idealist approach to the protection of human rights, despite the fact that there is no written constitution.[133] To elevate the written nature of the constitutional approaches prevalent in North America and Germany to become a *necessary* element of idealist constitutional developments underestimates these contrary examples, as well as the example of the United Kingdom's reception (though not incorporation) of the European Convention on Human Rights. Indeed, it could be plausibly argued that the pragmatic nature of the British constitution has enabled European Community law, for example, to be received more easily and effectively in Britain than in several other countries.

Nor does it follow that, when a community's values include fundamental individual rights, that should necessarily lead to judicial enforcement of those rights, though some argue otherwise.[134] As Calabresi has argued, 'the definition of fundamental rights and the judicial enforcement of those rights are two very different inquiries'.[135] (Or, as Waldron has argued, there is 'no necessary inference from a right-based position in political philosophy to a commitment to a Bill of Rights as a political institution along with an American-style practice of judicial review'.[136]) Calabresi continues:

In any given polity, with its own peculiar traditions, demography, and history, rights could be enforced by judges . . . or by the political process generally, that is by majoritarian legislatures, elected executives, administrative bureaucrats, or referenda voters. Of course the choice is not exclusive: rights could be enforced by all of these concurrently or by each of them at particular times and under particular circumstances.[137]

Conclusion

We have viewed our role as drawing some conclusions from the experience of the past forty years, rather than making yet more proposals for the future. Clearly, however, we hope that the perspectives to be found

---

[133] See A. Maoz, 'Defending Civil Liberties without a Constitution: The Israeli Experience' (1988) 16 *Melbourne Univ. LR* 815, and more generally D. Dyzenhaus, *Hard Cases in Wicked Legal Systems* (1991).

[134] An example is to be found in Robertson, *Freedom, Individual*, 6th edn. (1989), 400: 'The courts remain the *only* place where oppressive Government action against individuals may be checked' (emphasis added).

[135] G. Calabresi, 'Foreword: Antidiscrimination and Constitutional Accountability (What the Bork–Brennan Debate Ignores)' (1991) 105 *Harv. LR* 80.

[136] J. Waldron, 'A Right-Based Critique of Constitutional Rights', 13 *Oxf. J. Legal Studies* (1993), 18. See further, for comparisons between the USA and the UK, D. G. Barnum, 'Constitutional Organization and the Protection of Human Rights in Britain and the United States', in J. R. Schmidhauser (ed.), *Comparative Judicial Systems* (1987), 183; S. Lee, 'Bicentennial Bork, Tercentennial Spycatcher: Do the British Need a Bill of Rights', 49 *Univ. Pittsburg LR* (1988), 777.

[137] Calabresi, 'Foreword', 84.

here will contribute to the continuing debates on how best to secure the optimal protection of human rights in Britain. We can say, on the basis of the previous chapters and this chapter, that in the British context each of the three theories of the underpinnings of human rights (citizen's values, élite values, and constitutional mechanisms) has an important element of truth embedded in it. None of the three, however, should be supposed to be entirely reliable in the absence of the other two. To rely solely on one to the exclusion of the others, or to increase the importance of one to the significant weakening of the others, would be rash in the extreme. The same principle applies with respect to the variety of constitutional mechanisms and institutions whose role has been reviewed above. Reforms should be considered not only on the basis of whether they are desirable when considered by themselves, or whether they work in other countries, but also by taking into account what impact they are likely to have on existing mechanisms, imperfect as these inevitably are. If there is one general conclusion which can be drawn, it is that what is needed is a diversity and variety of techniques for the advancement of human rights. Techniques are needed which have the potential to create a multiplicity of risks for those who would seek to violate human rights, and a multiplicity of opportunities for those concerned with human rights to have their voices heard.

APPENDIX A

Significant Events in British and International Politics since 1950 and Key Statutory Developments

Year	Domestic events	Key statutes	International events
1949			Council of Europe established
1950	End of petrol rationing		Korean War begins
			McCarthy launches anti-communist crusade in USA
			First kidney transplant operation takes place in USA
			China invades Tibet
1951	Conservative government elected		First H-Bomb tested
	Festival of Britain		Iran nationalizes UK oil fields
			Libya becomes independent nation
1952	Britain tests first atomic bomb	Prison Act	
		Defamation Act	

1953	Relaxation of ban on civil servants' political activities	Death of Stalin Workers uprising in East Berlin
1954		Division of Vietnam along 17th parallel Racial segregation outlawed in US schools
1955	Conservatives re-elected ITV begins transmissions	Warsaw Pact signed West Germany becomes a sovereign state
1956	IRA begins border campaigns in NI	Hungarian uprising Suez crisis British troops sent to Cyprus to quell independence movement
1957	Publication of Wolfenden Report	Treaty of Rome creating EEC signed by six nations
1958	Macmillan's 'Wind of change' speech Notting Hill and Nottingham race riots First CND London to Aldermaston march 'Justice' founded	

Year	Domestic events	Key statutes	International events
1959	Conservatives re-elected	Obscene Publications Act	
1960	Ending of National Service		Sharpville killings
1961	Contraceptive pill goes on sale in UK		South Africa quits the Commonwealth and becomes a Republic Berlin Wall erected
1962		Commonwealth Immigrants Act	Cuban missile crisis Algeria becomes independent from France
1963	Profumo affair		Assassination of John Kennedy First nuclear test ban treaty
1964	Mods and rockers clash in Margate Labour government elected	Police Act	Nelson Mandela sentenced to life imprisonment Civil Rights Act signed in USA US attacks air bases in N. Vietnam
1965	Death Penalty abolished First woman High Court judge appointed	Race Relations Act	US sends Marines to Vietnam Civil rights marches in Selma, Ala. Race riots in Watts, LA Rhodesia declares UDI
1966	Labour re-elected UK grants right of petition under ECHR Timothy Evans, hanged in 1950, granted pardon		Mao proclaims Cultural Revolution

1967	Partial decriminalization of male homosexuality	Sexual Offences Act Matrimonial Homes Act Abortion Act Criminal Law Act Welsh Language Act	Six day Arab–Israeli war Colonels seize power in Greece Biafran rebellion in Nigeria Che Guevara shot in Bolivia
1968	Civil rights marches in NI Grosvenor Square anti-Vietnam war demonstrations results in 300 arrests Enoch Powell makes 'rivers of blood' speech *Last Exit to Brooklyn* obscenity conviction quashed	Commonwealth Immigrants Act Race Relations Act Caravan Sites Act Criminal Appeal Act Criminal Justice Act	Martin Luther King assassinated Student revolts in Paris and Prague Spring
1969	British troops sent to NI Bernadette Devlin elected to UK parliament from NI Anti-apartheid protest disrupts Springbok rugby tour	Children and Young Persons Act Family Law Reform Act	Yasser Arafat elected leader of PLO Death of Ho Chi Minh
1970	Conservative government elected	Equal Pay Act	Anti-war protesters shot by National Guard at Kent State University
1971	Internment introduced in NI Angry Brigade bomb the house of the Employment Secretary First British soldier shot in NI Post Office tower bombed by IRA Oz trial	Prevention of Terrorism Act Industrial Relations Act Immigration Act Divorce Reform Act Misuse of Drugs Act	Idi Amin seizes power in Uganda Civil War between East and West Pakistan US Supreme Court supports publication of 'Pentagon Papers'

Year	Domestic events	Key statutes	International events
1972	'Bloody Sunday' killings Direct rule by Westminster imposed on NI	Matrimonial Proceedings (Polygamous Marriages) Act	Black September attack at Munich Olympics Watergate burglary Nixon visits China Idi Amin expels 50,000 Asians from Uganda
1973	UK joins the EEC	Domicile and Matrimonial Proceedings Act	Ceasefire declared in Vietnam war President Allende of Chile dies in military coup Yom Kippur offensive against Israel
1974	'3 day' week Grunwick dispute IRA Birmingham pub bombing Ban on IRA membership introduced Labour government elected	Trades Union and Labour Relations Act Prevention of Terrorism (Temporary Provisions) Act Northern Ireland (Emergency Provisions) Act Consumer Credit Act Rehabilitation of Offenders Act	
1975	Referendum on EEC membership Margaret Thatcher elected leader of Conservative Party	Sex Discrimination Act	Trevi Group formed in Rome Helsinki Agreement Death of Franco

	IRA siege at Balcombe Street Internment without trial in NI suspended		Angolan independence from Portugal Khmer Rouge take control in Cambodia Fall of Saigon to N. Vietnam Civil War in Beirut Sakharov wins Nobel Peace Prize Gough Whitlam dismissed by Australian Governor-General
1976		Race Relations Act Domestic Violence and Matrimonial Proceedings Act Bail Act Motor-Cycle Crash Helmets (Religious Exemptions) Act Police Act	Death of Mao Tse Tung
1977	*Gay News* trial Anti-fascist riots in Lewisham Founders of NI peace movement win Nobel Peace Prize		Charter 77 Human Rights group formed in Czechoslovakia
1978	First British 'test-tube' baby	Employment Protection (Consolidation) Act Suppression of Terrorism Act	Aldo Moro kidnapped by Red Brigade terrorists
1979	Conservative government elected		Islamic Revolution in Iran

Year	Domestic events	Key statutes	International events
1979	Southall disturbances Devolution referendum in Scotland and Wales IRA kill Airey Neave in car bomb attack First direct elections in UK to European Parliament Lord Mountbatten killed by IRA bomb Anthony Blunt named as 'the fourth man'		Idi Amin deposed Brezhnev and Carter sign SALT-2 treaty Somoza ousted by Sandinistas in Nicaragua Soviet troops invade Afghanistan
1980	For first time since 1935 UK unemployment reaches 2 million Peter Sutcliffe, the Yorkshire Ripper, charged with murder	Employment Act	Zimbabwe becomes independent Death of Yugoslavian President Tito 'Solidarity' independent trade union movement recognized by Polish government Fascist bomb blast at Bologna station kills 84 Iran–Iraq war Military coup in Turkey John Lennon shot dead in New York
1981	Brixton, Toxteth riots Rupert Murdoch acquires *The Times* and the *Sunday Times* Formation of the Social Democratic Party	British Nationality Act Broadcasting Act Contempt of Court Act Education Act	Europe-wide demonstration against siting of cruise missiles Martial law declared in Poland First cases of AIDS reported in the USA

Year	Events	Legislation	World events
	Bobby Sands, IRA hunger striker, wins by-election in NI; UK unemployment exceeds 3 million		
1982	Falklands war; Women's peace camp set up at Greenham Common; Launch of Channel 4	Administration of Justice Act; Employment Act	Death of Brezhnev
1983	Conservatives re-elected; Cruise missiles arrive in Britain	Equal Pay (Amendment) Regulations	Soviets shoot down Korean airliner; US troops invade Grenada; Lech Walesa wins Nobel Peace Prize
1984	Freedom of Information campaign founded; IRA Brighton bomb attack on PM; Miners' strike; Sarah Tisdall jailed for leaking Foreign Office documents about cruise missiles	Police and Criminal Evidence Act; Data Protection Act; Matrimonial and Family Proceeding Act; Trade Union Act	'Schengen' treaty signed; Discovery of HIV, the virus that is said to cause AIDS; Indira Gandhi assassinated; Gas leak at Bhopal in India kills more than 2,000
1985	WPC Fletcher shot outside Libyan Embassy; Trade unions banned at GCHQ; Clive Ponting trial; 'Battle of the Beanfield'; Broadwater Farm riots; Anglo-Irish agreement signed by Thatcher and Fitzgerald	Prosecution of Offences Act; Interception of Communications Act; Single European Act; Prohibition of Female Circumcision Act	Gorbachev becomes Sec.-Gen. of CPSU; Heysel stadium disaster kills 41 soccer fans; French secret agents sink Greenpeace ship *Rainbow Warrior*

Appendix A

Year	Domestic events	Key statutes	International events
1986	Stalker suspended from 'shoot-to-kill' inquiry GLC abolished	Public Order Act Sex Discrimination Act	Olaf Palme assassinated President Marcos deposed in Philippines US launches air strikes on Libya Chernobyl nuclear disaster Iran–Contra arms-for-hostages affair Andrei Sakharov released from internal exile in the Soviet Union
1987	News International/Wapping dispute *Spycatcher* banned Conservatives re-elected IRA bomb blast at Enniskillen kills 11	Immigration (Carriers Liability) Act Access to Personal Files Act Criminal Justice Act Family Law Reform Act Local Government Act	Terry Waite kidnapped in Beirut
1988	Lockerbie bombing Broadcasting ban on BBC/IBA re IRA spokespersons	Access to Medical Reports Act Criminal Justice Act Employment Act Housing Act Local Government Act	

1989	Release of Guildford Four Publication of *Satanic Verses*	Official Secrets Act Fair Employment (NI) Act Children Act Education Reform Act Employment Act Extradition Act Security Service Act Social Security Act	Berlin Wall removed Revolutions in Eastern Europe Tianenmen Square uprising quashed
1990	Renewed IRA bombing campaign in Britain Poll Tax riots	Access to Health Records Act Courts and Legal Services Act Employment Act Human Fertilization and Embryology Act	Release of Nelson Mandela Iraq invades Kuwait Unification of Germany
1991	Release of Birmingham Six	Child Support Act Criminal Justice Act Legal Aid Act	Gulf war Soviet coup fails Rajiv Gandhi assassinated
1992	Conservative government re-elected for fourth term Ulster Defence Association banned Isle of Man legalizes homosexuality Judith Ward freed by Court of Appeal	Trade Union and Labour Relations (Consolidation) Act	Bill Clinton elected as US president Danish referendum rejects Maastricht treaty Collapse of the Soviet Union Slovenia and Croatia become independent republics 58 people killed in Los Angeles riots

APPENDIX B

International Human Rights Agreements*

MAJOR HUMAN RIGHTS AGREEMENTS CONCLUDED
UNDER UNITED NATIONS AUSPICES

Convention on the Prevention and Punishment of the Crime of Genocide, 9 December 1948, 78 UNTS 277, in force 12 January 1951, ratified by the United Kingdom.

Convention Relating to the Status of Refugees, 28 July 1951, 189 UNTS 137, in force 22 April 1954, ratified by the United Kingdom.

Protocol to the Convention Relating to the Status of Refugees, 31 January 1966, 606 UNTS 267, in force 4 October 1967, ratified by the United Kingdom.

Convention on the Political Rights of Women, 20 December 1952, 193 UNTS 135, in force 7 July 1954, ratified by the United Kingdom.

Convention on the Elimination of All Forms of Racial Discrimination, 21 December 1965, 660 UNTS 195, in force 4 January 1969, ratified by the United Kingdom. Declaration regarding Article 14 (competence of the CERD to receive communications from individuals), entry into force 3 December 1982, not accepted by the United Kingdom.

International Covenant on Economic, Social, and Cultural Rights, 16 December 1966, 993 UNTS 3, in force 3 January 1976, ratified by the United Kingdom.

International Covenant on Civil and Political Rights, 16 December 1966, 999 UNTS 171, in force 23 March 1976, ratified by the United Kingdom.

Optional protocol to the International Covenant on Civil and Political Rights, 16 December 1966, 999 UNTS 171, in force 23 March 1976, not ratified by the United Kingdom.

International Convention on the Elimination of All Forms of Discrimination Against Women, 18 December 1979, GA Res 34/180, UN Doc A/34/46, in force 3 September 1981, ratified by the United Kingdom.

Convention Against Torture and other Cruel, Inhuman or Degrading Treatment or Punishment, 10 December 1984, GA Res 39/46, Doc. A/39/51, in force 26 June 1987, ratified by the United Kingdom. Declaration regarding Article 21 of the Convention (competence of the Committee Against Torture to receive communications by a State Party against another State Party), in force 26 June 26 June 1987, accepted by the United Kingdom. Declaration regarding Article 22 of the Convention (competence of the Committee Against Torture to receive communications from individuals), in force 26 June 1987, not accepted by the United Kingdom.

* *Source*: J.-B. Marie, 'International Instruments Relating to Human Rights: Classification and Status of Ratifications as of 1 January 1992' (1992) 13 *HRLJ* 55.

Convention on the Rights of the Child, 20 November 1989, GA Res 44/25, in force 2 September 1990, ratified by the United Kingdom.

MAJOR INTERNATIONAL LABOUR ORGANIZATION CONVENTIONS

Convention (No. 87) concerning Freedom of Association and Protection of the Right to Organize, 9 July 1948, 68 UNTS 17, in force 4 July 1950, ratified by the United Kingdom.

Convention (No. 98) concerning the Application of the Principle of the Right to Organize and to Bargain Colectively, 1 July 1949, 96 UNTS 257, in force 18 July 1951, ratified by the United Kingdom.

Convention (No. 100) concerning Equal Remuneration for Men and Women Workers for Work of Equal Value, 29 June 1951, 165 UNTS 303, in force 23 May 1953, ratified by the United Kingdom.

Convention (No. 111) concerning Discrimination in Respect of Employment and Ocupation, 25 June 1958, 362 UNTS 31, in force 15 June 1960, not ratified by the United Kingdom.

MAJOR COUNCIL OF EUROPE CONVENTIONS AND OTHER TEXTS RELEVANT TO HUMAN RIGHTS

Council of Europe Convention for the Protection of Human Rights and Fundamental Freedoms, 4 November 1950, 5 ETS, in force 3 September 1953, ratified by the United Kingdom.

Protocol No. 1, 20 March 1952, 9 ETS, in force 18 May 1954, ratified by the United Kingdom.

Protocol No. 2, 6 May 1963, 44 ETS, in force 21 September 1970, ratified by the United Kingdom.

Protocol No. 3, 6 May 1963, 45 ETS, in force 21 September 1970, ratified by the United Kingdom.

Protocol No. 4, 16 September 1963, 46 ETS, in force 2 May 1968, not ratified by the United Kingdom.

Protocol No. 5, 20 January 1966, 55 ETS, in force 20 December 1971, ratified by the United Kingdom.

Protocol No. 6, 28 January 1983, 114 ETS, in force 1 March 1975, not ratified by the United Kingdom.

Protocol No. 7, 22 November 1984, 117 ETS, in force 1 November 1988, not ratified by the United Kingdom.

Protocol No. 8, 19 March 1985, 118 ETS, in force 1 January 1990, ratified by the United Kingdom.

Protocol No. 9, 6 November 1990, 140 ETS, not in force.

Protocol No. 10, 30 March 1993 (1992) 13 *Human Rights Law Journal* 182, not in force, not ratified by the United Kingdom.

Council of Europe, European Social Charter, 18 October 1961, 35 ETS, in force 26 February 1965, ratified by the United Kingdom.

Additional Protocol to the European Social Charter, 5 May 1988, 128 ETS, not in force, not ratified by the United Kingdom.

Protocol amending the European Social Charter, 21 October 1991, 142 ETS, not in force, not ratified by the United Kingdom.

European Convention on the Legal Status of Migrant Workers, 24 November 1977, 93 ETS, in force 1 May 1983, not ratified by the United Kingdom.

Convention for the Protection of Individuals with regard to Automatic Processing of Personal Data, 28 January 1981, 108 ETS, in force 1 October 1985, ratified by the United Kingdom.

European Convention for the Prevention of Torture and Inhuman or Degrading Treatment or Punishment, 26 November 1987, 126 ETS, in force 1 February 1989, ratified by the United Kingdom.

Human Rights Activity in the United Kingdom

1. HUMAN RIGHTS AND CIVIL LIBERTIES PRESSURE GROUPS AND LAW REFORM GROUPS
IN ENGLAND AND WALES

Organisation	Foundation date	Aims and objectives
Abortion Law Reform Association, 27–35 Mortimer Street, London W1N 7RJ	1936	'ALRS supports a woman's right to choose whether or not to continue with a pregnancy and campaigns for further liberalisation of the law and for improved NHS facilities.'
Amnesty International, 99 Rosebery Avenue, London EC1R 4RE	1961	'We stand for: prisoners of conscience who are being imprisoned for their beliefs but who have not applied violence; Fair and prompt trials for all political prisoners; We oppose the death penalty and torture or other cruel, inhuman or degrading treatment of all prisoners without reservation.'
Article 19, 90 Borough High Street, London SE1 1LL	1986	'Article 19's mandate is to promote and defend freedom of expression and to combat censorship in order to protect individuals and publications and to encourage action and awareness of censorship at local, national and international levels.'

Organisation	Founded	Aims
Board of Deputies of British Jews, Woburn House, Tavistock Square, London WC1H 0EP	1760	'. . . is the chief representative organisation of the Anglo-Jewish community on many issues, including those affecting the political and civil rights of anglo saxon and, in some cases, overseas Jews'.
Campaign for Press and Broadcasting Freedom, 96 Dalston Lane, London E8 1NG	1979	'. . . Campaigns for right of reply, legislation to break up the concentration of ownership in press and broadcasting and to promote a pluralistic media which adequately represents the range of views in our society.'
Charter 88, Exmouth House, 3–11 Pine Street, London EC1R 0JH	1988	'Through popular, peaceful change the Charter aims to bring about as soon as possible a modern written constitution in Britain, with human and political rights entrenched in law.'
Child Poverty Action Group, 1 Bath Street, London EC1V 9PY	1965	'A national charity which has been in the front line of the fight against poverty for over 25 years. CPAG seeks to ensure that families on low incomes get their full entitlements to welfare benefits and it campaigns for improvements in both benefits and other policies to eradicate the injustice of poverty.'
Children's Legal Centre, 20 Compton Terrace, London N1 2UN	1979	'We promote consultation with children. Our aim is to lobby and make decisions on behalf of children.'
Committee on the Administration of Justice (Northern Ireland), 45–7 Donegal Street, Belfast BT1 2FG	1981	'CAJ is an independent organisation which monitors civil liberties issues, provides information to the public and campaigns for change in the administration of justice in Northern Ireland.'

Organisation	Foundation date	Aims and objectives
European Women's Lobby, c/o National Association of Women's Organisations, 279–81 Whitechapel Road London E1 1BY	1990	'The aims of the lobby are to promote the interests of women living in the European Community including marginalised and ethnic minority women and to promote and extend the achievement of equal rights and equal opportunities for women. Comprises four UK organisations: the National Association of Women's Organisations, Women's Forum Scotland, Wales Women's Euro-Network, Northern Ireland Women's European Platform'.
Families Need Fathers, 134 Curtain Road, London EC2A 3AR	1974	'. . . a national charity based on regional branches. Founded . . . by a group of parents who, following divorce or separation had found themselves unwillingly losing contact with their children . . . primarily concerned with assisting parents in maintaining full relationships with children during and following divorce/separation but also offers support and assistance in other problems arising from family breakdown. The society also contributes to Governmental, academic and media reviews on the problems of family breakdown.'
Family Rights Group, The Print House, 18 Ashwin Street, London E8 3DL	1972	'A charitable organisation working throughout England and Wales. The group's main activities are to advise families whose children are involved in child protection procedures, in need of local authority services or looked after by the local authority.'
Fawcett Society, 46 Harleyford Road, London SE11 5AY	1866	'Campaign for equality between women and men.'
Freedom of Information Campaign, 88 Old Street, London EC1B 9AR	1984	'. . . aims to eliminate unnecessary official secrecy and to give people legal rights to information which affects their lives or which they need to hold public authorities properly accountable'.

Organisation	Date	Description
Gay London Policing Group, 36 Old Queen Street, London SW18 9JF	1982	'An organisation set up to combat homophobic violence and to uphold the fair treatment of the London lesbian and gay community by the police and court systems.'
Gingerbread, 35 Wellington Street, London WC2E 7BN	1970	'An Association which provides advice, support and friendship for lone parents through around 300 self-help groups and a national advice line.'
Haldane Society, 11 Doughty Street, London WC1N 2PG	1930	'An organisation which provides a forum for the discussion and analysis of law and the legal system, both nationally and internationally, from a socialist perspective. It is independent of any political party. Its membership consists of individuals who are lawyers, academics or students and legal workers, and it also has trade union and labour movement affiliates.'
Howard League for Penal Reform, 708 Holloway Road, London N19 3NL	1866	'The Howard League campaigns for a more humane criminal justice system.'
Index on Censorship, 32 Queen Victoria Street, London EC4N 4SS	1972	'"Index on Censorship" and "Writers and Scholars Educational Trust" protect and promote freedom of expression worldwide by circulating factual information on examples of banned work and analysis concerned with problems of free speech.'
Indian Workers Association, 112A The Green, Southall, Middlesex UB2 4BQ	1956	'The Association was set up for the purposes of providing for its members and the public at large the means of social intercourse, mutual helpfulness, mental and moral improvement and rational recreation.'
Inquest, 330, Seven Sisters Road, Finsbury Park, London N4 2PG	1981	'An organisation which gives advice information and support on death in police and prison custody, death in psychiatric hospitals and other coroners' inquests.'
Interights, 5–15 Cromer Street, London WC1H 8LS	1983	The organization 'pursues and promotes existing legal remedies to protect human rights. It provides an information and support service to practising lawyers engaged in human rights litigation before national courts. It also assists lawyers in bringing cases before international and regional tribunals.'

Organisation	Foundation date	Aims and objectives
Joint Council for the Welfare of Immigrants, 115 Old Street, London EC1V 9JR	1967	'We aim to get rid of the racism and injustices that are built into the country's immigration law and the way that it is administered'.
Justice, British Section of the International Commission of Jurists, London WC2A 1DT	1957	'Justice is a law reform society. It believes there must be continuous reform of the law if it is to remain just. To devise workable reforms requires thought, research and practical experience'.
Legal Action Group, 42 Pentonville Road London N1 9UN	1972	'The Legal Action Group works to improve legal services for the community.'
Lesbian & Gay Employment Rights, St Margaret's House, 21 Old Ford Road, Bethnal Green, London E2 9PL	1984	'LAGER is a multicultural project fighting the many forms of discrimination experienced by lesbians and gay men in and out of work. *Lesbian Employment Rights* is a self-organised group which deals with issues of specific relevance to lesbians. Both sections provide information advice and support to individuals, trade unions, local authorities, voluntary organisations and community groups.'
MIND, 22 Harley Street, London W1N 2ED	1946	'. . . the leading mental health charity in England and Wales. It works for a better life for people diagnosed, labelled or treated as mentally ill and campaigns for their right to lead an active and valued life in the community.'
Minority Rights Group, 79 Brixton Road, London SW9 7DE	1965	'MRG has as its principal aim to secure justice for minority (and non-dominant majority) groups suffering discrimination.'

National Abortion Campaign, The Print House 18 Ashwin Street, London E8 3DL	1975	'The NAC fights for a woman's right to choose over the question of abortion and to improve access to contraception.'
National Association for the Care and Resettlement of Offenders, 169 Clapham Road, London SW9 OPU	1966	'NACRO help people who have been in trouble with the law, and those at risk of becoming so, to deal with the problems they face and to be reaccepted into society without stigma. NACRO exists to promote a more humane and constructive criminal justice system'.
National Gypsy Council, Green Gate Street, Oldham, Greater Manchester OL4 1DQ	1966	'To obtain adequate accommodation for gypsies as defined in the Caravan Sites Act 1968 and an integrated state school education for gypsy children'.
National Viewers and Listeners Association, Ardleigh, Colchester, Essex CO7 7RH	1964	'Monitors programme content and publishes regular reports which have helped to stimulate public and Parliamentary concern, especially as to the cumulative effects of violent [and obscene] entertainment upon society.'
Outrage! c/o The London Lesbian/Gay Centre, 69 Cowcross Street, London EC1M 6BP	1990	'We are here to fight homophobia.'
Prison Reform Trust, 59 Caledonian Road, London N1 9BN	1981	'The Prison Reform Trust is a national charity which campaigns for better conditions in prison and the greater use of alternatives to custody.'
British Refugee Council, Bondway House, Bondway, London SW8 1SJ	1981	'The BRC is the focal point for UK-based voluntary organisations whose work concerns refugee policy in the UK and overseas, and the settlement of asylum seekers in this country.'

Organisation	Foundation date	Aims and objectives
Runnymede Trust, 11 Princelett Street, London E1 6QH	1968	'. . . an independent charity concerned with issues of racial equality and justice. It was set up . . . to help strengthen and shape policies and projects which: work towards eliminating all forms of racial discrimination; promote mutual respect, appreciation and learning between different traditions and values; and, release and develop the resources, talents and skills of all members of society.'
Society for the Protection of the Unborn Child, 7 Tufton Street, Westminster, London SW1P 3QN	1967	Aim is to 'uphold the right to life of all people from the time of conception onwards. Our primary concern is with the issues of abortion and human embryo experimentation (test tube babies), but we have also been active in lobbying in legal cases relating to euthanasia, surrogacy and the rights of the handicapped.'
Stonewall, 2 Greycoat Place, Westminster, London SW1P 1SB	1989	'To achieve social equality and legal justice for lesbians and gays in Great Britain'.
Union of Muslim Organisations, 109 Campden Hill Road, London W8 7TL	1970	'. . . is the only national umbrella organisation . . . which represents over two million Muslims residing in the United Kingdom and Ireland. Main objectives: (1) to promote unity amongst Muslims in the UK and Ireland, (2) to safeguard Islamic values and its heritage, (3) to facilitate the upbringing of Muslim Children in accordance with the Holy Qur'an and Sunnah, (4) to operate as the Spokesman on behalf of all Muslims in UK and Ireland on all matters relating to their religious, cultural, social, educational and economic issues, (5) to strengthen the bonds of brotherhood between Muslims throughout the world, (6) to promote Da'wah Islamiyya.'

| United Kingdom Immigrants Advisory Service (UKIAS): 2nd Floor, County House, 190 Great Dover Street, London SE1 4YB | 1970 | 'UKIAS offers advice and representation at court to people who have applied to come into Britain as visitors, students or partners. We also offer advice to people who are in this country as students or visitors who would like to vary that situation. We take up cases by applying to the Home Office and request appeals if we feel the case merits our support.' |
| 300 Group, 36–7 Charterhouse Square, London EC1M 6EA | 1980 | 'An all-party campaign for more women in Parliament, local government and public life.' |

2. ROYAL COMMISSIONS AND COMMITTEES OF INQUIRY OF RELEVANCE TO CIVIL LIBERTIES AND HUMAN RIGHTS ISSUES 1950–1992

1947 Royal Commission on the Press

1948 Committee on the Law of Defamation (the Porter Committee)

1957 Committee on Homosexual Offences and Prostitution (the Wolfenden Committee)

1961 Royal Commission on the Press

1962 Royal Commission on the Police

1966 Departmental Committee on Legal Aid in Criminal Court Proceedings (the Widgery Committee)

1966 Royal Commission on Trade Unions and Employers' Associations

1971 Enquiry into Allegations against the Security Forces of Physical Brutality in Northern Ireland arising out of the events on 9 August 1971 (the Compton Enquiry)

1971 Departmental Committee on Section 2 of the Official Secrets Act 1911 (the Franks Committee)

1972 Committee on Privacy (the Younger Committee)

1972 Criminal Law Review Committee, 11th report

1972 Commission to Consider Legal Procedures to deal with Terrorist Activities in Northern Ireland (the Diplock Commission)

1972 Tribunal Appointed to Inquire into the Events on Sunday 30 January which led to loss of life in connection with the Procession in Londonderry on that day (Lord Widgery)

1974 Inquiry into the Events of 15 June 1974 in Red Lion Square (the Scarman Inquiry)

1974 Committee on Contempt of Court (the Phillimore Committee)

1975 Committee on Defamation (the Faulks Committee)

1975 Committee to Consider in the Context of Civil Liberties and Human Rights, Measures to deal with Terrorism in Northern Ireland (the Gardiner Committee)

1977 Committee on the Future of Broadcasting (the Annan Committee)

1977 Inquiry into the Circumstances Leading to the Trial of Three Persons on Charges Arising out of the Death of Maxwell Confait and the Fire at 27 Doggett Road, London, SE5 (the Fisher Inquiry)

1977 Royal Commission on the Press

1977 Royal Commission on Legal Services

1978 Committee on Data Protection (the Lindop Committee)

1978 Review of the Operation of the Prevention of Terrorism (Temporary Provisions) Acts 1974 and 1976 (Lord Shackleton)

1979 Committee on Obscenity and Film Censorship (the Williams Committee)

1979 Committee of Inquiry into Police Interrogation Procedures in Northern Ireland (the Bennett Committee)

1979 Royal Commission on Criminal Procedure

1981 Inquiry into the Brixton Disorders, 10–12 April (the Scarman Inquiry)

1981 Law Commission Report on Breach of Confidence
1983 Review of the Operation of the Prevention of Terrorism (Temporary Provisions) Act 1976 (Earl Jellicoe)
1983 Law Commission Report on Offences Relating to Public Order
1984 Departmental Committee on the Prison Disciplinary System
1985 Committee of Inquiry into the Education of Children from Ethnic Minority Groups (the Swann Inquiry)
1985 Law Commission Report on Offences Against Religion and Public Worship
1986 Committee on Fraud Trials (the Roskill Committee)
1986 The Prevention of Terrorism (Temporary Provisions) Act 1984, Review of the Year 1985 (the Philips Review)
1987 Review of the Operation of the Prevention of Terrorism (Temporary Provisions) Act (Lord Colville)
1990 Review of the Northern Ireland (Emergency Provisions) Acts of 1978 and 1987 (Lord Colville)
1990 Committee on Privacy and related Matters (the Calcutt Committee)
1990 Inquiry into Prison Disturbances, April 1990 (the Woolf Inquiry)
1991 Royal Commission on Criminal Justice

APPENDIX D

Tables Concerning European Convention on Human Rights Litigation

Table D1. *European Commission on Human Rights— A Comparative Table of Cases by Country, 1990–1991*

	Provisional files		Applications registered		Applications inadmissible or struck off		Applications admissible		Applications referred to government		Friendly settlement reports		Reports on merits		Reports striking case off	
	1990	1991	1990	1991	1990	1991	1990	1991	1990	1991	1990	1991	1990	1991	1990	1991
Austria	254	195	159	135	79	129	9	24	23	45	1	1	7	8	—	1
Belgium	202	158	94	67	60	90	5	4	7	4	—	1	6	3	—	—
Cyprus	4	8	3	8	1	5	1	—	1	—	—	—	—	1	—	—
Denmark	72	78	14	22	14	24	—	—	1	1	—	—	—	—	—	—
Finland	63	69	13	38	2	14	—	—	1	1	—	—	—	—	—	—
France	1,017	1,728	248	400	135	247	16	25	32	70	2	3	12	16	1	—
Germany	618	620	146	139	124	154	4	5	7	10	—	—	1	6	—	—
Greece	51	88	26	29	17	24	5	5	13	2	—	—	3	4	—	—
Iceland	4	—	2	2	—	1	1	1	1	—	1	—	1	—	—	—
Ireland	21	27	13	7	8	12	2	—	3	1	—	—	1	2	—	—
Italy	353	463	154	133	72	74	62	64	134	79	—	—	14	51	—	1
Liechtenstein	3	—	1	—	—	1	—	—	—	1	—	—	—	—	—	—
Luxembourg	12	17	3	3	3	2	—	—	—	—	—	—	—	—	—	—

	1	2	3	4	5	6	7	8	9	10	11	12	13	14	15	16
Malta	4	8	2	7	1	5	—	—	—	1	—	—	1	—	—	2
Netherlands	178	165	109	98	58	116	3	20	28	19	1	2	3	12	—	—
Norway	34	27	18	11	5	12	—	—	2	1	—	—	1	—	—	—
Portugal	81	81	30	36	11	3	23	1	24	1	2	14	1	1	—	—
Spain	216	159	69	75	70	80	1	1	2	8	—	—	—	—	—	—
Sweden	247	225	125	90	85	143	6	13	19	8	3	9	9	3	—	—
Switzerland	184	227	107	113	97	103	6	5	6	16	1	1	4	3	—	—
Turkey	104	90	85	33	18	23	—	12	21	16	—	—	—	2	—	—
United Kingdom	1,067	843	236	202	205	178	7	38	29	22	1	1	12	32	1	—
Others	153	274	—	—	—	1	—	—	—	—	—	—	—	—	—	—
TOTAL	4,942	5,550	1,657	1,648	1,065	1,441	151	217	355	306	12	32	75	144	1	2

Source: European Commission of Human Rights, *Survey of Activities and Statistics 1991* (Council of Europe, Strasbourg, 1992).

Appendix D

Table D2. *Judgments of the European Court of Human Rights by Country 1959–1990 (excluding judgements under Art. 50 (remedies))*

State concerned	Cases which gave rise to a finding of:	
	at least one violation	non-violation
Austria	15	5
Belgium	14	4
Denmark	1	4
France	5	1
Germany	8	11
Greece	—	—
Iceland	—	—
Ireland	3	1
Italy	13	1
Malta	—	—
Netherlands	13	2
Norway	1	—
Portugal	5	—
Spain	2	—
Sweden	14	3
Switzerland	9	5
United Kingdom	27	10
TOTAL	130	47

Source: European Court of Human Rights, *Survey of Activities 1959–1990* (Council of Europe, Strasbourg, 1991), table viii.

Table D3. *UK cases before the European Court of Human Rights by outcome 1959–1992*[a]

Name	Case No.[b]	Subject	Series A No.	Date[c]	Outcome	Effect: UK Response
Golder	4451/70	Prisoner: access to a solicitor	18	21 Feb. 75	Breach of Arts. 6(1) and 8.	Access permitted after exhaustion of internal complaints procedure ('prior ventilation rule')
Handyside	5493/72	Little Red Book: obscene publication	24	7 Dec. 76	No breach of Arts. 10, 14, 18, or Protocol 1. Art. 1.	
Ireland	5310/72	Northern Ireland: internment and interrogation methods	25	18 Jan. 78	Breach of Art. 3. No breach of Arts. 5, 6, 14, or 15.	Administrative action in form of formal renunciation of 'five techniques'. Directives to security forces forbidding assaults on detainees. Change in medical procedures. New complaints investigation procedures.
Tyrer	5856/72	Isle of Man: judicial corporal punishment	26	25 Apr. 78	Breach of Art. 3.	Letter to Lieutenant Governor informing him that Manx law is in breach of the Convention
Sunday Times	6538/74	Freedom of the press: contempt of court in reporting Thalidomide case	30	26 Apr. 79	Breach of Art. 10.	Contempt of Court Act 1981
Young, James and Webster	7601/76	Closed shop	44	13 Aug. 81	Breach of Art. 11.	Employment Acts 1980 and 1982
Dudgeon	7527/76	Homosexuality: private life	45	22 Oct. 81	Breach of Art. 8.	Homosexual Offences (Northern Ireland) Order 1982 (in force 9 Dec. 82) decriminalising private homosexual acts between consenting adults, subject to certain exceptions.
X	6998/75	Mental patient: right to have detention reviewed	46	5 Nov. 81	No breach of Art. 5(1). Breach of Art. 5(4).	Mental Health (Amendment) Act 1982, s. 28(4) and Sch. 1. In Northern Ireland Mental Health (NI) Order 1986, SI 1986, No. 595, Part V.
Campbell and Cosans	7511/76 7743/76	Corporal punishment in state school: respect for parent's philosophical convictions	48	25 Feb. 82	No breach of Art. 3. Breach of Protocol 1. Art. 1.	Education (No. 2) Act 1986, ss. 47 & 48. Education (Corporal Punishment) (Northern Ireland) Order 1987
Silver and others	5947/72 6205/73 7152/75 7061/75 7107/75 7113/75 7136/75	Censorship of prisoners' correspondence, access to courts	61	25 Mar. 83	Breach of Arts. 6(1), 8, and 13.	Revised Prison Standing Order 5, into force December 1981. Simultaneous ventilation rule replaced prior ventilation rules re correspondence with legal adviser (see *Golder* above). Restrictions on content and correspondents eased. Standing Order published.
Campbell and Fell	7819/77 7878/77	Prison visitors: conduct of disciplinary proceedings	80	28 June 84	Breach of Art. 6 in two respects and Arts. 8 and 13.	Change in prison standing order in 1981. Letter to chairmen of Boards of Visitors (12 July 84), followed by amendment of *Manual on Conduct of Adjudications in Prison*

Table D3. *Continued*

Name	Case No.[b]	Subject	Series A No.	Date[c]	Outcome	Effect: UK Response
Malone	8691/79	Telephone tapping	82	2 Aug. 84	Breach of Art. 8.	Interception of Communications Act 1985 (in force 10 Apr. 86). Established machinery to regulate and monitor telephone tapping and metering
Ashingdane	8225/78	Detention of mental patient	93	28 May 85	No breach of Arts. 5(1), 5(4), or 6.	
Abdulaziz, Cabales, and Balkanadali	9214/80	Immigration: discrimination on grounds of sex	94	28 May 85	Breach of Arts. 13 and 14 in one respect. No breach of Arts. 3 or 8.	Immigration Rules amended August 1985. Further discrimination removed by Immigration Act 1988 s. 1 and Immigration Rules 1989.
James and others	8793/79	Leasehold reform	98	21 Feb. 86	No breach of Protocol 1, Art. 1, or Arts. 6(1) and 13.	
Lithgow and others	9006/80	Aircraft and shipbuilding nationalization	102	8 July 86	No breach of Protocol 1. Art. 1, or Arts. 14, 6(1), and 13.	
Rees	9532/81	Transsexual: reissue of birth certificate, right to marry	106	17 Oct. 86	No breach of Arts. 8 or 12.	
Agosi	9118/80	Forfeiture by customs of gold coins belonging to a third party	108	24 Oct. 86	No breach of Protocol 1, Art. 1.	
*Gillow	9063/80	Residence qualifications in Guernsey	109	24 Nov. 86	Breach of Art. 8 in one respect. No breach of Art. 8 on other.	Compensation to applicants
*Weeks	9787/82	Prisoner released on licence, lawfulness of redetention	114	2 Mar. 87	No breach of Art. 5(1). Breach of Art. 5(4).	Weeks pardoned, compensation paid. No general reform of the law until Criminal Justice Act 1991
Monnell and Morris	9562/81	Prisoner: period of appeal additional to sentence	115	2 Mar. 87	No breach of Arts. 5(1), 6(1), 6(3)(c), or 14.	
*O.	9276/81	Child-care: access to children in care of local authority, delay in court	120	8 July 87	Breach of Art. 6(1) for all 5 applicants. No breach of Art. 8 for O. Breach of Art. 8 for H., W., B., and R.	Partial anticipatory compliance in 1983 amendments to Child Care Act 1980. S. 34 of Children Act 1989 (in force Oct. 91) provides for all questions concerning parental contact with children in local authority care to be determinable by a court. S. 22 of Act provides for greater parental involvement in local authority decision-making
*H.	10496/83		120			
*W.	9480/81		121			
*B.	9749/82		121			
*R.	9840/82		121			
Boyle and Rice	9659/82 9658/82	Prisoners' correspondence and visits	131	27 Apr. 88	No breach of Art. 13. Breach of Art. 8 in respect of one letter.	Violation a result of erroneous application of Prison Rules. Measures taken to ensure Rules are correctly applied in future.
*Brogan and others	11209/84 11234/84 11266/84	Prevention of terrorism: length of detention	145	29 Nov. 88	No breach of Art. 5(1). Breach of Art. 5(3) for all parties. No breach of Art. 5(4). Breach of 5(5) for all.	Derogation made under Art. 15

Case	Application no.[b]	Description	No.[a]	Date[c]	Decision	Action taken
Chappell	10461/83	Search of premises in the execution of a court order in civil proceedings	152	30 Mar. 89	No breach of Art. 8	
Gaskin	10454/83	Access to personal records held by a local authority	160	7 July 89	Breach of Art. 8. No breach of Art. 10.	
Soering	14038/88	Extradition to USA to face a murder charge	161	7 July 89	Breach of Art. 3 if UK decide to extradite. No breach of Arts. 6(3)(c) or 13.	S. extradited to USA on condition that he would not be tried on any charges for which penalty was death
Powell and Rayner	9310/81	Heathrow: aircraft noise	172	22 Feb. 90	No breach of Art. 13.	
Granger	11932/86	Prisoner: legal aid for appeal	174	28 Mar. 90	Breach of Art. 6(3)(c) taken together with 6(1).	Practice Note circulated by the Scottish Lord Justice General to all appeal court chairmen and clerks in Dec. 1990.
*Fox, Campbell, and Hartley	12244/86 12245/86 12383/86	Northern Ireland: arrest and detention	182	30 Aug. 90	Breach of Art. 5(1)(c). No breach of Art. 5(2). Breach of Art. 5(5).	Section 11 of the Northern Ireland (Emergency Provisions) Act 1978, replaced by section 6 of the Northern Ireland (Emergency Provisions) Act 1987. No further action taken.
McCallum	9511/81	Prisoner's correspondence	183	30 Aug. 90	Breach of Art. 8	See above *Silver*.
Cossey	10843/84	Transsexual: reissue of birth certificate, right to marry		27 Sept. 90	No breach of Arts. 8 or 12.	
Thynne, Wilson, and Gunnell	11787/85 11978/86 12009/86	Prisoners: discretionary life sentences, lawfulness of detention and redetention	190	25 Oct. 90	Breach of Art. 5(4) for all applicants. Breach of Art. 5(5) in respect of Mr Wilson.	
Vilvarajah and others	13163/87 13164/87 13165/87 13447/87 13488/87	Asylum: Tamils removed to Sri Lanka		30 Oct. 91	No breach of Arts. 3 or 13.	
Observer and others and Guardian Newspaper and others	13585/88	Freedom of expression: *Spycatcher*	216	26 Nov. 91	Partial breach of Art. 10. No breach of Arts. 13 or 14.	
Sunday Times (No. 2)	13166/87	Freedom of expression: *Spycatcher*	217	26 Nov. 91	Breach of Art. 10. No breach of Arts. 13 or 14.	
Campbell	13590/88	Censorship of prisoner's correspondence with solicitor	233	25 Mar. 92	Breach of Art. 8.	

[a] Excluding judgments under Art. 50 (Remedies).

[b] The number after the stroke indicates the year the complaint was lodged with the European Commission.

[c] This column indicates the date of the judgment by the European Court.

Sources: Liberty, *A People's Charter: Liberty's Bill of Rights. A Consultative Document* (Liberty, London, 1991), appendix ii, table 1; European Court of Human Rights, *Survey of Activities 1959–1990* (Council of Europe, Strasbourg, 1991); R. R. Churchill, in J. P. Gardner (ed.), *The European Convention on Human Rights* (1992). 'Table showing cases where the United Kingdom has been found to have violated the Convention and the United Kingdom's response, 1975–1989'.

Bibliography

Aaron, B., 'The Labor-Management Reporting and Disclosure Act of 1959' (1960) 73 *Harvard Law Review* 851.

Abram, M. B., 'Affirmative Action: Fair Shakers and Social Engineers' (1986) 99 *Harvard Law Review* 1312.

Airey, C., 'Social and Moral Values', in R. Jowell and C. Airey (eds.), *British Social Attitudes: The 1984 Report* (SCPR, Gower, 1984).

Akhtar S., *Be Careful with Muhammad* (Bellew, London, 1989).

Albeda, W., *Disabled People and their Employment* (Commission of the European Communities, Brussels, 1984).

Alcock, A., *History of the International Labour Organisation* (Macmillan, London, 1971).

Alderman, G., *The Jewish Community in British Politics* (Clarendon Press, Oxford, 1983).

Allen, R. (ed.), 'Discretion in Law Enforcement' (1984) 47 Special Issue of *Law and Contemporary Problems* 4.

—— 'The Police and Substantive Rule-Making: Reconciling Principle and Expediency' (1976) *University of Pennsylvania Law Review* 62.

Alston, P. (ed.), *The United Nations and Human Rights: A Critical Appraisal* (Clarendon Press, Oxford, 1992).

—— and Parker, S., and Seymour, J. (eds.), *Children, Rights and the Law* (Clarendon Press, Oxford, 1992).

—— Amnesty International, *Allegations of Human Rights Abuses in Northern Ireland* (Amnesty International, London, 1991).

—— *Northern Ireland: Killings by the Security Forces and 'Supergrass' Trials* (Amnesty International, London, 1984).

—— *United Kingdom: Human Rights Concerns* (Amnesty International Publications, London, June 1991).

—— *Annual Report* (Amnesty International, London, 1992).

Amodio, E., and Selvaggi, E., 'An Accusatorial System in a Civil Law Country: The Italian Code of Criminal Procedure' (1989) *Temple Law Review* 1211.

Andrew, C., *Secret Service* (Dunton Green, London, 1986).

Appignanesi, L., and Maitland, S. (eds.), *The Rushdie File* (Fourth Estate, London, 1989).

Appleby, G., and Ellis, E., 'Formal Investigations: The Commission for Racial Equality and the Equal Opportunities Commission as Law Enforcement Agencies' (1984) *Public Law* 236.

Article 19, *Information, Freedom and Censorship: World Report 1991* (Library Association Publishing, London, 1991).

—— *No Comment: Censorship, Secrecy and the Irish Troubles* (Article 19, London, 1989).

Ashford, M., *Detention Without Trial* (Joint Council for the Welfare of Immigrants, London, 1993).

Ashworth, A., 'Concepts of Criminal Justice' (1979) *Crim. LR* 412.

Ashworth, A., 'Excluding Evidence as Protecting Rights' (1977) *Crim. LR* 723–35.
—— 'The "Public Interest" Element in Prosecutions' (1987) *Crim. LR* 595.
Atkins, S., and Hoggett, B., *Women and the Law* (Basil Blackwell, Oxford, 1984).
—— and Luckhaus, L., 'The Social Security Directive and UK Law', in C. McCrudden (ed.), *Women, Employment and European Equality Law* (Eclipse Publications, London, 1987).
Audit Commission for Local Authorities and the National Health Service, in England and Wales, *Effective Policing: Performance Review in Police Forces*, Police Paper 8 (HMSO, London, 1990).
Aust, S., *The Baader-Meinhof Group: The Inside Story of a Phenomenon* (The Bodley Head, London, 1987).
Austin, C., and Ditchfield, J., 'Internal Ventilation of Grievances: Applications and Petitions', in M. Maguire, J. Vagg, and R. Morgan (eds.), *Accountability and Prisons* (Tavistock, London, 1985).
Australian Parliament, Senate, Standing Committee of Constitutional and Legal Affairs, *A Bill of Rights for Australia?* (Australian Government Publishing Service, Canberra, 1985).
Bacchi, C., 'Pregnancy, the Law and the Meaning of Equality', in E. Meehan and S. Sevenhuijsen (eds.), *Equality Politics and Gender* (Sage, London, 1991).
Bailey, M., and Pillard, R., 'A Genetic Study of Male Sexual Orientation' (1991) *Archives of General Psychiatry* 1089.
Bailey, S. H., Harris, D. J., and Jones, B. L., *Civil Liberties: Cases and Materials*, 2nd edn. (Butterworths, London, 1985).
Bainham, A., *Children, Parents and the State* (Sweet & Maxwell, London, 1988).
—— 'The Privatisation of the Public Interest in Children' (1990) 53 *Modern Law Review* 206.
—— 'When is a Parent not a Parent? Reflections on the Unmarried Father and his Child in English Law' (1989) 3 *International Journal of Law and the Family* 208–39.
Baldwin, J. R., *Pre-Trial Justice: A Study of Case Settlement in Magistrates Courts* (Blackwell, Oxford, 1985).
—— and McConville, M., *Jury Trials* (Clarendon Press, Oxford, 1979).
—— —— *Negotiated Justice* (Martin Robertson, London, 1977).
Ball, W., and Solomos, J. (eds.), *Race and Local Politics* (Macmillan, Basingstoke, 1990).
Bankowski, Z., 'The Value of Truth' (1981) 1 *Legal Studies* 257–66.
Banton, M., *The Policeman in the Community* (Tavistock, London, 1964).
Barnum, D. G., 'Constitutional Organization and the Protection of Human Rights in Britain and the United States', in J. R. Schmidhauser (ed.), *Comparative Judicial Systems* (Butterworths, London, 1987).
—— Sullivan, J. L., and Sunkin, M., 'Constitutional and Cultural Underpinnings of Political Freedom in Britain and the United States' (1992) 12 *Oxford Journal of Legal Studies* 362–79.
Baxter, J., and Koffman, L., 'The Confait Case: Forgotten Lessons?' (1983) 14 *Cambrian Law Review* 11.
—— —— (eds.), *Police: The Constitution and the Community* (Professional Books, London, 1985).

Beddard, R., and Hill, D. M. (eds.), *Economic, Social and Cultural Rights: Progress and Achievement* (Macmillan, London, 1992).

Beer, C., Jeffery, R., and Munyard, T., *Gay Workers: Trade Unions and the Law*, 2nd edn. (National Council for Civil Liberties, London, 1983).

Bennett, Judge H. G., *Report of the Committee of Inquiry into Police Interrogation Procedures in Northern Ireland*, Cmnd. 7497 (HMSO, London, 1979).

Ben-Tovim, G., and Gabriel, J. G., 'The Politics of Race in Britain, 1962–79: A Review of the Major Trends and of Recent Debates', in C. Husband (ed.), *'Race' in Britain* (1982).

Bennett, T., 'The Social Distribution of Criminal Labels' (1979) 19 *British Journal of Criminology* 134.

Bennion, F., 'The Crown Prosecution Service' (1986) *Criminal Law Review* 3.

Benyon, J., and Bourne, C. (eds.), *The Police: Powers, Procedures and Proprieties* (Pergamon Press, Oxford, 1986).

Bercusson, B., *Fair Wages Resolutions: Studies in Labour and Social Law*, ii (Mansell, London, 1978).

—— 'Fundamental Social Rights and Economic Rights in the European Community', in A. Cassese, A. Clapham, and J. Weiler (eds.), *Human Rights and the European Community: Methods of Protection* (Nomos, Baden-Baden, 1991).

Berger, V., *Case Law of the European Court of Human Rights* (Round Hall Press, Dublin, 1989–).

Betten, L., Harris, D., and Jaspers, T. (eds.), *The Future of European Social Policy* (Kluwer Law and Taxation, Deventer, 1989).

Bevan, V., and Lidstone, K., *A Guide to the Police and Criminal Evidence Act 1984* (Butterworths, London, 1985, 2nd ed., 1992).

Birch, D., 'The Criminal Justice Act 1988: Documentary Evidence' (1989) *Criminal Law Review* 15.

—— 'Hunting the Snark: The Elusive Exception' (1988) *Criminal Law Review* 221.

—— 'The PACE Hots Up: Confessions and Confusions under the 1984 Act' (1989) *Criminal Law Review* 95.

Birkinshaw, P., *Freedom of Information: The Law, the Practice and the Ideal* (Weidenfeld & Nicolson, London, 1989).

Blackburn, R., and Taylor, J., *Human Rights for the 1990s: Legal, Political and Ethical Issues* (Mansell, London, 1991).

Blacke, C., 'Immigration Appeals: The Need for Reform' in A. Dummett (ed.), *Towards a Just Immigration Policy* (Cobden Trust, London 1986).

Bloed, A., and van Dijk, P. (eds.), *Essays on Human Rights in the Helsinki Process* (Martinus Nijhoff, Dordrecht, 1985).

—— —— (eds.), *The Human Dimension of the Helsinki Process: The Vienna Follow-up Meeting and its Aftermath* (Martinus Nijhoff, Dordrecht, 1991).

Bok, S., *Secrets* (Oxford University Press, Oxford, 1983).

Bollen, K. A., 'Political Rights and Political Liberties in Nations: An Evaluation of Human Rights Measures, 1950 to 1984' (1986) 8 *Human Rights Quarterly* 567.

Bottomley, K., and Pease, K., *Crime and Punishment: Interpreting the Data* (Open University Press, Milton Keynes, 1985).

Bottoms, A., and McClean, J. D., *Defendants in the Criminal Process* (Routledge

& Kegan Paul, London, 1976).

Bouchier, D., *The Feminist Challenge: The Movement for Women's Liberation in Britain and the United States* (Macmillan Education, London, 1983).

Bourn, C., and Whitmore, J., *Discrimination and Equal Pay*. (Sweet & Maxwell, London 1989).

Box, S., and Russell, K., 'The Politics of Discreditability: Disarming Complaints against the Police' (1975) 23 *Sociological Review* 2.

Boyle, K., Hadden, T., and Hillyard, P., *Ten Years On in Northern Ireland: The Legal Control of Political Violence* (Cobden Trust, London, 1980).

Bradley, A. W., 'Protecting Government Decisions from Legal Challenge' (1988) *Public Law* 1.

—— 'The United Kingdom before the Strasbourg Court 1975–1990', in W. Finnie, C. Himsworth, and N. Walker (eds.), *Edinburgh Essays in Public Law* (Edinburgh University Press, Edinburgh, 1991).

Bradley, D., 'The Development of a Legal Status for Unmarried Cohabitants in Sweden' (1989) 18 *Anglo-American Law Review* 322.

—— 'Homosexuality and Child Custody in English Law' (1987) 1 *International Journal for Law and the Family* 155.

Braithwaite, J., and Pettit, P., *Not Just Deserts: A Republican Theory of Criminal Justice* (Clarendon Press, Oxford, 1990).

Bratza, N., 'The Treatment and Interpretation of the European Convention on Human Rights: Aspects of Incorporation'. in J. P. Gardner (ed.), *European Convention on Human Rights: Aspects of Incorporation* (British Institute of Comparative Law and British Institute on Human Rights, London, forthcoming).

Brazier, R., *Constitutional Reform: Reshaping the British Political System* (Oxford University Press, Oxford, 1991).

Brest, P., 'In Defense of the Antidiscrimination Principle' (1976) 90 *Harvard Law Review* 1–54.

Brett, R., *The Development of the Human Dimension Mechanism of the Conference on Security and Co-operation in Europe* (CSCE), *Papers in the Theory and Practice of Human Rights*, No. 1 (Univ. of Essex, Colchester, 1992).

Brewster, C., and Teague, P., *European Community Social Policy: Its Impact on the UK* (Institute of Personnel Management, London, 1989).

Brogden, A., 'Sus is Dead, but what about Sas?' (1981) 9 *New Community*.

Brogden, M., and Brogden, A., 'From Henry VIII to Liverpool 8: The Complex Unity of Police Street Powers' (1984) *International Journal of Sociology of Law* 37.

—— *On the Mersey Beat: Policing Liverpool between the Wars* (Oxford University Press, Oxford, 1991).

—— *The Police: Autonomy and Consent* (Academic Press, London, 1982).

—— Jefferson, T., and Walklate, S., *Introducing Policework* (Unwin, London, 1988).

Bromley, P. M., *Family Law*, 1st edn. (Butterworths, London, 1957).

Brook, L., and Cape, E., 'Interim Report: Civil Liberties', in R. Jowell (ed.) *British Social Attitudes, the 8th Report* (Dartmouth, Aldershot, 1991).

Brophy, J., 'Custody Law, Child Care and Inequality in Britain' in C. Smart and

S. Sevenhuijsen (eds.), *Child Custody and the Politics of Gender* (Routledge, London, 1989).

Brown, C., *Black and White Britain: The Third PSI Survey* (Policy Studies Institute, London, 1984).

—— and Gay, P., *Racial Discrimination: 17 Years after the Act* (Policy Studies Institute, London, 1986).

Brown, D. *Detention at the Police Station under the Police and Criminal Evidence Act*, Home Office Research Study 104 (HMSO, London, 1989).

Brown, D., Ellis, T., and Larcombe, K., *Changing the Code: Police Detention Under the Revised PACE Codes of Practice*, Home Office Research Study No. 129 (HMSO, London, 1992).

Brown, M., *Working the Street: Police Discretion and the Dilemmas of Reform* (Russell Sage, New York, 1981).

Burke, E., *Reflections on the Revolution in France* (1790): (Penguin edn., Harmondsworth, 1986).

Buxton, R., 'Challenging and Discharging Jurors: 1' (1990) *Criminal Law Review* 225.

Bynoe, I., Oliver, M., and Barnes, C., *Equal Rights for Disabled People: The Case for a New Law* (Institute for Public Policy Research, London, 1991).

Byre, A., *Indirect Discrimination* (Equal Opportunities Commission, Manchester, 1987).

Byrne, A., and Lovenduski, J., 'The Equal Opportunities Commission' (1978) 1 *Women's Studies International Quarterly* 131.

Cain, M., 'On the Beat', in S. Cohen (ed.), *Images of Deviance* (Penguin, Harmondsworth, 1971).

—— and Sadigh, S., 'Racism, the Police and Community Policing' (1982) *Journal of Law and Society* 87.

Calabresi, G., 'Antidiscrimination and Constitutional Accountability (What the Bork–Brennan Debate Ignores)' (1991) 105 *Harvard Law Review* 80.

Callendar, C., 'The Development of the Sex Discrimination Act, 1971–75' (unpublished dissertation presented at Bristol University, 1978).

Campbell, T. D., *The Left and Rights: A Conceptual Analysis of the Idea of Socialist Rights* (Routledge & Kegan Paul, London, 1983).

Capers, B., 'Sex(ual Orientation) and Title VII' (1991) 91 *Columbia Law Review* 1158.

Carens, J. H., 'Aliens and Citizens: The Case for Open Borders' (1987) 49 *Review of Politics* 251.

Carlen, P., *Magistrates' Justice* (Martin Robertson, London, 1976).

Carr, J., *New Roads to Equality: Contract compliance for the UK?* (Fabian Society, London, 1987).

Carter, A., *The Politics of Women's Rights* (Longman, London, 1988).

Casale, S., *Minimum Standards for Prison Establishments* (NACRO, 1984).

Cassese, A., Clapham, A., and Weiler, J. (eds.), *Human Rights and the European Community: Methods of Protection* (Nomos, Baden-Baden, 1991).

—— —— —— (eds.), *Human Rights and the European Community*, iii. *The Substantive Law* (Nomos, Baden-Baden, 1991).

Central Community Relations Unit, *Race Relations in Northern Ireland* (CCRU, Belfast, 1992).

Chambers, G., and Horton, C., *Promoting Sex Equality: The Role of Industrial Tribunals* (Policy Studies Institute, London, 1990).

Charlton, R., and Kaye, R., 'The Politics of Religious Slaughter: an Ethno-religious Case Study' (1985–6) 12 *New Community* 490.

Chatterton, M., 'The Supervision of Patrol Work under the Fixed Points System', in S. Holdaway (ed.), *The British Police* (Edward Arnold, London, 1979).

Children Come First: The Government's Proposals on the Maintenance of Children, Cmnd. 1263 (HMSO, London, 1990).

Choo, A., 'Improperly Obtained Evidence: A Reconsideration' (1989) 9 *Legal Studies* 261–83.

Christian, T. J., and Ewing, K. D., 'Labouring under the Canadian Constitution' (1988) 17 *Industrial Law Journal* 73.

Churchill, R. R., 'Aspects of Compliance with Findings of the Committee of Ministers and Judgments of the Court with Reference to the United Kingdom', in J. P. Gardner (ed.), *European Convention on Human Rights: Aspects of Incorporation* (British Institute of Comparative Law and British Institute of Human Rights, London, 1992).

—— and Young, J., 'Compliance with Judgements of the European Court of Human Rights and Decisions of the Committee of Ministers: The Experience of the United Kingdom, 1975–1987' (1991) 62 *British Yearbook of International Law* 283–346.

Clapham, A., *Human Rights and the European Community: A Critical Overview* (Nomos, Baden-Baden, 1991).

Claude, R. B., and Jabine, T. B., 'Statistical Issues in the Field of Human Rights', editors' introduction (1986) 8 *Human Rights Quarterly* 563.

Clayton, R., and Tomlinson, H., *Civil Actions against the Police* (Sweet & Maxwell, London, 1987).

Cockburn, C., *In the Way of Women: Men's Resistance to Sex Equality in Organizations* (Macmillan, Basingstoke, 1991).

Cohen, B., and Staunton, M., 'In Pursuit of a Legal Strategy: The National Council for Civil Liberties', in J. Cooper and R. Dhavan (eds.), *Public Interest Law* (Blackwells, Oxford, 1986).

Colvin, M., and Hawksley J., *Section 28: A Practical Guide to the Law and its Implications* (Liberty, London, 1989).

Colville, Lord, *Review of the Northern Ireland (Emergency Provisions) Acts 1978 and 1987*, Cm. 1115 (HMSO, London, 1990).

—— *Review of the Operation of the Prevention of Terrorism (Temporary Provisions) Act 1984*, Cm. 264 (HMSO, London, 1987).

Commission for Racial Equality, *A Plural Society* (CRE, London, 1990).

—— *Annual Report 1991* (CRE, London, 1992).

—— *Free Speech* (CRE, London, 1990).

—— *Immigration Control Procedures: Report of a Formal Investigation* (CRE, London, 1985).

—— *Law, Blasphemy, and the Multi-Faith Society* (CRE, London, 1990).

—— *Schools of Faith: Religious Schools in a Multicultural Society* (CRE, London, 1990).

—— *Second Review of the Race Relations Act 1976: A Consultative Paper* (CRE, London, June 1991).

—— *Second Review of the Race Relations Act 1976* (CRE, London, 1992).

—— *Sorry its Gone: Testing for Racial Discrimination in the Private Rented Housing Sector* (CRE, London, 1990).

Commission of the European Communities, *Commission Communication on Community Accession to the European Convention for the Protection of Human Rights and Fundamental Freedoms and Some of its Protocols,* SEC (990) 2087 final, 19 Nov. 1990.

—— *Eurobarometer: Racism and Xenophobia* (European Commission, Brussels, 1989).

—— *Legal Instruments to Combat Racism and Xenophobia* (Luxembourg, Office of Official Publications, 1993).

Committee on the Administration of Justice, *Racism in Northern Ireland* (CAJ Pamphlet, No. 20, CAJ, Belfast, 1992).

Compton, Sir Edward, *Report of the Enquiry into Allegations against the Security Forces of Physical Brutality in Northern Ireland Arising out of the Events on 9 August 1971*, Cmnd. 4823 (HMSO, London, 1971).

Conference on Security and Co-operation in Europe, 'Concluding Document from the Vienna Meeting' (1989) 28 *International Legal Materials* 527.

Cooper, J., and Dhavan, R. (eds.), *Public Interest Law* (Blackwells, Oxford, 1986).

Corbin, A. L., 'Legal Analysis and Terminology' (1919) 29 *Yale Law Journal* 163.

Cornish, W. R., and Sealy, A. P., 'Jurors and the Rules of Evidence' (1973) *Criminal Law Review* 208.

Council of Europe, *The Situation of Women in the Political Process in Europe*, ii. *Women in the Political World in Europe* (Council of Europe, Directorate of Human Rights, Strasbourg, 1984).

Council of Europe Directorate of Human Rights, *Human Rights of Aliens in Europe* (Dordrecht, 1985).

Craies on Statute Law, 7th edn. by S. G. Edgar (Sweet & Maxwell, London, 1971).

Crane, P., *Gays and the Law* (Pluto Press, London, 1982).

Cray, E., *The Enemy in the Streets* (Anchor, New York, 1972).

Crenshaw, K. W., 'Race, Reform and Retrenchment: Transformation and Legitimation in Anti-Discrimination Law' (1988) 101 *Harvard Law Review* 1357.

Cretney, S. M., 'Defining the Limits of State Intervention: The Child and the Courts', in H. K. Bevan and D. Freestone (eds.), *Children and the Law* (Hull University Press, Hull, 1990).

—— and Masson, J. M., *Principles of Family Law* (Sweet & Maxwell, London, 1990).

Crime, Justice and Protecting the Public, Cmnd. 965 (HMSO, London, 1990).

Criminal Law Review, 'The New Prosecution Arrangements' (Jan. 1986) special edn.

Criminal Law Review Committee, *11th Report*, Cmnd. 4491 (HMSO, London, 1972).

Critchley, T. A., *A History of Police in England and Wales* (Constable, London, 1978).

Cruz, A., 'Carrier Sanctions in Five Community Countries', pamphlet issued to subscribers to *Migration News Sheet* (Brussels, 1991).

Damaska, M., 'Evidentiary Barriers to Conviction and Two Models of Criminal Procedure: A Comparative Study' (1973) *University of Pennsylvania Law Review* 506.

—— *The Faces of Justice and State Authority: A Comparative Approach to the Legal Process* (Yale University Press, New Haven, Conn., 1986).

Darbishire, P., 'The Lamp that Shows that Freedom Lives: Is it Worth the Candle?' (1991) *Crim. LR* 740.

Dauses, M. A., 'The Protection of Fundamental Rights in the Community Legal Order' (1985) 10 *European Law Review* 398.

Davies, G., and Murch, M., *Grounds for Divorce* (Clarendon Press, Oxford, 1988).

Davis, K. C., *Discretionary Justice* (University of Illinois Press, Urbana, Ill., 1969).

—— *Police Discretion* (West Publishing, St. Paul, Minn., 1977).

de Tocqueville, A., *Democracy in America* (Oxford University Press, Oxford, 1945 edn.).

De Smith, S., and Brazier, R., *Constitutional and Administrative Law*, 6th edn. by R. Brazier (Pelican Books, London, 1989).

Deakin, S., 'Equality under a Market Order: The Employment Act 1989' (1990) *Industrial Law Journal* 1.

Deech, R., 'The Case against Legal Recognition of Cohabitation', in J. Eekelaar and S. N. Katz (eds.), *Marriage and Cohabitation in Contemporary Societies* (Butterworths, London, 1980).

Delgado, R., 'The Ethereal Scholar: Does Critical Legal Studies have What Minorities Want' (1987) 22 *Harvard Civil Rights-Civil Liberties Law Review* 301.

Dennis, I. H., 'Corroboration Requirements Reconsidered' (1984) *Criminal Law Review* 316.

—— (ed.), *Criminal Law and Justice* (Sweet & Maxwell, London, 1987).

Devlin, Lord, 'The Conscience of the Jury' (1991) 107 *Law Quarterly Review* 398.

—— *The Judge* (Oxford University Press, Oxford, 1979).

Dicey, A. V., *An Introduction to the Study of the Law of the Constitution*, 10th edn., with an introduction by E. C. S. Wade (Macmillan, London, 1959).

A Digest of Information on the Criminal Justice System (Home Office Research and Statistics Dept., London, 1991).

Dingwall, R. W. J., and Eekelaar, J. M., 'Families and the State: An Historical Perspective on the Public Regulation of Private Conduct' (1988) 10 *Law and Policy* 341.

Diplock, Lord, *Report of the Commission to Consider Legal Procedures to Deal with Terrorist Activities in Northern Ireland*, Cmnd. 5185 (HMSO, London, 1972).

Dixon, D., *et al.*, 'Consent and the Legal Regulation of Policing' (1990) 17 *Journal of Law and Society* 345.

—— 'Reality and Rules in the Construction and Regulation of Police Suspicion' (1989) 17 *International Journal of the Sociology of Law* 185.

—— 'Safeguarding the Rights of Suspects in Police Custody' (1990) 1/2 *Policing and Society* 115.

Docksey, C., 'The Promotion of Equality', in C. McCrudden (ed.), *Women, Employment and European Equality Law* (Eclipse Publications, London, 1987).

Douglas, G., 'The Family and the State under the European Convention on Human Rights' (1988) 2 *International Journal of Law and the Family* 76.

—— 'Family Law under the Thatcher Government' (1990) 17 *Journal of Law and Society* 411.

—— *Law, Fertility and Reproduction* (Sweet and Maxwell, London, 1991).

—— and Lowe, N. V., 'Becoming a Parent in English Law' (1992) 108 *Law Quarterly Review* 414.

Dowty, A., *Closed Borders: The Contemporary Assault on Freedom of Movement* (Yale University Press, New Haven, Conn., 1987).

Drzemczewski, A., 'The need for a radical overhaul', *New Law Journal*, January 29 1993, 126.

Duff, R. A., *Trials and Punishments* (Cambridge University Press, Cambridge, 1986).

Dummett, A. (ed.), *Towards a Just Immigration Policy* (Cobden Trust, London, 1986).

—— and Nicol, A., *Subjects, Citizens, Aliens and Others: Nationality and Immigration Law* (Weidenfeld & Nicolson, London, 1990).

Duncan, W., 'The Divorce Referendum in the Republic of Ireland: the Tide' (1988) 2 *International Journal of Law and the Family* 62.

Dworkin, R., *A Bill of Rights for Britain* (Chatto Counterblasts No. 16, London, 1990).

—— 'Policy, Principle and Procedure', in C. Tapper (ed.), *Crime, Proof and Punishment: Essays in Memory of Sir Rupert Cross* (Butterworths, London, 1981).

—— *Taking Rights Seriously* (Duckworth, London, 1977).

Dyzenhaus, D., *Hard Cases in Wicked Legal Systems* (Clarendon Press, Oxford, 1991).

Eekelaar, J. M., 'The Eclipse of Parental Rights' (1986) 102 *Law Quarterly Review* 4.

—— 'The Emergence of Children's Rights' (1986) 6 *Oxford Journal of Legal Studies* 161.

—— 'The Importance of Thinking that Children have Rights' (1992) 6 *International Journal of Law and the Family* 221–35.

—— 'Parental Responsibility: State of Nature or Nature of the State?' (1991) *Journal of Social Welfare and Family Law* 37.

—— *Regulating Divorce* (Clarendon Press, Oxford, 1991).

Eekelaar, J. M., 'What are Parental Rights?' (1973) 89 *Law Quarterly Review* 210.

—— 'Parenthood, Social Engineering and Rights' (1994) *Zeitschrift für Rechts und Sozialphilosophie* (forthcoming).

—— 'The Interests of the Child and the Child's Wishes: the Role of Dynamic Self-Determinism' (1994) 8 *International Journal of Law and the Family* (forthcoming).

—— and Maclean, M., *Maintenance after Divorce* (Clarendon Press, Oxford, 1986).

—— Dingwall, R. W. J., and Murray, T., 'Victims or Threats? Children in Care Proceedings' (1982) *Journal of Social Welfare Law* 68.

Elias, P., and Ewing, K. D., *Trade Union Democracy, Members' Rights and the Law* (Mansell, London, 1988).

Emerson, T. I., 'Freedom of Association and Freedom of Expression' (1964) 74 *Yale Law Journal* 1.

Emmins, C. J., *A Practical Approach to Criminal Procedure*, 4th edn. (Blackstone, London, 1989).

—— 'Why No Advance Disclosure for Summary Offences?' (1987) *Criminal Law Review* 608.

Emsley, C., *The English Police: A Political and Social History* (Wheatsheaf, Hemel Hempstead, 1991).

—— '"The Thump of Wood on a Swede Turnip": Police Violence in 19th Century England', in L. A. Knafla (ed.), *Crime, Police and the Courts in British History* (Meckler, London, 1990).

Enright, S., 'Multi-racial Juries' (1991) 141 *New Law Journal* 992.

—— and Morton, J., *Taking Liberties: The Criminal Jury in the 1990s* (Weidenfeld & Nicolson, London, 1990).

Equal Opportunities Commission, *Equal Treatment for Men and Women: Strengthening the Acts* (Equal Opportunities Commission, Manchester, 1988).

—— *Health and Safety Legislation: Should We Distinguish between Men and Women?* (Equal Opportunities Commission, Manchester, 1979).

Ericson, R., *Making Crime: A Study of Detective Work* (Butterworths, Toronto, 1981).

—— *Reproducing Order: A Study of Police Patrol Work* (University of Toronto Press, Toronto, 1981).

European Commission for Democracy through Law, *European Convention for the Protection of Minorities*, repr. in (1991) 12 *Human Rights Law Journal* 2.

European Committee for the Prevention of Torture and Inhuman and Degrading Treatment or Punishment, *First General Report*, Jan. 1991 repr. in (1991) 12 *Human Rights Law Journal* 206.

Evans, J., *Immigration Law*, 2nd edn. (Sweet & Maxwell, London, 1983).

Ewing, K. D., *Britain and the ILO* (Institute of Employment Rights, London, 1989).

—— 'Freedom of Association in Canada' (1987) 25 *Alberta Law Review* 437.

—— *The Funding of Political Parties in Britain* (Cambridge University Press, Cambridge, 1987).

—— 'Trade Union Recognition: A Framework for Discussion' (1990) 19 *Industrial Law Journal* 111.

—— and Gearty, C. A., *Freedom under Thatcher: Civil Liberties in Modern Britain* (Clarendon Press, Oxford, 1990).

Fallon, R. H., and Weiler, P. C., 'Firefighters v. Stott: Conflicting Models of Racial Justice' (1984) *The Supreme Court Review* 1.

Farer, T. J., 'The UN and Human Rights: More than a Whimper, Less than a Roar', in A. Roberts and B. Kingsbury, *United Nations, Divided World: The UN's Roles in International Relations* (Clarendon Press, Oxford, 1988).

Fawcett, J. E. S., *The Application of the European Convention on Human Rights*, 2nd edn. (Clarendon Press, Oxford, 1987).

Feldman, D., 'Regulating Treatment of Suspects in Police Stations: Judicial Interpretations of Detention Provisions in the Police and Criminal Evidence Act 1984' (1990) *Criminal Law Review* 452–71.

Field, M. A., *Surrogate Motherhood: The Legal and Human Issues* (Harvard University Press, Cambridge, Mass. 1990).

Field, S., and Southgate, P., *Public Disorder* (Home Office Research Unit, London, 1982).

Final Report of the Royal Commission on Legal Services, Cmnd. 7648 (HMSO, London, 1979).

Findlay, M., and Duff, P. (eds.), *The Jury under Attack* (Butterworths, London, 1988).

Fine, B., and Millar, R. (eds.), *Policing the Miners' Strike* (London, Lawrence & Wishart, 1985).

Fisher, C., and Mawby, R., 'Juvenile Delinquency and Police Discretion in an Inner-City Area' (1982) 22 *British Journal of Criminology* 63.

Fiss, O. M., 'Groups and the Equal Protection Clause' (1976) 5 *Philosophy and Public Affairs* 107.

Fitzgerald, M. R., and Sim, J., *British Prisons* (Basil Blackwell, Oxford, 1982).

Fitzpatrick, P., 'Racism and the Innocence of Law', in P. Fitzpatrick and A. Hunt, *Critical Legal Studies* (Blackwell, Oxford, 1987).

Flaherty, D. H., Donohue, T. J., and Harte, P. J. (eds.), *Privacy and Data Protection: An International Bibliography* (Mansell, London, 1984).

Flanders, A., 'The Tradition of Voluntarism' (1974) 12 *British Journal of Industrial Relations* 352.

Flanz, G. H., *Comparative Women's Rights and Political Participation in Europe* (Transnational Publishers, Dobbs Ferry, NY, 1982).

Fletcher, G. P., 'Some Unwise Reflections about Discretion', in R. J. Allen (ed.), *Discretion in Law Enforcement*, special issue (1984) 47 *Law and Contemporary Problems* 282.

Foot, P., *Immigration and Race in British Politics* (Penguin, Harmondsworth, 1965).

Forbes, I., and Mead, G., *Measure for Measure: A Comparative Analysis of Measures to Combat Racial Discrimination in the Member Countries of the European Community* (Employment Dept., Research Series No. 1, Mar. 1992).

Forde, M., 'The European Convention on Human Rights and Labor Law' (1983) 31 *American Journal of Comparative Law* 9.

Forde, M., 'The Closed Shop Case' (1982) 11 *Industrial Law Journal* 1.

Forder, C., 'Constitutional Principle and the Establishment of the Legal Relationship between the Child and the Non-Marital Father: a Study of Germany, the Netherlands and England' (1993) 7 *International Journal of Law and the Family* 40.

Fransman, L., *Fransman's British Nationality Law* (Fourmat, London, 1989).

Frazier, I., *Great Plains* (Faber & Faber, London, 1990).

Fredman, S., 'Note: Crown Employment, Prerogative Powers, Consultation and National Security' (1985) 14 *Industrial Law Journal* 42.

—— and Morris, G. S., *The State as Employer: Labour Law in the Public Services* (Mansell, London, 1989).

Freeman, A., 'Legitimating Racial Discrimination through Anti-Discrimination Law' (1992) 62 *Minnesota Law Review* 96.

—— 'Racism Rights and the Quest for Equality of Opportunity: A Critical Legal Essay' (1988) 23 *Harvard Civil Rights—Civil Liberties Law Review* 295.

Freeman, M. D. A., 'Freedom and the Welfare State: Child-rearing, Parental Autonomy and State Intervention' (1983) *Journal of Social Welfare Law* 70.

—— 'The Jury on Trial' (1981) 34 *Current Legal Problems* 65.

—— 'Law and Order in 1984' (1984) 37 *Current Legal Problems* 175.

—— *The Police and Criminal Evidence Act 1984* (Sweet & Maxwell, London, 1985).

—— *The Rights and Wrongs of Children* (Frances Pinter, London, 1983).

Galligan, D. J., *Discretionary Powers: A Legal Study of Official Discretion* (Clarendon Press, Oxford, 1986).

—— 'More Scepticism about Scepticism' (1988) 8 *Oxford Journal of Legal Studies* 249–65.

—— 'Regulating Pre-Trial Decisions', in I. H. Dennis (ed.), *Criminal Law and Justice* (Sweet & Maxwell, London, 1987).

Gallivan, T., and Warbrick, C., 'Jury Vetting and the European Convention on Human Rights' (1980) 5 *Human Rights Review* 176.

Gardner, J., 'Liberals and Unlawful Discrimination' (1989) 9 *Oxford Journal of Legal Studies* 1.

Garland, D., *Punishment and Modern Society: A Study in Social Theory* (Clarendon Press, Oxford, 1990).

—— and Young, P. (eds.), *The Power to Punish* (Heineman, London, 1983).

Gastil, R. D. (ed.), *Freedom in the World: Political Rights and Civil Liberties 1989–90* (Freedom House, 1990).

Geach, H., and Szwed, E., *Providing Civil Justice for Children* (Edward Arnold, London, 1983).

Gearty, C., *Terror* (Faber & Faber, London, 1991).

Gee, D. J., and Mason, J. K., *The Courts and the Doctor* (Oxford University Press, Oxford, 1990).

Genders, E., and Player, E., *Race Relations in Prisons* (Clarendon Press, Oxford, 1989).

Gerwith, A., 'The Epistemology of Human Rights' (1984) 1 *Social Philosophy and Policy* 1.

Ghandhi, P. R., and Macnamee, E., 'The Family in UK Law and the International Covenant on Civil and Political Rights 1966' (1990) 5 *International Journal of Law and the Family* 104.

Glendon, M. A., *Abortion and Divorce in Western Law* (Harvard University Press, Cambridge, Mass., 1987).

Goldsmith, A., *Complaints against the Police* (Oxford University Press, Oxford, 1991).

Goldstein, H., 'Police Discretion: The Ideal vs. The Real,' (1963) 23 *Public Administration Review* 161.

Goldstein, J., 'Police Discretion Not to Invoke the Criminal Process: Low Visibility Decisions in the Administration of Justice' (1960) 69 *Yale Law Journal* 543.

—— Freud, A., and Solnit, A., *Before the Best Interests of the Child* (The Free Press, London, 1979).

Gomez del Prado, J. L., 'United Nations Conventions on Human Rights: The Practice of the Human Rights Committee and the Committee on the Elimination of Racial Discrimination in Dealing with Reporting Obligations of State Parties' (1991) 7 *Human Rights Quarterly* 492–513.

Goodwin-Gill, G., *The Refugee in International Law* (Clarendon Press, Oxford, 1983).

Goman, R. A., *Basic Text on Labour Law* (West Publishing Co, St Paul, Minn, 1976).

Gostin, L. O. (ed.), *Civil Liberties in Conflict* (Routledge, London, 1988).

—— *A Human Condition*, ii (MIND, London, 1977).

—— 'Professional Pressure Groups and Party Politics', in J. Cooper and R. Dhavan, *Public Interest Law* (Blackwells, Oxford, 1986).

Goulbourne, H., 'Varieties of Pluralism: The Notion of a Pluralist, Post-Imperial Britain' (1991) 17 *New Community* 211.

Grahl-Madsen, A., *The Status of Refugees in International Law* (Sijthoff, Leiden, 1972).

Grant, A., 'Video Taping Police Interviews' (1987) *Criminal Law Review* 375.

Grant, W., *Pressure Groups, Politics and Democracy in Britain* (Philip Allan, Hemel Hempstead, 1989).

Graveson, R. H., and Crane, F. R. (eds.), *A Century of Family Law* (Sweet & Maxwell, London, 1957).

Gray, K., *The Re-allocation of Property on Divorce* (Professional Books, Abingdon, 1977).

Greasley, P., and Williams, M., *Gay Men at Work: A Report on Discrimination against Gay Men in Employment in London* (Lesbian and Gay Employment Rights, London, 1986).

Greer, S., 'The Right to Silence: A Review of the Current Debate' (1990) 53 *Modern Law Review* 709.

—— and Morgan, R. (eds.), *The Right to Silence Debate* (Centre for Criminal Justice, Bristol University, 1990).

Gregory, J., *Sex, Race and the Law: Legislating for Equality* (Sage Publications, London, 1987).

—— *Trial by Ordeal: A Study of People Who Lost Equal Pay and Sex Discrimination Cases in the Industrial Tribunals during 1985 and 1986* (Equal Opportunities Commission, Manchester, 1989).

Grey, A., *Quest for Justice: Towards Homosexual Emancipation* (Sindair-

Stevenson, London, 1992).

Griffith, J. A. G., *The Politics of the Judiciary*, 3rd edn. (Fontana Press, Manchester, 1985); 4th edn. (Fontana, London, 1991).

—— and Ryle, M., *Parliament: Functions, Practice and Procedures* (Sweet & Maxwell, London, 1989).

—— *et al.*, *Coloured Immigrants in Britain* (OUP-IRR, London, 1960).

—— and Ryle, M., *Parliament: Functions, Practice and Procedures* (Sweet & Maxwell, London, 1989).

Grimshaw, R., and Jefferson, T., *Interpreting Policework* (Unwin, London, 1987).

Grosz, S., and Hulton, S., 'Using the European Convention on Human Rights', in J. Cooper and R. Dhavan, *Public Interest Law* (Blackwells, Oxford, 1986).

Gryk, W., 'AIDS and Insurance', in D. Harris and R. Haigh (eds.), *AIDS: A Guide to the Law* (Routledge, London, 1990).

Gudonsson, G. H., *The Psychology of Interrogations, Confessions and Testimony* (Wiley, Chichester, 1992).

Hampson, F. J., 'The United Kingdom before the European Court of Human Rights' (1989) 9 *Yearbook of European Law* 121.

Harden, I., and Lewis, N., *The Noble Lie* (Hutchinson, London, 1986).

Harding, R., 'Jury Performance in Complex Cases', in M. Findlay and P. Duff, *The Jury under Attack* (Butterworths, London, 1988).

Harlow, C., 'Public Interest Litigation in England: The State of the Art', in J. Cooper and R. Dhavan, *Public Interest Law* (Blackwells, Oxford, 1986).

Harris, D. J., *The European Social Charter* (University Press of Virginia, Charlottesville, Va., 1984).

—— 'The System of Supervision of the European Social Charter: Problems and Options for the Furture', in L. Betten, D. Harris, and T. Jaspers (eds.), *The Future of European Social Policy* (Kluwer Law and Taxation, Deventer, 1989).

Hart, H. L. A., *Law, Liberty, and Morality* (Oxford University Press, Oxford, 1963).

—— *Punishment and Responsibility: Essays in the Philosophy of Law* (Clarendon Press, Oxford, 1968).

Haskey, J., 'The Ethnic Minority Populations Resident in Private Households: Estimates by County and Metropolitan District of England and Wales' (Spring 1991) 63 *Population Trends* 22.

—— and Kiernan, K., 'Cohabitation in Great Britain: Characteristics and Estimated Numbers of Cohabiting Partners' (1989) 58 *Population Studies* 23.

Hayek, F. A., *The Constitution of Liberty* (Routledge & Kegan Paul, London, 1960).

Heinemann, B., *The Politics of the Powerless* (OUP-IRR, London, 1972).

Helfer, L., 'Finding a Consensus on Equality: The Homosexual Age of Consent and the European Convention on Human Rights' (1990) 65 *New York University Law Review* 1044.

—— 'Lesbian and Gay Rights as Human Rights: Strategies for a United Europe' (1991) 32 *Virginia Journal of International Law* 157.

Henkin, L. (ed.), *The International Bill of Rights: the Covenant on Civil and*

Political Rights (Columbia University Press, New York, 1981).

Hepple, B., 'Discrimination and Equality of Opportunity: Northern Ireland Lessons' (1990) 10 *Oxford Journal of Legal Studies* 408.

—— 'Implementation of the Charter of Social Rights' (1990) 53 *Modern Law Review* 643.

—— *Race, Jobs and the Law in Britain*, 2nd edn. (Penguin, Harmondsworth, 1970).

Hill, P., Young, M., and Sargant, T., *More Rough Justice* (Penguin, Harmondsworth, 1985).

Hindley, C., 'The Age of Consent for Male Homosexuals' (1986) *Criminal Law Review* 595.

Hiro, D., *Black British, White British*, 2nd edn. (Grafton Books, London, 1991).

HM Chief Inspector of Prisons, *Prison Suicides*, Cm. 1383 (HMSO, London, 1990).

—— *A Review of Prisoners' Complaints* (Home Office, London, 1987).

—— *A Review of Segregation of Prisoners under Rule 43* (Home Office, London, 1986).

Hogan, G., and Walker, C., *Political Violence and the Law in Ireland* (Manchester University Press, Manchester, 1989).

Hohfeld, W. N., *Fundamental Legal Conceptions* (Yale University Press, New Haven, Conn., 1919).

Holdaway, S., *Inside the British Police* (Blackwell, Oxford, 1983).

Home Office, *An Independent Prosecution Service for England and Wales*, Cmnd. 9074 (HMSO, London, 1983).

—— *Custody, Care and Justice*, Cm. 1647 (HMSO, London, 1991).

—— *Report on the Work of the Immigration and Nationality Department* (HMSO, London, 1984–   ).

—— *Statistics of the Criminal Justice System in England and Wales* (HMSO, London, 1980).

Honeyford, R., *Integration or Disintegration* (Claridge, London, 1988).

Hoskyns, C., 'The European Women's Lobby' (1991) 38 *Feminist Review* 67–9.

—— 'Women, European Law and Transnational Politics' (1986) *International Journal of the Sociology of Law* (special issue).

House of Commons, Home Affairs Committee, *The Forensic Source Service* (HC 26, 1988–89) (HMSO, London, 1989).

—— —— *The Law Relating to Public Order* (HC 756, 1979–80) (HMSO, London, 1980).

—— —— *Policing Football Hooliganism* (HC 1, 1990–91) (HMSO, London, 1991).

Howard, M. N., 'The Neutral Expert: A Plausible Threat to Justice' (1991) *Criminal Law Review* 98.

Humana, C., *World Human Rights Guide* (Economist Publications, London, 1986); 3rd edn. (Oxford University Press, New York, 1992).

Humphry, D., and Ward, M., *Passports and Politics* (Penguin, Harmondsworth, 1974).

Industrial Relations in the 1990s: Proposals for Further Reform of Industrial

Relations and Trade Union Law, Cmnd. 1602 (HMSO, London, 1991).

Institute for Public Policy Research, *A British Bill of Rights*, Constitution Paper No. 1 (IPPR, London, 1990).

Institute of Personnel Management, *Contract Compliance: the UK Experience* (IPM, London, 1987).

Interim Report on Joint Standing Industrial Councils by the Sub-Committee on Relations between Employers and Employed of the Committee of Reconstruction, Cmnd. 8606 (HMSO, London, 1971).

Inter-Parliamentary Union, *Participation of Women in Political Life and in the Decision-Making Process*, International Centre for Parliamentary Documentation series 'Reports and Documents', 15 (CIDP, Geneva, 1988).

Irving, B., *Police Interrogation: A Case Study of Current Practice*, Royal Commission Research Study 2 (HMSO, London, 1980).

—— and McKenzie, I., 'Interrogating in a Legal Framework', in R. Morgan and D. Smith (eds.), *Coming to Terms with Policing* (Routledge, London, 1989).

—— —— *Police Interrogation* (Police Foundation, London, 1989).

Jackson, J., 'In Defence of a Voluntariness Doctrine for Confessions: *The Queen v. Johnston* Revisited' (1986) NS 21 *Irish Jurist* 208.

—— 'The Insufficiency of Identification Evidence Based on Personal Impression' (1986) *Criminal Law Review* 203.

—— 'Recent Developments in Criminal Evidence' (1989) 40 *Northern Ireland Legal Quarterly* 105.

—— 'Theories of Truth Finding in Criminal Procedure: An Evolutionary Approach' (1988) *Cardozo Law Review* 475.

—— 'Two Methods of Proof in Criminal Procedure' (1988) 51 *Modern Law Review* 549.

—— *et al.*, *Called to Court: A Public View of Criminal Justice in Northern Ireland* (Belfast, 1991).

Jacobs, F. G., *The European Convention on Human Rights* (Clarendon Press, Oxford, 1980).

Jamieson, A., *The Heart Attacked: Terrorism and Conflict in the Italian State* (Marion Boyars, London, 1989).

Jarman, J., 'Equality or Marginalisation: The Repeal of Protective Legislation', in E. Meehan and S. Sevenhuijsen (eds.), *Equality Politics and Gender* (Sage, London, 1991).

Jefferson, T., *The Case against Paramilitary Policing* (Open University Press, Milton Keynes, 1990).

—— 'Race, Crime and Policing: Empirical, Theoretical and Methodological Issues' (1988) 16 *International Journal of the Sociology of Law* 521.

—— and Grimshaw, R., *Controlling the Constable* (Muller, London, 1984).

Jeffery-Poulter, S., *Peers, Queers and Commons: The Struggle for Gay Law Reform from 1950 to the Present* (Routledge, London, 1991).

Jellicoe, Earl, *Review of the Operation of the Prevention of Terrorism (Temporary Provisions) Act 1976*, Cmnd. 8803 (HMSO, London, 1983).

Jenkins, R., *Essays and Speeches* (Collins, London, 1967).

—— *Life at the Centre* (Macmillan, London, 1991).

Jenkins, R., and J. Solomos (eds.), *Racism and Equal Opportunity Policies in the 1980s*, 2nd edn. (Cambridge University Press, Cambridge, 1989).

Jenks, C. W., *Human Rights and International Labour* (Stevens, London, 1960).

Jennings, A. (ed.), *Justice under Fire: The Abuse of Civil Liberties in Northern Ireland* (Pluto Press, London, 1988).

—— 'Note' (1990) 140 *New Law Journal* 633.

Jennings, I., *The Law and the Constitution*, 5th edn. (University of London Press, London, 1959).

Joint Council for the Welfare of Immigrants, *Annual Reports* (1989/90–   ).

Jones, S., 'The Ascertainable Wishes and Feelings of the Child' (1992) 4 *Journal of Child Law* 181.

Jones, T., *Britains' Ethnic Minorities* (Policy Studies Institute, London, 1993).

Joshi, S., and Carter, B., 'The Role of Labour in the Creation of a Racist Britain' (1984) 25 *Race and Class* 53.

Jowell, J., 'Administrative Enforcement of Laws against Discrimination' (1965) *Public Law* 119.

Justice, *Contempt of Court: A Report by Justice* (Justice, London, 1959).

—— *Miscarriages of Justice* (Justice, London, 1989).

—— *The Prosecution Process in England and Wales* (Justice, London, 1970).

—— *Witnesses in the Criminal Court* (Justice, London, 1986).

Kadish, S. H., 'Methodology and Criteria in Due Process Adjudication: A Survey and Criticism' (1956) 66 *Yale Law Journal* 319.

Karst, K., 'The Freedom of Intimate Association' (1980) 89 *Yale Law Journal* 624.

—— 'Woman's Constitution' (1984) *Duke Law Journal* 447.

Katznelson, I., *Black Men, White Cities* (OUP-IRR, London, 1973).

Kennedy, I., 'Transsexualism and Single Sex Marriage' (1973) 2 *Anglo-American Law Review* 112.

—— and Grubb, A., 'HIV and AIDS: Discrimination and the Challenge for Human Rights' in A. Grubb (ed.), *Challenges in Medical Care* (John Wiley, London, 1992).

Kenrick, D., and Bakewell, S., *On the Verge: The Gypsies of England* (Runnymede Trust, London, 1990).

Keown J., *Abortion, Doctors and the Law* (Cambridge University Press, Cambridge, 1988).

Kiernan, K., and Wicks, M., *Family Change and Future Policy* (Family Policy Studies Centre, Joseph Rowntree Memorial Trust, York, 1990).

Kilbrandon, A., 'The Law of Privacy in Scotland' (1971) 2 *Cambrian Law Review* 35.

King, M., and May, C., *Black Magistrates: A Study of Selection and Appointment* (Cobden Trust, London, 1985).

King, R. D., and McDermott, K., 'British Prisons 1970–87: The Ever Deepening Crisis' (1989) 29 *British Journal of Criminology* 107.

—— and Morgan, R., *The Future of the Prison Service* (Gower, Farnborough, 1980).

Kinsey, R., Lea, J., and Young, J., *Losing the Fight against Crime* (Blackwell,

Oxford, 1986).

Knight, M., *Criminal Appeals* (Stevens, London, 1975).

Koppelman, A., 'The Miscegenation Analogy: Sodomy Laws as Sex Discrimination' (1988) 98 *Yale Law Journal* 145.

Kruger, H. C., 'Does the Convention Machinery Distinguish between States which have and have not Incorporated Textually, or Statistically, or Substantively?' in J. P. Gardner (ed.), *The European Convention on Human Rights: Aspects of Incorporation* (British Institute of Comparative Law and British Institute of Human Rights, London, 1992).

Kymlicka, W., *Liberalism, Community and Culture* (Clarendon Press, Oxford, 1989).

Lacey, N., 'Discretion and Due Process at Post-Conviction Stage' in I. H. Dennis (ed.), *Criminal Law and Justice* (Sweet & Maxwell, London, 1987).

—— 'Note: Dismissal by Reason of Pregnancy' (1986) 15 *Industrial Law Journal* 43.

—— *State Punishment: Political Principles and Community Values* (Routledge, London, 1988).

LaFave, W., *Arrest* (Little, Brown, Boston, Mass., 1965).

—— 'The Police and Non-Enforcement of the Law' (1962) *Wisconsin Law Review* 104; 179.

Lambert, J., *Crime, Police and Race Relations* (Oxford University Press, Oxford, 1970).

—— 'The Police Can Choose' (1969) 14 *New Society* 352.

Landon, S. F., and Nathan, G., 'Selecting Delinquents for Cautioning in the London Metropolitan Area' (1983) 23 *British Journal of Criminology* 128.

Laqueur, W., *The Age of Terrorism* (Weidenfeld & Nicolson, London, 1987).

Law, S., 'Rethinking Sex and the Constitution' (1984) 132 *University of Pennsylvania Law Review* 955.

Law Commission, *Breach of Confidence*, Law Comm. No. 110, Cmnd. 8388 (HMSO, London, 1981).

—— *Distribution on Intestacy*, Law Comm. No. 187 (HMSO, London, 1989).

—— *Family Law: Domestic Violence and Occupation of the Family Home*, Law Comm. No. 207 (HMSO, London, 1992).

—— *Family Law: Matrimonial Property*, Law Comm. No. 175 (HMSO, London, 1988).

—— *Family Law: Third Report on Family Property. The Matrimonial Home (Co-Ownership and Occupation Rights) and Household Goods*, Law Comm. No. 86 (HMSO, London, 1978).

—— *Offences against Religion and Public Worship*, Law Comm. No. 145 (HMSO, London, 1985).

—— *Polygamous Marriages*, Law Comm. No. 42 (HMSO, London, 1971).

—— *Polygamous Marriages*, Law Comm. No. 146 (HMSO, London, 1985).

—— *Criminal Law: Report on Offences Relating to Public Order*, Law Comm. No. 123 (HMSO, London, 1983).

Law Commission of Scotland, *Breach of Confidence*, Cmnd. 9385 (HMSO, Edinburgh, 1984).

Laws, J. G., 'The Ghost in the Machine: Principle in Public Law' (1989) *Public Law* 27.

—— 'Is the High Court the Guardian of Fundamental Constitutional Rights?' (1993) *Public Law* 59.

Lawson, E., *Encyclopedia of Human Rights* (Taylor & Francis, London, 1989).

Lea, J., and Young, J., *What is to be Done about Law and Order?* (Penguin, London, 1984).

Leader, S., 'The European Convention on Human Rights, the Employment Act of 1988 and the Right to Refuse to Strike' (19) 20 *Industrial Law Journal* 39.

Lee, J. J., *Ireland 1912–85: Politics and Society* (Cambridge University Press, Cambridge 1989).

Lee, J. A., 'Some Structural Aspects of Police Deviance in Relations with Minority Groups', in C. D. Shearing (ed.), *Organisational Police Deviance* (Butterworths, Toronto, 1981).

Lee, S., 'Bicentennial Bork, Tercentennial *Spycatcher*: Do the British Need a Bill of Rights?' (1988) 49 *University of Pittsburgh Law Review* 797.

—— *The Cost of Free Speech* (Faber & Faber, London, 1990).

Leigh, L. H., 'Liberty and Efficiency in the Criminal Process: The Significance of Models' (1977) 26 *International and Comparative Law Quarterly* 516.

—— *Police Powers in England and Wales*, 2nd edn. (Butterworths, London, 1985).

—— 'La Procedure penale anglaise à la lumiere de la Convention européenne des droits de l'homme' (1988) 3 *Revue de Science Criminelle* 453.

—— 'The Royal Commission on Criminal Procedure' (1981) 44 *Modern Law Review*.

—— 'Le Royaume Uni', in M. Delmas-Marty (ed.), *Raisonner la raison d'état* (PUF, Paris, 1989).

—— 'Some Observations on the Parliamentary History of the Police and Criminal Evidence Act 1984', in C. Harlow (ed.), *Public Law and Politics* (Sweet & Maxwell, London, 1986).

Leng, R., *The Right to Silence in Police Interrogation: a study of some of the issues underlying the debate*, Research Study No. 10 for the Royal Commission on Criminal Justice (HMSO, London, 1993).

Leonard, A., *Pyrrhic Victories: Winning Sex Discrimination and Equal Pay Cases in the Industrial Tribunals, 1980–1984* (HMSO, London, 1987).

Leopold, P., 'Freedom of Speech in Parliament: Its Misuse and Proposals for Reform' (1981) *Public Law* 30.

—— 'Incitement to Hatred: The History of a Controversial Criminal Offence' (1976) *Public Law* 389.

Lerner, N., *Group Rights and Discrimination in International Law* (Martinus Nijhoff, Dordrecht, 1991).

—— *The UN Convention on the Elimination of All Forms of Racial Discrimination*, 2nd edn. (Sijthoff & Noordhoff, Alphen aan den Rijn, 1980).

Lester, A., *Democracy and Individual Rights*, Fabian Tract, 390 (Fabian Society, London, 1968).

—— 'Fundamental Rights in the United Kingdom: The Law and the British

Constitution' (1976) 125 *University of Pennsylvania Law Review* 337.

—— 'Fundamental Rights: The United Kingdom Isolated?' (1984) *Public Law* 46.

—— and Bindman, G., *Race and the Law* (Penguin, Harmondsworth, 1972).

—— and Jowell, J., 'Beyond Wednesbury: Substantive Principles of Administrative Law' (1987) *Public Law* 368.

Levay, S., 'A Difference in Hypothalamic Structure between Heterosexual and Homosexual Men' (30 Aug. 1991) *Science* 1034.

Levi, M., 'Reforming the Criminal Fraud Trial' (1986) 13 *Journal of Law and Society* 117–30.

Lewis, R., *Anti-Racism: A Mania Exposed* (Quartet Books, London, 1988).

Liberty, *A People's Charter: Liberty's Bill of Rights: A Consultation Document* (Liberty, London, 1991).

—— (NCCL), *Standing Up for Your Rights!* (Liberty, London, 1991).

Lidstone, K., 'Magistrates, the Police and Search Warrants' (1984) *Criminal Law Review* 454.

—— 'Powers of Entry, Search and Seizure' (1989) 40 *Northern Ireland Legal Quarterly* 333–62.

Lind, E. A., and Tyler, T. R., *The Social Psychology of Procedural Justice* (Plenum, New York, 1988).

Lloyd-Bostock, S., and Clifford, B. R. (eds.), *Evaluating Witness Evidence* (Wiley, Chichester, 1983).

—— *Law in Practice*, Psychology in Action Series (British Psychological Society and Routledge, London, 1988).

London Equal Value Steering Group, *Job Evaluation and Equal Value: A Study of White Collar Job Evaluation in London Local Authorities* (LEVEL, London, 1987).

—— *A Question of Earnings: A Study of the Earnings of Blue Collar Employees in London Local Authorities* (LEVEL, London, 1987).

Loveday, B., 'The New Police Authorities in the Metropolitan Counties' (1991) 1/3 *Policing and Society* 193.

Lovenduski, J., and Randall, V., *Contemporary Feminist Politics: Women and Power in Britain* (Oxford University Press, Oxford, 1993).

Lustgarten, L., *The Governance of the Police* (Sweet & Maxwell, London, 1986).

—— 'The New Meaning of Discrimination' (1978) *Public Law* 178.

—— 'Racial Inequality and the Limits of Law' (1986) 49 *Modern Law Review* 68.

McBarnet, D., 'Arrest: The Legal Context of Policing', in S. Holdaway (ed.), *The British Police* (Edward Arnold, London, 1979).

—— *Conviction: Law, the State and the Construction of Justice* (Macmillan, London, 1981).

—— 'False Dichotomies in Criminal Justice Research' in Baldwin, J., and Bottomley, A. K. (eds.), *Criminal Justice* (Martin Robertson, London, 1978).

—— 'Fisher Report on the Confait Case: Four Issues', (1978) 41 *Modern Law Review* 455.

—— 'The Royal Commission and the Judges' Rules' (1981) 8 *British Journal of Law and Society* 109.

Maclean, M., and Eekelaar, J., 'Child Support: the British Solution' (1993) 7 *International Journal of Law and the Family* 205.

McCabe, S., and Purves, P., *The Jury at Work* (Blackwell, Oxford, 1972).

—— —— *The Shadow Jury at Work* (Blackwell, Oxford, 1974).

McConville, M., 'Search of Persons and Premises: New Data from London' (1983) *Criminal Law Review* 605–14.

—— *Corroboration and Confessions: the impact of a rule requiring that no conviction can be sustained on the basis of confession evidence alone*, Research Study No. 13 for the Royal Commission on Criminal Justice (HMSO, London, 1993).

—— Sanders, A., and Leng, R., *The Case for the Prosecution* (Routledge, London, 1991).

MacCormick, D. N., 'Delegated Legislation and Civil Liberty' (1970) 86 *Law Quarterly Review* 171.

McCrudden, C. (ed.), *Anti-Discrimination Law* (International Library of Essays in Law and Legal Theory, Dartmouth, 1991).

—— (ed.), *Fair Employment Handbook*, 2nd edn. (Eclipse, 1991). .

—— 'Between Legality and Reality: The Implementation of Equal Pay for Work of Equal Value in Great Britain' (1991) 3 *International Review of Comparative Public Policy* 177.

—— 'Codes in a Cold Climate: Administrative Rule-Making by the Commission for Racial Equality' (1988) 51 *Modern Law Review* 409.

—— 'Institutional Discrimination' (1981) 2 *Oxford Journal of Legal Studies* 303.

—— 'Northern Ireland and the British Constitution', in J. Jowell and D. Oliver, *The Changing Constitution*, 2nd edn. (Clarendon Press, Oxford, 1989), 297.

—— 'Options for Amending the Equal Pay Legislation', in *Equal Pay for Work of Equal Value: Conference Report* (Equal Opportunities Commission for Northern Ireland, Belfast, 1990).

—— 'Rethinking Positive Action' (1986) 15 *Industrial Law Journal* 219.

—— (ed.), *Women, Employment and European Equality Law* (Eclipse Publications, London, 1987).

—— Smith, D., and Brown, C. (and with the assistance of Jim Knox), *Racial Justice at Work: The Enforcement of the Race Relations Act 1976 in Employment* (Policy Studies Institute, London, 1991).

Macdonald, I., and Blake, N., *Immigration Law and Practice in the United Kingdom*, 3rd edn. (Butterworths, London, 1991).

MacEwen, M., *Housing, Race and Law: The British Experience* (Routledge, London, 1990).

McGoldrick, D., *The Human Rights Committee: Its Role in the Development of the International Covenant on Civil and Political Rights* (Clarendon Press, Oxford, 1991).

—— 'The United Nations Convention on the Rights of the Child' (1991) 5 *International Journal of Law and the Family* 132.

McKean, W. B., *Equality and Discrimination under International Law* (Clarendon Press, Oxford, 1983).

McKenzie, I., Morgan, R., and Reiner, R., 'Helping the Police with their Inquiries: The Necessity Principle and Voluntary Attendance at the Police Station' (1990) *Criminal Law Review* 22.

MacKinnon, C., *Sexual Harassment of Working Women: A Case of Sex Discrimination* (Yale University Press, New Haven, Conn., 1979).

MacKinnon, C., *Feminism Unmodified: Discourses on Life and Law* (Harvard University Press, Cambridge, Mass., 1987).

McManus, J. J., *Visiting Committees in Scottish Penal Establishments* (Scottish Office, Edinburgh, 1985).

McNee, D., *McNee's Law* (Collins, London, 1983).

McNitt, A. D., 'Some Thoughts on the Systematic Measurement of the Abuse of Human Rights', in D. L. Cingranelli, *Human Rights: Theory and Measurement* (Macmillan, Basingstoke, 1988).

McNorrie, K., 'Abortion in Great Britain: One Act, Two Laws' (1985) *Criminal Law Review* 475.

—— 'Symbolic and Meaningless Legislation' (1988) *Journal of the Law Society of Scotland* 310.

Maguire, M., 'Complaints against the Police: The British Experience' in A. Goldsmith (ed.), *Complaints against the Police* (Oxford University Press, Oxford, 1991).

—— 'Effects of the PACE Provisions on Detention and Questioning: Some Preliminary Findings' (1988) 28 *British Journal of Criminology* 19–43.

—— 'Prisoner's Grievances: The Role of the Board of Visitors', in M. Maguire, J. Vagg, and R. Morgan (eds.), *Accountability and Prisons* (Tavistock, London, 1985).

—— and Corbett, C., *Complaints against the Police* (HMSO, London, 1991).

—— and Pointing, J., *Victims of Crime: A New Deal?* (Open University Press, Milton Keynes, 1988).

—— and Vagg, J., 'Who are the Prison Watchdogs? The Membership and Appointment of Boards of Visitors' (1983) *Criminal Law Review* 238.

—— Vagg, J., and Morgan, R., *Accountability and Prisons* (Tavistock, London, 1985).

Maidment, S., *Child Custody and Divorce* (Croom Helm, London, 1984).

Malleson, K., 'Miscarriages of Justice: The Accessibility of the Court of Appeal' (1991) *Criminal Law Review* 323–32.

Manchester City Council, *Report of the Independent Inquiry Panel into Leon Britton's Visit to Manchester University Students' Union* (Manchester City Council, Manchester, 1985).

Manning, P., 'The Social Control of Police Work', in S. Holdaway (ed.), *The British Police* (Edward Arnold, London, 1979).

Mansfield, G., and Peay, J., *The Director of Public Prosecution: Principles and Practices for the Crown Prosecutor* (Tavistock, London, 1987).

Maoz, A., 'Defending Civil Liberties without a Constitution: The Israeli Experience' (1988) 16 *Melbourne University Law Review* 815–36.

Mark, R., *In the Office of Constable* (Collins, London, 1978).

Marshall, G., *Police and Government* (Methuen, London, 1965).

Martin, I., 'Combining Casework and Strategy: The Joint Council for the Welfare of Immigrants', in J. Cooper and R. Dhavan, *Public Interest Law* (Blackwell, Oxford, 1986).

Martin, J., *Women and Employment* (Report of Office of Population, Censuses and Surveys, HMSO, London, 1984).

May, R., *Criminal Evidence*, 2nd edn. (Sweet & Maxwell, London, 1991).

—— 'Fair Play at Trial: An Interim Assessment of s. 78 of the Police and Criminal Evidence Act 1984' (1988) *Criminal Law Review* 722–30.

Mayhew, P., Elliott, D., and Dowds, L., *The 1988 British Crime Survey*, Home Office Research Study 3 (HMSO, London, 1989).

Mayhew, P. and Maung, N. A., *Surveying Crime: Findings from the 1992 British Crime Survey* (Home Office Research and Statistics Department, London, 1992).

Mazey, S. P., and Richardson, J. J., 'British Pressure Groups in the European Community: The Challenge of Brussels' (1992) 45 *Parliamentary Affairs* 92.

Meehan, E., 'British Feminism from the 1960s to the 1990s', in H. Smith, *British Feminism in the Twentieth Century* (Edward Elgar, Aldershot, 1991).

—— *Women's Rights at Work: Campaigns and Policy in Britain and the United States* (Macmillan, London, 1985).

—— and Sevenhuijsen, S. (eds.), *Equality Politics and Gender.* (Sage, London, 1991).

Meijers, H., *et al.*, *Schengen: Internationalisation of Central Chapters of the Law on Aliens, Refugees, Security and the Police* 2nd edn. (Stichting NJCM-Boekerij, Leiden, 1992).

Meiklejohn, A., 'Free Speech and its Relation to Self-Government', in A. Meikel-john, *Political Freedom: The Constitutional Powers of the People* (Harper, New York, 1965).

Meldrum, J., *Attacks on Gay People* (Campaign for Homosexual Equality, London, 1980).

Meron, T., *Human Rights Law-making in the United Nations* (Clarendon Press, Oxford, 1986).

Metropolitan Police Commissioner, *Annual Report 1989*, Cm. 1070 (HMSO, London, 1990).

Meulders-Klein, M.-T. 'The Position of the Father in European Legislation' (1990) 4 *International Journal of Law and the Family* 131.

Miller, C. J., *Contempt of Court* (Clarendon Press, Oxford, 1990).

Miller, H., 'An Argument for the Application of Equal Protection Heightened Scrutiny to Classifications Based on Homosexuality' (1984) 57 *Southern California Law Review* 797.

Mirfield, P., *Confessions* (Sweet & Maxwell, London, 1985).

—— 'The Legacy of Hunt' (1989) *Criminal Law Review* 19–30.

Mitchell, A., *Children in the Middle: Living through Divorce* (Tavistock, London, 1985).

Modood, T., 'British Asian Muslims and the Rushdie Affair' (1990) 61 *Political Quarterly* 143.

—— 'The Indian Economic Success: A Challenge to Some Race Relations Assumptions' (1991) 19 *Policy and Politics* 177.

Montgomery, J., 'Legislating for a Multi-faith Society: Some Problems of Special Treatment', in B. Hepple and E. Szyszczak (eds.), *Discrimination: The Limits of Law* (Mansell, London, 1992).

Moody, S., and Tombs, J., *Prosecution in the Public Interest* (Scottish Academic

Press, Edinburgh, 1982).

Moran, L., 'The Uses of Homosexuality: Homosexuality for National Security' (1991) 19 *International Journal of the Sociology of Law* 149.

Morgan, J. and Zedner, L., *Child Victims: Crime, Impact and Criminal Justice* (Clarendon Press, Oxford, 1992).

Morgan, R., 'Policing by Consent: Legitimating the Doctrine' in R. Morgan and D. J. Smith (eds), *Coming to Terms with Policing* (Routledge, London, 1989).

—— 'Talking About Policing' in D. Downes (ed.), *Unravelling Criminal Justice* (Macmillan, London, 1992).

—— 'Woolf: In Retrospect and Prospect' (forthcoming) *Modern Law Review*.

—— and Bronstein, A.J., 'Prisoners and the Courts: The U.S. Experience', in M. Maguire, J. Vagg, and R. Morgan (eds.), *Accountability and Prisons* (Tavistock, London, 1985).

—— and Smith, D. (eds.), *Coming to Terms with Policing* (Routledge, London, 1989).

—— Reiner, R., and McKenzie, I., *Police Powers and Policy: A Study of the Work of Custody Officers*, End of Award Report (ESRC, Swindon, 1990).

Moss, P., *Childcare and Equality of Opportunity, Consolidated Report*, V/746/88 (European Commission, Brussels, 1988).

Mowbray, A. P., 'Administrative Law and Human Rights' (1991) 141 *New Law Journal* 1079.

Munday, R., 'Name Suppression: An Adjunct to the Presumption of Innocence and to Mitigation of Sentence' (1991) *Criminal Law Review* 680.

Munroe, M., 'The Prestige Case: Putting the Lid on the Commission for Racial Equality' (1985) 14 *Anglo-American Law Review* 187.

Nagra, J., 'Asian Supplementary Schools' (1981–2) *New Community* 431.

National Council for Civil Liberties, *Southall, 23 April 1979. The Report of the Unofficial Committee of Enquiry* (NCCL, London, 1980).

—— *Stonehenge: A Report into the Civil Liberties Implications of the Events Relating to the Convoys of Summer 1985 and 1986* (NCCL, London, 1986).

New Zealand Department of Justice, *A Bill of Rights for New Zealand: A White Paper* (Government Printer, Wellington, 1985).

NiBhrolchain, M., 'The Ethnicity Question for the 1991 Census: Background and Issues' (1990) 13 *Ethnic and Racial Studies* 542.

Nielsen, J., *A Survey of British Local Authority Response to Muslim Needs*, Research Paper 30/31, Centre for the Study of Islam and Christian–Muslim Relations (Birmingham, 1986), summarized in T. Gerholm and Y. G. Lithman (eds.), *The New Islamic Presence in Western Europe* (Mansell, London, 1988).

Nielsen, L., 'Family Rights and the "Registered Partnership" in Denmark' (1990) 4 *International Journal of Law and the Family* 297.

Norris, P., *Politics and Sex Equality* (Reinner, Boulder, Colo., 1987).

Northam, G., *Shooting in the Dark* (Faber & Faber, London, 1988).

Norton, P., *The Constitution in Flux* (Blackwell, Oxford, 1982).

Note, 'The Constitutional Status of Sexual Orientation: Homosexuality as a Suspect Classification' (1985) 98 *Harvard Law Review* 1285.

Note, 'Custody Denials to Parents in Same-Sex Relationships: An Equal Protection

Analysis' (1989) 102 *Harvard Law Review* 617.

Nozick, R., *Anarchy, State and Utopia* (Basic Books, New York, 1974).

O'Connor, P., 'The Court of Appeal: Trials and Tribulations' (1990) *Criminal Law Review* 615.

O'Donovan, K., 'A Right to Know one's Parentage?' (1988) 2 *International Journal of Law and the Family* 27.

—— *Sexual Divisions in Law* (Weidenfeld & Nicholson, London, 1985).

—— and Szyszczak, E., *Equality and Sex Discrimination Law* (Blackwell, Oxford, 1988).

O'Higgins, P., 'The European Social Charter', in R. Blackburn and J. Taylor, *Human Rights for the 1990s* (Mansell, London, 1991).

Packer, H., *The Limits of the Criminal Sanction* (Stanford University Press, Stanford, Calif., 1968).

Pais, M. S., 'The Committee on the Rights of the Child' (1991) 47 *ICJ Review* 36.

Palley, C., *The United Kingdom and Human Rights* (Sweet & Maxwell, London, 1991).

Pannick, D., *Sex Discrimination Law* (Clarendon Press, Oxford, 1985).

Parekh, B., 'Britain and the Social Logic of Pluralism', in *Britain: A Plural Society* (Commission for Racial Equality, London, 1990).

—— (ed.), *Colour, Culture and Consciousness* (Allen & Unwin, London, 1974).

Parker, S., 'Rights and Utility in Anglo-Australian Family Law' (1992) 55 *Modern Law Review* 311.

Parliamentary Assembly of the Council of Europe, 'The Geographical Enlargement of the Council of Europe: Policy Options and Consequences', repr. in (1992) 13/5–6 *Human Rights Law Journal* 230.

Parton, N., *Governing the Family: Child Care, Child Protection and the State* (Macmillan, London, 1991).

Pattenden, R., 'Abuse of Process in Criminal Litigation' (1989) 53 *Journal of Criminal Law* 341.

—— 'Should Confessions be Corroborated?' (1991) 107 *Law Quarterly Review* 317.

Peach, C., 'The Muslim Population of Great Britain' (1990) 13 *Ethnic and Racial Studies* 414.

Peay, J., *Tribunals on Trial: A Study of Decision Making under the Mental Health Act 1983* (Clarendon Press, Oxford, 1989).

Pennock, J. R., and Chapman, J. W., *Human Rights: Nomos 23* (New York University Press, New York, 1981).

Philips, Sir Cyril, *The Prevention of Terrorism (Temporary Provisions) Act 1984. Review of the Year 1985* (Home Office, London, 1986).

Pizzey, E., *Scream Quietly or the Neighbours will Hear* (Penguin Books. London, 1974).

Plender, R., 'Human Rights of Aliens in Europe', in Council of Europe Directorate of Human Rights, *Human Rights of Aliens in Europe* (Martinus Nijhoff, Dordrecht, 1985).

—— *International Migration Law* (Sijthoff, Leiden, 1972).

Polack, K., and Corsellis, A., 'Non-English Speakers and the Criminal Justice

System' (1990) 140 *New Law Journal* 1634.

Political and Economic Planning, *Racial Discrimination* (PEP, London, 1967).

Pollock, L. H., *Discrimination in Employment: The American Response* (Runnymede Trust, London, 1974).

Poncet, D., *La Protection de l'accusé par la Convention européenne des droits de l'homme* (Geneva, 1977).

Poulter, S., *Asian Traditions and English Law* (Runnymede Trust, Stoke-on-Trent, 1990).

—— 'The Claim to a Separate Islamic System of Personal Law for British Muslims', in C. Mallat and J. Connors (eds.), *Islamic Family Law* (Graham & Trotman, London, 1990).

—— *English Law and Ethnic Minority Customs* (Butterworths, London, 1986).

—— 'Ethnic Minority Customs, English Law and Human Rights' (1987) 36 *International and Comparative Law Quarterly* 589.

—— 'The Religious Education Provisions of the Education Reform Act 1988' (1990) 2 *Education and the Law* 1.

—— 'The Scattering of Cremated Ashes, River Pollution and the Law' (1989) 1 *Land Management and Environmental Law Report* 82.

—— 'Towards Legislative Reform of the Blasphemy and Racial Hatred Laws' (1991) *Public Law* 371.

Prashar, U., 'The Need for Positive Action', in J. Benyon, *Scarman and After* (Pergamon Press, Oxford, 1984).

Prosser, W. I., 'Privacy' (1960) 48 *California Law Review* 383.

Public Law Symposium on PACE, Autumn 1985 issue.

Radzinowicz, L., and Hood, R., *The Emergence of Penal Policy in Victorian and Edwardian England* (Clarendon Press, Oxford, 1990).

Raine, J. W., *Local Justice: Ideals and Realities* (T. & T. Clark, Edinburgh, 1989).

Raison, T., 'Cultural Diversity, Adaptation and Participation' (1980) 8 *New Community* 96.

Randall, V., *Women and Politics*, 2nd edn. (Macmillan Education, London, 1987).

Rawlings, H. F., *Law and Electoral Process* (Sweet & Maxwell, London, 1988).

Raymond, B., 'Not Guilty Verdict on the Jury', *Guardian*, 18 Nov. 1988.

Raz, J., 'Free Expression and Personal Identification' (1991) 11 *Oxford Journal of Legal Studies* 303.

Reform of Section 2 of the Official Secrets Act 1911, Cm. 408 (HMSO, London, 1988).

Reiner, R., *The Blue-Coated Worker* (Cambridge University Press, Cambridge, 1978).

—— *Chief Constables: Bobbies, Bosses or Bureaucrats* (Oxford University Press, Oxford, 1991).

—— 'Fuzzy Thoughts: The Police and "Law and Order" Politics' (1980) 28 *Sociological Review* 2.

—— 'The Politics of the Act' (1985) *Public Law* 394.

—— *The Politics of the Police* (Wheatsheaf, Brighton, 1985; 2nd edn., 1992).

—— 'Race and Criminal Justice' (1989) 16 *New Community*, special issue on 'Race, Criminal Justice and the Legal System'.

—— 'Policing a Postmodern Society' 55 *Modern Law Review* 6 (1992).

—— and Cross, M. (ed.), *Beyond Law and Order* (Macmillan, London, 1991).

—— and Spencer, S., *Accountable Policing: Effectiveness, Empowerment and Equity* (Institute for Public Policy Research, London, 1993).

Report of a Committee to Consider, in the Context of Civil Liberties and Human Rights, Measures to Deal with Terrorism in Northern Ireland, Cmnd. 5847 (HMSO, London, 1975).

Report of an Efficiency Scrutiny of the Prison Medical Service (Home Office, London, 1990).

Report of an Inquiry by the Hon. Sir Henry Fisher into the Circumstances Leading to the Trial of Three Persons on Charges Arising out of the Death of Maxwell Confait and the Fire at 27 Doggett Road, London SE 5, HC 90 (HMSO, London, 1977).

Report of an Inquiry by the Rt. Hon. Lord Justice Scarman, OBE: The Red Lion Square Disorders of 15 June 1974, Cmnd. 5919 (HMSO, London, 1975).

Report of an Inquiry by the Rt. Hon. Lord Justice Scarman, OBE: The Brixton Disorders, 10–12 April 1981, Cmnd. 8427 (HMSO, London, 1981).

Report of HM Chief Inspector of Constabulary 1989 (HMSO, London, 1990).

Report of HM Chief Inspector of Prisons 1988 (HC 491, 1988–89) (HMSO, London, 1989).

Report of the Commission to Consider Legal Procedures to Deal with Terrorist Activities in Northern Ireland. Cmnd. 5185 (HMSO, London, 1972).

Report of the Committee of Financial Aid to Political Parties, Cmnd. 6601 (HMSO, London, 1976).

Report of the Committee of Inquiry into the Education of Children from Ethnic Minority Groups (Chairman, Lord Swann), Cmnd. 9453 (HMSO, London, 1985).

Report of the Committee of Privy Councillors Appointed to Consider Procedures for the Interrogation of Persons Suspected of Terrorism (Chairman, Lord Parker of Waddingham), Cmnd. 4901 (HMSO, London, 1972).

Report of the Committee on Children and Young Persons (Chairman, Viscount Ingleby), Cmnd. 1191 (HMSO, London, 1960).

Report of the Committee on Contempt of Court (Chairman, Phillimore LJ), Cmnd. 5794 (HMSO, London, 1974).

Report of the Committee on Data Protection (Chairman, Sir Norman Lindop), Cmnd. 7341 (HMSO, London, 1978).

Report of the Committee on Defamation, Cmnd. 5909 (HMSO, London, 1975).

Report of the Committee on Fraud Trials (Chairman, Lord Roskill) (HMSO, London, 1986).

Report of the Committee on Homosexual Offences and Prostitution (Chairman: Lord Wolfenden), Cmnd. 247 (HMSO, London, 1957).

Report of the Committee on Obscenity and Film Censorship, Cmnd. 7772 (HMSO, London, 1979).

Report of the Committee on Privacy (Chairman, Kenneth Younger), Cmnd. 5012

(HMSO, London, 1972).

Report of the Committee on Privacy and Related Matters (Chairman, David Calcutt, QC), Cmnd. 1102 (HMSO, London, 1990).

Report of the Committee on the Future of Broadcasting, Cmnd. 6753 (HMSO, London, 1977).

Report of the Committee on the Law of Defamation, Cmnd. 7536 (HMSO, London, 1948).

Report of the Committee on the Prison Disciplinary System, Cmnd. 9641 (HMSO, London, 1985).

Report of the Departmental Committee on Legal Aid in Criminal Court Proceedings, Cmnd. 2934 (HMSO, London, 1966).

Report of the Departmental Committee on Section 2 of the Official Secrets Act 1911, Cmnd. 5104 (HMSO, London, 1972).

Report of the House of Lords Select Committee on Murder and Life Imprisonment, HL 78-I, 1988–9 (HMSO, London, 1989).

Report of the Interdepartmental Committee on the Distribution of Criminal Business between the Crown Court and Magistrates' Court, Cmnd. 6323 (HMSO, London, 1975).

Report of the Review Committee: The Parole System in England and Wales, Cmnd. 532 (HMSO, London, 1988).

Report of the Royal Commission on Criminal Procedure, Cmnd. 8092 (HMSO, London, 1981).

Report of the Royal Commission on the Police, Cmnd. 1728 (HMSO, London, 1962).

Report of the Tribunal Appointed to Inquire into the Events on Sunday 30 January 1972 which Led to Loss of Life in Connection with the Procession in Londonderry on that Day (Chairman: Lord Widgery), HL 101, HC 220 (HMSO, London, 1972).

Report of the Woolf Inquiry: Prison Disturbances April 1990, Cmnd. 1465 (HMSO, London, 1991).

Report of the Working Party on Disclosure of Information on Trials on Indictment (HMSO, London, 1979).

Report on the Work of the Prison Service April 1991–March 1992, Cm. 2087 (HMSO, London, 1992).

Report to the Secretary of State for the Home Department of the Departmental Committee on Evidence of Identification in Criminal Cases, HC 338 (HMSO, London, 1976).

Resnick, D., 'Due Process and Procedural Justice', in J. R. Pennock and J. W. Chapman (eds.), *Due Process* (New York University Press, New York, 1977).

Return to an Address of the Honourable House of Commons dated 12 July 1990 for the Inquiry into the Circumstances Surrounding the Convictions Arising out of the Bomb Attacks in Guildford and Woolwich in 1974, HC 556 (HMSO, London, 1990).

Review of Public Order Law, Cmnd. 9510 (HMSO, London, 1985).

Review of the Public Order Act 1936 and Related Legislation, Cmnd. 7891 (HMSO, London, 1980).

Rex, J., 'The Concept of a Multi-cultural Society' (1987) 14 *New Community* 218.

Richardson, J. J. and Jordan, A. G., *British Politics and the Policy Process: An Arena Approach* (Allen and Univin, London, 1987).

Richardson, G., 'The Case for Prisoners' Rights', in M. Maguire, J. Vagg, and R. Morgan, (eds.), *Accountability and Prisons* (Tavistock, London, 1985).

—— 'The Duty to Give Reasons: Potential and Practice' (1986) *Public Law* 437.

—— 'Judicial Intervention in Prison Life', in M. Maguire, J. Vagg, and R. Morgan (eds.), *Accountability and Prisons* (Tavistock, London, 1985).

—— *Law, Process and Custody: Prisoners and Patients* (Weidenfeld and Nicolson, London, 1993).

Riley, D., and Vennard, J., *Triable Either-Way Cases: Crown Court or Magistrates' Court* (HMSO, London, 1988).

Roberts, P., Willmore, C. and Davis, G., *The Role of Forensic Science Evidence in Criminal Proceedings*, Research Study No. 11 for the Royal Commission on Criminal Justice (HMSO, London, 1993).

Robertson, G., *Freedom, the Individual and the Law*, 6th edn.; earlier ends. by H. Street (Penguin, Harmondsworth, 1989).

Robilliard, S., *Religion and the Law* (Manchester University Press, Manchester, 1984).

Roche, P. M., 'The United Kingdom's Obligation to Balance Human Rights and its Anti-Terrorism Legislation: The Case of *Brogan* and Others' (1989–90) 13 *Fordham International Law Journal* 328.

Royal Commission on Criminal Justice, Home Office Memoranda (Home Office, London, 1991).

Royal Commission on Criminal Procedure, The Investigation and Prosecution of Criminal Offences in England and Wales: The Law and Procedure, Cmnd. 8091–2 (HMSO, London, 1981).

Royal Commission on Equal Pay, 1944–46 Report, Cmnd. 6937 (HMSO, London, 1946).

Royal Commission on the Press 1947–1949, Cmnd. 7700 (HMSO, London, 1949).

Royal Commission on the Press 1961–1962, Cmnd. 1811 (HMSO, London, 1962).

Royal Commission on the Press, Final Report, Cmnd. 6810 (HMSO, London, 1977).

Rubenstein, M., *The Dignity of Women and Men in the Workplace: The Problem of Sexual Harassment in the Member States of the European Community* (Office for Official Publications of the European Community, Luxembourg, 1988).

—— 'The Equal Treatment Directive and UK Law', in C. McCrudden (ed.), *Women, Employment and European Equality Law* (Eclipse Publications, London, 1987).

—— 'Sexual Harassment Recommendation and Code' (Jan.–Feb. 1992) *Equal Opportunities Review* 27.

Runnymede Trust, *Race and Immigration*, monthly bulletin (London, 1970–).

Russell, P. H., 'Canada's Charter of Rights and Freedoms: A Political Report' (1988) *Public Law* 385.

Rutherford, A. F., *Prisons and the Process of Justice* (Heinemann, London, 1984).

Sachs, V., 'The Equal Opportunities Commission: Ten Years On' (1986) 49 *Modern Law Review* 560.

Sanders, A., 'Class Bias in Prosecutions' (1985) 24 *The Howard Journal* 176.

—— 'Constructing the Case for the Prosecution' (1987) 14 *Journal of Law and Society* 229.

—— 'An Independent Prosecution Service?' (1986) *Criminal Law Review* 16–27.

—— 'Rights, Remedies and the Police and Criminal Evidence Act' (1988) *Criminal Law Review* 802–81.

—— 'Some Dangers of Policy Oriented Research: The Case of Prosecutions', in I. H. Dennis (ed.), *Criminal Law and Justice* (Sweet & Maxwell, London, 1987).

—— and Bridges, L., 'Access to Legal Advice and Police Malpractice' (1990) *Criminal Law Review* 494–509.

—— —— Mulvaney, A., and Crozier, G., *Advice and Assistance at Police Stations and the 24 Hour Duty Solicitor Scheme* (Lord Chancellors Dept., London, 1989).

Sargant, T., and Hill, P., *Criminal Trials: The Search for Truth* (Fabian Society, London, 1986).

Scanlon, T. M., 'A Theory of Freedom of Expression' (1972) 1 *Philosophy and Public Affairs* 204.

Scarman, Lord, *English Law: The New Dimension* (Stevens, London, 1974).

—— *The Scarman Report. The Brixton Disorders, 10–12 April 1981* (Penguin Books, London, 1986).

Schauer, F., *Free Speech: A Philosophical Enquiry* (Cambridge University Press, Cambridge, 1982).

Schwelb, E., 'The International Convention on the Elimination of All Forms of Racial Discrimination' (1966) 15 *International and Comparative Law Quarterly* 996.

Scorer, C., and Hewitt, P., *The Prevention of Terrorism Act: The Case for Repeal* (National Council for Civil Liberties, London, 1981).

Scottish Ethnic Minorities Research Unit, *Report on Anti-discrimination Law on the Grounds of Race: a Comparative Literature Survey of Provisions in Australia, New Zealand, Canada and the U.S.A.* (Edinburgh, S.E.M.R.U., 1992).

Scottish Law Commission, *Report on Family Law*, Scot. Law Comm. No. 135 (HMSO, London, 1992).

Scraton, P., and Gordon, P. (eds.), *Causes for Concern* (Penguin, London, 1984).

Shackleton, Lord, *Review of the Operation of the Prevention of Terrorism (Temporary Provisions) Acts 1974 and 1976*, Cmnd. 7324 (HMSO, London, 1978).

Shapland, J., *et al.*, *Victims in the Criminal Justice System* (Gower, Aldershot, 1985).

Sherr, A., *Freedom of Protest: Public Order and the Law* (Basil Blackwell, Oxford, 1989).

Sieghart, P., *The International Law of Human Rights* (Clarendon Press, Oxford, 1983).

—— *Privacy and Computers* (Latimer, London, 1976).

—— 'Sanctions against Abuse of Police Powers' (1985) *Public Law* 440.

Sim, J., *Medical Power in Prisons* (Open University Press, Milton Keynes, 1990).

Simitis, S., 'Reviewing Privacy in an Information Society' (1987) 135 *University of Pennsylvania Law Review* 707.

Sivanandan, A., 'Race, Class and the State: The Black Experience in Britain' (1976) 17 *Race and Class* 347.

Skogan, W. G., *The Police and the Public in England and Wales: A British Crime Survey Report*, Home Office Research Study 117 (HMSO, London, 1990).

Skolnick, J., *Justice Without Trial* (Wiley, New York, 1966).

Smith, A., *The Ethnic Revival* (Cambridge University Press, Cambridge, 1981).

Smith, A. T. H., *The Offences against Public Order Including the Public Order Act 1986* (Sweet & Maxwell, London, 1987).

Smith, D. J., *et al.*, *Police and People in London*, 4 vols. (Policy Studies Institute, London, 1983).

—— *Racial Disadvantage in Britain* (Penguin, Harmondsworth, 1977).

Smith, R., 'How Good are Test Cases?', in J. Cooper and R. Dhavan, *Public Interest Law* (Blackwell, Oxford, 1986).

Solomos, J., 'Equal Opportunities Policies and Racial Inequality: The Role of Public Policy' (1989) 67 *Public Administration* 79.

—— 'The Politics of Race and Housing' (1991) 19 *Policy and Politics* 147.

Sondhi, R., *Divided Families: British Immigration Control in the Indian Subcontinent* (Runnymede Trust, London, 1987).

Southgate, P., and Ekblom, P., *Police: Public Encounters* (Home Office Research Unit, London, 1984).

Spencer, J. R. (ed.), *Jackson's Machinery of Justice in England* (Cambridge University Press, Cambridge, 1989).

Spencer, J., 'The Neutral Expert: An Implausible Bogey' (1991) *Criminal Law Review* 106–10.

Spencer, M., *1992 and All That: Civil Liberties in the Balance* (Civil Liberties Trust, London, 1990).

Standing Advisory Commission on Human Rights, *Bill of Rights: A Discussion Paper* (1976).

—— *The Protection of Human Rights by Law in Northern Ireland*, Cmnd. 7009 (HMSO, London, 1977).

—— *Religious and Political Discrimination and Equality of Opportunity in Northern Ireland: Second Report* (Cn. 1107, 1990, Belfast).

Stern, V., *Bricks of Shame: Britain's Prisons*, 2nd edn. (Penguin, London, 1987).

Stevens, P., and Willis, C., *Ethnic Minorities and Complaints against the Police* (Home Office Research Unit, London, 1981).

—— —— *Race, Crime and Arrests* (Home Office Research Unit, London, 1979).

Stevenson, B., *The Ability to Challenge DNA Evidence*, Research Study No. 9 for the Royal Commission on Criminal Justice (HMSO, London, 1993).

Stevenson, S., and Bottoms, A., 'The Politics of the Police: A Royal Commission in a Decade of Transition', in R. Morgan (ed.), *Policing, Organised Crime and Crime Prevention* (Centre for Criminal Justice, Bristol, 1990).

Stockdale, R., 'Running with the Hounds' (1991) 141 *New Law Journal* 772.

Street, H., *Freedom, the Individual and the Law* (Penguin, Harmondsworth, 1963).

—— Howe, G., and Bindman, G., *Street Report on Anti-Discrimination Legislation* (PEP, London, 1967).

Sufian, J., 'DNA in the Courtroom' (Feb. 1991) *Legal Action* 7.

Supperstone, M., *Brownlie's Law of Public Order and National Security*, 2nd edn. (Butterworths, London, 1981).

Symmonds, C., 'The Effect of the European Convention on Human Rights on the Preparation and Amendment of Legislation, Delegated Legislation and Administrative Rules in the United Kingdom', in M. P. Furmston, R. Kerridge and B. E. Sufrin (eds.), *The Effect on English Domestic Law of Membership of the European Communities and of Ratification of the European Convention on Human Rights* (Mortinus Nijhoff, 1983).

Szyszczak, E., 'L' Espace sociale européenne: Reality, Dreams or Nightmares?' (1990) 33 *German Yearbook of International Law* 284.

Taitz, J., 'A Transsexual's Nightmare: The Determination of Sexual Identity in English Law' (1988) 2 *International Journal of Law and the Family* 139.

Tatchell, P., *Europe in the Pink: Lesbian and Gay Equality in the New Europe* (GMP Publishers, London, 1992).

Taylor, I., 'The Law and Order Issue in the British General Election and Canadian Federal Election of 1979' (1980) 5 *Canadian Journal of Sociology* 3.

Tefft, S. (ed.), *Secrecy: A Crosscultural Perspective* (Human Sciences Press, New York, 1980).

Temkin, J., *Rape and the Legal Process* (Sweet & Maxwell, London, 1987).

Thomas, D. A., *Current Sentencing Practice* (Sweet & Maxwell, London, 1982).

Thomas, P., and Costigan, R., *Promoting Homosexuality: Section 28 of the Local Government Act 1988* (Cardiff Law School, Cardiff, 1990).

Thornberry, P., *International Law and the Rights of Minorities* (Clarendon Press, Oxford, 1991).

Tolley, H., jun., 'Popular Sovereignty and International Law: ICJ Strategies for Human Rights Standard Setting' (1989) 11 *Human Rights Quarterly* 563.

Tregilgas-Davey, M., 'Miscarriages of Justice within the English Legal System' (1991) 141 *New Law Journal* 608.

Tribe, L. H., *American Constitutional Law* (Foundation, New York, 1978).

—— 'The Puzzling Persistence of Process-Based Constitutional Theories' (1980) 89 *Yale Law Journal* 1063.

Tuck, M., and Southgate, P., *Ethnic Minorities, Crime and Policing* (Home Office Research Unit, London, 1981).

Turpin, C., *British Government and the Constitution: Text, Cases and Materials*, 2nd edn. (Weidenfeld & Nicolson, London, 1990).

Tzannatos, P. Z., and Zabalza, A., 'The Anatomy of the Rise of British Female Relative Wages in the 1970s: Evidence from the New Earnings Survey' (1984) 22 *British Journal of Industrial Relations* 177.

United Kingdom Immigrants' Advisory Service, *Annual Reports* (UKIAS, London, 1983/4).

United Nations, *Report on National Legislation for the Equalization of Opportunities for People with Disabilities: Examples from 22 Countries and Areas* (United Nations, New York, 1989).

United Nations Committee on Economic, Social and Cultural Rights' (1989) 42 *ICJ Review* 33.

United Nations Educational, Scientific, and Cultural Organization, *Meeting of Experts (Cat IV) for the Europe Region to Examine Ways in Which Women may Exert a More Effective Influence on the Action of Public Authorities and Decision-Making Processes, Oslo, Norway, 5–9 February 1990: Final Report* (Paris, UNESCO, SHS-90/Conf. 610/14, 11 May 1990).

United States Dept. of State, *Country Reports on Human Rights Practices* (US Government Printing Office, Washington, DC, annual).

van Dijk, P., and van Hoof, G. J. H., *Theory and Practice of the European Convention on Human Rights*, 2nd edn. (Kluwer Law and Taxation, Deventer, 1990).

Vennard, J., *Contested Trials in Magistrates' Courts* (HMSO, London, 1982).

—— 'The Outcome of Contested Trials', in D. Moxon, *Managing Criminal Justice* (HMSO, London, 1985).

—— and Riley, D., 'The Use of Peremptory Challenge and Stand by of Jurors and their Relationship to Trial Outcome' (1988) *Criminal Law Review* 731.

Von Hirsch, A., *Doing Justice: The Choice of Punishments* (Northeastern University Press, Boston, Mass., 1986).

Waaldijk, K., *Tip of an Iceberg: Anti-Lesbian and Anti-Gay Discrimination in Europe* (Department of Gay and Lesbian Studies, University of Utrecht, 1991).

—— and Clapham, A., *Homosexuality: A European Community Issue* (Martinus Nijhoff, Dordrecht, 1993).

Wacks, R., *Personal Information: Privacy and the Law* (Clarendon Press, Oxford, 1989).

—— *The Protection of Privacy* (Sweet & Maxwell, London, 1980).

Waddington, D., Jones, K., and Critcher, C., *Flashpoints: Studies in Public Disorder* (Routledge, London, 1989).

Waddington, P. A. J., *The Strong Arm of the Law: Armed and Public Order Policing* (Clarendon Press, Oxford, 1991).

Wade, E. C. S., and Bradley, A. W., *Constitutional and Administrative Law*, 10th edn. (Longman, London, 1985).

Waldron, J., 'A Right-Based Critique of Constitutional Rights' 13 *Oxford Journal of Legal Studies* 18 (1993).

Walker, C., *The Prevention of Terrorism in British Law* 2nd edn. (Manchester University Press, Manchester, 1992).

Wallace, L., 'A Change in PACE' (1991) 155 *Justice of the Peace* 325.

Wallington, P., 'Policing the Miners' Strike' (1985) 14 *Industrial Law Journal* 145.

Walmsley, R., 'Indecency between Males and the Sexual Offences Act 1967' [1978] *Criminal Law Review* 400.

Walsh, B., 'The United Nations Convention on the Rights of the Child: A British View' (1991) 5 *International Journal of Law and the Family* 170.

Walzer, M., *Spheres of Justice: A Defence of Pluralism and Equality* (Robertson, Oxford, 1983).

Warren, S. D., and Brandeis, L. D., 'The Right to Privacy' (1890) 4 *Harvard Law Review* 193.

Wasserstrom, R. S., 'Racism, Sexism and Preferential Treatment: An Approach to the Topics' (1977) 24 *UCLA Law Review* 581.

Weber, M., *Law, Economy and Society* (Harvard University Press, Cambridge, Mass. 1954).

Webster, R., *A Brief History of Blasphemy: Liberalism, Censorship and 'The Satanic Verses'* (Orwell, Southwold, 1990).

Wedderburn, Lord, 'Freedom of Association and Philosophies of Labour Law' (1989) 18 *Industrial Law Journal* 1.

—— *The Social Charter, European Company and Employment Rights: An Outline Agenda* (Institute of Employment Rights, London, 1990).

Weeks, J., *Coming Out: Homosexual Politics in Britain from the Nineteenth Century to the Present*, revised edn. (Quartet Books, London, 1990).

Weis, P., *Nationality and Statelessness in International Law*, 2nd edn. (Sijthoff & Noordhoff, Alphen aan den Rijn, 1979).

Wells, C., Morgan, D., and Leat, D., 'Fetuses and Burials' (1991) 141 *New Law Journal* 1046.

Westin, F., *Privacy and Freedom* (Atheneum, New York, 1968; Bodley Head, London, 1970).

White, R. C., *The Administration of Justice*, 2nd edn. (Blackwell, Oxford, 1991).

Williams, D. G. T., *Keeping the Peace: The Police and Public Order* (Hutchinson, London, 1967).

Williams, G. H., *The Law and Politics of Police Discretion* (Greenwood Publishing, Westport, Conn., 1984).

Williams, G. L., 'Statutory Powers of Arrest without Warrant' (1958) *Criminal Law Review* 72.

Willis, C., *The Use, Effectiveness and Impact of Police Stop and Search Powers* (Home Office Research Unit, London, 1983).

—— Macleod, J., Naish, P., *The Tape Recording of Police Interviews with Suspects: A Second Interim Report*, Home Office Research Study 97 (HMSO, London, 1988).

Wilson, J. Q., *Varieties of Police Behaviour* (Harvard University Press, Cambridge, Mass., 1968).

Wiseberg, L., 'Protecting Human Rights Activists and NGOs: What More can be Done' (1991) 13 *Human Rights Quarterly* 528–9.

Woffinden, B., *Miscarriages of Justice* (Coronet, London, 1989).

Wood, J., and Crawford, A., *The Right of Silence: The Case for Retention* (Civil Liberties Trust, London, 1989).

Woolf, Sir H., *Protection of the Public: A New Challenge* (Stevens & Sons, London, 1989).

Young, J., 'The Role of the Police as Amplifiers of Deviancy', in S. Cohen (ed.), *Images of Deviance* (Penguin, Harmondsworth, 1971).

Young, M., and Hill, P., *Rough Justice* (British Broadcasting Corporation, London, 1983).

Zander, M., 'Access to a Solicitor in the Police Station' (1979) *Criminal Law Review* 342.

—— 'Are Too Many Professional Criminals Avoiding Conviction? A Study in Britain's Two Busiest Courts' (1974) 37 *Modern Law Review* 28–61.

—— *A Bill of Rights?* 3rd edn. (Oxford University Press, Oxford, 1985).

—— *A Matter of Justice* (Oxford University Press, Oxford, 1988).

—— *The Police and Criminal Evidence Act 1984*, 2nd edn. (Sweet & Maxwell, London, 1991).

—— 'Police and Criminal Evidence Bill' (1983) 133 *New Law Journal* 197, 220, 245, 269, 318, 339, 365, 389.

—— 'Police Powers' (1982) 53 *Political Quarterly* 153.

Zuckerman, A. A. S., 'Illegally Obtained Evidence: Discretion as a Guardian of Legitimacy' (1987) *Current Legal Problems* 55.

—— *The Principles of Criminal Evidence* (Clarendon Press, Oxford, 1989).

Zuijdwijk, T. J. M., *Petitioning the United Nations: A Study in Human Rights* (Gower, Aldershot, 1982).

Index